FIELD GUIDE TO THE
Birds of East Asia

Eastern China, Taiwan, Korea, Japan and Eastern Russia

Mark Brazil

D0771778

Illustrated by

Dave Nurney

Per Alström, Carl D'Silva, Martin Elliott, Kim Franklin, Alan Harris,
Ren Hathway, Hans Larsson, Derek Onley, Christopher Schmidt,
Brian Small, Laurel Tucker, Tim Worfolk and Bill Zetterström

CHRISTOPHER HELM
LONDON

for

Hannah

Published 2009 by Christopher Helm, an imprint of A&C Black Publishers Ltd.,
36 Soho Square, London, W1D 3QY

www.acblack.com

Reprinted with amendments 2009

ISBN 978-0-7136-7040-0

A CIP catalogue record for this book is available from the British Library

Commissioning Editor: Nigel Redman
Project Editor: Jim Martin

Design by Fluke Art, Cornwall

Maps by Fergus Crystal, Julie Dando and Mark Brazil

Printed in China by 1010 Printing International Ltd

10 9 8 7 6 5 4 3 2

Front cover: Steller's Sea Eagle (Alan Harris)
Back cover (top to bottom): Narcissus Flycatcher, Pryer's Woodpecker, Elegant Bunting, Tufted Puffin (Dave Nurney)

CONTENTS

ACKNOWLEDGEMENTS

Very many people have contributed to the success of this project, some with their field experience of the region, some with their experience of Asian vagrants in Europe or North America, and some by providing relevant literature, valued insights into taxonomy, or by critically reviewing drafts – I thank them all.

It was my fervent desire to have birders and ornithologists from both within and outside the region involved in this project from the outset. Of those I approached initially, the level of support was heartening and tremendous, with input ultimately coming from nationals of Australia, China, Estonia, Finland, Japan, Korea, New Zealand, Norway, Russia, Sweden, Taiwan, the UK and the United States. Behind every writer stand hidden ranks of vitally important readers and checkers of facts, grammar and spelling. I am indebted to the many friends and colleagues who have supported me by giving their time and consideration to labouring through complete drafts (sometimes several) of the manuscript. In particular, I would like to single out for my deepest thanks two special contributors, who gave so generously of their time and experience: Fergus Crystal (for reviewing repeated drafts, for significant contributions to the identification, voice and status code sections, and for helping with the maps) and Yoshiki Watabe (who not only reviewed drafts, but also took considerable time to help trace records, photographs and publications pertaining to recent records, and for responding to my innumerable questions). Others who helped very significantly and invaluably by painstakingly reviewing several earlier drafts of the entire text and whose efforts are deeply appreciated were: Neil Davidson, Will Duckworth, Ted Kenefick, Nick Lethaby and Peter Newbound. Paul Lehman gave considerable help with species also found in North America, whilst Norbert Bahr (compiler of *The Howard and Moore Complete Checklist of the Birds of the World*) provided invaluable help with recent ornithological literature pertaining to taxonomy, and along with Richard Klim made great efforts to ensure that species taxonomy was as up to date as possible. I thank James Lambert, Angus Macindoe and Guy Kirwan for carefully reading the manuscript.

Others who proffered invaluable insights and considerable time to the success of this work were: Peter Clement (thrushes, flycatchers and warblers), Greg Gillson (seabirds), Peter Harrison (seabirds), Peter Kennerley, Sean Minns, Dr Atle Ivar Olsen, Jevgeni Shergalin (Russian ornithological literature), and Thede Tobish (North American species).

Reviewers and supporters from various parts of the region have graciously helped this project towards completion by giving freely of their time and knowledge.

P. R. China Lei Jinyu and Liu Yang provided information concerning the status and distribution of birds, particularly recent records, in eastern and north-eastern China, whilst Paul Holt and Jesper Hornskov offered help with the checklist, and species recorded at Beidaihe, eastern China.

Taiwan, R. O. C For considerable input with the status of birds in Taiwan I thank: Dr Ruey-shin Lin, Cheng-te Yao, Dr Tzung-su Ding, Huan-Chang Liao, Steve Mulkeen, Dr Woei-horng Fang, Kuen-Dar Chiang and Wayne Hsu.

Korea For their invaluable input concerning birds in Korea, I thank: Will Duckworth, Tim Edelsten, Dr Kisup Lee, Robert Newlin, and particularly Yong-chang Jang, Chang-wan Kang and Eun-mi Kim.

Japan My special thanks go to: Yosuke Amano, Haruhiko Asuka, Chris Cook, Dr Yuzo Fujimaki, Masao Fukagawa, Atsunori Fukuda, Takashi Hiraoka, Hiroshi and Marie-Jo Ikawa, Hiroshi Ikenaga, Himaru Iozawa, Norio Kawano, Kazunori Kimura, Shinji Koyama, Dr Shigeru Matsuoka, Akiyo Nakamichi, Yuji Nishimura, Toshikazu Onishi, Shigeo Ozawa, Akiko Sasaki, Yuko Sasaki, Fumio Sato, Tomokazu Tanigawa, Satoshi Tokorozaki, Masahiro Toyama, Osao and Michiaki Ujihara, Dr Takeshi Yamasaki and Mike Yough, all of whom proffered considerable information. Also thanks to members of the *Kantori* e-mail group for stimulating discussions on the identification of several confusion species.

Russia For providing pertinent literature and invaluable input in reviewing the status and distribution of species occurring in the eastern Russian portion of this book's range, I thank: Dr Vladimir Yu. Arkhipov, Dr Vladimir Babenko, Dr Yuri Gerasimov, Dr Evgeny Koblik, Dr Yaroslav Red'kin, Dr Evgeny Syroechkovski and Dr Pavel S. Tomkovich.

Some who provided information relevant at earlier stages of this project included: Axel Bräunlich, Jesper Hornskov, Tim Inskipp and Dominic Mitchell. Others, who gave advice on particular subjects, single species, species groups, records or regions, were: Desmond Allen, Phil Battley, Dr Nigel Collar, Dr Árni Einarsson, Jon Hornbuckle, Heikki Karhu, Richard Millington, Pete Morris, Dr Urban Olsson, Dr Lars Svensson and Dr Angela Turner.

I thank the entire team of artists, Per Alström, Carl D'Silva, Martin Elliott, Kim Franklin, Alan Harris, Ren Hathway, Hans Larsson, Derek Onley, Christopher Schmidt, Brian Small, the late Laurel Tucker and Tim Worfolk, and especially Dave Nurney, for their sterling work in making this project a success. At A&C Black, I thank Nigel Redman for helping to launch the project and Jim Martin for navigating it to its successful conclusion, while Julie Dando tackled the complex task of designing the plates and laying out the book with tremendous skill.

Whilst my friends and colleagues have done their utmost to rescue me from glaring errors and mistakes, any inaccuracies that remain are entirely mine. To all who have been involved – and to any I may have inadvertently overlooked – my deepest thanks.

Finally, I thank Mayumi Kanamura for her understanding, support and encouragement throughout this project.

PREFACE

Having cut my ornithological teeth on a diet of bird books dedicated to the British avifauna, my introduction to Peterson, Mountfort and Hollom's *Field Guide to the Birds of Britain and Europe* was a revelation. For the first time I could feast my eyes on the exciting birds living just beyond the confines of my home country. I could read about them, and in particular dream of encountering them; in some measure I was able to prepare myself for birds that I had not seen, but that I *might* see. A generation later, Svensson, Grant, Mullarney and Zetterström in the *Collins Bird Guide* (1999) set a new standard in Europe that few will ever match, let alone supersede. The significance of such regional guidebooks is that they open the eyes of local birdwatchers to the possibilities beyond their own national borders, and provide information on species that may one day appear on their own local or national patch. These have been my aims in *Birds of East Asia*; to describe and illustrate all species that have occurred in the region (*c.*985), and to suggest those extralimital species and vagrants most likely to appear in the near future, so as to encourage birdwatchers living in the region to think of birds from beyond the immediate horizon. In so doing I hope to bring some of those birdwatchers together in a fraternity of observing and conserving the avifauna of this fascinating region. *Birds of East Asia* has also been designed with the particular interests of birders in Europe and North America in mind, in the hope of facilitating the identification of vagrants to those areas from the region.

East Asia has languished in a parochial ornithological sense. Yes, there have been numerous field guides published in recent decades, but these have all been national in their scope, and almost all of them photographic, with the obvious pitfalls of that medium. A focus on photography, rather than on critical bird observation, has created a local tradition of guides that serve more the goals of the photographer than the needs of the ornithologist/birder. Countries in the region do not have standing rarities committees soliciting, collating or assessing the acceptability of new records of rare birds for these regions. There are no equivalents of annual county, regional or national bird reports that birders from Western Europe and North America are so familiar with; national checklists for the region tend to follow outdated taxonomies, some even ignore numerous records of birds that have been convincingly photographed in the wild and even published in popular magazines or field guides. Furthermore, with little or no ornithological investigation of the cagebird trade (which is rampant in the region) and without critical assessment of rarities records, it is difficult to ascertain the likelihood of the wild occurrence of certain species. The lack of such baseline reference material has made assessing the species for inclusion here somewhat difficult. I have adopted the latest published taxonomic decisions available and have opted to err on the side of inclusiveness, on the basis that several species surely will appear in the region one day (if they have not already done so reliably), and that as a number of probably escaped species have bred in the wild, they may be on their way to becoming permanent features of the region's avifauna.

I first conceived the scope of the present book in the early 1980s, but it was a project that long languished as others came and went. My goal has not changed, however, and that is to expose birders within and outside the countries of East Asia to the extraordinary avian diversity of the region, to enable them to identify the birds they see and to be prepared for birds they may one day find.

Field guides are at best works in progress. Additional species are recorded in East Asia each year. Our ornithological knowledge of the region is growing steadily as research and publications continue unabated; new discoveries are made, new understandings are reached; taxonomy shifts and changes. Each project must, however, have its cut-off date; mine was at the end of 2006. I hope that you, the reader, will enjoy using this book in the field to enhance your birding, and that you will also be critical and tease out its insufficiencies, ambiguities and inaccuracies, whether in text, illustrations or maps. Please contribute your knowledge, updated information and experience to help improve this guide for the future, by contacting me care of A&C Black Publishers.

Through the process of writing we discover how little we know. As with my previous books, this one has been written, not because I consider myself a specialist or expert on the subject, but because I have wanted so much to have a single volume containing all of the relevant information that I could refer to when in the field. I dedicate this book to my daughter, Hannah, in the hope that she and her generation too will delight in the pleasure of observing, learning about and conserving wild birds.

Mark Brazil
Ebetsu, Hokkaido, June 2008

INTRODUCTION

Aims of this field guide

This book describes 985 species of birds found in East Asia and 19 extralimital species that may occur. Birds of East Asia evolved from a more limited original concept (including only the birds of the Japanese archipelago and the Korean Peninsula) to one that covers a much larger and more significant portion of the East Asian migratory flyway. The countries of this region share many similarities and many migrants, and are also the likely sources of a number of future vagrants to regions further west and north-east. I have tried to include all species (including established introductions and vagrants) known to have occurred in this region based on published records prior to the end of 2006. Species are described, illustrated and, if regular in the region, mapped. Their status in each part of the region is presented in Appendix 1 and a list of 46 species thought likely to occur as vagrants in the future is presented in Appendix 2.

Figure 1. The region.

Geographical scope

The East Asian region, as defined here, includes the entire Japanese archipelago, Taiwan, the Korean Peninsula and parts of eastern China and Russia. The western border of the region was chosen to incorporate all of the coastal provinces of eastern China from Fujian north, north-east China (eastern Nei Mongol, Liaoning, Jilin and Heilongjiang), and north to the Russian Arctic coast from the western border of Hebei province, bringing the western border loosely to 116°E, thereby including all of the immense region of north-eastern Russia (Siberia and the Russian Far East) from *c*.116°E (between the Taimyr Peninsula and the Lena Delta) to the Bering Strait. The eastern boundary comprises the western Pacific and Bering Sea coasts, from the Bering Strait, Kamchatka and the Commander Islands of Russia south to the Ogasawara and Izu islands of Japan. As many of the regions, island groups and island names may be unfamiliar, these are shown in Figure 1.

Taxonomy

The species recorded in East Asia are described and illustrated following *The Howard and Moore Complete Checklist of the Birds of the World* (Dickinson 2003), with some minor adjustments to give balance to the plates. I have strayed from Dickinson (2003) only where I am aware of more recently published papers (in peer-reviewed journals; largely with advice or suggestions from Norbert Bahr and Richard Klim) describing newly accepted species or taxonomic arrangements. Recent taxonomic reviews published by the British and American Ornithologists' Unions have been particularly helpful, as ornithological societies in the region tend to follow rather than lead taxonomic trends. Subspecific information is also given, particularly where it is helpful in the field, but also because the taxonomy of birds in East Asia is ripe for further study and, as Collar (2003) implied, the Asian avifauna is considerably over-lumped relative to other regions of the world. The arrangement of subspecies entirely follows Dickinson (2003).

While the region's avifauna urgently requires rigorous taxonomic research using the latest techniques, this book is not the place for such work, and currently there is much that is contentious and contradictory regarding the status of various taxa. Taxonomic changes are continuously altering the playing field and it is difficult to reach agreement over which taxa are accepted at the specific or subspecific levels at a given time; no doubt changes will occur even as the book goes to press. I make no pretensions to be an authority on taxonomy. Nevertheless, in compiling this work I have been forced to make decisions, often on the basis of limited information. Although personally tending towards the conservative, I have generally argued in favour of splitting the more recognisable 'suspect' taxa as species, or I have flagged potential splits by using distinct English names (e.g. Owston's Varied Tit). Because the East Asian region remains under-watched and under-studied, there are quite a number of subspecies that will no doubt one day be recognised specifically, perhaps including even common taxa such as the Oriental Crow (as *Corvus orientalis*), a close relative of the Carrion Crow *C. corone*. To encourage observation of subspecies (and hence of future species), I have at least mentioned those that occur, and where field criteria exist for separation I have included such details as space permits.

Whereas in Europe and North America a long history of ornithological research and amateur birding have led to the avifaunas there being relatively well understood, in the decades ahead it is likely that we will see considerable advances in our understanding of species limits in East Asia, as field work increases. Field identification criteria for several species are still evolving, and in some cases are still very poorly known (such as the recently split *Parus major*, *P. minor* and *P. cinereus*); those with time to critically assess these birds in the field, museum and photographic collections may yet prove some to be more readily distinguished than present knowledge permits. My hope is that this guide will stimulate research into the identification, ecology, distribution, vocalisations and taxonomy of such taxa, and by so doing serve also the conservation community, by providing a basis from which to work.

Nomenclature

For each species, the English name is followed by the scientific name. By convention, the generic name begins in upper case and the specific name is entirely lower case, thus Olive-backed Pipit is *Anthus hodgsoni*. Breaking with the normal practice of presenting only generic and specific names in field guides, I have also included subspecific names (where known) for those forms occurring or likely to occur in the region, as these distinctions should appeal to readers interested in species identification and species limits, and are also of particular interest to birders in Europe and North America seeking to identify birds out of range that may differ subtly from subspecies in those areas. Thus, here, both subspecies of Olive-backed Pipit are listed, as *Anthus hodgsoni hodgsoni* and *A. h. yunnanensis*, though it should be noted that it is not always possible to identify an individual to subspecies in the field.

Vernacular and scientific names are based generally on Dickinson (2003) and Gill & Wright (2006), but with minor variations where I do not agree with those authors. Scientific names are generally slow to change, although a number of significant revisions have been adopted here, especially where previous genera are now recognised as polyphyletic, and with respect to gender endings (following David & Gosselin 2002a,b). English names vary considerably from one book to the next, and while convention may sway first one way then another, it is ultimately the author's responsibility to choose. Aware that we achieve a measure of immortality through our children, our works, our discoveries and our eponymous appearances, where people have long been commemorated in bird names I have generally opted for these, in honour of their historical contribution. I feel that such names greatly enrich ornithology, rather than the (to my mind) drabber, less personal names that have become in vogue. Thus, for example, I favour retaining the colourful bush robin of Taiwan as Johnstone's, rather than the uninspiring and unenlightening Collared Bush Robin, and the enigmatic and endangered woodpecker of the Ryukyu Islands as Pryer's Woodpecker rather than the rather boring Okinawa Woodpecker, albeit correct (although, to be truly accurate, surely Yambaru Woodpecker would be more geographically accurate, as it is known only from the northern Yambaru district of Okinawa). Certain English names are less than appropriate given their known ranges, thus I have used Temminck's Cormorant and Black Woodpigeon rather than 'Japanese' in each case; conversely, for the grosbeaks, Japanese and Chinese are better descriptors than, for example, Yellow-billed (for the latter) since the former has the more yellow bill. Following precedents among owls and trogons, I consider Elegant Bunting a far better name for *Emberiza elegans* than the cumbersome Yellow-throated Bunting. In each case the common alternatives in use are also given, so that, while you may humour me, you are not forced to follow my own personal hobby-horse.

Bird identification

With almost a thousand species of birds recorded in the East Asian region, identification may seem at first daunting, but as with the development of any skill, a combination of pleasant effort and practice bring about competence; ultimately 'time in the field' is what counts, both for enjoyment of birding and for developing proficiency.

Keen field observation and writing field notes, supplemented by study of a field guide, will help you to hone your identification skills through processes of recognition, comparison and elimination. First it helps to be mindful of the season and time of day, to recognise the habitat in which a bird is seen, whether it is tundra or deciduous forest, wet mudflat or rocky shore, then it helps to recognise prominent and distinctive behaviours – ways of foraging or flying, for example. All can help significantly in assigning a sighting to a particular species.

Learning the families to which birds belong is a major step. My hope is that the family accounts (pp. 14–29) and their accompanying illustrations will help you, if you are a novice, to begin to familiarise yourself with the major groupings of birds, whilst for more experienced birders these will serve as a quick guide to finding the appropriate plates. Placing a bird in the correct family is the first step to its identification. Seasonality, habitat, habits and range may quickly eliminate many unlikely species. A quick eye, a sharp ear and attention to detail will all help in teasing apart the differences between a bird that hops or one that runs, a bird that forages from leaves and another that feeds on tree trunks or branches. The ways in which a bird flies, swims, walks, flicks its wings, bobs its tail, or sings can all significantly aid identification.

Having learned the main families, take time to become familiar with the commonest species of your region, a duck, a crow, a dove, a starling or myna and a sparrow. Knowing these, their sizes, patterns and behaviours will provide you with an instant benchmark against which to compare all subsequent sightings. When faced with a mystery bird, first determine its size in relation to your familiar five, then consider its body shape – is it elongated and slender, like a wagtail, or is it plump like a grosbeak? Starting at the head, examine the bill size and shape; is it short and blunt (finch-like) or short and hooked (shrike-like)? Is it short but fine, perhaps an insectivorous flycatcher, or warbler? Is it long, thin, thick, straight or curved, and how does the bird use its bill? Having confirmed the bill shape and size, consider the pattern of the head; crown-stripes, eye-stripes, ear patches, or the lack of these features are all relevant. Then look to the upperparts. Can you see any distinctive or conspicuous patches of contrasting colour on the back, rump or wings? Are there any wingbars? Observe the underparts carefully too, to see how they compare with the upperparts, and note whether they are plain, barred or streaked, whether they are pale, dark or distinctively patterned. Note also any details of the tail, in particular its length relative to the wings. If the bird is flying note its wing shape, whether they are long or short, narrow or broad, rounded or pointed, whether they move in a steady blur, a series of distinct slow beats, or several rapid beats interspersed with glides. At the same time try to note whether the tail is forked, notched, square-tipped, rounded or wedge-shaped, and whether there are distinctive or contrasting patterns on the wings, rump and tail. Where and how a bird flies, forages and perches are also useful keys to identification.

As your skills develop, you will learn to recognise an increasing number of species against which anything new can be referenced mentally. You will begin to recognise that whereas some species have only one plumage, the males, females, adults and juveniles of others differ. Over time you will build up a recognition pattern of features that make most species distinctive, even when seen in silhouette or in poor light – this combination of characters/features is known as the bird's 'jizz', a kind of subliminal *gestalt* of what the bird looks like and how it moves. With time, you may want to learn more about moult, about how and when birds change their plumage in a steady sequence from juvenile to first-winter, to first-summer, then to non-breeding adult and breeding adult. Some species achieve this sequence in less than a year from hatching, and breed at barely one year old. Other species take a number of years to reach adult plumage and may not breed until they are several years old. Understanding bird behaviour, whether it is flight, foraging, displaying or nesting, may all add to your pleasure of birdwatching.

A very important means of locating and identifying birds is by their vocalisations or other sounds. Sounds of walking over dry leaf litter, whistling wings, the hard tapping of bill on wood, all these and many more incidental or deliberate sounds can help locate a bird you may not otherwise notice and may even be vital clues to its identity. Birds essentially have two types of vocalisations – calls and songs. Calls may be used for contact, recognition, to indicate changing behaviour (such as taking flight) or alarm. Songs are a more prolonged series of notes, some of which may sound particularly melodious to our ears, and are particularly used by passerines to define territories and to attract mates. Tracking down each and every call or song to its maker is time-consuming, requiring a painstaking and patient approach. The availability of high-quality recordings of bird songs in particular makes learning bird sounds an easier and more enjoyable pastime these days. One can now learn relevant vocalisations prior to going into the field thereby facilitating more rapid recognition.

Bird habitats

The East Asian region spans a wide range of ecosystems. In the far north and north-east of Russia there is the high Arctic tundra *beyond* the treeline, whilst on the highest mountain ranges and isolated volcanoes of Kamchatka, Japan and Taiwan the ecologically similar high alpine tundra-like habitat occurs *above* the treeline. Both are dominated by prolonged periods of freezing (with permafrost in the north), with only brief, intense summers. In the far south, the southern islands of Japan, Taiwan, and southeast China are subtropical, dominated by high average annual temperatures, high humidity and high rainfall. In between lie the vast taiga (boreal forests) and swamps spanning most of northern Russia south of the tundra. Large areas are essentially temperate with well-defined seasonality (much of northern Japan, for example), with northern regions dominated by mainly deciduous forest, south of which occurs more extensive broadleaf evergreen forest in the warm-temperate zone, whilst in certain regions there are restricted areas of grassland, steppe and semi-desert (the extreme west of northeast China, for example).

Northern parts of the region experience extremes of climate, with very prolonged winters, locally heavy snowfalls, frozen lakes and rivers, and extensive sea-ice (that in the Sea of Okhotsk reaching Hokkaido is the southernmost in the Northern Hemisphere). In contrast, southern regions are annually affected by a rapidly advancing warm spring wind, followed by a spring/early summer monsoon or rainy season, hot summers and a powerful late summer/autumn typhoon season.

The main terrestrial habitats span high-altitude peaks and towering volcanoes (the highest in the region being Klyuchevskaya Sopka in Kamchatka, at 4,750m) to low coastal tundras in the Arctic and to seashores lapped by warm tropical seas in the south and by chilly Arctic waters in the north. Freshwater habitats range from fresh to brackish lakes and coastal lagoons, montane streams, and some of the world's largest rivers. Lake Biwa in Japan is one of the world's oldest extant lakes, while the Chiangjiang River of China is one of the ten largest in the world. Offshore, in the south there are coral reefs and seagrass beds, whilst in the north there are kelp beds and sea-ice. Warm currents flow north along the southern coasts of the region, whilst a major cold current flows south out of the Bering Sea along the north-east coast of the region. These, other minor currents, and regions of mixing of cold and warm currents all influence the distribution of marine planktonic organisms and hence of the fish and birds that are part of the marine food web.

Migration

A major bird migration route, the East Asian flyway, extends along the continental coast of Asia linking regions as far apart as southeast Australia and New Zealand with Yakutia, Chukotka and Kamchatka. Many regular summer and winter visitors and passage migrants to the region move along this flyway or its branches – these follow the line of the Japanese archipelago branching in Kyushu up the Korean Peninsula or through the main Japanese islands, branching again in Hokkaido, with some species migrating by way of the Kuril Islands and others via Sakhalin into northeast Russia (see Brazil 1991). Where these flyways cross water, such as the Yellow Sea between Shandong, China and Korea, headlands and capes or offshore islands are regular places of landfall for tired migrants, and so make excellent places to search for common migrants and vagrants.

In addition to being on the main Asian flyway, northeast Russia lies at the end of an extended North American flyway, along which many summer visitor shorebirds, and even warblers, reach as far as northwest Alaska or cross the Bering Strait to reach Russia. In this way, vagrants such as Rufous Hummingbird, Northern Waterthrush and Wilson's Warbler have reached the region, and more, such as Yellow Warbler, Blackpoll Warbler and American Tree Sparrow, may be expected in the future. Northeast Russia is probably the least watched area within the region of this guide; observations from the Bering Sea islands and St Lawrence Island to the east indicate that it offers the greatest opportunity for finding species new to the region. Vagrants may also arrive in the region from the west, straying on migration or being blown in by storms. Furthermore, young birds of a number of species are known for their 'reverse migrations', in which they follow the opposite compass direction from the norm, some ending up in East Asia. Then there are several species from ranges so far west that even though they have been recorded from the region it is tempting to speculate that they may have been escapees – part of the region's enormous bird trade. The arrival of such resident species as the North African Moussier's Redstart on Japan's Hegura-jima is an extreme example; it and several other species presumed to have been escapes have been omitted.

Vagrancy

Whilst the annual movements of millions of birds through the region provides seasonal excitement for birders, the arrival of vagrants adds spice to the mix. The region is generally under-watched so I have opted to include all vagrants known to have occurred up to 2006 in the main part of the book, as our true understanding of their status is poorly known, and because they are far more likely to be recorded again if birders in the region know what to expect. Unfortunately, the huge trade in captured wild birds, of a wide range of species, in China and, to a lesser extent, in Korea and Japan, must bring out-of-range records, especially of passerines, into question; there is simply no way of knowing how many such passerine 'vagrants' are truly wild, or escaped cagebirds. Nevertheless, the birder's skill is identification, and as a means to that end most are included here.

HOW TO USE THIS BOOK

The family accounts

Families are introduced briefly (pp. 14–29), in the order that they appear in Dickinson (2003), providing general information that applies to each of, or several of, the species in the family. A representative species from each family is illustrated alongside the family account, making this section a quick reference to the plates. Becoming familiar with the larger divisions of birds, the orders and families, is a considerable aid to developing field identification skills.

The species accounts

The majority of the book comprises the species accounts with their associated distribution maps, and facing them the 236 plates (pp. 30–499). The final plate draws attention to the taxa presumed or known to be extinct from the region, and highlights the need for continued conservation of habitats and species in East Asia. The species accounts and layouts are standardised and, I hope, largely self-explanatory. However to further explain some of the detail I have outlined their structure below.

The introductory line includes the English name, the scientific name and an abbreviated list of the regions in which the species has been recorded (C = Chinese coastal provinces; T = Taiwan and associated islands; J = the Japanese archipelago; K = the Korean Peninsula and its associated islands; R = Russia east of 116°E, including its associated Arctic Ocean, Bering and Okhotsk Seas islands). The global status of species of concern commences each account, following BirdLife International's (2001) *Threatened Birds of Asia* (either as Vulnerable, Endangered or Critical). Information on the status of each species, e.g. resident, migrant or visitor, within the region is included as Appendix 1 (pp. 502–512).

The length (L) of each species is followed, in some cases, by wingspan (WS) and mass (WT). Size ranges are derived from the major available published sources, principally *Handbook of the Birds of the World*, except where these volumes have been superseded by newer works (for example Svensson *et al.* 1999, Sibley 2000, Ferguson-Lees & Christie 2005), separating those for males and females where relevant. Whilst the accuracy of such measurements may be questionable they are invaluable for comparative purposes in the field.

Each species account comprises several sections (where information is relevant or available). Status and Distribution (**SD**) describes the species' global and regional range, and provides information for each subspecies known from the region. If no subspecies exist, then the species is defined as 'Monotypic'. Indication is also given of where the various accidentals/vagrants have been recorded. Some terms, such as 'resident', though unavoidable for brevity, may be somewhat misleading, as they do not imply 'sedentary' – for while birds may be present year-round in a given area, they may be represented by different populations, even different subspecies, at different seasons. Habitat and Habits (**HH**) provides information on which typical habitats a species prefers when breeding, on migration or in winter, and any uniquely distinguishing habits or unusual behaviours are described. In the Identification (**ID**) section, a summary is given of current knowledge of field identification criteria, beginning with information relevant to all individuals, followed by that for males, females, recognisably different winter and summer, and age-related, plumages. Particular attention is paid to the field separation of subspecies – the first time this has been attempted in any guide to East Asia. The Bare Parts (**BP**) section includes information on bill shape, size and coloration, eye colour and any associated features, and the structure, coloration and any special features of the legs. Voice (**Vo**) includes summarised descriptions and transliterations of common calls and songs, as well as any physical sounds produced, such as wing-whirring or drumming, where relevant. Transcribing calls and songs is not easy, especially with information coming via several languages. Cultural and individual differences render the ways we hear and transcribe sounds unique; however, some attempt seems worthwhile. Though largely based on personal experience, with so many species to describe I cannot lay claim to personal experience of all. Call and song descriptions are based on simple English transliterations. I have made no attempt to define differences in stress etc. Transcriptions of vocalisations, however good they are, are only really easy to recognise *after* one has heard the bird, not before; nevertheless they serve as a guide for retrospective identification. For species that I am not familiar with in the field, I have relied on data from many published sources (e.g. Mild 1987, Svensson *et al.* 1999, Mackinnon & Phillipps 2000, Robson 2000, Sibley 2000, Rasmussen & Anderton 2005, and especially the superb later volumes of *HBW*, as well as the many recordings of Tsuruhiko Kabaya, Toshio Oda and Hideo Ueda, and, to a lesser extent, on other guides pertaining to the region). I have also benefited greatly from the input of various reviewers, in particular on the helpful field notes of Fergus Crystal and Yoshiki Watabe.

The plates

The layout of the 236 plates with illustrations of 985 species essentially follows the currently accepted taxonomic order, with minor adjustments. The unifying aim of each was to stand alongside the text in presenting the maximum amount of information in the space available. Just as the text has often had to rely on abbreviation to achieve this aim, so many images have been reproduced quite small. Some change of scale within plates is inevitable, when including flight images and behavioural vignettes, but these changes should be self-evident. The artists have endeavoured to illustrate each of the main variations in plumage relating to gender, age and subspecies. Where a species has many subspecies, those more widely differing have been selected for illustration. A range of poses has been chosen to indicate important behavioural characteristics and within groups/plates, species have been illustrated at a comparable scale. The great advantage of plates over photographs is that species may be illustrated in poses that more clearly reveal plumage characters; thus although many species fold their wings such that the lower back, rump, upper- or undertail-coverts are concealed, briefly held poses revealing these characters have often been chosen to draw attention to diagnostic features.

Distribution and distribution maps

Within the text, for brevity's sake, short forms are commonly used for geographic regions. Thus for example, Russian Far East is used to refer to continental south-east Russia (Vladivostok north), to distinguish it from the contiguous regions of Yakutia, Chukotka and Kamchatka lying further north and east (see Figure 1).

Many vagrants have been recorded in Japan (which has the highest number of birders in the region), where they have been found only, or mostly, on islands in the Sea of Japan, such as Tobishima, Hegura-jima, Tsushima and so on. Similarly, in Korea many vagrants have occurred only on islands in the West or Yellow Sea. These are, respectively, abbreviated to 'offshore Japan' and 'offshore Korea', to distinguish them from the Japanese archipelago and Korean Peninsula, which refer to the larger islands or main landmasses of those countries.

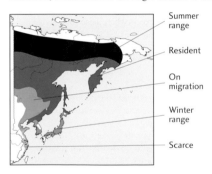

Summer range

Resident

On migration

Winter range

Scarce

A distribution map is provided for each species regularly occurring in the region. The shape of the region does not lend itself easily to mapping, so maps of varying scales have been used; a small-scale map for species ranging throughout the region, a slightly larger-scale map for species with ranges centring on Japan and the Korean Peninsula, and a further larger-scale map for species whose ranges centre on eastern China or Taiwan. Necessities of space and scale have rendered these maps small; however, they should provide a visual overview of where to look for each of these species. In some cases, arrows have been used to highlight presence on the smaller islands of the region. Because of the scale of the maps, no attempt has been made to map the occurrence of accidentals or rare migrants to the region.

AVIAN TOPOGRAPHY AND TERMINOLOGY

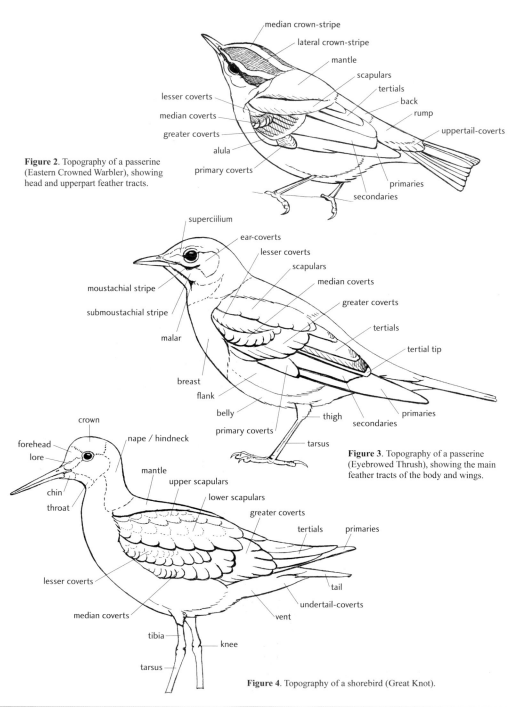

median crown-stripe
lateral crown-stripe
mantle
scapulars
tertials
back
rump
uppertail-coverts
lesser coverts
median coverts
greater coverts
alula
primary coverts
primaries
secondaries

Figure 2. Topography of a passerine (Eastern Crowned Warbler), showing head and upperpart feather tracts.

superciilium
ear-coverts
lesser coverts
scapulars
median coverts
greater coverts
tertials
tertial tip
moustachial stripe
submoustachial stripe
malar
breast
flank
belly
primary coverts
thigh
secondaries
primaries
tarsus

Figure 3. Topography of a passerine (Eyebrowed Thrush), showing the main feather tracts of the body and wings.

crown
forehead
lore
nape / hindneck
mantle
upper scapulars
lower scapulars
greater coverts
tertials
primaries
chin
throat
lesser coverts
median coverts
tibia
knee
tarsus
tail
undertail-coverts
vent

Figure 4. Topography of a shorebird (Great Knot).

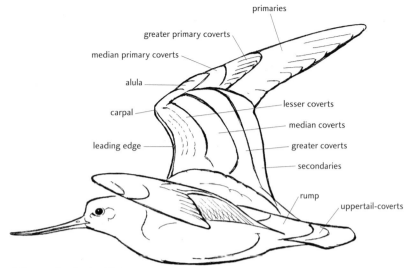

Figure 5. Topography of a shorebird in flight (Terek Sandpiper).

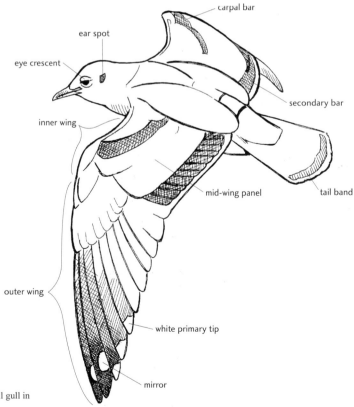

Figure 6. Topography of a small gull in flight (Saunders's Gull).

KEY TO FAMILIES

The numbers after the family headings indicate the number of species occurring in the region (based on this book) and the number of species in the family worldwide (following Dickinson 2003).

NON-PASSERINES   Plates 1–130

Grouse, pheasants, partridges and quails (Phasianidae) 23/180   1–6

A large, worldwide family well represented in the region. Most are essentially active on the ground, typically (though not always) ground feeding, often involving surface scratching, and always ground nesting, but some rest or roost in trees. They have small heads, with short, stout decurved bills, short rounded wings, though often longer tails. Among some species there is extreme sexual dimorphism: males are highly colourful, often with spectacular tails, whereas females and young are commonly cryptically plumaged; young precocial. Some species lek. Male displays commonly involve wing-whirring or beating, and calling. They run well and typically choose only to fly short distances when disturbed; flight is typically rapid and low, often audible, on stiff, arched, fast-whirring wings. Often highly vocal, especially at dawn and dusk.

Swans, geese and ducks (Anatidae) 64/158   7–21

A worldwide family closely associated with wetlands. They range in size from small to large, but all share webbed feet and a specialised short, broad, rather flattened bill covered with soft skin, often with lamellae along the edges and a nail at the tip. They have short legs, many have long necks, and powerful wings – many are long-distance migrants – and they fly with continuous fast wing beats, often audible. All of the long-necked geese and swans fly with their necks extended, and do not glide or soar. Larger waterfowl (swans, geese and shelducks) exhibit little sexual dimorphism. Swans and geese are essentially monogamous, and social, following traditional migration routes in family groups and large flocks in wavering lines and Vs; geese in particular may gather spectacularly at massed roosts. Geese typically walk and graze and have strong bills often with serrated edges; they have strong legs, walk strongly and can take off easily from land. Beware size variation within species, especially in varying light conditions; males are larger than females and adults bigger than juveniles, in addition subspecies vary in size. Ageing of individuals helps with species identification; adult geese typically have strong, deep dark 'grooves' down neck, and neat pale bars on upperparts formed by pale squarer edges to mantle and wing feathers. In contrast, young birds have indistinct neck 'grooves' and less distinct, pale rounder edges forming indistinct bars on upperparts. Smaller waterfowl (dabbling *Anas* and diving *Aythya*) are typically strongly dimorphic, and the bright plumaged males usually assume a post-breeding eclipse plumage, often retained into early winter. Juvenile birds and eclipse males more closely resemble the cryptic females. Many species perform pair formation displays in late winter while still on wintering grounds. Out-of-range individuals are often attracted to large flocks of locally common species.

Dabbling ducks 28/42   12–15

Small to medium-sized 'typical' ducks of a large range of freshwater habitats, but often also wintering on brackish or salt water. Surface feeders in shallow water that may upend, but rarely if ever dive. Some species also graze on land. Fairly agile and able to spring into the air for take off; flight is generally fast, during which they reveal a brightly-coloured speculum (which often has a metallic sheen). In most species sexual dimorphism is extreme, although after spring court-ship males quickly adopt a cryptic 'eclipse' plumage in which they resemble females, moulting back into full plumage in winter. During summer adults moult flight feathers simultaneously as in swans and geese, thus becoming flightless for several weeks. Unusually plumaged individuals resembling rarities should be scrutinised carefully. Dabbling ducks, and to a lesser extent diving ducks, are known to hybridise relatively frequently, complicating field identification (the subject of hybrid identification is beyond the scope of this guide – consult specialist literature). Also, beware of birds showing variably orange or rusty-coloured underparts, heads and necks – the result of iron oxide staining.

Diving ducks 19/63 16–20

'Diving duck' is a useful but rather loosely defined term, applied to a range of ducks that habitually dive deeply for food (animal or vegetable) rather than dabble at the surface. Found on large, deep fresh water bodies, in bays, and at sea, they are generally heavy-bodied with short wings, often giving audible wing beats and requiring a take-off run. Strongly sexually dimorphic, with male plumage commonly involving large blocks of colour; generally lack a speculum, and *Aythya* in particular have broad white wingbars. In eclipse plumage males are less bright than usual, but do not closely resemble females. In winter, different species are often found together, and may also forage around larger waterfowl such as swans, benefiting from their disturbance of bottom food. *Aythya* species are particularly prone to hybridising and these hybrids may resemble rarities.

Sawbills 5/6 21

Specialised piscivorous diving ducks ranging in size from small (Smew) to large (Goosander) with narrow, hook-tipped beaks with tooth-like lamellae along their cutting edges, enabling them to grasp their fish prey. Three larger species are long-bodied with rather long necks and bills; the smallest (Smew), is compact with a short bill. Shy during breeding season, social during winter, they may form large flocks of up to several thousand birds, sometimes mixing together. Goosander in particular (others occasionally) may engage in group-fishing, with a flock of birds cooperating to drive fish shoals. Float low in the water, recalling an elongated grebe or diver; frequently submerge head to look down into shallow water while foraging. Nest in natural tree holes and cavities, occasionally on the ground.

Divers (Gaviidae) 5/5 22

A small monogeneric family, all five species of which have been recorded in the region, four of which breed. Divers are foot-propelled deep-diving pursuit predators of fish; they have long bodies and long necks, their feet are placed well back, and they appear tailless. Larger and thicker-necked than grebes, they float low in the water like cormorants. Feeding dives, often following snorkelling, are smooth. Flight is rapid, with the wing-beats shallow and the head carried low. Widely dispersed during the breeding season on tundra or taiga pools and lakes, but on migration and during winter (on the sea) may be seen in loose gatherings; only rarely occurs inland in winter. On migration (usually over the sea) may occur in very loose, widely-spaced groups. Vocal during breeding season, but silent in winter. Summer plumages strikingly beautiful, with velvet appearance. Winter identification should focus on size, light and dark areas, bill colour and shape; but all species may carry bill at a raised angle.

Albatrosses (Diomedeidae) 4/13 23

Three species breed in the region, and a fourth has occurred accidentally. The taxonomy of this group is the subject of dispute. Highly specialised, large, long-lived, long-winged, pelagic seabirds, which at times fly with stiff wing beats, but more typically 'sail' effortlessly on strong winds. They use dynamic soaring to travel at high speed in arcing flight, dipping close to ocean surface then rising high. They are web-footed, and have peculiar tubular nostrils (divided, unlike other tubenoses) atop the bill. They forage from surface waters for squid/fish/carrion and are susceptible to hooking on long-lines. They visit land (isolated oceanic islands where they nest in colonies) only to breed during winter; the remainder of the year is spent at sea. The three breeding species, which range widely around the north Pacific Ocean and into the Bering Sea, are best seen from ships, but may be seen close to northern and eastern headlands, especially after stormy weather and typhoons.

Petrels and shearwaters (Procellariidae) 22/74 24–28

The northwest Pacific is particularly rich in seabirds; many breed and many more visit from Southern Hemisphere breeding grounds. Twenty-two species have been recorded so far, mostly pelagic, but with some visiting offshore islands and coasts to breed, and others appearing inshore only after storms and typhoons. Of particular note are the enormous numbers of Short-tailed and Sooty Shearwaters that are regularly seen off Japanese coasts, in the Sea of Okhotsk and off the Kuril Islands and Kamchatkan coast. These highly specialised, pelagic birds vary between the long-winged shearwaters and larger petrels, which fly like small albatrosses, sailing on the wind in high arcing flight or gliding close to the water's surface appearing to cut it with their reflections (hence 'shear'-water), and the smaller storm-

petrels, which are more active fliers, jinking, turning and fluttering over the sea surface (though they can shear too). Like the albatrosses, they are all web-footed, and have peculiar tubular nostrils atop the bill, from which concentrated salt solution seeps extracted by the salt gland. They forage from surface waters for small animals and carrion, usually while swimming buoyantly. They visit land (generally isolated oceanic islands, but also rocky offshore islands and capes, where they nest in colonies) only to breed; the remainder of the year is spent at sea. The *Pterodroma* petrels, all of which have short, stubby, black bills, are commonly referred to as 'gadfly' petrels, because of their fluttering flight style. Though best seen from ships, the commoner shearwaters and petrels may also be seen during coastal sea-watches and occasionally even appear inland on lakes after storms, typhoons, or in fog. They are typically noisily vocal at their nesting colonies, but mostly silent at other times, although on take-off from calm seas flocks of shearwaters may make an audible slapping sound as webbed feet and wing tips strike the water.

Storm-petrels (Hydrobatidae) 7/21   29–30

Storm-petrels are small or very small highly pelagic seabirds, rarely encountered close to land except during the breeding season. They visit their nesting colonies at capes, headlands and offshore islands at night to enter their nesting burrows, where they are very vocal. They are readily attracted to lights of ships at night and occasionally to lighthouses. During the day at sea they take tiny planktonic prey from water's surface while pattering and picking during butterfly-like flight. They swim buoyantly and may float in rafts. North Pacific species appear rather delicate but have rather large, chunky heads, with fine, hooked bills, short necks and bodies, long wings and short legs. All have short black bills with prominent nostril bulge (nostril tubes joined), most have black legs; all except Wilson's have black webs. Though best seen from ships, during fog, rain and after typhoons they occasionally (especially Leach's and Fork-tailed) appear in harbours or even at lakes inland. In particular, note presence or absence and shape of rump-patch, tail shape, foot extension and flight.

Grebes (Podicipedidae) 5/22   31

Five species of grebes breed in the region, several of which are migrants. Specialised, small to medium-sized aquatic birds (freshwater and inshore), they are streamlined with their feet placed well back on the body, and are very effective foot-propelled diving birds. They are smaller, thinner-necked and thinner-billed than divers, have rather small wings, and are essentially tailless. Unlike most aquatic birds they lack webbed feet, having lobed toes instead. They occur at well-vegetated inland waters during summer in the boreal and northern temperate areas of east Asia, and along adjacent or more southerly coasts and in shallow bays (also at inland lakes and reservoirs), during migration and winter. Nests are of anchored floating vegetation. Plumage is monomorphic, but males are larger than females. Young typically have striped heads and/or necks and may ride on the back of parents. They feed, by diving, on small fish and aquatic invertebrates. In breeding plumage they are attractive, typically with bright head plumes. In winter plumage they are more monochromatic, when identification should focus on size, head shape, and areas of contrast on face and neck, and of white on wings.

Storks (Ciconiidae) 4/19   32

A worldwide family of large wetland birds with long legs, long necks and long, powerful beaks. They stalk through wetland margins and wet grasslands in search of frogs and fish which they snatch or spear. Strong fliers on broad wings, they are long distance migrants, travelling by soaring and gliding using thermals. Wings are long and broad, the tail is short; neck and legs are invariably extended in flight. They breed in colonies in large trees, accumulating large stick nests. Although generally silent, pairs engage in loud bill-clattering displays at nest.

Ibises and spoonbills (Threskiornithidae) 6/32   33–34

Medium to large wetland birds with long legs, long necks, and long beaks adapted for probing into or sieving water and soft mud. They are smaller than storks but occur in similar habitats. Mostly social and colonial, they are strong fliers, but typically glide for brief periods between bouts of slow flaps. Flocks typically form ragged or messy lines in flight. Wings are long and broad, tail short; neck and legs invariably extended in flight. Ibises have long decurved bills, resembling curlews, used for probing. Spoonbills have long, broad bills with very broad spatulate tips; they have a characteristic feeding motion, during which they swing their head and sweep their bill from side to side while walking through shallow water.

Herons, bitterns and egrets (Ardeidae) 23/65   34–39

Small to large wading birds with long legs, long kinked necks (retracted in flight to form a bulge) and long, straight bills. They forage at inland and coastal wetlands, generally pacing slowly through shallow water, or waiting motionless, and spearing or snatching prey. Males are larger than females, but most species are monomorphic, some developing plumes or showing colour changes in soft parts for breeding season. They breed in colonies, mostly in large nests on trees, but some nest solitarily on ground. Most species are diurnal, but some are at least crepuscular if not nocturnal. Wings broad, rather rounded, tails short (and covered by tertials at rest). Flight is on arched wings with slow wing beats. All except Black Bittern and Black-crowned Night Heron have yellow iris.

Tropicbirds (Phaethontidae) 2/3   40

Tropicbirds occur in warm tropical and subtropical waters, breeding on remote islands, typically nesting on the ground in crevices, cavities and under overhangs among rocks or cliffs. Within the region they are confined to extreme southern areas except after typhoons. They are mainly white, and fly high and strongly, with a mechanical rowing action like a large tern but less buoyant and with faster, stiffer wing beats; sometimes glide, may hover before plunging to take food from the water's surface. Often rest on the water. Bill and tail colour, and distribution of black feathering, distinguish the species. Each has a short wedge-shaped tail, but with two greatly elongated central tail streamers.

Frigatebirds (Fregatidae) 3/5   40

A small family of large, extremely aerial seabirds with long, narrow rakish wings, short neck, prominent bill, long deeply forked tail (though often held closed), and small webbed feet. Strongly sexually dimorphic; the females are larger, males have a distensible, red gular pouch. Wing beats slow interspersed with prolonged glides, carpals held forwards, wing tips back. Occasionally soar with a cross-like silhouette. Piratical, pursuing other seabirds to force them to disgorge food, also snatch food from the surface with strongly hooked beak. Frigatebirds do not land on water. Breeding areas are on remote tropical oceanic islands south of the region, but they appear annually after typhoons, along coasts, at river mouths and even inland, occasionally occuring as far north as Hokkaido and the Russian Far East. Gradual transition from juvenile to adult plumage makes them confusingly difficult to identify to species when sub-adult; careful observation of detailed patterns on head, neck and belly is essential.

Pelicans (Pelecanidae) 3/7   41

A small family of large water birds. Unmistakeably long bill with prominent elastic gular pouch extending to bill tip, with which they scoop up fish. Pelicans have a long thin neck, large heavy body, short stout legs and webbed feet. They swim buoyantly and fly gracefully on long broad wings, often gliding and soaring. Will forage socially, with groups or lines of swimming birds herding shoals of fish. Sexes are alike in plumage. Accidental to the region from southern breeding grounds, but separation from escapees difficult. May occur at coasts, major lakes and rivers, where birds fish from a swimming position.

Gannets and boobies (Sulidae) 3/10   42

A small family of large coastal and oceanic seabirds, with a distinctive streamlined outline; the head, neck, wings and tail are all long and pointed. The prominently pointed heavy-based bill has serrated cutting edges and is unusual among seabirds in lacking obvious nostrils. These birds generally fly well above the sea surface and plunge-dive spectacularly, frequently in large groups, to catch fish beneath the surface. At the breeding grounds, Brown and Red-footed usually perch on vegetation, whereas Masked perches on bare ground or rocks.

Cormorants (Phalacrocoracidae) 5/36   43

A worldwide family of medium to large, mostly dark-plumaged aquatic birds with long, loose-skinned necks and long bodies. Some Asian species are exclusively coastal, one (Great) typically occurs inland. Conspicuous colonial nesters on cliffs, crags, or trees depending on species. Long body, and long neck lends superficial similarity to divers, but tail (not feet) long and prominent. Neck extended in flight. Bill generally long, strong and prominently hook-tipped, often with bare skin around base. Feet are totipalmate (the hind-toe is included in the webbing). They swim low in the water like divers, with the head and bill uptilted, and fish from the surface of the sea, lakes or rivers, by duck-diving, but this involves a slight upwards jump first, easily distinguishing them at distance from divers or large ducks; they do not normally snorkel before diving as divers do. After fishing may stand with wings spread to dry. Take-off requires a pattering run from water surface, but once airborne flight is strong and direct. Often forms lines or skeins in flight and may soar.

Falcons (Falconidae) 9/64   44–46

Small to medium-sized diurnal raptors, which are extreme aerialists. They are fast, accomplished flyers, with rakish, pointed wings and long tails; generally hunt aerial prey in open country ranging in size from large insects to waterfowl. Some perform spectacular dives or 'stoops' onto prey. Plumages usually similar, but females are typically larger than males and may take different-sized prey. Usually aggressively territorial and though usually silent can be vociferous near nest. Falconets are the smallest; these somewhat shrike-like raptors perch high and sally out after insect prey. The extent of falconry in the region is not clear, however, faced with a rare, out-of-range falcon look for evidence such as jesses and abnormal moult.

Buzzards, eagles, vultures and allies (Accipitridae) 41/233   46–56

A broad group containing Osprey, Old World vultures, hawks, kites, buzzards and eagles, most of which are broad-winged and capable of gliding and soaring. Many of them make short to long-distance migrations, and large numbers can be seen passing along traditional routes (flyways) at headlands, capes, and mountain passes. They are well-adapted for a hunting lifestyle, with often strongly hooked beaks for tearing prey, powerful eyesight, and strong claws and talons. Their prey consists largely of living animals, although some are scavengers and carrion feeders. The accipiters range from the highly specialised fish-hunting Osprey, to slow, low flying harriers and powerful eagles. The hawks are typically birds of woodland and forest with short, broad wings and long, rather broad tails; females are larger than males;. These birds are frequently seen soaring over their territory or in dashing pursuit of prey. The buzzards are larger than hawks with long, round-tipped wings and broad tails. Eagles are larger still with more massive heads, longer, broader wings, broad tails, powerful tarsi and toes; they may be encountered soaring in search of prey or in pursuit flights. Females are larger than males, but plumages are essentially similar; in the sea eagles large areas of white relieve the generally dark plumage. The Osprey is the most specialised, with dense oily plumage, a reversible outer toe and spiky scales on underside of toes to facilitate holding slippery fish; soaring flight gull-like, migratory. Kites and harriers are both slim, light, long-winged and long-tailed, the former usually with forked or notched tail; fly on slow, shallow wing beats or soar or glide; harriers have owl-like facial disks and strongly sexually dimorphic plumage that matures over several years. Vultures are large, long-winged raptors adapted to soaring in search of carrion prey. The larger raptors take several years to mature, passing through several identifiable age-related plumages. Identification may often depend on structural features and terms such as "arm" for the inner portion of the wing, "hand" for the outer portion beyond the outer wing joint, and "fingers" for the distinctly spread primaries, are used to describe important aspects of structural proportions or feather groups.

Bustards (Otidae) 2/26   57

A small family of terrestrial birds adapted to dry grassland, semi-desert and desert habitats. Rather large with long legs, heavy bodies and long necks held extended during flight. Wings are broad and long. They walk erect and slowly through open habitats, searching for large invertebrates and vegetable matter. They are shy and rely on cryptic plumage to avoid detection, but may flush at some distance, flying with slow deep wingbeats.

Rails and crakes (Rallidae) 15/141   57–60

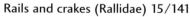

A large, cosmopolitan family of typically secretive species adapted to life amongst dense wetland vegetation. They appear plump, short-winged and long-legged, and are generally reluctant fliers. Often highly vocal (yet difficult to observe) with sharp, high-pitched and trilling calls, sometimes given at night. Typically elusive, foraging while pushing through dense vegetation, but at dawn and dusk may forage in open around wetland or field margins. Many species bob the head back and forth and flick tail while walking. When flushed commonly fly short distances, with legs dangling, dropping quickly back into vegetation, but some species surprisingly strong fliers and make long-distance migrations.

Cranes (Gruidae) 7/15   61–62

A small family of highly distinctive birds, all of which are large and long-legged, with long necks and rather long bills. In flight, both legs and neck are extended; flocks form lines and V formations. They breed in remote marshes but gather in winter, often in large flocks generally in open country, when they can be highly vocal. Commonly feed on small fish, frogs and other aquatic life during summer, but some species specialise on plant roots and bulbs during winter. Courting pairs duet and perform elaborate dances in late winter and spring before nesting in remote wetlands, where they build large mound nests.

Buttonquails (Turnicidae) 3/16   63

An odd group of small terrestrial birds, related to the cranes but showing reversed sexual dimorphism. Females are not only larger than males, but also more colourful; they call (usually low rhythmic hoots or cooing sounds) to attract males, and commonly mate with several males leaving them to incubate the eggs and care for the young, which are highly precocial. Walk and crouch secretively, and flush suddenly at point-blank range. Strangely lack a hind-toe, like some shorebirds.

Oystercatchers (Haematopodidae) 2/11   63

Large, stocky shorebirds with long thick bills and shortish, stout legs. Large-headed, broad-winged and short-tailed they have a distinctive silhouette on land or in the air. They feed on combination of shellfish and mud worms, developing specialised feeding habits and either slender or chisel-tipped bills depending on dietary adaptation. Sociable, usually in pairs but rarely in flocks, and often highly vocal; on breeding grounds give loud piping calls in slow display flight or while walking with bill pointed downwards.

Ibisbill (Ibidorhynchidae) 1/1   64

A monotypic family closely related to stilts and avocets, with which it is sometimes included. Long-billed and short-legged, the Ibisbill lives along gravel-bedded rivers.

Stilts and avocets (Recurvirostridae) 2/7   64

A small group of wading birds characterized by extremely long, slender legs and long, fine bills; in the avocets this is remarkably upturned. Stilts and avocets are typically gregarious, vociferous and aggressive, often driving other species away from their breeding territories. Graceful and elegant when feeding, either delicately picking from water or mud (stilts) or swinging neck, head and bill from side-to-side through the water's surface (avocets).

Plovers (Charadriidae) 16/66   64–68

A large and widespread family of birds adapted to the margins of wetlands and drier habitats from tundra to semi-deserts. Most are long distance migrants. The plovers are short-billed, with large eyes and strong legs lacking a hind toe. The larger plovers or lapwings may be found away from water in open or grassland habitats; they have broader, more rounded wings, longer necks and larger heads typically with crests, or facial wattles, and may have wing spurs. The smaller *Charadrius* plovers have distinctive head patterns and breast-bands, and bob nervously when disturbed, usually running some distance before taking flight.

Painted-snipe (Rostratulidae) 1/2   69

Only two species make up this unusual family of snipe-like shorebirds. Cryptic and inconspicuous amongst wet grasslands and wetlands, they have long bills, large eyes (and are largely crepuscular/nocturnal), broad, rounded wings and short tails. Flight is low, slow and rail-like. In common with certain shorebirds, jacanas and buttonquails, but with few other birds, the sex roles are reversed; females are brighter than males, and call and defend territories to attract males (sometimes several) to mate. Females lay the clutch but they leave the male to incubate and raise the young.

Jacanas (Jacanidae) 1/8 68

A small family of water birds resembling rails but with longer legs. They range throughout tropical regions, and all share the uniquely elongated toes that allow them to walk across floating vegetation and earn them their alternative name of 'lily-trotters'. Generally weak fliers, but vagrants have occurred north of breeding range. Their breeding system (polyandry) is rare among birds, with individual females (larger than males) courting several mates simultaneously. Often gregarious in winter.

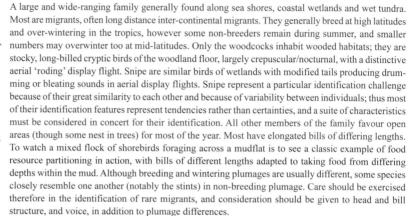

Sandpipers, woodcock and snipe (Scolopacidae) 58/92 69–84

A large and wide-ranging family generally found along sea shores, coastal wetlands and wet tundra. Most are migrants, often long distance inter-continental migrants. They generally breed at high latitudes and over-wintering in the tropics, however some non-breeders remain during summer, and smaller numbers may overwinter too at mid-latitudes. Only the woodcocks inhabit wooded habitats; they are stocky, long-billed cryptic birds of the woodland floor, largely crepuscular/nocturnal, with a distinctive aerial 'roding' display flight. Snipe are similar birds of wetlands with modified tails producing drumming or bleating sounds in aerial display flights. Snipe represent a particular identification challenge because of their great similarity to each other and because of variability between individuals; thus most of their identification features represent tendencies rather than certainties, and a suite of characteristics must be considered in concert for their identification. All other members of the family favour open areas (though some nest in trees) for most of the year. Most have elongated bills of differing lengths. To watch a mixed flock of shorebirds foraging across a mudflat is to see a classic example of food resource partitioning in action, with bills of different lengths adapted to taking food from differing depths within the mud. Although breeding and wintering plumages are usually different, some species closely resemble one another (notably the stints) in non-breeding plumage. Care should be exercised therefore in the identification of rare migrants, and consideration should be given to head and bill structure, and voice, in addition to plumage differences.

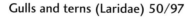

Phalaropes 3/3 85

Three species of slenderly proportioned, long-necked and fine-billed 'shorebirds' that exhibit reversed sexual dimorphism and form a distinct subfamily within the Scolopacidae. Females are larger and brighter than males and polyandrous, laying several clutches for several males, each of which incubates and rears a family. The two species breeding in the region breed on the tundra and winter at sea. Usually highly confiding; they float buoyantly on water, delicately picking insects from the surface, but walk awkwardly on land. Red-necked Phalarope in particular can be extremely common on migration, occurring in large flocks passing along coasts or in rafts at sea.

Coursers and pratincoles (Glareolidae) 1/18 85

Pratincoles are long-winged, short-legged aerialist 'shorebirds' that are somewhat tern-like or swallow-like in habits and appearance. Agile on the wing, they chase insect food in flight. Sociable and colonial; on the ground they appear large-headed, and have short arched bills. Only one representative of this group occurs in the region.

Gulls and terns (Laridae) 50/97 86–100

A diverse and numerous group of long-winged, short-tailed water birds. Large-headed, short-necked with short bills, their plumage is largely white (though tropical terns include all-dark species) with grey upperwing surfaces usually with black wing tips; juvenile plumage is typically brown. Smaller species take 1–2 years to reach adult plumage, with larger species taking 3–4 years, making identification difficult. Among larger gulls adult females are often noticeably smaller and smaller-billed than males, further complicating the identification of a range of species for which taxonomy is still in debate. Larids forage by picking, plunge-diving or scavenging on wide range of prey. They range in size from small (Little Tern) to large (Slaty-backed Gull). Some are active fishers, such as the terns, others are omnivores or scavengers such as the gulls, while active predators include birds such as Glaucous Gull. Bills range from stout and powerful among the gulls (deepest near the tip often with a pronounced gonydeal angle) to fine and delicate in terns. All have webbing between the three forward-facing toes. They breed widely across the northern tundra, at isolated islands or on rocky shoals and sandy beaches, migrating down the east Asian coast. Some of them winter well south of the region. They may be very vocal during the breeding season; their calls tend to be harsh. There is little or no sexual dimorphism, except in size, and they are typically monogamous. Identification of the complex of species previously treated as Herring or Lesser Black-backed Gull is challenging and further changes to the taxonomy are anticipated. Considerable overlap in size, colour-shade and jizz render field identification fraught with difficulty and some individuals, ages, or plumages defy accurate designation.

Skuas (Stercorariidae) 4/7 101–102

These medium to large seabirds are gull-like in structure, with short, strong bills (with hooked tips), long wings and short tails (with slightly to very, elongated central feathers), but recent evidence indicates they are more closely related to the auks than to the Laridae. Largely dark brown, but with prominent white 'flashes' in wings, they are powerful fliers and aggressive pursuit 'scavengers' (kleptoparasites, commonly chasing birds such as kittiwakes or terns and forcing them to disgorge their hard-won food) and sometimes active predators, such as the rodent-hunting Long-tailed Skua. They are long distance migrants mostly breeding on northern tundra, migrating past northern coasts and wintering out at sea, with the exception of South Polar Skua which is a Southern Hemisphere breeder that migrates north into the Pacific. Skua identification in the field is not easy, and requires careful consideration of species, age and morph. They typically appear darker than young gulls, though young Black-tailed Gull is potentially confusing at long range. Pay attention to a combination of factors such as bill, wing, tail proportions and flight style.

Auks (Alcidae) 18/23 103–107

The puffins, murres, murrelets, guillemots, auks and auklets make up a specialised small group of diving seabirds. They use their wings for underwater propulsion in pursuit dives after prey ranging from macroplankton to small fish. Their adaptations for underwater 'flight' make them poorly manoeuvrable in the air; their wings are short, and they fly with fast whirring wing-beats. They visit land only to nest. Some nest colonially, others more solitarily, in a range of locations from rock tumbles at the base of cliffs to talus slopes, cliff ledges, rock cracks, and burrows in deep cliff-top turf.

Sandgrouse (Pteroclididae) 1/16 107

An unusual group of seed-eating pigeon-like birds specialised for life in dry areas, with short bills, short feathered legs and long, pointed wings. Ground-nesters, these cryptic birds are typically encountered when flushed (with audible wing clattering), when seen flying high over breeding grounds calling noisily in small parties, or gathering at watering sites. Unpredictable in their irruptive behaviour, only one species has occurred so far in the region.

Pigeons and doves (Columbidae) 17/308 108–111

A diverse group of mostly forest or woodland species, ranging from large (most pigeons) to small (most doves), sharing the appearance of a small head which is nodded when walking, and the habit of foraging on fruits or seeds. All have delicate bills, often with fleshy ceres and bright, bare eyerings. They can drink directly from water without tipping their heads back to swallow like most other birds. Strong, direct flight can be noisy (clattering at take off) and larger species may appear crow- or hawk-like; often have aerial display flights involving wing clapping and undulating flight. Typically very vocal producing a range of soft, repetitive cooing or lowing sounds. Outside the breeding season they may be social to highly gregarious, gathering at food sources, freshwater or even salt water (Japanese Green Pigeon). Mostly monomorphic, tree nesting, and laying only one or two eggs. The soft-billed young (squabs) are fed on a milk-like secretion from the crop of the parents. Some species are widely domesticated and kept as cage birds.

Cockatoos (Cacatuidae) and parrots (Psittacidae) 8/364 112

A large and diverse family characterized by large heads, short bills, and short, thick tongues. They feed on a wide range of seeds, fruits, buds and even nectar. Their feet are zygodactylic (the fourth toe is reversible) giving them considerable agility among tree branches, including the ability to hang upside-down, and they use the bill (abruptly curved to sharp tip) as a third 'foot' to aid in reaching for and clambering after fruit. Typically sociable, often highly gregarious, gathering at food sources, water, or roost sites; very vocal, but calls are harsh, even raucous, or chattering. Cavity nesters. Common in trade, a wide range of species has been released or has escaped in the region, a number of which have established either self-sustaining breeding populations locally, or which are regularly supplemented by new releases or escapees.

Cuckoos (Cuculidae) 12/138   113–115

A group of medium-sized rather raptor-like birds with short, rather pointed wings (*Cuculus*) or short, rounded wings (*Hierococcyx*), short legs, long tails, largish heads and small bills. The hawk cuckoos even have a raptor-like cere. All species are insectivorous and highly vocal and either secretive forest species or conspicuous grassland/open woodland species. Those in the region are nest parasites (except for the coucals), using a wide range of small woodland and grassland bird hosts to raise their young for them. Identity of host is helpful in identifying cuckoo nestlings. Mostly monomorphic, but brown (hepatic) phases of several species exist.

Barn owls (Tytonidae) 1/15   115

A small group of medium to large nocturnal predators resembling true owls, with large heads, dark eyes, a laterally flattened bill and a distinctly heart-shaped facial disk with stiff-feathering around the margin. Hearing is acute, and small terrestrial mammal prey is located by sound. Most members of the family are cavity nesters, though some (including the grass owls) nest on the ground.

True owls (Strigidae) 21/180   116–120

A widespread group of small to very large predators, most of which are crepuscular or nocturnal. They range from the subtropical forests of the far south to the northern Russian tundra. Owls share certain features in common; a relatively large flat-faced head with forward-pointing eyes, almost silent flight, very broad wings and a short tail, a facial disk, and keen hearing. They tend to be very vocal. Several species have a 'false face' pattern on nape. Most are cavity nesters, though some typically and others occasionally nest on the ground. The largest owls (the eagle and fish owls) are massive and powerful with large ear tufts; the fish owls have yellow eyes, reduced facial disks, bare tarsi or toes (though not in Blakiston's) and spiny under surfaces to feet to aid in catching fish. Highly vocal with far-carrying calls. The wood owls (*Strix*) typically take mammalian prey, while smaller owls (*Otus, Ninox*) may take small birds, reptiles or large insects. The taxonomy of the smaller owls is complicated and subject to change.

Nightjars (Caprimulgidae) 2/89   121

Nocturnal or crepuscular insectivores. Often roost on ground or along branch; nest is on the ground. Large-headed, large, dark-eyed (reflective under direct light), short-billed, but with long rictal bristles forming net of whiskers facilitating catching of flying insects. Long-winged, and with long tails, they fly silently, erratically and rather slowly and are seen most commonly at or soon after dusk. Males typically have white patches near wing tips and tail corners, reduced or lacking in females and young; helpful in identification. Tarsi and toes are greatly reduced; instead of walking they hop with raised wings; elongated and pectinate middle toe is used for preening. Calls are typically monotonous and often repeated churring or deep knocking sounds.

Swifts (Apodidae) 6/94   121–122

A group of extreme aerialist insectivores, with long, pointed, arced wings, and either short or forked tails; small-bodied with extremely short legs and small toes. They have short bills, but with extremely wide gapes. Flight is fast and stiff-winged and often very high, though they descend to lower elevations to avoid rain storms. They feed, drink and copulate in flight, landing only during the breeding season at the nest site; take off from a horizontal surface is impossible. Nesting is often colonial in buildings, tree cavities, caves or cliff crevices.

Hummingbirds (Trochilidae) 1/331   122

A group of small to medium-sized, long-billed and short-tailed extreme aerialists foraging largely for nectar and to a lesser extent for insects. Unusual wing development with short 'arm' and highly-developed 'hand' with especially long primaries; capable of hovering flight, while foraging. Tarsi and toes are extremely short. Range is entirely within the Americas, with the exception of vagrancy of one species to extreme northeast Asia.

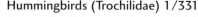

Trogons (Trogonidae) 1/39   122

A specialised, small group of tropical forest insectivores and frugivores. Large-headed, long-tailed, and extremely short-legged, they sit very upright in shady areas beneath the canopy. Despite their bright colours they are inconspicuous and are usually located only when flushed or when giving their deep resonant songs, which consist of a monotonous series of identical notes.

Rollers (Coraciidae) 1/12   123

A small group of colourful, large-headed, short-billed birds with long wings and tails. They typically perch conspicuously in open or semi-open country and drop onto invertebrate or reptilian prey, or sally for large aerial insects. Territorial, so usually encountered singly or in pairs, except on migration when Dollarbird, for example, may be encountered in flocks.

Kingfishers (Alcedinidae) 8/91   123–124

Kingfishers are typically colourful, with disproportionaly large heads and beaks for their body size; they have extremely short tarsi and partly fused front toes. Some are piscivorous, and many take insects and small vertebrates, particularly amphibians and reptiles. All are burrow-nesters, some excavating horizontally into river banks or cliffs, others, such as Ruddy Kingfisher, into rotting timber including tree trunks. Their flight is typically fast and direct. Larger species are usually vocal and conspicuous, often perching in the open; smaller species are generally more inconspicuous and usually call only in flight.

Bee-eaters (Meropidae) 3/25   125

An Old World family of slender, colourful insect-hunters. Most are graceful, with curved bills, short legs, and long tails. Sociable, vocal aerialists, they are typically found in open country sallying from perches to catch prey (which if wasps or bees they frequently batter to de-sting before consuming), or foraging in flight that is undulating and interspersed with prolonged glides. Some species breed socially in colonies of excavated burrows.

Hoopoe (Upupidae) 1/1   125

The taxonomy of this family remains unclear. Three subspecies (or one, two, or three species, one confined to Africa, another on Madagascar, and the third wide-ranging, including east Asia). Colourful, with a long curved bill, erectile crest, and black, white and pink plumage, broad rounded wings square-ended tail and rather short legs. They typically forage for invertebrates from soft ground, grass areas, and lawns; usually solitary or in pairs.

Toucans and barbets (Ramphastidae) 2/120   126

A family represented in the region only by the Asian barbets (Megalaiminae), a group of small forest frugivores with short, stout beaks. Inconspicuous, often sitting motionless in canopy, but commonly vocal, giving long series of repeated notes. Related to woodpeckers, which they resemble in excavating nesting cavities in trees, in having zygodactylic feet and in flight; they differ considerably, however, in their diet of fruit supplemented with large insects and small reptiles. Flight is direct, rapid, with audible flapping interspersed with glides, hence deeply undulating.

Woodpeckers (Picidae) 19/210   126–130

A large, widespread family well represented in the region. They have strong feet and rigid tails, and are adapted to foraging over vertical surfaces. Typically they have powerful, chisel-like bills, with which they excavate cavities in wood in search of invertebrates, and for nesting. The bill is also used for 'drumming' on wood, providing a means of long-range communication. The 'drum roll' varies in length and pattern from species to species. Adaptations to their arboreal life-style include stiffened tail feathers, which act as a brace during tree-climbing, a strengthened skull, an elongated tongue, and zygodactylic feet. Flight is rapid, wings sometimes whirring audibly, but usually deeply undulating. Most are forest residents, highly territorial in summer, but often joining mixed species flocks outside the breeding season.

Pittas (Pittidae) 2/30   131

A distinctive group of colourful small to medium-sized terrestrial birds of tropical forests, where they forage, thrush-like, for invertebrates beneath the leaf-litter. They have large heads, very short tails, long legs and upright posture. Generally shy, but very vocal (with simple, repetitive whistled songs), particularly at dawn and dusk, when may call from ground or from beneath the canopy.

Shrike-flycatchers and allies (Platysteiridae) 1/32   131

This family is represented in the region by the woodshrikes, an oddly shrike-like group with strongly hooked bills, large heads and squarish tails. They generally do not adopt the vertical posture of shrikes.

Woodswallows (Artamidae) 1/10   131

A group of aerial insectivores found usually in tropical regions around forests and open areas. With their large heads, broad-based, rather pointed wings and square-ended tail they share a passing resemblance in flight to starlings, but when perching appear more swallow-like. Flight is strong, graceful and includes soaring; insect food is caught in flight.

Cuckoo-shrikes (Campephagidae) 6/81   131–133

A large family of small to medium-sized insectivorous birds with several representatives in the region from two rather different genera, the cuckoo-shrikes (*Coracina*) and the minivets (*Pericrocotus*). They share short bills (often with rictal bristles), largish heads, elongated bodies, shortish wings, tarsi and toes and longish tails. Flight is strong, but undulating. Cuckoo-shrikes are typically unobtrusive, essentially monomorphic with drab pale or dark grey plumage. Minivets are usually sexually dimorphic, colourful (reds, yellows and blacks), gregarious and often highly active.

Shrikes (Laniidae) 7/30   133–135

A group of active predators, with long tails, large heads, hooked beaks with notched upper mandibles, occurring widely in the Old World and North America. Wings rather short and rounded, flight strongly undulating; typically drops from perch, flies low then swoops down to snatch prey or up to next perch often on top of a bush or tree; commonly at woodland or forest edge, or in open farmland with posts, poles or wires on which to perch. Plumage simply patterned in browns, blacks, greys and white, with bandit-like black face masks; males and females differ subtly, juveniles generally paler with barring. The status of several taxa is unclear. Tarsi and toes strong, claws sharply curved. Perch upright in open and swoop on large insects, small reptiles, amphibians and mammals, sometimes impaling them on thorns for later consumption; often conspicuously cast pellets of indigestible prey remains.

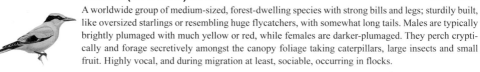

Orioles (Oriolidae) 2/29   135–136

A worldwide group of medium-sized, forest-dwelling species with strong bills and legs; sturdily built, like oversized starlings or resembling huge flycatchers, with somewhat long tails. Males are typically brightly plumaged with much yellow or red, while females are darker-plumaged. They perch cryptically and forage secretively amongst the canopy foliage taking caterpillars, large insects and small fruit. Highly vocal, and during migration at least, sociable, occurring in flocks.

Drongos (Dicruridae) 4/22   136

A group of generally dark-plumaged forest- or forest-edge-dwelling insectivorous species. They have strong hook-tipped bills, prominent rictal bristles, slender bodies, long wings and long, forked tails; they perch prominently and sally out aerobatically to catch large insects. Renowned as bold and vociferous they readily mob raptors, and lead mixed bird flocks. Calls are extremely varied and include considerable mimicry.

Monarchs (Monarchidae) 3/87 137

Members of this large group of tropical Old World flycatchers typically have long or very long tails and broad, flat bills. They inhabit the mid-level to canopy levels of forest, where they forage busily making short flights after medium to large insects amongst branches. Although usually solitary they will also join mixed bird flocks. Sometimes considered as a subfamily (Monarchinae), within the dicrurids.

Crows, jays and magpies (Corvidae) 19/117 138–141

A varied group of medium to very large passerines, often social, intelligent. Either predominantly black, or colourful. Ranging in habitats from open rocky tundra and alpine zone to tropical forest. Omnivorous; usually with powerful, straight bills. Largely black *Corvus* species are long-winged and strong fliers (even aerobatic in the case of the Raven), and may soar, whereas the colourful jays and magpies have shorter wings and longer tails and are weaker fliers, fluttering or flopping between trees, although jays in northeast Asia do undertake quite long seasonal movements.

Waxwings (Bombycillidae) 2/8 142

A small, specialised group of species, largely insectivorous in summer and frugivorous in winter. Distinctively crested with silky-smooth plumage. Sexes similar. Secondaries have unique fused wax-like tips. Starling-like in flight, but with big-headed appearance. Commonly flock outside the breeding season. Nomadic; wintering numbers, even presence, is erratic, with many present in some winters, few or none in others, and renowned for wandering during the winter. Commonly found around fruit- or berry-bearing trees, particularly rowans, or any trees with mistletoe; may remain in one locality, stripping shrubs and trees of their berries, before suddenly disappearing.

Tits (Paridae) 13/54 142–145

A group of small inhabitants of scrub, woodland edge and forest. They have rather large heads, very short, stubby bills, short rounded wings and medium to long tails. Mostly territorial during the breeding season (they nest in cavities) and gregarious in non-breeding period when they commonly associate in mixed flocks. They forage actively and acrobatically, constantly on the move gleaning insects from trunks, twigs and leaves. Despite apparently weak flight, several species undertake migrations in the east Asian region.

Swallows and martins (Hirundinidae) 11/84 146–148

A large, almost global family of small aerial insectivores. Extremely active, with long pointed wings, often with distinctly forked tails. Flight is rapid and agile but more relaxed, and not stiff-winged as in swifts. The bill is small (although the gape is large), and the legs so short that they appear virtually legless when perched in rather upright posture. Mostly long-distance migrants leaving colder northern regions entirely for the winter. Nests may be in pre-existing or excavated holes or of mud glued to buildings or rocks. Some breed colonially or semi-colonially, some in tree cavities or rock crevices, and they are often encountered in flocks outside the breeding season.

Long-tailed tits (Aegithalidae) 2/11 149

A small group of tiny-bodied, short-winged, long-tailed birds with very short, stubby bills. Outside the breeding season they are commonly encountered in highly active bands, foraging acrobatically in the outer twigs and leaves of shrubs and trees. They feed on small insects, and keep contact with constant churring or high-pitched calls.

Larks (Alaudidae) 8/92 149–151

A large, widespread family of terrestrial passerines, well represented in the region. They occur in a range of rather dry and open habitats. They are typically cryptically coloured, crouch when first disturbed, then burst into flight. Many are renowned songsters. The song is rich and varied, and often delivered during a prolonged, hovering or circling display flight, during which notes seem to cascade down. Nesting is on the ground. Outside the breeding season they commonly flock. Although they bear a superficial resemblance to pipits, their bills are heavier and more conical, and they often have crests.

Cisticolas and prinias (Cisticolidae) 8/110   151–153

A rather drably-coloured, mainly African and Asian group. They are typically tiny or small, with short rounded wings, rounded (*Cisticola*) or long graduated (*Prinia*) tails, and rather strong legs and feet. They are active, but usually skulk in dry grassland or scrub habitats, making observation and hence identification a challenge.

Bulbuls (Pycnonotidae) 10/118   153–155

A very large group of mainly African and Asian tropical and subtropical species, though the Brown-eared Bulbul occurs well north into the region. Medium-sized, somewhat thrush-like birds but entirely arboreal and with a very upright posture. Generally long-tailed and short-winged; sexes similar. Commonly highly vocal, sometimes even annoyingly so! Some form monospecific flocks outside the breeding season. They occur in a wide range of habitats from gardens and scrubland to secondary and primary forests, where they feed on a very wide range of berries and fruits. Research in Japan indicates that bulbuls are crucially important dispersers of shrub and tree seeds. Forest species may be secretive, though rather vocal, whereas open-country birds tend to perch conspicuously, and regularly flick their wings and tail.

Old World warblers (Sylviidae) 52/265   156–169

A diverse group of small insectivorous birds found arboreally or in scrub, and grasslands (the latter may be partly terrestrial in their habits). They vary in proportions depending on genus, but most have rather short, slender bills, short rounded or pointed wings, and medium to long tails. They pose extreme identification challenges when details of head pattern, wing pattern and structure and voice may all be critical. Songs may be beautiful and calls diagnostic, but many species are notoriously skulking, giving only tantalising glimpses from cover, or while dashing between patches of vegetation.

Babblers and parrotbills (Timaliidae) 34/273   169–178

A large, heterogeneous group of largely Asian species, many of which are sociable and noisy. Typically weak fliers with short, rounded wings and long tails, but with strong tarsi and toes. They occupy a range of niches from the ground to the canopy, from grassland and scrub to tropical forest, and while most are insectivorous some also take nectar, fruit or seeds. They are commonly noisy but skulking in dense vegetation and though often in social groups they can be difficult to see.

White-eyes (Zosteropidae) 3/95   179

A distinctive, homogeneous group, but one that is closely related to the yuhinas and may be better placed within the Timaliidae. Typically small to medium-sized, arboreal insectivores with green upperparts and off-white underparts; they have sharply-pointed, slightly arched bills, usually a white eyering (though the aberrant Bonin Honeyeater has a black mask), short, rounded wings and a short tail. They are highly sociable, maintain constant vocal communication and forage and migrate in flocks.

Crests (Regulidae) 3/5   180

Tiny arboreal insectivores with fine bills, rounded bodies, short rounded wings and tail and very high-pitched vocalisations. These tiny birds are, amazingly, resident at altitudes and latitudes that experience extremely cold winters. Resembling tiny *Phylloscopus* warblers, they are readily distinguished by their face patterns and their bright yellow or red crown stripes.

Wrens (Troglodytidae) 1/76   180

This large New World group has just one, very widespread, Eurasian representative. This active insectivore skulks in dense vegetation close to water courses, but reveals itself with extraordinarily loud song and hard calls. Its wings are very short and rounded, its tail somewhat short and usually held cocked upright.

Nuthatches (Sittidae) 3/25   181

A group of small, highly arboreal passerines. They have compact bodies with short tails, strong tarsi and toes, a large head with a stout chisel-like bill, and a dark facial mask. They forage over trunks and branches for insects during spring and summer, and for seeds, nuts and fruits during autumn and winter sometimes descending to the ground to forage after spring snow melt. Strongly territorial and highly vocal in summer, outside the breeding season they join mixed species flocks with tits and small woodpeckers. The aberrant Wallcreeper, often treated in its own subfamily, is closely related to the nuthatches, but has a very different bill structure. This spectacular bird has a long, slender, decurved bill and rather long but weak legs and toes. It forages creeper-like over rock faces, cliffs and outcrops in search of invertebrates among the cracks and fissures.

Treecreepers (Certhiidae) 1/8   181

A small group of arboreal insectivores that resemble shuffling tree trunk-loving mice. They spiral jerkily from tree base upwards probing bark crevices with their long fine bills for insects, their eggs and larvae. Their highly cryptic plumage, inconspicuous habits and very high-pitched sibilant calls make them difficult to locate. Outside the breeding season they may associate loosely with mixed bird flocks.

Starlings (Sturnidae) 14/115   181–184

A large group of small to medium-sized birds with stout bodies, strong tarsi and toes, pointed bills, and often glossy plumage. They are usually highly gregarious outside the breeding season, when they may gather into large flocks and large (even spectacular) roosts. They are extremely vocal, often with a wide range of harsh calls. They range in habitat from open grasslands, and suburban gardens to scrub and forest, where they take a wide range of food types. This group is commonly traded, with temporarily or permanently established populations occurring in out-of-range areas, especially on Taiwan and around major cities in Japan.

Thrushes (Turdidae) 28/165   184–191

A large, almost global group of medium-sized birds, most of which feed terrestrially at least part of the time, fossicking for invertebrates amongst the leaf litter, but they may also feed arboreally when seeds and fruit are seasonally available. Often appear plump, with long legs and a cocked tail. Many species are long distance migrants. Some are sexually dimorphic, many are excellent songsters; song is often given from a prominent, high perch. The larger thrushes have straight, stout bills and long pointed wings and medium-length tails; their flight is typically direct and undulating, and outside the breeding season they commonly occur in flocks. The *Zoothera* thrushes are heavily spotted, often crepuscular, and are rather more terrestrial than the *Turdus* thrushes. The latter in particular share genus-specific alarm and flight notes, which may be difficult to recognise to species; however their songs are richly varied and distinctive.

Chats, robins and flycatchers (Muscicapidae) 56/275   192–207

A diverse group; the exact relationships of the various genera remain unclear, and some may belong in other families. Generally small to medium-sized with large heads, rather short wings and longish tails, these essentially insectivorous birds share a rather perky, upright posture. The chats and robins are largely terrestrial or active in lower layers of the forest, while the longer-winged flycatchers are arboreal hunting for insects by sallying aerially at different levels in woodland or more open habitats. Some are long-distance migrants, some are accomplished songsters, but there is great variation within the group and some flycatchers have very undistinguished songs that are easily missed.

Dippers (Cinclidae) 1/5   208

A small, very specialised group of semi-aquatic species inhabiting fast-flowing rivers, often in mountainous areas, with riparian forest. Plump, stout-legged, strong-footed, with short rounded wings and with short cocked tails (resembling a large *Troglodytes* wren), they forage underwater for insect larvae along rocky river beds. Song is loud and often heard during winter as well as spring. Their flight is fast, whirring low over the water, sometimes taking short cuts where rivers curve.

Leafbirds (Chloropseidae) 1/8   208

A small group of highly arboreal species, typically found in the canopy. Smallish with slender, arched bills, they feed largely on insects but also on fruits and nectar. They are sociable and are sometimes found in mixed bird flocks.

Flowerpeckers (Dicaeidae) 3/44 208

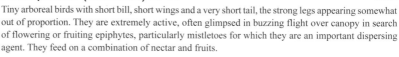

Tiny arboreal birds with short bill, short wings and a very short tail, the strong legs appearing somewhat out of proportion. They are extremely active, often glimpsed in buzzing flight over canopy in search of flowering or fruiting epiphytes, particularly mistletoes for which they are an important dispersing agent. They feed on a combination of nectar and fruits.

Sunbirds (Nectariniidae) 1/127 208

A large group of colourful, active nectarivores. These smallish, arboreal birds have sharp, down-curved bills with which they probe flowers. Flight fast, whirring and direct, and individuals travel considerable distances to find flowering plants, much like New World hummingbirds. Males are usually brightly coloured, often with iridescent sheen to plumage; females are typically drabber, but males assume a female-like eclipse plumage after breeding.

Sparrows and snowfinches (Passeridae) 4/40 209

A large, almost cosmopolitan group of small to medium-sized seed-eating birds. They typically have short, stout bills, rather large, rounded heads and strong legs/feet and a hopping gait. They are found in open habitats and urban areas where they feed largely on seeds and agricultural waste, or season-ally in forests where they are largely insectivorous. Highly vocal. Generally nest in tree cavities, or holes in buildings.

Weavers (Ploceidae) 2/40 210

A large, group of sparrow- or finch-like birds with stocky proportions and short stocky bills. Birds of open grassland or open woodland habitats where they commonly feed on seeds. Breeding males typically have patches of bright coloured plumage and distinctive display flights; non-breeding males and females are brown and cryptic. Colonial when breeding and highly gregarious in the non-breeding season, congregating to feed, drink and roost.

Waxbills and munias (Estrildidae) 9/130 210–212

A large group of small sparrow- or finch-like seed-eating birds, with short conical bills. Typically found in open dry habitats including grasslands, reedbeds, scrubland or secondary forest edge, where they feed largely on seeds. Many species in this family are traded locally and internationally, leading to repeated escapes or releases and temporary or long-established naturalised populations. Only those considered to be well-established in the region are treated here.

Indigobirds and whydahs (Viduidae) 1/20 212

A small Old World group of specific brood parasites dependent on particular waxbills as hosts. Males reach breeding plumage during the rainy season, when they have enormously elongated tails used prominently in display.

Accentors (Prunellidae) 4/13 213

This Old World group of small thrush-like birds is typically found in mountainous regions, where they feed on the ground. Colouring is usually subtle and rather cryptic. Song is often given from a prominent, rock, tree or bush-top perch.

Pipits and wagtails (Motacillidae) 17/64 214–218

A large, almost global, family of slender, terrestrial insectivores, mostly with long tails and long legs. They typically nod their heads like pigeons; many species bob or wag the tail, and have deeply undu-lating flight during which they often give distinctive calls. While wagtails are typically colourful or at least boldly marked with both sexual and age-related plumage variation, pipits are more cryptically coloured, are typically streaked and show little sex or age variation. Many are long-distance migrants, and commonly occur in small flocks outside the breeding season, or even in large flocks at communal roosts. Calls are typically uttered at take off and in flight and are an important means of identification, especially on migration. Walking gait is variable, steady and deliberate interspersed with rapid bursts of speed, all the while bobbing the tail.

Finches (Fringillidae) 24/168   219–224

A large, almost worldwide group of small to medium-sized seed-eating birds. They are typically large-headed with conical (often stout or stout-based) bills. Territorial when breeding, they are highly social and gregarious outside the breeding season, forming large flocks at food sources or roost sites. They are somewhat akin to buntings, but their bodies are generally stockier and their tails shorter, often with a clear notch at the tip. The wings are longish and pointed, and flight is undulating.

New World warblers (Parulidae) 4/112   225

A large family of small, active birds confined to the New World (with the exception of vagrancy to Europe and northeast Asia). With fine bills they are largely insectivorous; some species specialise in canopy niches and are generally brightly coloured, others prefer the shrub layer or are semi-terrestrial (and are generally drably coloured). They show considerable sexual dimorphism and age-related plumage variation.

New World blackbirds (Icteridae) 2/98   226

A varied group of medium to large birds, including blackbirds (un-related to Eurasian blackbirds, which are members of the thrush family), grackles and cowbirds. Rather heterogeneous, but all share rather slender, pointed bills. Meadowlarks and blackbirds (both of which have occurred in the region) are the most starling-like in shape, and are typically found in open habitats and agricultural areas.

Buntings and New World sparrows (Emberizidae) 34/308   227–235

A large family of terrestrial and rather secretive finch-like seed-eaters. They have pointed, conical bills, typically slighter than finches, and have a notch at the base of the upper mandible. Their wings are short to medium-long, and they typically have long, dark notched tails (longer than most finches), usually with contrasting white outer feathers; flight is undulating. Territorial when breeding, males are often brightly coloured, singing from high perches; females are duller and difficult to separate. Commonly found flocking in winter, often in mixed-species groups, when they are attracted to food sources on the ground. A number of North American sparrows are rare migrants or vagrants to the region.

REFERENCES

BirdLife International 2001. *Threatened Birds of Asia: The BirdLife International Red Data Book*. BirdLife International, Cambridge, UK.

Brazil, M. A. 1991. *The Birds of Japan*. Christopher Helm, London.

Collar, N. J. 2003. How many bird species are there in Asia? *Oriental Bird Club Bulletin* 38: 20–30.

David, N. & Gosselin, M. 2002a. Gender agreement of avian species. *Bulletin of the British Ornithologists' Club* 122 (1): 14–49.

David, N. & Gosselin, M. 2002b. The grammatical gender of avian genera. *Bulletin of the British Ornithologists' Club* 122 (4): 257–282.

del Hoyo, J., Elliot, A. & Sargatal, J. (eds). 1992–2008. *Handbook of the Birds of the World*. Volumes 1 to 13. Lynx Edicions, Barcelona.

Dickinson, E. C. (ed.). 2003. *The Howard and Moore Complete Checklist of the Birds of the World*. Third Edition. Christopher Helm, London.

Ferguson-Lees, J. & Christie, D. A. 2005 *Raptors of the World. A Field Guide*. Christopher Helm, London.

Gill, F. & Wright, M. 2006. *Birds of the World. Recommended English Names*. Christopher Helm, London.

MacKinnon, J. & Phillipps, K. 2000. *A Field Guide to the Birds of China*. Oxford University Press, Oxford.

Mild, K. 1987. *Soviet Bird Songs*. Krister Mild, Stockholm.

Peterson, R. T., Mountfort, G. & Hollom, P. A. D. 1966. *Field Guide to the Birds of Britain and Europe*. Collins, London.

Rasmussen, P. C. & Anderton, J. 2005. *Birds of South Asia: The Ripley Guide*. *Volumes 1 and 2*. Smithsonian Institution and Lynx Edicions, Washington, D.C. and Barcelona.

Robson, C. 2000. *A Guide to the Birds of Southeast Asia*. New Holland, London.

Sibley, D. A. 2000. *The North American Bird Guide*. Christopher Helm, London.

Svensson, L. Grant, P. J., Mullarney, K. & Zetterström, D. 1999. *Collins Bird Guide*. HarperCollins, London.

Extensive glossary and bibliography sections are available online at http://sites.google.com/site/birdsofeastasia

HAZEL GROUSE · CKJR
Tetrastes bonasia

L 34–39cm; WS 48–54cm; WT 370–430g. **SD** Wide range in taiga, from Scandinavia across Siberia to Sakhalin. In E Asia: *T. b. kolymensis* from Lena R through Yakutia, Magadan to Chukotka; *T. b. amurensis* south through Russian Far East and NE China including Hinggan Mountains; *T. b. sibiricus* Korea and Transbaikalia; *T. b. yamashinai* on Sakhalin; *T. b. vicinitas* widespread on Hokkaido, Japan. **HH** Quite common in evergreen coniferous, mixed and, sometimes, broadleaf deciduous forests with dense undergrowth; lowland and lower montane regions, to 2,100m in NE China, 1,900m in Korea, and c.1,100m on Hokkaido. Forages on ground and on buds and berries in trees. Often secretive but can be very confiding; flies readily but usually not far when disturbed. Roosts beneath snow in winter. **ID** Small, plump and small-headed. Grey-brown, finely barred white and grey, chestnut on scapulars, and chestnut and white on flanks. ♂ has prominent black chin and throat bordered white, small erectile crest, and small red eyebrow wattle; white scapular lines visible at rest and in flight, and grey-brown tail has broad black subterminal band (except central feathers), with white tips. ♀ lacks crest and has brown throat diffusely marked white. **BP** Bill short with arched culmen, blackish-grey; narrow white ring, eyes dark brown; legs feathered (buff), toes greyish-pink. **Vo** Calls from trees, remarkably sibilant and high-pitched, and recalling an exceptionally loud passerine: *tsi tsi tsi tseee* or *tsst-tsst, tse tsssssssssss*, also softer murmuring *pishururu shururu* from ground. Loud wing-whirring on take-off. **TN** Formerly *Bonasa bonasia*. **AN** Hazelhen.

SIBERIAN GROUSE · CR
Falcipennis falcipennis

L 38–43cm; WT c.700g. **SD** Rare E Asian endemic restricted to E Transbaikalia, S Yakutia, W Okhotsk, Sakhalin, and Amurland; perhaps also adjacent NE China, where status unclear. Monotypic. **HH** Moist shady areas in mossy taiga forests of spruce, fir and larch with dense undergrowth, especially lush berry bushes; 200–1,500m. **ID** Medium-sized, rather dark forest grouse, slightly larger than Hazel Grouse. ♂ rather blackish overall, with black face outlined white, prominent red eyebrow wattle and short white supercilium. Upperparts dark brown mottled grey, with white spots on back and scapulars; wings brown, outer primaries particularly narrow and pointed. Breast black, belly, flanks and vent grey, broadly scalloped white and slate-grey or black. Tail longer and fuller than Hazel, blackish with prominent white terminal band. ♀ paler with stronger russet tones; upperparts dark brown with ochre scales and spots, whilst underparts have whiter spots and scaling. **BP** Bill short with arched culmen, black; eyes dark brown; legs feathered (brown), toes dark pinkish-brown. **Vo** ♂ in courtship display (involving neck-stretching and tail-fanning) gives hard *ka-cha ka-cha* and rolling *u-u-u-rrr* calls. **TN** Formerly *Dendragapus falcipennis*. **AN** Siberian Spruce Grouse; Sharp-winged Grouse.

BLACK-BILLED CAPERCAILLIE · CR
Tetrao parvirostris

L 68–97cm; WT ♂ 3.35–4.58kg, ♀ 1.7–2.2kg. **SD** Uncommon regional endemic; *T. p. parvirostris* from Lake Baikal east to Chukotka (Anadyr), Russian Far East, Sakhalin and Amur basin; rare in adjacent NE China (Heilongjiang) and Hinggan Mountains; *T. p. kamschaticus* throughout Kamchatka. **HH** Larch and pine forests in taiga; lowland and montane areas to 1,000m. Also birch forest in Kamchatka. Typically terrestrial. **ID** Largest grouse in region. Very large, stocky, with rather large head, thick neck, 'beard' and full tail which may be raised and fanned. *T. p. parvirostris* ♂ has black head, neck and breast, glossed purple above and green below; prominent red eyebrow wattles; mantle, back and rump dark earth brown; scapulars and wings warm dark brown, scapulars and wing-coverts spotted white; long brown uppertail-coverts broadly tipped white contrast strongly with black tail, especially when raised and spread in display. Belly, flanks and vent dark earth-brown. ♀ smaller, dark brown, barred black and buff, white tips to coverts form wingbars; tail grey-tipped. *T. p. kamschaticus* ♂ differs from *parvirostris* in having far more extensive and larger spots on scapulars and wing-coverts, with white-tipped secondaries and fine white spots on flanks. **BP** Bill black; eyes dark brown; tarsi feathered (brown), toes grey to brownish-pink. **Vo** Display calls, given from ground before sunrise, involve strange clicks which become a short trill. **AN** Spotted Capercaillie.

EURASIAN BLACK GROUSE · CKR
Lyrurus tetrix

L ♂ 49–58cm, ♀ 40–45cm; WT ♂ 1.1–1.8kg, ♀ 0.75–1.1kg. **SD** W and N Europe, e. across Siberia to Amur basin. In E Asia found in boreal region, with *L. t. baikalensis* in Transbaikalia to NE Inner Mongolia and W Amurland, and *L. t. ussuriensis* east to Amur R and Sikhote Alin Mountains, and south to adjacent NE China and N Korea. **HH** Forest edge, clearings or open areas in pine and larch forests usually in mountains. Forms flocks in winter and spectacular leks in spring. Forages on ground and in trees on leaf buds. **ID** Medium-sized plump-bodied grouse, with a rather small head and long tail (c.15cm) held erect in display. ♂ black with bluish-green gloss and prominent red eyebrow wattles. Wings black with prominent white wingbar, white underwing and vent. Tail long with hooked, hammer-shaped tip, and long white undertail-coverts, raised and fanned prominently in display. ♀ smaller, browner, greyish-brown barred black, including underparts; rounded tail moderately long with many black bars; narrow whitish wingbar. **BP** Bill short, with arched culmen, black or blackish-grey; eyes dark brown; legs feathered (buff to brown), toes grey. **Vo** ♂ at lek utters extraordinarily ventriloquial communal bubbling or crooning known as rookooing; *kru-kru-kru-kru...*, which swells into general sound in which individual ♂s are indistinguishable. Also a strange air-pump-like *chooeesh* or *chuffshee*. ♀ gives fast cackling *kakakakakakeah*. Wings whirr audibly on take-off. **TN** Formerly *Tetrao tetrix*.

HAZEL GROUSE

vicinitas

ad ♂

ad ♀

SIBERIAN GROUSE

ad ♂

ad ♀

ad ♂

BLACK BILLED CAPERCAILLIE

(not to scale)

ad ♂ *kamschaticus*

ad ♀ *kamschaticus*

ad ♂ *parvirostris*

ad ♀ *parvirostris*

EURASIAN BLACK GROUSE

ussuriensis

ad ♂

ad ♀

ad ♂

ad ♀

ROCK PTARMIGAN   CJR
Lagopus muta

L 33–38cm; WS 58cm; WT ♂ 470–740g, ♀ 430–700g. **SD** Very wide range in N Holarctic, from Iceland across N Eurasia and in montane regions further south, east to Chukotka and Kamchatka. In N America, from Aleutian Is to Greenland. In E Asia, *L. m. pleskei* from Lena R to Bering Strait and south to Okhotsk Sea and Kamchatka; *L. m. kurilensis* on N & C Kuril Is; *L. m. japonica* in Japanese Alps, Honshu. **HH** Rocky tundra, coastal tundra, and alpine zone above treeline (e.g. Japan, where local and only above 2,500m), where excellently camouflaged as it moves slowly amongst lichen-covered rocks and dwarf vegetation; often very confiding. In winter forms flocks and seeks shelter at treeline. **ID** Stocky, with white wings (primary shafts black), white belly and legs year-round; slightly smaller than Willow Ptarmigan, but generally greyer and darker, lores black. ♂ (breeding) grey-brown, with black and white vermiculations giving overall greyish appearance; tail black; belly, wing-coverts, vent, outertail-feather tips all white. ♀ warm tawny-brown with fine black, grey and white bars and spots; wings mainly white; lacks black lores. Non-breeding ad. all white with black tail, though long white uppertail-coverts conceal it; ♂ has distinctive black eyestripe and lores. Red eyebrow wattles prominent in winter and summer. Confusable with Willow Ptarmigan, though habitat generally separates them. In winter, ♂ distinguished by black lores, in summer by lack of rufous-brown tones. **BP** Bill slender, even delicate (less arched than Willow), black; eyes black; legs and toes feathered white year-round, claws blackish. **Vo** Throaty series of croaks, almost coughing or snoring, *kuh kuh kwa guwaa*, from ground, but also when flushed, and sometimes by ♀. ♂ has aerial display flight, rising on rapidly beating wings, then descending stiff-winged and giving throaty rattled cackle, *ahrrrr-ka-ka-ka-ka-ka*. ♀ has softer *gweeaa* call.

WILLOW PTARMIGAN   CR
Lagopus lagopus

L 35–43cm; WS 61cm; WT ♂ 535–700g, ♀ 525–650g. **SD** Wide-ranging N Palearctic species, from N Eurasia to Mongolia and Okhotsk Sea. In E Asia *L. l. birulai* on New Siberian I.; *L. l. koreni* across taiga east to N Okhotsk Sea; *L. l. sserebrowskii* in Transbaikalia to W Okhotsk Sea and Khingan and Sikhote Alin Mountains; *L. l. kamtschatkensis* in Kamchatka and N Kuril Is; *L. l. okadai* on Sakhalin. **HH** Widespread across Arctic tundra and forest-tundra to taiga and montane areas with birch, conifer and willow scrub. Winters in more sheltered forested areas. **ID** Stocky and slightly larger than Rock Ptarmigan, but shares year-round white wings, belly and legs; generally more rufous; lores not black. ♂ (breeding) dark rufous-brown with black spots and vermiculations; tail black; belly, wing-coverts, legs and feathered toes, vent, outertail-feather tips all white. ♀ resembles ♀ Rock but larger

with larger bill; warm tawny-brown with fine black, grey and white bars and spots. Non-breeding ad. all white with black tail, though long white uppertail-coverts conceal it. Red eyebrow wattles prominent in winter and summer. **BP** Bill stout, with white or brown lores. Bill shape important, as Willow has larger bill with more arched culmen than Rock; eyes large, black, with narrow white ring; legs and toes feathered white year-round, claws whitish. **Vo** ♂ produces various harsh grating calls, and displays from ground or on elevated rock. Most distinctive call on ground or in flight is loud, guttural barking, *go-back go-back go back grrrrr*, and an accelerating, repetitive *ka ke ke-ke-ke-ke kekeke krrrr*. **AN** Willow Grouse.

CHUKAR PARTRIDGE   C
Alectoris chukar

L 32–39cm; WS 47–52cm; WT ♂ 510–800g, ♀ 450–680g. **SD** Ranges from SE Europe across C Asia to NE China; introduced in Europe, N America. In E Asia *A. c. pubescens* across NE China to Liaoning on W Bohai Gulf. **HH** Shy, occurs in dry grasslands, steppes, plateaux, rocky outcrops and open montane areas with dry vegetation. **ID** Medium-sized grey-brown partridge, with bold face and flank markings. Sexes similar: face, chin and throat creamy white, with black border from forehead through eyes to upper breast encircling throat; upperparts pinkish-grey; breast grades from grey to buff on belly; flanks boldly barred black, chestnut and white; vent and undertail-coverts dull orange-brown. **BP** Bill very short, arched, bright red; eye-ring red, irides black; tarsi dull red, ♂ has spurs. **Vo** Highly vocal, ♂'s may duet or chorus, head held high, a harsh series of accelerating calls: *ka-ka ka-ka kaka kaka kakerakakera*, and simpler chuckling *chukar-chukar-chukar*.

CHINESE FRANCOLIN   C
Francolinus pintadeanus

L 30–33cm; WT ♂ 347–388g, ♀ c.310g. **SD** NE India to SE Asia, and S and E China. *F. p. pintadeanus* occurs on Chinese coast north to Zhejiang and Fujian. **HH** Dry forest, woodland, grassland and secondary scrub to 1,600m. **ID** Medium-sized black and grey partridge. ♂ has long, broad rusty-orange supercilium, black eyestripe, and black moustachial stripe separating white cheeks from white chin and throat. Upperparts black with fine white spots on mantle and narrow white bars on rump and tail; scapulars dark brown. Underparts largely black, with fine white spots on neck and breast, and broader white fringes to belly and flanks, which appear largely white; vent rusty-orange. ♀ browner rather than black and barred rather than spotted on underparts. In flight, appears finely barred black and white, with rufous scapulars and short black tail. **BP** Bill black; eyes black; tarsi dull orange-red. **Vo** Highly vocal in late winter/spring breeding season. Harsh metallic grating calls: *come to the peak ha-ha*, often in chorus at dawn or dusk.

ROCK PTARMIGAN

ad sum ♂
ridgwayi

ad sum ♂
pleskei

ad sum ♂
pleskei

ad sum ♂
pleskei

ad sum ♀
pleskei

ad win ♂

WILLOW PTARMIGAN

ad sum ♂
koreni

ad sum ♂
koreni

ad sum ♂
koreni

ad sum ♂
koreni

ad sum ♀ *koreni*

ad sum ♂
okadai

ad win ♂

CHUKAR PARTRIDGE

pubescens

ad

CHINESE FRANCOLIN

pintadeanus

ad ♂

DAURIAN PARTRIDGE   CR
Perdix dauurica

L *c.*30cm; WT 200–340g. **SD** C Asia e. to Mongolia, Transbaikalia and NE China. In E Asia *P. d. suschkini* is fairly common in NE China and S Ussuriland. **HH** Dry rocky areas with sparse vegetation in steppe and low mountains to 3,000m. **ID** Small grey-brown partridge with orange-rufous face, very similar to extralimital Grey Partridge *P. perdix*. Orange-rufous of ♂ face extends on throat and belly, frames upper part of large black inverted U-shaped belly patch. Upperparts and neck largely grey, with fine brown vermiculations on back, rump and tail. Flanks barred black. In winter, sports orange-rufous 'whiskers' and 'beard'. ♀ paler, face and breast markings less clearly defined; has 'whiskers' and only a small black belly patch. Flanks barred brown. **BP** Bill short, arched, grey; eyes mid-brown; legs dull yellow-ochre. **Vo** Poorly documented. Harsh metallic creaking calls and accelerating series of harsh, rasping notes: *rex rex rex.*

JAPANESE QUAIL   CTKJR
Coturnix japonica

L 17–19cm; WT *c.*90g. **SD** Breeds from Lake Baikal to Japan, wintering south to NE India and S China. In E Asia, summers across NE China, adjacent N Korea, Russian Far East to Sakhalin, and N Japan. Winters in southern half of Japan, Korea, E China, Taiwan (?) and northern SE Asia. Once common but now scarce in much of range. Monotypic. **HH** Wet and dry meadows, dry grassland and agricultural land. Secretive, exploding into flight at close range. **ID** Small, plump, short-tailed, round gamebird with long creamy supercilia (both sexes). ♂ has plain reddish-brown face, chin and throat, whereas ♀ is mottled grey-brown; overall plumage mid- to dark-brown with cream streaks and broken black bars; well camouflaged. ♂ washed rufous on chest, white on belly, streaked rufous and cream on flanks; ♀ spotted on upper breast, white on belly. In flight, pale brown and rather uniform, with darker flight-feathers and tail barred paler brown. Differs from extralimital Common Quail *C. coturnix* (a likely vagrant to region) in throat colour and voice of ♂; ♀s probably indistinguishable, though Japanese may have redder brown centres to flank-feathers and heavier spotting on breast. Compare with much smaller buttonquails. **BP** Bill short, grey; eyes dark brown; legs dull flesh. **Vo** ♂ calls loud and rasping, variously transcribed as: *kextsu, kera keh, gwa-kuro, guwaguwaa* and *guku kr-r-r* (very different from liquid, whistled *wet-my-lips* of Common Quail); ♀ gives softer *pipipipii*; also a muffled *pirrr* on take-off.

BLUE-BREASTED QUAIL   CT
Coturnix chinensis

L 12–15cm; WS 25cm; WT 20–57g. **SD** India to Australia. In E Asia *C. c. chinensis* resident but uncommon in Fujian (also Guangdong), and Taiwan (rare), where perhaps concentrated in SW. **HH** Singly and in small coveys in both dry and wet dense grasslands, scrub, stubble and pineapple fields in lowlands, where

scarce and very secretive. **ID** Very small; smaller than Japanese Quail. ♂ very dark with distinctive pied face (white lores, malar stripe and throat crescent all bordered black), and black chin/throat; slate-blue extends from forehead to breast and flanks; belly, vent and undertail-coverts chestnut. Upperparts dark brown with faint white streaks (feather shafts). ♀ resembles small Japanese Quail, lacks ♂'s pied face, has rufous supercilia and dark brown ear-coverts, pale buff chin and sandy-brown underparts barred blackish on breast-sides and flanks. ♀ may be confused with buttonquail, but lacks contrast between black flight-feathers and pale wing panel. **BP** Bill short, black; eyes dark brown; thick tarsi and toes dull orange or yellow. **Vo** Explosive whistle *tee-yu, tee-yu...* and long, rasping hiss. **AN** Painted Quail; King Quail.

TAIWAN HILL PARTRIDGE   T
Arborophila crudigularis

L *c.*28cm; WT ♂ 311g, ♀ 212g. **SD** Endemic resident on Taiwan. Monotypic. **HH** Vocal and widespread in broadleaf montane forests from 100–2,300m. **ID** Small, grey-brown partridge with distinctively pied face. ♂ has long supercilia, white forehead, chin and cheeks contrasting with black eye patch, eyestripe and neck-sides. Olive-grey upperparts (crown to rump and tail) barred black; rufous wings have three broad grey bars. Chin to vent dark blue-grey, paler on belly, with prominent white lanceolate flank streaks. Wings and tail short, rounded. ♀ very similar but has less black streaking on throat and more white streaking on flanks. **BP** Bill short, blue-grey; eye has narrow bare red ring, irides black; stout tarsi and toes dull orange-red. **Vo** Series of rolling *gurru gurru gurru* calls rising in pitch to crescendo, then falling; often in duets or choruses. Single calls can be confused with those of Black-necklaced Scimitar Babbler.

WHITE-NECKLACED PARTRIDGE   C
Arborophila gingica

Vulnerable. L 25–30cm; WT 253g. **SD** Endemic to restricted range in southeast China; in region scarce resident only very locally in NW Fujian. Monotypic. **HH** Occurs in dense broadleaved and mixed hill forest (500–1,700m). **ID** Plump, medium-sized, grey and brown bird of the forest floor. White forehead extends into grey supercilium reaching to nape; crown chestnut, lores black, face orange-buff with black streaking on sides of neck. Upperparts grey-brown with black spotting across lower back, rump and short tail. Wings rufous with black terminal spots to wing-coverts. Broad black and chestnut bands across lower neck divided by narrow white band; breast and flanks grey, shading to white on belly, with chestnut streaks on flanks and black spots on undertail-coverts. **BP** Bill short, with strongly arched culmen, grey; eyes dark brown with red eye-ring; legs and feet red. **Vo** A loud, plaintive, two-toned whistle. **AN** Collared Hill Partridge.

DAURIAN PARTRIDGE

suschkini

ad ♂

ad ♀

JAPANESE QUAIL

ad ♂

ad ♀

BLUE-BREASTED
QUAIL

ad ♀

ad ♂

TAIWAN HILL
PARTRIDGE

ad ♀

ad ♂

WHITE-NECKLACED
PARTRIDGE

ad

CHINESE BAMBOO PARTRIDGE CTJ
Bambusicola thoracicus

L *c.*31 cm; WT 200–342g. **SD** *B.* (*t.*) *thoracicus*, restricted to SE China, and *B.* (*t.*) *sonorivox* (**Taiwan Bamboo Partridge**) endemic to Taiwan, are probably separate species. *B.* (*t.*) *thoracicus* introduced and now widespread and common in Japan (south of Hokkaido). **HH** Evergreen broadleaf forests with dense underbrush, and woodlands and parks; lowlands and low hills. Sociable, ♂'s often chorus at dawn and dusk, and families form small flocks in winter. **ID** Large, plump, reddish-brown partridge; highly vocal, but secretive. Chinese has distinctive pale blue-grey forehead, supercilium and upper breast; chin, face and neck-sides bright rufous-chestnut. Upperparts mid-brown with chestnut, black and white spots on mantle and wing-coverts, rump greyish-brown, tail finely barred, outertail-feathers rufous-brown, conspicuous in flight. Blue-grey lower neck, buff on lower breast, belly, flanks and vent, heavily spotted black on flanks. Taiwan Bamboo Partridge generally darker, with more extensive and darker grey on face, neck and breast, with rufous restricted to chin, and deeper orange-brown underparts with chestnut, not black, flank spots. **BP** Bill grey-black; eyes dark brown; tarsi greenish-grey, ♂ has tarsal spurs. **Vo** Highly vocal. Chinese pairs duet, loud *chottokoi chottokoi*, *pippyu kwai pippyu kwai* or *people pray people pray* calls winding to crescendo. Taiwan Bamboo Partridge has more complex and sonorous vocalisations.

CABOT'S TRAGOPAN C
Tragopan caboti

Vulnerable. L ♂ 61cm, ♀ 50cm; WT ♂1,400g, ♀ 900g. **SD** Endemic to SE China; *T. c. caboti* is rare and very local in NW Fujian and S Zhejiang. **HH** Subtropical evergreen and deciduous hill forest (800–1,400m). **ID** Large, short-tailed forest bird. ♂ has black head with orange facial skin and throat; golden-orange stripe behind eye, front collar and sides of neck scarlet. Throat and skin above eye inflate as wattle and horns for display. Upperparts chestnut with black-ringed buff spots on mantle and back, and larger buff spots on wing-coverts and rump; tail brown with broad dark band across tip. Underparts plain creamy-buff, from lower neck to belly, flanks and undertail coverts. ♀ dark grey with white flecking on upperparts including wings and tail, and white shaft streaks to feathers of underparts, especially on belly and flanks. **BP** Bill short, with arched culmen, grey; eyes brown; legs and feet pale pink. **Vo** Both sexes give harsh *ga-ra ga-ra* and *wear-wear-ar-ga-ga-ga* calls.

KOKLASS PHEASANT C
Pucrasia macrolopha

L ♂ 58–64cm, ♀ 52.5–56cm; WT 0.93–1.135kg. **SD** From Himalayas to E China, with *P. m. xanthospila* NE to Bohai Gulf coast, and *P. m. darwini* NE to E coast (Zhejiang and Fujian). **HH** Steep slopes in forested (coniferous and mixed) mountains at 600–1,900m in temperate zone. **ID** Large, short-tailed pheasant with bottle green head, white neck patch and prominent ear-tufts

(lacks bare facial skin or wattles). ♂ has crown and long nuchal crest grey-buff. Body generally mid-brown to buff- or grey-brown, very heavily streaked (grey plumes have black shaft streaks) except on breast, which is chestnut. ♀ smaller and lacks tufts (or crest). *P. m. darwini* has more buff on underparts; *P. m. xanthospila* more heavily striped on upperparts, whilst ♀ greyer. **BP** Bill reddish-black; eyes dark brown; tarsi and toes grey. **Vo** Varied. Calls from ground and trees include a loud *kok-kok-kok...* audible over long distances, a rasping *gur-gzee gze-gzeer* and weak *wut-wut wutt-trrh-trh.*

SILVER PHEASANT C
Lophura nycthemera

L ♂ 120–125cm, ♀ 70–71 cm; WT ♂ 1.13–2kg, ♀ 1.15–1.3kg. **SD** Primarily tropical, from S China to mainland SE Asia. In E Asia, only occurs in E China north to Fujian and Zhejiang (*L. n. fokiensis*). **HH** Lush evergreen and secondary forest, bamboo thickets and scrub; to 2,150m. **ID** ♂ large, long-tailed (60–75cm), with black crown and crest, broad red wattle across forehead to rear of eyes, cheeks and chin. Upperparts and full tail white, outertail feathers and wings white with fine black chevrons or streaks. Black from chin to vent. ♀ smaller and mostly plain olive-brown with short crest, bare red eye patch; underparts streaked and barred grey-buff; tail shorter (24–32cm), tail-sides vermiculated black and white. **BP** Bill pale horn; eyes dark brown; tarsi and toes dull red. **Vo** ♂ display includes wing-whirring and a weak, soft trill, *lo-lo-lo-lo*; also a high-pitched *ji-go-go-go* when disturbed.

SWINHOE'S PHEASANT T
Lophura swinhoii

L ♂ *c.*79cm, ♀ *c.*50cm; WT *c.*1.1kg. **SD** Endemic to mountains of Taiwan. Monotypic. **HH** Low to high elevations (200–2,500m) in damp evergreen and mixed forest. **ID** ♂ large, dark, steel-blue with white nuchal crest, large red facial wattle extending on to bill and in points above and below eye, large white upper-back patch, and long white central rectrices (41–50cm). Scapulars maroon and wing-coverts fringed metallic blue-green; blue-black back and rump scaled silver-blue, whilst neck has silver-grey streaks. ♀ smaller, shorter-tailed (20–22cm) and dark ash-brown with pale brown bars on wing, and prominent pale brown chevrons on neck, mantle and scapulars, dark-edged chevrons on foreneck and breast, and rufous outertail-feathers; underparts finely vermiculated black; bare red patch (smaller than ♂) across much of face. **BP** Bill arched, pointed, pale horn (♂) or grey (♀); eyes dark brown; tarsi bright coral red in both sexes. **Vo** ♂ gives a breathy *hus hus hus hus*, while the ♀ gives a guttural *ge ge ge* or a subdued *ku ku ku* and a high-pitched squeak. ♂ adopts erect posture in display, opening and shivering wings very briefly.

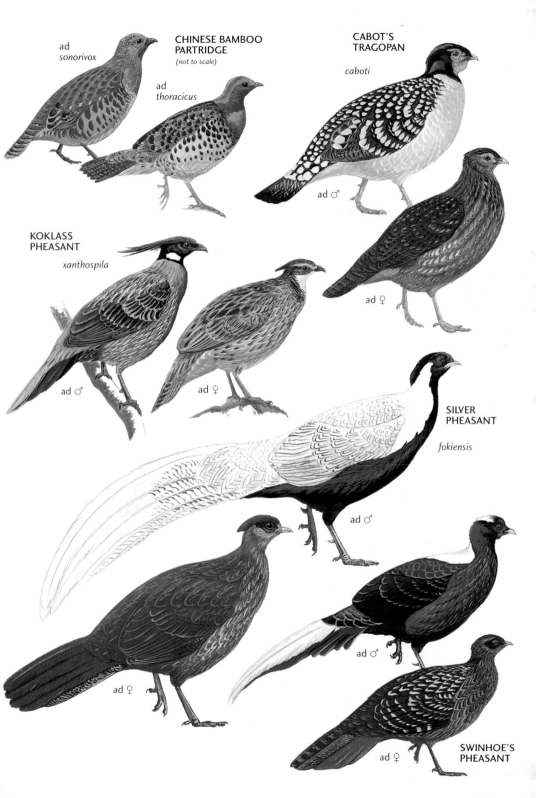

ad
sonorivox

**CHINESE BAMBOO
PARTRIDGE**
(not to scale)

ad
thoracicus

**CABOT'S
TRAGOPAN**

caboti

ad ♂

**KOKLASS
PHEASANT**

xanthospila

ad ♀

ad ♂

ad ♀

**SILVER
PHEASANT**

fokiensis

ad ♂

ad ♀

ad ♂

ad ♀

**SWINHOE'S
PHEASANT**

BROWN EARED PHEASANT   C
Crossoptilon mantchuricum

Vulnerable L 96–100cm; WT ♂ 1.65–2.475kg, ♀ 1.45–2.025kg. **SD** Endemic to N China, in region only in Hebei. Monotypic. **HH** Very rare in stunted montane coniferous or mixed forest with scrub and clearings; to 2,600m, descends in winter to 1,100m. Usually in small family parties. **ID** Large black and grey-brown pheasant with long (54.5cm) plumose tail. Bare facial skin red, crown velvet black and short white moustachial extends in white band below face and extends as pointed white ear-tufts. Entire body very dark grey-brown; lower back, rump and loose uppertail-covert plumes white, tail black, drooping. ♀ slightly smaller. **BP** Bill arched, pink; eyes orange; tarsi red, ♂ has tarsal spurs. **Vo** Displaying ♂ gives deep guttural *gu-gu gu-gu* and *gu-ji gu-ji* calls when foraging, also a high-pitched, raucous *trip-c-r-r-r-r-r-ah!*

ELLIOT'S PHEASANT   C
Syrmaticus ellioti

Vulnerable L ♂ *c.*80cm, ♀ *c.*50cm; WT ♂ 1.044–1.317kg, ♀ 0.726–1.09kg. **SD** Endemic to SE China. In region occurs in coastal Fujian and Zhejiang. Monotypic. **HH** Forested hills (evergreen broadleaf and coniferous) with bamboo thickets and dense undergrowth; to 1,500m. **ID** ♂ large, dark chestnut-brown with bare red facial skin, long, brown-banded silver-grey tail (39–44cm) and white belly. Head and nape grey-white, chin black, broad brown collar, mantle and wings largely dark chestnut-brown with black scaling, wings have two white bands, black rump scaled white. Chest dark blackish-chestnut, belly white, with flanks scaled dark brown. ♀ smaller, short-tailed (17–19.5cm), dark grey-brown with pale bars on wing, black chin, whitish belly, grey head-sides and bare red eyebrow. **BP** Bill pale horn (♂) or grey (♀); eyes dark brown; legs grey. **Vo** Deep guttural calls usually given in morning, also low clucks, chuckles and a shrill squeal. Audible wing-whirring display.

MIKADO PHEASANT   T
Syrmaticus mikado

L ♂ *c.*87.5cm, ♀ *c.*53 cm; WT ♂ *c.*1.3kg, ♀ 1.015kg. **SD** Endemic to Taiwan. Monotypic. **HH** Only in high mountains, 1,600–3,300m. Prefers shady, dense forest (evergreen broadleaf and coniferous), but also secondary forest, scrub and bamboo. **ID** ♂ elegant, long-tailed (49–53cm). Almost entirely steely blue-black with deep blue scallops on mantle, scapulars and rump, and prominent white wingbar on otherwise blue-black

wings, and narrower white tips to secondaries and tertials. Mantle, breast and rump have bluish-purple metallic sheen. Very dark blue-black tail has well-spaced narrow silver-grey bands. Large bare red patch above and below eye, somewhat pointed behind. ♀ smaller, shorter-tailed (17–22.5cm); dark earth brown with narrow red ring of bare skin around eye, black centres to most of mantle and scapulars, fine silver shaft-streaks to mantle, silver arrowheads on chest and bars on belly; tail brown with dark and pale bands, recalling ♀ Copper Pheasant. **BP** Bill sharply pointed and arched, grey; eyes dark brown; tarsi dark blue-grey or greenish-grey. **Vo** In breeding season ♂ gives shrill whistle; alarm a high *wok wok wok*. ♂ display involves abrupt wing-drumming or throbbing, audible up to 30m.

COPPER PHEASANT   J
Syrmaticus soemmerringii

L ♂ 87.5–136cm, ♀ 51–54cm; WT 907g. **SD** Endemic to Japan (Honshu, Shikoku and Kyushu Is), with five subspecies: *S. s. scintillans* in N & W Honshu; *S. s. subrufus* on Pacific side of C & SW Honshu and Shikoku, *S. s. intermedius* on northern side of SW Honshu, also Shikoku; *S. s. soemmerringii* in N & C Kyushu; and *S. s. ijimae* in S Kyushu. **HH** Mixed broadleaf evergreen and coniferous forests (even fairly undisturbed cedar plantations), and common in shady areas from low to mid elevations in mountains. **ID** ♂ elegant, chestnut with extremely long tail (48–98cm). Head and upper neck plain dark rufous-chestnut, whilst remaining plumage dark to mid chestnut with pale shafts and fawn fringes producing scaled effect; fringes to mantle and rump whiter. Very long cinnamon tail has dark brown and fawn bands. Bare facial skin red; does not extend as wattles. ♀ smaller, shorter-tailed (14–19cm), and dark brown with large pale scales on flanks, belly and vent; upperparts dark earth-brown, many feathers with blackish centres and white tip to tail. *S. s. scintillans* has broad white fringes to belly and flanks, and white fringes to rump and uppertail-coverts. *S. s. subrufus* is deeper chestnut than nominate, lacks white fringes to rump and uppertail-coverts, and has more golden rump. *S. s. soemmerringii* has amber fringes to rump and uppertail-covert tips. *S. s. ijimae* is deeper chestnut, lacks white fringes to feathers, and has large white rump and uppertail-covert patch. **BP** Bill grey-black; eyes dark brown; legs grey. **Vo** ♂ gives guttural *guru guru* or indistinct hoarse *ko-ko ko* calls. During late winter/early spring breeding season, ♂ produces deep drumming sounds with its wings, like distant deep motorcycle engine starting up ('wing-throbbing'). ♂ also gives high-pitched *heese!* in aggression or if flushed, and both sexes give quiet *kyup* calls while feeding.

BROWN EARED PHEASANT
(not to scale)

ad ♂

ELLIOT'S PHEASANT

ad ♀

ad ♂

MIKADO PHEASANT

ad ♀

ad ♂

COPPER PHEASANT

ad ♀
soemmerringii

ad ♂
soemmerringii

ad ♂
ijimae

REEVES'S PHEASANT   C
Syrmaticus reevesii

L ♂ c.210cm, ♀ c.75cm; WT c.949g. **SD** Endemic to C & E China, where declining due to overhunting. In region, S Shandong population very rare or extinct. Monotypic. **HH** Deciduous and mixed forests in wooded hills and steep valleys; 200–2,600m. **ID** The longest-tailed pheasant. ♂ has extremely long tail (100–160cm) with white central rectrices barred black, rendering it unmistakable and highly vulnerable to feather trade. Crown and collar white; black mask extends from bill to nape. Upperparts golden; feathers of neck, mantle and rump have black fringes. Underparts golden, scaled black on neck and breast-sides, centre of breast and belly dark chestnut spotted white. ♀ smaller, shorter-tailed (36–45cm) and pale-faced with rufous-brown underparts; tail grey-brown banded darker. Underparts dark earth brown with white shafts to neck and upper breast, buff shafts and fringes to belly and flanks. **BP** Bill pale horn (♂) or grey (♀); eyes black; legs grey. **Vo** Calls rarely. A fast, guttural gu-gu-gu-gu (♂) or ge-ge-ge.

RING-NECKED PHEASANT   CTKJR
Phasianus colchicus

L♂ 75–89cm, ♀ 53–62cm; WS 70–90cm; WT ♂ 0.77–1.99kg ♀ 0.545–1.453kg. **SD** Most familiar pheasant, introduced in Europe, N America, Australia, New Zealand and Japan (mainly Tsushima and Hokkaido), but natural range is from C Asia to E & NE China, adjacent S Russian Far East, patchily in Taiwan (mostly E lowlands) and widespread in Korea. Various subspecies in region: P. c. kiangsuensis in W Hebei and SE Inner Mongolia; P. c. karpowi in S Manchuria (Liaoning and Hebei), Korea, Cheju-do and Tsushima (possibly native), and introduced Japan; P. c. pallasi in SE Siberia, N & E Manchuria (Heilongjiang/Inner Mongolia) and NE Korea, and introduced Hokkaido; and P. c. formosanus on Taiwan. **HH** Dry scrub and woodland edge, open wood and farmland, from lowlands to mountains. Sexes typically encountered separately. Bursts into flight from cover with noisy wing-whirring. **ID** ♂ has greenish-black hood, with blue, green or purple sheen, greenish-black ear-tufts, and extensive warty red facial wattle extending to forehead, chin and cheeks. All subspecies in E Asia have complete or partial white collar. Brown to copper-bronze mantle scaled white, wings have large powder-grey panel, lower back and rump grey-green or brown, the loose feathers puffed and drooping somewhat; tail long, pointed, dark grey-brown with prominent transverse black bands. Underparts dark blackish-chestnut with purple sheen on breast, flanks browner and spotted white above and black below. ♀ smaller, much shorter-tailed and cryptically attired mid and dark brown, heavily scaled blackish-brown on mantle, wing-coverts and flanks, whereas tail has dark brown bands. Underparts mid or buffy-brown with some dark scaling on flanks (compare Japanese Green Pheasant). P. c. pallasi has green, rather than purple, gloss to chest. P. c. formosanus (currently threatened due to hybridisation with introduced subspecies, mainly P. c. karpowi), has very pale whitish-straw or buff flanks spotted black or dark chestnut, pale greyish-green uppertail-coverts,

pale grey rump, and particularly broad black tail bars. **BP** Bill short, arched, horn-coloured; eyes yellow (♂) or dull orange (♀); tarsi greyish-green. **Vo** ♂'s loud, guttural hok-kok hok-kok hok-kok display call given from prominent perch on ground or log is typically followed by noisy wing-drumming. A loud, harsh crowing kerrch-krch! When flushed gives screechy, rattling kr-krk-krk krk krk. **AN** Common Pheasant.

JAPANESE GREEN PHEASANT   J
Phasianus versicolor

L ♂ 81.5cm, ♀ 58cm; WT ♂ 0.9–1.4kg, ♀ 692–970g. **SD** Endemic to Japan, where found throughout Honshu, Shikoku and Kyushu, also Sado I., Izu Is, Yaku-shima and Tanega-shima. P. v. robustipes occurs NW Honshu and Sado; P. v. tohkaidi C & W Honshu and Shikoku; P. v. tanensis southern peninsulas and islands south of Honshu and Kyushu; and P. v. versicolor extreme W Honshu, Kyushu and islands west of Kyushu. **HH** Woodland and forest edge, brush, grassland and parkland. Indigenous Green Pheasant of Japan differs markedly from so-called 'green' pheasant hybrids in various parts of introduced range (e.g. Europe). **ID** Darker, more compact with shorter, broader tail (27–42.5cm) commonly held more cocked than Ring-necked. ♂ has bluish-purple hood with prominent ear-tufts, more rounded red wattles on forehead and face, and no white collar. Neck, mantle, breast and flanks deep bottle green. Wing-coverts, lower back and rump pale powder grey; tail long (but shorter with more prominent outertail-feathers than Ring-necked), pale grey-brown banded dark. ♀ smaller, shorter tailed (21–27.5cm) and cryptically coloured, with dark brown feathers fringed pale brown, affording darker heavily scaled pattern to entire body and wings than ♀ Ring-necked. White crescent below eye, scales on body and bands on tail more prominent and more uniform than in slightly smaller ♀ Copper, and lacks latter's white-tipped tail. **BP** Bill pale horn (♂) or grey (♀);large red wattles (with dark feather quills), eyes pale yellow (♂) or dark brown (♀); legs pinkish-grey. **Vo** ♂ ko-kyok calls (higher and hoarser than Ring-necked) are commonly followed by noisy wingbeats. **TN** Formerly within Ring-necked Pheasant.

INDIAN PEAFOWL   TJ
Pavo cristatus

L ♂ 180–230cm, ♀ 90–100cm; WS ♂ 130–160cm, ♀ 80–130cm; WT ♂ 4–6kg, ♀ 2.75–4kg. **SD** S Asian species (Pakistan to Sri Lanka) introduced to S Japan and established on Yaeyama Is, southern Nansei Shoto; escapes occur elsewhere. Monotypic. **HH** Social and polygamous; inhabits rural farmland and forest edge. **ID** ♂ enormous with largely bluish-purple head, neck and breast, small fan crest, and enormously long green 'train' (elongated uppertail-coverts; 140–160cm) with large multi-coloured eyespots. ♀ shares fan crest, but is smaller, drabber dark grey-brown, with white face, green neck, grey breast, white underparts and short (32.5–37.5cm) dark-grey tail lacking 'train'. In flight, short broad wings with chestnut (♂) or brown (♀) primaries. **BP** Bill short, blackish-grey; eyes dark brown; tarsi grey. **Vo** Very loud nasal wailing erWAAH!, oft-repeated in breeding season; may be accompanied by loud rustling made by rattling tail quills. **AN** Common Peafowl; Peacock (Peahen).

REEVES'S PHEASANT
(not to scale)

ad ♀

ad ♂

RING-NECKED PHEASANT

ad ♂
formosanus

ad ♀
pallasi

ad ♂
pallasi

JAPANESE GREEN PHEASANT

versicolor

ad ♂

ad ♀

ad ♀

INDIAN PEAFOWL
(not to scale)

ad ♂

LESSER WHISTLING DUCK CTJ
Dendrocygna javanica

L 38–40cm; WT 450–600g. **SD** Resident from Pakistan to SE Indonesia and S China. Occasional summer visitor to S & E China, accidental to Taiwan and Nansei Shoto, Japan (where formerly resident). Monotypic. **HH** Small to very large flocks at swamps, mangroves and wet ricefields; crepuscular. Perches in trees and walks well on land. **ID** Slightly larger than Eurasian Teal, but long neck and upright stance vey different from *Anas*. Greyish-buff face and neck, somewhat darker on crown. Upperparts dark blackish-brown with warmer rufous-brown fringes to coverts and mantle affording scalloped appearance. Underparts paler rufous-brown, flanks and belly warmer. Dark wings with bright chestnut patch on lesser and median coverts, and uppertail-coverts, and short blackish-brown tail. **BP** Bill small, dark grey; inconspicuous yellow eye-ring, eyes black; long tarsi and toes blackish-grey. **Vo** In flight gives sibilant, shrill, oft-repeated whistled *sweesik*, often in chorus.

SWAN GOOSE CTKJR
Anser cygnoides

Endangered L 81–94cm; WS 165–185cm; WT 2.85–3.5kg. **SD** E Asian endemic breeding from Mongolia, Transbaikalia, to NE China and S Russian Far East. Winters (Sep–Apr) south to E China (Changjiang Valley and L Poyang) and Korea (scarce). Very rare (not annual) winter visitor Japan and Taiwan. Rare and declining. Monotypic. **HH** Breeds in reeds and marshy meadows. On migration and in winter visits wetlands, especially those surrounded by grasslands or flooded vegetation, occasionally wet ricefields and tidal mudflats. **ID** Longest-bodied and most slender goose in region, with long, uniquely bicoloured neck. Long bill and head profile recall Whooper Swan, and feeds on land in similar fashion. Neck appears longer than other geese because of habit of craning neck upwards. Crown and stripe on nape dark chestnut, with buffy cheeks and breast, and white foreneck. Upperparts mid to dark brown, feathers fringed buff. Breast to belly grades from buff to dark brown, also with pale buff fringes; flanks, vent, rump, tail tip and undertail white. **BP** Bill long, black with narrow white border at base; eyes brown; tarsi and toes pinkish-orange. **Vo** Honking and cackling *gahan gahan gagagaga* in flight, like domestic geese ('Chinese Goose' derives from this species).

BEAN GOOSE CTKJR
Anser fabalis

A. (f.) middendorffii L 90–100cm; WS 180–200cm; WT ♂ 5.1kg, ♀ 4.6kg. *A. (f.) serrirostris* L 78–89cm; WS 140–175cm; WT ♂ 3.2kg, ♀ 2.8kg. **SD** Widespread N Palearctic species breeding in taiga and forest-tundra from Scandinavia across Russia north to Arctic, south to Lake Baikal and east to inland Kamchatka and Chukotka. Two taxa, probably full species, in region: larger **Taiga Bean Goose** *A. (f.) middendorffii* (split as *A. fabalis* including race *middendorffii*), and smaller **Tundra Bean Goose** *A. (f.) serrirostris* (split as *A. serrirostris*). Taiga breeds E Transbaikalia, Siberia and Kamchatka, Tundra in north NE Siberia and E Chukotka

south of tundra, and W Kamchatka. Both winter to E China, Korea and Japan. In E Asia, breeds south to W Okhotsk Sea and Amur River; winters largely in Japan (Honshu), these migrating via Sakhalin and Hokkaido, also Korea and locally in E & SE China; rare Taiwan. **HH** Large lakes, marshy wetlands and agricultural land; rivers when staging. **ID** Large and long-winged, with very dark head and neck contrasting with paler underparts. Pale buff fringes to scapulars and wings. Tail dark grey-brown with crescentic white base and narrow white tips. Lower neck, breast and belly paler grey-brown with darker scalloping on flanks; vent, uppertail base and undertail white. Taiga slightly larger, with longer, more slender bill and more Whooper Swan-like head and bill shape. Tundra has shorter thicker neck, rounder head, and stubbier bill, giving strongly sloping profile; appears 'glum-faced'. In flight, mantle, upperwing, rump and underwing all dark (cf. Greylag), whilst white fringes to tertials and scapulars make back more patterned than Greater White-fronted. Whilst Taiga and Tundra clearly differ in proportions from Greater White-fronted, in flight they might be confused; the ad. Beans have pale unbarred bellies. Juv. Greater White-fronted (lacks white facial blaze) best separated by plain drab blackish-yellow bill and voice. **BP** Bill mostly black, with yellow-orange patch near tip (unlike young of other *Anser*). Tundra bill is deep-based, short, with prominent 'grin' line. Taiga bill is longer, more slender. Eyes hazelnut or dark brown; tarsi orange. **Vo** Generally lower, slower and hoarser than other geese, but Tundra, though more similar to Greater White-fronted, gives a more metallic *gyahahaan* (more muffled or hollow than Taiga), whereas Taiga has deeper, more buzzy honking *gahahaan* or *gangh-gangh*, and *kakako* contact call in flight.

GREYLAG GOOSE CTKJR
Anser anser

L 76–89cm; WS 147–180cm; WT 2.5–4.1kg. **SD** From NW Europe to Amur R and NE China, wintering further south in Europe (some resident), India, northern SE Asia and China. In E Asia, *A. a. rubrirostris* migratory, a scarce to accidental winter visitor (Oct–Apr) to Taiwan, Korea and Japan. **HH** Lakes, rivers and wet ricefields in coastal Japan, Korea and SE China. **ID** Large, heavy, rather uniform pale grey goose, similar in size to Taiga Bean, but neck shorter and thicker, head larger and paler, bill shorter and stouter. Plain grey-brown, slightly darker on neck than head, with little contrast between head/neck and body. Broad pale tips to mantle and coverts form barred pattern. Deep dark grooves on neck; dark bars on flanks and some dark spots on belly, but never as extensive as Greater White-fronted. Young very similar but have scalloped rather than neatly barred upperparts. In flight, pale grey primary and lesser coverts contrast with dark grey-brown median coverts and back; rump noticeably pale grey, and grey-brown tail has white base and broad white band at tip; underwing-coverts also pale, contrasting strongly with darker remiges (see Bean and Greater White-fronted geese). **BP** Bill dull pink (orange in western nominate race); weak pink eye-ring, irides black; tarsi and toes dull pink. **Vo** Varied shrill, high-pitched notes and deep raucous honks, *gwen gwen* or *ahng ahng ahng* (deeper and more barked than Bean). Frequently gives clanging calls in flight.

SWAN GOOSE

ad

LESSER WHISTLING DUCK
(not to scale)

ad *serrirostris*

BEAN GOOSE

ad *middendorffii*

ad *serrirostris*

GREYLAG GOOSE

rubrirostris

ad

GREATER WHITE-FRONTED GOOSE   CTKJR
Anser albifrons

L 65–86cm; WS 130–165cm; WT 1.7–3kg. **SD** Breeds in Arctic, from NW Russia to E Chukotka, and Alaska to Greenland, wintering south to Europe, E Asia and SW USA, also in river valleys and other wetlands of NW USA. In E Asia *A. a. frontalis* is typical subspecies (but birds in region sometimes called *albicans*). Winters south to Japan (Honshu, Oct–Mar), passing through Hokkaido in Sep and Apr/May; also Korea and locally in E China (migrant through NE China), accidental Taiwan. Occasional larger individuals reported in Japan perhaps of N American origin. **HH** Breeds on open tundra near coast or inland near marshes, lakes and pools. Highly gregarious on migration and in winter, at large lakes, reservoirs, marshy wetlands especially those surrounded by grassland, or flooded vegetation and agricultural land; typically roosts (often in huge numbers) at lakes and forages on agricultural land in winter. **ID** Compact, mid-sized brownish-grey goose with rather short neck, prominent white face patch and strong flank line; pale breast and belly, with black transverse bars forming individually variable pattern like barcode on belly (often lacking in young). Note that young Lesser and Greater White-fronted and Greylag, have darker grey faces than ads. For separation of imm. from Bean, see latter. **BP** Relatively chunky deep-based bill, dull pink, bright pink or orange with white nail (dull yellow with dark nail in young, paler than Bean), surrounded at base and onto forehead by white blaze (cf. Lesser), developing late in first winter; eyes black; tarsi and toes orange. **Vo** Very vocal in flocks: repetitive, abrupt yapping or musical honking, *kyow-yow* or *kyow-yow-yow*, rapid, high-pitched jerky laughing *widawink widawink…* in flight, and deeper *kuwahahan kuwahahan* or *guwawawan guwawawan* when flushed.

LESSER WHITE-FRONTED GOOSE   CTKJR
Anser erythropus

Vulnerable. L 53–66cm; WS 115–135cm; WT 1.3–2.3kg. **SD** N Palearctic east to eastern Chukotka, in narrow habitat band of Arctic tundra marshes and bogs in uplands; winters in Europe, C Asia and E Asia. Breeds at high latitudes in Chukotka (possibly NE China), in marshes and bogs with willow and birch scrub, and winters (Oct–Mar) in Chiangjiang valley, Japan and Korea, but rare, local and in very small numbers in latter countries; accidental Taiwan. Monotypic. **HH** On migration and in winter typically occurs singly or in small family groups amongst flocks of other grey geese (often Greater White-fronted) at wetland roosts and on agricultural land, but foraging movements (walking and pecking) generally noticeably faster than other geese, thus often near leading edge of moving flocks. **ID** Smallest, most delicate grey goose, the largest birds just reach size of smallest Greater, but with smaller head, higher forehead, smaller bill and shorter (thicker) neck, more prominent white blaze and distinctive yellow eye-ring. Plumage like Greater, but rather darker-backed, belly bars fewer, less distinct and not extending as far onto sides as Greater; head rounder, wings longer and narrower, primaries extend well beyond tail tip. 1st-winter noticeably darker brown and more compact than young Greater. **BP** Bill short, rather delicate and bright pink; narrow white blaze extends on forehead of ad. as triangular wedge when seen front on (reduced or absent in juv. and bordered by narrow black line), higher on crown than Greater and more neatly rounded; prominent narrow yellow orbital in ad. and juv., eyes black; tarsi and toes yellowish-orange. Beware – some Greater show more prominent eye-rings and more extensive white shields. **Vo** Calls higher pitched, sharp and squeaky, more yelping than Greater: *kyuru-kyu kyu kyukyu*; *kyu-u-kyu-u* or *plewewew-whew*!

BAR-HEADED GOOSE   CKJR
Anser indicus

L 71–76cm; WS 140–160cm; WT 2–3kg. **SD** Breeds at high altitudes in C Asia, and migrates over Himalayas to N India and Myanmar for winter. Has straggled to E China, Japan on several occasions (Sep–May), Korea and Russia, though records may include escapes. Monotypic. **HH** Freshwater swamps, lakes and rivers. **ID** Large, very pale grey goose, with distinctive white head and face, white stripe on sides of plain dark grey neck and two black crescents, one across rear crown to eyes, the other shorter, around upper nape; hindneck dark greyish-black. Body pale grey with dark, blackish-grey patch on flanks. Young have greyer neck, white face and solid black rear crown (no bars). In flight, long wings largely pale grey, contrasting with black secondaries and primary tips, otherwise pale, but dark flank patch visible. **BP** Bill dull yellow-orange with black nail; eyes black; tarsi and toes dull yellowish-orange. **Vo** Low, nasal honking *gaaaa gaaaa gaaaa* or *guaa guaa guaa*.

EMPEROR GOOSE   KJR
Anser canagicus

L 66–89cm; WS 119cm; WT 2.766–3.129kg. **SD** Beringian endemic, breeding in coastal regions of extreme NE Chukotka and NW Alaska, and wintering to Kamchatka and Commander and Aleutian Is. Strays regularly down Pacific N American coast to California, but in Asia extremely rare south of Kamchatka; accidental Japan and Korea. Monotypic. **HH** Breeds on low coastal tundra, near lagoons, lakes or pools; on migration and in winter along rocky shores, sandy coasts and coastal marshes/wetlands. **ID** Uniquely patterned. Small, stocky goose, largely dark silver-grey, with black chin, throat and foreneck, white head, face, hindneck and tail. Head rather rounded with high forecrown; white head plumage may become stained rusty-orange in summer. Young have grey neck and head and entire plumage has rounded pale fringes, affording scalloped pattern as in young grey geese, though entirely blue-grey. In flight, all-dark, grey wings, blackish rump and white tail unique. **BP** Bill small, pink in ad., greyish-pink in juv.; eyes black; legs and feet bright orange. **Vo** Low grunting sounds on ground, and rapid, high-pitched double- or triple-note honks *kurahha kurahha* or *kla-ga kla-ga* on take-off and in flight; also *urru guup rugu* in alarm. **TN** Formerly *Chen canagicus*.

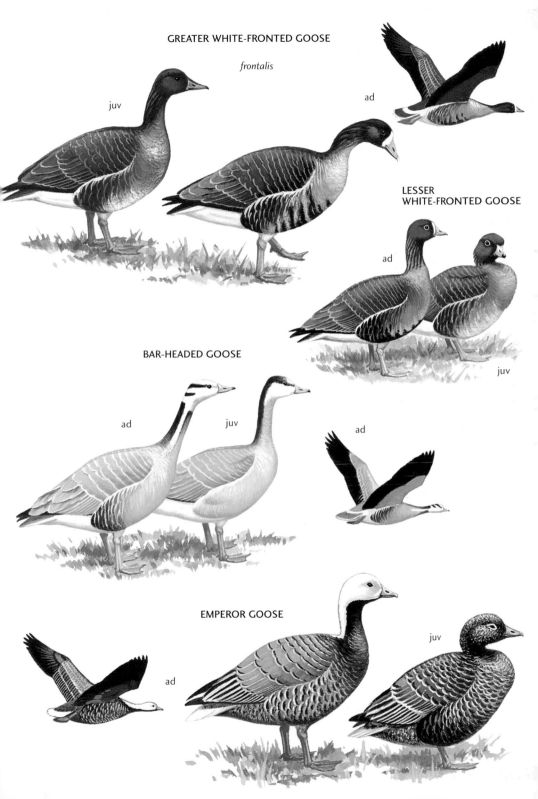

GREATER WHITE-FRONTED GOOSE

frontalis

juv

ad

LESSER WHITE-FRONTED GOOSE

ad

juv

BAR-HEADED GOOSE

ad

juv

ad

EMPEROR GOOSE

ad

juv

SNOW GOOSE CKJR
Anser caerulescens

L 66–84cm; WS 132–165cm; WT 2.42–3.4kg. **SD** Alaska to NE Canadian Arctic, wintering south to California and Texas, but in Asia (where once abundant) now breeds only on Wrangel I., and N & E Chukotka. Individuals, or family groups, reach Japan and, more rarely, E China coast and Korea (Oct–Mar). Most records *A. c. caerulescens* (**Lesser Snow Goose**), but *A. c. atlanticus* (**Greater Snow Goose**), of Canada and Greenland, vagrant to Japan. **HH** Breeds colonially on open tundra. On migration and in winter, lakes, marshes and agricultural land. **ID** Adult a small all-white goose with black primaries. 1st-winter has dingy grey upperparts and neck. See Ross's Goose. N American Lesser Snow has dark morph ('Blue Goose'; accidental N Japan). Largely dark blue-grey with white head and wing-coverts, but range of intermediates between 'blue' and white extremes. 'Blue Goose' recalls Emperor, but has clear-cut white hindneck, dark foreneck and much larger bill. *A. c. atlanticus* larger, heavier, with longer, deeper bill and obvious 'grin'. Beware Snow/Greater White-fronted hybrids. **BP** Bill pink with prominent black 'grin' (cutting edge) between gape and tip; eyes black; tarsi pink. **Vo** Various calls, a deep *angk-ak-ak-ak* in alarm to more typical repetitive, soft nasal monosyllabic *whouk* or upward-inflected heron-like *heenk*; low-pitched grunts when foraging. **TN** Formerly *Chen caerulescens*.

ROSS'S GOOSE R
Anser rossii

L 53–66cm; WS 115–130cm; WT 1.224–1.633kg. **SD** Arctic Canada, wintering south (some mixing with Lesser Snow Geese) to California, Texas and Mexico. Vagrant on Wrangel I., Russia. Monotypic. **HH** Tundra and short-grass areas. **ID** A very small all-white goose with black primaries. 1st-winter has dingy brownish-grey upperparts, neck and head, with greyer flight-feathers. Like Snow Goose, but has more rounded head, smaller bill (recalls Lesser White-fronted) and line separating bill from facial feathering is straight (convex in Snow). **BP** Bill short, delicate, blackish-pink at base, bright pink at tip, lacks prominent black 'grin' of Snow; eyes black; tarsi pink. **Vo** Higher pitched and faster than Snow, a low grunting *kowk* or higher, squealing *keek-keek*, but generally quiet. **TN** Formerly *Chen rossii*.

LESSER CANADA GOOSE CTKJR
Branta hutchinsii

L 55–66cm; WS 115–130cm; WT 2.3kg. **SD** Polymorphic Canada Goose of N America recently split into two species, of which Lesser Canada Goose occurs in our region. Most records in NE Asia pertain to **Aleutian Goose** *B. h. leucopareia*, though **Cackling Goose** *B. h. minima* recorded Japan, and possibly **Taverner's Goose** *B. h. taverneri*. Aleutian increasing, following eradication of foxes on breeding islands. Formerly bred NE Russia, but now accidental only in Yakutia, Chukotka and Commander Is; very rare but increasing winter visitor to Japan (Oct–Apr); accidental Korea and Taiwan. **HH** On migration and in winter, lakes, marshes and agricultural land. **ID** Small, dark grey-brown goose, slightly larger than Brent. Head (rather square) and neck black with white chin and cheeks, and white ring at base of neck (sometimes distinct, or incomplete and occasionally

lacking), with blackish feathers below. Upperparts dark ashy grey-brown with buff fringes to mantle, scapulars and coverts. Slightly paler on breast and belly, white on vent. In flight very dark, only vent and rump white. Wingbeats rather fast and wings held forwards. *B. h. minima* slightly smaller, with darker brown breast (purplish cast) and back, short neck, barely visible collar (sometimes absent), more rounded head, small bill and relatively long legs. Larger individuals, possibly *B. h. taverneri*, seen in Japan. **BP** Small black bill tapers to narrow tip; eyes black; tarsi black. **Vo** Rather high-pitched, even squeaky *yeek* or *uriik*, and deeper nasal *guwa guwa*.

GREATER CANADA GOOSE CJ
Branta canadensis

L 68–114cm; WS 119–152cm; WT 2.3–4.5kg. **SD** Alaska and Canada, with well-established introductions elsewhere in USA, W Europe, New Zealand and, locally, E China (Beijing). A large subspecies, established in Yamanashi, Shizuoka and Aichi prefectures, Honshu, Japan, is expected to spread. A very large, apparently wild goose in Hokkaido in 2006 was perhaps *B. c. parvipes*. **HH** Wetlands from grassy tundra to city parks. **ID** Medium to large, black-necked, white-cheeked grey-brown goose, with paler and greyer lower neck/breast. Upperparts mid to dark ashy grey-brown with narrow buff/grey fringes to mantle, scapulars and coverts. Slightly paler on breast and belly, white on vent. In flight, contrast between black neck and paler body/wings obvious, as is white rump against black tail. Often hybridises with Snow and Greater White-fronted geese. **BP** Short to long black bill; eyes black; tarsi black. **Vo** Repetitive deep nasal honks and distinctive bisyllabic *gah-hut*, with clear pitch change from low to high.

BRENT GOOSE CTKJR
Branta bernicla

L 55–66cm; WS 110–120cm; WT 1.2–2.25kg. **SD** Holarctic on tundra, wintering south on coasts. In E Asia, *B. b. nigricans* (**Black Brant**) breeds on coasts of N & E Chukotka, and Wrangel I. Winters (mid Oct–late Apr) south to E China (Yellow Sea south to Hebei), Korea (local E coast), and Japan (mainly N, rare in S); accidental Taiwan.

HH Breeds on lowland, usually coastal, tundra. On migration and in winter forages in shallow bays, coastal lagoons, estuaries, sandy shores; very rare inland, joins other geese at lakes or on agricultural land. **ID** Smallest common goose (same size as Lesser White-fronted and Ross's). Dark, with strongly contrasting white on flanks, tail and vent. Head and neck black, as are upperparts and breast, albeit fringed grey, giving frosted appearance. Ad. *nigricans* has prominent broad white necklace on throat, broadest at front, with dark flecks. Tail-sides, vent and uppertail-coverts white (mostly conceal blackish tail). Extensive white barring on flanks. Juv. lacks white necklace and has broad pale tips to upperwing-coverts. In flight, only white vent and uppertail-coverts prominent. Wings pointed, slightly swept back; flight fast, often low, and usually in large straggling flocks, less linear than larger geese. **BP** Bill very short, dark blackish-grey; eyes black; short tarsi black. **Vo** Rolling, guttural bark-like *gururu* or *guwawa*, deeper than other geese. Flocks maintain constant low murmuring or gargling sounds. **AN** Brant.

SNOW GOOSE

ad *atlanticus*
blue morph

ad *caerulescens*

ad *caerulescens*

juv *caerulescens*

ad

ROSS'S GOOSE

LESSER CANADA GOOSE

ad *leucopareia*

juv

ad *minima*

ad

GREATER CANADA GOOSE

parvipes

BRENT GOOSE

nigricans

ad

ad

MUTE SWAN *Cygnus olor* CTKJR

L 125–160cm; WS 200–240cm; WT 6.6–16kg. **SD** E Europe and C Asia; introduced W Europe, N America and elsewhere. Breeds locally NE China. Scarce winter visitor to Yellow R delta, S China, and Korea; accidental elsewhere, with well-established introduced, partly migratory, population in N Japan (W Hokkaido, C & W Honshu). Monotypic. **HH** Breeds on shallow lakes; moves s. to winter on lakes and rivers. Introduced birds confiding but can be aggressive. **ID** Probably heaviest species breeding in region. Graceful S-shaped neck. Ad. white, sometimes tinged rusty-orange on head and neck. Juv. dark grey-brown, whitening with age, but introduced population has high percentage of all-white ('Polish') cygnets. On water, wings form high cushion-like dome, tail longer than congeners, more pointed and typically cocked. Runs heavily and noisily across water to take-off, once airborne wingbeats laborious. **BP** Deep orange-red bill of ad. has black tip, nostril, base and knob (larger in ♂), black facial skin reaches eye. Juv. bill black, fading to dull grey-pink, lacks knob; eyes black; legs and feet black. **Vo** On territory, snake-like hissing, grunts and explosive snorting *nyeeorrr*. In flight wings make loud rhythmic throbbing, quite unlike other swans.

BLACK SWAN *Cygnus atratus* J

L 110–140cm; WS 160–200cm; WT 3.7–8.75kg. **SD** Widely introduced (from Australia) to ornamental waters; has perhaps bred Japan. **ID** All black, with extremely long, slender neck, raised 'crinkly' wing-feathers and white primaries in flight. **BP** Bill and bare skin extending to eyes waxy red, with white subterminal band; eyes red; tarsi dark grey or pinkish-grey.

WHOOPER SWAN *Cygnus cygnus* CTKJR

L 140–165cm; WS 205–243cm; 7.5–12.7kg. **SD** Palearctic breeder, from Iceland to E Russia. In E Asia, breeds from Lake Baikal to E Chukotka and Kamchatka; winters (Oct–mid April) south to E Hokkaido and N Honshu, also Korea and E China; accidental Taiwan. Monotypic. **HH** When breeding, pairs typically shy and solitary, on isolated pools, lakes and marshes in taiga, and at riverine wetlands; on migration and in winter highly social, at large coastal bays, large rivers and ice-free lakes, less often farmland. **ID** Large and long-necked, upper neck held rather vertical, head horizontally, crown peaks towards rear of head. Long straight bill gives imperious profile. Ad. white, often tinged rusty-orange on head, neck and breast. Juv./1st-winter dark grey, whitening with age. On water, folded primaries cover most of rather rounded tail, which lies horizontally. More agile on land than Mute, and runs for take-off more easily. In flight forms lines and Vs. **BP** Bill (ad.) largely black with extensive deep yellow wedge from feathers, on culmen and sides to at least nostrils, often beyond. Some have black line at feathers. Bill initially pink (juv.), whitens, then slowly becomes yellow in late winter. Eyes brown, but blue and blue-grey not uncommon; tarsi and feet black. **Vo** Calls loudly and frequently on water and in flight. Bugling similar to Bewick's but typically louder and deeper, and given in threes and fours *a-hoo a-hoo a-hoo*. Also frequent soft *hu-hu* or *koo koo* from winter flocks. During social interactions in winter and displays in summer, strong bugling calls in duet.

TRUMPETER SWAN *Cygnus buccinator* JR

L 150–180cm; WS 230–260cm; WT 7.3–12.5kg. **SD** N American species increasing and expanding after earlier population collapse. Vagrant Chukotka, perhaps summering, and has wintered N Japan (Oct–Apr). Monotypic. **HH** Shallow ice-free lakes, lagoons and rivers, with other swans; on migration also forages on farmland. **ID** Large, long-necked swan resembling Whooper. Ad. pure white. Young retain grey-brown on mantle, wings, head and neck through first winter. Feathering on forehead extends forward to point (cf. dark-billed Whistling). **BP** Bill long, straight and rather angular, black with dark red wedge at base and on lower cutting edge, exposed when beak open (though red 'lips' or gape line sometimes lacking); bare black facial skin around eye, juv. has black-based and black-tipped dull pink bill; eyes black; tarsi olive-brown in young, black in ad. **Vo** Softer nasal honking is less barking than Whooper.

TUNDRA SWAN *Cygnus columbianus* CTKJR

C. c. columbianus L 120–150cm; WS 167–225cm; WT 4.3–9.6kg. *C. c. bewickii* L 115–140cm; WS 167–225cm; WT 3.4–7.8kg. **SD** Holarctic tundra breeder; moves to temperate zone for winter. In E Asia, **Bewick's Swan** *C. c. bewickii* breeds largely to west of our region, and winters (Oct–mid April) in large numbers in Japan (migrating via Sakhalin and Hokkaido to Honshu), in Korea, and uncommonly in NE & E China; accidental Taiwan. **Whistling Swan** *C. c. columbianus* breeds NE Chukotka; most winter in N America, though some reach Japan. Some difficult to assign to race, presumably due to mixed pairings. **HH** Nests dispersed across tundra; on migration and in winter highly social at ice-free lakes; on migration also farmland. **ID** Smaller and more compact than Whooper, with shorter neck, proportionally larger, rounder head and gentler profile. Head peaks at mid crown, and feathering on forehead rounded. All-white plumage may be stained rusty-orange, especially on head, neck and breast. Young dark grey, whitening with age. On water, folded primaries cover most of tail, which lies horizontally. On land, more agile than Whooper or Mute. In flight, travels in lines and Vs. Whistling closely resembles Bewick's, but generally larger, with heavier, deeper based bill and different bill pattern. **BP** Bill largely black with some yellow; pattern subspecifically and individually variable. Bewick's shorter, slightly more concave, and proportionally broader at tip than Whooper's. Culmen often partly, or entirely, black to feathers, but may have rounded deep yellow patch from eyes towards nostril. Yellow patch smaller, more rounded, or squarer, less wedge-shaped than Whooper's. Commonly has narrow yellow wedge at base of lower mandible. Young initially deep reddish-pink, then black, yellow portion initially white until first winter. Whistling has larger, less concave bill (more Trumpeter-like in shape), all-black culmen, yellow patch generally limited to pre-orbital teardrop. Entirely black-billed individuals (scarce) resemble Trumpeter, but overall smaller size, rounder head shape, and rounded feather line on forehead distinguish it. Hybrid Bewick's x Whistling has intermediate pattern or resembles either parent. Eyes typically brown, occasionally blue or blue-grey; tarsi and toes black. **Vo** Bewick's has higher-pitched, more yapping calls than Whooper (some overlap). Calls in winter and on migration usually given singly or doubled, a nasal *koho* or *koho koho*. Whistling has higher pitched more musical *klooo* or *kwooo* call than Bewick's.

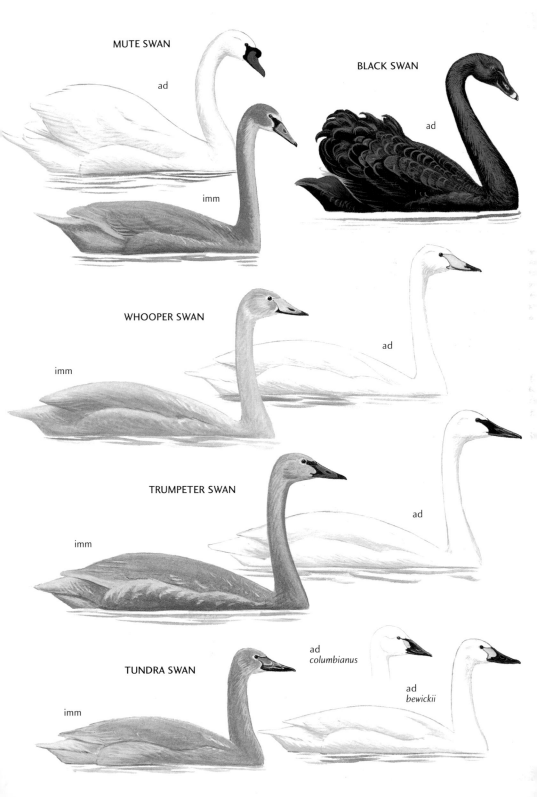

MUTE SWAN

ad

imm

BLACK SWAN

ad

WHOOPER SWAN

imm

ad

TRUMPETER SWAN

imm

ad

TUNDRA SWAN

imm

ad
columbianus

ad
bewickii

PLATE 11 : SHELDUCKS, MANDARIN DUCK AND PYGMY-GOOSE

COMMON SHELDUCK CTKJR
Tadorna tadorna

L 55–65cm; WS 100–133cm; WT 0.8–1.45kg. **SD** W Europe to C Asia and NE China; winters south to N Africa, India and coastal SE China. Winters (Oct–Apr) commonly in SW Korea, and locally in S Japan (especially Kyushu); scarce winter visitor Taiwan, accidental S Russian Far East. Monotypic. **HH** Breeds in dry habitats (burrow-nesting) often near saline lakes; winters at coastal mudflats and wetlands, rarely on land. Commonly seen walking across mudflats or in shallow water, sieving food from water and wet mud. **ID** Large, plump, black, white and chestnut duck, with goose-like proportions, though head larger, neck shorter and legs longer. Ad. has blackish-green head and scapulars, and white body with broad dark chestnut band on chest and upper back. Flight-feathers, scapulars, belly stripe and tail black, secondaries with green gloss, tertials chestnut. Juv./1st-winter has grey head, neck and upperparts, and white-tipped flight-feathers forming trailing edge to wing. In flight, clearly pied; wings long, pointed, white coverts contrast with black remiges, beats strong and fast. **BP** Bill short, broad (especially at tip), deep red in ad. ♂ (with prominent knob on forehead), dull pinkish-red in ♀ (knob lacking), which has white flecking on face; eyes black; tarsi and toes pale pink. **Vo** Unlike other ducks; ♂ gives sibilant single- or double-noted whistled *suwees-suwees* on water or in aerial pursuit, also a rapid, somewhat goose-like, guttural growling; ♀ gives nasal quacking *gaggagagaga...* Wings emit dull whistle in flight.

RUDDY SHELDUCK CTKJR
Tadorna ferruginea

L 58–70cm; WS 110–135cm; WT 0.925–1.64kg. **SD** Mainly C Asia, ranging east of Lake Baikal in S Russia and across N China, but not to Asian coasts. Winters in SE China, northern SE Asia and S Asia; stragglers occur east to Asian coast; locally common (mid Oct–Apr) Korea, rare Japan and Taiwan. Monotypic. **HH** Breeds (burrow-nesting) at inland wetlands, rivers and streams in arid regions, including steppe, and at high altitude. Generally winters on agricultural land (where walks well and grazes like a goose) or at coastal wetlands, where may join other waterfowl. **ID** Large orange-brown duck, larger than smallest geese. Body bright orange-chestnut, paler peach on upper neck and head; sometimes white around eyes or bill base. ♂ (summer) has narrow black collar. ♀ has paler head and lacks collar. Rump and tail black. Dark flight-feathers (black primaries, green-glossed secondaries) contrast strongly with white forewing, and white underwing-coverts. Young have greyer forewing. **BP** Bill blackish-grey; eyes black; tarsi black. **Vo** Calls both day and night; rather loud, nasal honking, somewhat goose-like *aakh* or *aang-aang*, an abrupt *pok-pok-pok* and rolling *ahrrrr* or *porrr*.

MANDARIN DUCK CTKJR
Aix galericulata

L 41–51cm; WS 65–75cm; WT 444–500g. **SD** Endemic to E Asia, where relatively scarce except in Japan; introduced into Europe and parts of N America, where apparently increasing. Breeds S Russian Far East, NE China, N Korea, C & N Honshu and Hokkaido, Japan, and Taiwan. Winters from C Honshu and S Korea south to SE China, occasionally northern SE Asia. Monotypic. **HH** In summer, favours secluded forested rivers and streams, where nests in tree cavities. On migration and in winter forms flocks; winters at ponds and lakes with wooded margins, or fast-flowing rivers, roosts in trees and often feeds onshore; on migration may appear at almost any waterbody, preferably with well-vegetated margins. **ID** Medium-sized duck, slightly smaller than Eurasian Wigeon, with large, maned head and long tail. Breeding ♂ gaudy; largely orange-brown, forehead green, hindcrown orange; white stripe from eye arcs back over head-sides with prominent orange 'whiskers' and 'mane' making head appear even larger, and sweeping creamy white band from eyes to tip of mane. Enlarged orange tertials form distinctive 'sails' rising vertically over lower back. Vertical black and white stripes on breast-sides contrast with purplish-black breast and orange-brown flanks. ♀ mostly grey and olive-brown, with neat white spectacles around eyes and white line behind eyes on plain grey head, white also at bill base. Lower breast and flanks have pale round spots. Eclipse ♂ resembles ♀, but has bright pink bill. In flight, appears large-headed, with plain dark grey-brown wings, blue-green speculum with white trailing edge to secondaries; ♂'s chestnut tertials visible. **BP** Bill reddish-pink (♂ all year) or grey (♀) with white nail; eyes black; tarsi orange. **Vo** ♂ gives rather varied high, melodious, upwardly inflected whistles: *kehp kehp, wippu,* or *pyui,* and rather high, hoarse *vhett* or *ghett* calls, whereas ♀ gives deeper clucking *kyu* or *kwa* calls.

COTTON PYGMY-GOOSE CTJ
Nettapus coromandelianus

L 31–38cm; WT 380–403g. **SD** Resident from India to E Australia, and summer breeder north to SE & E China. *N. c. coromandelianus* accidental to Hebei, Taiwan, and Nansei Shoto, Japan. **HH** Freshwater ponds, well-vegetated streams and wet ricefields. **ID** Tiny, very pale goose-like duck with black cap. Breeding ♂ white and green, with plain white face, black crown, black band on breast and lower neck. ♀ less contrasting, browner instead of blackish-green, and lacks black breast- and white wing-band. ♀, young and eclipse ♂ all have black stripe from bill to ear-coverts. In flight, wings largely bottle green with broad white band on remiges (♂) or greyish-black with white trailing edge to secondaries (♀). **BP** Black bill is tiny; eyes dark reddish-brown (♂) or brown (♀); tarsi black. **Vo** ♂ gives rather low, nasal *wuk wirrarakwuk,* ♀ gives a weak *quack*; groups utter babbling *nyar-nyar-nyar*. **AN** Cotton Teal.

COMMON SHELDUCK

juv

ad sum ♂

ad ♂

RUDDY SHELDUCK

ad

ad ♂

ad ♀

ad ♀

ad ♀

COTTON PYGMY-GOOSE

coromandelianus

ad ♀

MANDARIN DUCK

ad ♂

ad ♂

ad ♀

ad ♂

ad ♂

GADWALL
Anas strepera

CTKJR

L 46–58cm; WS 78–95cm; WT 850–999g. **SD** Common Holarctic duck. *A. s. strepera* breeds in our region in NE China, Russian Far East, possibly northern Korea, locally in E Hokkaido and Kamchatka, wintering (mid Sep–Apr) south to E China, Korea and Japan (C Honshu to Kyushu); occasional Taiwan. **HH** Shallow freshwater lakes, coastal lagoons and rivers on migration and in winter; prefers secluded, well-vegetated wetlands for breeding. **ID** Fairly large, rather dark dabbling duck, slightly smaller and noticeably slimmer than Mallard, with more rounded head and smaller bill. Breeding ♂ has dull brownish-grey head, dark grey body with pale breast scalloping, black tail-sides, vent, rump and tail-coverts. Chestnut patch and white square visible on closed wing. Eclipse ♂ resembles ♀, but has plain grey tertials and retains wing pattern. ♀ dull blackish-brown like ♀ Falcated (but has white patch on closed wing), with dark tertials; tail and vent brown, not black. Floats higher than Mallard and feeds more like a wigeon. In flight, wings narrower than Mallard; ♂ has prominent white speculum, black and chestnut patches on upperwing; ♀ also has white speculum (less conspicuous, even absent), but lacks chestnut patch; belly white but less clean-cut than Eurasian Wigeon. **BP** Bill dark greyish-black (♂), or has grey culmen with orange sides speckled grey (♀); eyes black; tarsi reddish-orange. **Vo** ♂ gives high-pitched whistle and low croaking *kua*; ♀ a repetitive rather high-pitched quacking *gaa-gaa-gaa* or *ge-ge-ge* recalling Mallard, but more nasal. In flight, harsh *gerssh gerssh*.

FALCATED DUCK
Anas falcata

CTKJR

L 46–54cm; WS 78–82cm; WT ♂ 590–770g, ♀ 422–700g. **SD** E Asia west to Lake Baikal in summer and south to N India and S China in winter. In region, breeds from NE China to Russian Far East, Sakhalin, Hokkaido, Kuril Is, and Kamchatka. Winters (Oct–mid Apr) in E China, Korea, occasionally Taiwan, and fairly commonly in Japan south of Hokkaido. Monotypic. **HH** Breeds at well-vegetated lakes and marshes. Post-breeding commonly associates with other *Anas* wintering at lakes, rivers and coastal wetlands. **ID** Large and unmistakable Asian speciality. Breeding ♂ largely pale grey, with large head, forehead and crown all dark metallic chestnut, facesides, 'whiskers' and extensive 'mane' glossy green, contrasting with white chin and cheek bar, and black and white neck bars. Small white spot at base of culmen unique; breast and flanks finely vermiculated black and white, tail black but sides yellowish-white, largely obscured by long elegant drooping (falcated) black tertials with pale grey fringes. ♀ like blackish-brown ♀ Eurasian Wigeon (without white belly), with (squarer) plain grey head and blackish-grey bill; underparts scalloped chestnut and dull brown, exposed tertials grey-fringed. Eclipse ♂ resembles ♀, but has darker head, pale grey forewing, and may retain long tertials. In flight, ♂ has prominent pale grey forewing and blackish secondaries; ♀ has grey forewing and whitish greater covert bar; short rounded tail. **BP** Both sexes have blackish-grey bill, black eyes and blackish-grey tarsi. **Vo** ♂ gives disyllabic *foo-ee* and whistle-and-buzz *foo-ee-brurururu*, ♀ a hoarse quacking.

EURASIAN WIGEON
Anas penelope

CTKJR

L 42–50cm; WS 71–85cm; WT 415–970g. **SD** Europe to extreme NE Russia. In region, breeds NE China and Russian Far East, Chukotka and Kamchatka, wintering (mid Sep–late Apr) south to E China, Taiwan, Korea and Japan in large numbers. Monotypic. **HH** Summers at tundra pools and taiga forest ponds, wintering on lakes, coastal lagoons and large rivers, especially near their mouths; often grazes onshore. Forms large flocks, frequently with other *Anas*. **ID** Medium-sized, compact, with rather short neck, large rounded head and small bill. Breeding ♂ has bright chestnut head and creamy buff blaze from forehead to crown. Breast vinaceous, upperparts and flanks pale to mid grey finely vermiculated white; rear flanks white contrasting with pointed black tail. Ad. ♀ rather uniform dark rufous-brown, best identified by shape. Eclipse ♂ like ♀, but has white forewing. In flight (fast and agile) wings pointed, slightly swept back at tips; large head on narrow neck and pointed tail visible; forewing of adult ♂ has large white patch and dark green speculum; both sexes have a prominent white belly and dusky-grey axillaries and underwing-coverts (unlike American Wigeon). **BP** Both sexes have pale blue-grey bill with black tip, lacking black at gape; eyes black; tarsi dark grey. **Vo** ♂ gives high, whistled *fee-oo* or *pee-you*, often preceded by brief low *wu-fee-oo* and more subdued whistles; ♀ gives crackling, quacking *gwa gwa gurrerr*.

AMERICAN WIGEON
Anas americana

TKJR

L 45–56cm; WS 76–89cm; WT 680–770g. **SD** Common N American breeder, wintering south to C America, but also breeds (rare) in NE Russia, mainly Anadyr valley. Rare but annual winter visitor in small numbers (Oct–mid Apr, occasionally May) to Korea and Japan; accidental Taiwan. Monotypic. **HH** Commonly joins Eurasian Wigeon at lakes, large rivers, coastal wetlands and, occasionally, on sea. Hybrids rare NE Russia, but frequent in winter in Japan. **ID** Size and shape as Eurasian Wigeon, but has larger head with steeper forehead and more bulging nape. Breeding ♂ has greyer head and neck than Eurasian; speckled grey face, prominent glossy dark green crescent from eyes to nape, and white or buff (not cream/yellow) blaze on forecrown. Breast, flanks and back pinkish-brown, lacking grey of Eurasian. ♀ has paler grey head, especially forehead and lores, and darker blackish-grey patch around eye, affording more masked appearance than Eurasian, with darker rufous flanks. In flight, both sexes have prominent white belly and ♂ a white forewing, but axillaries and underwing-coverts white (grey in Eurasian), and ♀ has white bar on greater coverts. Hybrids occur fairly frequently, confusing the issue. **BP** Bill short, pale blue-grey with black tip; eyes black; tarsi dark grey. Essentially as Eurasian, but narrow black border at gape lacking in Eurasian. **Vo** ♂ gives distinctive bi- or trisyllabic whistle resembling Eurasian: *wiwhew*; ♀ a low growling *warr warr warr*.

GADWALL

ad ♂

ad ♂

ad ♀

strepera

ad ♀

FALCATED DUCK

ad ♂

ad ♂

ad ♀

ad ♀

EURASIAN WIGEON

ad ♂

ad ♀

ad ♂

ad ♀

AMERICAN WIGEON

ad ♀

ad ♂

ad ♂

ad ♀

BLACK DUCK KJ
Anas rubripes

L 53–61cm; WS 89cm; WT 1.15–1.35kg. **SD** N American species largely confined to E of continent, but has strayed to Alaska (though introductions in W USA cloud vagrancy picture) and to Korea and Japan (though some doubt attached to these records, despite Korean one apparently being of a ringed bird). Monotypic. **ID** Sexes alike. Size and structure as Mallard, recalls ♀ Mallard but much darker with very dark, cold brown body, wings and tail; dark-centred feathers have paler fringes; head paler grey-brown, streaked greyish on throat, crown and eyestripe dark blackish-brown, offering marked contrast between head/neck and darker body. Often hybridises with Mallard. In flight, speculum dark blue or purple, lacking obvious white borders of Mallard, though a faint rear border may be present. **BP** Bill greenish-yellow in ad. ♂, olive-green in ♀ and juv.; eyes black; tarsi red-orange (♂) or brownish-orange (♀). **Vo** Like ♀ Mallard, but quack lower pitched.

MALLARD CTKJR
Anas platyrhynchos

L 50–60cm; WS 81–95cm; WT 0.75–1.575kg. **SD** Commonest large Holarctic dabbling duck, generally the most familiar, and origin of many domesticated forms. In region, *A. p. platyrhynchos* breeds from NE China, northern Korea northeast through Russian Far East and Hokkaido to Kamchatka and Commander Is; small population also in C Honshu. Winters (mainly Oct–Apr) south to E China, Taiwan (uncommon), Korea and throughout most of Japan south of Hokkaido (scarce in winter; some in summer). **HH** Breeds at wetlands; on migration and in winter occurs in flocks at ponds, lakes, rivers, wet ricefields, estuaries and sometimes on sea; often in large numbers with other *Anas*. **ID** Large, stocky, with large head and bill and short tail. Breeding ♂ has glossy dark bottle green head, separated from purplish-chestnut chest by narrow white collar. Body pale grey, 'stern' black, tail white with upcurled black central uppertail-coverts; white-bordered blue speculum visible on closed wing. Eclipse ♂ like ♀, but retains ochre bill and black rump. ♀ brown, streaked, dark crown and eyestripe darker than face, belly pale brown. In flight, wings broad and blunt, wingbeats rather slow, heavy and shallow (slower than most other dabbling ducks, but compare Eastern Spot-billed Duck); blue speculum with white borders prominent in both sexes; underwings pale, greyish-white, belly lacks white patch. **BP** Bill dull yellow-ochre in ♂, orange with black culmen and centre in ♀; eyes black; tarsi orange. **Vo** ♂ gives soft rasping, nasal *kwehp* and sometimes soft teal-like whistled *piu*. ♀ gives archetypal loud quack, *gwaa kuwa kuwa* and laughing, descending series, *quek quek quek quek quak quak*.

PHILIPPINE DUCK TJ
Anas luzonica

Vulnerable. L 48–58cm; WS 84cm; WT 725–977g. **SD** Rare resident in Philippines; accidental to Taiwan and S Japan. Monotypic. **HH** Lowland lakes, mangrove-lined estuaries and marshes, with other waterfowl. **ID** Medium-sized dabbling duck (sexes alike) with black crown, nape and eyestripe, rufous-brown supercilium, face and upper neck, dark grey-brown upperparts, pale greyish-buff breast and dark grey-brown underparts with some pale feather fringes affording scalloped appearance to flanks; rump and vent black, tail dark grey. In flight, pale underwing with dark trailing edge to grey secondaries, and bright green speculum bordered above and below with black and white (also visible on closed wing). **BP** Bill bluish-grey; eyes dark brown; tarsi dark greenish-brown. **Vo** Resembles harsh Mallard, similar to quacks of Eastern Spot-billed Duck.

EASTERN SPOT-BILLED DUCK CTKJR
Anas zonorhyncha

L 58–63cm; WS 83–91cm; WT 0.75–1.5kg. **SD** Quite common, largely resident and widespread in E China, Korea and most of Japan, but summer visitor (May–Sep) further north to NE China, northern Korea, Russian Far East, Hokkaido, S Sakhalin and the southern Kuril Is. **HH** Wetlands, from lakes and ricefields, agricultural ditches and streams to smallest urban ponds, even in city centres. In winter, with other *Anas* at ponds, lakes, rivers, and wet ricefields. **ID** The largest and heaviest dabbling duck; stocky, with large head and long bill. Sexes similar; very dark above and below, with pale face and black 'stern'. Crown, eyestripe and lower cheek bar dark ashy-brown; supercilium whitish. Closed wing shows small area of blue speculum and white tertials. In flight, brown wings have dark blue or purple speculum with narrow white borders, usually only hind border prominent; underwings white. Flight heavy with slower beats than most dabbling ducks except Mallard. **BP** Black bill is long, broad, especially at tip, which has large yellow spot; eyes dark reddish-brown; tarsi bright orange. **Vo** Mallard-like, but stronger. ♀'s series of quacking *gwe gwe* or *guwa guwa* calls often louder and in descending series. **TN** Formerly considered a race of Spot-billed Duck *A. poecilorhyncha*. **AN** Chinese Spot-billed Duck.

BLACK DUCK

ad ♂

MALLARD

✓

platyrhynchos

Tokyo 2/16/19

ad ♂

ad ♀

PHILIPPINE DUCK

ad

EASTERN SPOT-BILLED DUCK

ad

BLUE-WINGED TEAL J
Anas discors

L 35–41cm; WS 58–69cm; WT 266–410g. **SD** Widespread in N America northwest to Alaska; winters south to northern S America. Accidental winter visitor to Japan. Monotypic. **HH** Winters on brackish or saline wetlands, usually at or near coast. **ID** Size of Garganey, with similar long bill. Breeding ♂ head dark slaty blue-grey with prominent white facial crescent; breast and underparts warm orange-brown with black spots and vermiculations; upperparts darker brown, elongated tertials fringed orange-brown, rear flanks white, undertail-coverts and tail black. ♀ (and eclipse ♂) duller grey-brown with pale buff fringes to most body-feathers and upperparts; dark eyestripe splits white eye-ring, white loral spot connected to white chin. Eclipse ♂ has more distinct ghost of facial crescent. In flight, ♂ has prominent blue forewing, contrasting with dark grey primaries, and dark green speculum bordered in front by bold white bar; underwing white with a broad black leading edge and white axillaries (compare Garganey). ♀ has similar wing pattern, but blue forewing less bright and white bar narrower. Similar Garganey ♂ has silver, not blue, forewing, narrow green speculum bordered above and below with white; ♀ has grey forewing and similar speculum pattern to ♂. ♀ Garganey has darker eyestripe, and pale loral spot separated from whitish chin. **BP** Bill rather long, dark blackish-grey; eyes black; tarsi dull ochre. **Vo** ♂ gives thin whistled *peeew* or *tui*, whilst ♀ utters coarse, nasal quacking recalling Northern Shoveler; also *tui* or *tsi tsi* calls.

NORTHERN SHOVELER CTKJR
Anas clypeata

L 44–52cm; WS 73–82cm; WT 0.41–1.1kg. **SD** Wide-ranging Holarctic breeder wintering south to Africa, S & SE Asia and C America. In region, breeds in NE China and Russian Far East, Chukotka and Kamchatka, also probably Hokkaido; winters (mid Sep–late Apr) in E China, Korea, Japan and Taiwan. Monotypic. **HH** Breeds in wetlands, typically shallow with plentiful vegetation; on migration and in winter usually found in small numbers amongst larger flocks of *Anas* at lakes and coastal lagoons. **ID** Unmistakable, rather large dabbling duck with larger, more spatulate, bill than any other duck. Short neck and large bill give it front-heavy look on water and in flight. Breeding ♂ dark bottle green head, white breast and chestnut flanks and belly distinctive. ♀ resembles ♀ Mallard, but smaller, with much larger bill, and green speculum on closed wing. Eclipse ♂ resembles ♀, but has grey head with

pale crescent in front of eye, and retains wing pattern. In flight, ♂ has prominent blue forewing and green speculum with bold wedge of white in front; white underwing-coverts contrast with blackish flight-feathers. ♀ in flight has grey forewing, no white trailing edge to dull green speculum, darker belly than Mallard and prominent bill. **BP** Bill deep-based, long, broadly spatulate at tip, black in ♂, greyer with orange cutting edges in ♀ and eclipse ♂; eyes bright yellow in ad. ♂, dull orange-yellow in ♀ and brown in juvenile; legs bright orange. **Vo** ♂ generally quiet, but gives strange nasal rattling *took took* on take-off and quiet, hoarse *kue* or *kusu* in display; ♀ gives soft, low, staccato descending, somewhat Mallard-like quacking, *gaa-gaa-gaa-gaa...* or *kwe-kwe-kwe-kwe...* and a wheezy *kerr-aesh*.

NORTHERN PINTAIL CTKJR
Anas acuta

L ♂ 61–76cm, ♀ 51–57cm; WS 80–95 cm; WT 850g. **SD** Very common Holarctic dabbling duck wintering south to Africa, S Asia and C America. In region breeds in Russian Far East, Sakhalin, throughout Chukotka and Kamchatka, wintering (mid Sep–mid Apr) in large numbers south to E China, Taiwan, Korea and widely throughout Japan. Monotypic. **HH** Breeds at wetlands from tundra pools to taiga lakes; on migration and in winter mixes with other *Anas* in large flocks at ponds, lakes, rivers, wet ricefields, shallow bays and coastal lagoons. **ID** Medium-sized, but much slimmer and more elegant than other dabbling ducks of its size, with long slender neck and long narrow tail. Breeding ♂ has dark chocolate head and nape, white foreneck and narrow stripe up neck-side, creamy white breast and belly, grey flanks, creamy rear flanks to tail, black vent and tail with extremely long (c.10cm), pointed central rectrices, tail-sides white. ♀ greyer brown, much more slender and longer necked than ♀ Mallard, with grey bill. Eclipse ♂ resembles ♀, but with ad. ♂ upperwing. Wings long and narrow with wingtips swept back; flight agile and fast, often in lines. Grey wings have green speculum with broad hind border in ♂, dark speculum with broad white trailing edge in ♀. Underwing of both sexes grey with pale bars. **BP** Bill slender, pale blue-grey with black culmen, tip, cutting edges and band at base (♂), or plain grey (♀); eyes black; tarsi dark grey. **Vo** Generally rather quiet, but displaying ♂s give various calls, in particular a low, Eurasian Teal-like whistle, a rolling *furrr-furrr*, and a quieter *kishiin*, audible only at close range. ♀ gives somewhat crow-like *cr-r-r-rah*, a soft, hoarse, somewhat Mallard-like quacking *guegue kuwa kuwa* or low *kwuk*; and in flight a four-note descending series, *keersh-kursh-kurrh-kurrh*.

BLUE-WINGED TEAL

ad ♂

ad ♀

ad ♀

ad ♂

NORTHERN SHOVELER

2-16 19
Tokyo

ad ♂

ad ♂

ad ♀

ad ♀

NORTHERN PINTAIL

ad ♂

ad ♀

ad ♀

GARGANEY
Anas querquedula
CTKJR

L 37–41cm; WS 59–67cm; WT 290–480g. **SD** Ranges across temperate and boreal regions of W Europe to E Asia, wintering in sub-Saharan Africa, India and SE Asia. In region, rather uncommon, breeds in NE China and Russian Far East, Sakhalin, Okhotsk coast and Kamchatka; winters E China and Taiwan; uncommon migrant (Mar–May, Sep/Oct) through Taiwan, Korea and Japan. Monotypic. **HH** Well-vegetated small ponds and pools in mixed forest, forest-steppe and steppe. Lakes and coastal lagoons on migration and in winter. **ID** Slightly larger than Eurasian Teal, with more prominent, heavier bill, squarer head and slightly longer tail. Breeding ♂ head, dark at distance, is purplish-brown with silvery stripe from just in front of eyes down neck-sides. Dark brown, finely speckled breast contrasts with pale grey flanks vermiculated black. Tertials elongated, grey, black and white, droop over flanks; vent and tail brown. ♀ like Eurasian Teal, but pale supercilium contrasts more strongly with dark crown and very dark eyestripe, and pale loral spot separated from buffy-white chin. Eclipse ♂ resembles ♀, but has whiter supercilium, darker eyestripe, whiter lores and dark line on cheeks. In flight (faster, more direct than Eurasian Teal), ♂ has pale silvery-grey forewing and primary bases, dark green speculum bordered front and aft by bold white bars, also pale grey belly contrasts markedly with dark breast, as does very dark leading edge to underwing with pale grey axillaries and rest of underwing. ♀ in flight: note face pattern, broad white trailing edge to secondaries, recalling Northern Pintail, and paler 'hand' than Eurasian Teal. **BP** Bill rather long, dark blackish-grey; eyes orange-brown to dark brown; tarsi dark grey or black. **Vo** ♂ gives hard, wooden clicking or rattling croak *kar-r-r...* on water or in flight; ♀ a soft croaking *kwak* or *ke*, and erratic series of harsh, high-pitched quacks: *graash-graash-graash-graash*.

BAIKAL TEAL
Anas formosa
CTKJR

Vulnerable. L 39–43cm; WS 65–75cm; WT 360–520g. **SD** Breeds across N & E Siberia to Okhotsk Sea, W Chukotka and N Kamchatka. Winters (Oct–Apr) south to E China, Taiwan (occasional), Korea and Japan (once abundant, now scarce). Currently winters mainly in southern S Korea where 400,000+ congregate at a few localities; large concentrations recently also found in Shanghai, China. Monotypic. **HH** Tundra, forest-tundra and taiga pools; winters at lakes and large ponds, feeding on wet ricefields and shallow wetlands (sometimes at night). Makes mass aerial manoeuvres on approaching and leaving roost. **ID** Noticeably larger and longer tailed than Eurasian Teal, and smaller billed than Garganey, with more peaked hindcrown than other teals. Breeding ♂ highly distinctive. Head uniquely patterned with creamy yellow, bottle green, black and white stripes and swirls; breast pinkish-brown with black spots and grey flanks with vertical white bar (longer and narrower than Green-winged). Upperparts brown, vent black, long drooping tertials striped orange, black and white. ♀ has distinctive pale round loral spot, often with dark border, dark crown and eyestripe

(from rear of eye) contrasting with paler brown supercilium and head-sides. May have pale vertical bar or wedge extending onto cheeks from pale chin and throat. Eclipse ♂ resembles ♀, but generally more rufous-brown. In flight, forewing grey, ♂'s greenish-black speculum has prominent rufous frontal border and broad white rear border; ♀ has only white rear border and no midwing bar (cf. Eurasian). **BP** Bill small, black or dark grey; eyes black; tarsi greenish-grey (♂) or greyish-black (♀). **Vo** ♂ gives various deep clucking calls: *klo-klo-klo*; *wot-wot-wot*, or *proop*, and ♀ a harsh, jerky, low *quack*.

EURASIAN TEAL
Anas crecca
CTKJR

L 34–38cm; WS 53–59cm; WT 340–360g. **SD** The commonest small Palearctic dabbling duck. In region, breeds from NE China across Russian Far East, Sakhalin, Chukotka and Kamchatka, winters (Sep–May) south to E China, Taiwan, Korea and Japan (except Hokkaido). **HH** Breeds in secluded and well-vegetated small wetlands in taiga forest. On migration and in winter, visits ponds, lakes, rivers and wet ricefields; often in large numbers and mixed with other *Anas*. **ID** Smallest dabbling duck; appears compact with rounded head and small bill. Breeding ♂ has chestnut head with long, broad dark green eye patch bordered by buff lines, creamy breast spotted black, grey neck and flanks finely vermiculated black and white, tertials pale grey-brown and hang loosely slightly over tail-sides, which are cream, vent black; broad horizontal bars, white above black, at edge of closed wing. ♀ is a plain, dark brown duck, heavily scalloped on flanks, face generally clean with eyestripe. Eclipse ♂ resembles ♀. Flight rapid and agile, leaps from water, turns and twists frequently even in dense flocks. Grey wings with green speculum which has prominent white borders; at distance appears dark with short, broad white upperwing bar. **BP** Bill small, dark grey; eyes black; tarsi dark grey or black. **Vo** ♂ has sharp, rattled *kyireek*, and fluty *piri piri*; ♀ gives short, gruff *graurk* and descending series of raspy nasal quacks *gwee gwe gwe gwe* or *peeht pat pat pat*.

✓GREEN-WINGED TEAL
Anas carolinensis
CTKJR

L 34–38cm; WS 58cm; WT 350g. **SD** Common Nearctic duck that breeds across northern N America to Alaska, wintering in S. In region rare, typically found in winter among Eurasian Teal and in mixed flocks with other *Anas*. Because of difficulty in separating ♀ from Eurasian, most records relate to ♂s. **HH** Rivers, pools and lakes. **ID** Closely resembles Eurasian, but breeding ♂ has less distinct buffy 'frame' to eye patch, lacks bold white horizontal bar at sides but has prominent vertical white bar on breast-sides, and breast darker and buffier. ♀ virtually identical to Eurasian, but may show more contrast between crown, eyestripe, dark cheek patch and general ground coloration of face. In flight, grey wings with green speculum, but borders usually more rusty, less white than Eurasian. **BP** Bill small, dark grey; eyes black; tarsi dark grey or black. Occasional (♂) hybrids may have vertical *and* horizontal white bars. **Vo** Seemingly as Eurasian Teal. **TN** Formerly within Common Teal *A. crecca*, now split into Eurasian and Green-winged.

GARGANEY

ad ♂

ad ♀

ad ♂

ad ♀

BAIKAL TEAL

ad ♂

ad ♀

ad ♂

ad ♀

EURASIAN TEAL

ad ♀

ad ♂

ad ♀

ad ♂

GREEN-WINGED TEAL

2/12/19
Osaka

ad ♂

RED-CRESTED POCHARD   CTKJ
Netta rufina

L 53–57cm; WS 85–90cm; WT 0.83–1.32kg. **SD** Europe and C Asia east towards Lake Baikal. Rare winter visitor (Nov–Mar) to E China, Taiwan, Korea and Japan. Monotypic. **HH** Reed-fringed lakes and reservoirs, lagoons, coastal marshes and bays, with other diving ducks. **ID** Large and bulky with large head. Breeding ♂ has bright rusty-orange head, crown typically paler. Lower neck, breast, central belly, tail and vent black, contrasting with large pale, buffy-grey flanks and brown back. ♀ has pale grey-brown crown and nape, and whitish face (recalling smaller Black Scoter ♀); rest of plumage pale, plain grey-brown. Eclipse ♂ like ♀, but has red eyes and bill and larger, redder brown crown. Floats rather higher than other diving ducks. In flight, both sexes have long, prominent white wingbar contrasting with rather black tips to remiges, and grey-brown or blackish upperwing-coverts; underwing rather pale greyish-white; ♂ has chestnut axillaries, white flanks and black belly stripe. **BP** Bill slender, waxy red with yellow nail (♂) or grey with buff tip (♀); eyes red (♂) or brown (♀); tarsi orange-red, webs black. **Vo** ♂ has rasping, sneezing *gyi* or *byiii* and subdued *rerr-rerr* in display, whilst ♀ calls sound like distant barking, *wrah-wrah-wrah…* or growling *keurr-keurr*; otherwise silent.

CANVASBACK   CTKJR
Aythya valisineria

L 48–61cm; WS 74–90cm; WT 0.85–1.6kg. **SD** N America, breeding west to C Alaska. Rare, but probably annual, winter visitor (Sep–May) to Japan (mainly C & N), mostly alone or in pairs but occasionally small flocks; accidental in NE Russia, Korea, Taiwan and NE China. Monotypic. **HH** Large lakes, reservoirs, coastal lagoons and bays, and on sea, often with other *Aythya*. **ID** Larger, longer necked and longer billed than Common Pochard, with flat-headed profile, long sloping forehead and long black bill. Breeding ♂ has dark, blackish-chestnut head and neck, black breast and 'stern', and extremely pale grey wings and body (can appear white). ♀ has pale brown head, breast and tail, body and wings very pale grey-brown. In flight, wings very pale grey (♂), or mid to dark grey (♀), underwing very pale grey to white, axillaries white; neck appears longer and thicker than other *Aythya*. **BP** Bill long, deep at base, slender at tip, black or dark grey (lacks Common Pochard's pale subterminal band); eyes red (♂) or dark brown (♀); tarsi dark grey. **Vo** ♂ gives *wikku wikku wikku*, ♀ a low coarse growling *kururu kururu kururu* or *grrrt grrrt grrrt*, but both generally silent.

REDHEAD   KJR
Aythya americana

L 40–56cm; WS 74–85cm; WT 1.03–1.08kg. **SD** N America, nearest breeding grounds to region in Alaska. Vagrant in winter to New Siberian Is, Russia, S Korea and Japan. Monotypic. **HH** Ponds, lakes, and bays with other *Aythya*. **ID** Mid-sized with rather rotund appearance, body short, back high and rounded; like Common Pochard, but rounder, redder brown head. Breeding ♂ has bright, rufous-brown head and neck, black breast and 'stern', mid to dark grey upperparts and body. ♀ plain brown overall, greyer on upperparts, paler on chin. In flight, slow shallow beats, wings broad with some contrast between pale grey flight-feathers and dark to very dark forewing; underwing very

pale-grey to white, axillaries white. **BP** Bill blue-grey with white subterminal band and black tip; eyes dull yellow (♂; red/orange in Common Pochard), dark brown (♀/juv.); tarsi dark grey. **Vo** Generally silent. ♂ gives cat-like *myuuou* or *waow*, whilst ♀ has harsh *squak* or softer, nasal *grehp*.

COMMON POCHARD   CTKJR
Aythya ferina

L 42–49cm; WS 67–75cm; WT 0.9–1.1kg. **SD** W Europe to Transbaikalia in summer, south into Africa and S & E Asia in winter. Breeds very locally in SE Hokkaido. Winters (Oct–mid Apr) widely, in Japan, Korea and E China; locally common. Scarce in Taiwan. Monotypic. **HH** Ponds, lakes, coastal lagoons and bays with other *Aythya*, where it feeds by dabbling, up-ending and diving. **ID** Mid-sized with high domed crown, sloping forehead and concave culmen, and short tail. Breeding ♂ has dark chestnut head and neck, black breast and 'stern', mid to dark grey upperparts and body. Eclipse ♂ has duller brown head and dark brownish-grey breast and stern. ♀ has grey-brown head with pale stripe behind eye, pale lores and chin, brownish breast and 'stern', and brownish-grey upperparts and body. Flight strong, direct, rather whirring; wings differ from other *Aythya* in grey wingbar with little or no contrast between pale grey flight-feathers and grey forewing, though primary and secondary tips blackish. **BP** Bill black at tip and base with blue band between (♂), or grey at base, blue in middle and black at tip (♀); eyes red/orange (♂) or brown (♀); tarsi grey. **Vo** ♂ has soft chattering and series of wheezy whistles, while ♀ gives harsh, rattling growl *grrr-grrr* or *kururu kururu*, often in flight.

BAER'S POCHARD   CTKJR
Aythya baeri

Vulnerable L 46–47cm; WS 70–79cm; WT 680–880g. **SD** Endemic to E Asia: breeds NE China, S Russian Far East and probably (at least formerly) northernmost Korea; winters to E China south of Changjiang River, and south to S & SE Asia. Scarce visitor (mid Oct–late Mar) to Taiwan; accidental Korea and Japan. Once common, now rare. Monotypic. **HH** Shy on breeding grounds, at well-vegetated ponds, lakes and slow-flowing rivers. In winter favours larger lakes, ponds or marshes with other *Aythya*. **ID** Small, compact and very dark, resembling Tufted Duck, though rounded head and bill recall Common Pochard. Breeding ♂ has green-glossed black head and neck, blackish-brown upperparts, dark chestnut breast, and paler chestnut and white flanks. Small area of white on vent-sides recalls Ferruginous Duck. Tail short, held low. ♀/1st-winter and eclipse ♂ are less clean-cut, blackish-brown head and neck blend into chestnut-brown breast and flanks. In flight, white wing-stripe contrasts strongly with blackish upperwing-coverts and flight-feathers tips; underwing largely white, as are belly and vent. **BP** Bill dark grey, bluer towards tip, black nail; eyes white or pale yellow (♂), or brown (♀); tarsi dark grey. **Vo** Generally silent, but both sexes give harsh *graaaak* in courtship, and *koro koro* (♂) or *kura kura kura* (♀) at other times.

RED-CRESTED POCHARD

ad ♀

ad ♂

ad ♂

ad ♀

CANVASBACK

ad ♀

ad ♂

ad ♀

ad ♂

REDHEAD

ad ♂

02|12|19
Osaka, JP

ad ♀

2-16-19
Tokyo

COMMON
POCHARD

ad ♂

ad ♀

ad ♀

BAER'S POCHARD

ad ♂

ad ♂

ad ♂

ad ♀

FERRUGINOUS DUCK CTKJ
Aythya nyroca

L 38–42cm; WS 60–67cm; WT 410–650g. **SD** E Europe to C Asia, wintering south to Africa and S Asia. Winter vagrant (Nov–Mar) to E China, Taiwan, Korea and Japan. Monotypic. **HH** Marshes, freshwater lakes and coastal lagoons. **ID** Compact, rather uniformly dark, with high domed head (less rounded than Baer's) and conspicuous white undertail-coverts (many Tufted also have white undertail). Tail short, held low. Breeding ♂ deep, rich chestnut head and breast, and browner flanks (lacks white on flanks of Baer's), blackish upperparts, rump and tail. Eclipse ♂ duller but retains white eye. ♀ browner, less chestnut. In flight, very prominent white wingbars obviously contrast with blackish upperwing-coverts and wingtips; wingbar longer and more striking than Tufted or Baer's; underwing white, as are belly and vent. **BP** Bill dark grey, with blue-grey towards tip and black nail; eyes white (ad. ♂) or dark brown (♀/juv.); tarsi grey. **Vo** Generally silent except in courtship when ♂ whistles (*wee-few*) or gives hard nasal *chk-chk-chk.* ♀ has low, snoring *ka ka ka ka* or *errr errr errr*. **AN** White-eyed Pochard.

RING-NECKED DUCK J
Aythya collaris

L 37–46cm; WS 61–75cm; WT 690–790g. **SD** N America. Winter vagrant (Nov–Mar) to Japan (Hokkaido to C Honshu). Monotypic. **HH** Ponds and lakes. **ID** Superficially resembles Tufted Duck but has larger head with high, peaked hindcrown, longer neck, different bill and wing patterns, and more prominent tail. Breeding ♂ has glossy black head (with suggestion of rounded crest on rear crown), neck, breast, upperparts and 'stern', pale grey flanks with prominent white spur on breast-sides (cf. Tufted). Purplish-brown neck ring indistinct. ♀ has less contrasting plumage, with dark brown cap, grey cheeks and narrow, broken white eye-ring and lores. In flight (unusually erratic for *Aythya*), wingbar grey not white; grey flight-feathers have blackish tips and offer little contrast with black (♂) or dull grey-brown (♀) forewing; underwing pale grey (greyer than scaups). **BP** Bill dark grey, with white at base (♂), white subterminal band and black tip, and narrow white bands at bill base and nostrils; eyes orange (juv. brown); tarsi dark grey. **Vo** ♂ typically silent except in display; ♀ utters guttural growling *kerp kerp …* and both sexes a throaty *kua* or *kwa kwa*.

TUFTED DUCK CTKJR
Aythya fuligula

L 40–47cm; WS 65–72cm; WT 1–1.4kg. **SD** Breeds from Iceland to Kamchatka, wintering Africa, S & E Asia. In E Asia, breeds NE China and most of Russia south of tundra, east to Kamchatka, Commander and Kuril Is, and Sakhalin. Winters (Oct–late Mar), often in large numbers, in E China, Taiwan, Korea and in Japan (mostly south of Hokkaido). Monotypic. **HH** Forested lakes and marshes; on migration/in winter at bays, coastal marshes, lagoons, large inland lakes and major rivers, often with Common Pochard. **ID** Small, compact, boldly pied *Aythya* with rather rounded, crested head, high forehead and conspicuous yellow eyes. Breeding ♂ black with white flanks, head sometimes tinged purple; rear crown-feathers hang in long, dense crest. Tail short, held low. Eclipse ♂ short crest, grey flanks. ♀ browner, with brown flanks, rear crown has very

short tuft, limited white at bill base (rarely as extensive as Greater Scaup), and may show inconspicuous, sometimes more prominent, white undertail-coverts (cf. Ferruginous Duck, Baer's Pochard). Resembles ♀ Greater Scaup, but darker brown, smaller, with different head shape, pale band and broad black tip to bill. In flight, wingbar white but outer primaries mostly dark grey, thus bar is short; underwing, axillaries and belly white. **BP** Bill blue-grey, paler distally, tip and nail black; eyes pale yellow (♂) or dull orange-yellow (juv./♀); tarsi grey. **Vo** Usually silent. ♂ gives soft whistled *wheeoo* in breeding season, and low *kyu* and *gagaa*. ♀ (like other *Aythya*) has abrupt *krr-krr-krr*… Both sexes utter low *gurrrr gurrrr* on take-off and in flight, when wings whistle.

GREATER SCAUP CTKJR
Aythya marila

L 42–51cm; WS 71–80cm; WT 0.9–1.25kg. **SD** Arctic and subarctic breeder across Eurasia and N America. In E Asia *A. m. nearctica* breeds from Lena River to Bering Strait, Kamchatka and Commander Is. Winters (Oct–Apr) in Kuril Is, Japan, Korea, and Taiwan, but uncommon in continental Asia. Locally in very large numbers in Japan, with some in summer. **HH** Montane or tundra marshes and pools, or on coast; on migration/in winter in bays, lagoons, large freshwater lakes and major rivers, often with other *Aythya*. **ID** Heavy, compact, boldly marked *Aythya* with uncrested head. Breeding ♂ has black head with dark green gloss, black breast and 'stern', white flanks, grey back. Tail short, held low. ♀ browner, with grey-vermiculated flanks and mantle, and broad white area at bill base; many also show pale patch on ear-coverts in autumn/winter; confusable with Tufted Duck, but larger, head rounder and only bill nail is black. Eclipse ♂ like washed-out, slightly browner, breeding ♂. In flight, ♂ has dark grey forewing, long white wingbar, dark grey outer primaries and black tips to flight-feathers; underwing, axillaries and belly white; ♀ has brownish-grey forewing and back. **BP** Bill uniform blue-grey (♂) or dark grey (♀), with large black nail; eyes yellow; tarsi grey. **Vo** Generally silent, but displaying ♂ gives low, hollow hooting; ♀ a more prolonged, deeper growl than Tufted.

LESSER SCAUP JR
Aythya affinis

L 38–48cm; WS 64–74cm; WT 800–850g. **SD** N America. Winter (mid Oct–Mar) vagrant to Japan and NE Russia, e.g. Kamchatka. Monotypic. **HH** Ponds, lakes and large rivers, preferring more freshwater habitats than Greater Scaup, though they overlap. **ID** Very similar to Greater, but smaller, with taller, narrower head, and shorter, narrower-tipped bill. Crown peak further back, thus head appears more angular than Greater. Breeding ♂ head gloss more purplish than green, mantle has broader black barring, white flanks appear partly grey due to fine vermiculations. ♀ browner, with grey-vermiculated flanks and mantle, and slightly less white at bill base than ♀ Greater; typically lacks white auriculars of many ♀ Greater in autumn/winter, but often has them in spring/summer. In flight like Greater, but wingbar confined to secondaries. Underwing distinctive; lesser and median coverts white, contrasting with rest of grey underwing. **BP** Bill uniform pale blue-grey, with small black nail; eyes yellow (♂) or orange (juv./♀); tarsi dark blackish-grey. **Vo** ♂ gives husky whistling in display; ♀ a guttural *karr karr* or *garf garf*.

FERRUGINOUS DUCK

ad ♀

ad ♂

ad ♀

ad ♂

ad ♂

ad ♀

RING-NECKED DUCK

ad ♂

ad ♀ win

TUFTED DUCK 2/15/19 Tokyo

ad ♂

ad ♀

ad ♂

ad ♀

ad ♂

ad ♀

GREATER SCAUP

nearctica

ad ♀ win

ad ♀

ad ♂

ad ♂

ad ♀ sum

LESSER SCAUP

ad ♀ win

ad ♂

STELLER'S EIDER CJR
Polystica stelleri

L 42–48cm; WS 68–77cm; WT 860g. **SD** High Arctic tundra across E Siberia to Bering Strait and Alaska (but population declining seriously there, and perhaps more widely). In winter reaches NW Europe, and south to Commander and Aleutian Is, Kamchatka, Kuril Is and, rarely, Hokkaido; very rare Heilongjiang and Hebei, China. Monotypic. **HH** Breeds in coastal tundra at freshwater pools, also sheltered lagoons and bays; at other seasons off rocky coasts, commonly near weed-covered shores and capes; typically in small groups. **ID** Smallest eider, with proportionately longer body, smaller, squarer head and much shorter bill; tail longer and often held cocked. Unusual in resembling *Anas* in form and bill. Flattish crown, steep forehead and vertical nape. Breeding ♂ white with creamy orange-buff breast, belly and sides, black eye patch, chin, collar, spot on breast-sides, upperparts, 'stern' and tail; rear crown has greenish-black patch forming slight crest. Black and white tertials droop. Eclipse ♂ dark blackish-brown with white scapulars; 1st-year ♂ recalls washed-out ad. ♀ has same angular head shape, all dark, mottled sooty-brown, plumage with broad white borders to blackish speculum. Wings longer and narrower than other eiders, thus take-off easier, flight faster, sometimes in tight flocks; ♂ black with white shoulders, wing-coverts and tips of secondaries; secondaries, primaries and primary-coverts black. ♀ dark with broad white borders to speculum, and whitish axillaries and underwing-coverts. **BP** Bill lacks lobes, is grey, blunt and has wedge-shaped tip; eyes black (breeding ♂) or has pale eye-ring (eclipse ♂ and ♀); tarsi dark ochre-grey. **Vo** ♀ gives guttural quacking *gaa gaa geah* and loud *cooay*; wings whistle loudly in flight.

SPECTACLED EIDER R
Somateria fischeri

L 51–58cm; WS 84cm; WT 1.63kg. **SD** Narrow Arctic range, from Lena Delta to Chukotka and NW Alaska. Monotypic. Declining. **HH** Breeds on marshy coastal tundra; moult migrations to remote coasts; winters in Bering Sea. **ID** Similar size to King Eider, smaller than Common, with plumage resembling latter, but has white 'goggles', dark forehead and blackish-grey breast; bill, though long and wedge-shaped, appears small (feathered to nostril), tail longer and more pointed than Common. Breeding ♂ has green head with white black-bordered 'goggles'; chin, neck, back, wing-coverts, tertials and patch on tail-sides white. Breast very dark grey, flanks, 'stern', tail and flight-feathers black. White tertials droop over flanks. Eclipse ♂ dark greyish-brown with buffy 'goggles'. ♀ has pale 'goggles', dark brown forehead and boldly barred mid brown plumage. In flight, ♂ largely white with black flight-feathers; ♀ very similar to King, but 'goggles' may be visible and axillaries pale grey not white (cf. Common). **BP** Orange bill without lobes and heavily feathered (green and white) to nostrils (♂); ♀/juv. has dark grey bill, also feathered (brown) to nostrils; tarsi dull ochre. **Vo** Typically silent. ♂ gives weak *ho-hoo* or *ah-hoo* display call and ♀ a gargling *gogogo*….. Quiet *cro cro ko ko* and hoarser, raven-like croaking, *krro*.

KING EIDER JR
Somateria spectabilis

L 55–63cm; WS 87–100cm; WT 1.5–2.01kg. **SD** N shores of Holarctic. In E Asia, NE Siberia from Lena Delta to Bering Strait, including south Chukotski Peninsula. Winters south to Kamchatka and Aleutians; accidental further south. Monotypic. **HH** Pools on coastal tundra; moults along coasts; winters at sea in deepwater areas. **ID** Large with unusual head and bill shape. Slightly smaller than Common, with smaller, more angular head, bulbous forehead and shorter bill. ♂ mainly black with white band on upper back and breast, breast salmon-pink (darker than Common), white flank bar and large white patch at tail-sides; crown and nape powder blue, face greenish, bordered black. Eclipse ♂ dark brown with reduced orange shield and bill. 1st-winter ♂ like eclipse, but breast white. ♀ warmer brown than Common, with bold dark crescents on flanks (Common barred), distinctive head and bill shape. Flight heavy, slow, often low over water; appears deep-bellied and very short-necked; ♂ has white upper back separated from white wing-coverts by black shoulders. **BP** Bill orange (♂) with variable-sized shield, blackish-grey (♀/young), with lobe near eye, feathering extending towards nostril and upturned black gape line afford smiling expression; eyes black; tarsi dull ochre. **Vo** ♂ gives deep, hollow crooning *yooo hruru ruru* or *broo broooo brooooo broo* and rolling *arr-arr-arr*; ♀ deep quacking *gwaaku gwaaku* and deep, hoarse *gogogogo*….

COMMON EIDER JR
Somateria mollissima

L 60–70cm; WS 95–105cm; WT 1.915–2.218kg. **SD** Widespread in northern N America and NW Europe; in region *S. (m.) v-nigra* (**Pacific Eider**; potentially distinct species) in NE Siberia from Lena Delta to Bering Strait, south to N Kamchatka and NE Okhotsk Sea. Wintering areas unknown (perhaps ice-free areas near breeding range), but accidental N Japan. **HH** Tundra pools and rocky coasts. In winter off rocky coasts. **ID** Very large, with distinctive long profile, large head, long pointed bill with lobes extending towards eye, and feathering along sides to nostrils. Breeding ♂ white above, black below, rear crown, rear head-sides and upper neck tinged pale grey-green, breast washed peach. Cap black with point extending nearly to nostrils; flanks, tail, vent and flight-feathers black, white back, drooping tertials and round patch near tail base also white. 1st-winter ♂ blackish-brown with white breast/lower neck. ♀ greyish-brown with finely barred breast, flanks and broader pale brown fringes to darker mantle. Flight heavy, slow, on short broad wings, rather goose-like and often low over water; appears deep-bellied; ♂ has white back continuous with white wing-coverts, and black flight-feathers; ♀ has brown speculum bordered white and pale grey axillaries, wings otherwise brown. **BP** Bill orange-yellow, paler at nail, with feathering extending in rounded lobe to nostrils (**Atlantic Eider** has greenish-grey bill with pointed feathering on bill); eyes black; tarsi dull grey. **Vo** Generally silent but highly vocal when breeding: ♂ has far-carrying, deep mournful cooing *ah-oooh-eh oh oooh oo*; whereas ♀ gives deep, guttural *gag-ag-ag*….

STELLER'S EIDER

ad ♂

ad ♀

ad ♂

ad ♀

ad ♀

ad ♂

SPECTACLED EIDER

ad ♂

ad ♀

ad ♂

ad ♀

KING EIDER

ad ♀

ad ♂

ad ♀

ad ♂

ad ♀

ad ♂

ad ♂

COMMON EIDER

v-nigra

ad ♀

HARLEQUIN DUCK CKJR
Histrionicus histrionicus

L 38–45cm; WS 63–70cm; WT 540–680g. **SD** Disjunct range in NW Atlantic and N Pacific. Breeds across Russian Far East, around Sea of Okhotsk, Sakhalin, Chukotka, Kamchatka, Kuril Is and N Japan; winters (Oct–mid Apr) south to Korea, N Kyushu to C Japan (year-round in N), and very rarely in NE China (Heilongjiang to Shandong). Monotypic. **HH** Favours turbulent water; breeds along fast-flowing cold rivers with white water and visits rocky coasts; moults in groups away from breeding grounds; winters off rocky coasts in small flocks. **ID** Uniquely patterned small, dark sea duck, with tiny bill, rounded head, steep forehead, and pointed, often cocked tail. Breeding ♂ essentially steel blue with chestnut flanks, and attractive white markings outlined black on face, ear-coverts, neck, collar, breast-sides and mantle; white line fringing crown filled chestnut from eyes back. Despite flamboyant plumage, can look surprisingly dark/monochrome in poor light or at distance. ♀ uniform sooty-brown, with round white spot on ear-coverts, indistinct whitish supraloral and whitish patch below eye. 1st-year ♂ shows elements of ad. ♂ facial pattern in otherwise ♀-like plumage, but perhaps not until midwinter. Flight fast and agile on all-dark wings. **BP** Bill pale to dark grey; eyes black; tarsi pinkish-grey. **Vo** On breeding grounds ♂ gives high whistled *tiiv* and ♀ soft quacking *koa koa koa* and nasal *ekekekek…*; in winter ♂ gives whistled *feee* and *fee-ah* and ♀ deeper *guwa guwa* calls.

SURF SCOTER JR
Melanitta perspicillata

L 45–56cm; WS 76–92cm; WT 0.9–1kg. **SD** N American vagrant, usually singly, occasionally in small parties, to Chukotka, Commander Is, Hokkaido and N Honshu (mostly Jan–Mar, exceptionally May). Monotypic. **HH** In region on coasts, often with Black and White-winged Scoters. **ID** Size intermediate between Black and White-winged with squarer head and strangely swollen bill with slightly convex profile. ♂ glossy black, with small white forehead and large white nape patch; no white in wing. ♀ dark blackish-brown with dark crown, diffuse pale patch near bill base, another behind eye, and some have pale nape patch. In flight, wings all-dark like Black, but has darker primaries. **BP** Bill large, triangular: ♂ orange with yellow nail, but black on swollen base of culmen, black and white on sides and basal cutting edge, and pinkish-orange from nostril to tip; in ♀, slightly swollen-based bill is blackish-grey; eyes white or dark brown (young); tarsi orange-red. **Vo** Usually silent, but ♂ gives low, whistled *puk puk* and ♀ moaning *aa aa aa* or *krrraak krrraak*. Wings whistle in flight.

WHITE-WINGED SCOTER CKJR
Melanitta deglandi

L 51–58cm; WS 86–99cm; WT 1.67kg. **SD** C & E Siberia and N America. In E Asia, *M. d. stejnegeri* (which possibly merits specific status) breeds east of Lena River across Chukotka and south to Kamchatka and Sea of Okhotsk (**Velvet Scoter** *M. fusca* breeds to west); winters (Oct–Mar) on N Pacific coast south to E China (uncommon), commonly in Korea and very commonly off N Japan, S Sea of Okhotsk and Kuril Is.

HH Wetlands of tundra, forest-tundra and taiga, but in winter favours rocky and sandy coasts, occasionally harbours or inland lakes and rivers. Less common than Black Scoter, with which it often mixes. **ID** Largest scoter. Dumpier than other black sea ducks, with larger head and longer bill than Black. ♂ glossy black, with neat white crescent below and rear of eye, and large white wing patch visible even on closed wing. Flanks glossy black in breeding ♂, perhaps tinged brown in non-breeder. ♀ dark blackish-brown with pale oval patches near bill base and behind eye; also has white wing patch. Deep base to bill gives distinctive straight forehead and bill profile. Feathering reaches almost to nostril in all ages and sexes. Flight fast and direct though heavier and slower than Black, often in tight groups and straggling lines low over sea; all-white secondaries form rectangular patch in flight. **BP** Bill of ♂ deep pink, even red proximally, yellower distally, black on sides at base, and ad. has distinct black knob on base of upper mandible and large, round 'see-through' nostrils. 1st-winter ♂ lacks knob, deep red or pink areas duller, more orange; ♀'s deep-based bill is dark grey to black; eyes pale blue-grey (ad. ♂), brown (imm. ♂) or dark brown (♀); legs bright pink. **Vo** Usually silent, but displaying ♂ gives whistled *fee-er* and low nasal *aah-er*; ♀ a gruff croaking *kraa-ah kraa-ah kraa*. Wings whistle in flight. **TN** Formerly within Velvet Scoter *M. fusca.*

BLACK SCOTER CKJR
Melanitta americana

L 43–54cm; WS 70–90cm; WT 950g. **SD** Widespread and common in NE Siberia and northwest N America. Breeds east of Lena River (**Common Scoter** *M. nigra* breeds to west) across Chukotka to Bering Sea coast and Kamchatka; winters (Oct–Mar) south to E China, Korea and, very commonly, off N Japan, S Sea of Okhotsk and Kuril Is. **HH** Wetlands (marshy bogs and rivers) of taiga and tundra. In winter, favours shallow coastal waters, harbours, even visits inland lakes. Gregarious. **ID** Dumpy, with small rounded head and rather prominent pointed tail. Ad. ♂ all black. Imm. ♂ resembles ♀ but acquires some yellow to bill base. ♀ dark grey-brown with pale greyish-white cheeks contrasting with dark brown crown and nape. Flight fast and direct on somewhat broad, black wings, with paler greyish-brown primaries; often in bunched groups and straggling lines low over sea. **BP** Bill of ♂ partly black, with swollen base to upper mandible yellow or orange, nostrils small and oval, near front edge of yellow 'shield'; imm. ♂ and ♀ all black; eyes black; tarsi dark grey. **Vo** ♂ in winter may be highly vocal, uttering haunting, fluting *pyuuu*, *pyeee* or *pyi-feeee* sounds; ♀ gives harsh rasping or growling *urururu* or *kaarrr*. Wings of ♂ whistle in flight. **TN** Formerly within Common Scoter.

COMMON SCOTER Extralimital
Melanitta nigra

Ranges from Iceland and British Isles almost to Lena River, in Vilyuy basin, so likely to stray to far N of region. Closely resembles Black, but ad. ♂ has largely black bill with swollen basal knob, and yellow restricted to culmen ridge and around nostrils.

HARLEQUIN DUCK

ad ♀

ad ♂

ad ♂

ad ♀

SURF SCOTER

ad ♂

ad ♂

ad ♀

ad ♀

WHITE-WINGED SCOTER

ad ♂

ad ♀

ad ♂

stejnegeri

ad ♀

BLACK SCOTER

ad ♂

ad ♂

ad ♀

COMMON SCOTER

ad ♀

ad ♂

LONG-TAILED DUCK CKJR
Clangula hyemalis

L ♂ 51–60cm; ♀ 37–47cm; WS 65–82cm; WT 650–800g. **SD** Widespread in Holarctic. In region, breeds mainly above Arctic Circle, from Lena River to Bering Strait south to NE Sea of Okhotsk and N Kamchatka. Winters (Nov–Apr) north to pack-ice leads in N Bering Sea and south on coasts, commonly to N Japan, rarely to Korea and very rarely in E China south to Fujian and Guangdong. Monotypic. **HH** Isolated tundra pools and marshy tundra, but in winter favours waters off rocky and sandy shores, around capes, and occasionally harbours or inland lakes and rivers. Gregarious. **ID** Rather small, dumpy, with long neck, small head and small bill, and dark wings in all plumages. Sexually and seasonally dimorphic. ♂ has distinctive extremely long (10–15cm) central rectrices. ♂ (ad. winter/spring) has grey cheeks, dark chestnut ear and breast patches, back and wings, with white head, neck, shoulders, tertials, underparts and most of tail; elongated plumes black. In early summer mainly blackish-brown with white face patch, flanks and 'stern'. ♀ in winter has white face with dark forehead, crown and ear patch, dark brown breast and upperparts, 'stern' white; in spring resembles ♂ but lacks long tail, head and neck become duskier. Flight fast and agile on all-dark, narrow, pointed wings, beats only rising to horizontal; elegant in flight, though quite pot-bellied. **BP** Small bill, black with large pink spot near tip in ad. ♂, dark grey in ♀/juv. with black tip; eyes dull orange (♂), brownish-orange (♀) or dark brown (juv.); tarsi dark grey. **Vo** Very vocal late winter. ♂ gives deep yodelling *ow ow a ow-na a ow-na*; ♀ a weak high quacking *kuwaa* or soft *kak kak kak kak*; pleasant piping chorus maintained by flocks. **AN** Oldsquaw.

BUFFLEHEAD JR
Bucephala albeola

L 33–40cm; WS 53–61cm; WT 330–450g. **SD** N America. Winter vagrant (Nov–Mar) to NE Russia (e.g. Commander Is) and Japan (Hokkaido). Monotypic. **HH** Singly or in pairs, offshore or in bays and coastal lagoons. **ID** Smallest diving duck; compact with relatively large head but small bill. Breeding ♂ has black head with purple and green gloss and huge white band from eyes back over head. Upperparts black; breast, scapulars and body white. Eclipse ♂ resembles ♀ but has large white oval on black head. ♀ has sooty-brown head with white ovals below and behind eye; body dark grey. Juv. like ♀, but face patch less distinct. In flight, head appears raised, with body angled upwards; ad. ♂ very white, with large white patch from speculum across forewing; ♀ has small white speculum and white belly patch; flight rapid on whirring wings. **BP** Bill pale grey; eyes black; tarsi dark grey. **Vo** Typically silent, but ♂s sometimes squeal or growl and ♀s give soft grunting *gururu gururu* or a low *prrk prrk*.

COMMON GOLDENEYE CTKJR
Bucephala clangula

L 40–48cm; WS 62–77cm; WT 770–996g. **SD** Widespread in Holarctic. In E Asia *B. c. clangula* breeds from Transbaikalia and N Heilongjiang through Russian Far East and N Sakhalin, Koryakia and Kamchatka. On migration and in winter (mainly Nov–Mar), in Kamchatka, Kuril Is, throughout Japan and shores of Sea of Japan, Korea, and E & S China. **HH** Rivers and lakes in taiga (nests in tree cavities, such as old Black Woodpecker holes). Winters on rivers, lakes and coasts in pairs to fairly large flocks. **ID** Medium-sized diving duck, with a large triangular head, high-peaked crown, short neck and compact body. Tail held very low on water. Breeding ♂ has black head with green gloss and white oval on lores. Upperparts and 'stern' black; breast and flanks pure white; scapulars (white with narrow black fringes) droop over flanks. ♀ brown-headed and grey-bodied, with diffuse white collar and whitish belly; white speculum visible on closed wing. Eclipse ♂ like ♀ but has ♂ wing pattern. Flight fast with deep wingbeats and whirring wings which produce distinct musical whistle. In flight, ad. ♂ wings black, with white speculum, forewing and scapulars; ♀ wings greyer with white speculum and smaller forewing patch divided by black line; underwing of both dark. **BP** Bill short, somewhat broad, dark blackish-grey (♂) or mostly black with yellow band near tip (♀); eyes bright yellow (♂), pale yellow (♀) or brown (juv.); tarsi dull yellow/orange. **Vo** Generally quiet but ♂ gives forced *bee-beeech*, and dry, grating rattle *drrrr*, recalling Garganey, and hoarse, buzzy whistle (*kyi riiku kyi riiku*) in display (common in winter quarters from Mar), whilst ♀ gives *Aythya*-like low guttural *arr arr arr* and harsh, dry staccato quacks: *grak grak grak* or *kuwa kuwa kuwa*. In flight, stiff wings produce loud musical whistle.

BARROW'S GOLDENEYE KJR
Bucephala islandica

L 42–53cm; WS 67–84cm; WT 0.737–1.3kg. **SD** Iceland and N America west to Alaska. Vagrant in winter to Russia, Japan and Korea. Monotypic. **HH** Lakes and coasts. **ID** Slightly larger than Common Goldeneye, with larger head, steeper forehead, flatter crown, crown peak further forwards, and 'mane' at rear. Breeding ♂ has black head with purple gloss (visible only in good light), and large white crescent on lores forming point higher than eye. Upperparts like Common, but black more extensive, including black spur on breast-sides, and row of white windows on black scapulars. ♀ like Common, but shares ♂'s head-shape. Eclipse ♂ like ♀ but has ♂ wing pattern. Juv. distinguishable from Common only by head shape and wing pattern. In flight, ad. ♂ wings blacker, with white speculum, divided from white forewing by black line, scapulars have only white spots; ♀ wings also have less white; underwing of both dark; wings emit low whistle, quieter, less musical than Common. **BP** Bill black in ♂, mostly orange with grey base in ♀; eyes bright yellow (♂), pale yellow (♀), or brown (juv.); tarsi dull yellow/orange with dark webs. **Vo** ♂ gives gruff *kakaa* in breeding season and ♀ a low guttural *arr arr arr* and dry staccato quacks: *grak grak grak* (deeper than Common).

LONG-TAILED DUCK

ad ♀ win

ad ♂ win

ad ♂ sum

ad ♂ win

ad ♀ win

ad ♀ sum

BUFFLEHEAD

ad ♀

ad ♂

ad ♂

ad ♀

COMMON GOLDENEYE

ad ♂

ad ♀

clangula

ad ♂

ad ♀

BARROW'S GOLDENEYE

ad ♀

ad ♂

ad ♂

ad ♀

SMEW CTKJR
Mergellus albellus

L 38–44cm; WS 56–69cm; WT 515–935g. **SD** Scandinavia to Kamchatka, rarely Aleutians and mainland N America. In E Asia breeds in NE Russia east of the Lena River, Yakutia, Chukotka, Kamchatka, locally in NE China. Winters Japan, Korea and NE & E China; accidental Taiwan. Monotypic. **HH** Taiga swamps, pools, lakes and rivers; also fresh or brackish coastal lagoons. May form dense mixed sawbill flocks. **ID** Compact with slightly tufted head and small bill. Breeding ♂ mostly white, black 'panda' mask around eyes and on lores, fine black lines on rear crown and breast-sides; mantle/back black, white flanks finely vermiculated grey, 'stern' dark grey. ♀ grey with dark chestnut head, white cheeks and chin. Eclipse ♂ brown-headed, grey-bodied, but has white crest and breast. In flight, ad. ♂ distinctly pied, black flight-feathers and mantle contrasting with white scapulars and forewing; ♀ greyer, with smaller white forewing and mostly white underparts. **BP** Bill dark grey; eyes black; legs grey. **Vo** Breeding ♂ gives frog-like husky *eruru eruru eruru ukuu* and ♀ gruff, crackling *krrr* or *grrr*. **TN** Formerly placed in *Mergus*.

HOODED MERGANSER J
Lophodytes cucullatus

L 42–50cm; WS 56–70cm; WT 453–879g. **SD** N American vagrant to Hokkaido. Monotypic. **ID** Larger and longer billed than Smew with steep forehead and unique erectile crest; long tail. Breeding ♂ largely black with white chest, chestnut flanks and large white fanned crest with black margin; tertials black with white fringes. ♀ plain grey-brown, but size, head shape and tertial pattern distinctive (cf. Red-breasted). Wingbeats fast and shallow, wings appear narrow and dark with small white area on inner secondaries; underwing and axillaries pale grey. **BP** Bill black (breeding ♂), dark grey (eclipse), or dull ochre-yellow with grey tip and culmen (juv./♀); eyes bright yellow (♂), or orange to dull yellow (♀/juv.); legs dull pinkish-orange. **TN** Formerly in *Mergus*.

GOOSANDER CTKJR
Mergus merganser

L 58–68cm; WS 78–94cm; WT 0.89–2.16kg. **SD** Wide Holarctic range. Two races in the region: *M. m. merganser* from the Lena to Chukotka, Kamchatka and Commander Is; *M. m. orientalis* south through Russian Far East, Sakhalin, Hokkaido and NE China. Winters south to E China, Korea and N & C Japan; accidental Taiwan. **HH** Forest with rivers and lakes; winters mainly on fresh water, sometimes in very large flocks. **ID** Largest sawbill, with smooth crest. Breeding ♂ has black head and upper neck glossed green, black mantle, white neck, breast and underparts flushed pink or peach; tail grey. ♀ (and eclipse ♂) like large ♀ Red-breasted, but has pale chin and neat demarcation between dark brown head and whitish neck. Wingbeats shallow, neck held straight, recalling grebe or diver. Wing pattern distinctive: ♂ has extensive, almost undivided white innerwing patch, and white scapulars, ♀ has large white undivided speculum. **BP** Bill long, thick-based with darker well-hooked tip, red (ad.), or orange (juv.); eyes dark brown (ad.) or

white (juv.); tarsi reddish-orange. **Vo** ♂ gives deep, muffled *krroo-kraa*, ♀ short, deep *kar-r-r kar-r-r*; clearer notes given by both sexes in wide, circular display flights. In normal flight, gives fast chuckling *chakerak-ak-ak-ak*. **AN** Common Merganser.

RED-BREASTED MERGANSER CTKJR
Mergus serrator

L 52–58cm; WS 67–82cm; WT 0.78–1.35kg. **SD** Wide Holarctic range. In E Asia breeds from the Lena to Bering Sea, Chukotka, Kamchatka, Commander and Kuril Is, south through Russian Far East, Sakhalin, to N Heilongjiang, China, and at least formerly N Korea. Winters coastal E & SE China, Korea and Japan. Monotypic. **HH** Clear, freshwater pools, lakes and rivers in tundra and taiga, and on coasts; nests on ground. Winters at sea, or in smaller numbers on lakes and coastal lagoons. Gregarious. **ID** Slim, elegant, with long thin neck, rather vertical forehead, and wispy crest. Floats low in water with neck erect, recalling grebe or diver. Breeding ♂ has bottle green head, white neck ring, ginger-brown neck to black chest, sides black with unique white spots, black mantle and grey/white flanks, and grey 'stern'. ♀ has mid brown head grading into grey neck and body (unlike Goosander). In flight, note long slim neck; ad. ♂ has large white patch on innerwing divided by two fine black lines. ♀ has rather dark forewing, white patch divided by single black line, and white belly. **BP** Bill long, very slender, deep red with dark culmen (ad. ♂) or orange-red (juv./♀), small terminal nail hook; eyes red (♂), orange (♀) or orange-brown (juv.); tarsi dull orange. **Vo** Displaying ♂ gives hiccupping and sneezing *chika pitcheew*, purring *ja-aah* or *kwa kwa* and more metallic *koroo* or *yeow*. ♀'s harsh grating calls resemble Goosander, but higher pitched, *prrek prrek* or *grak grak*. Both sexes also call in wide, circular display flights in spring.

SCALY-SIDED MERGANSER CTKJR
Mergus squamatus

Vulnerable. L 52–62cm; WS 70–86cm. **SD** Rare E Asian endemic. Breeds S Russian Far East, NE China (Jilin, Heilongjiang) and N Korea. Winters in Korea, E China, Taiwan (accidental), and very rare, but annual, in C & S Japan. Monotypic. **HH** Secretive along montane rivers in mixed coniferous/broadleaf forests below 900m. Winters on fast-flowing rivers, lakes and reservoirs, usually singly or in groups of 2–3, on migration rarely up to 10 together. **ID** Large, boldly marked sawbill, with long, double, wispy hindcrest. Breeding ♂ has longer glossy green head than Red-breasted, flatter crown; neck and mantle black, breast and flanks white, rump and tail grey; flank feathers have grey fringes forming large scales, finer at rear, and grey vermiculations on lower back and rump. ♀ like large ♀ Red-breasted, but neater neck pattern (more like Goosander) and fewer grey scales on flanks. Wing pattern like Red-breasted; ♂ has large white patch on innerwing divided by two fine black lines, ♀ has smaller white patch divided by single black line, thus differing from otherwise similar Goosander. **BP** Bill long, thick, bright red with yellow tip; eyes dark brown; tarsi reddish-orange. **Vo** Like Red-breasted, but also makes deep, hoarse hiss. **AN** Chinese Merganser.

ad ♂

ad ♀

ad ♂

SMEW

HOODED MERGANSER

ad ♂

ad ♀

ad ♀

GOOSANDER

ad ♂

ad ♂

merganser

ad ♀

ad ♀

SCALY-SIDED
MERGANSER

RED-BREASTED
MERGANSER

ad ♂

ad ♂

ad ♀

ad ♀

ad ♀

ad ♀

ad ♂

ad ♂

RED-THROATED DIVER *Gavia stellata*   CTKJR

L 53–69cm; WS 91–110cm; WT 0.988–2.46kg. **SD** Commonest diver. Higher latitudes (above 50°N) of N America, and Eurasia south to Lake Baikal, Amur River mouth, Sakhalin, Chukotka, Kamchatka. Winters Kamchatka and Kuril Is to Japan, Korea, Yellow Sea and S China (rare), accidental Taiwan (but regular Kinmen and Matsu). Monotypic. **HH** Breeds at fishless tundra/taiga pools, commuting elsewhere to feed. Migration and winter coastal, may enter bays and harbours; rare inland. **ID** Smallest, slimmest diver; flat-chested, slim vertical neck, rather flat crown, bill typically slightly uptilted. Breeding ad. has dark rufous throat contrasting with grey face and neck. Much whiter in winter. In flight, slim-necked, head often low slung, appears hunch-backed, toes rather small; faster, deeper wing beats than other divers; narrow wings angled back. **BP** Bill slender, slightly upturned to sharp tip, black (breeding) or dark grey; eyes deep red; tarsi dark grey. **Vo** Prolonged mournful moaning *aaroooah* by breeding ♂; ♀ gives higher *arro-arro-arro*. In territorial flight, short nasal barking *gwaa-gwaa-gwaa-gwaa...* or more prolonged goose-like cackling *gark gark gark gargark gaagarag*. **AN** Red-throated Loon.

BLACK-THROATED DIVER   CTKJR
Gavia arctica

L 62–75cm; WS 100–130cm; WT 2.6g. **SD** British Isles to Siberia. *G. a. viridigularis* from Transbaikalia and the Lena east to Chukotka, Kamchatka (and extreme NW Alaska), south to Amur River and N Sakhalin. Winters Kuril Is, Japan, continental coast of Sea of Japan, Korea, Yellow Sea and E China (rare); accidental Taiwan. **HH** Mostly coastal, but breeds at clear lakes and sea bays. Migration/winter essentially coastal, sometimes bays and harbours; uncommon inland. **ID** Larger, heavier than Red-throated (RTD), very like Pacific (PD), but larger. Low in water, full-chested, neck thicker than RTD, longer than PD, typically held in S-curve, head more angular with steep forehead. Distinctive white flank patch separates it from very similar PD in all plumages. Summer adult has black throat patch with green gloss (purple in PD), and clear black and white stripes on neck and breast-sides (bolder than PD). In winter, head and neck darker than RTD, dark cap extends to eye, cheeks white, but back and neck-sides mid grey (border darkest), forming strong contrast between front and back of neck, which is only half white. Juv./1st-winter: browner upperparts, with pale, scaled pattern on back and scapulars. In flight, toes rather prominent; slower, shallower wing beats than RTD. **BP** Bill dagger-shaped, horizontal, rather thick, dark blackish-grey, darkest at tip; eyes deep red (breeding) or black; tarsi dark grey. **Vo** Growling *aaah-oh*, raven-like *kraaw* and croaking snore *knarr-knorr*, and repetitive whistling *owiiil-ka owiiil-ka owiiil-ka*, on territory. Usually silent in flight and in winter. **AN** Arctic Loon.

PACIFIC DIVER *Gavia pacifica*   CKJR

L 63–66cm; WS 91–112cm; WT 1.7kg. **SD** Alaska, northern Canada, N Siberia and E Chukotka. Winters south down Pacific coasts (Alaska to Baja California; Kamchatka to Japan, Korea and N Yellow Sea). Monotypic. **HH** As Black-throated Diver (BTD). **ID**

Very similar to BTD, but slightly smaller, head less angular, more rounded, bill thinner. Tends to hold head more level than BTD. Ad. sum crown/nape noticeably paler than BTD, white stripes on neck and breast-sides narrower; black throat patch glossed purple (though hard to see). Ad. win as BTD, but upperparts darker grey and typically narrow dark 'choker' on throat contrasts with white chin and neck, and sometimes absent. Flanks typically dark, though occasionally shows small area of white. Juv./1st-winter: usually have dark brown 'choker'. **BP** Bill black (ad. breeding) or grey with blackish culmen and tip (juv./non-breeding); eyes deep red; tarsi dark grey. **Vo** Repetitive mournful *guaan* or *gwain*, raven-like *kowk*, plaintive yodelling *o-lo-lee*, and a mournful long call on territory. **TN** Formerly within Black-throated Diver. **AN** Pacific Loon.

GREAT NORTHERN DIVER   JR
Gavia immer

L 69–91cm; WS 122–148cm; WT 2.78–4.48kg. **SD** N America and Iceland. Accidental in winter to Commander Is and N Japan. Monotypic. **HH** Winters on coasts. **ID** Large and heavily built. This and White-billed much larger and thicker-necked than other divers. Head angular with steep forecrown and forecrown bump. Ad. sum head and neck black, with white-striped chin bar and neck patch. Upperparts blackish-grey with extensive white chequers on shoulders and back, elsewhere fine white spots. Ad. win large size and pale patch around eyes distinguishes it from BTD and PD. Has partial white collar on neck-sides in all except breeding adults. White chin and fore throat contrast strongly with dark border to rear neck. Juv./1st-winter: similar to ad., but has light scaly pattern on upperparts. Flight heavy, slow, and goose-like, with thick neck, thick bill and large toes. **BP** Bill, heavy, dagger-shaped, black (ad. breeding) or blue-grey with dark culmen and tip (juv./non-breeding); eyes blood red (breeding) or dark brown; tarsi dark grey. **AN** Common Loon.

WHITE-BILLED DIVER *Gavia adamsii*   CKJR

L 76–91cm; WS 135–152cm; WT 4.05–6.4kg. **SD** Breeds along N Russian coast to NE Chukotka, and high arctic Alaska and Canada; winters on Pacific coasts to Yellow Sea and Vancouver. Winter: scarce Japan, Korea, accidental E China. Monotypic. **HH** Tundra pools and lakes, also low-lying Arctic coasts; winters on sheltered inshore waters. **ID** Largest, most heavily built diver, with thick neck and bill, which is usually held upwards. Head shape like Great Northern (GND). Ad. sum neck patch larger and white markings on upperparts larger and cleaner than GND. Paler in winter, particularly around eyes and ear-coverts, with little contrast on neck. Frequently has mottled look to face, even at distance. Juv./1st-winter similar to ad., but light scaly pattern on upperparts, pale head and neck, dark ear-coverts. In flight, much like GND. **BP** Bill dagger-shaped with wedge-shaped tip, ivory or yellowish, less yellow in winter (ad.) or pale horn with grey basal half of culmen (juv.); eyes blood red (breeding) or dark reddish-brown; tarsi black. **Vo** On breeding grounds gives yodelling screams *aaa aaa ...* and loud, harsh calls, lower and slower than GND. **AN** Yellow-billed Loon.

RED-THROATED DIVER

ad sum

ad win

ad sum

juv

ad win

BLACK-THROATED DIVER

viridigularis

ad sum

juv

ad sum

ad win

ad win

PACIFIC DIVER

ad sum

characteristic 'snorkelling' posture

ad sum

ad win

juv

juv

GREAT NORTHERN DIVER

ad sum

juv

ad sum

ad win

ad win

WHITE-BILLED DIVER

ad sum

juv

ad sum

ad win

ad win

LAYSAN ALBATROSS   CTJR
Phoebastria immutabilis

L 79–81cm; WS 195–203cm; WT 2.3–2.8kg. **SD** N Pacific, breeding only on some Hawaiian islands (400,000 pairs on Midway represent 50–70% of population), islands off Mexico, and on remote southern Japanese islands (Ogasawara chain) in winter. Commonest albatross in E Asian waters. Monotypic. **HH** In non-breeding season common off E Japan, Kuril Is, and in Kamchatkan Pacific Ocean and S Bering Sea. Occasionally seen from shore, especially after storms. **ID** Rather heavily built; pied plumage superficially resembles that of large gull; does not change with age. Head, neck, rump and underparts white, otherwise dark. Dark smudge below and around eye. Upperwing, mantle, upper rump and tail very dark blackish-brown. In flight, white shafts to upper primaries visible; underwing pattern variable, but broad leading edge, trailing edge and primaries always black, with small to large black patch on carpal and closer to body. **BP** Bill large, dark pink or deep yellow with dark grey tip in ad., all dark in juv.; eyes black; legs greyish-pink. **Vo** Mostly silent, but on breeding grounds emits groaning sounds during slow, elaborate courtship dance. **TN** Formerly placed in *Diomedea*.

BLACK-FOOTED ALBATROSS   CTJR
Phoebastria nigripes

L 68–74cm; WS 193–213cm; WT 3–3.6kg. **SD** N Pacific, breeds on Leeward, Marshall and Johnston Is (western Hawaii) and on certain remote Japanese islands (Torishima, Ogasawara and Senkaku; Oct–May). One-third (20,000 pairs) breeds on Midway and one-third on Laysan. Monotypic. **HH** Pelagic, but occasionally seen from shore, especially after storms. In non-breeding season occurs off Japan, Kuril Is and in Kamchatkan Pacific Ocean and Bering Sea, also in Taiwan Strait, but rare off SE China and accidental in Taiwan. **ID** Darkest of the three breeding albatrosses in the region (except imm. Short-tailed Albatross), appearing longer winged. Ad. dark sooty-brown, slightly paler on underparts, with white around bill base and below eye. Uppertail-coverts, vent and undertail-coverts white. Some have pale on belly and breast, and may be confused with imm. Short-tailed, but have dark collar and dusky-pink bill. Juv. also all dark, with all-dark tail, rump and vent. In flight, note white shafts to primaries at all ages. Pale ad./juv. differs from juv./imm. Short-tailed in lacking pale patches on inner secondaries and by black, or at most dusky-pink, bill. **BP** Bill long, thick, black or dusky pinkish-grey with dark tip; eyes dark greenish-brown; tarsi blackish-grey. **Vo** Generally silent, but at sea and breeding colonies gives long, extended groaning *uuwoouu* or short, continuous *uuuu*. At sea bickers over food with sheep-like bleating and shrill whistles. Simple courtship display dance also incorporates bill-snapping and groaning calls. **TN** Formerly placed in *Diomedea*.

SHORT-TAILED ALBATROSS   CTKJR
Phoebastria albatrus

Vulnerable. L 84–94cm; WS 213–229cm; WT Av 5.3kg. **SD** N Pacific albatross recovering from near extinction in early 20th century; breeds (Oct–May) only on remote Izu (Torishima) and Senkaku Is (East China Sea). In non-breeding season disperses north in Yellow Sea as far as Shandong, and in NW Pacific and S Bering Sea historically as far as Bering Strait. Seen off Japan, Kuril Is and Kamchatka; accidental Taiwan. Monotypic. **HH** Breeds on open flats or scree slopes with grass on oceanic islands. Rest of year exclusively pelagic, dispersing more widely in N Pacific, often near or over continental shelf zones of Bering Sea and between Commander and Aleutian Is. **ID** Largest, rarest and most spectacular of the region's breeding albatrosses. Wings and tail broader, and bill larger and more obviously pink than other species. Ad. plumage (acquired after 12+ years) largely white with golden crown and face, and dusky grey-brown nape; black primaries (with white shafts), secondaries, tertials and outer wing-coverts; shoulder and inner secondaries white; tail black. White mantle, rump and uppertail coverts readily distinguishes ad. from other albatrosses. Juv. entirely very dark brown for several years, then becomes increasingly pale. As subad. look for large pink bill, and pale or white patch on inner secondaries. **BP** Bill pink with grey tip; eyes black; tarsi dull greyish-pink. **Vo** At breeding colonies gives loud braying *bwaaaaa* or *bwiiaaaa* or *u uuu u* (lower than Black-footed), and a loud clattering of mandibles in display (lower and slower than Black-footed). Otherwise largely silent. **TN** Formerly placed in *Diomedea*. **AN** Steller's Albatross.

WANDERING ALBATROSS   J
Diomedea exulans

Vulnerable. L 107–135cm; WS 254–351cm; WT 6.25–11.3kg. **SD** Largest albatross. S Hemisphere species accidental at Senkaku Is, Japan (two in Nov 1970); subspecies unconfirmed. **ID** Extremely large, wings longer than northern albatrosses. Ad. largely white, with black only on wings; may appear dusted with white on coverts and tail. White mantle, rump and uppertail-coverts. Takes 8–10 years to reach full adult plumage. Juv. initially dark brown with white only on face and on underwing which increases with age; white on upperwing spreads from body along centre, increasing in width; body all white at 4–5 years, but extensive white wings only with maturity. Black on primaries and trailing edge; juv. tail white with black terminal band (all dark in early years), subsequently black-tipped or all white; ♀ (and some subspecies) less white than fully ad. ♂. **BP** Bill massive, pink with yellow tip; eyes black; tarsi pale flesh. **Vo** Usually silent at sea, but croaks, yaps, brays, shrieks and bill-claps at colonies, and may call when squabbling over food.

LAYSAN ALBATROSS

ad lighter

ad darker

ad

SHORT-TAILED ALBATROSS

imm

imm

ad

young imm

young imm

BLACK-FOOTED ALBATROSS

ad

ad

ad

imm

imm

imm

ad

imm

imm

WANDERING ALBATROSS

NORTHERN FULMAR CKJR
Fulmarus glacialis

L 45–50cm; WS 102–112cm; WT ♂ *c*.700g, ♀ *c*.835g. **SD** N Atlantic and N Pacific. *F. g. rodgersii* is largest and commonest of large petrels in Bering Sea, NW Pacific and NE Okhotsk Sea. Post-breeding dispersal reaches NE Japan and Kamchatka, the Russian Bering Sea coast and E Pacific; rare off Korea and NE China. **HH** Breeds in colonies, visited diurnally, in Kuril Is, and Sea of Okhotsk (e.g. *c*.1,000,000 on Yamskiye Is). In non-breeding season exclusively pelagic but occasionally seen from shore. An opportunistic and versatile feeder, taking carrion, fish offal from fishing fleets and macro-plankton. **ID** Superficially gull-like; compact, with large head, short, thick neck, broad, rounded wings and stubby bill. Dark form sometimes mistaken for Flesh-footed Shearwater (see p.84). Ad. may be dark, light or very light (the former 'blue' form predominates in N Pacific and S Bering Sea, and the pale forms in N Bering Sea), but all three have dark tail (pale grey, concolorous with rump in Atlantic birds) and dark smudge around eye. The darkest are plain grey-brown with paler inner primaries. Light ad. has white head, neck, upper tail and underparts; wings and uppertail-coverts mid-grey with pale and dark speckling; primaries darker, with white patch on innermost. Very light ad. mostly white, with a little grey on upperwing. Colour morphs vary in percentage between colonies, but darkest birds commonest in S Bering Sea. Ungainly on land; floats high on water. Takes off by jumping from cliffs or pattering across water. Glides on broad stiff wings, interspersed with rapid shallow beats. **BP** Tubenose bill is yellowish-horn, stubby and deep with prominent nostrils; eyes black; tarsi pinkish-grey. **Vo** Highly vocal on breeding grounds, gives deep, guttural cackling *aark aak aak*, accelerating in duet, and also squabbles noisily, grunting and cackling in flocks at sea or when feeding.

CAPE PETREL J
Daption capense

L 38–40cm; WS 81–91cm; WT 340–480g. **SD** S Hemisphere species accidental in Japan (*D. c. capense*). **HH** Pelagic. Commonly follows ships. **ID** Northern Fulmar-like in proportions, but distinctively pied. Hood, mantle and tail-band black; mantle and rump white, heavily spotted black. Upperwing black with large white patches on inner primaries and inner secondaries. **BP** Bill short, tube-nosed, black; eyes black; tarsi black. **Vo** Gives short, sharp whistles and harsh rattling calls at sea.

PROVIDENCE PETREL JR
Pterodroma solandri

Vulnerable. L 40cm; WS 95–105cm; WT 500g. **SD** Migrant to N Pacific from breeding grounds off E & NE Australia; scarce, encountered only in very small numbers in deep waters east of Japan, north to Hokkaido and southern Kuril Is; Jun–Oct. Monotypic. **HH** Pelagic. **ID** Large, thickset gadfly petrel. Rather dark, grey and brown; head darker, often giving hooded appearance, but paler around base of stubby bill and on chin, and underparts slightly paler than breast and head. M across upperwing, and dark blackish-grey tail (broadly tipped darker), both contrast with dull upperparts, though M can be inconspicuous at distance. Mantle and undersides of primaries and secondaries may be mottled or entirely pale; white bases to primaries and greater primary-coverts form two neat crescents on otherwise uniform dull grey underwing, creating skua-like white flash at base of primaries. **BP** Bill short, thick, black; eyes black; tarsi grey. **Vo** Generally silent at sea. **AN** Solander's Petrel.

JUAN FERNÁNDEZ PETREL J
Pterodroma externa

Vulnerable. L 43cm; WS 95–97cm; WT 500g. **SD** Migrant to N Pacific from breeding grounds off western S America; a post-typhoon accidental to Japan. Monotypic. **HH** Pelagic. **ID** Large grey and white gadfly petrel, with rather typical grey upperparts and prominent black M on upperwings. Black cap extends below eyes and is continuous with back, lacking white collar of White-necked Petrel (see p.78), but bordered at sides by white extending from throat. Tail grey, but sometimes has basal off-white horseshoe. Underparts entirely white. Underwing largely white, with black tips and leading edge to otherwise white primaries, a narrow black trailing edge to wing, and very short black bar from carpal towards body (cf. Stejneger's, p.80). **BP** Bill short, black; eyes black; tarsi grey. **Vo** Generally silent at sea.

NORTHERN FULMAR

rodgersii

dark morph

light morph

dark morph

darkest variant

CAPE PETREL

capense

PROVIDENCE PETREL

fresh

worn

JUAN FERNÁNDEZ PETREL

WHITE-NECKED PETREL J
Pterodroma cervicalis

Vulnerable. L 43cm; WS 95–100cm; WT 380–545g. **SD** Migrant to warmer waters of N Pacific from breeding grounds northeast of New Zealand; rare in Japanese subtropical waters, e.g. off Ogasawara Is (mid Jul–late Nov). Monotypic. **HH** Pelagic. **ID** Large grey and white gadfly petrel, resembling Juan Fernández Petrel (see p. 76), but appears heavier and shorter winged at sea, and upperparts more brown-toned. Dark grey or black cap, pale grey mantle and innerwing, and dark grey tail. Neck and underparts white; the white hind-collar, separating black cap from grey upperparts, further distinguishes it from Juan Fernández. Underwing mostly white, with black border to forewing and on secondary and primary tips. **BP** Bill short, black; eyes black; tarsi grey. **Vo** Generally silent at sea.

KERMADEC PETREL TJ
Pterodroma neglecta

L 38–39cm; WS 92cm; WT 509g. **SD** *P. n. neglecta* is a scarce migrant to N Pacific from breeding grounds on oceanic islands across SW & C Pacific; very small numbers in deep waters east of Honshu north to Hokkaido (most frequent Sep). Accidental Taiwan. **HH** Pelagic. **ID** Polymorphic gadfly petrel with obvious skua-like wing flashes (bases to inner webs of primaries) on upper and underside of wing in pale and dark morphs, and rather short, square tail. Dark morph mostly plain dark grey-brown, recalls similar Providence Petrel (see p. 76), but is brown rather than grey, and lacks white underwing greater primary-coverts crescent. Pale morph is grey-brown with paler head, and very pale grey or off-white lower breast and belly. Typically a fleck of white between base of stubby black bill and eye. Flight (in moderately calm conditions) leisurely, with deep wingbeats followed by long glides, rising and banking in broad arcs. **BP** Bill stubby, black; eyes black; tarsi pinkish-orange. **Vo** Generally silent at sea.

HAWAIIAN PETREL J
Pterodroma sandwichensis

Vulnerable. L 40–43cm; WS 90–91cm; WT 434g. **SD** Separate races of former Dark-rumped Petrel *P. phaeopygia* on Galápagos and Hawaii migrate to C Pacific, but now recognised as separate species. Hawaiian has strayed to Honshu, Japan, on a few occasions. **HH** Pelagic. **ID** Rather large, long-winged, dark greyish-brown and white gadfly petrel, similar to Juan Fernández Petrel (see p. 76). Forehead white, crown and nape black, the cap extending below eyes and onto neck/upper-breast-sides; mantle and back pale grey to greyish-brown (slightly paler than head and wings), wings and tail dark slaty-grey. Slight M mark on wings. Underparts largely white, with broad black markings close to leading edge (carpal and primary-coverts), rather broad black tips to white-based primaries, and narrow black trailing edge; axillaries white, only rarely showing any dark markings. **BP** Bill short, black; eyes black; tarsi bluish-flesh. **Vo** Generally silent at sea.

MOTTLED PETREL JR
Pterodroma inexpectata

L 33–35cm; WS 74–82cm; WT 247–441g. **SD** Common migrant from breeding grounds off New Zealand to deeper waters of N Pacific, around Aleutian Is and north into Bering Sea, where occurs in summer. Rare in Japanese waters, but common between Kamchatka and Commander Is, numerous between there and Attu I (USA), and east to C Aleutians. Should be looked for during any summer pelagic off E & NE Japan, the Kuril Is and Kamchatka. Monotypic. **HH** Pelagic. **ID** Very distinctive, rather heavy gadfly petrel. Almost entirely pale or mid grey on upperparts, with a strong dark grey M on upperwings. Underwing white with strong black bar from carpal almost to the axillaries. The most obvious feature is the dark grey belly patch, which contrasts well with white underwing and vent (though can be surprisingly difficult to see even at moderate range); neck- and breast-sides also grey. **BP** Short, stout black bill contrasts with white lores and chin; black patch around black eye; tarsi flesh-coloured with distal webs black. **Vo** Generally silent at sea.

WHITE-NECKED PETREL

fresh

worn

KERMADEC PETREL

pale morph

intermediate morph

dark morph

intermediate morph

dark morph

worn

MOTTLED PETREL

HAWAIIAN PETREL

fresh

fresh

worn

BLACK-WINGED PETREL   J
Pterodroma nigripennis

L 28–30cm; WS 63–71cm; WT 140–200g. **SD** Migrant to NW Pacific (Jun–Nov). So far recorded only once from Japan (Hakodate, Hokkaido, after a typhoon), but should be looked for during summer pelagic trips off E & NE Japan, and perhaps even Kuril Is and Kamchatka. Monotypic. **HH** Pelagic. **ID** Grey, black and white gadfly petrel with strong underwing pattern. Upperparts entirely pale or mid grey with strong dark grey M on upperwings recalling Mottled Petrel (see p. 78), but lacks grey belly patch of that species, and is smaller and flight is less towering. Underwing closely resembles Bonin Petrel, white with very strong black bar from carpal almost to axillaries (but has much narrower black primary-coverts patch), and prominent black trailing edge to underwing (not noticeable in Mottled), but cap less black and head and neck-sides paler, greyer than Bonin. **BP** Bill stout and black, lores and chin white; eyes black with white eyebrow and small black mask; tarsi black. **Vo** Generally silent at sea.

BONIN PETREL   CTKJR
Pterodroma hypoleuca

L 30–31cm; WS 63–71cm; WT *c*.182 g. **SD** Only gadfly petrel that breeds in region, on outer Japanese islands south from the Ogasawara (Bonin) Is (also breeds Hawaii). Outside winter/ spring breeding season (Nov–May) disperses across NW Pacific, occasionally being reported in Japanese waters north to Russian Kuril Is; accidental Korea, Taiwan and Fujian coast, China. Monotypic. **HH** Pelagic. **ID** Small, graceful gadfly petrel, similar to Black-winged; dark grey upperparts with darker, blackish-grey wings. Mask and neck-sides blacker than Black-winged. Lores, chin, foreneck and underparts pure white. Distinctive underwing pattern with broad black primary-coverts patch and strong black carpal bar extending towards, but not reaching, flanks; undertail-coverts white. **BP** Bill short, black; eyes black; tarsi pink, toes blackish-pink. **Vo** Generally silent at sea.

STEJNEGER'S PETREL   J
Pterodroma longirostris

Vulnerable. L 26–31cm; WS 53–66cm. **SD** Though only rarely recorded off Japanese coast, from Okinawa to N Honshu, this migrant to NW Pacific occurs in very large numbers just east of Japanese waters (May–Sep/Oct), and should be looked for after typhoons during any summer or autumn pelagic off E Japan. Monotypic. **HH** Pelagic. **ID** Small grey, black and white gadfly petrel with very strong M mark on upperparts and very pale trailing edge to inner wing. Very similar to Black-winged Petrel, but eye patch less obvious, paler grey on upperparts and has more prominent dark grey M on upperwings. Underwing white, with only small black carpal patch and inconspicuous black trailing edge. **BP** Bill black; eyes black; tarsi dark grey. **Vo** Generally silent at sea.

TAHITI PETREL   T
Pseudobulweria rostrata

L 38–40cm; WS 84cm. **SD** Ranges across tropical and subtropical W Pacific Ocean; accidental in Taiwan, though subspecies uncertain: *P. r. rostrata* of Society and Marquesas Is and *P. r. trouessarti* of New Caledonia not seperable at sea. **HH** Pelagic. **ID** Two-toned petrel, with blackish-brown upperparts including head, wings and tail. Jizz distinctive, appears small-headed with long wings and straight leading edge. Hood, to lower neck black, underwing including axillaries dark blackish-brown, but breast, belly and vent clean white. In some lights dark underwing has pale central stripe. Very similar extralimital **Beck's Petrel** *P. becki* could reach Japan; same plumage but smaller with thinner bill. **BP** Bill short and black; eyes black; tarsi pink with black webs. **Vo** Generally silent at sea. **TN** Formerly placed in *Pterodroma*.

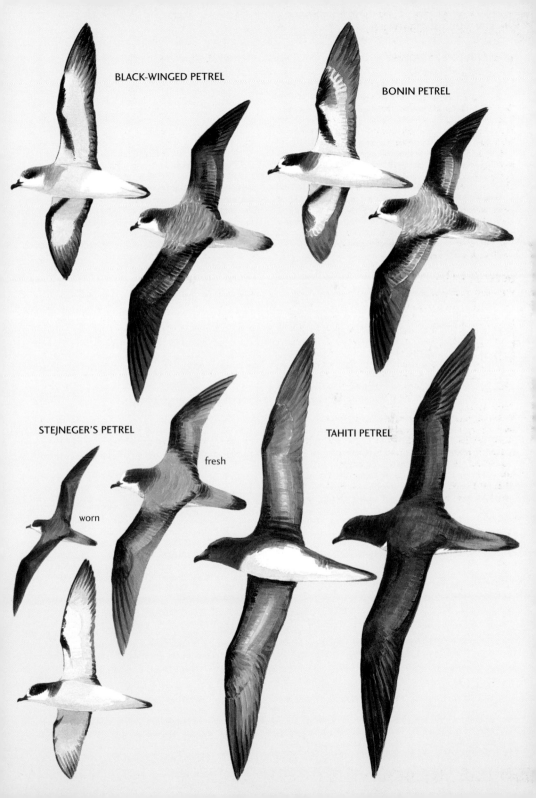

BLACK-WINGED PETREL

BONIN PETREL

STEJNEGER'S PETREL

TAHITI PETREL

fresh

worn

STREAKED SHEARWATER CTKJR
Calonectris leucomelas

L 48–49cm; WS 122cm; WT 440–545g. **SD** Breeding endemic primarily in warm-water regions of NW Pacific, around Japan, Sea of Japan and Yellow Sea. Most frequent Mar–Nov when breeds colonially on islands, particularly off Japanese coast south from Hokkaido to the Nansei Shoto, but also off S Russian Far East, Korea, Shandong, E China, and Taiwan. Disperses north, uncommonly as far as Sea of Okhotsk and Kuril Is and, more especially, south to Australian waters when not breeding. Should be looked for anywhere from northern Kuril Is to Fujian (China). Monotypic. **HH** Pelagic. The commonest shearwater in sight of land, and after typhoons; during fog and heavy rain sometimes recorded inland. Burrow-nester on forested offshore islands, visited only at night. **ID** Large, long-winged shearwater (largest in region), with broad wings, white underparts and underwings, and prominent pale or white head at distance. Plain mid brown hindneck, mantle, wings, rump and tail, with pale fringes to secondaries, some wing-coverts and mantle, producing impression of dark M across paler wings and mantle, and pale V on uppertail-coverts in worn plumage. Head and neck flecked white to almost entirely white (age-related?); face and underparts all white. Flight steady with slow flaps, gull-like on angled wings, interspersed with long glides; pale underwing, dark carpals, white body, long thick neck and pale head diagnostic. **BP** Bill grey/pink with dark tip; eyes dark brown or black; tarsi pinkish-grey, feet do not reach tail tip. **Vo** Noisy growling at colony, with returning ♂ giving *pee wee pee wee* and ♀ deeper, more vigorous *guwaae guwaae* calls.

KIRITIMATI SHEARWATER J
Puffinus nativitatis

L 35–38cm; WS 71–81cm; WT 324–340g. **SD** C Pacific (Hawaii, and Kiritimati to Marquesas), but reported from Pacific coast of Honshu, Japan, on several occasions. Monotypic. **HH** Pelagic. **ID** Small, slender, uniformly dark-brown shearwater. Resembles Wedge-tailed Shearwater but smaller, with rounded tail. Upperparts and underparts uniform dark brown, like small Short-tailed Shearwater (p.84), but long narrow wings lack silvery-grey underwing. **BP** Bill black; eyes black; tarsi dark brown. **Vo** Generally silent at sea. **AN** Christmas Island Shearwater.

WEDGE-TAILED SHEARWATER CTJ
Puffinus pacificus

L 38–46cm; WS 97–105cm; WT 320–510g. **SD** Tropical Indian and Pacific Oceans, its range extending north to the Ogasawara and Iwo Is, Japan (breeds late Mar–early Oct), where replaces Streaked Shearwater; scarce further north in Izu Is and very rare off main Japanese islands, Nansei Shoto, and Chinese coast, but a common migrant past Taiwan. Monotypic. **HH** Pelagic. **ID** Mid-sized shearwater with pale and dark morphs, both have long wedge-shaped tail. Dark morph entirely dark brown (including underwing); pale morph resembles Streaked, mid brown above with pale underparts, but lacks Streaked's pale head and face. Underwing-coverts all white,

leaving only primaries and secondaries dark. Many Japanese birds are pale morphs, but dark morphs numerous off Taiwan. Flight rather lazy with short glides on bowed wings held forwards, interspersed by slow flaps; less graceful than other shearwaters. **BP** Bill dark grey; eyes black; tarsi pink. **Vo** At breeding sites gives deep moaning *uu-oo uu-oo* or *vuu oo vuu oo*. Silent at sea. **TN** May change to *Ardenna pacifica* in the future.

BULLER'S SHEARWATER JR
Puffinus bulleri

Vulnerable. L 46–47cm; WS 97–99cm; WT 342–425g. **SD** New Zealand endemic breeder which migrates to N & E Pacific outside austral summer breeding season. Rare and most commonly encountered in Japanese waters off E Honshu (Aug–Oct); has strayed north to Kamchatkan coast. Monotypic. **HH** Pelagic. **ID** Medium to large, slender-bodied grey, black, and white shearwater with striking *Pterodroma*-like pattern to upperparts. Between Short-tailed and Sooty in size (see p.84), but more slender with longer tail, broader based wings and neatly patterned plumage. Upperparts have striking M pattern, with blackish-grey crown, nape and tail, and dark grey flight-feathers; very dark wing-coverts contrast with pale grey hindneck, mantle, shoulders and greater coverts, rump and uppertail-coverts. Underparts clean white. Flight is graceful, languid, and buoyant, recalling smaller albatrosses, with long effortless glides low over water even in completely calm conditions. **BP** Bill grey with dark culmen and tip; eyes black; tarsi dark grey. **Vo** Generally silent at sea. **TN** May change to *Ardenna bulleri* in the future. **AN** New Zealand Shearwater.

TROPICAL SHEARWATER J
Puffinus bailloni

L 27–33cm; WS 64–74cm; WT 150–230g. **SD** Tropical Indian and Pacific Oceans. *P. b. bannermani* (sometimes considered specifically as **Bannerman's Shearwater**) a local and uncommon endemic, seemingly restricted to vicinity of its remote breeding grounds, on Ogasawara and Iwo Is, where occurs year-round, though has reached Nansei Shoto. **HH** Pelagic. **ID** Smallest shearwater in region. Forehead rather vertical like Short-tailed Shearwater (see p.84), but crown appears flatter. Bill plain and two-toned, with slate-black upperparts (mantle, rump and upperwing-coverts may show paler fringes) and white chin to undertail-coverts; dusky blackish-grey neck- and breast-sides. White at bill base below lores, white spot in front of eyes suggesting supercilium, pale grey mottling around eyes and ear-coverts, white face below eye, white flanks and dark undertail-coverts. At rest, wingtips fall close to or short of tail tip. Wings rather broad and round-tipped, underwing white with broad dusky margin at rear and tip, whilst leading edge also dark and broad especially on inner wing. Undertail dark with white, or at least pale, central rectrices. Flight rapid, with fast wingbeats and short glides. Previously considered conspecific with Atlantic **Audubon's Shearwater** *P. lherminieri*, but has paler, greyer head and hind-collar, contrasting with black mantle and whiter face, mottled blackish-grey underwing-coverts, neck- and breast-sides, and paler fringes to upperparts. **BP** Bill dark grey with blue base; eyes black; tarsi and toes dark blue-grey, webs pink. **Vo** Generally silent at sea. **TN** Formerly part of Audubon's Shearwater *P. lherminieri*. **AN** Baillon's Shearwater.

STREAKED SHEARWATER

KIRITIMATI SHEARWATER

BULLER'S SHEARWATER

fresh

worn

WEDGE-TAILED SHEARWATER

dark morph

dark morph

pale morph

pale morph

intermediate
morph

bannermani

TROPICAL SHEARWATER

MANX SHEARWATER   J
Puffinus puffinus

L 30–38cm; WS 76–89cm; WT 350–575g. **SD** Migrant between N and S Atlantic, but has strayed to Pacific, regularly to W USA north to Alaska, and suspected of breeding in NE Pacific (possibly Triangle I., British Columbia). Claimed Japan. Monotypic. **ID** Medium-sized shearwater with uniform sooty-black upperparts (can appear brownish-black in strong light), extending below eyes to lores and auriculars, and onto neck- and breast-sides, and in some to axillaries. Underparts white including undertail-coverts. On water shows clean white flanks; wingtips extend beyond tail. In flight, tends to flap more (and glide less) with extremely rapid, stiff, almost auklet-like wingbeats interspersed by longer glides, unlike congeners; shears and banks in stronger winds. Differs from Tropical (see p.82) in being larger, with white undertail-coverts and black lores. **BP** Bill grey with dark culmen and tip; eyes black; tarsi dark grey. **Vo** Generally silent at sea.

FLESH-FOOTED SHEARWATER
Puffinus carneipes   KJR

L 40–45cm; WS 99–107cm; WT 580–765g. **SD** Ranges north to W, N & E Pacific, and Yellow Sea (small numbers) from breeding grounds off Australia and New Zealand. Occurs year-round, but most May/Jun, and relatively scarce compared to Sooty and Short-tailed Shearwaters, particularly in autumn. Monotypic. **HH** Pelagic. **ID** Large, stocky, dark brown shearwater with rather large head and full chest. Most easily distinguished, other than by size, by uniform dark upperwing and underwing-coverts, slightly pale grey flight-feathers (below) and leg and bill colour. Darker and noticeably broader-winged than Sooty. At rest resembles Wedge-tailed (see p.82), but larger with paler bill, and wingtips extend beyond tail. Flight more laboured, almost gull-like, beats deeper and slower than Sooty. Dark-morph Northern Fulmar (see p.76) often mistaken for this species in Japanese waters, but Flesh-footed much darker brown (not milky grey), with longer, narrower wings, smaller head and slender-bodied jizz, and longer, more slender bill with obvious dark tip. Wings lack prominent pale panel at base of primaries of Northern Fulmar. **BP** Bill nearly as long as head, straw-coloured to pink with dark grey tip (bill conspicuously bicoloured even at distance); eyes black; tarsi pinkish-grey often obvious against dark tail. **Vo** At breeding sites gives *kwi kwi aa* calls; though generally silent at sea, gives sharp, high squeals during feeding frenzies. **TN** May change to *Ardenna carneipes* in future. **AN** Pale-footed Shearwater.

SOOTY SHEARWATER   CTKJR
Puffinus griseus

L 40–51cm; WS 94–109cm; WT 650–978g. **SD** Abundant, breeding primarily off New Zealand and Australia during austral summer and performing loop migration around N Pacific post-breeding; also breeds off Chile and ranges north into N Atlantic. Millions migrate past Japan to Sea of Okhotsk and Bering Sea in summer (Apr–Aug), with smaller numbers passing between Japan and Asian mainland to N Hokkaido. Accidental E China and Taiwan, reported Korea. Monotypic. **HH** Pelagic. Follows only fishing boats. **ID** Medium-sized, essentially uniform sooty-grey or blackish-brown, becoming browner with wear. Long, narrow,

pointed wings and obvious silvery-grey to white feathering (individually variable in extent) on underwing-coverts contrasts strongly with dark brown body, axillaries and dark grey-brown flight-feathers. Broadest and brightest white is on primary-coverts contrasting with dark primaries. Whitish secondary-coverts form bar that gradually narrows towards body. Separation from Short-tailed Shearwater often impossible, except at close range, thus enormous numbers of shearwaters off N Japanese and Russian coasts in summer may involve either or both species. On water appears evenly dark. Crown rather flat, the forehead gently sloping to bill which is nearly as long as head. Neck short and wingtips extend beyond short tail. Bill longer than Short-tailed and head less rounded. Flight powerful, smooth and rather direct, banking strongly; wings angled back in strong wind; body stocky, more uniformly tubular than Short-tailed, thus looks dumpy at rear. In calm weather, gives 3–7 quick, stiff-winged flaps, followed by glide of 3–5 seconds. In moderate breeze flight lacks sudden changes in direction; flaps briefly then arcs up in longer glide with ventral side facing into wind. In strong winds, glides without flapping, arcing up very high followed by long downward glide. **BP** Bill long (almost as long as head), fairly stout at base with 'pinched-in' middle and heavy hooked tip, dark-grey with black culmen and tip; eyes black; tarsi dark grey. **Vo** Feeding groups can be noisy, giving a raucous, nasal *aaaa*.

SHORT-TAILED SHEARWATER   TKJR
Puffinus tenuirostris

L 40–45cm; WS 95–100cm; WT 480–800g. **SD** Abundant, but breeds only off New Zealand and Australia in austral summer, and makes loop migration around N Pacific. Millions migrate past Japan into Sea of Okhotsk, northern N Pacific and Bering Sea in summer (May–Oct), with smaller numbers passing between Japan and continental mainland to N Hokkaido. Scarce at other seasons. Smaller numbers pass Korea and Taiwan. Monotypic. **HH** Pelagic. Follows fishing boats. **ID** Dark brownish-grey shearwater closely resembling Sooty. Ad. generally as Sooty, but underwing-coverts, though very pale grey or brownish-grey with silvery sheen (light-dependent, and usually brightest on median secondary-coverts not median primary-coverts), usually darker and more uniform than Sooty (rarely all dark). In flight appears duller than Sooty (though some, confusingly, have uniform silvery wing linings, including most underwing-coverts and bases to flight-feathers), with toes extending noticeably beyond tail. Juv. (late autumn/winter) has entirely pale grey underwings and brown underparts, especially pale on breast, neck, and throat, contrasting with dark crown and ear-coverts. Usually shows pale throat and short neck; bill appears shorter and more slender than Sooty. Wings shorter and narrower than Sooty, with more even width, and wingtips appear slightly more rounded. Flight action highly variable (Sooty rather consistent), faster, more erratic, with more mechanical wingbeats and longer periods of flapping, usually less arcing and gliding, but frequent changes of direction. Characteristically rocks side-to-side while flapping. **BP** Bill short (clearly shorter than head), dark grey with black culmen and tip; eyes black; tarsi dark grey. **Vo** Generally silent at sea. **AN** Slender-billed Shearwater.

MANX SHEARWATER

worn

fresh

FLESH-FOOTED SHEARWATER

SOOTY SHEARWATER

SHORT-TAILED SHEARWATER

fresh

fresh

BULWER'S PETREL   CTJ
Bulweria bulwerii

L 26–28cm; WS 68–73cm; WT 78–130g. **SD** Ranges in subtropical Atlantic and Pacific Oceans. Breeds in region only on remote islands off S Japan, Taiwan and China's Fujian coast, visiting colonies only at night. Reasonably common around outer Japanese islands south from Ogasawara Is and the Nansei Shoto in Apr–early Oct, but recorded from mainland coasts or inland only after typhoons. Monotypic. **HH** Pelagic, dispersing widely across W & C Pacific. **ID** Long-winged and long-tailed dark petrel with small head, intermediate between shearwaters and storm-petrels. Entirely blackish-brown, with only slightly paler diagonal upperwing-coverts bar. Slightly larger and longer-tailed than Tristram's Storm-Petrel (see p.88), with which it may be confused, though wings broader, tail pointed when closed, wedge-shaped when open (not forked). Similar all-dark shearwaters (Sooty and Short-tailed, p.84) larger, with larger heads, grey or silvery underwings, longer, slender bills, and different flight actions. Flight also differs from Tristram's, being graceful, somewhat shearwater-like, but lighter, erratic, buoyant, even tern-like, with sudden changes of speed and direction, employing loose, deep wingbeats, and often gliding somewhat skua-like on arched wings, but head and bill point down; tail is fanned rarely. **BP** Bill short, rather heavy, black; eyes black; tarsi pale pink, webs black distally. **Vo** Generally silent at sea, but at nest gives hoarse barking *hroo hroo hroo*, guttural *krsh krsh* notes and moans.

WILSON'S STORM-PETREL   J
Oceanites oceanicus

L 15–19cm; WS 38–42cm; WT 34–45g. **SD** S Hemisphere breeder; *O. o. exasperatus* is occasional off Japan's Pacific coast. **HH** Pelagic. **ID** Dark blackish-brown plumage relieved only by pale panel on greater upperwing-coverts, variably narrow to broad, and by broad square white rump and extensively white undertail-coverts. Tail square, but may appear notched. As storm-petrel plumages largely similar, note wing/'hand' shape and flight action. Flight bounding, wings outstretched, skipping, pattering, walking on water with legs extended, occasionally showing diagnostic yellow webs; flight also often slow and fluttering, with very prolonged glides on flat, depressed wings. Wings short, with indistinct carpal bend, 'arm' short and broad, 'hand' longer, creating broadly triangular wing shape with rounded tip (compare Leach's, p.88). Centre of otherwise dark underwing is pale brown, and toes extend well beyond tail tip. More frequently foot-patters sea surface whilst feeding than other storm-petrels, with wings and tail raised, revealing white lateral undertail-coverts. **BP** Bill slender, short, black; eyes black; long legs black, toes dark grey, webs yellow. **Vo** Generally silent, but foraging groups (and ad. at colonies) give harsh, stuttering, chattering *kerr kerr kerr kerr* calls.

BAND-RUMPED STORM-PETREL   JR
Oceanodroma castro

L 19–21cm; WS 44–46cm; WT 29–56g. **SD** Atlantic and Pacific. In region, breeds only Pacific coast of Japan (mostly seen May–Oct). Monotypic. **HH** Pelagic. Most frequently encountered storm-petrel from NW Pacific ferries. **ID** Largely brownish-black to black, with broad white rump (wider than long), extending to rump-sides (but not vent), forming house martin-like patch. In flight, dark brown carpal bar weak and does not reach leading edge, not as pale as Wilson's or as long as Leach's. Most similar to Leach's (see p.88), but has broader wings, wing-shape intermediate between Leach's and Wilson's. Wings longer (especially 'arm') than Wilson's; tail shorter and squarer, only slightly forked, and toes do not extend to tip. Flight erratic, with many twists and turns, often turning back on its course, less bounding than Leach's, with quicker, shallower beats; glides on bowed wings. **BP** At close range, black bill is stubbier, heavier than other species; eyes black; short tarsi black (not visible in flight). **Vo** Highly vocal around colonies: squeaky *chiwee* and repetitive, deep purring *kuwa kuwa gyururu* or *kerr wheecha wheecha wheecha wheeeechuh* (deeper and hoarser than Leach's). **TN** May become *Thalobata castro* in the future. Recent studies indicate that species may be polytypic. **AN** Madeiran Storm-Petrel.

SWINHOE'S STORM-PETREL   CTKJR
Oceanodroma monorhis

L 19–20cm; WS 44–46cm; WT 38–40g. **SD** Breeds (May–Sep) mainly in NE Asia, on a few remote islands around Japan, S Russian Far East, Yellow Sea, Korea, Shandong (China) and Taiwan. Disperses to south and west, typically reaching N Indian Ocean, even N Atlantic (though records in W Europe perhaps more suggestive of undiscovered breeding colonies in that region). Scarce, vulnerable and enigmatic; most frequently seen from offshore ferries. Monotypic. **HH** Pelagic. **ID** Small all-black storm-petrel, with no white rump patch (much smaller than Tristram's and Matsudaira's). Similar in size, structure and wing pattern to Leach's (p.88), but blacker, with all-dark rump and only slightly forked tail (note that some Leach's show little white on rump). In flight, dark brown carpal bar less prominent than other species, and 2–3 pale, outer primary shafts may be inconspicuous, except at very close range (cf. Matsudaira's, see p.88). Wings have long angular shape typical of *Oceanodroma* species. Flight, like Leach's, rather tern- or pratincole-like, powerful, graceful, bounding but erratic, with sudden banks and arcs; also rather fluttering flight interspersed with glides. **BP** Black stubby bill; eyes black; short tarsi and toes black. **Vo** Generally silent at sea. **TN** May become *Cymochorea monorhis* in the future.

BULWER'S PETREL

exasperatus

WILSON'S
STORM-PETREL

BAND-RUMPED STORM-PETREL

SWINHOE'S STORM-PETREL

LEACH'S STORM-PETREL CJR
Oceanodroma leucorhoa

L 19–22cm; WS 45–48cm; WT 45g. **SD** Widespread in N Atlantic and N Pacific. *O. l. leucorhoa* breeds (May–Aug) in north of region, on Commander and Kuril Is, in Sea of Okhotsk and off SE Hokkaido. In non-breeding season encountered anywhere at sea off Pacific coast of Honshu north to Sea of Okhotsk and Kamchatka; accidental NE China. **HH** Breeds in burrows. Commonly attracted to ships' lights and even onshore lights near breeding grounds. May even appear inland during heavy rain and fog. Pelagic outside breeding season. **ID** Slender, fairly large, long-winged storm-petrel, largely blackish-brown (slightly paler than Band-rumped or Wilson's, see p.86), with prominent, long V-shaped white rump, sometimes with dark central feathers, dark vent and all-dark undertail-coverts, thus less bright than Band-rumped. Some in N Pacific are dark-rumped, but no such birds reported in E Asia. In flight, long wings are pointed, narrower in 'arm' than Band-rumped or Wilson's, with strong carpal angle and very prominent pale-brown carpal bar extending diagonally across entire wing. Flight strong and bounding, wingbeats deep, sometimes glides on bowed wings with carpals held forward, or wings raised; deeply notched tail; some birds show white primary shafts. **BP** Bill short, black; eyes black; tarsi black. **Vo** Silent at sea, but highly vocal at and near colonies when gives a crooning *uooooo u* from inside burrow and a very rapid, highly repetitive chattering cooing: *oteke-te-toto* or *totte ketto tep top top* and more alarm-sounding *ki ki ki ki kyururuuru...* calls in flight over colony. **TN** May become *Cymochorea leucorhoa* in the future.

TRISTRAM'S STORM-PETREL
Oceanodroma tristrami TJ

L 24–25cm; WS 56cm, WT 70–112g. **SD** In region breeds (Oct–Apr) only in Izu, Ogasawara, Iwo chains (also in Hawaii), but occurs in warm waters around and between breeding areas all year. Monotypic. **HH** Breeds on volcanic islands, otherwise pelagic. **ID** Large, long-winged storm-petrel, mostly black with blue-grey to brown tinge, and lacks white rump. Similarly all-dark Swinhoe's is smaller. Approaches Bulwer's Petrel (p.86) in size, but wings longer and narrower in relation to body. Wings long and angled at carpal. In flight, brown carpal bar long and prominent, lacks distinctive white primary bases of Matsudaira's Storm-Petrel. Tail long and deeply forked. Flight strong, wings held more stiffly and less angled than smaller storm-petrels, and much longer than Swinhoe's (p.86), which glides less frequently. **BP** Bill longer, more prominent than smaller storm-petrels; eyes black; tarsi black. **Vo** Breeding colonies are noisy at night; birds in burrows give crooning *auooo koo* notes, and those in flight *keekoo kyukukuku*. **TN** May become *Cymochorea tristrami* in the future.

MATSUDAIRA'S STORM-PETREL TJ
Oceanodroma matsudairae

L 24–25cm; WS 56cm; WT 62g. **SD** Endemic breeder on the Ogasawara and Iwo Is, Japan, where mostly occurs mid Jan–mid Jun; seen several times off Taiwan. In non-breeding season disperses south as far as Indonesia and west to E Africa. Monotypic. **HH** Breeds on volcanic islands, otherwise pelagic. **ID** Large and long-winged, mostly black with brownish tinge, lacks white rump patch. Carpal bar short and brown; tail long and deeply forked. Much larger than similarly patterned Swinhoe's (p.86), closely resembles Tristram's but carpal bar less distinct and has prominent white bases to outermost 4–5 primaries. Flight strong, purposeful, but slow with wings held more stiffly and less angled than smaller storm-petrels; heavier in flight than Swinhoe's or Bulwer's (p.86). **BP** Bill black, longer, more prominent than smaller storm-petrels; eyes black; black tarsi very long, toes black. **Vo** Generally silent at sea. **TN** May become *Loomelania matsudairae* in the future.

FORK-TAILED STORM-PETREL TJR
Oceanodroma furcata

L 20–23cm; WS 46cm; WT 59g. **SD** N Pacific. *O. f. furcata* breeds in N Kuril Is, Commander and Aleutian Is, with *O. f. plumbea* ranging from Alaska along American W coast to California; both races winter in N Pacific. Common, especially at night, around ships near Kuril Is. In non-breeding season confined to Bering Sea (vagrant in N Bering Sea), N Pacific and Sea of Okhotsk. Regular in small numbers off Hokkaido and N Honshu in all months except May. **HH** Breeds on remote islands, otherwise pelagic. Occasionally visits harbours after storms. **ID** The only pale grey storm-petrel. Head appears over-large, with steep bulging forehead and flat crown, with blackish eye patch. Relatively stocky and broad-winged, with sooty black underwing- and upperwing-coverts. On upperwing, appears to have double bar, black and pale grey, on inner wing. Rump and longish forked tail are pale grey and tail tip is dark grey to black. In flight, wingbeats fairly shallow, fluttering with erratic zigzags, speed changes and interspersed by short glides. Field separation of the subspecies probably impossible, but *O. f. plumbea* is smaller and darker. **BP** Bill short, black; eye patch and eyes black; tarsi dark grey. **Vo** Generally silent at sea, but at colonies gives soft twittering, chirping and high rasping *skveeee skew skwe*. **TN** May become *Hydrobates furcatus* in the future. **AN** Grey Storm-petrel.

LEACH'S STORM-PETREL

leucorhoa

MATSUDAIRA'S STORM-PETREL

worn

fresh

TRISTRAM'S STORM-PETREL

FORK-TAILED STORM-PETREL

plumbea

✓LITTLE GREBE *Tachybaptus ruficollis*   CTKJR

L 23–29cm; WS 40–45cm; WT 130–236g. **SD** Europe, Africa to S, SE & E Asia. Year-round E China, Taiwan, Korea, Japan south of Hokkaido; summer visitor NE China, Russian Far East and Hokkaido. *T. r. poggei* in E China, Russia, Korea, Japan; *T. r. philippensis* Taiwan. **HH** Small well-vegetated ponds. Winters at lakes, ponds and rivers, not coasts, sometimes in small groups, and with other waterfowl. **ID** Smallest Asian grebe; short-necked, with rounded body and tail-less 'bottom'. Ad. sum has chestnut cheeks and foreneck, dark chocolate-brown crown, nape and upperparts. Ad win pale brown and buff. Wings small; flight generally low and laboured. Uniformly dark upperwing brown (may show narrow white trailing edge), contrasts with white underwing. **BP** Bill small, black with bright yellow base/gape and pale tip (breeding), or pale horn with yellow base (non-breeding); eyes white or pale-yellow; tarsi dark grey. **Vo** ♂ gives rapid series of high-pitched *ke-ke-ke* notes, rising to long wavering far-carrying trill *kiri kiri kiri kirirìrìrì,* commonly given during courtship and chases. Abbreviated trills in winter, and short, sharp clicks in alarm. **AN** Dabchick.

RED-NECKED GREBE *Podiceps grisegena*   CKJR

L 40–50cm; WS 77–85cm; WT 806–925g. **SD** Alaska, Canada, and NW Europe to Sakhalin, Chukotka, Kamchatka, also Hokkaido (rare) and NE China. Winters along Pacific coast south to Japan, Korea and SE China. Race in the region is *P. g. holboellii*. **HH** Well-vegetated ponds and lakes; winters along coasts, bays, occasionally inland lakes and rivers. **ID** Slightly smaller, more compact and darker than Great Crested (GCG), stockier with squarer, broader head, and shorter, thicker neck; 'bottom' more rounded than GCG. Winter ad. duskier than GCG, especially head, face and neck; dark cap extends to eyes and lores; cheeks dusky-grey; rear ear-coverts pale. Juv./1st-winter like winter ad. with black and white face bars. In flight, similar to GCG, but darker neck shorter and thicker. **BP** Bill long, straight, often angled below horizontal, yellow base with dark grey culmen and tip (breeding ad.) or dull yellow (non-breeding ad./juv.); eyes black; tarsi dark grey. **Vo** On territory: combination of gull-like wailing howls, loud harsh squeals, and Little Grebe-like continuous neighing *kerekerekere...* Pairs give deep mournful *uwaa uwaa.*

GREAT CRESTED GREBE   CTKJR
Podiceps cristatus

L 46–51cm; WS 59–73cm; WT 0.596–1.49kg. **SD** W Europe, Africa, to Australia and New Zealand. *P. c. cristatus* is summer visitor N & NE China, adjacent Russian Far East, and locally N Honshu. Winters through Japan south of Hokkaido to Kyushu, Korea, and E China south of Changjiang; rare migrant Hokkaido and Taiwan (commoner Kinmen and perhaps Matsu). **HH** Breeds at large reed-fringed inland lakes; courtship displays elaborate and noisy. Migration and winter on coastal or inland lakes, reservoirs and large rivers. **ID** Largest E Asian grebe; elegant, with long thin neck, striking white face and eyebrows

contrasting with black and orange head plumes. Winter ad. paler buff-brown without crown- and ear-tufts; black of crown does not extend to eyes (cf Red-necked), leaving distinct narrow, black loral line. In flight, rapid flickering wingbeats, long neck and large feet with white on forewing and secondaries. **BP** Bill dagger-like, long, straight, pinkish-horn; eyes dark red; tarsi grey. **Vo** Noisy, far-carrying deep guttural *kuwaa*, rolling *crra-ahrr* and slow series of wooden *breck-breck-breck* calls on breeding grounds, with distinctive head-waving and nodding display.

SLAVONIAN GREBE   CTKJR
Podiceps auritus

L 31–38cm; WS 46–55cm; WT 300–470g. **SD** Holarctic. *P. a. auritus* NW Europe to Lake Baikal, Russian Far East, Sakhalin, Kamchatka, S Chukotka. Winters south to coasts of N Japan, S Korea, NE China (migrant), SE China (rare in winter) and Taiwan (accidental). **HH** Shallow, well-vegetated boreal lakes. Migration/winter less dependent on freshwater than Black-necked (BNG), often inland (lakes/rivers), but more often in shallow coastal waters. **ID** Medium-sized grebe, smaller, more compact than Red-necked. Only likely to be confused with slightly smaller BNG. Head appears larger, flatter (than BNG) and triangular with crown peak at rear. Ad. sum colourful with attractive head plumes and broad golden yellow band from bill base to rear of head. Ad. win black cap extends down to eyes and contrasts with clean white cheeks and neck-sides. Juv./1st-winter like winter ad. with pale bill. In flight, upperwing pattern similar to Great Crested and Red-necked. **BP** Bill short but strong, black with grey-white tip (ad. breeding) or grey with white tip (juv./ad. win), diagnostic bare pink loral stripe between bill and eye in all plumages; eyes deep red; tarsi grey. **Vo** On territory various hoarse, guttural rattling calls, *way-urrr* or *hee ahrr*, and deep, pulsating Little Grebe-like whinnying trill *piii piii piii*. **AN** Horned Grebe.

BLACK-NECKED GREBE   CTKJR
Podiceps nigricollis

L 28–34cm; WS 41–60cm; WT 265–450g. **SD** Widespread Holarctic and southern Africa. *P. n. nigricollis*: Europe east to restricted area of NE China and SE Russian Far East. Winters N Japan to S Korea, Chinese coasts; accidental Taiwan. **HH** Loose colonies at heavily vegetated lakes and ponds. Migration/winter commonly inland on lakes and rivers, also estuaries, less often on sea; foraging flocks of 10s to 100s. **ID** Mid-sized, rather dark grebe, slightly smaller and thinner-necked than Slavonian, with loose fan of golden-yellow ear plumes. Dumpy, rounded body, especially 'bottom', recalls Little. Head rounded, crown peak central; forehead steep. In winter, black crown extends as bulge below eyes to cheeks, face less clean than Slavonian; neck dusky. Juv./1st-winter like winter ad. with browner ear-coverts and neck. In flight, lacks Slavonian's white shoulder patch. **BP** Bill slender, pointed, uptilted to tip, black (breeding) or pale grey (juv.), with dark tip and culmen (non-breeding); eyes bright red; tarsi dark grey. **Vo** On territory: plaintive *pew-ee*, a short, hard, emphatic whistle *fuuuweech* and low hard chittering trill, otherwise silent. **AN** Eared Grebe.

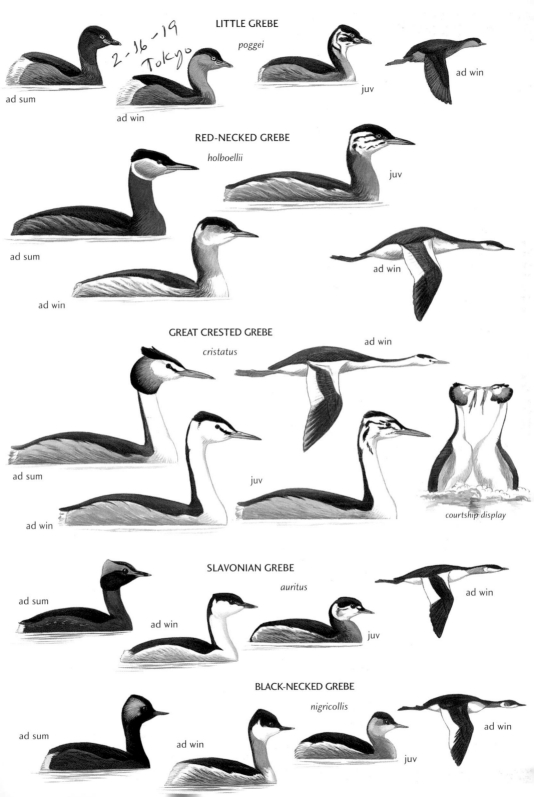

LITTLE GREBE

poggei

2-16-19
Tokyo

ad sum

ad win

juv

ad win

RED-NECKED GREBE

holboellii

juv

ad sum

ad win

ad win

GREAT CRESTED GREBE

cristatus

ad win

ad sum

juv

ad win

courtship display

SLAVONIAN GREBE

auritus

ad sum

ad win

juv

ad win

BLACK-NECKED GREBE

nigricollis

ad sum

ad win

juv

ad win

PAINTED STORK   C
Mycteria leucocephala

L 93–102cm; WT 2–3.5kg. **SD** Ranges across Indian subcontinent to SE Asia and SE China north to Changjiang (where possibly extinct). Monotypic. **HH** Rivers, ponds and lakes with marshy margins, breeds colonially in waterside trees. **ID** Very tall, between Grey Heron (see p.102) and Oriental Stork in size. Ad. largely white with black scapulars and wing-coverts narrowly barred white, tertials pink. Underparts white with black and white barring on lower breast. Non-breeding plumage and bare parts duller. Juv. drab brown, somewhat whiter on wings. In flight, broad wings and tail black, upperwing and underwing-coverts black with white barring; from below has prominent black breast-band. **BP** Face bare red; bill deep-based, long, heavy and slightly droop-tipped, pale yellow; eyes brown; tarsi dull reddish-pink. **Vo** Silent except at colonies where moans and bill-clatters.

BLACK STORK   CTKJR
Ciconia nigra

L 90–105cm; WS 173–205cm; WT *c.*3kg. **SD** Wide-ranging summer visitor to Palearctic, wintering in Africa, S & E Asia. In region, summer visitor to N & NE China, Russia (Transbaikalia east to W Sea of Okhotsk), S Russian Far East and Korea, but scarce, shy and declining; winters S China and northern SE Asia; rare in wetlands and farmland areas of Korea, rare/accidental in Taiwan and Japan. Monotypic. **HH** Nests in remote swamp forests, nesting on tall trees and foraging at wetlands. On migration, sometimes travels in small groups. May perch on electricity poles. **ID** Very tall, between Grey Heron (see p.102) and Oriental Stork in size. Ad. glossy black (with metallic green/purple sheen), with white belly, thighs, vent and axillaries. Young dark brown where ad. black. Flight strong, with slow shallow beats, but prefers to soar. In flight, white belly and axillaries prominent on otherwise all-black bird. **BP** Bill, lores and extensive eye-ring deep red (ad.) or grey-green (juv.); eyes black; tarsi dull reddish-pink (ad.) or grey-green (juv.). **Vo** In display, raises and lowers head while emitting raptor-like whistles (*pyuuree pyuuree*), or lowers head in threat posture (*fee fee*), but unlike Oriental Stork only rarely bill-clatters, and does so quietly.

WHITE STORK   C
Ciconia ciconia

L 95–110cm; WS 180–218cm; WT 2.3–4.4kg. **SD** Wide-ranging summer visitor to Europe, N Africa and C Asia, wintering in Africa and India. *C. c. asiatica* has, apparently, strayed to NE Inner Mongolia, though record may actually refer to Oriental Stork. **HH** In summer in farmland, marshes and wet river valleys. **ID** Large, cleanly marked black and white stork with contrasting black flight-feathers and white wing-coverts. Distinguished from Oriental Stork by shorter all-red bill, black eye-ring and all-black secondaries. **BP** Long, deep-based bill is waxy red (ad.) or dull red with dark tip (juv.); eyes black, surrounded by narrow black patch; tarsi red (ad.) or dull pink (juv.). **Vo** Silent except during breeding season when bill-clatters loudly.

ORIENTAL STORK   CTKJR
Ciconia boyciana

Endangered. L 110–115cm; WS 195–200cm. **SD** Endemic to E Asia, breeds NE China, Korea (extinct but there have been recent attempts) and S Russian Far East. Captive breeding and reintroduction being attempted in NW Japan, where became extinct. Scarce winter visitor to Korea, rare migrant and winter visitor to Japan, and winters south to SE China, rare Taiwan (breeding has been attempted). Monotypic. **HH** Nests in loose colonies in tall trees, feeds at shallow wetlands such as wet ricefields. May travel in small groups on migration. **ID** Very large, between White-naped Crane and Hooded (see p.150–152) in size. White, with black flight-feathers and primary-coverts. Stands tall with neck erect; lower neck and upper breast-feathers elongated and may blow loosely. Secondaries have greyish-white outer webs, prominent in flight, but also visible on folded wing. Flight strong, with slow shallow beats but prefers to soar. In flight, wings distinctly pied. **BP** Black bill is long, deep at base, sharply pointed, lower mandible upturned at tip; narrow eye-ring, small loral patch, base of lower mandible and gular region red; irides whitish-yellow; long tarsi red. **Vo** On breeding grounds gives weak *shuu* or *hyuu* prior to loud bill-clattering display when mandibles are pointed upwards, waved and clattered together loudly like castanets.

PAINTED STORK

ad

juv

ad

ad

BLACK STORK

ad

juv

ad

WHITE STORK

asiatica

ORIENTAL STORK

ad

SACRED IBIS
Threskiornis aethiopicus

T

L 65–89cm; WS 112–124cm; WT c.1.5kg. **SD** Natural range sub-Saharan Africa, but *T. a. aethiopicus* well established in Taiwan, mainly in coastal NW, following accidental release, now numbers 300–350 pairs and increasing, so may eventually spread beyond Taiwan. **HH** Forages in damp grasslands and well-vegetated swamps, also agricultural fields, where walks purposefully and probes vigorously for food. Flocks wander widely, often found alongside egrets and herons, and occasionally attracting rare wintering Black-headed Ibis. **ID** Approaches size of spoonbills. Ad. white plumage contrasts with naked black skin of head and neck; wings largely white, but has black tips to flight-feathers forming black rim to wing. Tertials long, loose and black, covering tail at rest. In breeding season, thighs turn yellow and naked skin of underwing 'arm' scarlet. Juv. greyer, with feathering on much of neck to chin, and has shorter black tertials. **BP** Bill thick-based, long and arched, black; eyes black; tarsi dark blackish-grey. **Vo** Gives loud croaking calls at colonies.

BLACK-HEADED IBIS
Threskiornis melanocephalus

CTKJR

L 65–76cm. **SD** Resident from Pakistan through SE Asia to Indonesia. In E Asia, migratory with records of wintering birds (Oct–Apr) from E & S coasts of China, Taiwan, Korea and Japan; no breeding grounds known in region, thus origin of these birds unclear. Monotypic. **HH** Scarce; solitary or in small groups, foraging in flooded grasslands and well-vegetated swamps, where walks purposefully and probes vigorously for food. **ID** Approaches size of spoonbills. Ad. all white bar naked black skin on head and upper neck; wings white. Breeding ad. has loose ruff of white feathers on lower neck, grey scapulars, elongated grey tertials and variable yellow wash on breast and mantle; non-breeding ad. loses grey coloration and neck ruff. Juv. has grey feathering on head and white-feathered neck. In flight all-white, with line of bare, pink skin on 'arm' visible on underwing at base of greater wing-coverts (juv. has black tips to outermost primaries); black head and neck contrasting with white body and large bill distinguish it from egrets, and decurved bill from spoonbills (but see Sacred Ibis). **BP** Ad. has strong, black, decurved bill; eyes dark red-brown to black; strong black tarsi. **Vo** Subdued grunting *guwaa* or *kuwaa*, and mellow squeals at nesting colonies, otherwise silent.

CRESTED IBIS
Nipponia nippon

CTKJR

Endangered. L 55–60cm. **SD** Very rare E Asian endemic. Once bred across E & NE China to Russian Far East, Korea and Japan, wintering south to SE China, but now extirpated over most of range due to habitat loss, agricultural intensification, use of agro-chemicals and hunting. Lingers on outside region in southern Shaanxi where sedentary. Captive populations in Beijing and on Sado I., Japan where being introduced to wild. Formerly, occasionally straggled during winter to Taiwan (pre-1950) and S Korea. Monotypic. **HH** Habitually social, nesting in small colonies in deciduous treetops, foraging at nearby shallow wetlands and wet ricefields, where probes purposefully for food. **ID** Large ibis (size of Black-faced Spoonbill), with unique plumage. Ad. white with delicate peach or salmon-pink cast to flight-feathers, neck and breast, long loose-feathered nuchal crest; head, neck and mantle become deep grey when breeding, stained by oily secretion. Young grey. In flight, wings essentially white, with strong pink cast, especially to undersides of flight-feathers. **BP** Bill strong, decurved, black with red tip and base; bare skin of face and forecrown vermilion; eyes yellow; short tarsi bright red. **Vo** Calls: *taaa* or *aaa*.

GLOSSY IBIS
Plegadis falcinellus

CTJ

L 55–65cm; WS 88–105cm; WT 485–580g. **SD** Ranges from S Europe and Africa to S Asia and Australasia; also SE USA, Middle America and northern S America. In region, has strayed to SE China, Taiwan and S Japan. Monotypic. **HH** Shallow freshwater wetlands. **ID** Large dark ibis. Distinguished by all-dark, chocolate-brown or chestnut plumage, with purple and green gloss; wings metallic green; lores blue with white border above and below; non-breeder duller with white streaks on head and neck. In flight, recalls dark curlew, with long thin neck, distinctive bill and toes protruding well beyond tail tip. **BP** Bill long, relatively narrow-based and curved, curlew-like, grey-green to dark brown; eyes black; long tarsi dark grey or brown (bill and legs dull yellow in non-breeding season). **Vo** Generally silent away from nest, but occasionally gives deep grunting *grrr*, a muffled, nasal moaning, *urnn urnn urnn*, and deep quacking sounds.

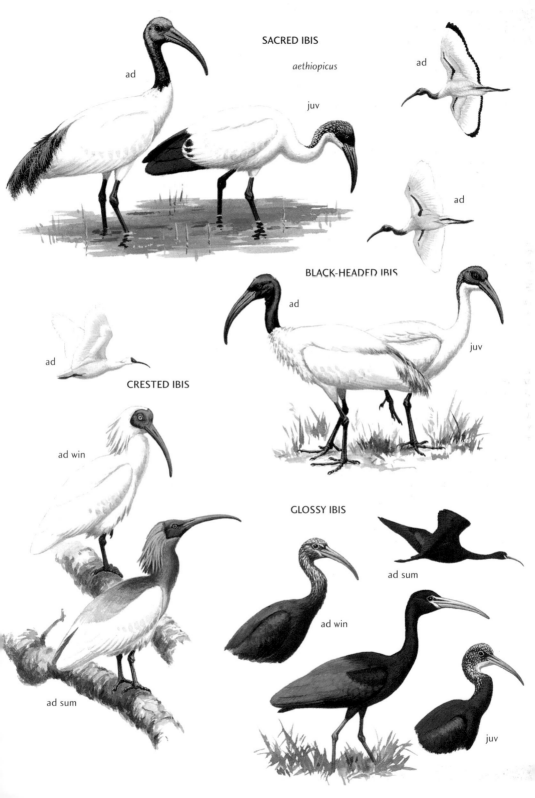

SACRED IBIS

aethiopicus

ad

ad

juv

ad

ad

BLACK-HEADED IBIS

ad

ad

juv

CRESTED IBIS

ad win

GLOSSY IBIS

ad sum

ad win

ad sum

juv

EURASIAN SPOONBILL CTKJR
Platalea leucorodia

L 80–93cm; WS 120–135cm; WT 1.13–1.96kg. **SD** Wide-ranging summer visitor across Palearctic, wintering south to Africa, resident in India, and migrant in E Asia. In region, *P. l. leucorodia* uncommon and very local. Summer visitor to NE China and SE Russian Far East, wintering (Oct–Apr) SE China, with small numbers in Korea and S Japan; accidental Taiwan. **HH** Loose colonies in trees or reeds within extensive swamps. On migration/in winter, singly or in small groups at pools, lakes or muddy wetlands, where wades slowly, swinging head and bill from side-to-side to sieve food. Frequently occurs with Black-faced Spoonbill in winter. Partly nocturnal, thus often seen roosting during day. **ID** Large, white, with all-white wings. Ad. breeding has yellow 'spoon', yellowish throat and band across lower neck/upper breast, and prominent white nape plumes. Non-breeding lacks crest and yellow. Juv. also lacks crest and has paler bill and legs. Roosting individuals with bill concealed separable from rare Black-faced by eye outside black mask. In flight, neck and legs outstretched, ad. all white (juv. has black tips to outermost primaries); extended neck, large bill and quicker wingbeats interspersed with glides distinguish it from egrets. **BP** Very long dark grey to black bill has transverse ridges with spatulate tip (yellow in summer), bill pink in young; facial feathering extends in front of eye to bill base; gular patch pinkish. Unlike egrets, eyes dark red-orange, not pale yellow; legs dark-grey to black, brownish-pink in juv. **Vo** Generally silent, but calls include guttural *fuu*, *ufuu* or *puu*, and gives various grumbling sounds at nest.

BLACK-FACED SPOONBILL CTKJR
Platalea minor

Endangered. L 60–78.5cm; WS 110cm. **SD** E Asian endemic breeder. Summer visitor to islands off NE Korea and Shandong coast, China. Winters Taiwan and S China to N Vietnam, with small numbers in Kyushu and Nansei Shoto, Japan (occasionally summering). Rare but gradually increasing; world population 2,065 (Mar 2008), of which *c.*50% winters in Taiwan. Accidental S Russian Far East. Monotypic. **HH** Breeds on coastal islands, foraging singly or in small groups (sometimes with Eurasian Spoonbill in winter), at coastal mudflats, muddy wetlands and lakes, where swings head and bill from side-to-side to sieve food while wading slowly, or by more actively chasing and lunging for prey; partly nocturnal, thus often seen roosting by day. **ID** Smaller than Eurasian, but similarly white, with all-white wings. Ad. has black bill with black 'spoon', bare black face, and when breeding has yellowish or rusty band on lower neck/upper breast, and prominent nape plumes may also be tinged yellow. In flight, ad. all white (juv. has black tips to outermost primaries and black shafts to all primaries), extended neck, large bill and quicker wingbeats interspersed by glides distinguish it from egrets. **BP** Very long dark grey to black bill has transverse ridges with spatulate tip; bill pink in juv. Facial skin black, extending across forehead, back to eye and surrounds bill base, enclosing small yellow pre-orbital patch; chin/gular region feathered white; eyes dark red-orange; legs dark grey to black. **Vo** Generally silent but utters guttural *ubuu ubuu* in breeding season.

EURASIAN BITTERN CTKJR
Botaurus stellaris

L 69–81cm; WS 100–130cm; WT 0.867–1.94kg. **SD** Widespread, from W Europe to E Asia and south to Africa and S Asia. *B. s. stellaris* breeds from Lake Baikal east to Sea of Okhotsk, S Russian Far East, Sakhalin, Hokkaido, N & C Honshu, and NE China. In winter, mainly E China south of Changjiang River to SE Asia, but small numbers also winter Korea and Japan. **HH** Extensive wetlands, generally in reedbeds, but almost any large wetland further south in winter, from Honshu to Taiwan and Fujian, though highly secretive. **ID** Large, buff-brown, heavily streaked heron. Neck thicker and shorter than Grey Heron (see p.102), and short-necked appearance exacerbated by hunched posture. Legs shorter and toes larger than Grey Heron. Occasionally appears at reedbed edge or in ricefields where finely patterned, cryptic ginger-and-buff plumage, combined with blackish-brown crown, nape and moustachial stripe distinctive. Abandons hunched posture if disturbed and stretches erect, revealing vertical dark brown stripes on throat and neck. Rarely flies, but when does so resembles large owl with broad rounded wings (though all flight-feathers and primary-coverts narrowly barred blackish and dark ginger), with huge trailing toes. **BP** Bill long, strong, sharply pointed, yellow or green at base/lores; eyes reddish-brown; rather short legs greenish-yellow, very large greenish-yellow toes protrude well beyond tail in flight. **Vo** In breeding season ♂ gives very deep, far-carrying resonant booming call, often repeated slowly three times *bwooo bwooo bwooo*, or from close range breathy inhalation audible *uh-buwooo uh-buwooo uh-buwooo*; in non-breeding season also gives gruff *kau* at night, and when flushed a short, gruff *grek grek…*. **AN** Great Bittern.

YELLOW BITTERN CTKJR
Ixobrychus sinensis

L 30–40cm; WS 53cm. **SD** Resident India to SE Asia, Philippines, Indonesia, S China and Taiwan. Summer visitor (May–mid Oct) across China, northeast to Korea, NE China and S Russian Far East, Japan and Sakhalin; has wintered SW Kyushu. Monotypic. **HH** Inland wetlands, lake margins, reedbeds, lotus and ricefields. **ID** Small, rather plain buff-brown bittern. ♂ has black cap, pale tawny-brown hindneck and mantle, paler foreneck and buff belly. ♀ duller with dark grey crown with some streaking, pale streaking on back and more distinct streaking on foreneck than ♂. Juv. like diminutive Eurasian Bittern, though dark and brown streaks vertical. Bursts from cover, flops across reeds and drops again quickly. In flight, pale buff wing-coverts contrast strongly with blackish-brown flight-feathers and primary-coverts. Tail short, blackish, toes protrude beyond tip. **BP** Bill orange-yellow with dark culmen and tip, lores yellow; eye-ring yellow, eyes yellow; tarsi yellow. **Vo** In breeding season a repetitive, pumping *oo oo*, *whoa whoa* or *wooo wooo* call by day or night. Flight call is a dry *kik-kik-kik* or raspy *tschek*, lower pitched but similar to Whiskered Tern (p.226).

EURASIAN SPOONBILL

leucorodia

juv

ad

ad Eurasian

ad Black-faced

BLACK-FACED SPOONBILL

juv

ad

EURASIAN BITTERN

stellaris

ad

YELLOW BITTERN

ad

ad

juv

SCHRENCK'S BITTERN CTKJR
Ixobrychus eurhythmus

L 33–39cm; WS 55–59cm. **SD** Summer visitor (May–Sep) from S China to NE China, Korea (now apparently rare), S Russian Far East and Sakhalin, and from C Honshu north through Japan (now rare). Winters in Philippines, Borneo and peninsular SE Asia; scarce migrant Taiwan. Monotypic. **HH** Reedbeds and marshes, occasionally ricefields; solitary and secretive, thus perhaps overlooked, but seemingly in decline having become scarce or disappearing from various breeding areas in Japan in last 25 years. **ID** Small two-tone bittern (slightly larger than Yellow Bittern, see p.96). ♂ has dark chestnut face, hindneck, mantle, rump and innerwing-coverts, blackish crown, tail and flight-feathers, with prominent pale buff patch on larger wing-coverts. Underparts warm whitish-buff with dark gular stripe extending to breast. ♀/young differ strikingly: crown dark grey, neck, mantle, wings and tail mid-brown spotted white, and neck, breast and flanks streaked mid brown on pale buff. Occasionally bursts from cover, flies across reeds and drops again quickly. In flight, ♂'s pale buff wing-coverts contrast strongly with blackish flight-feathers and primary-coverts, whilst ♀'s white spotted wing-coverts are noticeable. Tail short, blackish, toes and part of legs protrude beyond tip. **BP** Bill dark in ♂ with pink facial skin, ♀/young have black culmen, yellow at base of upper and all-yellow lower mandible, and yellow lores; eyes yellow but black at rear, forming distinct 'C' shape, similar to Cinnamon but distinct from Yellow; tarsi dull yellow. **Vo** Repetitive, rhythmic *oo oo oo* or *gup-gup-gup* calls at night in breeding season recall Yellow Bittern, but tempo is twice as fast; also gives a low, gruff *wek* in flight. **AN** Von Schrenk's Bittern.

CINNAMON BITTERN CTKJR
Ixobrychus cinnamomeus

L 40–41cm. **SD** Mostly resident, from India to Sulawesi north to SE China, Taiwan and Nansei Shoto, Japan, but summer visitor from S China to N Bohai Gulf. Accidental Korea to Ussuriland. Monotypic. **HH** Solitary; wetlands including wet rice and rush fields, swamps and wet grasslands, where may be mango orchards in Amami and Yaeyama Is, Japan, and Luzon, Philippines. **ID** Small and stocky, plain but rich cinnamon-orange bittern. ♂ entirely rich cinnamon, somewhat paler on underparts. ♀ and young resemble ♀ Schrenck's, with dark brown crown, rear neck and mantle, foreneck and breast heavily streaked dark brown, but upperparts dark and pale brown, flecks not white, and flight-feathers cinnamon-rust. Flight typically low and slow. ♂ plain cinnamon, tail very short, toes and much of legs protrude beyond tail tip. **BP** Bill yellow with dark culmen, bare lores yellow; eye-ring and eyes yellow, but black at rear, forming distinct 'C' shape, similar to Schrenck's but distinct from Yellow; tarsi yellow. **Vo** Flight call, given especially on take-off, deep croaking *kwe kwe* or stuttering *kik-kuk kuk kik kik*. At night and at dawn on breeding grounds gives an even series of deep *aaa aaa* notes, and mellow *kokokokoko* notes. **AN** Chestnut Bittern.

BLACK BITTERN CTKJ
Dupetor flavicollis

L 54–66cm; WS 80cm; WT 300–420g. **SD** Mostly resident from Pakistan to S China and through SE Asia to Australia. Summer visitor to E China north to latitude of Changjiang River mouth. *D. f. flavicollis* is accidental (mostly spring: Mar–May) north and east to Taiwan, Korea, and Nansei Shoto and even Honshu, Japan. **HH** Rice and rush fields, swamps and wet grasslands, where may be relatively conspicuous. **ID** Mid-sized, very dark bittern. ♂ almost entirely black from crown and face to wings and tail, but has yellow throat and foreneck to upper breast with two broken lines of black and chestnut streaks on upper neck to breast. ♀ dark brown where ♂ is black. In flight, appears uniformly dark. **BP** Bill rather deep-based, yellow with grey culmen and dark tip; eyes yellow or pale green; rather short tarsi dull green. **Vo** Series of staccato *nyuk-nyuk-nyuk* or deep croaking calls in flight, and a mellow, descending booming on territory, *kwoouh kwoouh* or *booouhh*. **AN** Mangrove Bittern.

STRIATED HERON CTKJR
Butorides striata

L 35–48cm; WS 52–60cm; WT 135–250g. **SD** Range includes Africa, India to Australia and S America. *B. s. actophila* resident north into S China. *B. s. javanica* north to Taiwan and possibly S Nansei Shoto, Japan. *B. s. amurensis* summer visitor (mid Apr–late Sep) E & NE China, Korea, S Russian Far East and Japan south of Hokkaido. **HH** Shy and usually solitary, at streams, ponds, wet fields, swamps and mangroves, usually near dense cover. May adopt vertical 'bittern-like' posture if disturbed. **ID** Stocky, dark grey heron, with short legs but relatively long straight bill. Ad. has black crown and loose nuchal crest. Face to underparts mostly pale to dark grey, with white at edge of crown, on chin and in streaks on mid neck to belly. Mantle feathers long, pointed, loose, dark grey with green gloss; tail short and blackish-grey. ♂ generally grey-toned whilst ♀ browner. Juv. greyish-brown with heavily streaked and spotted underparts. In flight, rather sluggish, with long, narrow wings. Flight-feathers all dark grey or blackish, but upperwing-coverts and secondaries have pale grey or off-white fringes. **BP** Bill black, lores and eye-ring greenish-yellow; eyes yellow; tarsi yellow. **Vo** Mainly at night in breeding season, but on migration after dark, a sharp, loud *kyeowp, qewee, pyuu* or *tiuu.* 'Song' on territory is a hard, low *chi-kwowp*, repeated at long intervals. **AN** Little Green Heron; Green-backed Heron.

SCHRENCK'S BITTERN

ad ♂ sum

ad ♀

juv

ad ♂

CINNAMON BITTERN

Juv

ad ♀

ad ♂ sum

ad ♂

BLACK BITTERN

flavicollis

ad ♀

ad ♂ sum

STRIATED HERON

amurensis

ad ♂

juv

WHITE-EARED NIGHT HERON
Gorsachius magnificus   C

Endangered. L 54–58cm. **SD** A rare and very poorly known species. Range seemingly restricted to Hainan and adjacent S China, and parts of Zhejiang and Fujian. Northern birds believed to winter in Hainan, but may be resident. Monotypic. **HH** Presumably nocturnal and breeds very locally in bamboo and dense subtropical and tropical forest, in montane regions, where they forage at marshy streams within forest. **ID** Larger and more distinctively marked than other *Gorsachius* in region. ♂ generally blackish-brown with black crown, nuchal crest, cheeks and ear-coverts; lores, eye patch and curving postocular stripe, chin and throat all white; neck-sides chestnut. Underparts largely dark brown, but heavily flecked white/silver on lower neck and breast. ♀ resembles ♂, but has less distinct head pattern, shorter crest and pale streaks or spots on upperparts. **BP** Bill short, dagger-like, black, with bare yellow lores and base of lower mandible; eyes large, yellow-orange; short tarsi green. **Vo** From tree perch and in flight, a brief deep, throaty, raspy *whoaa* repeated at 5–15-second intervals, typically in early evening.

JAPANESE NIGHT HERON
Gorsachius goisagi   CTKJR

Endangered. L 49cm; WS 87cm. **SD** Secretive summer visitor (mid Apr–mid-Sep) and endemic breeder to S Japan (Kyushu north to C Honshu). Scarce but perhaps overlooked. Winters Philippines and Sulawesi, perhaps in Nansei Shoto and Taiwan; rare migrant to E China, Taiwan and Korea; accidental north to Ussuriland. On migration also small islands in Sea of Japan, e.g. Hegura-jima. Monotypic. **HH** Mature forest, but may forage crepuscularly at woodland edge or grassland. **ID** Small, stocky night heron with short thick neck, often hunched. Ad. very similar to Malayan. Upperparts dull chestnut-brown with fine black vermiculations, head and neck-sides rufous-brown to chestnut. Underparts pale buff with dark streaks. Young have black crown, upperparts finely vermiculated dark and mid brown, more extensively streaked neck, and off-white lores and eye-ring. Very broad-winged, blackish flight-feathers brown-tipped, alula and secondary fringes white. **BP** Bill stout, dark grey with greenish-yellow lores (blue when breeding); shorter than Malayan, has same decurved culmen; eye-ring yellowish-green, irides pale yellow; tarsi dark yellowish-green. **Vo** Slow, repetitive, reverberating owl-like hooting on territory (mainly early evening) and sometimes on migration. First part of deep *o-bwoo* only audible close by, thus typically sounds like *bwoo bwoo bwoo*.

MALAYAN NIGHT HERON
Gorsachius melanolophus   TJ

L 47–51cm; WT 417–450g. **SD** SE Asia to Borneo and Philippines. Resident Taiwan and S Nansei Shoto, Japan. Monotypic. **HH** Secretive and nocturnal or crepuscular; mature subtropical forest, but may forage at forest edge at dawn and dusk. Regular by day in Taiwan (where now common in lowlands, even in urban areas) and occasionally on Ishigaki, Japan. Many are migrants. Roosts in camphor trees in palm plantations. Tolerates cedar plantations.

Sea level to 800m. Prefers short grass or weeds in shade; avoids tangles and dense understorey when feeding. **ID** Stocky with short thick neck, often hunched, like Japanese but has blacker crown and more prominent nuchal crest. Ad. face and neck-sides more rufous-brown than Japanese. Juv. paler than Japanese with more prominent white spotting on crown, face and nape (see imm. Black-crowned); takes *c.*2 years to achieve ad. plumage (appears first on ear-coverts). In flight, very broad-winged, blackish secondaries brown-tipped, whereas primaries white-tipped. **BP** Bill stout, arched, black, blue base to lower mandible, blue lores, gape and eye-ring all more prominent than Japanese; irides greenish-yellow; tarsi dull, dark yellowish-green. **Vo** Calls most frequently at dawn/dusk, in breeding season only, deep, owl-like hoots, *oongh oongh* or *whoop whoop* at close range and *toob toob* at distance. Generally repeated in pulsing series of 6–20 notes, at *c.*1 per second; several may call simultaneously.

BLACK-CROWNED NIGHT HERON
Nycticorax nycticorax   CTKJ

L 58–65cm; WS 90–100cm; WT 525–800g. **SD** Virtually cosmopolitan. *N. n. nycticorax* resident SE & S China, Taiwan, Nansei Shoto north to C Honshu, Japan, and Korea; summer visitor north to NE China and N Japan. **HH** Noisy colonies in trees above water, at range of wetlands (including urban parks with ponds). Secretive, mostly nocturnal, but occasionally diurnal, sometimes confiding. Disperses from roosts at dusk to feed at wetlands. **ID** Mid-sized, very stocky, with large head, short neck and legs, and stout bill. Ad. has black crown, nape, mantle and back, with long white nuchal plumes. Supraloral stripe and forehead, and underparts from chin to vent white or very pale grey. Wing- and tail-feathers pale to mid grey. Young dull brown with indistinct, broad pale streaks on mantle/underparts, prominent white spots on wing-coverts, and fine white streaks on crown. In flight, compact; wings broad, short, rounded and strongly bowed; slow floppy wingbeats. **BP** Bill black with grey lores in breeding ad., yellow-green base with black tip and grey culmen, and green lores in imm.; irides dark reddish-orange; tarsi range from dull yellow to orange and pink. **Vo** Deep croaking calls at colonies, and deep nasal barking, *kwok*, *goa* or *guap guap*, in flight.

RUFOUS NIGHT HERON
Nycticorax caledonicus   TJ

L 55–59cm; WS 95–110cm; WT 550–990g. **SD** Philippines to Tasmania, formerly Ogasawara Is, Japan, where endemic *N. c. crassirostris* extinct *c.*1889. However, *N. c. manillensis* from Philippines has strayed to Taiwan and could reach S China. **HH** Inland and coastal wetlands; urban/suburban gardens and parks to estuaries, mangroves and damp forest edges. Mainly crepuscular or nocturnal. Essentially sedentary, but considerable post-breeding dispersal. **ID** Ad. small with distinctive black cap and generally warm cinnamon-brown plumage, darker on upperparts, and long nuchal plumes when breeding. Young heavily spotted/streaked brown and white. **BP** Bill more slender, more pointed than Black-crowned, culmen less arched, but black; eyes large with green lores, irides yellow-orange (bright red when breeding); short tarsi bright yellow to dusky-ochre. **Vo** Deep croaks.

WHITE-EARED
NIGHT HERON

ad

ad

juv

ad

JAPANESE
NIGHT HERON

juv

ad

MALAYAN
NIGHT HERON

juv

ad

ad

juv

RUFOUS
NIGHT HERON

ad

BLACK-CROWNED
NIGHT HERON

ad

CHINESE POND HERON CTKJR
Ardeola bacchus

L 42–52cm; WS 79–90cm. **SD** Bang-ladesh to SE Asia. Resident S China, summer visitor to N & E China and adjacent Russian Far East; winters in peninsular SE Asia and Borneo. Regular in small numbers in S Nansei Shoto in winter (Oct–Apr), and uncommon spring overshoot elsewhere, e.g. Korea (may breed), offshore Japan and Taiwan. Has bred Kyushu. Monotypic. **HH** Solitary or small loose flocks feed at wetlands. Nests among other egrets and herons; on migration at wet meadows, woodland edges, streams and ditches. **ID** Small heron, dark and bittern-like when foraging, but like dark-backed cattle egret in flight. Breeding ad. has rich chestnut head, breast and prominent nape plumes, whilst mantle is charcoal-grey with elongated loose plumes that cover most of wings at rest. Wing-feathers, rump and tail white. In winter, mantle is plain, dull grey-brown, whilst head, neck and breast are whitish, heavily streaked grey-brown. **BP** Relatively long straight bill yellowish with black tip in breeding season, duller with dark upper mandible in winter; bare pale eye-ring, irides yellow; short tarsi yellow. **Vo** Call a gruff *kwa*.

EASTERN CATTLE EGRET CTKJR
Bubulcus coromandus

L 46–56cm; WS 88–96cm; WT 340–390g. **SD** Widespread, mostly resident, in warmer regions from Pakistan to Australia. Year-round in SE Asia and S China. Summer visitor (Apr–Sep) to E China, Taiwan, Korea and Japan north to N Honshu with spring over-shoots reaching north to Hokkaido and even Sakhalin. Some winter in S Japan. Monotypic. **HH** Nests colonially, roosts socially and often encountered in groups around stock animals, in grasslands or drier margins of wetlands, rice and fallow fields. Walks boldly with pigeon-like head-bobbing. Often locally common. **ID** Essentially all-white egret, but breeding ad. has white face and puffy, bright rusty-orange head, neck, upper breast and mantle plumes. Non-breeding ad. all white. Smallest of the region's white egrets, with shorter neck, legs and shorter bill than other species. In flight, wings slightly more rounded and arched, and wingbeats deeper than other white egrets. **BP** Bill yellow with deep reddish-pink base in breeding season, plain yellow when non-breeding, rather blunt, lores yellow; feathered chin extends forwards of nostril; eyes yellow to red; legs short, rather thick, pinkish-orange in breeding season, dark greyish-green or blackish in winter. **Vo** Rather vocal at colonies, but also calls occasionally away from them: typically a short, gruff, barking *goaa* or *guwaa*. **TN** Formerly within Cattle Egret *B. ibis*, which has now been split into two species. Extralimital **Western Cattle Egret** *B. ibis*, from Europe, Middle East and Africa (and self-introduced to New World) closely resembles Eastern, but is smaller and stockier with shorter bill, neck and legs. Breeding plumage less rusty, and confined to shaggy crest on crown and plumes of chest and lower back. Legs yellowish-or greyish-olive, never black. Possible vagrant to region though as yet unrecorded.

GREY HERON CTKJR
Ardea cinerea

L 84–102cm; WS 155–175cm; WT 1.02–2.073kg. **SD** Widespread Old World species ranging to Philippines and Indonesia. In E Asia resident over most of China (probably *A. c. cinerea*), whilst *A. c. jouyi* is summer visitor to NE China, Korea, S Russian Far East and Sa-khalin, and is resident on main Japanese islands. Previously only a summer visi-tor on Hokkaido, now year-round, and has reached Kamchatka. **HH** Large, conspicuous colonies in tall trees near lakes; also roosts in trees. Forages at lakes, large rivers and shallow coastal lagoons, on mudflats and exposed rocky shallows. Outside breed-ing season may gather in large loose flocks and forage in loose associations at wetlands. Increasingly common at urban wetlands where reasonably confiding. **ID** Largest regularly occurring heron in region. Ad. largely pale to mid grey, with white crown, face, foreneck and underparts; neck- and breast-sides buffy-grey. Black 'bandana' around crown and small nuchal crest. Neck and breast streaked black. Mantle and tail grey, inner thighs black. Young greyer on neck and blacker on head than ad. Tall but often stands with neck hunched into shoulders. Spends long periods motion-less waiting patiently for prey. In flight, retracted neck forms conspicuous rounded bulge and appears whitish; wings markedly two-tone, with black flight-feathers and primary-coverts contrast-ing with grey upper- and underwing-coverts. Flight slow, on deeply bowed wings. **BP** Bill large, stout, dagger-like and varies from pinkish-orange (breeding) to yellow (non-breeding); imm. has dull-grey bill. Eyes pale to deep yellow. Very long legs flush bright pinkish-orange when breeding, otherwise dull greyish-flesh. **Vo** Flight call a deep, hoarse long *fraaank* or bisyllabic *frah-aank* typically given on take-off; also vocal at colonies when gives throaty rattling and croaking *kuwaa* or *gua*.

GREAT BLUE HERON R
Ardea herodias

L 97–137cm; WS 180–183cm; WT 2.268–3.629g. **SD** N America, wintering especially in temperate regions of W USA and Mexico, some migrating as far as Canada to breed and others reaching Alaska. Accidental on Chukotski Peninsula, Russia (presumably *A. h. fannini*, which normally reaches SE Alaska). **ID** Largest heron recorded in region, tall and bulky, with a very heavy bill. Ad. resembles Grey Heron, but larger and has chestnut 'shoul-ders' and thighs. Face and crown white, with black crown-sides extending as long plumes; neck greyish-buff with narrow black-and-white streaks on foreneck; upperparts and underparts largely mid to dark grey. After take-off retracts long neck; flight-feathers black contrasting with grey upperparts, wing-coverts and tail. **BP** Bill orange-yellow with dark grey culmen; eyes white; long tarsi grey. **Vo** Very vocal when flushed or in flight, a hoarse trumpeting *fraaahnk* or *braak*.

ad sum

ad sum

EASTERN CATTLE EGRET

CHINESE
POND HERON

ad sum

ad win

ad win

GREY HERON
2/17/19 *jouyi* Tokyo

juv

ad

ad sum

ad sum

juv

GREAT BLUE HERON

ad

PURPLE HERON CTKJR
Ardea purpurea

L 70–90cm; WS 110–145cm; WT 0.525–1.345kg. **SD** Widespread Old World species; resident in S & SE Asia to Philippines and Indonesia. In E Asia *A. p. manilensis* resident through S China and southernmost Nansei Shoto, Japan. Summer visitor across China to S Russian Far East, and scarce migrant to Korea and offshore Japan, accidental mainland Honshu and Hokkaido. Migrant (some winter) in Taiwan. **HH** Nests in reeds or trees. Solitary, rather secretive, crepuscular, foraging in well-vegetated swamps, reedbeds and lake margins; much less common than Grey Heron. **ID** Tall rakish heron with long, slender neck and bill; often leans forward or sideways with neck curved and head cocked, ready to lunge at prey. Ad. is mixture of chestnut, black and grey. Prominent black and chestnut stripes from face to chest; crown and nape blackish, mantle and tail dark grey, thighs and lower back plumes also chestnut. Juv. browner overall. Flight, jerky, more buoyant than heavier Grey Heron (see p.102), on deeply arched wings. In flight, neck bulge very conspicuous, more angular than Grey and wings less markedly two-tone than latter, as upper and underwing-coverts tinged brown in ad. and all brown in young. **BP** Long, narrow, bayonet-like bill has dark culmen and yellowish sides; eyes yellow; long legs yellowish-brown, but flush pinkish-brown when breeding; large, bunched toes very obvious in flight. **Vo** Flight call, a guttural *guwaa*, or emphatic *skrech* is shorter, less resonant than Grey Heron; also a growling *graaaau* or nasal gurgling *gurrh*.

GREAT WHITE EGRET CTKJR
Casmerodius albus

L 80–104cm; WS 140–170cm; WT 0.7–1.5kg. **SD** Widespread in warmer regions of both New and Old World, including S & SE Asia. In E Asia, *C. (a.) albus* (perhaps specifically distinct) is summer visitor to NE China (Heilongjiang), S Russian Far East, Japan (north to Honshu) with spring migrants increasingly overshooting to Hokkaido. Winters increasingly from Hokkaido south, but mainly from C Honshu to Kyushu and in Korea. *C. (a.) modestus* (probably deserving full species status, **Eastern Great White Egret**) breeds in E Asia, in E China from Jilin to Fujian, to S Japan and Korea, wintering in Taiwan and S China. **HH** Locally common at lakes, rivers and large wetlands, scarce elsewhere, often solitary or with other herons/egrets. **ID** Largest egret, comparable in size to Grey Heron (see p.102). Large, long-necked (with characteristic kink), large stout bill, very long legs. Appears tall, often with hunched shoulders. Commonly leans forward, craning neck while foraging. Breeding ad. has extensive, lacy lower-back plumes, but no head or breast plumes. The two races are very similar, but *albus* almost same size or slightly larger than Grey Heron and larger than *modestus*, which is slightly smaller than Grey Heron. Non-breeding *albus* has pale or bright yellow tarsi and toes, and pink or yellow tibia and tarsus when breeding, whereas non-breeding *modestus* has black tarsi and toes, and deep pink tibia when breeding. In flight, legs extend well beyond tail, toes very large; wings well forward of mid line, neck bulges deeply. **BP** Bill changes from black with blue-green lores (breeding) to yellow with greenish lores (non-breeding); gape line extends rear of eye (see Intermediate); eyes yellow; tarsi largely blackish, but flush pink when breeding; in *modestus* legs are black during non-breeding, in slightly larger *albus* legs are yellowish or all yellow outside breeding season. **Vo** Commonly gives loud, hoarse *crrack* or *guwaa* and a dry, wooden, rather grating *grrraah* on take-off. **TN** Has formerly been placed in *Egretta* and *Ardea*.

INTERMEDIATE EGRET CTKJR
Mesophoyx intermedia

L 65–72cm; WS 105–115cm. **SD** Resident in sub-Saharan Africa, S & SE Asia to Australasia. In E Asia, *M. i. intermedia* is summer visitor (Apr–Oct) across S & E China to Changjiang, S Korea and Japan, north to N Honshu (scarce spring overshoots reach Hokkaido). Winters south through Taiwan, Philippines, SE Asia and Indonesia. **HH** Inland and coastal wetlands, from wet ricefields to mangroves and mudflats. **ID** All-white egret, between Little (p.106) and Great White in size. Separated from both by rather short bill, relatively small rounded head and long slender neck often held in curved S, lacking angular kink of Great White (neck also proportionately thicker than Great White), and from Little by lack of head plumes. Note crown peaked at centre, giving characteristic rounded triangular head shape from side. Breeding ad. has extensive lacy lower-breast and back plumes, but no head plumes. In flight, long legs extend well beyond tail with toes forming prominent clump even at long range; wings appear placed forward of mid-line. **BP** Bill short, yellow with black tip and yellow lores (non-breeding), or largely black with yellow base and greenish-yellow lores (breeding); gape line extends level with or to rear of eye (compare Great White Egret); eyes yellow; long tarsi black. **Vo** Hoarse barking *goah-goah* or *graak graak* and more prolonged rasping *graaarsh*. **TN** Formerly *Egretta intermedia*. **AN** Yellow-billed Egret.

PIED HERON T
Egretta picata

L 43–55cm; WT 210–372g. **SD** Resident of Indonesia and Australasia, which has strayed to Taiwan. Monotypic. **HH** Like Pacific Reef Egret (p.106), favours rocky shores. **ID** Small egret with unique pied plumage. Mask, crown and nape plumes slate-grey, chin, throat, lower face, neck and breast plumes pure white, whilst rest of plumage dark slate-grey. **BP** Bill yellow, lores blue; eyes pale yellow; long tarsi deep ochre-yellow. **Vo** Harsh croaks.

PURPLE HERON

manilensis

juv

ad sum

ad sum

GREAT WHITE EGRET

ad win *albus*

ad win *modestus*

ad sum *albus*

ad win

intermedia

**INTERMEDIATE
EGRET**

ad sum

PIED HERON

juv

ad

ad

WHITE-FACED HERON CT
Egretta novaehollandiae

L 65–69cm; WS 106cm; WT 550g. **SD** Resident of Indonesia to New Zealand that has strayed to Xiamen (Amoy), Fujian and Taiwan. Monotypic. **HH** In native range, ponds and streams, grassy fields, mudflats and mangrove edges, also urban wetlands where can be reasonably confiding. **ID** Graceful grey egret, size of Intermediate (p.104). Ad. forehead, face, chin and throat white, otherwise entirely pale to mid grey, with elongated pale grey back and pale chestnut lower-neck plumes. Juv. has less white on face and lacks back and neck plumes. In flight, wings distinctly two-toned, with paler grey coverts contrasting with dark grey primaries and secondaries. **BP** Bill dark greyish-black, lores grey; eyes yellow; long tarsi deep greenish-yellow. **Vo** Harsh croaks.

LITTLE EGRET CTKJ
Egretta garzetta

L 55–65cm; WS 88–106cm; WT 280–638g. **SD** Widespread in warmer regions of Europe, Africa, Asia and Australasia. *E. g. garzetta* resident north to S China and Taiwan, and summer visitor to E China and Korea. Also resident on main Japanese islands north to N Honshu. **HH** Nests colonially, roosts socially, in dense trees near marshes, lakes or rivers, but usually forages alone or in small groups, sometimes with other egrets. Occurs on wetlands, wet ricefields, streams and shallow coastal lagoons, but rarely on beaches or rocky shores. Reasonably confiding in urban areas. Sometimes stirs mud or water with feet to disturb prey. **ID** Mid-sized white egret, slightly taller and slimmer than Eastern Cattle (p.102), more elegant, with longer neck; smaller than Intermediate (p.104). All white; breeding ad. has two elongated nape plumes and elongated, lacy, breast and lower-back plumes, some of them erectile. Scarce dusky form has grey wash to breast, belly, wings and tail. In flight, legs extend moderately beyond tail; wings appear centrally set (see Intermediate and Great White, p.104). See also Pacific Reef and Swinhoe's. **BP** Bill long, slender, black, lores yellow to greyish-green (non-breeding), pink or orange (breeding); western (extralimital) birds have blue-grey lores. Young have yellowish lower mandible. Eyes yellow at all ages; long tarsi generally all black (sometimes greenish-black or patchily yellow), toes contrastingly yellow. **Vo** Guttural gargling croaks, *guwa* or *goa*, bleating *mmyaaaaw* and bizarre gargling *blublublublublub* at colonies, and grating, throaty, crow-like *aaah*, *ark-ark-ark*, or *kra a a ak* when flushed and in flight.

PACIFIC REEF EGRET CTKJ
Egretta sacra

L 58–66cm; WS 90–100cm; WT 330–700g. **SD** Resident of coastal Australasia and Pacific islands through Indonesia, SE Asia, Philippines, Taiwan, Nansei Shoto to C Japan and S Korea. Wanderers have reached SE China. Subspecies in region *E. s. sacra*. **HH** Rocky coasts, foraging at tideline and nesting on rock stacks, cliffs or amongst boulders; usually solitary. Forages

by crouching, creeping and lunging. Vagrants occasionally appear at other wetland habitats, e.g. river mouths and mudflats, and may be confused with other white egrets. **ID** Size of Little Egret, but oddly proportioned, with shorter thicker neck, shorter more rounded wings, and shorter legs – only feet protrude beyond tail. Dimorphic, blue-grey form commoner than white. Dark form: blue-grey, with small, inconspicuous white chin, nape rounded. Fewer plumes than other egrets when breeding, restricted to lower back. White form superficially like other white egrets, but stocky proportions and bill shape differ. Imm. white form may be flecked grey, especially on wing-coverts. Note creeping gait while feeding. **BP** Bill long, thicker based, broader at tip and less sharply pointed than Little, narrowing more continuously from base than Swinhoe's, culmen somewhat convex, varies from dark yellowish-grey (dark morph) to pale yellow with dark culmen (white morph); lores blue-grey to yellowish; eyes yellow; strong legs appear short (especially tibia; compare Swinhoe's), rather stout with chunky joints, yellowish-green with large yellow toes. **Vo** Distinctive, deep *gruk*, similar to Northern Raven, downslurred *nyarp* and, in flight, nasal *gyaaah*. **AN** Eastern Reef Egret.

SWINHOE'S EGRET CTKJR
Egretta eulophotes

Vulnerable. L 65–68cm. **SD** E Asian endemic breeder. Rare summer visitor, breeding locally only on Yellow Sea coastal islets, off E China (Shandong), Korea and coastal Russian Far East (Sea of Japan), and recently islets off Fujian. Winters in Philippines, Indonesia and Borneo. On migration also Taiwan, and scarce along E Chinese coast. Recorded north to Liaoning, also Hokkaido and Sakhalin. Monotypic. **HH** Coastal mudflats, lagoons, river mouths and rocky areas on offshore islands. Distinctive when foraging; hunched, even crouching, leaning posture, suddenly dashing forward to snatch prey; darting about excitedly, chasing and lunging at disturbed prey. **ID** Mid-sized white egret, usually larger (but sometimes smaller and slimmer) than Little Egret, smaller than Great White (p.104). Neck proportionately thicker and shorter than Little, legs also shorter. Erectile plumes on crown and nape far more extensive than other egrets, breast plumes and lower-back plumes lacy as in others. Proportions and behaviour separate it from Pacific Reef. **BP** Bill thicker than Little's, symmetrically dagger-shaped, narrowing only near tip, yellow/orange (breeding), blackish with yellow base to lower mandible (non-breeding); lores blue-grey to blue (summer), yellow (winter); eyes yellow; legs blackish, with variable amount of yellow; may have strong contrast between yellow-green toes and dark tarsi, or may grade between yellow-green and blackish, but tarsi clearly shorter and slightly thicker than Little. **Vo** Guttural croaking *gwa* sometimes given on taking flight. **AN** Chinese Egret.

WHITE-FACED HERON

ad sum

juv

LITTLE EGRET

garzetta

ad

ad win

ad sum

ad dark
morph

PACIFIC REEF EGRET

sacra

ad white
morph

juv

SWINHOE'S EGRET

ad sum

ad sum

ad win

RED-TAILED TROPICBIRD   TJ
Phaethon rubricauda

L 46–96cm; WS 104–119cm; WT 600–835g. **SD** Warm waters of Indian and Pacific oceans. *P. r. melanorhynchos* breeds north only to S Nansei Shoto and Ogasawara/Iwo Is. **HH** Seldom seen except at remote Japanese breeding islands (May–Aug), though accidental Taiwan. **ID** Ad. essentially pure white (the whitest tropicbird), with black loral crescent, streak behind eye, black shafts to outer primaries and small black area on tertials. White tail short and rounded, but has two very long (30–35cm) red central tail-feather shafts (often hard to detect at long range). Juv. largely white with wide-spaced, coarse blackish-brown bars on upperparts and wing-coverts, wings largely white with narrow black area on primary shafts to tips; lacks tail-streamers. Commonly flies high above sea, hovers and plunges to feed. Largest and heaviest of all tropicbirds; relatively short-winged. **BP** Bill prominent, arched and sharply pointed, red (ad.) or black (juv.), changing with age to yellow then orange; eyes black; tarsi blue/grey with black distal webs to toes. **Vo** Screams loudly at nest, giving high, nasal yapping *nyak* or *nyap*, also long rolling call, and in flight a rasping *skrawk* and ratchet-like *ki ki ki*.

WHITE-TAILED TROPICBIRD   TJ
Phaethon lepturus

L 37–82cm; WS 90–95cm; WT 300g. **SD** Pantropical. *P. l. dorotheae* rare on S Nansei Shoto, Ogasawara Is and, rarely, mainland Japan, usually after typhoons (April–Sep); accidental Taiwan. **HH** Pelagic. Typically flies high over sea, then swoops or plunges for food. **ID** Ad. essentially pure white, with broad black streak from lores to well behind eye, broad black outer primary bases contrasting with white primary-coverts, and black bar on innerwing-coverts and tertials. Tail short, rounded, with two very long (33–45cm) white central tail-feather shafts. Juv. mainly white, with wide-spaced, fine greyish-black bars on upperparts and wing-coverts; wings largely white, with broad black area in primary bases; lacks tail plumes. Smaller than Red-tailed with faster, more buoyant, flight. **BP** Bill short, orange-yellow or yellow; eyes black; short tarsi and toes black. **Vo** Harsh tern-like *ki-dit-kit* and squawking *skrech* at colony, and rattling *ki-di ki-di ki-di...* in flight.

RED-BILLED TROPICBIRD   Extralimital
Phaethon aethereus

L 45–106cm; WS 99–115cm; WT 700g. **SD** May be expected to stray to S of region (recorded S China). **ID** Distinguished by red bill, black eyestripe, white upperparts finely barred black in ad., white wings with black primaries and primary-coverts and very long (46–56cm), thin white central tail-feathers. **Vo** Generally silent away from colonies.

CHRISTMAS ISLAND FRIGATEBIRD   CJ
Fregata andrewsi

Critically Endangered. L 89–100cm; WS 205–230cm; WT ♂ *c.*1.4kg, ♀ *c.*1.55kg. **SD** Endemic breeder on Christmas I., Indian Ocean. Accidental SE China and Japan. Monotypic. **HH** Pelagic. **ID** Large, mostly black with strongly forked tail. Recalls Lesser Frigatebird, but much larger. Ad. ♂ largely black with green gloss to mantle; wrinkled red gular pouch (inflated only on breeding grounds) and white oval belly patch (not extending to axillaries). ♀ has black hood to lower neck, distinct white hind-collar, large white belly patch and white axillary 'spurs', with black of hood extending to point on chest, and black bar on breast-sides. Juv. has white or rusty-buff head, white collar, prominent pale alar bar, prominent black breast-band and white underparts; white belly patch hexagonal, extending as 'spurs' onto axillaries. **BP** Bill very long, slender, sharply hooked, dark grey (♂) or pink (♀); eyes black, surrounded by red (♂) or reddish-pink (♀) orbital ring almost joining gape; tarsi dark (♂) or orange (♀). **Vo** Silent away from breeding grounds.

GREAT FRIGATEBIRD   CTKJ
Fregata minor

L 85–105cm; WS 205–230cm; WT ♂ 1–1.4kg, ♀ 1.215–1.64kg. **SD** Widespread in tropical oceans. *F. m. minor* is rare visitor to region, most frequently after autumn typhoons (Aug–Nov) from Taiwan to Yellow Sea and Bohai Gulf north to Japan and S Russian Far East. **ID** Large black frigatebird with black axillaries. Ad. ♂ black with bronzy-green sheen to upperparts, alar bar slight or absent, small wrinkled red gular pouch (only inflated when breeding) and black underparts. ♀ has black cap and head-sides, indistinct brown hind-collar, diagnostic greyish-white throat, and large white, saddle-shaped chest patch extending to breast-sides but not onto axillaries; black point on belly broadly rounded. Juv. has combination of elliptical white belly patch, sandy-brown head and breast, and some have short axillary spurs angling outwards. **BP** Bill long, sharply hooked, dark grey (♂), pale blue-grey or pinkish-horn (♀/juv.); eyes black (♀ with pink eye-ring); tarsi dark red (♂) or grey (♀/juv.). **Vo** Silent away from breeding grounds.

LESSER FRIGATEBIRD   CTKJR
Fregata ariel

L 71–81cm; WS 175–193cm; WT ♂ 625–875g, ♀ 760–955g. **SD** Widespread in tropical seas. *F. a. ariel* is annual, albeit erratic and rare, during or after summer storms and autumn typhoons (May–Nov). Recorded regularly from coastal SE China and Taiwan to Japan, exceptionally even to Hokkaido, Ussuriland and Amur R mouth. **HH** Pelagic. **ID** Smaller, with slighter build than other frigatebirds. Ad. ♂ largely black with blue sheen to upperparts, and narrow white axillary patches or 'spurs' extending to flanks, but not joining across belly; alar bar slight or absent. ♀ has black hood extending to throat, white hind-collar, moderate off-white alar bar, and restricted white breast patch reaching neck- and breast-sides and onto axillaries (triangular and angled outwards), leaving large, pointed-oval black belly patch. Juv. has whitish-sand or pale rusty-cream hood, blackish-brown breast-band and small, triangular white belly patch extending to axillaries as 'spurs' angled outwards. **BP** Bill long, slim, sharply hooked, dark grey (♂), pale blue-grey or pinkish-horn (♀/young); eyes black, ♀ has pink eye-ring; tarsi dark red (ad) or grey (young). **Vo** Silent away from breeding grounds.

CHRISTMAS ISLAND FRIGATEBIRD

ad ♂

WHITE-TAILED
TROPICBIRD

ad   *dorotheae*

juv

RED-TAILED
TROPICBIRD

ad

melanorhynchos

ad ♀

CHRISTMAS
ISLAND

juv

juv

minor

GREAT
FRIGATEBIRD

ad ♀

GREAT

juv

ad ♂

ad ♂   ad ♀   LESSER

juv

ad ♂

ariel

ad ♀

LESSER FRIGATEBIRD

juv

*juv Lesser harasses a
Red-tailed Tropicbird*

PLATE 41 : PELICANS

GREAT WHITE PELICAN
CTJ
Pelecanus onocrotalus

L 140–175cm; WS 245–295cm; WT ♂ 9–15kg, ♀ 5.4–9kg. **SD** Main range includes Africa, E Europe, SC & S Asia. Rare visitor to SE China, Taiwan and Japan, but escapees from collections may remain for years associating with other waterbirds, e.g. Grey Herons at heronries. Straggles north as far as Hokkaido. Monotypic. **HH** Estuaries, large rivers and lakes. **ID** Unmistakable. Huge, white, short-legged, long-necked and long-billed waterbird. Ad. has yellow breast patch in summer. Flight-feathers black, forming striking contrast with remainder of white plumage in flight. Juv./1st-winter rather dark brownish-grey, but also have bare pink facial skin. In flight, at all ages has all-dark flight-feathers. **BP** Bill largely yellow with grey culmen and sides to lower mandible, gular or bill pouch yellow or yellowish-orange; bare facial skin extensive, pinkish-orange (only narrow point of feathering in centre of forehead), eyes black; short, stout legs and large feet pink or yellowish-orange. **Vo** Usually silent away from colonies, but gives deep grunting calls on taking flight. **AN** Rosy Pelican; Eastern White Pelican.

SPOT-BILLED PELICAN
CTKJ
Pelecanus philippensis

Vulnerable. L 127–152cm; WS 250cm; WT *c*.5kg. **SD** Formerly widespread in S & SE Asia, S & E China and Philippines, now primarily SE India and Sri Lanka. Possibly rare resident of E & S Chinese coasts (though confusion with Dalmatian Pelican likely), but probably extinct. Accidental Korea, Japan and Taiwan (but most historical records possibly reflect confusion with Dalmatian). Monotypic. **HH** Occurs on estuaries, large rivers and lakes. **ID** Smallest pelican of region. Overall plumage rather dull, greyer even than Dalmatian, with greyish-white head, neck, underparts and wings. Breeding ad. develops tufted crest on rear crown and hindneck, a dull pink rump and vent, cinnamon underwing-coverts and spotted bill pouch. Non-breeding ad. has shorter crest and paler, pinker bill pouch and face; lacks pink wash. In flight, ad. has flight-feathers and tail tip blackish-grey. Young have dark axillaries and underwing-coverts, dark flight-feathers and pale central underwing stripe. **BP** Face mostly feathered, upper mandible dull yellow or pinkish-orange with blue spots, bill rather short for a pelican, pouch pinkish-grey with blue spots and blotches, tip has small hook; bare patch around eye off-white, irides dark, lores bluish-grey; tarsi black. **Vo** Generally silent away from colonies.

DALMATIAN PELICAN
CTKJ
Pelecanus crispus

L 160–180cm; WS 270–320cm; WT 10–13kg. **SD** Main range E Europe, C Asia and S Asia to SE China. Rare and local winter visitor to SE China (Nov–Apr); accidental Taiwan and Japan. Monotypic. **HH** Estuaries, large rivers and lakes. **ID** Large, dusky-grey pelican. Similar to Spot-billed but larger. Ad. grey-tinged white. When breeding has a golden-yellow breast patch, unkempt, curly head plumes and feathered face and forehead. Non-breeding ad. lacks yellow breast patch and has paler, duller bill. Juv. similar to Great White, but face feathered and bill darker, greyish-pink. In flight, all flight-feathers are black above, but secondaries, especially, are pale grey below, black only at tips. This and Spot-billed show whitish medial band against otherwise dusky-grey underwing. **BP** Bill long and black with strongly hooked tip, pouch reddish-orange when breeding, bill pink and pouch yellowish-pink when non-breeding; bare skin around eye pink (less extensive than Great White with broader band of feathers on centre of forehead), irides pale yellow; short stout legs and large feet dark grey (breeding) or pink (non-breeding). **Vo** A rather muffled roaring sound when breeding and though generally silent away from colonies may call while fishing.

GREAT WHITE PELICAN

ad

ad sum

ad

juv

juv

ad sum

SPOT-BILLED PELICAN

ad

juv

ad

roosting
Spot-billed Pelicans

juv

ad

juv

ad

juv

ad sum

ad

DALMATIAN PELICAN

juv

MASKED BOOBY   TKJ
Sula dactylatra

L 81–92cm; WS 150–170cm; WT 1.55kg. **SD** Widespread in tropical seas. *S. d. personata* is very local summer visitor (mostly Jun–Oct) to S Japan and off Taiwan, with a few records off S Korea. **HH** Breeds on remote islands; otherwise pelagic. **ID** Larger, longer-winged and shorter-tailed than other boobies. Ad. mostly white with black face, flight-feathers, primary-coverts and rectrices. Underwing-coverts white. Juv. largely brown above with brown hood and white below, resembling Brown Booby, but upper chest white and white collar separates brown of head from mantle, and white rump separates brown of mantle from tail; underparts white, including underwing. **BP** Bill pale yellow (ad.), dusky greyish-yellow (juv.), with area of bare, black skin around eye and bill base; irides pale yellow (dark in juv.); short legs and feet dull grey or olive. **Vo** Usually silent except at colonies, where ♂ gives wheezy rising then falling *fuuwheeeoo* and ♀ a loud braying.

RED-FOOTED BOOBY   CTJR
Sula sula

L 66–77cm; WS 124–152cm; WT 0.9–1kg. **SD** Widespread in tropical seas, but *S. s. rubripes* only very local summer visitor to S Japan and Taiwan, mostly Jun–Oct; has strayed north to Tartar Strait (S Russian Far East) and, in winter, to SE China. **HH** Breeds on remote islands; forages at night far from shore. In non-breeding season pelagic. **ID** Smallest of the three boobies in the region, with proportionately longer tail, shorter slimmer neck, rounder head and slender bill. Polymorphic, with various white and brown morphs. In region, only white morph recorded. Ad. white morph all white with black primaries and secondaries, but white tertials (thus black trailing edge does not reach body), black primary-coverts and all-white tail (see Masked Booby). Underwing pattern same

as upperwing, the black under primary-coverts are diagnostic. Juv. brown above and below (though vent whitish-brown) with all-dark underwing (compare juv. Brown). Flight faster than other boobies. At all ages, slender jizz and smaller size of Red-footed is helpful, especially so in relation to juv. Brown. **BP** Bill pale blue-grey, with bare pink facial skin at base in ad., pink with dull-grey tip and grey/pink facial skin in juv.; large eyes black; feet dark grey (juv.) to bright red (ad.). **Vo** Both sexes give guttural rattling squawks and screeches, but only at breeding colonies.

BROWN BOOBY   CTKJ
Sula leucogaster

L 64–74cm; WS 132–150cm; WT 0.724–1.55kg. **SD** Ranges widely in tropical and subtropical oceans. In E Asia, *S. l. plotus* is locally common breeder on offshore islands of Taiwan, S Nansei Shoto, Kyushu, Izu, Ogasawara and Iwo Is. Commonly reaches coasts and bays of Kyushu (often in small flocks) and S Honshu, and may also be encountered off Pacific coast. Year-round in Japanese waters; rare off S Korea and SE China. **HH** Breeds on remote islands and forages over inshore waters. **ID** Ad. mostly dark, chocolate-brown, upper surfaces entirely so, head, neck and upper breast also chocolate-brown, strongly contrasting with white lower chest, belly and undertail-coverts, and white underwing-coverts. No white hind-collar (compare juv. Masked Booby). ♀ larger than ♂. Juv. like ad., but mottled brown and white on underparts and has white axillaries and underwing-coverts (see Red-footed). In flight, wings broad-'armed', with sharply pointed 'hand'; tail long, graduating to point, all dark. Upperwing all dark, underwing has broad dark leading and trailing edges, and all-dark primaries/primary-coverts; axillaries and underwing-coverts white. **BP** Bill long, strongly tapered and pointed, lacks obvious nostril; bare facial skin bluish/purple (♂) or yellow (♀); eyes dark brown; short legs and large webbed feet pale to bright yellow (ad.) or dark grey (juv.). **Vo** Usually silent at sea, but at colonies ♂ gives high, sibilant *schwee* and ♀ short, rather guttural grunting and honking *guwa guwa guwa* or *guu guu guu* calls.

MASKED BOOBY

ad

personata

2/9/19

MASKED

ad

juv

ad white
morph

RED-
FOOTED

ad intermediate
morph

ad brown
morph

2/6/19 RED-FOOTED BOOBY

rubripes

juv

ad white
morph

ad brown
morph

ad intermediate
morph

ad

BROWN

ad

plotus

juv

BROWN BOOBY

DOUBLE-CRESTED CORMORANT R
Phalacrocorax auritus

L 76–91cm; WS 137cm; WT 1.67–2.1kg. **SD** Widespread in N America. *P. a. cincinatus* in Aleutians. Vagrant Chukotka. **HH** Inland waters and coasts (rocky to estuarine). **ID** Large, dark cormorant, slightly smaller than Great. Ad. black with bronze/green sheen to mantle, back and scapulars. Off-white ear plumes when breeding (Mar–May). Juv. pale brown to blackish-brown, usually pale on neck and breast with dark belly. From Great by deeper orange, rounded bare facial skin; lacks white face patch; head smaller. **BP** Bill dull blue-grey with deep-orange facial skin; eyes dark blue-green; stout tarsi blackish-grey. **Vo** Grunts and clear *yaaa yaa ya* calls at colonies, but usually silent.

GREAT CORMORANT CTKJR
Phalacrocorax carbo

L 77–94cm; WS 121–149cm; WT 1.81–2.81kg. **SD** Europe to SE Asia, Australasia and Atlantic N America. *P. c. sinensis* resident north to Yellow Sea and Korea, with *P. c. hanedae* in Japan, but limits between them unclear. Summer visitor north to Russian Far East, Amurland, and N Honshu (also Hokkaido?). Mainly resident Japan, but may disperse post-breeding. Winters to SE Asia, SE China and Taiwan. **HH** Often assumed to occur on inland lakes and rivers, but habitat overlap and confusion with Temminck's likely, as visits estuaries and bays to forage. **ID** Largest cormorant in region. Ad. black with long thick neck, white facial patch extending to throat and variable narrow white plumes on crown, nape and head-sides. In late winter/early spring has large white flank patch. Upperwing-coverts and mantle have bronze sheen with black fringes (green sheen in Temminck's); however, beware effects of light and note flight-feathers have greenish sheen. Juv. like dull ad., but commonly white or mottled brown and white on underparts. Flight heavy, goose-like with broad wings, but neck short, thick and kinked. **BP** Bill long, thick, pale grey, darker at tip, yellow at base and gape, upper mandible dark grey, lower whitish, and yellow skin angles back from eye to point of gape, then forms prominent rounded gular patch on chin and below bill (see Temminck's); eyes dark green; strong legs and feet greyish-black. **Vo** At colonies makes gruff, guttural croaking sounds: *gock gock* or *guwaa*, sometimes repeated in growling series. Also croaking *kursh kursh kursh....*

TEMMINCK'S CORMORANT CTKJR
Phalacrocorax capillatus

L 81–92cm; WS 152cm. **SD** Endemic to E Asia, from S Kuril Is, Sakhalin and Russian Far East to C Japan, also locally Korea, and Liaoning, Hebei and Shandong. Winters locally from Hokkaido to N Taiwan and SE China. Monotypic. **HH** Favours rocky coasts, but may also visit coastal lagoons and estuaries, where confusion with Great Cormorant possible. **ID** Large (Great and Temminck's overlap), with plumage as Great but white facial patch more extensive. Ad. upperwing-coverts and mantle have deep green sheen with black margins (bronze sheen on Great), and flight-feathers also have dark green sheen. Facial skin and gular pouch yellow, but less extensive than Great. Yellow skin forms vertical border just behind eye then extends to sharp point at gape, and forms small rounded patch on chin and below bill (see Great). Juv. like dull ad., but commonly white or mottled brown and white on underparts, and has yellow lower mandible. **BP** Bill long, thick, dark grey darkest at tip, yellow at base, on gape and gular region; eyes dark green; strong tarsi greyish-black. **Vo** Gruff *guwaa*, deep rolling *guwawawa* or *gururu* at colonies. **AN** Japanese Cormorant.

PELAGIC CORMORANT CTKJR
Phalacrocorax pelagicus

L 63–76cm; WS 91–102cm; WT 1.474–2.438kg. **SD** *P. p. pelagicus* restricted to N Pacific, Bering and Okhotsk seas. Summer visitor from Bering Strait islands to Kamchatka and Sea of Okhotsk. Resident in Aleutians, Kurils and Hokkaido; also small islands off Liaoning. Winters (Nov–Apr) along continental coast of Sea of Japan, Japanese archipelago, Korea and E China coast to Fujian; accidental Taiwan. **HH** Breeds and roosts on coastal cliffs, rock stacks and offshore islands; typically forages inshore, wintering in similar habitat. **ID** Smallest and most slender cormorant in region, with long, thin neck, small rounded head and very thin bill. Overall jizz best separates it from other cormorants. Ad. all black with white flank patch and sparse white filoplumes on neck-sides in Mar–May, also purple neck gloss and wispy crests on crown. Lacks white facial patch of larger cormorants. Juv. like ad., but has black face. Wingbeats faster than larger species. **BP** Bill slender, pale (even yellow) to dark blackish-grey (can appear extremely pale in sunshine, when often mistaken for Red-faced Cormorant), and bare facial skin dark reddish and restricted to small area below eye, though when breeding red is brighter and lower mandible may be yellower; irides dark green; tarsi black. **Vo** At colonies gives grunting and groaning calls.

RED-FACED CORMORANT JR
Phalacrocorax urile

L 71–89cm; WS 110–122cm; WT 1.664–2.552kg. **SD** Uncommon breeder in coastal N Pacific, S Bering Sea, Aleutian and Commander Is, SE Kamchatka, at isolated colonies in Kuril Is, and rarely E Hokkaido. Some breed at mixed colonies with Pelagic Cormorant (their ranges overlap extensively). Rare winter visitor Hokkaido; accidental C & S Japan and reported Bohai Gulf. Monotypic. **HH** Breeds and roosts on cliffs, rock stacks and offshore islands, and typically forages inshore. Winters mainly near colonies, occasionally S & N of breeding areas. **ID** Slender, all-black cormorant. Slightly larger and appears stockier, larger headed and thicker in neck and bill than Pelagic, but there is overlap. Ad. as Pelagic, but in breeding season has more prominent, slightly bushier, crest. Bare facial skin bright red and extends from base of upper mandible over forehead, around eye to chin, covering much larger area than in Pelagic. Juv. lacks red face, instead has yellowish eye-ring and much paler face than Pelagic. In non-breeding season confusion easily possible with pale-billed Pelagic. Wingbeats faster than larger cormorants. **BP** Bill pale yellowish-grey, brighter yellow in ad., and breeding ad. has blue gape; eyes dark green; tarsi black. **Vo** At colonies gives low, weak guttural croaking *kwoon* or *gwoo*.

DOUBLE-CRESTED CORMORANT

ad win

juv

cincinatus

ad sum

juv

ad win

ad sum

GREAT CORMORANT

sinensis

juv

GREAT

juv

GREAT

TEMMINCK'S

juv

ad

TEMMINCK'S

ad

GREAT

PELAGIC

ad win

ad sum

juv

ad sum

Pelagic with two Temminck's

2-16-19
Tokyo Japan

TEMMINCK'S CORMORANT

TEMMINCK'S

ad win

juv

ad win

juv

ad win

PELAGIC CORMORANT

pelagicus

RED-FACED

juv

ad sum

juv

juv

ad

PELAGIC

ad

RED-FACED

ad win

PELAGIC

RED-FACED CORMORANT

PIED FALCONET   C
Microhierax melanoleucus

L 16–18cm; WS 33–37cm; WT 55–75g. **SD** NE India to E China, northern SE Asia. E China from Fujian to Jiangsu (very rare). Monotypic. **HH** Evergreen and deciduous woodland edge and clearings, in wooded foothills (to 1,500m) or open country; uses exposed perch (often a long-dead trunk or bough), makes dashing flights after large insects or small birds, or drops to prey on ground. **ID** Smallest raptor in region. Rather long-tailed, upperparts slate-black, face white with black patch to ear-coverts; underparts white with some black on flanks. In flight, compact, wings pointed, tail long, very dark, upperparts completely so, axillaries and underwing-coverts white, flight-feathers black with fine white bars, undertail black with several white bars. Flight fast with rapid beats and glides. **BP** Bill dark grey in ad., paler in juv., with blackish-slate cere; eyes large, dark brown, with black orbital skin; tarsi greyish-black. **Vo** Screams and sharp, high-pitched chattering whistles: *kip kip kip….*

LESSER KESTREL   CTKJ
Falco naumanni

Vulnerable. L 26–31cm; WS 62–73cm; WT ♂ 90–172 g, ♀ 128–216g. **SD** Europe and C Asia, moving to sub-Saharan Africa in winter; also breeds Mongolia and Hebei, China, and migrates along coast. Vagrant Japan, Taiwan, reported Korea. Monotypic. **HH** Colonial in open areas with wooded hills and cliffs, grasslands and low-intensity cultivation, steppe and semi-desert; winters on savannas; takes insects caught in air or on ground. **ID** Slim, slightly smaller and slimmer than Eurasian. ♂ has pale grey head and face, without dark moustachial. Upperparts and smaller upperwing-coverts plain rufous-brown, lacking black spots, greater coverts blue-grey forming panel on innerwing; flight-feathers black. Tail plain grey with broad black band at tip. Underparts deep buff-orange with small black spots on breast and flanks. ♀/juv. like Eurasian but paler, with no dark eyestripe or moustachial streak, and paler underparts. At rest, wingtips almost reach tail tip and more rounded than Eurasian. Flight fast and agile with quick beats, stiffer and shallower than Eurasian; hovers less than latter but circles and soars more. Compared to Eurasian, appears shorter-bodied, with more pointed wings and more wedge-shaped tail (central rectrices often protrude). Underwing of ♂ very pale, primaries almost unmarked white contrasting with black tips, underwing-coverts only slightly spotted; ♀ has pale grey underwing with darker grey barring on flight-feathers, spotting on axillaries and underwing-coverts, far less conspicuously marked than Eurasian. **BP** Bill very short, relatively deep, sharply hooked, yellow with dark grey tip; eye-ring yellow, irides dark brown; tarsi yellow with white claws. **Vo** Highly vocal when breeding. Shrill, trisyllabic flight call a scratchy, chattering *kyi kyi kyi, chay chay chay* or *jhee jhee jhet*, differing markedly from Eurasian, but also gives fast Eurasian-like notes.

EURASIAN KESTREL   CTKJR
Falco tinnunculus

L 27–35cm; WS 57–79cm; WT (*F. t. interstinctus*) 150–185g. **SD** From W Europe to Yakutia south to southern Africa and SE

Asia. Northern breeders migratory. *F. t. tinnunculus* through boreal and temperate regions east to the Lena; *F. t. perpallidus* east to Yakutia and Magadan, south through Russian Far East to NE China and Korea; *F. t. interstinctus* from Himalayas to Japan. Some northern birds winter in S Japan, Taiwan. **HH** From taiga and forest-tundra to open plains, steppe, agricultural and suburban areas; lowlands to mountains. **ID** Small to mid-sized falcon with long, narrow wings, with rather rounded tips, and long rather broad tail with rounded tip. ♂ has grey head flecked dark grey and diffuse dark eyestripe and single moustachial (which separates Eurasian from extralimital **American Kestrel** *F. sparverius*, a possible vagrant that has reached Aleutians). Upperparts rufous-brown with black 'anchor' spots on mantle and wing-coverts; primaries and primary-coverts black (no grey on upperwing). Tail plain grey with broad black band at tip. Underparts buff-orange with many black streaks. ♀/juv. have noticeably streaked rufous-brown crown and nape, dark eyestripe and moustachial; tail rufous with many narrow dark bars and bold band on tip. Occasionally much paler, sandy-brown birds recorded Korea and Japan, but whether a pale morph or *F. t. perpallidus* is unclear. At rest, usually perches very upright, wingtips reach just beyond mid tail. Flight fast and agile, graceful with shallow beats; soars on flat wings with tail fanned, and commonly hovers into wind or with fast shallow wingbeats when hunting, dropping to prey on ground, though also taken on wing. Underwing pale, heavily barred on underside of flight-feathers. **BP** Bill very short, relatively deep, sharply hooked, yellow with dark grey tip; eye-ring yellow, irides dark brown; tarsi yellow with blackish-grey claws. **Vo** When breeding noisy, uttering piercing, high-pitched screams, sharp *kyi kyi kyi* or *klee-klee-klee* notes, and chittering series of hard *stik stik stik* notes. Typically silent at other times.

AMUR FALCON   CTKJR
Falco amurensis

L 26–30cm; WS 63–71cm; WT ♂ 97–155g, ♀ 111–188g. **SD** E Asian endemic breeder which migrates via India to winter in southern Africa. Breeds in S Russian Far East, NE China and N Korea, scarce migrant S Korea, and E China, and rare S Japan, accidental Hokkaido and Taiwan. Monotypic. **HH** Open areas, woodland edge, wooded steppe, agricultural areas; perches freely on trees and wires; migrates in flocks and may also hunt socially, taking insect prey from ground or in air. **ID** Small to mid-sized dark falcon with long, narrow wings and tail. ♂ plain grey with darker eyestripe, moustachial and upperparts; thighs, vent and undertail-coverts rufous-orange; underwing-coverts bright white. ♀ (recalls Northern Hobby, p.118) is dark grey above, narrowly barred black, with white cheeks, short blunt moustachial, blackish spots on breast and bars on flanks; thighs and vent buff; tail grey barred black, with broad band at tip; young ♂ intermediate between ♀ and ad. ♂. Juv. strongly streaked blackish-brown below, with rufous-brown crown; upperparts have rufous-buff fringes. At rest, wingtips extend to or just beyond tail tip. Flight fast and dashing, in pursuit of large insects, and frequently hovers. **BP** Bill grey with reddish-orange cere; eye-ring reddish-orange, irides dark brown; tarsi orange-red (ad.) or yellow-orange (juv.), claws white. **Vo** Sharp chittering *kikikikik…* or *kew kew kew* at colonies and communal roosts. **AN** Eastern Red-footed Falcon.

LESSER KESTREL

ad ♂

ad ♀

imm ♂

ad ♂

imm ♂

ad ♀

ad ♀

ad ♀

EURASIAN KESTREL

tinnunculus

ad ♂

PIED FALCONET

ad

ad

ad ♂

ad ♂

ad ♂

ad ♀

ad ♀

ad ♂

AMUR FALCON

ad ♀

juv

ad ♂

imm ♂

ad ♂

imm ♂

ad ♀

MERLIN CTKJR
Falco columbarius

L 24–32cm; WS 53–73cm; WT (*F. c. insignis*) ♂ 164–190g, ♀ 155–205g. **SD** Boreal N America and Eurasia east to E Chukotka, winters Europe, Asia and C America. *F. c. insignis* breeds south of tundra from Yenisey to Kolyma, winters south to Sakhalin, E China, S Korea and Japan, on migration more widely in NE China and Korea. *F. c. pacificus* Russian Far East, Anadyr to Koryakia, Okhotsk and Sakhalin, winters to China, rarely N Japan. **HH** Tundra, wooded steppe, uplands, grasslands, marshes and agricultural land with copses; in winter coastal lowlands and wetlands. **ID** Small, compact, like tiny Peregrine with short, broad-based, pointed wings, and short tail, but has pale cheeks, narrow whitish supercilium, weak moustachial, and pale hind-collar. ♂ has pale blue-grey upperparts, rufous nape-sides, pale orange-buff underparts with dark streaks, plain grey tail with broad dark subterminal band, white at tip. Larger ♀ and juv., brown above, rather mottled, with pale supercilium, brown streaking on pale buff underparts, and narrow-barred brown uppertail with broad dark subterminal band and whitish tip. At rest, wingtips fall just short of tail tip. In flight, ♂ grey above with darker primaries and tail-band; ♀ brown with broadly banded tail. Flight in pursuit of prey extremely fast, changing direction rapidly; wingbeats fast, flickering, interspersed with short glides; soars on flat wings with tail partly spread. *F. c. insignis* and *F. c. pacificus* very similar, but *pacificus* larger and darker, and *insignis* generally paler than N American or western races. **BP** Bill black-tipped, yellow at base, cere yellow; eyes large, dark brown with narrow yellow eye-ring; tarsi yellow. **Vo** A shrill jerky series of fast *ki ki ki...* or *kui kui kui...* notes on breeding territory, and *quik-ik-ik-ik* notes in alarm. Otherwise silent.

NORTHERN HOBBY CTKJR
Falco subbuteo

L 28–34cm; WS 68–84cm; WT ♂ 131–232g, ♀ 141–340g. **SD** Iberia and N Africa to Kamchatka, winters Africa, India and S China. *F. s. subbuteo* in taiga east to Magadan, Chukotka, Kamchatka, S Russian Far East, Sakhalin and Hokkaido, N & C Korea, NE & E China; accidental Taiwan. Occurs widely on migration south of breeding range. **HH** Forested and open areas, agricultural land with trees, and locally (Hokkaido and Korea) urban areas with wooded parks; usually near water. Perches upright in open or canopy. **ID** Mid-sized dark falcon. Long, narrow, scythe-like pointed wings and mid-length dark-tipped tail. ♂ has slate head, white face, prominent slate moustachial, upperparts mid grey, flight-feathers blackish; underparts cream with bold black streaks from throat to flanks; thighs, vent and undertail-coverts rufous. ♀ has brown-tinged upperparts and dark-streaked thighs. Grey replaced by dark brown in juv.; thighs and vent buffy-orange. Extremely active low-level pursuit flight after large insects or small birds, on powerful stiff wingbeats interspersed with short glides, occasionally soars on flat wings angled back at carpal with tail spread. **BP** Bill dark grey with yellow cere; eye-ring yellow, irides dark brown; tarsi yellow. **Vo** On territory gives repetitive, agitated shrill *klee-klee-klee...* or scolding *kew-kew-kew*; vocal with young, post-breeding, and on autumn migration.

SAKER FALCON CK
Falco cherrug

L 47–57cm; WS 97–126cm; WT ♂ 750–990g, ♀ 0.975–1.15kg. **SD** E Europe to NE China. *F. c. milvipes* breeds N China, Mongolia, Transbaikalia, Inner Mongolia; on migration around Bohai Gulf (Liaoning, Hebei), vagrant Korea. **HH** Forest, steppe, semi-desert, desert with cliffs. Hunts in low flight or from perch. **ID** Large, *Buteo*-sized, with long, broad-based, blunt-tipped wings and long tail. Ad. variable, but generally head pale with whitish supercilium, poorly marked narrow moustachial stripe and somewhat darker crown; upperparts warm brown, with broad orange-buff tips giving tawny-barred appearance to back and innerwing; underparts buff with dark streaks/bars, typically broadest and blackest on sides/thighs (dark 'trousers'). Juv./1st-year like ad. but upperparts tawny-brown, moustachial more prominent, supercilium whiter, underparts boldly streaked with all-dark 'trousers'. At rest, tail extends just beyond wingtips. Flight powerful, beats slower, lazier than Peregrine, on blunter wings; soars and glides on flat wings, occasionally hovers; upperwing 'hand' dark blackish-brown, 'arm' paler brown, dark covert band on underwing contrasts with greyer remiges; upper tail has numerous pale oval spots. **BP** Bill pale grey, tip blackish-grey, base pale, cere yellow (dull greyish-green in juv.); eye-ring yellow, irides dark brown; feet yellow (ad.) or dull blue-grey (juv.). **Vo** Peregrine-like calls, a loud *kyak-kyak-kyak* or screaming *keek-keek-keek*, usually only when breeding.

GYR FALCON CJR
Falco rusticolus

L 50–63cm; WS 105–131cm; WT ♂ 0.8–1.32kg, ♀ 1.13–2.1kg. **SD** Boreal Eurasia and N America. In E Asia from Lena to Bering Strait, south through E Chukotka, Kamchatka and Commander Is, winters Sea of Okhotsk, Sakhalin, and rarely Hokkaido; vagrant NE China. Monotypic. **HH** Tundra, forest-tundra, rocky coasts, nesting on crags or coastal cliffs; in winter favours coasts, roosting on cliffs. Takes prey in air, on ground and also from water. **ID** Very large, barrel-chested, broad-bodied falcon with long, broad-based, blunt-tipped wings. Can recall large *Accipiter* or similar-sized *Buteo*. Lacks Peregrine's boldly marked face. Ad. variable, almost pure white to dark grey; mid-grey morph most common and widespread. White morph unmistakable, with black spots and bars on otherwise white upperparts, and black wingtips. Darkest has black or dark grey streaking on underparts and barring on upperparts, lacks distinguishing plumage features, but size distinctive. Juv./1st-year darker than ad., upperparts brown-tinged, underparts heavily streaked. At rest, third of tail extends beyond wingtips. Flight extremely powerful, apparently leisurely but deceptively fast, with stiff beats (slower, shallower than Peregrine), chasing down prey in level flight; also glides and soars on flat wings. In flight, underwing-coverts darker than flight-feathers in all but white morph, which has all-white underwing; tail long, broad-based, undertail-coverts full, giving impression of bulky rear end. **BP** Bill grey with yellow cere and gape (ad.) or blue-grey (juv); eye-ring yellow, irides dark brown; tarsi yellow (ad.) or blue-grey (juv.). **Vo** Vocal on territory, gruff calls recall Peregrine, but slower, more nasal and drawn-out *kak kak kak...* or *kwah kwah kwah...* and rattling *keeak keeak keeak*.

MERLIN

♀ pursuing Snow Buntings

insignis

ad ♂

ad ♀

ad ♂

ad ♀

NORTHERN HOBBY

subbuteo

ad

ad

juv

juv

hunting a dragonfly

SAKER FALCON

ad

milvipes

ad

GYR FALCON

ad white morph

ad grey morph

ad white morph

ad grey morph

white morph hunting Harlequin Duck

PEREGRINE FALCON CTKJR
Falco peregrinus

L 38–51cm; WS 84–120cm; WT (*F. p. calidus*) ♂ 588–740g, ♀ 0.825–1.33kg. **SD** Cosmopolitan; generally scarce; some populations resident, some nomadic, others migratory. *F. p. calidus* tundra/taiga Lena to Chukotka (*F. p. tundrius* replaces it in Alaska and may stray to region), Kamchatka, Russian Far East; winters NE & E China (also W Taiwan coast); *F. p. peregrinus* W Europe to NE China; *F. p. japonensis* NE Siberia, Japan, Korea, E China, but reaches SE China and has bred Taiwan; *F. p. peregrinator* Pakistan to E China. Some subspecies very local, e.g. *F. p. fruitii* of Iwo Is and *F. p. pealei* of Commander, Aleutian and Kuril Is (has reached Japan). **HH** Rocky coasts, cliffs, cities, tundra, mountains, desert and remote islands; open country, wetlands, rivers and coasts in winter. **ID** Large, compact falcon (♀ larger than ♂), dark above, pale below (e.g. *calidus*); large head with striking dark moustache (all ages/races); long, broad-based wings and mid-length, broad-based tail. Sexes/ages variable; subspecific ID difficult. Ad. upperparts dark grey, darkest on crown, nape and moustachial; rear of cheeks, chin and throat white, underparts whitish barred slate-grey, dark grey tail has many narrow paler bars. Juv. brown instead of grey, and buff rather than white on underparts. Some subspecies ad. much darker, e.g. *peregrinator* is smaller, darker above, with blackish hood and dark rufous underparts barred dark grey, and *pealei* is larger, heavier, generally browner with broader moustachial. At rest, wingtips extend nearly to tail tip. Flight powerful, fast; smooth, shallow beats interspersed with short glides on flat wings; soars on flat wings (leading edge angled, trailing edge straight); often makes spectacular stoops. In flight, head and rump appear broad, wings triangular (broad-based 'arm', 'hand' narrowly pointed), tail tapers slightly; upperwing lead grey, may contrast with paler rump and tail; uniformly barred underwing lacks contrast. **BP** Bill short, deep, mostly grey; cere yellow; eye-ring yellow, irides dark brown; tarsi yellow to yellowish-orange (ad.) or bluish-grey to bluish-green (juv.). **Vo** Generally silent away from territory, but can be very noisy giving harsh, scolding chatter or screaming *kyek-kyek-kyek* or *rehk rehk rehk* in alarm on territory, and whining *shri-shreee-shreeee*.

OSPREY *Pandion haliaetus* CTKJR

L 50–66cm; WS 127–174cm; WT ♂ 1.12–1.74kg, ♀ 1.21–2.05kg. **SD** Cosmopolitan. *P. h. haliaetus* east to Russian Far East, Sea of Okhotsk, Chukotka, Kamchatka and Hokkaido; summer visitor and possibly some resident Japan, also NE China; migrant and former breeder Korea. Winters from Honshu to E & SE China, S Korea and Taiwan. **HH** Rivers, lakes or coasts, nesting on treetops, crags and cliffs; in winter favours coasts. Rare away from water. Commonly soars, circles and hovers over water, stooping steeply then plunging feet first to catch fish. **ID** Mid-sized, long-winged, two-tone raptor. Head off-white with broad black mask; upperparts blackish-brown, underparts white with band of dark streaks across upper breast (heaviest in ♀). In flight, head prominent, tail short, wings narrow, long in 'hand' and typically angled at carpal (appears gull-like), with four primary tips separated. Underwing (ad.) largely white,

with black carpal patch and black fringes to coverts forming dark midwing bar, juv. has pale lesser and median coverts, and streaked greater coverts, thus less distinct underwing bar. **BP** Bill black with pale blue-grey base; eyes blue-grey (ad.), or orange (juv.); upper legs feathered white, lower tarsi blue-grey. **Vo** Short plaintive whistle *cheup cheup…* in display, or series of 4–20 clear notes, with whistles and chirps, e.g. *pyo pyo pyo*. Also sharp *kew-kew-kew…* or *kui kui kui…* in alarm, and longer slurred *teeeeaa* whistle.

BLACK BAZA *Aviceda leuphotes* CTK

L 28–35cm; WS 64–74cm; WT 169–224g. **SD** NE India to S China. *A. l. syama* has strayed to Taiwan, N Chinese coast (Beidaihe/Happy I.), reported Korea. **HH** Low forested mountains, broadleaf evergreen forest with streams and open areas. **ID** Largely dark, broad-winged, with unusually long erectile crest. Ad. has black upperparts with white and brown spots on back. White upper-breast-band contrasts with black hood; breast and flanks off-white with broad, orange-brown or chestnut bars, central belly and undertail-coverts black. Juv. has more chestnut barring on underparts and less white in wing. Flight crow-like; wings narrow at base, broad in 'hand', rather rounded, black above with white patch on inner wing; greyish-white undersides of primaries contrast with dark underwing-coverts and dark secondaries. **BP** Bill short, dark blue-grey; eyes blue-grey; eyes dark brown; tarsi black. **Vo** Rather vocal; 1–3 soft quavering squeals or whistles, also sibilant, downslurred whistles *chawee yuu* and more even *chu weep*.

ORIENTAL HONEY BUZZARD CTKJR
Pernis orientalis

L 54–65cm; WS 128–155cm; WT ♂ 0.75–1.28kg, ♀ 0.95–1.49kg. **SD** Transbaikalia to Russian Far East, NE China, Korea and Japan, common migrant Korea, E China, and via Taiwan to SE Asia, Philippines. Monotypic. **HH** Lowland and montane deciduous broadleaf or mixed broadleaf/coniferous forests with river valleys, to c.1,800m. Feeds on ground and at bee or wasp nests. **ID** Largish, dark, broad-winged, long-tailed, head small, neck long. Pale, intermediate and dark forms; all have small nuchal crest. Upperparts greyish-brown, face plain grey, wings and tail dark greyish-brown; underparts cream with finer dark barring on underside of flight-feathers, through rufous with chest barring, to almost entirely dark brown. Commonly has dark malar and dark gular stripe on pale throat. ♂ has grey face, ♀ brown. At rest, wingtips fall short of tail-tip. In flight, black wingtips very rounded, six primary tips distinct; no dark carpal spot; ♂ has three black lines on underside of remiges, broad dark trailing edge to underwing, tail dark with broad pale central band, ♀ has four bands on remiges, narrow trailing edge to underwing and pale tail with two narrow dark bands near base and broader subterminal band. Normal flight involves several flaps followed by long glide with wingtips depressed, soars on flat wings. **BP** Bill slender, small, blackish at tip, paler grey at base; eyes large, dark reddish-brown (♂) or yellow (♀); tarsi yellow. **Vo** High, flute-note whistle *wee hey wee hey* or *pii-yoo pii-ee*, or a whistled scream *kleeeur* during breeding season. **TN** Oriental Honey Buzzard formerly referred to *P. ptilorhynchus*, which has been split.

ad *pereginator*

PEREGRINE FALCON

ad *japonensis*

juv *pereginator*

ad *calidus*

juv *japonensis*

OSPREY

ad

haliaetus

juv

ad

ad

ad

BLACK BAZA

syama

♀ ad

ad ♂

ad

ad

ad

♀ ad

ORIENTAL
HONEY
BUZZARD

juv

ad ♂

ad plumage
variant

juv

BLACK-SHOULDERED KITE   CTJ
Elanus caeruleus

L 31–37cm; WS 77–92cm; WT ♂ 197–277g, ♀ 219–343g. **SD** Iberia south through Africa and east through India to Indonesia, Philippines and New Guinea. Previously bred in E China north to Zhejiang, currently local resident in S Fujian, on Kinmen I. (Taiwan), and increasingly recorded in E China north to Hebei; accidental S Japan. Subspecies in region *E. c. vociferus* (included by some in nominate). **HH** Tropical and subtropical regions, frequenting woodland edges, grassland, savanna, cultivation and scrub. **ID** Small, grey, white and black falcon-like kite, with rather large, somewhat owl-like head with large eyes, broad-based wings and long, notched tail. Ad. has eye patch, upperwing-coverts and underside of primaries black, contrasting strongly with pale silver-grey upperparts, clean white underparts and otherwise white underwing. Juv. appears less clean than ad., with rufous-tinged crown and breast, and white fringes to dark grey mantle, back and greater coverts, affording somewhat scaly appearance. At rest, wingtips extend beyond tail tip. In flight, wings commonly angled back at carpal, 'arm' long, 'hand' appears short. Flight light, buoyant, with fast beats, commonly hovering with legs dangling, or gliding, harrier-like, with wings raised. **BP** Bill small, black, broad-based, with yellow cere; eyes red (ad.) or greenish to dull orange (juv.); tarsi yellow. **Vo** Usually silent away from territory; weak calls include soft, high-pitched whistled *pee-oo* or *pyuu; fwit*; aggressive *kree-uk*, and *pweeip* or *wheep wheep* in alarm, and chattering *kek-kek-kek*. **AN** Black-winged Kite.

BLACK-EARED KITE   CTKJR
Milvus lineatus

L 58–66cm; WS 125–153cm; WT 0.75–1.08kg. **SD** Asia from east of Urals to Japan, Himalayas and N Indochina. Previously common and widespread throughout most of region, but absent from N Sea of Okhotsk eastwards; in local decline. Summer visitor to NE China and Russian Far East and Sakhalin, with northern birds moving short distances to winter. **HH** Lowlands and coasts with farmland, woodland and wetlands (reservoirs in Taiwan where scarce and local), but also in mountains during summer. Also near habitation where scavenges garbage. Large conspicuous tree nests mainly of large twigs and branches, but commonly incorporates plastic. **ID** Large, dark kite, with long, broad wings, and long tail with shallow fork. Ad. varies from dark, dull, blackish-brown to warmer rufous, but generally paler on head and vent, with prominent dark ear-covert patch. Underparts pale brown with heavy dark streaking. Pale buff spotting on mantle, wing-coverts and streaked underparts, especially in juv. In flight, wings held flat, often angled at carpal, but frequent adjustment to trim of wings and especially tail, often twisting and turning when soaring, tip barely notched; primary tips distinct; prominent pale bases to primaries form large, distinctive crescent on underwing (absent or reduced in smaller, extralimital **Black Kite** *Milvus migrans*). **BP** Bill weak, black tip, grey base; eyes dark to mid brown; legs feathered, feet yellow. **Vo** Highly vocal in spring when gives plaintive, tremulous, descending whistle: *pi-rrr* or *piihyorohyorohyoro pyiippippi*. **TN** Formerly placed within Black Kite.

BRAHMINY KITE   CT
Haliastur indus

L 44–52cm; WS 110–125cm; WT 520–700g. **SD** Indian subcontinent to S China and Australia; *H. i. indus* declining and probably extinct E China, as in inland SE Asia. Resident in Philippines N to Luzon; accidental to S Taiwan. **HH** Tropical and subtropical, near rivers, lakes and coasts, or forest, also towns. **ID** Mid-sized white and chestnut kite, smaller than Black-eared. Ad. has largely dark to pale chestnut plumage, except on head, neck and breast, which are white flecked grey; ♂ has bolder black shaft-streaks on hood than ♀. Imm. takes three years to reach ad. plumage; from brown with streaked chest and dark ear-covert patch in 1st-year (recalling young Black-eared, but note structural differences) to greyish-white in 2nd-year. In flight, wings broad, especially in 'arm', black primary tips contrast with chestnut wing-coverts and pale secondaries; cinnamon tail is broadly rounded, not shallowly notched, with pale tip. Juv. has square pale patch covering most of primaries (Black-eared has white crescent at base of primaries) and plain, grey tail. Glides with wings raised, soars in flat V. **BP** Bill dull greenish-yellow, cere bluish-white; eyes dark brown; tarsi dull yellow. **Vo** On territory gives plaintive mewing calls, moderately or strongly descending *nyaoww* and shorter *nyuk-nyuk*; in aggression a harsh, squealing *peeee-yah*.

WHITE-BELLIED SEA EAGLE   CT
Haliaeetus leucogaster

L 70–85cm; WS 178–218cm; WT ♂ 1.8–2.9kg, ♀ 2.5–3.9kg. **SD** Indian subcontinent to SE Asia and Australia. In E Asia, southernmost Chinese coast and offshore islands, and in Philippines to N Luzon. Accidental to Taiwan and Zhejiang. Monotypic. **HH** Islands, along coasts and estuaries, with adjacent forest. **ID** Large (albeit relatively small for sea eagle) with short diamond-shaped tail. Perches upright and prominently in trees or cliffs near water, wingtips reaching beyond tail tip. Ad. pied, with clean white head, neck, underparts and tail, and mid to dark grey upperparts, wings and rump. Juv. heavily streaked dark brown, with pale sandy-brown head and off-white belly/vent. Distinctive shape and wing action. Neck long, head small, tail short and wedge-shaped, with long broad wings bulging at innerwing, narrower at base and primaries. Ad. from below has black primaries and secondaries contrasting strongly with white underwing-coverts and body, and black base to white tail. Juv. has brown-streaked underparts and underwing-coverts, thus less contrasting, and large white panel on inner primaries, primary tips and outer primaries and secondaries brownish-black, tail greyish-white with broad dark subterminal band. Flight graceful and aerobatic, soaring and gliding easily with wings pressed forward and in pronounced dihedral, wingbeats slow and powerful, but makes spectacular dive into water after prey. **BP** Bill and cere grey; eyes dark brown; tarsi pale grey. **Vo** Loud, barking or honking *nyak nyak nyak…* or *ank ank ank*, a nasal *ka ka ka* and abrupt *kyit kyit*.

BLACK-SHOULDERED KITE

BLACK-EARED KITE

BRAHMINY KITE

WHITE-BELLIED
SEA EAGLE

PALLAS'S SEA EAGLE C
Haliaeetus leucoryphus

Vulnerable. L 72–84cm; WS 185–215cm; WT ♂ 2–3.3kg, ♀ 2.1–3.7kg. **SD** Iraq across C Asia, India, to NE China. Breeds Heilongjiang and Inner Mongolia, and migrant across NE China south to Jiangsu. Monotypic. **HH** Inland, often in dry steppe-like terrain, at lakes, marshes and large rivers. Perches for long periods immobile and upright on posts and trees. Flight sluggish. **ID** Large, pale-headed, broad-winged eagle with pied tail. Slightly smaller than White-tailed and much darker, with more protruding head and neck, slighter bill and longer tail. Ad. upperwing and underwing blackish-brown, head and mantle pale tawny, breast pale to dark brown; tail white with broad black terminal band and black base. Juv. pale to mid brown, heavily streaked, with dark ear-coverts patch; takes up to five years to acquire ad. plumage. In flight, wings long and broad, wingbeats measured, and glides on flat wings. Ad. has all-dark wings and prominently white-banded tail; juv. resembles young White-tailed but has all-dark tail, dark brown leading edge to underwing, prominent underwing-covert bar and large whitish panel in inner primaries; outer primaries and secondaries brownish-black. **BP** Bill and cere grey; eyes yellow; tarsi grey- to yellowish-white. **Vo** Quite vocal on breeding grounds, a hoarse, barking *kuok-kuok-kuok* or *kyow-kyow-kyow*, guttural *kha-kha-kha-kha* or *gao-gao-gao-gao* and nasal gull-like yelps and squeals. Usually silent in winter.

WHITE-TAILED SEA EAGLE CTKJR
Haliaeetus albicilla

L 74–92cm; WS 193–244cm; WT ♂ 3.1–5.4kg, ♀ 3.7–6.9kg. **SD** Most widespread sea eagle, from Iceland to NE Russia. *H. a. albicilla* throughout NE Russia, Chukotka, Kamchatka, Sakhalin, also Hokkaido, and locally in NE China and Korea, with some (especially young) wintering on Honshu, Korea, E China coast and Taiwan. **HH** Lakes, marshes and large rivers inland, shallow coastal lagoons and rocky coasts with cliffs; enormous stick nest is sited atop tree or crag. Perches upright, hunched, for long periods, scanning for prey, or horizontally on ground or ice. **ID** Large brown eagle. Ad. has pale yellowish-brown head, long, broadly rectangular wings, short, conspicuous baggy brown 'trousers', and short, white wedge-shaped tail. Neck, mantle and wings have pale fringes, giving 'frosted' appearance. Young more mottled; tail may have white base and dark tip. Becomes progressively 'cleaner' and tail whiter with age. At rest wingtips reach tail tip (see Steller's). Flight-feathers dark sooty-brown, underwing-coverts more chestnut-brown; young have pale band on underwing-coverts and paler axillaries. In flight, wingbeats shallow and laborious on long, broad vulture-like wings, with 5–6 'fingered' primaries, series of shallow beats occasionally interspersed with glides on flat or slightly bowed wings; frequently soars, sometimes at considerable height, with wings slightly raised above horizontal. **BP** Bill large, deep, horn to pale yellow with yellow cere (bill dark grey in imm.) and whitish lores; eyes pale yellow (ad.) or dark brown (imm.); legs feathered, feet yellow. **Vo** Rather vocal on territory, a series of gentle barks or yelps, *kyee-kyee-kyee*, or deeper stronger *kra-kra-kra* in alarm, but also calls in winter in alarm or when competing for food.

BALD EAGLE R
Haliaeetus leucocephalus

L 70–90cm; WS 180–230cm; WT 2.5–6.3kg. **SD** N America nw. to C Aleutians. *H. l. washingtoniensis* accidental Commander Is, Kamchatka Peninsula and Kolyma River. **HH** Coasts, large rivers and lakes. **ID** Large, blackish-brown, with large head. Ad. unmistakeable with white head, rump, tail and undertail-coverts. Juv. very dark, with white only in axillaries, underwing-coverts and base of tail; becomes whiter on belly, underwing, tail and upper back in 2nd-year. By 3rd-year head begins to whiten, but has dark band from ear to nape. Juv. and subad. resemble young Steller's and White-tailed, but structural differences important: neck shorter, tail longer, and has whitish-buff axillaries and contrasting buff and brown bars on underwing-coverts. Long, broad wings have pale grey panel at base of primaries on underside; younger birds have more prominent secondary bulge and extensive white mottling in brown underwing. **BP** Bill large, bright pale yellow (ad.), dark grey (juv.), with dull yellow base (2nd-year) and bright yellow base (3rd-year); eyes dark brown (juv.), dull brown (2nd-year) and off-white (3rd-year/ad.); tarsi yellow. **Vo** Calls variable but surprisingly weak; a flat whistled *kah-kah-hah* in flight or perched.

STELLER'S SEA EAGLE CTKJR
Haliaeetus pelagicus

Vulnerable. L 85–105cm; WS 195–230cm; WT ♂ 4.9–6kg, ♀ 6.8–9kg. **SD** Breeds coasts of Kamchatka, around the Sea of Okhotsk, Sakhalin, and inland short distances along major rivers. Winters S Kamchatka and Hokkaido, also to N (even C) Honshu, Sea of Japan, and S Korea; rare, declining visitor to Bohai Gulf, China; reported Taiwan. Monotypic. **HH** Rocky coasts, large coastal lagoons and near river mouths, nesting atop mature trees or rock crags; in winter ranges inland to large lakes and wetlands, also coastal lagoons; may congregate at rich food supplies (fish and waterfowl) and locally at roosts. **ID** Very large blackish-brown eagle, with unique wing and tail shape, rather long neck, large dark brown head frosted on crown and neck, with white blaze on forehead, massive yellow bill dwarfs that of White-tailed. Entire forewing, thighs, rump and large diamond-shaped tail of adult white; appears to have large baggy white 'trousers'. ♀ larger with deeper bill. Young variable, initially extremely dark with pale axillaries, underwing covert bar, and enormous pale horn-coloured bill. With age, tail, thighs and forewing whiten steadily, becoming progressively 'cleaner' and tail whiter; white forehead blaze is last to develop. Wings clearly 'fingered', with 'hand' and base both narrower than midwing, giving distinctive wing shape. In flight, beats often appear laborious, but can be extremely active, especially in aerial displays and chases, when 'rows' powerfully like a massive falcon. Soars frequently. **BP** Bill depth particularly impressive, strongly hooked, deep yellow including gape and lores; eye-ring yellow, irides pale yellow to white; legs massive, white feathered, large feet yellow. **Vo** On territory gives loud, throaty barking *kyow-kyow-kyow* or *gra-gra-gra* from perch or during aerial display; also in winter when squabbling over food or roosting site. Louder and deeper than White-tailed. **TN** Purported subspecies *H. p. 'niger'* (possibly morph, aberrant or mis-identified subadult) of Korea and S Russian Far East, all blackish-brown, lacking white except on tail. Only old specimens and records; perhaps extinct.

PALLAS'S SEA EAGLE

ad

imm

ad

juv

juv

ad

BALD EAGLE

juv

ad

ad

WHITE-TAILED SEA EAGLE

ad

ad

ad

juv

imm

juv

ad

STELLER'S SEA EAGLE

juv

ad

ad

imm

ad

juv

imm

BEARDED VULTURE CK
Gypaetus barbatus

L 94–125cm; WS 231–283cm; WT ♂ 4.5–7kg, ♀ 5.6–6.7kg. **SD** Africa, S Europe to C Asia, W & C China, Himalayas. *G. b. aureus* accidental E China and Korea. **HH** High mountains, cliffs and gorges; steppe with rocky outcrops; eats carrion and bones. **ID** Huge, unusually shaped, with long tapering wings and long, wedge-tipped tail. Ad. head whitish-buff; black band from eye to bill base extends in tuft of feathers (beard); upperparts blackish-grey with paler shaft streaks; neck and underparts cream to rusty-orange. Black hood of juv. contrasts with greyer underparts; blackish-brown with pale streaks on body. More aerobatic than other vultures; soars on flat or slightly down-turned wings. Black underwing-coverts of adult contrast with greyer flight-feathers, rusty-orange axillaries and body; juv. has ragged trailing wing-edge. **BP** Bill long, yellowish-horn with black lores; black around eyes, irides yellow (ad.) or orange (juv.); legs heavily feathered rusty-orange (ad.) or buff (juv.), feet grey. **AN** Lammergeier.

MONK VULTURE CTKJ
Aegypius monachus

L 100–120cm; WS 250–295cm; WT ♂ 7–11.5kg, ♀ 7.5–12.5kg. **SD** Iberia, Middle East, C Asia, Mongolia, NE China. Scarce, local winter visitor Korea, regular, rare winter visitor E China, accidental Taiwan, Japan. Monotypic. **HH** Mountains, forest-steppe, semi-desert; farmland with forest in winter. **ID** Enormous. Ad. head small, pale brown, with long brown neck ruff; bare grey skin (extent variable) on ear-coverts; eye patch, lores and throat dark brown. Juv. blacker, especially cap, face and ruff. In flight, long, broad wings with 6–7 long 'fingered' primaries recall *Aquila*; underwing dark; tail short, ragged, rounded or wedge-shaped; wingbeats deep, slow, soars on flat wings or with 'hand' drooping and angled back slightly. **BP** Bill heavy, tip black, dull bluish-pink base and gape (bright pink in juv.); eyes dark brown; legs heavily feathered; feet whitish-grey. **Vo** Hoarse cackling rattle on take-off, various croaks, grunting *kuwaa* or *kaa*, and hisses in gatherings at food. **AN** Eurasian Black Vulture.

SHORT-TOED SNAKE EAGLE C
Circaetus gallicus

L 62–70cm; WS 166–188cm; WT ♂ 1.2–2kg, ♀ 1.3–2.3kg. **SD** S Europe to S Asia and Mongolia. Accidental NE China. Monotypic. **HH** Dry plains and wooded hills. Perches conspicuously. **ID** Variable, generally pale, with large head. Ad. largely dark sandy- or grey-brown, mantle, scapulars and wing-coverts with dark shafts, and secondaries and tertials with pale brown fringes; underparts off-white with rusty-brown streaks on neck and chest. Commonly pale-breasted with darker hood; or, darker with dark greyish hood, narrow black bars on underparts/underwing; or, much paler, almost white-headed. Juv. similar. In flight *Buteo*-like, but body rather small, long broad wings narrow at base, underwings pale, no carpal patch, and only 3–4 wide-spaced dark bars on underside of flight-feathers instead of numerous fine bars; very pale grey tail rather long with narrow base and three dark bands; glides/soars on flat or slightly raised wings, also hovers. **BP** Bill short, weak, grey, tip black, cere yellow;

eyes orange-yellow (ad.) or yellow (juv.); tarsi pale blue-grey. **Vo** High-pitched, clear mewing whistle *kyew-lu keow-weow* recalls Mountain Hawk or Crested Serpent.

CRESTED SERPENT EAGLE CTK
Spilornis cheela

L 50–74cm; WS 109–169cm; WT 0.42–1.8kg. **SD** India to Indonesia, E China to latitude of Changjiang River (*S. c. ricketti*). *S. c. hoya* (perhaps full species) most abundant resident raptor on Taiwan; vagrant Korea. **HH** Well-wooded hills to 1,900m (>2,500m Taiwan), also lowlands, margins of agricultural land and mangroves. Often soars, but also perches openly on trees/poles. **ID** Large, dark, *Buteo*-sized with large flat-crowned, shaggy-hooded head; dark ear-coverts and crown contrast with bare yellow face. Ad. mottled grey-brown above, brown below, numerous silver-white spots form bars on flanks/belly. Juv. paler, underparts entirely cream, upperparts mottled. At rest, wingtips only reach mid tail. Wings short, broad, clearly 'fingered', flight-feathers have broad white band across bases and broad black tips; underwing-coverts orange- to dark-brown; short black tail has broad, white central band; juv. has much paler, less distinct underwing and tail pattern. Wingbeats generally slow, glides and soars, holding wings in shallow V. *S. c. hoya* smaller, much darker, with blackish cheeks and throat, plain breast and spotted underparts; juv. has dark (brown) and pale (rufous) morphs – both moult into identical ad. plumage. **BP** Bill weak, grey, tip dark, cere, lores and gape yellow; yellow eye-ring, irides yellow (ad.), white or grey-brown (juv.); tarsi dull yellow. **Vo** Gives quavering *fwee-ee-eeer* from perch. Pairs often give loud, shrill whistles and ringing *kleer-kleer-kleer*, broken *kleyeep* or highly distinctive *pi-pi-wheeeah-wheeeah* over territory.

RYUKYU SERPENT EAGLE J
Spilornis perplexus

L 50–56cm; WS 110–123cm. **SD** Endemic to Ishigaki and Iriomote Is, Japan; accidental Yonaguni I. Monotypic. **HH** Subtropical evergreen forest <500m, coastal lowlands and farmland. Often soars; perches openly on trees and poles. **ID** Small, pale, with short bushy nuchal crest. Ad. crown and loose crest black, upperparts greyish-brown with fine silver-white spotting, wings and tail darkest, blackish-grey primaries white-tipped, tail almost black with narrow white tip and broad central band; underparts paler greyish-brown with light spotting on breast and more extensive, broader grey-white ovals across lower breast, belly, flanks and thighs. Juv. pale cream with dark brown flight-feathers and tail, dark brown ear-coverts, cream supercilium, dark brown spots on crown and neck and blotches on mantle and wing-coverts, underparts largely cream with fine brown barring on thighs. Wings and tail generally as Crested; underwing-coverts greyish-brown heavily spotted greyish-white. Juv. in flight extremely pale, underwing-coverts mostly white, primaries/secondaries very pale with black tips and two narrow black bands. **BP** Bill and legs as Crested; yellow eye-ring, irides yellow (ad.) or white (juv.). **Vo** Loud, shrill *peeee piee fee feee*. **TN** Formerly within Crested Serpent Eagle.

BEARDED VULTURE

ad

aureus

ad

juv

juv

ad

MONK VULTURE

ad

juv

ad

CRESTED SERPENT EAGLE

ad *ricketti*

ad *hoya*

ad *hoya*

SHORT-TOED SNAKE EAGLE

ad pale

ad

SHORT-TOED

CRESTED

RYUKYU

RYUKYU SERPENT EAGLE

juv

ad

pale juv *hoya*

ad

WESTERN MARSH HARRIER CKJ
Circus aeruginosus

L 42–54cm; WS 115–145cm; WT ♂ 405–730g, ♀ 540–960g. **SD** Europe and C Asia, wintering in Africa and India; accidental E China, Japan and possibly Korea. Monotypic. **HH** Lakes and riverine wetlands with extensive reedbeds, and open grasslands. **ID** Large bulky harrier with long, broad, rather rounded wings; like Eastern Marsh Harrier, but ♂ has brown mantle and upperwing-coverts, grey secondaries and primary-coverts and black primaries (tricoloured upperwing); forewing white to cream, head typically very pale fawn and long tail is plain grey. ♀ extremely dark chocolate-brown with paler upperwing-coverts, creamy white crown and chin, dark eyestripe, and plain reddish-brown (unbarred) tail without white rump; underparts dark brown, except pale breast-band; axillaries and underwing-coverts dark brown, secondaries dark grey, primaries mid brown with blackish-brown tips. Juv. dark blackish-brown, with buttery-yellow crown and throat, pale brown to off-white underwing-coverts and dark brown tail, also without white rump. At rest, wingtips fall just short of tail tip. Flight as Eastern, wingbeats heavy, glides and soars on raised wings, often rocks from side to side; legs sometimes dangle and may hover. **BP** Bill short, slender, blue-grey with black tip, and yellow cere; eyes yellow (♂), dull orange (♀), or brown (juv.); long tarsi and toes yellow, visible in flight against dark underparts. **Vo** Typically silent outside breeding season.

EASTERN MARSH HARRIER CTKJR
Circus spilonotus

L 43–54cm; WS 119–145cm; WT ♂ 405–730g, ♀ 540–960g. **SD** Widespread in E Asia, in summer in NE China, Korea (at least formerly), SE Siberia, Russian Far East, Sakhalin, and Hokkaido, and wintering south through Japan, Korea and E China, passing through Taiwan to SE Asia. Monotypic. **HH** Grasslands or marshes with extensive reeds or tall grasses. Nests on ground. **ID** Heavy, bulky harrier, the size of Eastern Buzzard but slimmer with rather long tail and long, broad wings. ♂ variable but typically grey and black, somewhat resembling Pied Harrier but far less neatly marked; mantle and upperwing-coverts dark grey to black, head largely black with white streaks, black extending to neck/chest, or grey with dark face. Underparts white streaked grey or black on upper chest; underwing white, primary tips black, secondaries mostly plain pale grey sometimes with narrow black bars near tips; upperwing has black coverts (spotted white), primaries and secondaries mostly pale grey, outer primary tips black, others, and secondaries, have some faint black bars; tail unbarred grey with narrow, inconspicuous white rump. ♀ somewhat resembles Hen Harrier, generally dark grey-brown, paler on head, breast and underwing. ♀ and juv. show much individual variation. ♀ rather rufous-brown on underparts, cooler brown on face (lacks cap and dark eyestripe of Western Marsh), underwing-coverts heavily streaked mid to dark brown, flight-feathers grey with darker barring; tail has narrow grey bands, and narrow rump patch is white, though some have buff rump or no rump patch. Juv. warm brown with pale creamy brown head, breast and mantle; underwing largely dark but coverts and axillaries creamy white (see Western Marsh). In flight, wings black-tipped, with 3–4 visible 'fingers'. Wingbeats generally slow, glides frequently holding wings in pronounced V, sometimes hovers with legs dangling before dropping onto prey.

BP Bill has grey-black tip, yellow base and gape; eyes yellow (♂) or dull orange-brown (♀); long tarsi yellow. **Vo** Typically silent, but when breeding gives myuaa myuu kyuii and ke ke calls, also a high-pitched squealing pishee pishee (♂) and soft kyu-kyu (♀). **TN** Formerly considered conspecific with Western Marsh Harrier.

HEN HARRIER CTKJR
Circus cyaneus

L 42–50cm; WS 100–121cm; WT ♂ 300–400g, ♀ 370–700g. **SD** The most wide-ranging N Hemisphere harrier. Occurs across N Eurasia, wintering to S Europe, N Africa, Middle East, locally in northern SE Asia and E Asia. In E Asia, breeds NE China, E Siberia, Russian Far East to Chukotka and Kamchatka (?), and winters through Japan, Korea, E China south of Changjiang River and occasionally Taiwan. Monotypic. **HH** Bogs in taiga, moorland; open marshes, grassland and agricultural habitats in winter; habits similar to marsh harriers. **ID** Smaller and slighter than marsh harriers but larger than all others, with typical long, broad-tipped wings, but broader in 'arm' than 'hand', and long tail. ♂ mid grey above, with grey hood, broad black wingtips, pale grey primaries, secondaries and tail, broad white rump and white underparts below grey chest; underwing white with black primary tips and broad, dark grey trailing edge to inner primaries and secondaries. ♀ dark brown with heavily streaked underparts and pale-fringed upperparts, clearly barred wings and tail both above and below, prominent white rump and pale brown underparts streaked darker. Tail has broad dark grey subterminal band and 3–4 bands on mid tail. Juv. dark-headed, with somewhat orange-brown body and underwing. ♀ and juv. have owl-like facial disc and pale collar. In flight, long-winged rowing action is slow and smooth, glides on flat or slightly raised wings, frequently rocking from side to side revealing white rump. **BP** Bill has grey-black tip, yellow cere and gape; eyes orange-yellow (♂), brown to pale yellow (♀); long tarsi yellow. **Vo** Usually silent, but on territory gives whistled piiyo piiyo and chattering ke ke ki ki. ♂ performs aerial display flight over territory, during which it utters a deeper, rapid series of dry barks, chuk-uk-uk-uk, ♀ responds with softer, higher kek-ek-ek-ek.

NORTHERN HARRIER J
Circus hudsonius

L 41–50cm; WS 97–122cm; WT ♂ 290–390g, ♀ 390–600g. **SD** N American sister species of Hen Harrier that has reached Japan (Hokkaido and Honshu). Monotypic. **HH** Open habitats, marshes and boreal forest. **ID** Mid-sized harrier with rather broad, rounded wings and long narrow tail. Ad. ♂ essentially like Hen, but has five black primary tips (Hen has six) covering a smaller area, and trailing edge to secondaries black (not grey). Juv. dark-headed, with somewhat orange-brown body and underwing, recalling Montagu's, almost orange breast and upperparts (unstreaked), but lacks bold face markings of young Montagu's and Pallid (see p.130). **BP** Bill has grey-black tip, yellow cere and gape; eyes yellow (♂), or brown to yellow (♀); long legs (10% longer than Hen) and feet orange-yellow. **Vo** Usually silent away from breeding grounds. **TN** Formerly considered Nearctic race of Hen Harrier. **AN** American Harrier.

WESTERN MARSH HARRIER

ad ♂
ad ♂
ad ♂
ad ♀
ad ♀

EASTERN MARSH HARRIER

ad ♀
ad ♂
juv
juv
ad ♀
ad ♀
ad ♀

HEN HARRIER

ad ♂
ad ♂
ad ♀

NORTHERN HARRIER

ad ♂
ad ♀
ad ♂
imm ♂
ad ♀
ad ♀
juv

PALLID HARRIER C
Circus macrourus

L 40–50cm; WS 100–121cm; WT ♂ 235–415g, ♀ 400–550g. **SD** Largely extralimital, C Asia to W China north to Lake Baikal, but has strayed to Hebei and Jiangsu, China. Monotypic. **HH** Favours open country, dry grassland and steppe, but also cultivation and marshes; habits similar to other harriers. **ID** Elegant, very similar to Montagu's, or like small, slim, pale Hen (p.128); long wings narrow to rather pointed tip. ♂ very clean, pale grey, may appear white, head especially pale grey, throat and chest white (lacks Hen's hooded appearance). Longer legged than Montagu's; ♀ has paler tail and less clearly marked underwing. ♀ has bolder head pattern with dark eyestripe, dark ear-coverts and paler, more prominent supercilium, and more noticeable narrow hind-collar than Montagu's; flanks and 'trousers' have diffuse broad double rufous spots, whereas Montagu's has narrower, neater, rufous streaks. Juv. unstreaked orange-buff on underparts and underwing-coverts, with bolder face pattern (pale tawny collar contrasting with dark neck and white patch below eye). In flight, ♂ upperwing pale grey (lacks Montagu's black wingbar) with only four outer primaries solid black; underwing white with black outer primaries forming distinct dark wedge; tail pale grey with mid grey barring, white rump narrow and indistinct. ♀ underwing has pale, irregularly barred primaries, contrasting with darker secondaries which have very narrow grey bars; lacks Montagu's prominent coarse barring on axillaries, instead dark with pale spots; rump white. Flight buoyant, light with rapid beats and wavering glides (wings in shallow V). **BP** Bill yellow with black tip, yellow cere; eyes pale yellow (grey with brown ring in young); long tarsi yellow. **Vo** High-pitched tremulous *pirrr* in display and loud chattering *gik-gik-gik* in alarm, with ♂ also giving chuckling *chack-ack*; but generally silent away from breeding areas.

PIED HARRIER CTKJR
Circus melanoleucos

L 43–50cm; WS 110–125cm; WT ♂ 254–325g, ♀ 390–455g. **SD** Mainly restricted as breeder to NE Asia, winters south to India and Borneo. In E Asia, breeds NE China and adjacent Russian Far East only, especially Amur region; not uncommon migrant through E China and Korea, scarce Taiwan, and rare or vagrant elsewhere, e.g. Japan (has bred). Winter range includes SE China south of Changjiang River. Monotypic. **HH** Open dry steppe, damp river valleys, swamps, lakes and coastal marshes, also rice fields and reedbeds; habits as other harriers. **ID** Small, slightly built, slender-winged harrier. ♂ unmistakably elegant, with black head, upper breast and mantle, narrow black band across upperwing on upper median coverts and black outer primaries; shoulders white and rest of wing very pale grey; tail grey; underparts white, underwing white with black primaries, secondaries all white at tips (see Hen, p.128). Can recall larger Eastern Marsh Harrier ♂ (p.128), but far more cleanly marked. ♀ resembles ♀ Hen, but greyer, shares white rump, but is darker, more contrasting, with dark greyish-brown upperparts and wings, with pale forewing and dark grey, barred brown remiges; tail grey with 4–5 fine dark bars, the subterminal band being darkest; underwing much paler than ♀ Hen, with pale underwing-coverts and brown streaking, remiges pale with narrow dark grey barring. ♂ resembles ad. ♀, but much darker brown upperwings and tail and narrower white rump patch, underparts and underwing-coverts more rufous than ad. ♀. Juv. ♀ is largely dark brown with whitish nape, streaked head, buff rump, no grey in wings and dark, banded tail. Flight lighter than Eastern Marsh or Hen, but also glides and soars on V-shaped wings. **BP** Bill has grey-black tip, yellow base and gape; eyes yellow (♂ bright, ♀ dull) or dark brown (juv.); long tarsi yellow. **Vo** Usually silent except on territory where ♂ gives lapwing-like *kii bii* or *kiiy-veeee* and ♀ an abrupt *kee-kee*; also rapid chattering *wek-wek-wek* or *chak-chak-chak-chak* in alarm.

MONTAGU'S HARRIER CJ
Circus pygargus

L 39–49cm; WS 102–123cm; WT ♂ 227–305g, ♀ 254–445g. **SD** Vagrant in winter to E China from C Asian and European range, and reported Japan; normally winters in Africa and India. Monotypic. **HH** Open country, grassland, scrub, crops and young forest; habits as other harriers. **ID** Small, slender and very long, narrow-winged harrier; wings reach tail tip when perched (in most harriers wingtips fall nearly to or short of tail tip). ♂ resembles ♂ Hen Harrier (p.128), but black wingtip narrower, has unique black bar across secondaries, and distinct contrast between dark mantle and paler grey flight-feathers; upperwing darker than Pallid and more extensive black on primaries; underwing darker than Hen with more extensive black on underside of primaries and outer secondaries, two black bars on secondaries, and variable extent of rufous streaking on underwing-coverts and flanks. ♀ resembles ♀ Hen and Pallid, but much slighter than Hen with inconspicuous narrow white rump, black band on secondaries, underwing cleanly marked with dark trailing edge and clear blackish bands on secondaries, several narrower bands on underside of primaries. Juv. has unstreaked orange-buff underparts and underwing-coverts, with dark secondaries and trailing edge, pale primaries with several black bars, black outer primary tips, and paler face with broader supercilium, reduced dark ear-coverts and crucially lacks distinct dark rear neck and pale collar (distinguishing it from young Pallid). In flight, more agile than Hen, narrower wings (especially wingtips, with only four 'fingers') and more buoyant flight than Hen. Dark phase unknown in E Asia. **BP** Bill yellow with black tip; eyes pale orange (ad. ♂), yellow (♀) or brown (juv.); tarsi yellow, legs shorter than other harriers, more horizontal stance. **Vo** Thin downslurred whistle, *squeah*, in display, slow *chink chink ...* or more rapid *shik-shik-shik* and, in alarm, a sad *kek-kek-kek* or rapid *chit-er chit-er chit-it-it-it*, but usually silent away from territory.

PALLID HARRIER

ad ♂

ad ♂

imm ♂

juv

ad ♀

PIED HARRIER

ad ♂

ad ♂

juv ♂

ad ♀

ad ♀

MONTAGU'S HARRIER

ad ♂

ad ♂

juv

ad ♀

ad ♀

CRESTED GOSHAWK *Accipiter trivirgatus*   CT

L 30–46cm; WS 54–79cm; WT 224–450g. **SD** Primarily S & SE Asia to Indonesia and Philippines: *A. t. indicus* scarce resident E China (e.g. Zhejiang); *A. t. formosae* resident Taiwan. **HH** Forested (generally broadleaf) montane areas to <1,000m. In Taiwan, also secondary growth, urban parks, gardens and lowlands. Still-hunts from perch. **ID** Large, powerful, between Northern Goshawk and Eurasian Sparrowhawk (see p134); prominent dark mesial on white throat, small, grey nuchal crest and heavily barred underparts distinctive. Ad. has grey head and crest, brown mantle and wings; breast heavily streaked dark brown; chest, belly and flanks boldly barred brown on cream. Juv. browner, lacks crest; broad brown spots on neck and breast, narrower brown bars on thighs. At rest wingtips reach just beyond tail base. Flight agile; most often seen soaring. Wings appear broader and rounder than congeners, and narrower at base. In flight, underwing-coverts pale with brown spots, remiges clearly barred, with 4+ dark bars on pale background; tail from above or below has three broad dark bars on grey background; vent feathers fluffy white. *A. t. formosae* averages marginally larger than *A. t. indicus*. **BP** Bill heavy, grey with yellow cere; eyes orange (ad. ♂), yellow (♀) or brown (juv.); legs short, sturdy, yellow. **Vo** Usually silent, even in aerial display, occasionally gives shrill, screaming *he he hehehe* and yelping whistles *fyeew-fyeew-fyeew*.

CHINESE SPARROWHAWK   CTKJ
Accipiter soloensis

L 25–30cm; WS 52–62cm; WT 140–204g. **SD** Breeds across E China, Korea and S Russian Far East; northeastern birds migrate via Korea, Kyushu and Nansei Shoto, S Japan, and Taiwan, to SE Asia and Philippines, others cross Yellow Sea from N Korea via Shandong Peninsula to E & S China, thence south to winter range. Monotypic. **HH** Forested/wooded country, and open fields near wooded hills. Locally common migrant (May, Sep/Oct); often in flocks. **ID** Pale, mid-sized *Accipiter* with plain grey head (no supercilium) and upperparts, darker wings, mantle and tail, wingtip rather pointed and tail rather short. Underparts white with pale rufous wash or bars. Young similar to Japanese Sparrowhawk, but darker brown with blackish-brown crown, grey face lacking supercilium, and distinct underwing pattern. Flight fast and agile; often soars. In flight, ad. has narrow, rather pointed black wingtip (unique in *Accipiter*), contrasting strongly with mainly white, unbarred underwing with grey trailing edge; undertail very pale with several narrow dark grey bars; tail proportionately short. Juv. in flight recalls juv. Eurasian Sparrowhawk, but breast and belly more rufous and underwing-coverts very pale buff and mostly plain; underside of flight-feathers pale with very narrow dark barring. **BP** Bill grey, tip black, swollen cere orange-yellow, reaching forehead and prominent on plain grey face; eyes dark brown to dark red (ad.) or yellow (juv.); tarsi bright yellow-orange. **Vo** On territory high-pitched *kyii ki ki*, lower pitched and less lilting than similar Japanese, also fussy, staccato *kiwit-kiwit-kiwit*, but typically silent away from breeding grounds. **AN** Chinese Goshawk; Horsfield's Sparrowhawk.

JAPANESE SPARROWHAWK   CTKJR
Accipiter gularis

L 23–30cm; WS 46–58cm; WT ♂ 92–142g, ♀ 111–193g. **SD** Summers west of Lake Baikal to Sakhalin, Japan, Korea and NE China. Migrates/winters E China, SE Asia, Philippines, Borneo and Sulawesi. *A. g. sibiricus* SE Yakutia to Sea of Okhotsk, migrates to SE Asia; *A. g. gularis* Russian Far East, NE China, Sakhalin and Japan, moves to SE China and Philippines; *A. g. iwasakii* resident S Nansei Shoto (Ishigaki and Iriomote). **HH** Breeds mainly in deciduous forests, also urban parks; migration/winter, subtropical and tropical forests. **ID** Smallest *Accipiter* in region. *A. g. gularis* resembles small, plain Eurasian Sparrowhawk, with long primary projection and four narrow dark bands on largely pale tail (terminal broadest). ♂ dark bluish-grey on face and upperparts; underparts whitish, breast, flanks and belly variably washed and barred brick red. ♀ browner above, white below with dark rufous barring, throat white with faint gular stripe. Juv. browner with rufous streaks on lower neck, barred flanks and breast. *A. g. iwasakii* darker grey on face and head, browner on wings and tail. At rest, primary projection long and tail has pale bars obviously broader than dark bars. Flight fast and agile, often soars; like larger Eurasian, but wings shorter, 'arm' broader, tips narrower and more pointed, tail shorter. Field ID of subspecies uncertain, but *iwasakii* averages smaller and darker above than *gularis*, underside browner; wings blunter and tail more densely barred. **BP** Bill blue-grey with black tip, cere greenish-yellow (young) to yellow (ad.); eye-ring yellow; eyes yellow (♀/young ♂), darkening to orange-yellow (*iwasakii*) or dark red (ad. ♂ *gularis*); tarsi greenish-yellow (young) to yellow (ad.). **Vo** High-pitched chattering *kyik kye-kye-kye*, often given on migration, also sharp, ringing *kyip-kyip*. **TN** Formerly part of Besra.

BESRA *Accipiter virgatus*   CT

L 24–36cm; WS 42–65cm; WT 112g. **SD** India and SE Asia to Indonesia and Philippines. *A. v. affinis* in SE China; *A. v. fuscipectus* in forests of Taiwan. **HH** Broadleaf evergreen montane forest and open woodland. **ID** Small *Accipiter*, recalls Crested Goshawk (CG), but some as small as Japanese Sparrowhawk (JS), others as large as Eurasian (ES), and lacks nuchal crest. Prominent dark gular on white throat distinguishes it from JS and ES; short primary projection and broader darker tail-bands from JS and underparts pattern – streaked lower neck, darkly barred breast and flanks – from ES. ♂ dark slate-grey above, heavily streaked and barred dark rufous below. ♀/juv. browner, but share prominent gular stripe (JS and CG have weak or no gular stripe). ♀ dark chestnut streaks on neck/breast and heavy chestnut barring on belly and flanks. Juv. looser rufous-brown streaks on breast and teardrop spots on flanks and thighs. *A. v. fuscipectus* larger, with darker, browner upperparts. Primary projection short; dark tail bars broader than pale bars. In flight, wing short, bluntly pointed at tip; heavily barred axillaries and underwing coverts, dark-barred flight-feathers, tail short with broader dark bars than congeners. **BP** Bill dark grey at tip, paler at base, cere greenish-yellow to yellow; eyes red (♂) or yellow-orange (♀); legs long, slender, yellow. **Vo** On territory noisy, piercing series of accelerating, descending *tchiew tchew-che-checheche*, squealing *ki-weeeer* or rapid *tchew-tchew...* notes.

ad ♂ *indicus*

CRESTED GOSHAWK

ad ♂ *formosae*

ad ♂ *indicus*

juv

juv

juv

CHINESE SPARROWHAWK

ad

juv

juv

ad ♀

juv

ad

JAPANESE SPARROWHAWK

gularis

ad ♂

ad ♂

juv

ad ♀

ad ♀

BESRA

ad ♂

affinis

ad ♂

juv

juv

ad ♀

ad ♂

EURASIAN SPARROWHAWK CTKJR
Accipiter nisus

L 28–40cm; WS 56–78cm; WT (*A. n. nisus*) ♂ 105–196g, ♀ 185–350g. **SD** Common from W Europe to Kamchatka. European, Korean and Japanese birds mostly resident, but those in taiga largely or partially migratory, wintering south to India. In E Asia, *A. n. nisosimilis* breeds across Russian Far East to Chukotka, Kamchatka, NE China, Korea and Japan, with northern birds migrating through region in large numbers, wintering in E & S China; migrant Taiwan. **HH** Well-forested montane areas and wooded lowlands, also wooded urban areas and parks, generally in forest or at edges. **ID** Medium-sized *Accipiter* with rather small head, short, broad, blunt wings and long tail (with four dark bands from above). ♂ small, with slate- to blue-grey upperparts, plain face, rufous cheeks and rufous-barred breast. ♀ large and resembles Northern Goshawk, but slimmer, with dark grey upperparts, white underparts barred grey (lacks gular stripe of Besra, Japanese and Chinese sparrowhawks). Large ♀ confusable with Northern Goshawk ♂, but has shorter 'arms' and longer 'hands', tail longer, narrower at base with squarer corners and even slight notch, relatively smaller head, less prominent white vent. Primary projection moderate. In flight, may appear pigeon-like, with gentle undulations on rapid, light flaps interspersed by short glides; less direct than Northern Goshawk, but agile in pursuit of prey, often very close to ground, but also soars over canopy and high on migration. Underwing, incl, coverts and axillaries, rather strongly barred, tail has three bold dark bars below, and is rather narrow, longer than width of wings, thus wings appear narrower, and tail longer compared with Japanese Sparrowhawk (p.132). Juv. has brown-barred underparts with little or no streaking. **BP** Bill dark grey with yellow cere; eyes yellow-orange to red (♂), or yellow to orange (♀); legs greenish-yellow to yellow, feet yellow. **Vo** Generally silent, except when breeding, when gives chattering or cackling *ki ki ki*; *kee kee kee* or *kekekekeke*, higher pitched than Northern Goshawk.

NORTHERN GOSHAWK CTKJR
Accipiter gentilis

L 46–63cm; WS 89–122cm; WT (*A. g. albidus*) ♂ 0.894–1.2kg, ♀ 1.3–1.4kg. **SD** Very wide range across Holarctic, largely resident, but northern birds reach C Asia and southern E Asia in winter. *A. g. schvedowi* resident S Urals to Transbaikalia, Kuril Is, Russian Far East, Sakhalin and NE China. Probably breeds Korea, but race uncertain; *A. g. albidus* from Lena across Chukotka and Kamchatka, and has straggled to Japan; *A. g. fujiyamae* in N & C Japan; winters in Korea and E China; accidental Taiwan and Attu (USA), presumably Commander Is. **HH** Forests and plantations. Rather shy, views often distant, but can be territorial and aggressive near nests; in winter may visit areas with concentrations of waterfowl. **ID** Very large, powerful *Accipiter*, reaching size of Eastern Buzzard (p.136). Deep chest, long but very broad wings, and long, broad tail, longer neck, more protruding head than Eurasian Sparrowhawk, lacks crest and mesial throat-stripe of several congeners, but has dark crown and ear-coverts, and very prominent white brows. Ad. upperparts rather dark grey (some brown tones), underparts

white with fine dark grey barring from breast to belly, undertail-coverts long, white, often conspicuous; grey tail has three broad dark bands; ♀ larger. Juv. brown upperparts and buff underparts heavily streaked brown. Rather long primary projection and long undertail-coverts. Flight aggressive, turns and accelerates with remarkable speed in pursuit of prey, also glides and soars over territory; differs from Eurasian Sparrowhawk – short series of relaxed flaps followed by straight glide without losing height. Wingbeats stiff, powerful and steady, wings bulge at secondaries, 'hand' somewhat narrower at tip and clearly 'fingered'. In flight, ad. broad-bodied and plain grey from below; juv./1st-year densely streaked dark brown on buff underparts, underwing narrowly barred, upperwing also barred and has pale bar on greater coverts. Tail broad-based and rounded at tip, long white undertail-coverts bulge and may wrap around rump-sides; tail commonly spread when soaring. *A. g. fujiyamae* smaller, darker above and more strongly barred below than *A. g. schvedowi*. *A. g. albidus* differs in being very pale, almost white, with pale brown markings. *A. g. schvedowi* typical of northeast of region, but larger *A. g. albidus* reaches NE China and, rarely, Japan in winter, and is separable by its extremely pale grey upperparts and almost white underparts; juv. buffy-white with pale to dark brown streaking, brown tail and remiges, and whitish underparts with dark brown vertical streaks. *A. g. fujiyamae* occurs in Taiwan in winter. **BP** Bill well hooked, grey tip, greenish-yellow cere; eyes orange-red (♂), bright orange-yellow (♀) or pale yellow (juv.); legs short and thick, yellow. **Vo** Generally silent, except when breeding, when gives stronger, deeper, chattering *kya kya kya kya*, *kee kee kee*, or *kyik-kyik-kyik* than Eurasian Sparrowhawk.

GREY-FACED BUZZARD CTKJR
Butastur indicus

L 41–48cm; WS 101–110cm; WT ♂ 375–433g. **SD** Common, but locally declining, migrant to E Asia, summering (Apr–Oct) in NE China, Korea (at least formerly), immediately adjacent Russian Far East and N Korea, and Japan south of Hokkaido. Winters to SE Asia, Philippines and Indonesia. Some winter in Nansei Shoto and Taiwan. Monotypic. **HH** Coniferous and mixed evergreen forests in mountains, at forest edge, also around agricultural land. On migration, common and often in flocks. **ID** Mid-sized, rather slender, buzzard-like hawk. Ad. varies from mid-grey to dark chocolate-brown, all but darkest (almost entirely black) forms have grey-brown head, grey cheeks and white chin, with black gular stripe; crown flat, white supercilium prominent. Upperparts and breast typically mid to dark brown, lower breast and flanks brown barred white. Juv. more heavily streaked and has more contrasting head pattern. Wings and tail longer and narrower than Eastern Buzzard (p.136), resembling large sparrowhawk, underwing pale with fine dark barring, tail has three dark bars and is square-cut; pale U on uppertail-coverts; flight generally rather laboured, but on migration often appears quite buoyant, with almost falcon-like glides. **BP** Bill reasonably long with prominent hook, basal half yellow, cere orange-yellow; eyes bright yellow (ad.) or brown to pale yellow (juv.); tarsi dull yellow. **Vo** Rather vocal year-round, especially prior to nesting. Call, while soaring or perched, a whistled *whick-awee* or tremulous, drawn-out, but strong, *pik-wee; pik-kwee* or *kin-mii kin-mii*, the second part prolonged and upslurred. **AN** Grey-faced Buzzard-eagle.

EURASIAN SPARROWHAWK

ad ♂

nisosimilis

ad ♂

ad ♀

juv ♀

ad ♀

juv ♀

NORTHERN GOSHAWK

ad ♀
schvedowi

juv ♀

ad ♀
albidus

ad ♂
albidus

ad ♂
schvedowi

ad ♂
fujiyamae

juv ♀

ad ♀ *schvedowi*

ad

juv

juv

ad

GREY-FACED
BUZZARD

EASTERN BUZZARD *Buteo japonicus*   CTKJR

L 40–52cm; WS 109–136cm; WT ♂ 630–810g, ♀ 515–970g. **SD** *B. j. japonicus* from Lena to Sea of Okhotsk, Russian Far East, Sakhalin, Japan, NE China; winters Japan, S Japan, E & S China. Some northern birds migrate; *B. j. toyoshimai* endemic Ogasawara Is; *B. j. oshiroi* endemic to Daito Is (status uncertain). **HH** Forest, well-wooded hills. **ID** Compact, broad-winged, typically dark brown above and very pale below, with small bill and feet. Ad. variable, but head typically paler than upperparts, underparts off-white with prominent dark brown patches on sides/thighs extending to lower chest and contrasting with pale belly and vent. Superficial resemblance to Rough-legged, especially in winter when reflection off snow makes Eastern appear whiter below, but upper tail uniform. Juv. more heavily streaked below. Wingbeats slow, stiff, glides on flat wings or with lowered 'hands', commonly soars with wings in shallow V, rides thermals, even hovers for short periods. Underwing flight-feathers pale, coverts pale brown, large carpal patch and primary tips black; undertail pale with fine grey bars. *B. j. toyoshimai* smaller, shorter winged and usually paler than *B. j. japonicus*, with only small brown patches on underside. *B. j. oshiroi* smaller, darker, redder and shorter winged. **BP** Bill small, weak, tip grey, cere yellow; eyes brown (ad.) or pale grey to pale brown (juv.); tarsi yellow. **Vo** Call, given in aerial display, a loud mewing *kiiii-kiiii* or *peeyou*; may be heard year-round.

ROUGH-LEGGED BUZZARD   CTKJR
Buteo lagopus

L 45–62cm; WS 120–153cm; WT (*B. l. menzbieri*) ♂ 1.29–1.38kg, ♀ 1.24–1.43kg. **SD** Across N Holarctic. In E Asia from Lena to Bering Strait, Kamchatka and Sea of Okhotsk coast; winters Russian Far East, south to Japan (scarce; mostly Hokkaido), Korea, and NE China, occasionally Taiwan. Exact borders of *B. l. menzbieri* and *B. l. kamtschatkensis* ranges are unclear; *kamtschatkensis* occurs around N Sea of Okhotsk, Kamchatka and on Kuril Is, with *menzbieri* further north and west. Most Japanese records probably *menzbieri*, though *kamtschatkensis* has also occurred, most NE Chinese winterers are *kamtschatkensis*. **HH** Montane and coastal tundra, sparse taiga, and in winter steppe, mountains, grassland, farmland, wetlands including coast. **ID** Large, long-winged, long-tailed buzzard; larger and paler than Eastern. Ad. grey-brown, with very pale head, variable brown streaking on throat and neck, dark flank patches may merge across belly; prominent black carpal contrasts with whitish primary bases, upperparts dark, contrasting with white tail with black subterminal band (juv. tail extensively grey-brown, whitish only at base); narrow black eyestripe. ♂ has darker head and pale breast. ♀/juv. have paler head, can appear bright white. Wings long, but broad-tipped (reaching tail tip at rest). Flight slow, often low, on shallow beats (resembles large harrier); hovers frequently; glides on flat wings but may soar on slight dihedral. Upperwing largely dark, but pale band across coverts, large pale area formed by most of primaries being grey-white, and most of uppertail white. Underwing pale, largely white, but prominent dark carpal, flight-feather tips (forming marginal band); undertail white with 2–3 prominent dark terminal bands (♂) or one (♀).

Larger *kamtschatkensis* very similar to *menzbieri*, but throat and upper breast darker, giving hooded appearance; flank patches paler, less extensive; pale panel on upper surface reduced, almost absent, with extensive white tips to body, mantle and scapulars. Dark morph unknown in E Asia. **BP** Bill has grey tip with yellow cere; eyes yellow or brown (ad.), or pale to brownish-grey (juv.); legs feathered, toes yellow. **Vo** On territory, gives strong mewing *piiyororo*, *piiee* and thin descending *vyeeeuuw* or *mee-oo*, louder than Eastern, but more downslurred. Alarm call loud and high-pitched *pi-i-aay*. **AN** Rough-legged Hawk.

UPLAND BUZZARD *Buteo hemilasius*   CTKJ

L 57–67cm; WS 143–161cm; WT ♂ 0.95–1.4kg, ♀ 0.97–2.05kg. **SD** Tibetan plateau, Mongolia, C & NE China, possibly N Korea and adjacent Russia. In E Asia breeds only in NE China; scarce in winter Korea, accidental Fujian, Taiwan and Japan. Monotypic. **HH** Steppe and desert in summer, hills, open country and farmland to sea level in winter. **ID** Larger with more eagle-like proportions than Eastern – wings and tail longer, broader. May also recall Rough-legged. Ad. pale morph has whitish head with dark moustachial streak, grey-brown upperparts; underparts whitish with large blackish-brown flank patches, thighs and large black carpal patch (resembling Eastern). Underwing-coverts pale brown, wingtips and trailing edge to underwing black; large grey-white patches at bases of primaries 'window' on both surfaces of wing, tail greyish-white finely barred near tip. Dark morph is blackish-brown, with pale underside to flight feathers. Tail pale grey-brown with fine barring (lacks subterminal band of Rough-legged); only distinguished from dark Eastern by size/structure. Juv. like dark-morph ad. Flight slow, wingbeats deep, glides with 'arms' raised and 'hands' flat, soars with wings in deep V, also hovers. **BP** Bill blue-grey, cere greenish-yellow; eyes whitish or yellowish to golden-yellow; tarsi greyish-yellow to yellow. **Vo** Mewing calls resemble Eastern, but longer and more nasal *piiyoo piiyoo*; also short, nasal yapping calls.

INDIAN BLACK EAGLE *Ictinaetus malayensis*   CT

L 65–80cm; WS 148–182cm; WT (*I. m. perniger*) 1–1.6kg. **SD** S India, Sri Lanka to SE Asia, Indonesia. *I. m. malayensis*: rare resident Taiwan, Fujian and N Guangdong. **HH** Montane evergreen forests (150–2,500m). **ID** Large, unspotted, very dark eagle; slighter, more elongated than *Aquila* eagles, somewhat kite-like with relatively small head and bill, and long wings and tail. Ad. all black except narrow pale grey tail-bands (sexes similar, though ♀ considerably larger). Juv. mostly dark blackish-brown, but has buffy-brown head, underparts and underwing-coverts, all streaked blackish; rump and tail barred. At rest, wings extend to or beyond tip of tail. Wings narrow in 'arm', broad in 'hand' with 7–8 distinctly 'fingered' widespread primaries. Flight light, slow with 'rowing' action, often with tail fanned, soars low over forest canopy, with wings flat or raised in very shallow V. In flight, underwing has pale bases to primaries. **BP** Bill rather small, grey with obvious yellow cere and long gape; eyes dark brown (ad.) or reddish-brown (juv.); legs long, feathered, yellow toes contrast with very dark plumage. **Vo** Occasionally gives plaintive *kleeee-kee* or *hee-lee-leeeuw*, and shrill *kip kip kip* in display.

EASTERN BUZZARD

ad

juv

japonicus

ad

ROUGH-LEGGED BUZZARD

kamtschatkensis

ad

juv

ad

ad

juv

UPLAND BUZZARD

ad

juv

ad

ad pale

ad dark

juv

INDIAN BLACK EAGLE

malayensis

juv

ad

ad

juv

GREATER SPOTTED EAGLE CTKJR
Aquila clanga

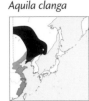

L 59–71cm; WS 157–179cm; WT ♂ 1.7–1.9kg, ♀ 1.8–2.5kg. **SD** Baltic to Sea of Okhotsk, winters Africa, S & SE Asia. In E Asia breeds NE China and S Russian Far East, winters SE China; scarce Taiwan, accidental Korea, S Japan. Monotypic. **HH** Taiga forest, forest-steppe with wetlands; winters at wetlands or farmland/forested areas. **ID** Large, dark, compact eagle; flat-crowned head with spiky nape-feathers; wings very broad, tail broad, shorter than Golden or Eastern Imperial with rounded tip. Ad. has whitish crescent on uppertail-coverts, and pale bases to primaries. Wings broad, dark, 'arm' short, secondaries bulge slightly, 'hand' broad with 6–7 long prominent 'fingers'. Juv. creamy white spots on upperparts, double wingbars, and off-white vent; rarer pale '*fulvescens*' imm. strongly contrasting dark flight feathers and pale tawny head, body and wing coverts. Juv. from below like ad., but remiges narrowly barred; '*fulvescens*' has pale buff underwing-coverts and dark midwing bar. At rest appears hunched; wingtips almost reach tail tip. Flight slow, soars with wings flat or slightly arched at carpal. **BP** Bill rather small, dark grey, cere and short gape line (to eye) yellow, with round nostrils (slit-like in other *Aquila*); eyes dark brown; thighs heavily feathered, legs closely feathered, toes yellow. **Vo** Gives staccato yelps: *kleep kleep kleep*; *kyui kyui* or *pyo hyohyohyo* on territory and in winter. During undulating and swooping aerial display gives repetitive plaintive, dog-like bark *kiak-kiak-kiak…* or *kyow kyow kyow…*

STEPPE EAGLE *Aquila nipalensis* CKJ

L 60–81cm; WS 165–214cm; WT (*A. n. nipalensis*) ♂ 2.5–3.5kg, ♀ 2.3–4.9kg. **SD** E Europe to SE Siberia, winters Africa, S & SW Asia. In E Asia *A. n. nipalensis* breeds Inner Mongolia, Hebei, SE Siberia, wintering south to Guang-dong. Has reached Korea and Japan. **HH** Steppe, semi-desert, wooded hills and open grasslands. **ID** Large mid to dark brown *Aquila*, with rather large head/neck. Five, age/plumage classes. Ad. has rufous nape and pale chin, otherwise dark with pale fringes to mantle and wing-coverts, and prominent baggy 'trousers'. Dark brown underwing-coverts; unlike Greater Spotted has dark blackish-brown trailing edge and fine pale-barred dark flight-feathers, all dark-tipped; tail also narrowly barred with dark terminal band. Juv. greyish-brown with broad white bars on tips of greater coverts and secondaries, underparts unstreaked; wing pattern distinctive, a broad white mid-underwing bar; two narrower white bands on upperwing-coverts; trailing edge to wing and terminal tail-band white; uppertail-coverts also have narrow white band; white underwing bar and trailing edge less prominent with age. At rest, wingtips reach or extend beyond tail tip. Flight slow, sluggish, wingbeats heavy; long broad wings straight and level in powered flight, but glides and soars on slightly bowed wings with well-spread 'fingers' (more prominent than Greater), drooping below 'wrist'. **BP** Bill large, dark grey, cere and long gape line yellow (extends to rear of eye); eye brown under heavy brow; thighs and legs feathered, large feet yellow. **Vo** Breeding birds give deep, hoarse, barking *kau kau*, muffled crow-like croaks, and loud whistle.

EASTERN IMPERIAL EAGLE CTKJ
Aquila heliaca

Vulnerable. L 68–84cm; WS 176–216cm; WT ♂ 2.45–2.72kg, ♀ 3.16–4.53kg. **SD** E & SE Europe, C Asia to Transbaikalia, winters NE Indian subcontinent, S & SE China. Rare migrant or accidental winter visitor to E & NE China, Korea, Taiwan and Japan. Monotypic. **HH** Lowland and montane forest (<1,800m), steppe and agricultural areas; favours wetlands in winter. **ID** Larger than Greater Spotted or Steppe, with prominent head and large deep-based bill. Ad. dark brown, with distinctive buff or straw-coloured crown, nape and head-sides, and small white patches on mantle-sides. Tail (shorter than Golden) grey with broad black band at tip. Juv. pale tawny-brown, head particularly pale, with dark-streaked mantle and wing-coverts, broad pale tips form double wingbar; rump and back plain buff. In flight, ad. wings appear long, plain and rather dark, broad to base with prominently 'fingered' primaries; tail long, rather square-ended and two-toned. Wings narrower than Golden, often parallel-edged. Flight slow and deliberate with heavy floppy beats; glides/soars on flat or slightly raised wings with primary tips upcurled. Juv. very different wing pattern and bulging secondaries; flight-feathers and tail blackish-brown with pale tips, inner primaries pale grey, forming panel ('window') from above and below, underwing and upperwing-coverts buff heavily streaked dark brown, and upperwing greater coverts blackish with white fringes forming narrow wingbar; uppertail-coverts creamy white; thighs and vent plain buff. **BP** Bill large, deep, grey at tip, yellow at deep base; eyes yellowish-grey to pale brown; rather short legs heavily feathered, small feet yellow. **Vo** Quite vocal, especially during aerial display flight; series of deep quacking barks *kuwa kuwa kuwa* or *owk-owk-owk…*, also a squealing whistled *skiwip*.

GOLDEN EAGLE *Aquila chrysaetos* CKJR

L 66–90cm; WS 180–234cm; WT (*A. c. canadensis*) ♂ 3–4.3kg, ♀ 3.6–6.4kg. **SD** Eurasia, N Africa and N America. *A. c. kamtschatica* (perhaps = *canadensis*) Lena River to Chukotka, Bering Strait, Kamchatka, Sea of Okhotsk, Russian Far East and NE Mongolia; unknown race (possibly *A. c. daphanea*) Hebei and Beijing; *A. c. japonica* Korea (rare) and Japan (local), vagrant Chinese E coast. **HH** Open habitats; taiga to temperate, semi-desert, high mountains with rocky crags and dense forests. Generally resident, but N populations migratory. **ID** Largest *Aquila*. Ad. has golden feathering on rear crown/nape that contrasts with overall dark blackish-brown plumage. At rest, wings reach tail tip. In flight, head and neck prominent; wingbeats slow and deliber-ate on long dark wings with broad 'hand', bulging secondaries and narrower bases, with prominently 'fingered' primaries and curved trailing edge to secondaries; tail long, broad but narrow-based. Soars high with wings flat or in shallow V and pressed forwards. Paler upperwing-coverts of ad.form panel on upperwing; juv. even darker, 1st-year has distinctive large, white wing patches and conspicuous white tail with black terminal band. *A. c. japonica* smaller and darker than *kamtschatica*. **BP** Bill grey with black tip, yellow cere and grey lores; eyes amber to red-brown (ad.) or dark brown (juv.) under heavy brows; legs heavily feathered, toes yellow. **Vo** Calls rather weak; *japonica* has range of whistled *pee-yep* or *pyo pyo* calls, deeper barking *kiek-kiek-kiek* or *kra kra*, and yapping *nyek nyek*.

GREATER SPOTTED EAGLE

ad

juv

ad

ad

juv

STEPPE EAGLE

ad

nipalensis

ad

ad

juv

imm

EASTERN
IMPERIAL EAGLE

ad

ad

ad

ad

juv

juv

GOLDEN EAGLE

kamtschatica

ad

ad

ad

juv

BONELLI'S EAGLE C
Aquila fasciata

L 55–67cm; WS 142–175cm; WT 1.56–2.49kg. **SD** N Africa and W Europe to India, E China and Indonesia. *A. f. fasciata* uncommon, breeding only in SE coastal China north to Fujian and Zhejiang (accidentally north to Hebei); essentially resident. **HH** Mountains (to 3,000m), hills and plains with forest or woodland, cliffs, crags and gorges, with some moving to lowlands, even wetlands, in winter. Generally in pairs. **ID** Mid-sized eagle, almost size of Greater Spotted Eagle, but more honey buzzard-like proportions, with protruding head, long rather narrow wings (broadest in secondaries, narrowing in primaries), and longer, narrow square-ended tail. Ad. dark brown above with pale mantle and underparts with black streaking. Juv. buff to rufous-brown. At rest, wingtips fall short of tail tip. In flight, somewhat large-headed with long neck; wingbeats fast, loose and shallow, in gliding flight wings arched, carpals project forwards, whilst rear edge is straight; soars on flat or slightly raised wings. Ad. from below has dark underwing contrasting with pale belly streaked dark; leading edge of innerwing and scapulars pale, contrasting with blackish underwing-coverts forming midwing bar; dark secondaries with broad dark trailing edge, slightly paler primaries (except dark tips) and pale grey tail with broad black terminal band. From above has pale, even white, triangle on mantle, contrasting with otherwise dark upperparts. Juv. in flight has pale buff to brown underwing-coverts, with dark midwing bar, narrow dark carpal crescent, barred flight-feathers (darker secondaries than primaries) and grey, barred tail. **BP** Bill large, dark grey, cere and short gape line yellow, tip black; eyes yellow to orange (ad.) or hazel-brown (juv.); thighs and long legs feathered, large toes yellow. **Vo** Silent, except on territory, where gives sibilant squeals *scheeuup scheeuup*, slowing series of *skur-skur…* notes, and various barking calls. **TN** Formerly *Hieraaetus fasciatus*.

BOOTED EAGLE CK
Aquila pennata

L 42–51cm; WS 113–138cm; WT ♂ 510–770g, ♀ 0.84–1.25kg. **SD** Summer visitor to SW Europe, NW Africa to Transbaikalia; winters Africa and Indian subcontinent. May breed extreme NW China; vagrant (perhaps overlooked) to coastal NE & E China and Korea. Monotypic. **HH** Desert, savanna and wooded hills. Generally solitary. **ID** Small, compact eagle, size of Eastern Buzzard (p.136), with long wings and long, square-ended tail, small white patches ('landing lights') at inner edge of forewing, pale brown panel on greater upperwing-coverts, and white crescent on uppertail-coverts. Ad. variable; head, neck, upperparts and underparts pale whitish-cream, or rufous-brown, depending on morph. In pale morph, underwing-coverts pale, contrasting with dark flight-feathers, except grey panel on inner primaries. Tail grey with dark tip and centre. Dark morph has brown to rufous-brown underwing-coverts, typically with darker greater coverts forming midwing bar, inner primaries pale. Juv. dark morph has more rufous-brown body-feathers. At rest, wingtips almost reach tail tip. In flight, six clearly 'fingered' primaries (five in Eastern Buzzard); white patch on forewing, pale brown panel on upper midwing; flight faster than buzzard, on loose, deep beats with longer glides; glides with wings arched, tips upcurled, and soars on flat wings pressed forward. Stoops spectacularly. **BP** Bill dark grey to black with yellow cere; eyes dark yellow-brown to red-brown; thighs and rather long legs well feathered ('booted'), toes yellow. **Vo** Generally silent, except on breeding grounds, where gives buzzard-like mewing *hiyaaah* and shrill chattering *kli kli kli* or *yug-yug-yug*. **TN** Formerly *Hieraaetus pennatus*.

MOUNTAIN HAWK EAGLE CTKJ
Nisaetus nipalensis

L 66–84cm; WS 134–175cm; WT ♂ 2.5kg, ♀ 3.5kg. **SD** Himalayas to Hokkaido (from where some may retreat in winter), and S India to SE Asia. **Japanese Hawk Eagle** *N. (n.) orientalis* (possibly distinct species) throughout Japan, possibly NE China and Russian Far East (and N Korea?). Vagrant S Korea, recorded Sakhalin. *N. n. nipalensis* SE China north to Zhejiang, and Taiwan. **HH** Heavily wooded or forested mountains to *c.*2,000m. Local movements by northern birds, and higher montane birds may descend in winter. **ID** Large, extremely broad-winged, rather pale eagle, with broad tail with rounded tip. Ad. dark brown with particularly dark ear-coverts, very narrow grey supercilium, heavy streaking on nape and neck-sides, partly erectile rear crown-feathers form short, broad rounded crest; mantle and wing-coverts dark brown with some pale fringes; white throat with dark mesial stripe and dark striping on upper neck; flanks, belly, thighs, feathered legs and undertail-coverts off-white heavily barred dark rufous-brown. Juv. paler tawny-buff, especially on underparts. At rest wingtips extend just beyond tail base. In flight, very broad wings with broadly rounded tips, brown finely barred underwing-coverts, and narrow dark barring across pale underside to flight-feathers, trailing edge strongly curved; tail has 4–5 dark bars on pale background. Juv. has plain buff-brown underwing-coverts and thighs, breast and belly pale buffy-white. Glides and soars on flat or shallow V-shaped wings. *N. n. orientalis* larger and paler than *N. n. nipalensis*, with much paler, less heavily marked underparts, much shorter, less erectile crest, and longer tail. **BP** Bill strong, deep, dark grey with dark grey cere in ad., dark grey with yellow basal half in juv.; eyes golden-yellow to orange (ad.) or pale bluish-grey to yellow (juv.); legs large, feathered, large feet dull yellow. **Vo** Usually silent, but may give various shrill screams and whistles (some likened to Green Sandpiper, others to Little Grebe, including *pippii pippii, pie pie, pyopyo* or *pyo pyii*) on territory, from perch and in flight. **TN** Formerly *Spizaetus nipalensis*. **AN** Hodgson's Hawk Eagle.

BONELLI'S EAGLE

fasciata

ad ♂

ad

juv ♀

ad

ad

juv

BOOTED EAGLE

ad dark

ad pale

ad pale

ad dark

ad pale

juv

MOUNTAIN HAWK EAGLE

orientalis

ad

ad

ad

ad

GREAT BUSTARD　　　　　　　　　　CKJR
Otis tarda

Vulnerable. L ♂ 90–105cm, ♀ 75–85cm; WS ♂ 210–260cm, ♀ 170–190cm; WT ♂ 5.8–18kg, ♀ 3.3–5.3kg. **SD** SW Europe to NE China. *O. t. dybowskii* breeds C Asia, from Altai to Transbaikalia, Mongolia and NE China (Inner Mongolia and Heilongjiang), moving south to winter in N China from Gansu to Shandong. Rare, much-reduced, winter visitor on Chinese coast to Fujian, east to Korea (none recently) and accidental Japan (Sep–Apr). **HH** Forest-steppe, steppe and desert; also in open croplands. Wary. **ID** Large, stout-bodied, goose-sized bird, with large head, long neck and long legs. Bulky, tail short. Head and neck pale grey, upperparts pale to mid orange-brown barred black, tail somewhat darker; underparts white. ♂ has elongated white facial whiskers on chin and broad chestnut chest band when breeding, loses whiskers and breast-band narrower at other times. ♀ much smaller, lacks chest-band, and is paler, sandier brown, neck thinner and less grey. Take-off requires lumbering run, but flight fast with deep beats, recalling *Aquila* eagles. Wings long, broad, rounded, with prominent 'fingers', largely white with black carpal crescent, primary tips and broad secondary band, secondary wing-coverts concolorous with back, but primary-coverts and flight-feathers mostly greyish-white, forming extensive pale panel contrasting with black primary tips and broad black band on secondary tips. **BP** Bill greyish-horn; eye large, irides dark brown; legs greenish-grey. **Vo** In breeding season ♂ makes deep hollow *umb*, and in alarm a bizarre short nasal bark; juv. gives haunting, melodious hooting; silent away from breeding grounds.

LITTLE BUSTARD　　　　　　　　　　　J
Tetrax tetrax

L 40–45cm; WS 83–91cm; WT ♂ 794–975g, ♀ 680–945g. **SD** SW Europe and N Africa, Middle East and C Asia. Vagrant Japan (Kyushu). Monotypic. **HH** Dry grasslands and short cropfields. **ID** Small bustard with bulky body, long legs, short tail, long neck and upright stance; size of ♀ pheasant, but proportions very different. ♂ (breeding) has grey head, black neck with two white collars, one narrow forms V on foreneck, one broad around lower neck, sandy to mid brown upperparts and white from upper breast to vent. ♀/non-breeding ♂ more uniform, lack grey, black and white on head and neck, body and upperparts sandy brown with dark mottling. Wings broad, rounded, wingtips arched and wingbeats rapid and shallow, recalling grouse, wings may whistle in flight. Broad white panel across secondaries and inner primaries (wings appear very white); primary tips and crescent on greater coverts black. **BP** Bill short, slightly arched, dull grey, blackish-grey at tip; eyes reddish-brown (breeding ♂), paler brown (♀); long tarsi dull greyish-green. **Vo** Essentially silent, except on breeding grounds, but when flushed may give low *ogh* and wings whistle weakly, but distinctively, in flight.

SWINHOE'S RAIL *Coturnicops exquisitus*　CKJR

L 13cm. **SD** Rare, secretive and poorly known. Endemic breeder in NE China and Khanka region of extreme S Russian Far East; in winter (Oct–Apr) or on migration, reaches Korea, Japan (has bred N Honshu) and SE China. Monotypic. **HH** Breeds in wet meadows and marshes, and winters in similar habitats and wet ricefields. **ID** Very small crake, which may be confused with Baillon's (see p.146). Upperparts mid to dark brown, with broad blackish-brown streaks and very fine transverse white bars. Warm brown face and breast finely speckled white, ear-coverts and breast dark brown, belly white. Wings short and rounded; coverts and tertials patterned like upperparts; flight-feathers plain greyish-brown, secondaries form clear white patch in flight. **BP** Bill short, grey with dull yellow base; legs grey. **Vo** Range of grunting, squealing and cackling calls, and distinctive hard note repeated like two stones tapped together. Also a fast series of continuous sounds described as *kyo kyoro ru....* **TN** Formerly part of Yellow Rail *C. noveboracensis*.

RED-LEGGED CRAKE *Rallina fasciata*　　T

L 23–25cm. **SD** SE Asia, from NE India to Indonesia and Philippines. Accidental at least twice in Taiwan, e.g. Lanyu I. Monotypic. **HH** Wet areas in forests to rivers, swamps and wet cropfields. **ID** Rufous-brown with broadly barred black and white wing-coverts, belly, flanks and undertail-coverts. Head and neck bright rufous-brown, mantle, rump and tail duller mid brown. Like Slaty-legged Crake, but white flank bars broader and bolder, and has white tips to wing-coverts, clearly visible on closed wing. **BP** Bill strong, blackish-grey; eyes and eye-ring dark red; tarsi pinkish-red. **Vo** In breeding season calls at dawn and dusk, either a slow descending trill or nasal *pek pek pek* or clucking *kunkunkunkunk...*; advertising calls *gogogogok* given at night.

SLATY-LEGGED CRAKE　　　　　　　　TJ
Rallina eurizonoides

L 21–25cm; WS 47.5cm; WT 90–180g. **SD** Pakistan through S & SE Asia to Sulawesi and Philippines with various races. *R. e. telmatophila* occurs north to SE China; *R. e. formosa* occurs on Taiwan and Lanyu, and *R. e. sepiaria* in S Nansei Shoto, Japan. **HH** Shy and retiring: damp forest, forest edge, scrub, marshes and wet fields. **ID** Large crake. Ad. has rufous-brown head, neck and breast, plain dark brown upperparts and wings; underparts to undertail-coverts dark blackish-brown or black, narrowly barred white. Resembles Red-legged Crake, but lacks black and white barring on wing-coverts and legs not red. Juv. duller, cooler brown on head and neck, with grey cheeks and white throat. **BP** Bill strong, blackish-grey; eyes red (ad.) or brownish-orange (juv.); tarsi greenish-grey. **Vo** Calls at night from tree perch, a distinctive double *beep-beep* or *kuwa, kuwa*; recalls Mountain Scops Owl (p.260). Also hollow, repetitive clucking *pok-pok pok-pok...* and downslurred, nasal *paau paau*. In winter also calls at dusk, a prolonged, but slowing, harsh growling *grgrgr ger ger-ker-ker-ker....* In circling display flight after dark, low over lowland forest, an evenly repeated *aow aow*. In alarm a harsh metallic *kik kik kik* or loud *kwek*. **AN** Banded Crake.

GREAT BUSTARD
(not to scale)

dybowskii

ad ♀

imm ♂

ad ♂

LITTLE BUSTARD
(not to scale)

ad ♂ win

ad ♀

ad ♂

SWINHOE'S RAIL

ad

ad

RED-LEGGED CRAKE

ad

ad

SLATY-LEGGED CRAKE

telmatophila

juv

OKINAWA RAIL
Gallirallus okinawae

J

Endangered. L 29–33cm; WS 48–50cm; WT 433g. **SD** Endemic to northern quarter of Okinawa I., Japan, where range contracting. Monotypic. **HH** Dense primary and secondary subtropical forest, generally seen at forest edge and at pools and streams; sociable and may live in loose 'troops'. **ID** Large, dark, essentially flightless rail; terrestrial, but roosts in trees. Ad. upperparts, tail and wings dark brown; face black with broad white band from eye to neck-sides; underparts slate-grey with extensive fine white barring. Juv. duller, less clearly marked. **BP** Bill rather thick-based, bright red with yellow-horn tip, duller red in immature; long, strong legs and feet red. **Vo** Typically heard early morning, late afternoon or evening, though also calls at night. Range of notes include guttural grunting sounds (kyo kyo kyo; ke ke; gu gu; kururu) and loud kwi kwi kwi kwi ki-kwee ki-kwee, rolling kikirr krarr followed by rising kweeee until becomes pig-like squeal followed by ki-kwee-ee ki-kweee-ee kwee ke-kwee, often in duet, followed by ki-ki-ki and kyip-kyip-kyip calls. Eerie calls at night sound like hyena yelps. One regular call is a frog-like, loud, chirping kurrrroiIT! (rising at end), and on ground regularly gives a deep gurgling bubbling, perhaps in agitation or alarm, gu-gu-gugugugu. **TN** Formerly Rallus okinawae.

SLATY-BREASTED RAIL
Gallirallus striatus

CT

L 25–30cm; WT 100–142g. **SD** Indian subcontinent to Greater Sundas, and China. Uncommon in E Asia: G. s. jouyi Hainan and SE China; G. s. taiwanus on Taiwan. **HH** Shy, partly nocturnal. Swamps, mangroves, marshes, wet grasslands and crops, to 1,000m. **ID** Like Eastern Water Rail but crown and nape dark chestnut; blackish-olive upperparts including tail and wings covered with fine grey bars; face, neck and breast grey, throat white; blackish-grey belly, flanks and undertail coarsely barred white. Flight weak. **BP** Bill thicker, straighter and slightly shorter than Eastern Water Rail, dull red at base and on lower mandible, greyer tip and culmen; eyes red (no eye-ring), tarsi grey. **Vo** Hard ter rek or series of metallic shtik notes, and buzzing, repeated kech kech kech notes rising and falling in strength. **TN** Formerly Rallus striatus. **AN** Blue-breasted Banded Rail.

EASTERN WATER RAIL
Rallus indicus

CTKJR

L 23–29cm; WS 38–45cm; WT 74–190g. **SD** S Asia, northern SE Asia and N Mongolia to E Asia. Summer visitor to SE Siberia, NE China, Transbaikalia to Amur Estuary, Sakhalin and N Japan. Winters (Oct–Apr) in S Japan, Korea, SE China and Taiwan (has bred Lanyu). Monotypic. **HH** Shy, in swamps, marshes, reedbeds, wet meadows and wet cropfields. **ID** Ad. upperparts dull olive-brown, the feathers with dark brown centres; face and neck-sides grey-blue, with white supraloral stripe and broad brown eyestripe, breast grey-blue, but with variable brown wash across lower neck/breast (face pattern and breast coloration differ from extralimital **Western Water Rail** R. aquaticus); flanks to undertail-coverts blackish with white barring. Juv. has buff, not grey, underparts, with black and buff flanks barring. Wings broad, relatively long, brown; flight surprisingly strong and fast. **BP** Bill long and slightly decurved, dark red at base and on lower mandible, tip and culmen blackish; eyes deep red; tarsi pinkish-grey. **Vo** Low grunts, high squeals and a softer jip jip jip, also an abrupt series of piping kyu kyu kyu notes (quite different from kdik of Western). In breeding season utters series of c.10 sharp skrink skrink or kyi kyi kyi… notes. **TN** Formerly considered race of Water Rail R. aquaticus.

BROWN CRAKE
Amaurornis akool

CT

L 26–28cm; WT ♂ 114–170 g, ♀ 110–140g. **SD** Resident from India to E China. In E Asia A. a. coccineipes from NE Vietnam to SE China, north to Jiangsu; also Matsu Is (Taiwan). **HH** A shy, skulking resident of reed and grass swamps, and wet fields, in lowlands and low hills. **ID** Slightly smaller than White-breasted Waterhen, structure similar, though duller and plainer. Dark olive-brown from crown to wings and short tail; face, foreneck and underparts grey, white on chin and browner on flanks, vent and undertail-coverts. **BP** Bill short, stout, greenish at base, grey at tip; eyes red-brown; legs brownish-pink to purple. **Vo** Calls at dawn and dusk; a tremulous descending whistle or musical whirring whr r r r r r, and short plaintive notes.

WHITE-BREASTED WATERHEN
Amaurornis phoenicurus

CTKJR

L 28–33cm; WS 49cm; WT ♂ 203–328g, ♀ 166–225g. **SD** Resident from Pakistan to the Sunda Is and S & SE China. In E Asia, A. p. phoenicurus on E Chinese coast north to Zhejiang, Taiwan, and S Nansei Shoto. Some move to north of Changjiang, and has strayed Kyushu, islands off C Honshu, Korea and Ussuriland. **HH** Common and often conspicuous, foraging in open areas amongst mangroves, along streams and rivers, wet marshes, grassy swamps, wet fields, and grasslands to 1,500m. **ID** Large, size of Common Moorhen (p.148). Ad. boldly pied; dark slate-grey from crown to tail and wings, contrasting strongly with white forehead, face and underparts; lower belly and undertail-coverts cinnamon. Juv. has grey-white face, foreneck and breast. Restless, flicks cocked tail. **BP** Bill stout, straw yellow, but red at base of upper mandible; eyes black; tarsi bright yellow. **Vo** Vociferous, wide range of loud calls, at dawn, dusk or at night, often from raised perch. A generally monotonous u-wok u-wok, long series of kru-ak kru-ak kru-ak-a-wak-wak notes for many minutes or short nasal wid repeated slowly or more gulping whigh in rapid series, but will also chorus at dawn and dusk with more varied grunts, chuckles and frog-like noises: kshorr kor kor… and korokorororowa; and cat-like gyaoo gyaoo. Also a Ruddy-breasted Crake-like (p.146) rapid descending trill kyegegegegegege, though deeper and more guttural.

OKINAWA RAIL

ad roosting in tree

ad

ad

jouyi

SLATY-BREASTED RAIL

EASTERN WATER RAIL

ad

BROWN CRAKE

coccineipes

ad

ad

WHITE-BREASTED WATERHEN

phoenicurus

juv

ad

BAILLON'S CRAKE CTKJR
Porzana pusilla

L 17–19cm; WS 33–37cm; WT ♂ 23–45g, ♀ 17–55g. **SD** SW Europe to Japan and Africa to New Zealand. In E Asia, *P. p. pusilla* (perhaps separate species, **Eastern Baillon's Crake**, requires further study) is fairly common summer visitor to NE China and Russian Far East, and scarce visitor (mid Apr–late Oct) to C & N Japan. Migrant Korea and most of E China; accidental Taiwan. **HH** Secretive, in swamps, marshes, wet grasslands and reedbeds. **ID** Tiny with very short wings. Ad. warm mid or rufous-brown above with irregular large black and narrow white streaks; face, foreneck and breast blue-grey, ear-coverts, hindneck and neck-sides brown; flanks, belly, vent and undertail-coverts narrowly barred black and white. Juv. drabber above, lacks blue-grey, instead face brown, chin, throat and belly off-white with brown wash on breast and pale grey-white barring on belly, flanks and undertail-coverts. **BP** Bill rather thick, dull green (ad.), or dull yellow with black tip (juv.); eyes deep red (ad.) or orange-brown (juv.); tarsi dull brownish-yellow. **Vo** Territorial song rather weak and does not carry far; a dry descending sputtering trill or rattle, *kokkokko* or *tou tou tou tou*, given at night. In alarm gives sharp *tac* or *tyuik* and a low, continuous growling or series of hard notes, *kraa-kraa-kraa-chachachacha*.

RUDDY-BREASTED CRAKE CTKJR
Porzana fusca

L 19–23cm; WS 37cm; WT *c*.60g. **SD** Resident from Pakistan to Sulawesi, Philippines, SE China, Taiwan and S Japan north to S Honshu; a summer visitor (Apr–Oct) further north to NE China, Korea, N Japan and Sakhalin. *P. f. erythrothorax* occurs NE China, Korea and Japan; *P. f. phaeopyga* occurs E China, Taiwan and Nansei Shoto, Japan. **HH** Around marshes, wet cropfields, streams and ditches, occasionally venturing into open. **ID** Medium-sized dark crake; mid to dark brown from crown to tail and wings, chestnut face and underparts to belly; chin/throat white; lower belly, rear flanks and undertail-coverts black with very narrow white barring. Juv. dull olive-brown, with no chestnut, barred like ad., but also has fine white barring on dull olive chest. **BP** Bill dark grey, short; eyes deep red; tarsi red. *P. f. phaeopyga* is slightly darker than *P. f. erythrothorax* but extremely difficult to separate in the field. **Vo** Crepuscular, gives brief descending trills with an increasing tempo, commencing with a single hard *kyot*, repeated and speeding up like someone knocking at a door, *kyot kyot kyokkyokyo burururu*; also quiet, clucking *puk* notes while foraging.

BAND-BELLIED CRAKE CTKJR
Porzana paykullii

L 20–22cm; WS 42cm; WT 96–132g. **SD** Restricted to E Asia, where uncommon. Breeds from Amur River and Ussuriland in Russian Far East across NE China and, at least formerly, N Korea. Recorded Sakhalin. Winters to SE Asia and Borneo; accidental Japan and Taiwan. Monotypic. **HH** Shy, retiring and little-known species of marshes, wet grasslands, meadows and wet cropfields. **ID** Medium-sized short-billed crake, slightly larger than Ruddy-breasted; crown, upperparts and wings dark grey-brown, some with fine barring on upperwing-coverts; flanks, lower breast, belly and undertail-coverts barred black and white. Distinguished from Ruddy-breasted and Red-legged (p.142) crakes by grey rather than deep brown upperparts and wings, and paler orange rather than red tarsi. **BP** Bill short, greenish-grey; eyes red; tarsi orange-red. **Vo** Rapid wooden purring or drumming call, *tototototo...*, blending into a trilled *urrrrr*, given at dusk, dawn and at night.

WHITE-BROWED CRAKE TJ
Porzana cinerea

L 15–20cm; WS 27cm; WT 40–62.5g. **SD** Philippines and SE Asia to NE Australia, including numerous tropical Pacific islands. In E Asia, known from Iwo Is, Japan (extinct 1911), and as an accidental to Taiwan. Monotypic. **HH** Freshwater and saline swamps, marshes, wet grasslands and wet cropfields, also lakes, particularly favouring areas with mats of floating vegetation. **ID** Grey and brown crake. Size of Ruddy-breasted, but foreparts grey, not rufous, lacks barring on underparts, off-white belly and buff-brown flanks and undertail. Most distinctive features are: straw-buff mantle streaked dark brown, and face pattern, with black eye patch, red and white lores, prominent white supercilium, and white from bill base below eye patch to ear-coverts. In flight, narrow pale fringes to trailing edge of wing. **BP** Bill slender, yellowish-orange; eyes deep red; tarsi pale yellowish-olive. **Vo** Noisy high-pitched piping *cutchi cutchi cutchi*. Call, when flushed or on landing, a rapid *kwekwekwekwekwek* with an electric, bubbling quality. While foraging utters a loud, sharp *kek-kro* and quiet *charr-r* in alarm. **TN** Formerly *Poliolimnas cinereus*. **AN** Ashy Crake.

BAILLON'S CRAKE

pusilla

ad

juv

RUDDY-BREASTED CRAKE

erythrothorax

ad

BAND-BELLIED CRAKE

ad

juv

WHITE-BROWED CRAKE

ad

WATERCOCK CTKJR
Gallicrex cinerea

L ♂ 42–43 cm, ♀ 35–36cm; WS 68–86cm; WT ♂ 300–650g, ♀ 200–434g. **SD** Largely resident from Pakistan to Indonesia, Philippines and China. Summer visitor to China, Korea (now very rare in some areas where once abundant) and S Nansei Shoto, Japan. Regular in spring on islands in Sea of Japan and has reached Hokkaido and even Kamchatka. Monotypic. **HH** Shy and largely nocturnal; normally occurs in marshes and wet fields, but on migration also on offshore islands with grassland and scrub. **ID** Large, oversized but slender version of Common Moorhen; tail typically cocked. Breeding ♂ mostly slate-blue or dark greyish-black (lacks white flank stripe), mantle and wing-feathers fringed brown. Non-breeding ♂ and smaller ♀ appear orange-buff; plain-faced with large dark eye; underparts have fine darker brown barring, the upperparts with buff fringes to dark-centred scapulars and wing-feathers. Flight strong, tarsi trailing obviously beyond short tail. **BP** Bill orange with yellow tip, and red shield extends to raised horn on forehead (breeding ♂), or yellowish-horn lacking shield (non-breeding ♂, ♀ and juv.); eyes deep red (breeding ♂), otherwise brown; tarsi dull red (breeding ♂), otherwise dull green or ochre, toes very long. **Vo** Rather noisy when breeding, giving rhythmic series of deep booming *ka-pon ka-pon ka-pon* or *ku-wa ku-wa*, also softer, hollow *youmb youmb youmb* and resonant gulping *tyokh* or *kluck* notes; silent in winter.

PURPLE SWAMPHEN T
Porphyrio porphyrio

L 43–50cm; WS 90–100 cm. **SD** S Europe and Africa to Australia and New Zealand. *P. p. poliocephalus* of S & SE Asia breeds north to Shantou, Guangdong, China (just outside region) and has strayed to Taiwan and possibly S Japan. Asian taxa probably deserve specific recognition (from European/African and Australasian), as **Grey-headed Swamphen** *P. poliocephalus* though further research required to establish species limits in E Asia. **HH** Freshwater and brackish swamps and wetland margins. **ID** Largest and most colourful of the rails/crakes; heavy, plump with deep, prominent bill. Deep purplish-blue plumage, with dark head, cerulean-blue face, throat, breast, scapulars and upperwing-coverts, greener wings, blackish-blue upperparts, and white vent. **BP** Bill deep, stout bright red with red frontal shield (ad.) or blackish-red (juv.); eyes reddish-brown to red; tarsi and enormous feet red. **Vo** Highly vocal with extremely varied vocabulary of moans, croaks, squawks and wails including *hnyaarh nyaaar nyaaar*, and harsh *kraarrk*. Song a series of powerful nasal rattles. **AN** Purple Gallinule.

COMMON MOORHEN CTKJR
Gallinula chloropus

L 30–38cm; WS 50–55cm; WT ♂ 249–493g, ♀ 192–343g. **SD** The most widespread rail, from southern N America to S America, across Eurasia, Africa and S & E Asia. In E Asia, *G. c. chloropus* is a summer visitor to S Russian Far East, Sakhalin, NE China, N Japan and Korea, wintering and resident from C Japan south through Nansei Shoto, Taiwan and SE coastal China. **HH** Wide range of freshwater wetlands, usually with extensive and dense fringe vegetation; at times secretive, at others bold and conspicuous, foraging in open on water or on land. **ID** A large dark crake/rail. Ad. mostly slate-grey, though upperparts tinged brown, with prominent broken white bar on flanks. Tail quite long, undertail-coverts white with vertical black central stripe, particularly noticeable due to constant flicking of tail; also jerks head back and forth like pigeon. Juv. dull grey-brown with whitish chin and throat, white flank bar and undertail-coverts, and same mannerisms as adult. Take-off conspicuous, with long splashing run across water; flight weak, feet trailing prominently. **BP** Ad. has stout, bright red bill with yellow tip, and red shield extending up forehead; juv. has dull, dark red bill; eyes chestnut; tarsi greenish-yellow with orange-red spot near feather edge, toes very long. **Vo** Wide vocabulary of various loud, harsh gargling sounds, *pruruk-pruuk-pruuk*, or series of brief guttural *ku ku ku...* notes, an explosive single *krrrruk*, creaking and clucking sounds, and a sharp *keek* or *kit i tit kit i tit* in alarm and in flight.

COMMON COOT CTKJR
Fulica atra

L 36–39cm; WS 70–80cm; WT ♂ 0.61–1.2kg, ♀ 0.61–1.15kg. **SD** Very widespread, from British Isles to New Zealand, resident in Europe, India, Australasia but migratory across swathe of Eurasia from Black Sea to Pacific. In E Asia, *F. a. atra* is a summer visitor to Transbaikalia, NE China, adjacent Russian Far East and from Lena to Sakhalin, Korea and C & N Japan. Winters in C & S Japan, probably C & S Korea, E China south of the Changjiang and Taiwan. **HH** Standing freshwater bodies, from ponds to large lakes with abundant vegetation; aggressive during breeding season when fights and chases conspicuous, but often gregarious at other times. **ID** Large, stocky, hunched, all-black 'crake', the most aquatic of its group, rarely seen far from water, though may graze onshore; commonly upends and dives for food; constantly nods head. Ad. all black with striking white bill and frontal shield. Juv. plain ash-brown with whitish throat and breast. Flight laboured, take-off requires running start, tarsi/toes often dangling, trailing edge of secondaries white. **BP** Ad. bill white with pinkish tinge, extending up forehead as distinctive shield; juv. bill grey; eyes deep dark red (ad.) or dark brown (juv.); tarsi dull green, feet grey with prominently lobed toes. **Vo** Highly vocal, often noisy, with rich repertoire, especially at night. Explosive *kik kik kik*, also loud *krek, tyok* and *kyururu* calls, and shorter guttural *gu-gu-gu* amongst commoner calls. At night and in flight gives nasal trumpeting *neee beeep*.

ad ♂

WATERCOCK

ad ♀

PURPLE SWAMPHEN

poliocephalus

ad

juv

COMMON MOORHEN

COMMON COOT

ad

atra

chloropus

juv

juv

juv

DEMOISELLE CRANE
CTKJ

Grus virgo

L *c*.90cm; WS 150–170cm; WT 2–3kg. **SD** From Turkey across C Asia to Mongolia and NE China. In E Asia breeds only NE China. Migrates over Himalayas to NW India, but accidental (mostly Oct–Feb, also May–Jul) Korea, Japan and Taiwan (Penghu). Monotypic. **HH** Steppe, semi-steppe and desert-edge; often at high altitudes (to 5,000m), on migration appears at wetlands but mainly departs region; mostly mixes with flocks of other cranes. **ID** Large, size of Great White Egret (see p.104), but smallest crane; slender, elongated appearance, but neck and bill actually shorter than other species, and head fully feathered. Ad. elegant, pale grey, with black head, neck and breast, extending as chest plumes (black chest separates from Common Crane at distance). Hindcrown pale grey; whitish ear-tufts extend from eye to nape (unique amongst cranes). Pale grey tertials sleek and particularly elongated, drooping very low. 1st-winter has contrast of head less developed and neck less black. In flight, primaries and primary-coverts, secondaries and tail, all dark grey contrasting with paler grey upperparts and tertials, as Common (p.152), but black of neck extends to centre of breast. Flight faster, more agile than larger cranes. **BP** Bill yellowish horn at tip, paler at base; eyes blood-red (♂) or orange (♀), dark in juv.; tarsi pinkish-grey; the only crane in region that lacks bare skin on head. **Vo** Vocal on breeding grounds and in winter, when trumpeting is shriller, higher than Common, but drier and more wooden, also a rolling trill, *kuu kururuu* or *kd r r r r r kd r r r r r*.... Young give high, thin whistle. **TN** Formerly *Anthropoides virgo*.

SIBERIAN CRANE
CKJR

Grus leucogeranus

Critically Endangered. L 140cm; WS 210–230cm; WT 4.9–8.62kg. **SD** Summer visitor to restricted area of Russian Arctic. In E Asia, local east of the Lena, wintering in freshwater marshes and lakes in E China, such as Poyang. Moves through NE China and singles stray occasionally in winter (mostly Oct–Feb) to Korea and Kyushu, joining other cranes, but has also occurred in Japan in Apr–Aug at, for example, wetlands in E Hokkaido. Monotypic. **HH** Wet tundra and forest-tundra with pools; on migration and in winter favours extensive wetlands including marshes, tidal flats and shallow lakes. **ID** Large all-white crane, larger than Common Crane (p.152). Ad. white, with red face and black wingtips. At rest like large white egret, as white tertials mostly cover black primaries. Young dark cinnamon, becoming mottled white with age; brown remaining on head, neck and flight-feathers through first winter. In flight, solid black wingtips contrast with otherwise white plumage, like Snow Goose (p.46); Oriental Stork (p.92) has black bill and black secondaries. Flight slow, measured, black primaries deeply 'fingered'; legs and neck extended. **BP** Bill yellow-orange; bare facial skin to eye and forecrown red; eyes staring yellow; long tarsi reddish-pink. **Vo** Rolling *kuru kuruu kuru kuruu* and soft musical *koonk koonk* in flight, higher pitched than other cranes. Young give thin, high, downslurred whistles *tchyu tchyu*.... **AN** Siberian White Crane.

SANDHILL CRANE
CKJR

Grus canadensis

L 88–120cm; WS 160–210cm; WT ♂ 3.75kg, ♀ 3.35kg. **SD** Widespread in N America but also breeds NE Russia. In E Asia, *G. c. canadensis* breeds from Kolyma River to Chukotski Peninsula south to Koryakia and NE Sea of Okhotsk. Most are assumed to winter in N America, but small numbers reach SW Kyushu (Arasaki) in Nov–late Mar; vagrant Korea and E China (e.g. Beidaihe, Jiangsu, Shanghai, Zhejiang). Joins large flocks of Hooded Crane (p.152). **HH** Breeds in grassy coastal tundra, in dry areas near water. Winters in open grasslands, wetlands and agricultural areas. **ID** A grey crane similar to or slightly larger than Hooded, but much paler. Ad. mostly grey, with variable warm rusty-brown on upperparts, especially wing-coverts and rump, but may extend to neck and back (though some have essentially none). Forehead and forecrown to rear of eye red (bare skin), appearing somewhat heart-shaped at front; head and neck pale grey, lacking black, and often paler, even white, below eye, on lores and chin. Young similar but lack bare red forehead and are browner. Flight straight, leisurely, slow, deep downbeats and quick upbeats, legs and neck extended; flight-feathers blackish, but little contrast with brown of upperwing coverts; underwing paler grey, only outer primaries blackish. **BP** Bill dark grey above, lower mandible yellowish-horn; eyes yellow; tarsi greyish-black. **Vo** A loud rolling, bugling *karr-roo karr-roo* or *kururuu* (fairly similar to Common Crane (p.152), but lower, more drawn-out), with juv. giving sibilant whistled *sweer* notes.

WHITE-NAPED CRANE
CTKJR

Grus vipio

Vulnerable. L 120–153cm; WS 200–210cm; WT 4.75–6.5kg. **SD** Breeds only in E Mongolia, NE China and closely adjacent Russian Far East (Amur and Ussuri basins), though historically bred Hokkaido, where has attempted to nest recently. Winters (mid Oct–late Mar) mainly in SW Kyushu (Arasaki; *c*.40–50% of population), with small numbers in Korea, though mainly a migrant there, also in E China (lower Changjiang valley). Vagrant to Fujian, Taiwan and Sakhalin. Monotypic. **HH** Swamps with extensive reedbeds along river valleys in wet forest-steppe. Gregarious in non-breeding season, family groups forming large flocks at wetlands and on fallow agricultural land. **ID** Large, elegant grey crane which forms spectacular gatherings with Hooded Crane. Ad. various shades of grey and white; head mostly white with dark grey lores and forehead, large red eye patch bordered dark grey, and dark grey ear spot; neck white with dark grey line forming point below head; breast, underparts and mantle dark grey, scapulars pale grey, coverts whitish-grey, tertials almost white. Flight straight, leisurely, slow, deep beats are even, legs and neck extended; flight-feathers blackish (primaries have white shafts), forming strong contrast with pale grey upperwing and underwing-coverts; rump and tail grey. Young similar but have rusty-brown on head and neck, and wings less white, more grey-brown. **BP** Bill yellowish-horn; eyes yellow; tarsi dull pink. **Vo** Various strong bugling *kururuu* or *guruu* calls, stronger and louder than other cranes. Sexes duet on wintering grounds, with ♂ giving deep *kururu* or *gyururu* and ♀ *ko ko ko* or *kuwa kuwa kuwa* calls.

DEMOISELLE CRANE

ad

juv

ad

ad

SIBERIAN CRANE

ad

juv

SANDHILL CRANE

ad

ad

canadensis

juv

WHITE-NAPED CRANE

ad

ad

juv

COMMON CRANE
CTKJR
Grus grus

L 96–125cm; WS 180–200cm; WT ♂ 5.1–6.1kg, ♀ 4.5–5.9kg. **SD** Very widespread, from Scandinavia to NE Siberia, wintering in Iberia, NE Africa, Himalayan region and E Asia. In E Asia, *G. g. lilfordi* breeds in taiga swamps and forest tundra, from the Lena to the Kolyma, also NE China, most moving southwest, but small numbers move south to winter (mid Oct–late Mar) in E & S China, Korea and S Japan (Kyushu). **HH** Wide variety of shallow wetlands from small bogs to extensive swamps. On migration and in winter occurs on agricultural land, grasslands and around lakes and wetlands. Typically gregarious, families remain together throughout winter; often join other cranes, like them roosts in shallow water. **ID** Very large, taller than Grey Heron (p.102) or Oriental Stork (p.92). A medium-sized grey crane. Ad. has black and white head, black foreneck, and broad dark-grey 'bustle' (loose tertials) with some black feathers obscuring tail. Forecrown red; broad white band extends from eye over sides and back of head and neck. Young have brown markings on scapulars/coverts, and buffy-grey head and neck; ad. head pattern acquired during first winter onwards; bustle is grey tinged brown. Flight straight, leisurely, slow, deep beats are even, legs and neck extended; flight-feathers blackish, very similar to Demoiselle (p.150), but has pale carpal patch on leading edge of wing at base of outer primaries. Groups/flocks commonly in lines or V formation. **BP** Bill dull, greenish-horn, dagger-shaped; eyes yellow; tarsi very long, black. **Vo** Breeding pairs utter bugling duet with various *kaw, karrroo, kleeeur* and *kluuer* notes, and a short rolling trill *kr r r reeech*. In flight and on migration gives deep trumpeting *krraw* and rattling *kururuu kururuu* or *k d d dew*. Lone adult may give wooden knocking sound and young a plaintive whistling *peerp peerp*. **AN** Eurasian Crane.

HOODED CRANE
CTKJR
Grus monacha

Vulnerable. L 91–100cm; WS 160–180cm; WT ♂ 3.28–4.87kg, ♀ 3.4–3.74kg. **SD** Range-restricted E Asian species, from Upper Lena, S Yakutia and Lake Baikal to Amur and Ussuriland, and NE China. Winters (mid Oct–early Apr) in large numbers in SW Kyushu (Arasaki has *c*.80% of population), with small numbers in SW Honshu, arriving via Korea and Tsushima; also in east China (Changjiang valley); vagrant Taiwan. Monotypic. **HH** Extensive mossy taiga swamps, upland bogs and swampy lakes and large river systems. On migration, and in winter, wetlands and grasslands including farmland, and around lakes. Extremely gregarious, families remain together throughout winter, forming large, dense flocks. **ID** Very small dark crane. Ad. mostly dark blackish-grey with white head and neck (hood). Forehead and lores black and small red forecrown is largely covered by black feathering. Young lack black and red of forehead, and have rusty off-white head and neck and somewhat browner plumage than adult. Flight straight, leisurely, slow deep beats are even, legs and neck extended; flight-feathers and coverts blackish with no contrast on upperwing. Hybrids between Hooded and Common reported most years in Japan; variable, but commonly lack black on nape, throat/foreneck more dark grey than black, and shorter-billed than Common. **BP** Bill greenish-horn; eyes deep red; long tarsi greyish-black. **Vo** Loud rolling *krrrk; kleeer k d d duuur;* or *kuururun*; juv. keeps contact with family using regularly uttered high-pitched *reeh* or *pyii pyii* in winter.

RED-CROWNED CRANE
CTKJR
Grus japonensis

Endangered. L 138–152cm; WS 220–250cm; WT 7–12kg. **SD** E Asian regional endemic, breeding in NE China, SE Russian Far East and Hokkaido. Japanese population recovering from near extinction (fewer than 20 birds) in early 20th century; now over 1,000 birds mainly concentrated in E Hokkaido, though range steadily expanding. Continental population migratory, breeding in Amurland and Ussuriland and adjacent NE China, moving south to Korea and E China, particularly to Jiangsu. Vagrant in winter to Kyushu and Taiwan. Japanese birds essentially resident, making short-distance movements to winter mainly at provisioned sites within its breeding range, principally Kushiro marsh. Significant in Korean and Japanese culture, symbolic of long life and happiness. Monotypic. **HH** Extensive wetlands with reedbeds, more aquatic than other cranes. On migration (also non-breeders in summer, and in winter) coastal lagoons, mudflats, lakes, swamps and open ricefields. In winter, family groups form flocks of 100+ at river roosts and on open agricultural land. **ID** The largest, most distinctive and arguably the most beautiful crane of the region. Ad. largely white with black head and neck, broad white band from eye to nape; forehead black, crown red (larger in ♂, but variable in size depending on state of arousal/excitement); rest of plumage snow white except black 'bustle' of loose tertials and black secondaries obscuring tail. Young similar but have dark rusty-brown, not black, head and neck and lack red crown, 'bustle' partly brown and wing-coverts have brown tips. Subad. commonly has black in tips of primaries. In flight, white primaries deeply 'fingered', black secondaries and tertials contrast with otherwise white wing, back and tail. Leisurely, slow, deep wingbeats are even; legs and neck extended. **BP** Bill dark horn, eyes reddish-brown (ad.) or black (juv.); tarsi blackish-grey to black. **Vo** Trumpeting and bugling calls on breeding grounds far-carrying; winter duet *ka, kaa-kaa, ka,* with ♂ giving short first and last notes, particularly in association with display dances in Feb–Apr; softer *krewip* notes before and during flight. Juv. uses plaintive high-pitched whistle as a contact call. **AN** Japanese Crane; Manchurian Crane.

COMMON CRANE

lilfordi

ad

juv

ad

ad

HOODED CRANE

ad

juv

ad

ad

RED-CROWNED CRANE

juv

display

SMALL BUTTONQUAIL   CT
Turnix sylvaticus

L 15–16cm; WT ♂ 32–34g, ♀ 39–54g. **SD** N Africa and S Eurasia to Philippines. In E Asia, *T. s. davidi* resident in S China and Taiwan. **HH** Favours dry grassland, heathland or cropland habitats, where secretive and hard to flush, best located by voice; watch for dust-bathing in clearings or at edge of dry vegetation. Polyandrous. **ID** Tiny, finch-sized bird, appearing almost tail-less (*T. s. davidi* is perhaps as small as 11cm). ♀ larger and brighter than ♂. Upperparts dark chestnut-brown, feathers finely vermiculated black and cinnamon, rufous mantle and coverts have buff fringes (scaling), those of scapulars forming lines; underparts paler, rusty-orange on breast of ♀, with fine black spotting on neck-sides and larger spots on flanks. Non-breeding has duller plumage. When flushed reveals short, rounded wings; pale wing-coverts contrast with dark flight-feathers and primary-coverts. **BP** Bill blue-grey; eyes dull yellowish-brown; tarsi yellowish-flesh or grey. **Vo** Territorial calls, given by ♀, mostly early morning but also at dusk, are soft, low, mooing, crooning or booming *hooon…hooon…hooon* or *fwooooaaah… fwooooaaah…* sounds at 1–3-second intervals, and gently reverberating. **AN** Andalusian Hemipode.

YELLOW-LEGGED BUTTONQUAIL   CTKR
Turnix tanki

L 17cm; WT ♂ 35–78g, ♀ 93–113g. **SD** Indian subcontinent to SE Asia. In E Asia, *T. t. blanfordii* resident throughout SE, E & NE China, with those in NE and adjacent S Russian Far East migrants; occurs in Korea on migration where scarce and secretive. **HH** Favours dry grassland, scrub, marshland and cropfields (especially rice stubble), to 2,000m; best located by voice. **ID** Small, but larger and plainer (with less scaling) than Small Buttonquail; nape and upper breast rufous-orange, dark grey-brown mantle and paler wing-coverts have dark spots, pale buff flanks also spotted black. ♀ larger and brighter than ♂, which has rufous-brown mantle heavily streaked black and white. Non-breeding duller. Rarely flies, but wing-coverts contrast less with flight-feathers than in Small. **BP** Bill yellowish-pink; eyes yellowish-white; tarsi yellow. **Vo** Louder, more booming than other buttonquail; a low-pitched hoot becoming stronger and more moaning, also *off-off-off* and *pook-pook*.

BARRED BUTTONQUAIL   CTJ
Turnix suscitator

L 15–17cm; WT ♂ 35–52g, ♀ 47–68g. **SD** Indian subcontinent to Sulawesi and Lesser Sundas. In E Asia, resident in SE China, lowland Taiwan (*T. s. rostratus*), and the Nansei Shoto, Japan (*T. s. okinavensis*). **HH** Dry grassland, scrub and cropfields, especially sugar cane; best located by voice. **ID** Small, size of Yellow-legged Buttonquail; warm rusty-brown upperparts, paler on wing-coverts. ♀ larger and brighter than ♂. ♀ chin, throat and upper breast black, the black extending almost to ear-coverts,

nape blackish-brown; crown and face speckled black and white, upperparts scaled white, breast-sides barred black. ♂ chin black, breast and sides barred black, face pattern more quail-like, with buff supercilium and malar. Belly and vent plain rufous-brown in both sexes. Non-breeding duller. Rarely flies, but wing-coverts contrast less with flight-feathers than in Small. **BP** Bill blue-grey; eyes grey-white; legs blue-grey. **Vo** During breeding season, females court males with a prolonged, deep purring *drr-r-r-r-r-r* and series of very deep rhythmic crooning or hooting notes: *hoon-hoon-hoon-hoon-hoon, uhuu uhuu uhuu* or *pwoo pwoo pwoo*.

AMERICAN BLACK OYSTERCATCHER   R
Haematopus bachmani

L 43–45cm; WS 81cm; WT ♂ 555–648g, ♀ 618–750g. **SD** W coast of N America from Baja California to Alaska including E Aleutians; limited seasonal movements, with some reaching C Aleutians and Pribilofs. Accidental on Chukotski and Kamchatka peninsulas. Monotypic. **HH** Found exclusively on rocky shores and jetties. **ID** Large, plump black shorebird with broad wings, short tail and prominent bright red bill. Ad. all black; juv. (Aug–Feb) has paler brown fringes to mantle, scapulars and wing-coverts, and paler underwing in flight. **BP** Bill thick, chisel-like, bright red (dull red with blackish tip in juv.); eyes yellow with bright red eye-ring; rather thick tarsi pale yellow in ad., dull pink in juv. **Vo** Very vocal; various loud piping calls include *queep* and *weeyo*, run together in alarm, *kleep kleep klidik-klideeew*.

EURASIAN OYSTERCATCHER   CTKJR
Haematopus ostralegus

L 40–47.5cm; WS 72–86cm; WT ♂ 425–805g, ♀ 445–820g. **SD** Widespread from NW Europe to NE Russia. Rather isolated *H. o. osculans* on shores of Kamchatka and Sea of Okhotsk to Amur River, and NE coastal China, including Liaoning, and Korea. Winters south on Asian continental coast to Korea, SE & S China, also Japan (uncommon Sep–Apr), rare Taiwan. **HH** Seacoasts with sand and mudflats, estuaries and large rivers, where takes molluscs and various worms by probing deeply. Alone or in small groups. **ID** Large, boldly pied shorebird with prominent bright red bill. Ad. head, neck and upperparts black. Underparts white, from shoulder and mid chest to undertail-coverts. Non-breeding may develop trace of white half-collar on throat. Juv./1st-winter mantle and scapulars have narrow brown barring, creating scalloped effect. In flight, broad white wingbar on primaries and secondaries, rump and tail mostly white with black terminal band; wingbeats deep, rather slow. *H. o. osculans* shows more extensive white on closed wing and has somewhat longer bill than nominate. **BP** Bill long, thick, largely red, but orange at extreme tip in ad., blackish at tip in imm.; eye-ring and irides red; legs red when breeding, otherwise flesh pink to orange-red. **Vo** Very vocal on ground and in flight: various sharp, shrill piping *pi pi pi*, *kip*, *kleep* and *klee-eep* notes, and on breeding grounds gives rolling *kr r r r r r r* and duet of *klee-eep* notes in display.

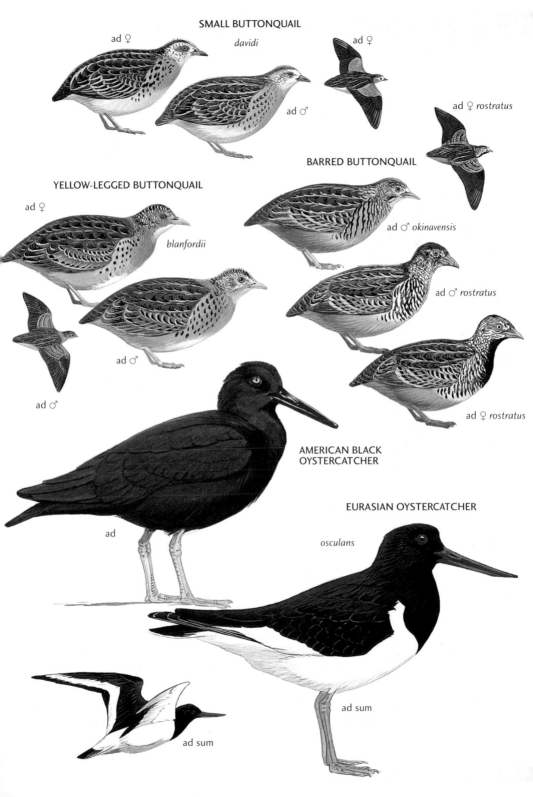

SMALL BUTTONQUAIL

ad ♀

davidi

ad ♀

ad ♀ *rostratus*

BARRED BUTTONQUAIL

ad ♂ *okinavensis*

ad ♂ *rostratus*

YELLOW-LEGGED BUTTONQUAIL

ad ♀

blanfordii

ad ♂

ad ♂

ad ♀ *rostratus*

AMERICAN BLACK OYSTERCATCHER

ad

EURASIAN OYSTERCATCHER

osculans

ad sum

ad sum

IBISBILL C
Ibidorhyncha struthersii

L 39–41cm; WT 270–320 g. **SD** Himalayas and SC Asia across W & C China; rare and local resident northeast to E Inner Mongolia, Hebei and Beijing; old records from Beidaihe. Monotypic. **HH** Fast-flowing, stony- or shingle-bedded rivers to 4,400m in summer, lower in winter. **ID** Very distinctive with long bright red bill. Ad. crown, face and chin black, narrowly bordered white; neck and upperparts grey, underparts bright white with black gorget on upper breast. Juv. browner, with less distinct pattern; brown fringes to mantle, back, scapulars and wings. In flight, wings and tail grey, former with white bases to primaries, tail square-cut with narrow grey and white barring on outer rectrices. **BP** Crimson bill is long, slender and decurved (greyish-pink in juv.); eyes red or reddish-brown; legs proportionately rather short, crimson (ad.) or grey (juv.). **Vo** Sandpiper-like piping *klew-klew*, rapid *tee-lee-tee-lee* or high-pitched, ringing *fwee-ki-ki-ki-ki*.

BLACK-WINGED STILT CTKJR
Himantopus himantopus

L 35–40cm; WS 67–83cm; WT 166–205g. **SD** Widespread: Americas, SW Europe and Africa to S & SE Asia, Australia and New Zealand, also China and SE Russia. *H. h. himantopus* W Europe to E Asia, where migrant E China (also breeds) and Korea; spring overshoots scarce north to Magadan and S Kamchatka, but regular Hokkaido and Sakhalin. Some winter in south, e.g. Taiwan (1,000+). Breeds coastal C Honshu (where resident), also Taiwan, has bred Korea. Australasian *H. h. leucocephalus* (sometimes considered specifically as **White-headed Stilt**) accidental to Japan. **HH** Alone or in small groups at shallow wetlands. **ID** Tall, slender, boldly pied shorebird. Ad. has all-black wings with narrow white trailing edge to secondaries. ♂ has glossy black back; ♀ has brown-tinged back. Tail pale grey. Underparts, rump and wedge extending up back white. Head and neck entirely white, or have variable amounts of black on crown, nape and ear-coverts (♂ usually with more black, ♀ grey). Juv. brown-backed, with brown wings and crown. In flight, white back and narrow white band at tips of secondaries; legs from tarsus extend beyond tail. Ad. *H. h. leucocephalus* marginally larger, with slightly longer, slightly uptilted bill tip, slightly shorter legs, and has glossy black (♂) or brown (♀) hindneck from rear crown to base of neck. **BP** Bill needle-like, all black (ad.), or with reddish base (juv.); eyes dark red (ad.) or brown (juv.); extremely long flesh pink legs. **Vo** Noisy, but only when breeding: tern-like *kik-kik-kik* or *skyip-skyip-skyip*, and softer *pyu* and *pyuii* notes.

PIED AVOCET CTKJR
Recurvirostra avosetta

L 42–45cm; WS 67–80cm; WT 225–397g. **SD** W Europe to E Asia, breeding north to Transbaikalia. Winters on Chinese coast; rare migrant/winter (Nov–mid April) Korea and Japan; accidental Taiwan. Monotypic. **HH** Shallow wetlands, including inland steppe or desert lakes; on migration alone or in small groups at coastal and inland wetlands; forms larger flocks in winter. Swings bill from side to side in shallow water; also swims and upends. **ID** Tall, elegant,

boldly pied, with long legs and uniquely upturned bill. Ad. mostly white, with glossy black forehead, crown and nape, scapulars, primaries and carpal bar. Juv. brown where adult is black, with extensively brown-mottled mantle and wings. In flight, tarsi extend well beyond tail tip, upperwing white with black scapulars, carpal band and primaries; from below, wings all white except black primaries; flight fast. **BP** Bill black; eyes dark brown; long legs pale greyish-blue. **Vo** Commonly utters series of clear, flute-like notes: *klee-ee-klee* or *kluiit kluiit kluiit*; alarm call from ground or in flight *kweep kweep*.

NORTHERN LAPWING CTKJR
Vanellus vanellus

L 28–31cm; WS 82–87cm; WT 128–330g. **SD** Breeds British Isles to E Russia. In E Asia across N China to S Russian Far East, formerly Korea, and winters to south of 32°N in E China, also Korea and Japan south of N Honshu; scarce winter visitor Taiwan. Monotypic. **HH** Marshland fringes or wet meadows. Flocks in winter occur on grasslands and fields. **ID** Large, sociable and vocal; unique in having crest, deep black breast-band, and broad, rounded wings. Ad. crown, face, chin, breast-band and long wispy crest black. Lores, behind eye, cheeks, belly to flanks white; undertail-coverts dull orange/buff. Upperparts dark metallic green, with purple sheen at bend of wing. In winter has less contrasting face pattern and pale buff fringes to mantle and wing-coverts. Juv.: has shorter crest and pale buff fringes to back and wings. In flight, broad black wings with white-tipped outer primaries, underwing-coverts and rump white, and tail black; feet do not extend beyond tail tip. **BP** Bill short, black; eyes black; legs short, dull brownish-pink. **Vo** Very vocal. From ground gives rather cat-like *myuu* and in flight rather nasal *chee-zik chee-zik*, and a higher pitched plaintive *pee-wit*. ♂'s display flight impressively aerobatic with various calls combined into more complex 'song'.

GREY-HEADED LAPWING CTKJR
Vanellus cinereus

L 34–37cm; WS 75cm; WT 236–296g. **SD** Breeds NE China, also Jiangsu, Fujian, and Honshu. Migrates through E Asia to S China, northern SE Asia and NE India. Resident C Japan. Has strayed to SE Russian Far East and Transbaikalia. Monotypic. **HH** Wet rice fields, also wet grasslands and marshes; in winter also at riversides. **ID** Large, vocal and boldly marked. Ad. striking, with grey head, neck and upper chest, broad black bar separating grey chest from white belly; brown upperparts; has tiny yellow loral spot. Winter ad. has less prominent chest-band. Juv./1st-winter head, upperparts and chest mid brown vermiculated grey. In flight, black primaries contrast with white secondaries, brown wing-coverts and back; white rump and tail with black subterminal band; feet extend beyond tail tip. **BP** Bill yellow with black tip; eyes reddish-orange, narrow eye-ring yellow; legs long, dull ochre to bright yellow. **Vo** Flight call a sharp *kik kik*, and on ground gives plaintive but insistent *chee-it chee-it* in alarm, and series of staccato *chyink-chyink-chyink…* notes.

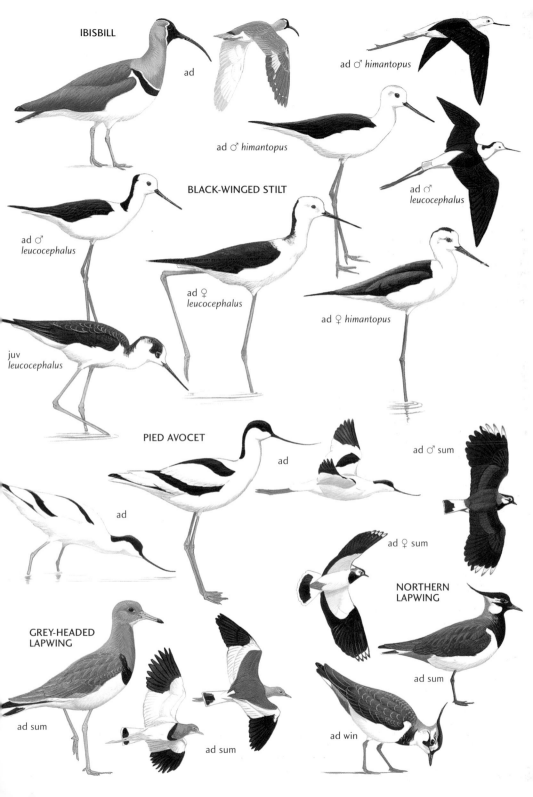

IBISBILL

ad

ad ♂ *himantopus*

ad ♂ *himantopus*

BLACK-WINGED STILT

ad ♂ *leucocephalus*

ad ♂ *leucocephalus*

ad ♀ *leucocephalus*

ad ♀ *himantopus*

juv *leucocephalus*

PIED AVOCET

ad

ad

ad ♂ sum

ad ♀ sum

NORTHERN LAPWING

GREY-HEADED LAPWING

ad sum

ad sum

ad sum

ad win

ad sum

SOCIABLE PLOVER   C
Vanellus gregarius

Critically Endangered. L 27–30cm; WS 70–76cm; WT 150–260g.
SD C Asia to SW Siberia, wintering from Africa to India, and
has strayed to Hebei, China. In serious decline. Monotypic.
HH Breeds on grassy steppes and winters on dry plains, sandy
and short-grass areas; migrants generally appear at agricultural
areas with short grass. **ID** Large, distinctly marked plover, with
prominent head pattern when breeding, and bold wing pattern. Ad.
crown and eyestripe black, supercilium and bar below eye clean
white, face and rear of supercilium creamy buff, neck, mantle,
back, scapulars and wing-coverts pale grey-buff. Underparts
grey-buff to lower breast, black on central belly, with chestnut
rear belly and white undertail-coverts. ♀ has less distinct black
areas mottled white. Winter ad. has broader supercilium, no black
or chestnut on belly and is scaled buff across grey breast. Juv.
paler, with Eurasian Dotterel-like head pattern (see p. 164), buff
fringes to scapulars and wing-coverts, and fine grey streaking
on neck. In flight has distinctive pattern of black primaries and
primary-coverts, white secondaries, grey-buff wing-coverts, and
white rump and tail with black band at tip. **BP** Bill short, slender,
black; eyes black; long, slender tarsi black. **Vo** Dry, harsh chattering *kretch* calls uttered singly or in series, *kretch-etch-etch*,
but usually silent outside breeding season.

PACIFIC GOLDEN PLOVER   CTKJR
Pluvialis fulva

L 23–26cm; WS 60–67cm; WT 100–192g.
SD The common large plover of the E
Palearctic, breeding across the E Russian
tundra, E Chukotka south to NE Sea of
Okhotsk and CW Kamchatka, and wintering in S Asia (including Nansei Shoto,
Ogasawara, Taiwan and S China), East
Africa and Australasia. Common migrant
through E China and Japan (April/May
and Aug–Oct), less common through Korea. Monotypic. **HH**
Open tundra. On migration, grasslands, fields, lakeshores, rivers
and coastal mudflats, in loose but often large flocks. **ID** Large, but
somewhat slender, long-legged plover, with rather upright stance
and small head. Ad. breeding face and underparts black, bordered
white from forehead, supercilia merging into broad neck-stripe,
which extends, though scalloped black, on flanks to vent. Upperparts
spangled gold, white and black. Non-breeding ad./juv. duller, but
with warmer buff or golden tones. Dark cap contrasts strongly with
supercilia, ear-coverts have prominent grey spot. At rest, wings
extend beyond tail tip when folded, but only 2–3 primary tips extend
beyond tertials, which are longer than in American Golden Plover. In
flight, faint white wing-stripe, dusky-grey underwing and axillaries;
toes project beyond tail tip. **BP** Bill black, longer than American or
Eurasian; eyes dark brown; legs long, greyish-black. **Vo** Plaintive,
somewhat soft, clear whistled *chu-vit* both in flight and on ground,
and more drawn-out *tu-ee, kyo-bee* and *chu-veee* (second syllable
clearly stressed); less commonly a trisyllabic *chu-ee-uh*. In display
flight, gives slow series of well-spaced slurred whistles *chuvee
chooeee…* **TN** Formerly considered conspecific with American
Golden Plover (as Lesser Golden Plover *P. dominica*).

AMERICAN GOLDEN PLOVER   JR
Pluvialis dominica

L 24–28cm; WS 65–72cm; WT 122–194g. **SD** Nearctic migrant
that regularly occurs in N Chukotka (bred there in 2006) and
is an accidental in Kamchatka, Japan (particularly Okinawa
in winter) and Hong Kong, but is probably overlooked among
the large numbers of Pacific Golden Plovers migrating through
E Asia. Monotypic. **HH** Dry tundra slopes. On migration on
mudflats and grasslands. **ID** Large plover with long wings and
legs, small head and rather upright stance; confusingly similar
to Pacific. Ad. breeding: ♂ has face and underparts solid black,
♀ slightly mottled on face and undertail-coverts; crown, nape
and upperparts black spangled white and gold. Prominent white
forehead and supercilia (more prominent than Pacific) merge
into very broad neck-stripe, which terminates and is broadest at
wing bend. Non-breeding ad. and young duller, with cool grey
tones (see Pacific). Dark cap contrasts strongly with broader
pale supercilium. At rest, wings long, extending well beyond tail
tip (by 12–22mm); tertials short, with usually 4–5 primary tips
extending beyond them (Pacific only 2–3). In flight, feet do not,
or very barely, project beyond tail tip (see Pacific), faint white
wing-stripe, grey underwing and axillaries. **BP** Bill black, smaller
than Pacific; eyes dark brown; legs greyish-black. **Vo** Mournful
whistled *kyuee* or *klee-i* (first syllable stressed), though often
sounds shorter, monosyllabic *klee or kleep*; occasionally trisyllabic *dlu-ee-oo* given in flight; usually higher pitched than Pacific.
On territory, display involves repeating single phrases over and
over, *koweedl koweedl…* or *wit wit weee wit wit weee…*

EURASIAN GOLDEN PLOVER   Extralimital
Pluvialis aricaria

L 26–29cm; WS 67–76cm; WT 160–280g. **SD** Breeds across
Russia almost to region, hence likely vagrant. **ID** Similar to Pacific and American Golden Plovers but larger and stockier, with
shorter wings barely extending past tail tip, broader wings with
prominent white wingbar in flight, and white (not grey) axillaries
and underwing. Breeding ♂ has white border to black of neck,
breast and flanks. **Vo** Call a plaintive whistled *peeuw* or *tüü*.

GREY PLOVER   CTKJR
Pluvialis squatarola

L 27–31cm; WS 71–83cm; WT 174–
320g. **SD** N Holarctic, wintering on most
southern continental coasts. In E Asia, *P.
s. squatarola* breeds across entire Russian
tundra to Chukotka, with *P. s. tomkovichi*
on Wrangel I. Migrates through E Asia
to SE Asia, including much of E China,
Korea and Japan, wintering on coast from
Jiangu south, with some in Japan from
C Honshu to Nansei Shoto, also S Korea and Taiwan. **HH** Open
tundra. On migration and in winter mainly on sand and mudflats.
May form dense flocks at roost, but usually alone or in wide-spaced
flocks. **ID** Largest plover, heavy-set, with very large eyes and stout
bill. Breeding ad. has black face and underparts (mottled white in
♀); white extends from forehead, via supercilia into broad neck-
stripe ending at wing bend. Vent and undertail-coverts white. Upperparts largely black, with very prominent white spots and feather
tips, affording silver-spangled appearance. Non-breeding ad. and
young duller, browner or grey, lack distinct supercilia, black eye
very prominent on rather plain face. In flight shows white rump
and tail (with grey tail barring), prominent white wing-stripe,
white underwing with black axillaries distinctive. **BP** Bill black;
eyes black; legs dark grey/black. **Vo** A mournful trisyllabic *pleee-
you-ee* or *tyu-eer-lee*. In display flight repeats melodious *kudiloo*
or *trillii*. **AN** Black-bellied Plover.

SOCIABLE PLOVER

ad sum

juv

ad win

ad sum

**PACIFIC
GOLDEN PLOVER**

ad ♂ sum

juv

ad win

ad win

AMERICAN GOLDEN PLOVER

ad ♂ sum

juv

ad win

ad win

ad win

EURASIAN GOLDEN PLOVER

ad win

ad ♂ sum

juv

GREY PLOVER

squatarola

juv

ad ♂ sum

ad win

COMMON RINGED PLOVER CTKJR
Charadrius hiaticula

L 18–20cm; WS 48–57cm; WT 42–78g. **SD** Wide-ranging Holarctic species of NE Arctic Canada, Greenland and NW Europe to Chukotka. Most winter well outside their breeding range, around Mediterranean, sub-Saharan Africa and Middle East. In E Asia, *C. h. tundrae* is a locally common breeder in NE Russia, from the Lena to E Chukotka south to Koryakia and NE Sea of Okhotsk. Generally scarce on migration through E Asia, and only accidental in NE China, vagrant Korea and Taiwan, and scarce migrant Japan (mostly Apr/May and Aug/Sep, occasional in winter). **HH** Breeds on drier coastal tundra, along coasts, lakeshores and rivers; on migration at coastal wetlands, mudflats and estuaries. **ID** Fairly large, plump plover, deeper-chested than Little Ringed Plover, with broad black breast-band, obvious white supercilium behind eye, white forehead, black forecrown, grey-brown crown and white collar from chin to nape. Winter ad./juv. have narrower, fainter, browner breast-bands and face markings, cheek patch rounded at bottom edge (see Semipalmated Plover). Flight fast and agile; shows prominent white bar on flight-feathers, tail plain, but darker at tip. **BP** Bill short, tip black, basal half orange in breeding season, mostly black with small yellow area at base in winter, juv. has largely dark bill; lores always black, extending to gape (see Semipalmated); eyes dark brown, lacking prominent eye-ring of Little Ringed, with distinct white supercilium in all plumages; legs dull ochre (juv.) to bright orange (breeding ad.), tiny web only between middle and outer toes. **Vo** Vocal on breeding grounds: a mellow, whistled *pyuui*; *too-li* or *tu-wheep* (emphasis on first syllable), usually in flight. When agitated, piping calls and sharp *skreeet*; and display involves low, slow bat-like flight and repetitive *tweeah-tweeah* calls.

SEMIPALMATED PLOVER JR
Charadrius semipalmatus

L 17–19cm; WS 43–52cm; WT 28–69g. **SD** Nearctic breeder, wintering south to S America. Just reaches E Asia in Chukotka, where a scarce breeder that presumably migrates across Bering Strait to winter in New World, but to be expected elsewhere in E Asia as vagrant. Reported three times Japan. Monotypic. **HH** Tundra; on migration at coastal wetlands and mudflats. **ID** Very similar to Common Ringed Plover, but slightly smaller and more compact. ♂ breeding lacks or shows only faint white supercilium, and black breast-band is narrow at front. ♀ has narrower breast-band than ♂; much overlap between two species. In winter ad. black replaces dark brown. Juv. like winter ad., but upperparts, coverts and tertials have dark subterminal bands and pale fringes. In flight, white wingbar slightly shorter than in Common Ringed Plover. **BP** Bill shorter, stubbier than Common, and more strongly tapered, black at tip, orange at base in ad., black with hint of orange at base in winter ad./juv., often with white on lores above gape; large eyes black, with very narrow yellow eye-ring (usually absent in Common); tarsi dull yellow-orange with webs between all three toes. **Vo** A clearly two-noted whistle recalling Spotted Redshank, *chu-wee* or *tew-it*; or longer *tu-eet* aids separation from Common. In display flight gives husky whistled *too-ee too-ee…*

LONG-BILLED PLOVER CTKJR
Charadrius placidus

L 19–21cm; WS 45cm; WT 41–70g. **SD** Restricted to NE Asia as a breeder, reaching SE Asia in winter. Occurs north to SE Russia, where scarce; summer visitor and resident NE & E China and Hokkaido; resident through main islands of Japan south of Hokkaido and Korea. Winters in SE China, scarce Nansei Shoto and rare Taiwan. Monotypic. **HH** Predominantly rivers with gravel or rocky bars and banks, also similar lakeside habitat, and on migration may appear at wetlands and wet fields. **ID** Large 'ringed' plover, averaging larger than Common Ringed, but appears noticeably less plump and more elongated because tail longer. Plumage very similar, but whereas forehead bar is black, eyestripe is brown rather than black. Black collar narrower than Common. Non-breeder has less contrast on face and only upper margin of collar black, the remainder brown. Flights usually short, shows narrow white wingbar (less prominent than Common); long tail has white tip and black subterminal band. Superficially similar Little Ringed much smaller, with small head, lacks wingbar, less attenuated at tail/wingtips, and has shorter bill (and prominent yellow eye-ring when breeding). **BP** Bill long, all black; narrow eye-ring pale yellow, irides dark brown; legs dull ochre to yellow. **Vo** Clear, piping *piwee* usually given in flight. Also *pyiu* or *byu*, lower and flatter than Little Ringed, and an explosive *sfreeit*. In display a duet of sharp *kip* notes in series, rising and falling in pitch, and a sharp, strong *pi pi pi pi…* in alarm.

LITTLE RINGED PLOVER CTKJR
Charadrius dubius

L 14–17cm; WS 42–48cm; WT 26–53g. **SD** Occurs from W Europe and N Africa to SE Asia and New Guinea. In E Asia, *C. d. curonicus* a common summer visitor (Apr–Oct) to E coastal China, Korea, the Russian Far East to western shores of Okhotsk Sea (scarce to the north), Sakhalin, and Japan. Year-round in Taiwan and on S China coast, and may also be found in S Japan in winter. **HH** Coastal, riverine and inland wetlands, often where substrate is muddy, sandy or shingle. Sociable, vocal, very active. **ID** Small, rather slim, sandy brown plover with prominent black collar. Appears more slender, more elongated, more long-legged than Common Ringed, thus structurally like Long-billed. Broad white and narrow black collars separate brown of nape from mantle. Upperparts including rear crown mid brown, flight-feathers darker, but very long tertials cover primaries. Face, lores and forecrown black (with white fringe above), and face patch extends to point below ear-coverts, forehead white. Juv. brown where ad. black, has less prominent eye-ring, but also has face patch extending in point below ear-coverts. Underparts white. Flight fast and agile, lacks (or has only very narrow) wingbar, tail plain, but paler at sides and darker at tip. **BP** Bill short, black with small orange base to lower mandible; eye-ring prominent, bright yellow, irides dark brown; legs dull ochre in summer, brown in winter. **Vo** In flight, occasionally on ground, utters soft, drawn-out downslurred *pee-oo* or more abrupt *peeu peeu*. On breeding grounds various notes: *pio pio, pyuu pyuu* and in slow bat-like flight a rapid, hard *pipipipipi*.

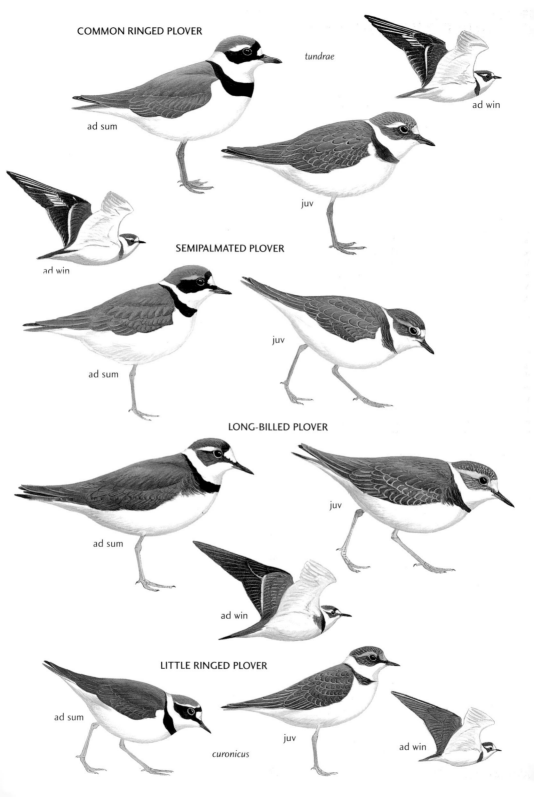

COMMON RINGED PLOVER

tundrae

ad win

ad sum

juv

ad win

SEMIPALMATED PLOVER

ad sum

juv

LONG-BILLED PLOVER

ad sum

juv

ad win

LITTLE RINGED PLOVER

ad sum

curonicus

juv

ad win

KILLDEER R
Charadrius vociferus

L 23–26cm; WS 59–63cm; WT 72–93g. **SD** N America and northern South America, with *C. v. vociferus* breeding as far north as Alaska; accidental to Alaskan Bering Sea coast and Chukotski Peninsula, Russia. **HH** Open areas with short vegetation, from savanna to agricultural land. **ID** Slender, very long-tailed plover, recalling Long-billed Plover, but has diagnostic double breast-band. Ad. upperparts mid to dark sandy brown crown, nape, ear-coverts and mantle, back, rump and uppertail coverts brighter, more orange-brown; forehead and supercilium from eye white. White of chin extends on neck-sides as white collar, below which are two black bands, narrowly separated by white; rest of underparts entirely white. Juv. (Jun–Sep) similar, but less clearly marked. In flight, which is buoyant and relaxed, wings long and slender, largely black with prominent broad white band on flight-feathers; very long tail and rump largely orange, with black subterminal band and white tips. **BP** Bill longer than lores, black; eyes large, black with narrow red ring; legs long, pale brownish-pink. **Vo** A strident piping *deee deeeyee tyeeeeeee deew deew tewddew*, but when agitated becomes even more strident, giving a rapid trilled *tttttttttttt*.

KENTISH PLOVER CTKJR
Charadrius alexandrinus

L 15–17.5cm; WS 42–45cm; WT 32–56g. **SD** Europe to Japan, and in N and S America. In E Asia, *C. a. nihonensis* breeds in E China north to Liaoning, Transbaikalia, and coasts of SE Russia and Hokkaido; year-round in much of Japan, Taiwan and E & SE China south of Changjiang. **HH** Beaches, coastal rivers and marshes, favouring drier areas but also mixes with other shorebirds on mudflats. Runs rapidly, stops, forages; runs again. **ID** Small, short-billed plover, size of Little Ringed Plover, with sandy brown upperparts. Forehead, supercilium and underparts white; forecrown, eyestripe and narrow band at upper-breast-sides black; chest band does not meet in middle. ♀ lacks black on crown and in non-breeding plumage both sexes have brown, not black, breast-sides. Flight rapid and agile, with prominent white wingbar and white tail-sides, central rectrices brown, darker at tip. **BP** Bill black, thicker and longer than extralimital nominate *C. a. alexandrinus*, which ranges east to N China; eyes dark brown, legs greyish-flesh. **Vo** Call, typically given in flight, a rattling trilled *trrrt*, soft *pi..pi..pi* or abrupt *pik*; from ground during breeding season gives *piru piru poi pirururu* and harder more rattling *geregeregeree* in bat-like display flight. **AN** Snowy Plover.

LESSER SAND PLOVER CTKJR
Charadrius mongolus

L 18–21cm; WS 45–58cm; WT 39–110g. **SD** Restricted as a breeder to NE Asia and C Asia, though winters to Africa, SE Asia and Philippines to Australasia. In region breeds E Chukotka, Kamchatka and NE Sea of Okhotsk. On migration occurs Japan, Korea, Chinese coast and Taiwan; often found in winter in S Japan, Taiwan and S China coast. Perhaps mostly *C. m. mongolus* on continental coasts, but more westerly *C. m. atrifrons* also noted on migration in Korea, Japan and S China, and *C. m. schaeferi* of E Tibet reported Japan. *C. m. stegmanni*

of NE Siberia, Kamchatka and Commander Is is common migrant through Japan, wintering in Taiwan, Philippines and, occasionally, Japan. Two groups, eastern *mongolus* and *stegmanni*, and western *atrifrons* and other subspecies, sometimes considered separate species (as **Mongolian** and **Lesser Sand Plovers** respectively). **HH** Tundra, particularly montane, in inland mountains. On migration, common on continental coasts, on sandy beaches, mudflats, estuaries and wetlands. **ID** Brightly coloured plover; proportions similar to Common Ringed Plover but darker above with rufous chest-band. The *mongolus* group has longer wings, shorter bill and tarsi than the *atrifrons* group; black mask, white forehead (more extensive and bisected by vertical black line in *C. m. stegmanni*), white chin and throat; broad rufous chest-band (with narrow black upper border) extends onto flanks and nape. The *atrifrons* group has shorter wings and longer bill and tarsi, a bolder mask, broader black forehead lacking white and breeding ♂ lacks black border to rufous chest-band. Non-breeders lack rufous band, face less contrasting, but with short narrow white supercilium, and are overall duller brown. In flight, narrow but noticeable white wingbar across flight-feathers (narrower on primaries than Greater Sand Plover); tail has white sides, rump and central rectrices uniform mid brown. **BP** *mongolus* group: bill short, rather thick, black and blunt-tipped, bill length from tip to base equals base to eye centre (see Greater); eyes black; legs greyish-green. **Vo** Call, generally in flight, a soft trilled *puriri* or *prrrp* (recalls Ruddy Turnstone) and a sharper *kip-ip*.

GREATER SAND PLOVER CTKJR
Charadrius leschenaultii

L 22–25cm; WS 53–60cm; WT 55–121g. **SD** Breeds from Turkey to C Asia and Mongolia, winters south to Africa, India and SE Asia to Australasia. *C. l. leschenaultii* moves through E China and Taiwan, less commonly Japan (mostly mid Apr–May, Aug/Sep; earlier than Lesser Sand in autumn), and rare in Korea. Winters uncommonly in S China, Taiwan and Nansei Shoto, accidental Transbaikalia and Ussuriland. **HH** Coastal mudflats, frequenting drier, sandier areas. **ID** Very like Lesser Sand Plover, with considerable variation and overlap, but generally larger, heavier, longer legged with thicker, longer bill. Ad. breeding has less black on head and narrower rufous chest-band (some with black upper margin) which does not extend far on flanks. Black mask mostly bordered above with rufous, and reduced white above eye; small forehead patch, chin, throat and underparts white. Non-breeding lacks rufous band, face is less contrasting and white supercilium more prominent. Juv. confusingly like Lesser Sand, best separated by general structure and bill size. In flight, faint narrow innerwing bar, conspicuous and broad on primaries; tail has white sides, rump and tail centre uniform mid brown; feet protrude clearly beyond tail tip. **BP** Bill black, longer from bill base to rear edge of eye, appearing stronger, with more prominent gonys and sharper tip than Lesser; eyes dark brown; legs longer than Lesser and dull pale yellow or green, not greyish. **Vo** Flight call a trilling *kuriri*, *trrrt* or *prrrirt* very similar to, but slightly deeper and drier than Lesser Sand.

KILLDEER

ad

vociferus

ad

ad win

KENTISH PLOVER

ad ♂ sum

ad ♀ sum

nihonensis

juv

ad ♀ sum

ad ♂ win

LESSER SAND PLOVER

ad win
mongolus

ad win

ad sum
atrifrons

ad sum
stegmanni

juv

GREATER SAND PLOVER

ad win

leschenaultii

ad win

ad sum

juv

ORIENTAL PLOVER CTKJ
Charadrius veredus

L 22–25cm; WS 46–53cm; WT 95g. **SD** Breeds Mongolia, N China and Transbaikalia, wintering in SE Asia and Australasia. Uncommon migrant (mostly Mar/Apr, Aug/Sep) through E China; rare Korea and Taiwan; very rare Japan. Monotypic. **HH** Dry grasslands, steppe and stone desert. **ID** Medium-sized elegant plover with long neck, long wings and long legs. Ad. breeding plumage striking, with head/face mostly greyish-white, lower neck and band across chest orange to chestnut, deepening to black on upper belly. Head has variable brown on rear crown/nape; upperparts, wings and tail mid brown; underparts white. Non-breeding ad./juv. like Lesser Sand Plover (p.162), with brown band on chest, rather plain face with pale forehead and supercilium, but juv. has prominent broad pale buff fringes to most of upperparts. In flight, lacks white wingbar; axillaries and underwing (coverts and flight-feathers) uniform brown (similar Caspian has white axillaries and underwing-coverts); tail brown with white outer rectrices and white tips to all but central feathers. **BP** Bill black, longer than from bill base to eye (like Greater Sand); eyes dark brown; legs longish, yellow to pink. **Vo** Flight call a loud, repetitive *chip-chip-chip*, also gives piping whistled *kwink*, a brief stony *dzhup* and piping *klink-klink-link*.

CASPIAN PLOVER Extralimital
Charadrius asiaticus

L 18–20cm; WS 55–61cm; WT 60–91g. **SD** Formerly considered conspecific with Oriental Plover; breeds in C Asia and migrates to Africa, but could stray eastwards. **ID** Similar to Oriental but slightly smaller, with slightly shorter neck and legs, dusky-white underwing-coverts and greenish-brown legs. **Vo** Call: a sharp downturned *tchup*.

EURASIAN DOTTEREL CKJR
Charadrius morinellus

L 20–22cm; WS 57–64cm; WT ♂ 86–116g, ♀ 99–142g. **SD** Breeds across Palearctic from Scotland to Chukotka. In E Asia, breeds in montane NE Siberia to Chukotski Peninsula, but migrates southwest to N Africa and Arabia. Scarce migrant through NE China (NE Inner Mongolia and Heilongjiang) and rare in Japan (Hokkaido to Okinawa) in spring, autumn and winter; accidental Korea. Monotypic. **HH** Breeds on montane and mossy tundra. On migration usually found on dry grasslands,

occasionally mudflats. **ID** Medium-sized compact plover with long legs, deep-chested, head small, with prominent white supercilia meeting on nape in distinctive V, also has white band across chest. Ad. breeding plumage attractive (♀ brighter), with dark blackish-grey crown contrasting with white supercilia, white throat, grey neck, narrow white chest-band bordered narrowly black above, with orange/chestnut breast, flanks and belly darkening to black; vent and undertail-coverts white. Non-breeding ad./juv. duller, greyish-brown with pale buff fringes to mantle, coverts and flight-feathers, underparts streaked greyish-brown, but long pale buff supercilium and white chest-band retained (albeit often faded). In flight, uniform upperparts lacking wingbar, but outermost primary shaft is white. **BP** Bill blackish-grey, slender; eyes large, black; legs greenish-yellow to ochre. **Vo** Calls, given in flight, a deep rolling *brroot* and, in alarm, a clear whistled *weet-weeh* or soft *pee-u-ee*. **AN** Mountain Dotterel.

PHEASANT-TAILED JACANA CTKJR
Hydrophasianus chirurgus

L 31–58cm; WT ♂ 126g, ♀ 231g. **SD** Pakistan to Philippines, E China and Taiwan, north in summer to the Chiangjiang River. Occasionally straggles to N China, Korea, S & C Japan (mostly Jun–Mar) and Ussuriland. Scarce local resident in S Taiwan and has bred in Nansei Shoto. Monotypic. **HH** Ponds (especially water chestnut ponds) and lakes with extensive surface vegetation, on which it forages. **ID** Rail-like, but has long legs and extremely long toes. Attractive, boldly patterned breeding adult is dark chocolate-brown above, black below with white head and throat, yellow nape bordered black, white wings, and extremely long (23–35cm) black tail. Non-breeding is white below, retaining only narrow chest-band, wing-coverts mottled brown, crown blackish with white supercilium, elongated tail-feathers lost. Juv. pale to mid brown above, whitish below, with brown crown, yellowish neck and dark brown bar on neck-sides. Wings long and broad. In flight, wings almost all white, with narrow black tips, long trailing tail (♂) and prominent bundle of toes extending well beyond tail (♀) are all distinctive. **BP** Bill grey, legs greenish-grey; eye large, irides dark in adult, yellowish in young. **Vo** Generally silent, but song consists of a bell-like *ku-wuuul*; calls include a purring *hrrrrrt*, a harder *chuu chuu* and a nasal mewing in alarm.

ORIENTAL PLOVER

ad ♀

juv

ad ♂ sum

CASPIAN PLOVER

ad ♀

juv

juv

ad ♂ sum

EURASIAN DOTTEREL

juv

ad win

ad ♂ sum

juv

PHEASANT-TAILED JACANA
(not to scale)

ad win

ad win

ad sum

juv

GREATER PAINTED-SNIPE
CTKJR
Rostratula benghalensis

L 23–28cm; WS 50–55cm; WT 90–200g. **SD** Widespread across Africa, India to E China, Taiwan and Japan. In E Asia, *R. b. benghalensis* resident in E China north to Changjiang R, but a summer visitor north to S Liaoning; also resident Taiwan (but most perhaps summer migrants) and Japan north to C Honshu. Rare on migration and in winter in Korea, where also a very rare breeder. Has strayed to N Japan and Russian Ussuriland. **HH** Wet grasslands, including ricefields and rush fields, from sea level to *c*.900m, Very secretive, crepuscular or nocturnal. **ID** Plump, short-tailed snipe-like shorebird with prominent 'spectacles' and 'braces', which exhibits reverse sexual dimorphism. ♀ bright and boldly marked, with white eye-ring extending back from eye, contrasting with chestnut or maroon head and chest, and white stripe curving up breast-sides ('braces') from white underparts and extending into sandy brown V on back. Mantle and wings greenish-brown. Often raises wings, revealing white underwing-coverts and pale brown spots on flight-feathers. ♂ considerably duller and smaller, mid brown above with irregular dark markings, sandy eye patch and stripes on neck and back, and whitish underparts. Wings broader and rounder at tip than typical shorebirds. Flight typically short on rounded and arched wings, with feet dangling well beyond tail tip. **BP** Bill longish with slightly drooping tip, ♀ pinkish-orange, ♂ pinkish-grey, with dark tip; eye large, irides dark brown; legs greenish-grey. **Vo** ♀ gives long series of low, slow, rhythmic, somewhat owl-like hoots, *hooo-hooo-hooo*, rising and falling slowly, or *koh koh koh uk uk* at night from ground or in circular flights. Also an explosive *tooick* and *twick-twick*.

EURASIAN WOODCOCK
CTKJR
Scolopax rusticola

L 33–35cm; WS 56–60cm; WT 144–420g. **SD** Widespread, ranging across forested regions of Russia to Sea of Okhotsk, Sakhalin, adjacent NE China, Hokkaido, and N & C Honshu. Migrates through E China. Winters (Nov–Apr) from C Honshu south including Nansei Shoto, Korea, SE China and Taiwan (scarce). Monotypic. **HH** Damp deciduous and mixed forest (and edges) of lowlands and hills, but in winter moves to lower altitudes/latitudes where often near streams or in damp woodland, and dry grasslands and sugarcane fields.

On passage in any thicket-type habitat. **ID** Rotund, long-billed, short-legged 'shorebird'. Larger than snipe, with broader, more rounded wings and occurs in different habitats. Forehead grey, crown and nape have broad dark brown and pale greyish-buff transverse bars. Upperparts warm brown, coverts, scapulars and particularly tertials with blackish-edged oval pale-centred spots and warm cinnamon-brown bands; underparts narrowly barred dark brown. Relies on cryptic plumage, thus freezes if disturbed. When flushed may run or fly; flight slow, straight on deeply bowed wings, head held high. In flight, squat and heavy, wings plain dark brown, rump and short tail orange-brown, tail has very dark brown subterminal band and grey terminal band. Twilight roding display flight over woodland canopy around margins of territory is slow, with bill pointing down. **BP** Bill long, straight, deep-based, pinkish-grey with dark tip; eye very large with split white eye-ring, irides black; short tarsi pinkish-grey. **Vo** On take-off gives abrupt *chiki chiki chiki*. During display flight gives series of soft grunting *buu buu, buu buu* calls followed by louder, sibilant *tswissick*.

AMAMI WOODCOCK
J
Scolopax mira

Vulnerable. L 34–36cm. **SD** Endemic to N and C Nansei Shoto of Japan, from Amami-Oshima to N Okinawa and associated smaller offshore islands. Monotypic. **HH** Damp subtropical evergreen broadleaf hill forest, where highly terrestrial. Eurasian Woodcock winters in same range, albeit in drier non-forested habitats. **ID** Very similar to Eurasian Woodcock, but upperparts darker, greyer brown, oval spots darker, lacking pale centres, with much narrower brown bands; distinctive triangular orange-brown patches separate dark spots at leading edge of flight-feathers, noticeable on closed wing. Displays on ground and, in March, during roding flight in 0.5–1km circles over territory. If disturbed most likely to run, but may fly up and land in trees, descending only later. Flight typically slow on deeply bowed wings, but can be fast, and direct. **BP** Bill long, straight, deep-based, pinkish-grey with dark tip; eye very large, irides black with pink and white crescents above and below eye, and bare pink skin immediately behind it (lacking in Eurasian); tarsi pinkish-grey. **Vo** On take-off gives sharp *je je* or *vett vet-vett*, snipe-like *jheet* and louder, duck-like *ghett!* in spring, as well as high *puu* or low *vuu* flight calls; when roding gives a low, burping *wart wart wart* similar to Eurasian, interspersed with the loud *ghett!* call. On ground, strong *gu* and softer *ku* calls given during display.

GREATER PAINTED-SNIPE

benghalensis

ad ♂

ad ♀

ad ♂

ad ♀ display

EURASIAN WOODCOCK

ad

AMAMI WOODCOCK

ad

ad

JACK SNIPE
CTKJR
Lymnocryptes minimus

L 17–19cm; WS 38–42cm; WT 28–106g.
SD Scandinavia to C & NE Siberia.
Breeds across Russia to the Chaun Gulf.
Rare migrant through NE China and E
China coast; rare or accidental in Japan
(Sep–Jan), vagrant Korea and Taiwan.
Monotypic. **HH** Marshes and water-
logged bogs in sparse taiga and tundra.
Conspicuous switchback display flight
over territory. In winter, solitary, favouring dense wet vegetation, marshes and fields. Probably overlooked on migration and
in winter due to its extremely skulking habits (largely nocturnal/
crepuscular and close flushing distance). **ID** Small, dark snipe,
with rather short bill. Crown dark (lacks pale central crown-stripe
of other snipe), bordered by broad, pale, split supercilium. Cheeks
pale, with dark crescent below eye. Upperparts dark brown with
distinct metallic purple-green gloss on mantle; two pale yellow-
brown mantle-stripes. Heavily streaked dark brown on breast-
sides, paler streaking on chest, belly and vent mostly unstreaked
buff. Juv. has white undertail-coverts. In flight, very dark wings
more rounded at tip than other snipe, with narrow white trailing
edge to secondaries, dark rump and black wedge-shaped tail
(lacks white and rufous bands of most snipe), underwing-coverts
and belly white; flight weak and slow, feet do not extend beyond
tail tip. Bobs head while feeding, but freezes rather than flushes,
until nearly trodden on, and if flushed flies less erratically than
other snipe before dropping again very quickly. **BP** Bill short,
straight (c.1.5 times head length), deep-based, mostly dark grey-
brown with pale, pinkish-brown culmen; eyes black; legs short,
greenish-yellow. **Vo** Typically silent, but may give single soft
gah, in series *gah gah gag gag*, or *gatch* when flushed. During
display flight gives short series of hollow *tok-tok-tok, tok-tok* or
ogogok-ogogok-ogogok notes, likened to cantering horse.

SOLITARY SNIPE
CKJR
Gallinago solitaria

L 29–31cm; WS 51–56cm; WT 126–
227g. **SD** Himalayas and C Asian mountains, e.g. Pamirs, Tien Shan and Altai,
northeast to Chukotka, wintering in N
India, Korea and Japan. *G. s. japonica*
from coastal SE Russia (Sikhote Alin;
breeding unconfirmed), and W Okhotsk,
possibly C Kamchatka, to Upper Anadyr
region in Chukotka; NE China. Winters
in sheltered areas of montane breeding range and at lower
altitudes elsewhere. Scarce migrant and winter visitor to Korea,
Changjiang Valley and SE China, but not yet in Taiwan despite
suitable habitat. **HH** Damp valleys, streams and alpine bogs. In
winter (mid Oct–late Apr) may remain at high altitudes in snowy
regions, provided wooded streams, springs or small marshes are
ice-free (e.g. Hokkaido and N Honshu), down to 150m, rarely
in muddy ricefields, occasionally overlapping with Common
Snipe (p.170) and Eurasian Woodcock (p.166). Typically alone.
Often bobs or rocks body steadily while feeding. **ID** Large, very
dark snipe unlikely to be confused with or seen alongside other
species due to habitat preferences. Dark, cold, milky blackish-
brown or greyish-brown, with heavy barring on flanks to belly.
Face appears whiter and back stripes broader and whiter than

other species. Unlike other snipe, barring whitish rather than buff;
well camouflaged. Primaries/tertials extend almost to tail tip; feet
do not project beyond tail in flight. Typically flies heavily and
more slowly than smaller snipe, dropping at streamside, but may
also escape high into forest; appears very dark and lacks white
trailing edge to secondaries of Common or Pin-tailed Snipes
(p.170), with small white belly patch and much rufous in graduated or wedge-shaped tail. **BP** Bill very long, straight, black at
tip becoming grey at base; eyes black; tarsi greenish-yellow. **Vo**
Harsh, Common Snipe-like *pench, kensh* or *jeht* when flushed,
and on breeding grounds a distinctive *chok-achock-a* in display,
augmented during display flight by a thrumming or 'bleating'
sound produced by tail-feathers.

LATHAM'S SNIPE
CTKJR
Gallinago hardwickii

L 23–33cm; WS 48–54cm; WT 95–277g.
SD E Asian breeding endemic that
winters in Australia. Breeds primarily
in Hokkaido and S Kuril Is, but also
highlands of C & N Honshu, occasionally
further south; also S Sakhalin and nearby
coastal mainland. Arrives mid Apr,
departs late Sep. Migrants pass through
W Honshu mid Jul–mid Aug. Status on
migration unclear due to difficulties of field identification; generally assumed that most migrate directly to Australia, but has been
recorded in Hebei and Liaoning, China; rare migrant through
Korea and Taiwan. Monotypic. **HH** Breeds in wet meadows,
grasslands with scrub, open woodland edge from sea level to
1,400m. Often perches prominently on treetops, fence posts or
utility poles. Display unique, involving high climbing flight followed by series of spectacular rushing dives and upward swoops.
On migration, dry grasslands and rice stubble. **ID** Largest snipe
in region. Closely resembles Swinhoe's Snipe (p.170), but larger.
Rather pale, tan or yellow-brown snipe; narrow black eyestripe to
base of bill, supercilium very broad between eye and bill; face
very pale. Mantle mid brown with broad buff fringes to scapulars;
wing-coverts largely buff. Distinctive tail pattern visible only in
hand or if exposed during preening: warmer orange tone (see
Swinhoe's cooler grey tone), central feathers pale orange-buff
to tip, though latter may appear whiter, narrow grey subterminal
band; outer feathers pale grey-brown with narrow grey and white
bars (see Swinhoe's, plain dark grey with white tips). Wings
long, but tail extends well beyond primaries/tertials; primaries
entirely obscured by tertials. Total length overlaps Swinhoe's,
but Latham's clearly longer-tailed than Swinhoe's, and both
are longer than Common and (especially) Pin-tailed (p.170).
Flushes heavily and flies more slowly than smaller snipe, usually more direct, less erratic than others; feet project beyond tail.
Underwing mainly dark; upperwing has prominent pale tawny
panel on median coverts and lacks obvious white trailing edge to
secondaries. **BP** Bill long, straight, black at tip becoming brownish at base; eyes black; tarsi greenish. **Vo** Calls less frequently
than Common or Pin-tailed but sequence of calls like Common.
Take-off call similar to other snipe, though usually more abrupt,
a deeper *geh geh* or *jek jek*. During display flight over territory
(May–Jun only) produces dramatic *tsupiyaku tsupiyaku tsupiyaku
gwo gwo gwo gwo-o-o*, a combination of vocalisations (first part)
and the thrumming of the stiff outertail feathers as it dives and
jinks. Otherwise generally silent. **AN** Japanese Snipe.

JACK SNIPE

ad

ad

SOLITARY SNIPE

japonica

ad

ad

ad

LATHAM'S SNIPE

ad

ad

PIN-TAILED SNIPE   CTKJR
Gallinago stenura

L 25–27cm; WS 44–47cm; WT 84–170g. **SD** NE Europe to Okhotsk Sea, N Yakutia, Chukotka, south to Altai, possibly N Mongolia; winters S China, SE Asia, Indonesia, Philippines. Common migrant E China, some winter Fujian, Taiwan, uncommon Korea, Okinawa (some winter S Japan). Monotypic. **HH** Taiga, forest-tundra and montane E Siberia. Migration/winter wetlands. **ID** Small to medium-sized, relatively dumpy snipe with pot belly. Smaller, more narrow-chested with shorter bill and tail than Common Snipe (CS) and more pointed wings. Head slightly larger relative to body than CS. Paler, cooler, greyer than similar CS and Swinhoe's, extensive white fringes to upperparts recall Solitary (p.168). Crown dark blackish-brown with pale ochre median stripe, supercilium, and face stripe below eye. Eyestripe narrows near bill base, where narrower than supercilium. Upperparts brown with two pale buff lines on mantle, and broad pale buff outer and narrow inner fringes (inner half width of outer) to more rounded scapulars, which have brighter, more rufous centres. Underparts dark brown, heavily barred on flanks, but belly white. Narrowly barred outer four tail feathers very narrow (less than half width of central feathers), but distinctive tail features of snipe only visible in hand or when preening. At rest, wings and tail short, wingtips more rounded than CS, primaries fall short of tail tip, tertials mostly cover primaries, whereas Swinhoe's tail extends well beyond tertials. CS's tail is intermediate in exposed length. Flushes fast but flight slower, less erratic than CS; may fly off high or drop quickly (more often than CS); feet protrude prominently beyond tip of short tail. Mostly dark underwing lacks central white bar; underwing-coverts black, very finely and extensively barred white; only very narrow white trailing edge to secondaries sometimes visible at close range when flushed. Paler coverts panel (like Latham's) differs from CS. **BP** Bill black at tip, brownish at base, deep-based and relatively shorter than others, tapering evenly from base to tip (CS's is parallel-sided); eyes black; legs greenish-yellow. **Vo** Usually calls when flushed, though seldom in sequence, similar to CS but less urgent and rasping; rarely gives high-pitched abrupt *jik* (unlike other snipe). Display flight involves steep dives accompanied by *chvin-chvin-chvin* calls and buzzing mechanical sound of air passing over tail-feathers, higher pitched than Swinhoe's.

SWINHOE'S SNIPE   CTKJR
Gallinago megala

L 27–29cm; WS 47–50cm; WT 82–164g. **SD** SC Russia (Altai to Transbaikalia) and S Russian Far East; winters south to S & SE Asia to N Australia. Common migrant E China, scarce Taiwan; uncommon migrant Japan (winters in extreme SW), Korea, Taiwan (some winter). Monotypic. **HH** Meadows or forest margins. Migration/winter on same wetland habitats as Common (CS) and Pin-tailed (PT). **ID** Large, with long bill and pointed wings. Very like CS and PT, but larger, heavier, closer in size but not coloration to Latham's. Rather uniform buff-brown, the saddle, wing-coverts and breast lack striking markings. Unlike Latham's, lacks contrast between dark saddle and pale coverts panel. Dark eyestripe extremely narrow at bill base, whereas supercilium very broad. Crown peak is to rear of eye, rather than forward of it like PT. Narrow outer tail feathers plain, broader than PT but narrower than CS. At rest, wings and tail short (much white on corners), primaries occasionally extend slightly beyond tertials, suggesting dark tip to tertials rather than obvious primary projection, but primaries fall short of tail tip. Like Latham's, tertial tips equal in length to, or just longer than, uppertail-coverts but Swinhoe's tail shorter and due to overall uniformity reddish-orange tail patch is most eye-catching plumage feature. Flushes heavily and more slowly, and flight less erratic than CS/PT; feet protrude less than latter's. Underwing mostly dark; lacks white trailing edge to secondaries. **BP** Bill black at tip, brownish at base; eyes black; legs thicker and yellower-olive than PT. **Vo** Often rises silently, rarely calls more than once, a rasping, unstressed, sneeze-like *jeht* or *chert*, very similar to CS but higher pitched, and repeated abrupt *sketch skretch*. Display flight involves repeated stooping from great height giving rasping *cheek-cheek-cheek...* and *chee-ua-chee* calls combined with rushing *ssseeeu* sound of air through tail.

COMMON SNIPE   CTKJR
Gallinago gallinago

L 25–27cm; WS 44–47cm; WT 72–181g. **SD** W Europe to Yakutia, Chukotka, Kamchatka and Commander Is, also NE China. *G. g. gallinago* common migrant E China and coastal archipelagos. Winters China south of the Chiangjiang, from C Honshu southwards, Korea, Taiwan. **HH** Damp, swampy areas from tundra to steppe. Migration/winter on wetlands. **ID** Medium to large, with long bill and pointed wings. Crown blackish-brown with pale ochre median stripe, supercilium and stripe below eye; eyestripe typically broader near bill base, where wider than supercilium. Upperparts brown, with two pale buff lines on mantle, and broad pale buff outer fringes to rather pointed scapulars which often taper to teardrop spot near tip of inner fringes, not continuous with outer web fringe (inner fringes darker than Pin-tailed's). Underparts dark brown, heavily barred on flanks, but belly white. Rusty-orange outer tail feathers almost as broad as central rectrices (see Pin-tailed and Swinhoe's). Juv. has neater buff fringes to upperwing-coverts. Wings short, tertials mostly cover primaries and primaries fall well short of tail tip. Flushes abruptly, with frequent, erratic zigzags, often high, though may drop quickly to cover. Mostly dark underwing has white bars along covert tips; secondaries have broad white tips. **BP** Bill black at tip, brownish-pink at base; eyes black; legs greenish-yellow. **Vo** Calls harshly on take-off and in flight, often in rapid sequence, a rasping *jaak* or *j'yak*, *jeht-jeht* or *ca-atch* and *skraaik*. Seemingly endless series of *chippa-chippa-chippa…* notes from ground or elevated perch when breeding; display involves circling, deeply undulating flight while 'bleating' or 'drumming' with outertail-feathers during short descent.

WILSON'S SNIPE *Gallinago delicata*   R

SD Nearctic species. **ID** Resembles Common (formerly considered conspecific) but has all-dark heavily barred underwing (lacks Common's white underwing bar), contrasting with white belly, and lacks broad white trailing edge to secondaries, thus resembles Pin-tailed in flight. Back broadly striped cream with almost black centres to scapulars and mantle, recalling Swinhoe's. Potential for vagrancy to E Asia compounds the difficulties of snipe identification.

PIN-TAILED SNIPE

ad

ad

SWINHOE'S SNIPE

ad

COMMON SNIPE

gallinago

ad

ad

WILSON'S SNIPE

ad

ad

SHORT-BILLED DOWITCHER J
Limnodromus griseus

L 25–29cm; WS 45–51cm; WT 65–154g. **SD** Nearctic shorebird, accidental in Japan (subspecies uncertain, presumably *L. g. caurinus* which breeds north to S Alaska). **HH** Coastal wetlands. **ID** Large, dark snipe- or godwit-like shorebird (very similar to Long-billed Dowitcher, from which best separated by flight call). Averages smaller and slimmer, with longer wings, shorter legs and bill than Long-billed, but considerable overlap in measurements. Eye located slightly higher on face and, combined with supercilium shape and bill structure, affords slightly different facial expression from Long-billed. Ad. breeding differs from Long-billed in having pale/white belly, dense dark spots on neck and barred flanks. Upperparts dark; scapulars have pale fringes. Non-breeding less plain than Long-billed, with arched supercilium, fine dark streaking on face, spotting on breast and paler grey flanks. Upperparts show more contrast between frostier pale grey and darker feather centres. Juv./1st-winter like juv. Long-billed, but warmer orange wash on breast and brighter upperparts also rufous fringes to, and bars across, dark tertials (Long-billed lacks bars). In flight, whitish secondary panel, dark tail and white wedge from rump onto lower back recall Spotted Redshank (p.178); axillaries and underwing-coverts darker than Long-billed; feet project just beyond tail. **BP** Bill long, thicker than snipe, shorter than Long-billed, with slightly drooping outer third (important distinction from Long-billed), blackish with slightly swollen tip; eyes black; tarsi toes greenish-yellow. **Vo** Distinctive flight call a low, slurred, double or triple series of rattling notes: *kew chu-chu* or *too-dulu*.

LONG-BILLED DOWITCHER CTKJR
Limnodromus scolopaceus

L 24–30cm; WS 46–52cm; WT 90–135g. **SD** Nearctic species that also breeds on Siberian tundra from E Chukotka almost to the Lena. Most winter southeast to S America, but also common in western N America, and small numbers appear in E Asia on migration and in winter (Sep–mid May). Rare migrant and winter visitor in Japan (Hokkaido to Nansei Shoto) and Taiwan; vagrant Korea and Chinese coast (Beidaihe). Monotypic. **HH** Boggy areas amidst tundra; in winter at muddy freshwater or brackish wetlands, occasionally coastal mudflats. **ID** Large, stocky, dark snipe- or godwit-like shorebird (see Short-billed Dowitcher for structural and other differences). Ad. breeding largely rufous-brown, including belly. Dark barring on breast-sides/flanks; scapulars dark rufous with white tips. Non-breeding pale grey, with clear division between grey breast and flanks and white belly; pale straight supercilium prominent; mantle/scapulars/coverts have dark shaft-streaks. Juv. resembles non-breeding ad., but has warm buff wash to grey breast, rufous fringes to dark scapulars/coverts, and rusty-fringed grey tertials. Non-breeding Long-billed and Short-billed almost impossible to separate using plumage alone (but tertials of juv./1st-winter dark-centred with neat pale fringes; see Short-billed); calls always distinctive. Winter birds in region all presumed to be this species. In flight, whitish secondary panel, dark tail and white wedge from rump onto lower back; white tail. **BP** Bill long, straight, thicker than snipe's, longer than Short-billed, blackish from base to swollen tip; eyes black; tarsi greenish-yellow. **Vo** Flight call a single high, sharp *keek* or sharp series, varying in length, of repeated chattering *keek-keek-keek* or *kyik-kyik-kyik-kyik*.

ASIAN DOWITCHER CTKJR
Limnodromus semipalmatus

L 33–36cm; WS 59cm; WT 127–245g. **SD** Rare, breeds in SW Siberia, Mongolia, NE China and SE Russia. Migrates (Apr/May; Aug–Oct) through E China, wintering to Hong Kong, SE Asia and N Australia. Rare migrant in Korea, Japan and Taiwan. Monotypic. **HH** Grassy swamps, where nests in small, loose colonies. On migration, coastal mudflats where feeds with distinctive rocking and rapid vertical probing action. **ID** The largest dowitcher, close to Bar-tailed Godwit in size. From smaller Long-billed and Short-billed by dark back (no white rump) and black legs; from godwits by bill size and shape. Ad. breeding largely rufous-brown, face plain rufous, lacking pale supercilium of other dowitchers or godwits, but has dark loral stripe with small white spot at base of lower mandible. Upperparts dark brown, all feathers with narrow to broad warm rufous margins. Non-breeding pale grey-brown, the upperparts with pale fringes to most feathers, underparts lightly barred, face very pale with dark lores and a pale supercilium. Juv. recalls non-breeding ad. but has warmer browner tones, buff fringes to upperparts and buff wash to breast, and bill extensively pinkish at base. In flight, wingtips rather black, inner primaries and secondaries paler, and has pale trailing edge to flight-feathers; pale, unmarked underwing (other dowitchers dark); rump and tail grey with dark bars (see Bar-tailed Godwit, p.174). Long legs trail beyond tail. **BP** All-black bill, long, straight and thicker than godwits, with broad base and prominent snipe-like swollen tip; eyes black; long tarsi blackish-grey. **Vo** A plaintive yelping *tye chu* or *chep chep*, and soft *kru-ru kru-ru*; on territory gives booming purrs from ground or in display flight.

SHORT-BILLED DOWITCHER

ad win

juv

ad sum

LONG-BILLED DOWITCHER

ad win

juv

ad win

ad sum

ASIAN DOWITCHER

ad win

juv

ad win

ad sum

EASTERN BLACK-TAILED GODWIT   CTKJR
Limosa melanuroides

L 36–44cm; WS 70–82cm; WT ♂ 160–440g, ♀ 244–500g. **SD** Widespread across E Palearctic. Breeds NE China and S Russia, from Lake Baikal and middle Lena River to Sakhalin, Kamchatka and Chukotka; migrates (Apr/May; Aug– Oct) through E China, Korea, Japan and Taiwan; winters southeast China, Taiwan, SE Asia and Philippines south to Australasia. Monotypic. **HH** Prefers fresh water. Breeds at wet meadows, swamps and lake margins in forested areas. On migration and in winter occurs at coastal wetlands and mudflats, margins of rivers and lakes, also wet fields; typically feeds in quite deep water. **ID** Large, tall, long-legged, long-necked and long-billed shorebird. Breeding ♂ has rufous or chestnut head, with dark crown, rufous/chestnut neck and breast, prominent white supercilium from bill to eye, pale chin, heavy black barring on lower breast, flanks and belly, vent whitish, barred. Mantle has rufous, grey and black spots; wing-coverts rounded with pale fringes. ♀ duller. Non-breeding much greyer, white on belly and vent. Juv. has warm peach/orange-buff wash to breast and neck. In flight, wings more rounded than either Bar-tailed or Hudsonian; reveals blackish-brown forewing (pale to mid grey in extralimital Western Black-tailed Godwit *L. limosa*), with little contrast, narrow white wingbar (much narrower than in Western), white rump, and black tail (legs extend well beyond). Underwing clean white (see Hudsonian). **BP** Very long bill straight, basal half pink, tip dark; eyes black; tarsi and toes black in breeding season, greenish-grey in non-breeder. **Vo** A high, strident yapping *ki ki ki* or *kek kek kek*. **TN** Formerly part of Black-tailed Godwit *Limosa limosa*.

HUDSONIAN GODWIT   J
Limosa haemastica

L 36–42cm; WS 74cm; WT ♂ 196–266g, ♂♀ 246–358g. **SD** Breeds Alaska, winters southeast S America. Annually reaches more southerly E Asian flyway in New Zealand; expected in E Asia on northward migration (reported Korea; accidental Japan). **ID** Breeding ad. has grey neck and dark rufous underparts, upperparts almost black; non-breeding ad. dark, plain grey, with almost black crown. In flight, wings more pointed than Black-tailed, and readily distinguished by black axillaries and underwing-coverts, and narrow white underwing bar. From above, has dark flight-feathers and narrow white upperwing bars, a largely black tail with a narrow white band across the uppertail-coverts; only the toes project beyond the tail-tip. **Vo** Call a high *kwidwid* or *kwehweh*.

BAR-TAILED GODWIT   CTKJR
Limosa lapponica

L 37–41cm; WS 70–80cm; WT ♂ 190–400g, ♀ 262–630g. **SD** Widespread Arctic breeder, primarily Palearctic, but also W Alaska; winters in Africa, S & SE Asia, Australia and New Zealand. *L. l. menzbieri* breeds NE Russia from Yana River to Chaun Gulf, whilst *L. l. baueri* (= *novaezealandiae*) breeds W Alaska. Taxonomy of small population in Anadyr uncertain. Moves through NE & E China, Korea, Japan (Apr/May, Aug/Sep), Taiwan, with some wintering in Taiwan and on SE China coast. **HH** Tundra and forest-tundra. On migration and in winter, coastal wetlands, mudflats, estuaries, where wades in shallow water and probes soft mud. *L. l. baueri* suspected of making longest non-stop migration of any bird – between Alaska and New Zealand – though en route pauses on E Asian coasts and in Japan. **ID** Large, tall, long-legged and long-billed shorebird, eastern races are slightly larger, but shorter-legged than Eastern Black-tailed Godwit, and stocky in comparison. Breeding ♂ has varying amounts of rufous on head, neck and breast, with pale supercilium, fine black streaking on neck, breast and flanks, and barring towards whitish vent. Mantle has dark brown, tawny and grey streaking; wing-coverts pointed, with very dark shaft-streaks. Larger ♀ typically has little rufous in breeding plumage. Subad. may migrate with little breeding plumage. Non-breeding duller, upperparts more curlew-like with pale fringes to grey-brown mantle, scapulars and wing-coverts. Juv. has dark brown upperparts notched pale buff, giving strongly patterned appearance to back. At rest, primaries extend beyond tail. In flight, grey-brown wings relieved only by darker primary-coverts and lack prominent wingbar; rump and lower back white, sparsely barred grey-brown in *L. l. menzbieri* and essentially appears grey-brown in *L. l. baueri* because heavily barred; tail variably barred grey-brown and white, but appears largely dark in both. **BP** Very long bill slightly upcurved, basal half pink, tip dark (culmen also dark in breeding season; bill noticeably longer in ♀); eyes black; tarsi black in breeding season, dark grey in non-breeder. **Vo** Slower and lower pitched than Eastern Black-tailed, a strident *ke ke ke, kek kek* or *kirrik*; *k-tek k-tek k-tek* heard prior to northbound migration.

EASTERN BLACK-TAILED GODWIT

ad win

juv

ad win

ad sum

ad win

ad win

ad sum
Western
Black-tailed Godwit

HUDSONIAN GODWIT

ad ♂ win

ad ♂ sum

ad ♀ win
baueri

ad win
menzbieri

BAR-TAILED GODWIT

ad win
baueri

juv
baueri

ad ♂ sum
baueri

LITTLE CURLEW
Numenius minutus

CTKJR

L 28–32cm; WS 68–71cm; WT 118–221g. **SD** E Palearctic, between Yenisei River and W Chukotka; winters New Guinea and Australia. Fairly common migrant E China; rare or scarce through Korea, Japan and Taiwan. Monotypic. **HH** Isolated subalpine areas with stunted forests. On migration and in winter, more commonly found in short grassland or dry cropfields. **ID** Smallest Asian curlew, with shortest, least-curved bill (see Whimbrel). Delicate; pale brown neck and breast with some dark streaking; upperparts darker with pale fringes; head prominently striped, like Whimbrel, with narrow, dark eye-stripe, broad pale supercilium and two dark brown lateral crown-stripes. At rest, primaries fall just short of tail tip. In flight, wings plain brown, underwing dark; rump and lower back pale brown, brown tail narrowly barred dark and white. Potential confusion with similar-sized Upland Sandpiper, a probable vagrant to region (see below). **BP** Bill very short, slender, mostly straight, slightly decurved at tip, brown with pink basal half of lower mandible; eyes appear large, irides black; tarsi greenish-grey. **Vo** Flight call a chattering three-note whistle *pipipi* or *te-te-te*; shorter, higher, more metallic than Whimbrel.

UPLAND SANDPIPER
Bartramia longicauda

Extralimital

L 26–32cm; WS 64–68cm; WT 98–226g. **SD** Breeds N America, winters southeast S America; migrates through C & N Alaska so likely to overshoot to Chukotka. **ID** Superficially resembles curlews, and same size as Little, with brown upperparts and whiter underparts, but small unstriped head on thin neck, very prominent large black eye, short straw-coloured bill with dark tip and yellow legs all distinctive. In flight, wings largely plain but has white outer primary shaft (like curlews) and pale innerwing contrasts with darker primaries. **Vo** Flight call a liquid *qui-di-di-du*.

WHIMBREL
Numenius phaeopus

CTKJR

L 40–46cm; WS 76–89cm; WT ♂ 268–550g, ♀ 315–600g. **SD** Widespread Holarctic breeder, wintering S Hemisphere. In E Asia, *N. p. variegatus* breeds Yakutia and Chukotka, winters south to S Asia and Australasia. Common migrant on coasts of Sea of Okhotsk, E China, Japan (Apr/May, Aug/Sep) and Korea, with some wintering S Japan, Taiwan and coastal SE China. N American *N. p. hudsonicus* (sometimes considered specifically distinct) recorded Japan. **HH** Forest-tundra, and some tundra and subalpine areas. On migration coastal mudflats, beaches, rocky shores, pastures, in flocks alone or with other species. **ID** Small curlew, with relatively short bill and somewhat short legs. Crown has two dark lateral stripes, separated by pale median stripe, prominent pale brown supercilium and narrow dark eyestripe. Upperparts dark brown; underparts buff. In flight, upperwing plain brown, axillaries and underwing-coverts white, barred dark grey-brown; rump and lower back pale greyish-brown, with narrow white stripe on lower back (recalling Spotted Redshank; see p.178); toes do not protrude beyond tail. *N. p. hudsonicus* differs from *N. p. variegatus* in being overall richer buff with brown, barred underwing and axillaries, dark back, rump and uppertail-coverts concolorous with wings and tail, lacking white stripe. **BP** Bill decurved at tip, brown with pink basal half to lower mandible; eyes black; tarsi bluish-grey. **Vo** Distinctive trilling seven-note whistle: *hwi pipipipipipi* or *hu-hu-hu-hu-hu-hu-hu*, given commonly in flight year-round; on breeding grounds only gives long bubbling trill in display.

BRISTLE-THIGHED CURLEW
Numenius tahitiensis

JR

Vulnerable. L 40–44cm; WS 82–90cm; WT ♂ 254–553g, ♀ 372–796g. **SD** Restricted as breeder to extreme W Alaska, wintering on C Pacific islands. Migrants have reached Chukotski Peninsula and Japan (Mar–May, Jul–Sep). Monotypic. **HH** Dwarf-shrub tundra. On migration/winter prefers drier habitats (including grassland) than Whimbrel, but also occurs in typical Whimbrel habitats. **ID** Similar to Whimbrel in size and shape, but distinguished by flatter crown, large pale cinnamon-buff spots on mantle and wing-coverts, finely streaked breast abruptly demarcated from plain pale belly, and paler buffy-brown supercilia. In flight, wings plain brown, axillaries and underwing-coverts cinnamon-brown barred dark brown; most noticeably, rump pale buff and tail pale cinnamon-buff with dark brown bars; toes do not protrude beyond tail. Only in close views are thigh bristles visible. **BP** Bill, like Whimbrel's, long, decurved at tip, mostly brown with pink basal half of lower mandible; eyes black; tarsi bluish-grey. **Vo** Calls include *chi-u-it; kwi; kuiiyo; piiyo*, whistled *whe-whe-whe-whe* and ringing *peeuu-pee* or *whee-wheeoo*.

SLENDER-BILLED CURLEW
Numenius tenuirostris

J

Critically Endangered. L 36–41cm; WS 80–92cm; WT 255–360g. **SD** Possibly extinct C Asian species wintering to Iraq and N Africa, which strayed to Japan in early 20th century. Monotypic (though some have recently debated status as a species). **ID** Similar to larger Eurasian Curlew in plumage (but close to Whimbrel size), from which distinguished by short black streaks on breast, large heart-shaped spots on lower breast/flanks, unmarked belly and vent, and in flight by inner primaries/secondaries being very pale. **BP** Bill slender, rather short and mostly straight, black with pink base to lower mandible, fine and slightly decurved at tip; eyes black; legs shorter than curlew, dark grey. **Vo** Higher pitched, sweeter *cour-lee* than Eurasian Curlew; alarm call a halting *bi-bi-bi*.

LITTLE CURLEW

ad

UPLAND SANDPIPER

ad

WHIMBREL

ad *hudsonicus*

ad *hudsonicus*

ad *variegatus*

ad *variegatus*

BRISTLE-THIGHED CURLEW

ad

ad

SLENDER-BILLED CURLEW

ad

ad

EURASIAN CURLEW   CTKJR
Numenius arquata

L 50–60cm; WS 80–100cm; WT ♂ 0.41–1.01kg, ♀ 0.475–1.36kg. **SD** Common and widespread from W Europe to SE Siberia, Transbaikalia and NE China. *N. a. orientalis* breeds NE Russia and migrates through E China, Korea, Japan and Taiwan, with some wintering in coastal C & SW Japan, S Korea, SE China and Taiwan. **HH** Peat bogs, wet grasslands and swamps in forested regions. On migration wet pastures and coastal grasslands, but most frequently at coastal wetlands and extensive mudflats; forages by deep probing. **ID** Large curlew with rather plain head and long bill. Overall mid to pale brown with grey tones, especially on closed wings; neck and breast buff, streaked dark brown, belly and vent white; at long distance appears pale overall. In flight, plain brown wings contrast with rather black outer primaries and primary-coverts, white patch extends to point on back, white rump and pale brown tail barred dark brown, and whitish axillaries and underwing-coverts separate it from similar Far Eastern Curlew; feet protrude beyond tail; wingbeats slow. **BP** Bill very long (♀ longer, ♂/juv. shorter), strongly arched, dark brownish-grey with pink basal half to lower mandible; eyes black; legs long, blue-grey. **Vo** Haunting, somewhat cracked, oft-repeated rising *couer-leuw couer-leuw*, or *curr-lee* given by foraging and flying birds; on territory rather vocal, has range of calls, also has a slow display flight on shivering wings interspersed with long descending glides when gives prolonged series of melancholic notes and whistling trills, culminating in frenzied series of churring whistles.

FAR EASTERN CURLEW   CTKJR
Numenius madagascariensis

L 53–66cm; WS 110cm; WT 0.39–1.35kg. **SD** E Asia, breeding in SE Russia north to upper Yana and Kamchatka; winters south to Australia. Uncommon migrant (mid Mar–early Jun, mid Jul–Oct) through E China, Korea, Japan and Taiwan, often with Eurasian Curlews. Occasionally winters in Taiwan. Monotypic. **HH** Wet swampy meadows, rivers and coasts in wet grasslands and swamps. On migration, coastal wetlands and mudflats. **ID** Largest curlew in region with very long curved bill. Overall brown with warmer orange-brown tones than Eurasian; neck, breast, belly and vent warm buff or brown, streaked darker to belly and flanks; at long distance appears dark overall. In flight, plain brown wings, brown lower back, brown rump and tail barred dark brown, and dark axillaries and underwing-coverts separate it from similar Eurasian; toes protrude beyond tail. **BP** Bill very long (♀longer, ♂/juv. shorter), strongly arched, dark brownish-grey with pink basal half to lower mandible; eyes black; legs long, blue-grey. **Vo** Similar to Eurasian, but flatter, deeper and longer *hoo ii nn* and on migration more moaning single notes; in alarm a guttural *gurrr-wheeuh*, and on territory calls from ground and in display flight like Eurasian, but more powerful and harsher.

SPOTTED REDSHANK   CTKJR
Tringa erythropus

L 29–32cm; WS 61–67cm; WT 97–230g. **SD** N Palearctic breeder, winters from S Europe and Africa to SE Asia. Breeds across N

Yakutia and Chukotka south to NE Sea of Okhotsk, winters from S China to S Asia. Common migrant (Mar/Apr, Aug/Sep) E China, Korea, Japan, scarce Taiwan; with some wintering in Taiwan and in coastal SE China. Monotypic. **HH** Wet tundra and forest-tundra. On migration, inland and coastal wetlands including wet fields, pools, rivers, marshes and mudflats. **ID** Tall, elegant, long-legged and long-billed shorebird. Breeding plumage unmistakable, black with fine white spots on mantle, scapulars, wing-coverts and flight-feathers, white crescents above and below eyes, fine white barring on black flanks and vent. In non-breeding plumage pale grey, with prominent white supercilium, grey mantle, grey upperparts with fine white speckling, and grey on neck and chest but white on throat and belly. Juv. like non-breeding ad. but browner above, more extensively grey below. At rest, wings relatively short, tertials overlay primaries and wingtips just reach tail tip. In flight, plain black or grey wings lack wingbar, but contrast with narrow white rump and lower-back stripe, tail black or grey with fine barring; axillaries and underwing-coverts white; tarsi protrude beyond tail. **BP** Bill long, fine, black with red base; eyes black with narrow white ring; long slender legs deep blackish-red in breeding plumage, red-orange in non-breeding and orange in juv. **Vo** Short *chip chip chip* in alarm but most distinctive is sharply whistled *chui* or *chew-ick* flight call (recalls Pacific Golden Plover but sharper). In display flight gives slow chattering and trilling *tyou-vee-veeyou tik-tik-tik tyou-vee-veeyou* song.

COMMON REDSHANK   CTKJR
Tringa totanus

L 27–29cm; WS 59–66cm; WT 85–155g. **SD** Breeds across N Europe, W & C Russia east across S Siberia to Hokkaido (though very local). *T. t. ussuriensis* breeds NE China, southern Russian Far East, fairly common migrant E China, Korea, Japan (scarce) Taiwan; some winter in Taiwan and coastal SE China. **HH** Favours wet marshes (fresh and brackish) and grassland pools; on migration, inland and coastal wetlands including wet fields, pools, rivers, marshes and mudflats. **ID** Medium-sized shorebird with long legs and mid-length bill, less elegant than Spotted Redshank, with shorter neck and shorter bill. Ad. breeding, brown above and on head; underparts white with prominent dark streaking/spotting on neck, breast and flanks. Non-breeding greyish-brown above, buff on breast-sides with faint streaking. Juv. like non-breeding ad. but browner upperparts have pale buff fringes; legs paler orange. In flight, prominent white panel on trailing edge of secondaries and inner primaries, white wedge from rump up back, tail pale grey with fine dark bars. *T. t. ussuriensis* larger and paler than races to west, and juv. has paler tones, rendering them confusable not only with Spotted but even with yellowlegs (p.180). **BP** Bill has black tip, orange base; eyes black with narrow white ring; long legs red in breeding plumage, reddish-orange at other seasons; orange toes protrude beyond tail. **Vo** Highly vocal, calling on ground, in flight and particularly during display flight when circles on stiffly beating wings; typically gives a whistled, rather anxious, *chew-chew chew-hu-hu*, but also a longer, more drawn-out and plaintive *tyooooo*, and excited *tyeeu*.

EURASIAN CURLEW

orientalis

ad Eurasian

ad Far Eastern

FAR EASTERN CURLEW

ad

ad

COMMON REDSHANK

SPOTTED REDSHANK

ussuriensis

ad sum

ad sum

juv

juv

ad win

ad win

MARSH SANDPIPER *Tringa stagnatilis*   CTKJR

L 22–26cm; WS 55–59cm; WT 43–120g. **SD** European Russia, Kazakhstan, NE China to Transbaikalia, C Yakutia, and Ussuriland. Migrant E China, Korea, Japan and Taiwan; some winter SW Japan, SE China and Taiwan, but most S Asia. Monotypic. **HH** Steppe and forest-steppe swamps; wetlands on migration. **ID** Graceful, pale grey above, white below, with small head and long neck, differing from Green and Wood sandpipers (p.182) by very long legs and white back, and from larger greenshanks in size. Breeding has pale brownish-grey mantle, scapulars and wing-coverts with black centres giving black-spotted appearance to back. Head and neck pale grey with pale supercilium; dark spotting extends to flanks, underparts otherwise white. Non-breeding is plain pale grey; face pale, though may show contrasting dark crown and ear-coverts, and pale supercilium, giving capped appearance. Juv. like non-breeding ad. but has darker fringes to upperparts, face plain white. In flight, long white wedge extends up back from rump; tail white with fine dark bars, tarsi protruding beyond tip. **BP** Bill mid length, needle-fine, black; eyes black; very long legs greenish-yellow. **Vo** High-pitched single *kiu* or *pyoo*, sometimes repeated in greenshank-like series *kiu-kiu-kiu*, but faster, higher; also a repeated abrupt *yup* when flushed, and series of slurred whistles *tu-ee-u* or *tuleeeoh* on breeding grounds.

COMMON GREENSHANK *Tringa nebularia* CTKJR

L 30–35cm; WS 68–70cm; WT 125–290g. **SD** Scandinavia across E Siberia to Yakutia, Chukotka and Kamchatka. Common migrant through E China, Korea, Japan and Taiwan; some winter SE coastal China, S Japan, Korea and Taiwan. Monotypic. **HH** Taiga swamps, lakes and streams. Wet fields and coastal wetlands on migration. **ID** Heavy-set, long-legged, pale grey above and white below; heavier looking than Marsh with long stout bill. Ad. breeding crown, face and neck finely streaked dark grey, upperparts grey-brown with many feathers having blackish centres and black marginal triangles. Lower neck and breast-sides heavily streaked grey. Non-breeding paler with less prominent streaking on underparts. Juv. browner than non-breeding ad., upperparts pale-fringed. In flight, wings dark, underwing dusky grey; long white wedge extends up back from rump, tail also white with fine dark bars; tarsi protrude beyond tail. **BP** Bill greyish at base, black tip slightly upturned; eyes black with narrow white ring; legs long, dull greenish-grey. **Vo** Calls frequently on ground and in flight, a loud, level, three- (or more) note whistle: *chew-chew-chew*. In display flight a prolonged series of loud *clu-wee…* or *tlee-tyouee* whistles. In alarm a series of shrill *skyip* notes.

NORDMANN'S GREENSHANK   CTKJR
Tringa guttifer

Endangered. L 29–32cm; WT 136–158g. **SD** Very rare E Asian endemic, breeds only near shores of W (and probably N) Sea of Okhotsk, also N Sakhalin, possibly W Kamchatka. Winters SE Asia. Rare migrant in Japan, Korea, Taiwan and SE coastal China. Monotypic. **HH** Tree-nesting, favours larch forests with swamps near wet coastal meadows and mudflats. Migration/winter on mudflats. Feeding diagnostic as chases hurriedly after

prey. **ID** Slightly smaller, stockier, with larger head, and shorter, thicker neck than Common Greenshank (CG). Ad. breeding like CG, but mantle, scapulars and wing-coverts blacker with pale fringes; head/neck heavily streaked, lower neck, breast-sides and flanks heavily spotted black; black spotting on crown, mantle and breast-sides may coalesce; under-parts white with isolated black spots on belly. Non-breeding upperparts more scaled and less streaked than CG, and pale grey mantle and wings have even paler grey fringes. Juv. darker, browner on crown, upperparts and wings than non-breeding ad. In flight, white rump extends in wedge up back, tail also white with few grey bars; underwing pure white. Toes just protrude beyond tail. **BP** Bill thicker (particularly at base) than CG, bicoloured (black tip, yellowish base) and slightly upturned; eyes black; legs thicker, shorter and yellower than CG. All three toes webbed (only two in CG). **Vo** Trilled *kee* or *kwee* and harsh *kwork* or *gwark* in flight. **AN** Spotted Greenshank.

GREATER YELLOWLEGS   KJR
Tringa melanoleuca

L 29–33cm; WS 70–74cm; WT 111–235g. **SD** Nearctic. Vagrant E Asia, on Wrangel I., Japan (has wintered) and Korea. Monotypic. **HH** Freshwater and coastal wetlands. **ID** Large, elegant, with plain face marked by white eye-ring and white supraloral. Similar to Common Greenshank (CG) but has longer legs, more spotted mantle and square white rump. Breeding ad. darker, more heavily streaked on neck and breast than CG; mantle, scapulars and wing-coverts almost black, with narrow tips giving distinctly spotted appearance; underparts largely white, with extensive dark barring on flanks and some on vent. Non-breeding like CG, but scapulars and wing-coverts more spotted than scaled, and head, neck and breast more heavily streaked. Juv. resembles breeding ad. though less black above, with clean, fine grey streaking on breast. At rest, primary tips reach just beyond tail, tertials just beyond tail base. In flight, wings uniform dark above, white or pale grey underwing-coverts; white rump and tail diagnostic (like Lesser lacks wedge on back); tail heavily barred dark grey; tarsi protrude beyond tail. **BP** Bill long, slightly upturned and more delicate than CG, thicker, more uptilted than Lesser, black tip, greenish-grey base; eyes black; long legs bright yellow. **Vo** Three or more loud CG-like *pyuu* or *chew* notes, the final syllable lower pitched; also single, double and multiple calls; in alarm repeats single *tew* call.

LESSER YELLOWLEGS *Tringa flavipes*   TJR

L 23–25cm; WS 59–64cm; WT 48–114g. **SD** Nearctic. Vagrant to E Asia: on Wrangel I, Chukotka Peninsula, most regularly Japan (migration/winter). Monotypic. **HH** Freshwater and coastal wetlands. **ID** Slightly larger, slimmer and longer winged than Wood Sandpiper, with only short, prominent supraloral (supercilium indistinct behind eye). Adult best separated from Greater by smaller size, longer legs and wings, and voice. Juv. greyer, with less streaking on lower neck and upper breast. At rest, primary tips extend beyond tail, tertials halfway along tail. In flight, square white rump resembles Greater; underwing-coverts white and pale grey; tail heavily barred dark grey; tarsi protrude beyond tail. **BP** Bill black at tip, greyish or brownish at base: shorter, thinner, straighter than Greater; longer, thinner, darker than Wood Sandpiper, p.182); eyes black; legs long, bright yellow. **Vo** Clear, high-pitched *tew* or *pyuu*, sometimes in quick series of 2–4 notes, rarely in distinct sets of three (typical of Greater). In alarm a rising *kleet*.

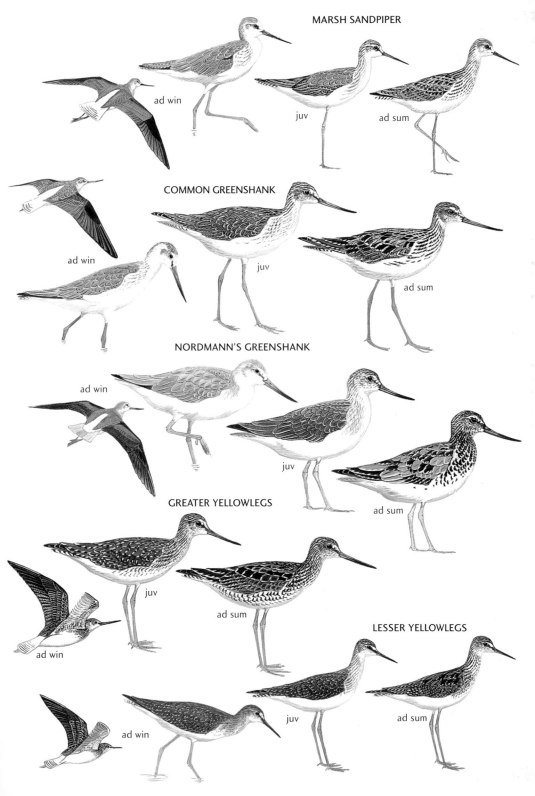

MARSH SANDPIPER

ad win

juv

ad sum

COMMON GREENSHANK

ad win

juv

ad sum

NORDMANN'S GREENSHANK

ad win

juv

ad sum

GREATER YELLOWLEGS

juv

ad sum

ad win

LESSER YELLOWLEGS

juv

ad sum

ad win

GREEN SANDPIPER *Tringa ochropus* CTKJR

L 21–24cm; WS 57–61cm; WT 53–119g. **SD** W Europe to Amur River, Yakutia and Sea of Okhotsk; winters Africa to SE Asia. Migrant Japan, Korea, E China and Taiwan; some winter south from C Honshu and Korea to coastal SE China. Monotypic. **HH** Forests with pools; migration/winter on streams, rivers, freshwater wetlands; rarely coasts. **ID** Stocky, with blackish-brown upperparts and prominent white rump and belly. Recalls smaller Wood and Solitary sandpipers. Ad. breeding greenish-brown upperparts with smattering of tiny white spots; head and neck streaked; heavy dark spotting on breast contrasts with clean white underparts; dark loral bar with white supraloral and narrow white eye-ring. Non-breeding plainer and greyer on upperparts and breast. Juv. above like breeding ad., underparts like non-breeding. At rest, wingtips reach tail tip; tertials fall just short of primary tips. In flight wings very dark (uniform with back); underwing almost black; square white rump very prominent; tail with 2–3 black bars; toes protrude just beyond tail. **BP** Bill just longer than head, fine-tipped, but somewhat thick basal half, black; eyes black; legs dark greenish. **Vo** Often calls explosively on flushing: *tluueet-veet-veet*. In alarm staccato series of *kwik* notes. In display flight gives complex, repetitive *tluuee-tewi…*

SOLITARY SANDPIPER Extralimital
Tringa solitaria

L 18–21cm; WS 55–59cm; WT 38–69g. **SD** Breeds northern N America; winters S America. Potential vagrant: reported Chukotski Peninsula, Russia (record not approved). **ID** Resembles Wood and Green, but has prominent white 'spectacles' and pale lores. Slighter than Green with longer, narrower wings, primaries extend beyond tail at rest; tertials much shorter. Rump and central tail all dark; white outertail-feathers barred black. Very dark underwing and uniform dark plumage with fine spots like Green. Breeding ad. darkest, with most prominent eye-ring; non-breeding slightly duller, juv. brownest. **BP** Bill longer than head, grey with blackish-grey tip; black eyes appear large in white ring; long tarsi dull yellowish-green. **Vo** Recalls Green, but less inflected: *peet-veet*, and when flushed *peet-weet-weet*.

WOOD SANDPIPER *Tringa glareola* CTKJR

L 19–23cm; WS 56–57cm; WT 34–98g. **SD** Scandinavia to E Asia; Yakutia, Chukotka, Kamchatka, Commander and N Kuril Is, Sakhalin, and NE China. Winters Africa, S & SE Asia to Australia. Common migrant Japan, Korea, E China and Taiwan, with some wintering Nansei Shoto, Taiwan and E coastal China. Monotypic. **HH** Swampy forest-tundra and taiga marshes. Migrants on inland freshwater and coastal brackish wetlands, wet fields, rivers and mudflats; often in small flocks. **ID** Medium-sized, rather slender sandpiper, similar to Green, but more elegant with longer legs. Ad. browner with prominent whitish supercilium, checkerboard back pattern, less contrasting rump and noticeably paler underwing. Breeding plumage brown, upperparts pale-fringed, giving overall spotted appearance. Non-breeding browner, less spotted. Juv. warm brown upperparts and grey-brown wash to breast. At rest, wingtips reach

tail tip; tertials fall just short of primary tips. In flight, wings and back brown; underwing pale grey; rump white; tail white with several narrow brown bars; toes protrude well beyond tail. **BP** Bill just longer than head, fine, straight, black; eyes black; legs pale, yellowish. **Vo** Common alarm or flight call a high *chiff iff-iff*. In display flight (typically above forest), gives lark-like torrent of *ti-lew ti-lew…* notes.

GREY-TAILED TATTLER *Tringa brevipes* CTKJR

L 24–27cm; WS 60–65cm; WT 80–162g. **SD** Breeds Siberia east of Yenisei across Yakutia, Chukotka and Kamchatka; winters south to Australasia. Fairly common coastal migrant in Korea and Japan. Some winter SW Japan and Taiwan. Monotypic. **HH** Stony rivers in montane forests, montane taiga; on migration coastal wetlands, lakes, rivers, beaches and, principally, mudflats and estuaries, occasionally rocky shores. **ID** Medium to large, rather uniform ash-grey sandpiper, with mid-length bill and relatively short legs, prominent dark loral stripe and white supercilium extending beyond eye. Ad. breeding has unmarked grey upperparts, distinctive facial pattern and narrow grey bars on lower neck, breast, flanks and vent-sides, though central belly and vent are unbarred (see Wandering). Non-breeding plain grey above, white below, lacking barring on underparts, but has dusky wash over breast and flanks. Juv. pale-fringed upperparts and grey wash over chest and flanks, supercilium broad, diffuse behind eye, contrasting with black eyestripe that is particularly obvious between bill and eye; cheeks largely white. At rest primary tips reach tail tip. In flight, upperwing, rump and tail grey; underwing uniform dark grey. **BP** Black bill, may be yellowish at base of lower mandible; eyes black; legs short, yellowish. **Vo** Clear double whistle, *pyu-ii* or *tuee-dee*, recalls Grey Plover, sometimes run in series: *tuee-dee-dee*. **TN** Formerly placed in genus *Heteroscelus*.

WANDERING TATTLER *Tringa incana* CTKJR

L 26–29cm; WS 66cm; WT 72–213g. **SD** Breeds Alaska, adjacent Canada, also locally in Chukotka. Most move to USA, Pacific islands. Very scarce or accidental visitor to coasts of E Russia, Kamchatka and Japan. Accidental E China, Taiwan, Korea and Sakhalin. Monotypic. **HH** Alpine zone, along fast-flowing streams. Migration/winter almost exclusively along rocky shores, sometimes with Ruddy Turnstone and Rock Sandpiper. **ID** Very similar to, but slightly larger and longer winged than, Grey-tailed. Ad. breeding has broader and more extensive dark barring on underparts. Non-breeding like Greytailed, but darker, best distinguished on habitat and voice. Juv. like non-breeding ad. but upperparts pale-fringed; supercilium short, narrow, only in front of eye, contrasting with diffuse black eyestripe between bill and eye; cheeks largely grey. At rest longer primary tips extend beyond tail. In flight, like Grey-tailed, but underwing blackish-grey, upperwing dark grey with blackish carpal and primaries. **BP** Bill dark grey, may be yellowish at base of lower mandible, lores black (bill shorter and darker than Grey-tailed); eyes black; legs short, yellowish. **Vo** Flight call a series of short clear piping notes on even pitch: *pi pi pi pi*. **TN** Formerly placed in genus *Heteroscelus*.

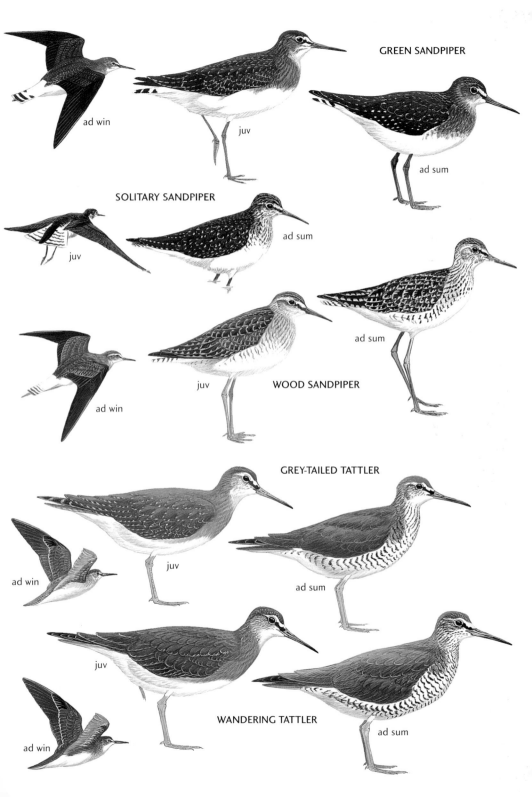

GREEN SANDPIPER

ad win

juv

ad sum

SOLITARY SANDPIPER

juv

ad sum

ad win

ad sum

juv

WOOD SANDPIPER

ad sum

GREY-TAILED TATTLER

juv

ad win

ad sum

juv

ad win

WANDERING TATTLER

ad sum

TEREK SANDPIPER
Xenus cinereus

CTKJR

L 22–25cm; WS 57–59cm; WT 50–126g. **SD** Breeds from Finland across Siberia to Yakutia, Chukotka and N Kamchatka. Winters on coasts of Africa, S & SE Asia to Australia. Fairly common migrant (Apr/May; Jul–Oct) with other shorebirds, but rarely numerous on its own, through Japan, Korea, E China and Taiwan, with some wintering in Taiwan. Monotypic. **HH** Breeds in forest-steppe to tundra, especially at rivers and lakes in lowland taiga. On migration mainly at brackish or saltwater wetlands. Appears hunched and leans forwards when running, forages typically close to water or tide edge on wettest mud. **ID** Medium-sized but rather oddly proportioned shorebird, appearing short-legged and stocky with distinctive long, upturned bill and steep forehead. Ad. breeding is brownish-grey with prominent black centres to coverts and scapulars, forming black bar on shoulder; head and neck grey with white streaking, underparts white. Non-breeding paler, greyer brown and lacks black scapular line. Juv. darker and browner than ad. with dark centres and pale buff fringes to coverts, creating scaled effect. At rest, blackish primary tips just visible beyond grey-brown tertials, and reach tail tip. In flight, grey-brown wing-coverts contrast with almost black leading edge, primaries and greater covert bar, and white secondaries form prominent white trailing edge; the rump and tail are both grey; toes do not protrude beyond tail. **BP** Bill long, thick and yellowish at base, with black tip; eyes black; legs short, yellowish-orange. **Vo** Various whistled calls, in short series of clear, liquid notes, *pwee-wee-wee* or clattering *wick-a-wick-a-wick*. On territory, song a repeated three-note phrase, *ka klee-rreee*, and in display flight a loud *kuved-ryouyou*.

COMMON SANDPIPER
Actitis hypoleucos

CTKJR

L 19–21cm; WS 38–41cm; WT 33–84g. **SD** Breeds at high and low latitudes from British Isles across Europe and Siberia to Chukotka and Kamchatka, south to Sakhalin, also NE China, Korea and C & N Japan. Fairly common migrant, but usually solitary or in very small groups, through Japan, Korea, E China and Taiwan. Winters from Africa to Australia and, in region, from C Japan south including E China south of the Chiangjiang, and Taiwan. Monotypic. **HH** Breeds around waterbodies, streams and rivers in various habitats from forest-tundra to steppe. On migration and in winter, generally solitary, at freshwater, brackish and coastal wetlands, often along lakeshores, streams and rivers, also beaches. **ID** Small brown and white sandpiper with rather short neck, bill and legs, distinctive horizontal stance and almost constant tail-bobbing habits. Ad. breeding mid brown above with brownish-grey breast patches and white wedge extending between closed wing and breast patch; white eye-ring and supercilium. Non-breeder plainer, breast patches less distinct. Juv. has scaled upperparts, the scapulars and wing-coverts have pale buff fringes. At rest, long-tailed, primaries fall well short of tail tip; tertial fringes have dark notches. Flight flickering, rapid wingbeats on down-angled wings interspersed with stiff-winged glides distinctive. Upperwing-coverts brown, primaries blackish-brown, with clear white wingbar on bases of flight-feathers to innerwing (also on underside); brown rump and tail has barred outer feathers; toes not visible beyond tail in flight. Spotted Sandpiper, accidental in region, very similar in non-breeding plumage and probably overlooked (see that species for separation). **BP** Bill short, dark grey; eyes black; legs short, olive or greyish-green to dull yellowish-brown. **Vo** Calls readily on taking flight and in alarm, a thin, high-pitched and plaintive *wee-wee-wee*, *hee-dee-dee*, or *chee-ree-ree* in quick series. Also a single, high-pitched, drawn-out *tweeeh* commonly given. Display flight, low on rapidly pumping wings, accompanied by fast twittering *sweedidee sweedidee ....*

SPOTTED SANDPIPER
Actitis macularius

JR

L 18–20cm; WS 37–40cm; WT 19–64g. **SD** Breeds across N America and winters south to S America. Accidental in region, to Chukotski Peninsula and Hokkaido, but probably overlooked in non-breeding or juv. plumages. Monotypic. **HH** Favours similar habitats to Common Sandpiper. **ID** Almost identical to Common in size and structure, but tail distinctly shorter, projecting less far beyond wingtips. Ad. breeding unmistakable due to heavy black spotting on white throat, neck and breast, and reduced spotting on flanks; underparts otherwise white; upperparts greyish-brown with bolder darker markings on mantle, scapulars and wing-coverts. Prominent blackish eyestripe, narrow white supercilium and eye-ring. Non-breeding like Common, lacks spotting but is greyer brown. Juv. also like Common, but has plainer, greyer lateral breast patches, and strongly barred median and lesser coverts (more prominent than in Common). Tail and wing proportions differ subtly from Common; tail shorter and projects less beyond wing tips. Feeding and flight habits identical. Tertial fringes lack faint dark notches of Common. Upperwing has similar pattern to Common, but wingbar short, narrower, reduced on innerwing, and underwing mostly white with dark flight-feathers and dark carpal bar; rump and tail like Common, but has less white on outer rectrices. **BP** Bill short, orange in breeding plumage, pale pinkish-horn in non-breeding, with dark tip and culmen (subtly thicker bill than Common, with tip slightly more drooping); eyes black; legs short, pale yellowish-flesh to bright yellow. **Vo** Though vagrants often silent, separation from Common Sandpiper possible by voice, being quieter, less resonant and lower pitched than Common, a single whistled *peet*, sometimes doubled, *peet-weet*, or in descending series like Green when flushed, *tueet-ueet-ueet*.

TEREK SANDPIPER

ad sum

ad win

ad win

COMMON SANDPIPER

juv

ad sum

ad win

SPOTTED SANDPIPER

1st-win

juv

ad sum

RUDDY TURNSTONE CTKJR
Arenaria interpres

L 21–26cm; WS 50–57cm; WT 84–190g. **SD** Largely coastal breeding range spans entire Arctic; winters in temperate and tropical regions from S America, Africa and Asia to Australasia. *A. i. interpres* breeds along Arctic Ocean east to Bering Strait and south to Anadyr. Common migrant in coastal E Russia, Japan, E China, Korea and Taiwan, with some wintering S Japan, Taiwan and coastal SE China. **HH** coastal tundra, on sandy or rocky shores, rarely inland. Outside breeding season typically on rocky shores, also beaches and occasionally coastal mudflats. May mix with other species, e.g. Sanderling, on beaches and shores, or may form dense flocks on mudflats, even foraging in short coastal grasslands. **ID** Small, stocky shorebird with complex breeding plumage, stubby bill and short legs. ♂ breeding has bright rufous-orange upperparts and wings, with white face and underparts, and various complex and individually variable black markings through eye, on neck, band across breast and on the scapulars. ♀ similar but more brown than orange on upperparts and less white, more brown on head. Non-breeding retains same black markings, but lacks orange tones on upperparts, being duller grey-brown. Juv. has blackish-brown upperparts with pale fringes, reduced face pattern and large blackish breast patch. At rest, primary tips just reach tail tip. In flight in all seasons has contrasting pattern of black, white and brown; back, scapular bar and bars at bases of flight-feathers white, rest of flight-feathers black, tail black with white base and tip; underwing white. **BP** Bill very short, black, wedge-shaped, and lower mandible slightly angled upwards; eyes black; legs short, orange when breeding, pale orange at other times. **Vo** Call, when flushed or in flight, a low staccato, rattling *tuk tuk-i-tuk-tuk.* On territory, gives rapid, rolling *chuvee-chuvee-viti-viti-vitvitititi.*

BLACK TURNSTONE JR
Arenaria melanocephala

L 23cm; WS 53cm; WT 120g. **SD** Beringian endemic, breeding only in W & NW Alaska. Most winter on Pacific coast of N America to Baja California, but breeding-season vagrant to Chukotski Peninsula, and reported Wrangel I. Accidental once on Hokkaido. Monotypic. **HH** Grass and dwarf-shrub meadows and marshes near coast. In winter, more exclusively favours rocky shores than Ruddy Turnstone, feeding in splash zone, particularly where barnacles proliferate. **ID** Small, stocky shorebird with stubby bill and short legs, closely resembling Ruddy in structure, but has very different plumage. Ad. breeding has almost entirely jet black upperparts, with black extending from head to upper breast; small areas of white below eye, forming large loral spot, and frosting on feathers of neck; rather long, loose scapulars and wing-coverts have narrow white fringes; underparts clear white. Non-breeding similar, but lacks white on face and is duller, slightly browner. Juv. more extensively blackish-brown on head and upperparts. In flight, shows distinctive white back patch, white band at tail base, white wing-coverts patch and wingbar, on otherwise black upperwing and tail. **BP** Bill black; eyes black; legs short, dark orange-brown. **Vo** Flight call a chattering *keerrt* (higher pitched than Ruddy). In alarm, a long low rattle or clear nasal *weepa weepa weepa…*

GREAT KNOT CTKJR
Calidris tenuirostris

L 26–28cm; WS 62–66cm; WT 115–248g. **SD** Uncommon to locally common range-restricted species of NE Siberia (Yakutia, Chukotka and Koryakia). Migrates (mostly Apr/May, Aug–Oct) to S & SE Asia and Australasia, through coastal Russian Far East, Japan, Korea (where one of commonest shorebirds) and China, with some wintering in Taiwan and S China. Monotypic. **HH** Subarctic montane tundra (300–1,600m). On migration and in winter forms dense flocks on intertidal wetlands. **ID** Large, grey sandpiper, with longish bill (longer than head); slightly larger, more attenuated rear, longer-necked and smaller-headed than Red Knot. Ad. breeding heavily streaked grey on head and neck, with prominent black spots/chevrons on breast/flanks, forming band of almost solid black on chest. Upperparts dark grey, feathers fringed pale grey except some scapulars, which are chestnut. Non-breeding ad./juv. plainer grey, lacking chest-band and with grey spots from neck to flanks. Belly and vent white at all seasons, but with scattered black spots in breeding plumage. Wings extend beyond tail tip, giving more elongated appearance than Red Knot. In flight, appears grey with narrow white wingbar and white rump contrasting with grey tail. **BP** Bill long (longer than Red), blackish, thicker and greyer at base, slightly decurved to finer tip; eyes black; legs dull grey-green. **Vo** Double whistle, *queet queet* or *nyut-nyut,* and *chucker-chucker-chucker.* On breeding grounds gives soft whistles.

RED KNOT *Calidris canutus* CTKJR

L 23–25cm; WS 45–54cm; WT 85–220g. **SD** Holarctic species of high-Arctic tundra, currently in serious decline; winters S America, Europe and Australasia. Scarce in region: *C. c. roselaari* breeds Wrangel I. and NW Alaska; *C. c. piersmai* (?= *C. c. canutus*) on New Siberian Is; *C. c. rogersi* in NE Chukotka. On migration (mostly Apr/May, Aug–Oct) occurs on entire E Asian coast, Japan, Korea (common) and Taiwan. Small numbers winter in Taiwan and S China. **HH** High-Arctic coastal and montane tundra. On migration and in winter forms close, cohesive flocks (often dense) or joins mixed flocks of other shorebirds, on coastal intertidal wetlands, particularly wet mudflats. **ID** Medium-sized, heavy-set grey sandpiper with long wings but shortish bill and short legs. Ad. breeding reddish-orange on face, chest and belly; crown grey, streaked darker, upperparts grey with black centres and tips to many feathers (some orange), may have blue-grey appearance. Non-breeding ad./juv. largely plain grey above, grey feathers of mantle and wings have dark grey and white fringes, breast has grey wash, finely streaked black, belly and vent white with some dark streaking on flanks. In flight, grey with noticeable white wingbar contrasting with black flight-feathers and grey coverts; mostly white rump barred grey, showing little contrast with pale grey tail. **BP** Bill short (equals head), rather thick, straight, blackish; eyes black; legs dull greenish-grey. **Vo** Flocks give steady chattering *kyo kyo kyo,* and in flight a low *knutt knutt.* On breeding grounds gives loud, whistled *plee-vee plee-vee* and fast *kuveet-kuveet* in display flight.

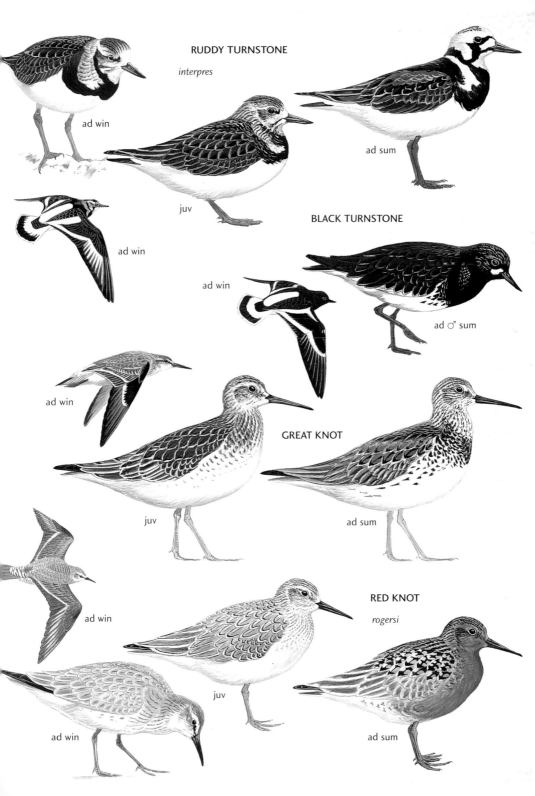

RUDDY TURNSTONE

interpres

ad win

ad sum

juv

ad win

BLACK TURNSTONE

ad win

ad win

ad ♂ sum

ad win

GREAT KNOT

juv

ad sum

ad win

RED KNOT

rogersi

juv

ad win

ad sum

SANDERLING *Calidris alba*   CTKJR

L 20–21cm; WS 35–39cm; WT 33–110g. **SD** Holarctic, in region *C. a. rubida* breeds in Lena Delta and Novosibirsk Is; winters south to Australia. Migrant over entire E Asia coast, and winters (Aug–May) in Japan, Korea, Taiwan and S China. **HH** High-Arctic stony tundra. On migration/in winter, close-knit flocks forage on sandy beaches (follows tide, runs with mechanical gait at speed); occasionally on mudflats with other waders. **ID** Small, stocky, but larger than stints. Ad. breeding mixture of rufous, black and white, but intensity variable. Head, face, neck, breast and upperparts largely rufous (recalling Red-necked Stint), mantle and wing-coverts grey with black shafts and tips (some with rusty-orange, others fringed white). Non-breeding is palest small Asian shorebird, with black lesser coverts forming prominent black patch on closed wing. Juv. white with black and silver-spangled upperparts. Wingtips reach tail tip or just beyond. In flight, broad white wingbar, black leading edge to wing with prominent carpal patch, black flight-feathers, black central tail with white sides. **BP** Bill black; eyes and tarsi black, foot lacks hind toe. **Vo** In flight, a short hard *cheep cheep*; excited *twick twick* or liquid *pleet*; in alarm *veek-veek*. Song a complex mix of churring, croaking and hissing, with hoarse, trilling *trrr-trrr*.

SEMIPALMATED SANDPIPER   TJR
Calidris pusilla

L 13–15cm; WS 34–37cm; WT 20–41g. **SD** High-Arctic Alaska and Canada, recently also Chukotski Peninsula, where rare to locally common; winters in coastal C & S America. Accidental Taiwan; reported Japan. Monotypic. **HH** Coastal plain tundra with dwarf shrubs. On migration/winter, fresh, brackish and saltwater areas; less fond of water edge than some stints. **ID** Small, rather drab, recalling Red-necked Stint but lacks rufous tones. Closely resembles Western. Ad. breeding uniform grey-brown, with weak supercilium, dark loral patch, pale forehead, grey-brown wing-coverts (lacking rufous), scapulars black with buff to orange-buff fringes; crown and cheeks darker than supercilium, grey-brown to warmer rufous-brown. Non-breeding ad. largely uniform grey, with fine black centres to mantle and black shaft-streaks to scapulars; underparts white, with grey-washed breast. Wingtips fall just short of tail tip, imparting plump look. Juv. dark-capped (cap richly coloured as scapulars), dark ear-coverts, 'scaly' upperparts, black mantle and back with pale buff fringes (no rufous), scapulars grey with black anchor marks and off-white fringes; recalls Little Stint (p.190) but lacks white V on mantle. In flight, long narrow white wingbar, and white underwing. **BP** Bill short, straight, black (shortest in juv., longest in ♀), eyes black; tarsi black, with webbing between middle and outer toes (like Western but unlike other stints). **Vo** Flight call brief, a rather low-pitched, husky *chrup* or coarse *krrit*; sometimes a higher pitched Western-like *kit* or *cheet*.

WESTERN SANDPIPER *Calidris mauri*   CTKJR

L 14–17cm; WS 28–37cm; WT 18–42g. **SD** Beringian endemic. Mostly breeds Alaska, but also easternmost Chukotka. Majority migrates to northern S America, but has appeared on both passages in Japan (especially Aug–Oct), at Beidaihe, China, in

Taiwan, and in spring in Korea. Monotypic. **HH** Coastal and mossy tundra. On migration: coastal mudflats, estuaries and wetlands. **ID** Small stint with long bill, commonly appearing droop-tipped; structurally recalls Semipalmated, but has longer legs. Ad. breeding heavily streaked dark grey on neck, grey triangles on breast-sides/flanks, but prominently rufous on crown-sides and ear-coverts, and rufous. Non-breeding ad. pale grey (usually paler on face and breast than Semipalmated), but has less prominent black shafts to mantle/coverts, and grey-streaked breast. Juv. has distinct rufous fringes to upper scapulars, forming stripe, also distinctive black anchors on grey lower scapulars; cap less richly coloured than scapulars. In flight, narrow white wingbar, tail has black centre and white sides; underwing mostly whitish, underside of flight-feathers grey. **BP** Bill length variable (♀ often has longer bill than ♂), but usually longer than Semipalmated, slightly decurved and fine-tipped, black; eyes black; legs short, black, front toes clearly webbed (like Semipalmated, unlike other stints). **Vo** Raspy, snipe-like *jeet* or *krreep* in flight, and during display a murmuring trilled *tzree*. Calls generally higher pitched and more drawn-out than Semipalmated.

RED-NECKED STINT   CTKJR
Calidris ruficollis

L 13–16cm; WS 29–33cm; WT 18–51g. **SD** Commonest NE Asian stint; breeds only C & E Siberia, from Taimyr and Lena Delta to Chukotski Peninsula and Koryakia; winters SE Asia to Australasia. Very common (Apr/May, Aug–Oct) along continental coast: Russia, Korea, China, and through Japan and Taiwan. Monotypic. **HH** Dry tundra areas, mainly low mountains. Migration/winter on mudflats, coastal wetlands, inland marshes. **ID** Small, dumpy (bill slightly thicker, legs slightly shorter, body subtly longer than Little (p.190), with longer primary projection lending slightly elongated, longer bodied appearance). Ad. breeding has variable rufous-orange to bright rusty-red wash over head and breast (throat not white; compare Little). Crown streaked; mantle/wing-coverts orange to chestnut with black centres, tertials paler, grey-brown centred (black-centred in Little). Scapulars are most rufous (mantle, scapulars and wings concolorous in Little). Central tail-feathers not fringed orange (unlike Little). Non-breeding has pale grey upperparts with narrow dark shafts (broader than Semipalmated), and grey wash restricted to breast-sides. Juv. has largely black mantle with rufous fringes, 'braces' on mantle not obvious, whilst wing-coverts and tertials grey-brown with buff fringes; drab grey-brown wash on breast-sides, but plumage overlaps with Little and Semipalmated, though dark cap typically has paler sides than either. At rest, tail tip typically projects slightly beyond wingtips, whilst primary projection is as long as Little. Beware moult, when tail/wing measurements overlap with Little. In flight, prominent narrow white wingbar; rump mostly white with narrow black centre and grey tail; underwing mostly whitish, but flight-feathers grey. **BP** Bill short, straight, blunt and black; eyes black; legs short, black. **Vo** Flight call a rasping *quiit*, dry *churi*, chattering *chrit-chrit* or *chreek*; shriller, higher pitched and more 'cracked' than Semipalmated, but lower and less drawn-out than Western.

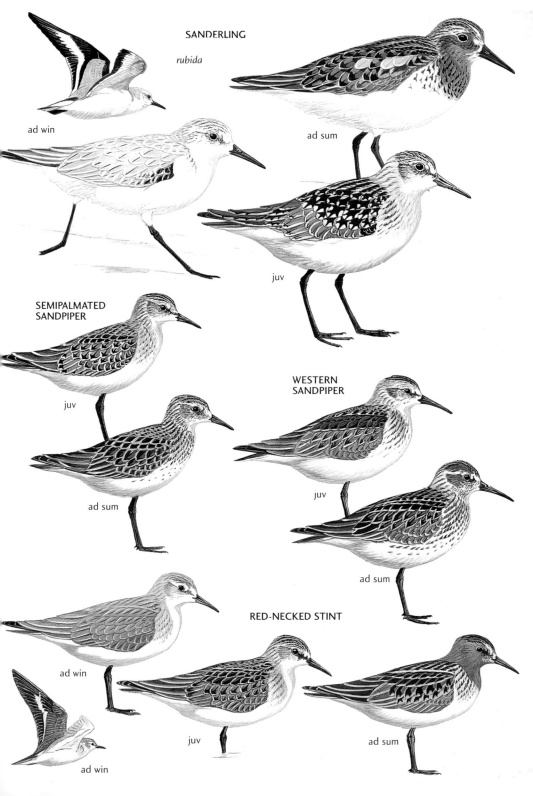

SANDERLING

rubida

ad win

ad sum

juv

SEMIPALMATED
SANDPIPER

juv

ad sum

WESTERN
SANDPIPER

juv

ad sum

ad win

RED-NECKED STINT

ad win

juv

ad sum

LITTLE STINT *Calidris minuta*   CTKJR

L 12–14cm; WS 28–31cm; WT 17–44g. **SD** Breeds from N Scandinavia east to Arctic coast of Yakutia. Rare in region except on breeding grounds west of Indigirka River. Winters Africa and S Asia. Accidental in Sakhalin, Korea, rare Japan (but has wintered), and rare migrant to E coast of China. Monotypic. **HH** Low-altitude high-Arctic tundra. On migration, small inland wetlands, and in winter typically with other waders at coastal mudflats and wetlands. **ID** Small, dumpy stint, slightly smaller than Red-necked (p.188) but confusingly similar in non-breeding plumage. Ad. breeding has reddish cheeks and reddish wash over dark-spotted breast, but chin and throat always white. Upperparts like Red-necked but more orange, with mantle, wing-coverts and tertials all fringed orange; prominent off-white or buff V on mantle. Central rectrices fringed orange (unlike Red-necked). Non-breeding pale grey like Red-necked, but has broader dark feather centres. Juv. has darker cap, streaked breast-sides, prominent white 'braces', and bright-edged dark grey-centred wing-coverts (but coloration overlaps both Red-necked Stint and Semipalmated Sandpiper). In flight, prominent white wingbar; rump white with narrow black centre, tail grey. Underwing mostly whitish, underside of flight-feathers grey. **BP** Bill short, straight, black with fine tip, but slightly longer and less blunt than Red-necked; eyes black; black legs slightly longer than Red-necked. **Vo** A sharp, incisive *chit* or *stit*, and weak *pi pi pi* in flight. On territory, displays aerially or from ground, giving repetitive weak *tswee-tswee-tswee-tswee*.

TEMMINCK'S STINT *Calidris temminckii*   CTKJR

L 13–15cm; WS 34–37cm; WT 15–36g. **SD** From Scandinavia across Russia to N Yakutia, Chukotka and Koryakia; winters primarily in Africa and S & SE Asia. Scarce migrant (Apr, Aug–Oct) through Korea, Japan, Taiwan and E China (wintering in small numbers in Japan, Taiwan and Fujian). Monotypic. **HH** Breeds in forest-tundra and on sparse grasslands such as those backing beaches, along lakes and rivers. On migration/winter sometimes found with other shorebirds at mudflats and coastal wetlands, but favours inland freshwater wetlands (preferring well-vegetated areas) and wet fields. **ID** Small, dumpy stint, with rather horizontal stance and quite long tail. Ad. breeding has brown upperparts lacking rusty tones of Red-necked, Long-toed and Little stints, and Western Sandpiper, with full, rather rounded grey breast patch with fine streaking. Dark-centred scapulars, wing-coverts and tertials fringed brown and grey. Non-breeder very plain pale grey with grey wash on neck and breast, resembling small Common Sandpiper (p.184). Juv. grey, but has clear dark shafts, dark submarginal edges with pale fringes. Creeps quietly on flexed legs and pecks more slowly than Red-necked. In flight shows white wingbar, prominent white sides to dark-centred rump and tail. **BP** Bill short, slightly arched, dull greenish at base, blacker towards fine tip; eyes black; pale legs may appear yellowish or greenish. **Vo** Song a distinctive, insect-like trilling *chirririri* or *tirrrrrrit*, more conspicuous because given while ♂ hovers. Flight call also a purring trilled *tirrr* or prolonged *tirrr-r-r* (also given at night on migration) and when flushed usually flies off fast, high and erratically.

LONG-TOED STINT *Calidris subminuta*   CTKJR

L 13–16cm; WS 26–31cm; WT 20–37g. **SD** Asian species with disjunct breeding range. In E Asia in E Chukotka, Koryakia, Magadan region, throughout Kamchatka, Commander Is, N Kuril Is and N Sakhalin. Migrant (mostly Apr/May, Aug–Oct) through Mongolia, Transbaikalia, Russian Far East, Japan, Korea and Taiwan. Winters Nansei Shoto, S Kyushu, S China, S & SE Asia and Australasia. Monotypic. **HH** Grassy swamps and montane tundra in sub-Arctic and taiga zones. On migration generally in small numbers with other shorebirds at well-vegetated freshwater wetlands, muddy pools, rivers, wet ricefields and occasionally brackish coastal wetlands. **ID** Small, slender, long-legged stint. Ad. breeding has dark-streaked rufous-brown cap extending forward to bill base and narrow dark lores, bordered by pale supercilium that often splits over eye. Cheeks, neck- and breast-sides finely streaked black. Upperparts largely blackish, but scapulars, coverts and tertials all fringed warm rufous. Non-breeding ad. greyer with clearly grey-fringed but solidly dark-centred scapulars. Juv. has dark ear patch, rufous-brown-fringed scapulars and tertials, but grey-fringed median/lesser coverts. Closed wing shows no primary projection; tertials almost as long as primaries. In flight, white wingbar and white rump-sides like other stints, but protruding toes diagnostic. Closely resembles smaller Least Sandpiper, but stands more upright and appears long-necked, like tiny Sharp-tailed Sandpiper (p.192). **BP** Bill short, straight or slightly curved, blackish with pale yellow (or orange-brown) base to lower mandible; eyes black; legs long (especially tibia) yellowish-brown or greenish-yellow. **Vo** In flight gives soft, low, slightly disyllabic purring *prrrp* and *chirrup*, recalling Curlew Sandpiper. Display involves purring series of call notes.

LEAST SANDPIPER *Calidris minutilla*   TJR

L 13–15cm; WS 32–34cm; WT 17–33g. **SD** Breeds across N America from Alaska to Newfoundland and winters S America; accidental NE Russia (though displaying ♂ reported, so breeding possible) and Taiwan, and reported Japan. Monotypic. **HH** To be looked for amongst other stints at freshwater wetlands and at brackish coastal wetlands and intertidal mudflats. **ID** The smallest sandpiper, appears hunched, small-headed and short-winged with very fine bill. Very similar to slightly larger Long-toed Stint, but crouches more, has shorter neck, appears more compact, overall drabber, less clean. Ad. breeding appears grey-brown, with cap less rufous than Long-toed and pale supercilia usually joining on forehead and contrasting with broader, darker lores; breast streaked black on grey, upperparts and tertials narrowly fringed rufous, though these wear off in summer. Non-breeding ad. grey with dark centres to grey-fringed scapulars less clearly marked than Long-toed. Juv. generally warm rufous above; resembles Little Stint in having narrow white V on scapulars, with rufous-brown fringes to scapulars, tertials and wing-coverts. Closed wing shows no primary projection and tertials almost as long as primaries. In flight, wings shorter and darker than other stints, white wingbar more prominent than Long-toed; toes do not extend beyond tail. **BP** Bill short, straight or droop-tipped, very fine, black; eyes black; legs yellow or yellow-ochre, but shorter than Long-toed, and toes short. **Vo** A soft purring *preet*, but more commonly a short series of shriller rising *kureee* or *brreeep* notes on take-off, thinner and higher pitched than Long-toed.

LITTLE STINT

juv

ad sum

ad sum

ad win

TEMMINCK'S STINT

juv

ad sum

ad sum

ad win

LONG-TOED STINT

ad sum

ad win

juv

LEAST SANDPIPER

juv

ad win

ad win

ad sum

WHITE-RUMPED SANDPIPER CJR
Calidris fuscicollis

L 15–18cm; WS 36–38cm; WT 30–60g. **SD** Breeds Canadian Arctic and adjacent tundra of N Alaska, wintering in S America. Accidental Chukotka, Baikal and Ussuriland, Russia, Beidaihe (China) and Japan. Perhaps overlooked elsewhere on Asian coasts. Monotypic. **HH** With other stints at coastal wetlands ranging from intertidal mudflats to saltmarshes and wet fields. **ID** Medium-sized, long-bodied, slender, long-winged sandpiper with distinctive narrow white rump, slightly smaller than Dunlin, but like Baird's Sandpiper appears longer and lower. Ad. breeding finely streaked on head, neck, chest and flanks; crown and ear-coverts tinged rufous; mantle has rufous fringes, scapulars and wing-coverts dark-centred with broad grey-brown fringes. Non-breeding ad. grey with dark centres to grey-fringed scapulars; obvious white supercilium. Juv. brightly patterned with prominent supercilium, dark loral bar, and bright mantle and scapulars fringed rufous and white forming prominent V. Closed wing is long, primaries project beyond tail, whilst tertials fall short of tail tip. In flight shows prominent white wingbar; entire rump white, contrasting with dark grey tail. **BP** Bill short, straight or droop-tipped, black with brown base to lower mandible; eyes black; short tarsi black. **Vo** In flight, gives thin, shrill insect-like or somewhat snipe-like *jeeeet* or *tzreet*, and shorter, clearer *tit* in rapid bat-like series.

BAIRD'S SANDPIPER *Calidris bairdii* CTJR

L 14–19cm; WS 36–46cm; WT 38g. **SD** Breeds more widely in Canadian Arctic and Alaska than White-rumped Sandpiper, also breeds on Chukotski Peninsula and Wrangel I. Most winter in S America, but occasional from Kuril Is (Aug–Oct) to Taiwan and Fujian. Monotypic. **HH** Rocky tundra, occasionally on sparsely vegetated coastal sandbars. In winter, marshes, sandy beaches and mudflats, generally preferring drier habitats than White-rumped, and often far from water, e.g. on open short grass and drier, upper shores. **ID** Medium-sized, very long-winged, long-bodied and short-legged sandpiper, similar to but longer-winged than White-rumped, and lacking white rump. Body has odd 'flattened oval' shape when viewed from head or rear. Ad. breeding duller than other sandpipers, rather plain greyish-buff on head, neck and breast, upperparts similar but silvery-grey with large black centres to pale scapulars. Non-breeding ad. grey-brown, warmer than White-rumped, with broader, more diffuse dark centres to scapulars/coverts. Juv. has most brightly patterned plumage, with buffy breast-band and pale scaly pattern to mantle, scapulars/wing-coverts, rather than neater, more restricted white V of White-rumped. At rest, primaries extend well beyond tail, whilst tertials fall well short of tip, accentuating elongated appearance. In flight, shows only weak white wingbar; rump dark brown with pale sides offering no contrast with dark brown tail. **BP** All-black bill finer at base and straighter than White-rumped; eyes black; short tarsi black. **Vo** Purring trill recalls Curlew Sandpiper but softer, *kyrrp, krrt* or *krreep*, and in alarm *veet-veet-veet*.

PECTORAL SANDPIPER *Calidris melanotos* CTKJR

L 19–23cm; WS 37–45cm; WT ♂ 45–126g, ♀ 31–97g. **SD** W Canadian Arctic, N & W Alaska and across N Siberia from Taimyr to E Chukotka. Winters mainly to S America, but also reaches

Australia and New Zealand. In E Asia, breeds from Lena River to Chukotski Peninsula. Surprisingly scarce south of Asian breeding range (due to its primarily easterly migration), but accidental in Transbaikalia, Russian Far East and Beidaihe (China), a rare migrant in Taiwan and Korea, and rare to scarce migrant in Japan (primarily Aug–Oct, occasionally in spring). Monotypic. **HH** Wet and well-vegetated Arctic tundra; on migration favours freshwater wetlands, wet ricefields, marshes and lakeshores, but also grassy coastal wetlands. **ID** Large sandpiper (♂ larger than ♀) with distinctive upright stance and slightly decurved two-tone bill, resembling small Ruff (p.196) in stance and proportions. Ad. breeding very similar to Sharp-tailed; warm rufous fringes to black-centred upperparts, cap, neck and breast streaked black on buff; breast streaks end abruptly and form clear division from otherwise white underparts, and form point in centre of chest. White supercilium weaker than in Sharp-tailed. Non-breeding ad. greyer, but retains clear demarcation between streaked chest and white belly. Juv. like warm ad. with narrower rufous fringes to upperparts and prominent narrow white V on mantle. Primaries of closed wing reach tail tip. In flight, narrow white wingbar, rump and tail blackish with white rump-sides recalling Ruff, but less distinctive. **BP** Slightly curved bill is dark at tip, grey or yellowish-brown at base; eyes black; legs longish, yellow or greenish-yellow. **Vo** On take-off a Curlew Sandpiper-like rich, liquid trilling *kureep; prrit, kirrp* or *churrk*, but deeper; on breeding grounds a muffled *do-do-do* from elevated ground.

SHARP-TAILED SANDPIPER CTKJR
Calidris acuminata

L 17–22cm; WS 36–43cm; WT ♂ 53–114g, ♀ 39–105g. **SD** NE Asia, from Taimyr across N Yakutia to Chaun Gulf. Fairly common migrant through E Asia (mostly Apr/May, Aug–Oct), through NE China and coastal areas, also common Korea and Japan. Most winter south to New Guinea, Australia and New Zealand, but some as far north as Taiwan. Monotypic. **HH** Low-Arctic and sub-Arctic tundra, among hillocks with damp vegetation and willow thickets. Winter habitats range from intertidal mudflats to wetlands (including ricefields). Only adults migrate inland, whilst juveniles occur only on coasts. **ID** Large sandpiper, similar in size and structure to Pectoral (PS). Ad. breeding has streaked rufous cap, more prominent face pattern than PS, with pale supercilium that tends to flare behind eye (unlike PS), narrow white eye-ring; neck and breast boldly spotted and streaked black on buff wash (lacks clear demarcation on underparts of PS), streaks extend as bold crescents/arrowheads on flanks to vent, and generally brighter rufous than PS. Non-breeding ad. greyer, but retains clear rufous cap. Juv. less boldly streaked than ad., the streaks forming a necklace, below which there is a warm orange-buff wash to breast, and narrower rufous-fringed upperparts. Closed wing shows no primary projection. In flight, pattern recalls PS and Ruff (p.196). **BP** Slightly curved bill is black, brown or pink at base of lower mandible; eyes black; legs longish, yellowish to green. **Vo** On breeding grounds a twittering *whit-whit whit-it-it* recalling Barn Swallow and a muffled, trilled *trrr*; in flight or when flushed a soft trilled *trrrp* or *purii*, and various low grunting noises, also a subdued, repetitive *ueep-ueep*.

WHITE-RUMPED SANDPIPER

ad win

ad sum

BAIRD'S SANDPIPER

ad win

juv

ad sum

ad win

PECTORAL SANDPIPER

juv

juv

ad ♂ sum

ad win

SHARP-TAILED SANDPIPER

ad win

juv

ad sum

CURLEW SANDPIPER *Calidris ferruginea* CTKJR

L 18–23cm; WS 38–41cm; WT 44–117g. **SD** Quite widespread from Yamal Peninsula to Arctic coast of Chukotka; winters W Africa to New Zealand. Uncommon migrant (mostly Apr/May, Aug–Oct) across NE China and on continental coast, Korea and Japan. Mainly winters in S Asia and Australasia, but some as far north as Taiwan. Monotypic. **HH** Relatively dry tundra in high-Arctic lowlands. On migration, muddy coastal wetlands, also wet ricefields inland. **ID** Slender, medium to large sandpiper, more elegantly proportioned than Dunlin, appearing longer necked and more upright, with longish, curved bill. Ad. breeding rusty-red on head, neck and underparts, white undertail; patterned on back and wings with black, white and orange-brown. Non-breeding ad. rather plain grey with prominent pale supercilium, lightly streaked grey on head and neck, and white underparts. Juv. appears warmer, browner than non-breeding with pale-scalloped fringes to mantle, scapulars and wing-coverts, unstreaked buff head, neck and breast, and narrow white supercilium. Closed wing long, primaries extending beyond tail and tertials almost to tail tip. In flight resembles Dunlin, but has distinctive white rump. **BP** Long curved bill black; eyes black; legs longish (toes extend beyond tail in flight, not in Dunlin), black. **Vo** Flight call a distinctive soft purring *prrrp*, *prrriit* or *chirrup*, quite different from Dunlin's rasping call. Song consists of ascending sweet whistles, *fuweeet fuweeet…* **TN** Cox's Sandpiper '*C. paramelanotos*', a rare hybrid of Curlew and Pectoral Sandpipers with intermediate characters of both, has been recorded several times in Japan.

ROCK SANDPIPER *Calidris ptilocnemis* CJR

L 20–23cm; WS 43cm; WT 70g. **SD** Beringian endemic, breeding in W Alaska, E Chukotka and islands of Bering Strait (*C. p. tschuktschorum*), the Commander Is, S Kamchatka and possibly N Kuril Is (*C. p. quarta* including *kurilensis*), with most wintering on N American W coast, but some on Pacific coast of Japan (Dec–Apr). Birds resembling both Chukotkan and Aleutian (*C. p. couesi*) races have reached Japan. **HH** Montane or coastal tundra often with streams. On migration/winter alone or in small groups, sometimes with Ruddy Turnstone, almost exclusively on rocky shores foraging in splash zone. **ID** Dumpy, dark grey sandpiper. Ad. breeding Dunlin-like, with extensive pale rufous fringes to mantle and scapulars; head pale, greyish, throat pale, upper breast streaked; bold black belly patch extends from lower breast to legs (in Dunlin extends between legs). Aleutian birds much darker. Non-breeding ad. dark grey on head, neck and breast, upperparts grey-fringed with dark brown shafts and centres. Juv. has chestnut and buff-fringed scapulars. Closed wing falls short of tail tip. In flight, mostly white underwing and conspicuous white wingbar. **BP** Bill short, black but yellow at base, drooping at tip; eyes black; legs short, ochre to green. **Vo** Flight call a husky *cherk*. Display flight accompanied by low growling trills in long series of *grreee* notes, followed by low *grrdee* notes.

STILT SANDPIPER *Calidris himantopus* TJR

L 18–23cm; WS 43–47cm; WT 40–68g. **SD** Breeds Alaskan and Canadian Arctic, wintering in S America. Accidental to Japan (Jul–Sep), Taiwan and Russia. Monotypic. **HH** Prefers freshwater wetlands in which wades deeply, leaning well forward, probing mud rapidly with 'sewing machine' action recalling dowitchers. **ID** Elegant, long-billed and long-legged sandpiper, resembling large, slender Wood Sandpiper, or small, slender dowitcher, but longer-legged. Ad. breeding is dark, heavily blotched black on mantle and wing-coverts, heavily streaked on head, neck and upper breast, and heavily barred on underparts; crown and cheeks rusty-orange, separated by prominent white supercilia. Non-breeding ad. paler and plainer grey above than winter dowitchers, retaining prominent supercilia. Juv. has scaly brown upperparts and lightly streaked head and neck. Closed wings reach just beyond tail, tertials almost to tail tip. In flight, underwing largely grey with white centre; upperwing dark to pale greyish-brown; rump white, tail grey with long tarsi trailing obviously. **BP** Bill long, black, rather heavy and droops slightly towards swollen tip, resembling Curlew Sandpiper in shape; eyes black; long tarsi greenish-yellow. **Vo** Usually silent, but in flight a rattling *kirrr* recalling Curlew Sandpiper, clearer whistled *whu* or *feu* and husky *toof*, a wheezy *keewf* and clearer *kooowi*. **TN** Formerly *Micropalama himantopus*.

DUNLIN *Calidris alpina* CTKJR

L 16–22cm; WS 33–40cm; WT 33–85g. **SD** Widespread in Holarctic. In region, breeds from the Lena to Chukotka and south through Kamchatka. Several subspecies: *C. a. centralis* of NE Siberia to Kolyma River migrates southwest probably to Middle East; *C. a. sakhalina* of Chukotka, from Kolyma to Chukotski Peninsula, winters to E China, Korea, Japan and Taiwan; *C. a. kistchinskii* of N Sea of Okhotsk, S Koryakia to Kamchatka and N Kuril Is; *C. a. actites* of Sakhalin and *C. a. arcticola* of Alaska and NW Canada winter to E China, Korea and Japan. Common migrant through E Asia, with some wintering (Aug–May) in Korea, C & S coastal Japan, Taiwan and coastal SE China. **HH** Arctic tundra, mostly in areas with wet or boggy ground. On migration and in winter common along beaches, coastal wetlands with muddy margins, and inland freshwater wetlands including wet ricefields. Massed flocks make spectacular aerobatic twists and turns with white underparts flashing. **ID** Stocky, somewhat large-headed sandpiper with long, droop-tipped bill. Races *C. a. centralis* and *C. a. sakhalina* recorded in China, but most Asian coastal migrants presumed to be *C. a. sakhalina*. Field identification of subspecies uncertain. Ad. breeding has grey head and neck, with fine dark streaks; upperparts extensively dark rufous, underparts dominated by long black belly patch from lower breast back between legs (see Rock Sandpiper); vent and undertail-coverts white. Non-breeding ad. plain brownish-grey above, white below, resembling Curlew Sandpiper, but has shorter legs and wings. Juv. buffer than winter ad. with faintly streaked underparts, blackish mantle and scapulars with buff fringes. At rest, closed wing reaches almost to tail tip. In flight, mostly white underwing and conspicuous white wingbar, rump has white sides but black centre. **BP** Bill black, but shorter, thicker and less curved than most Curlew Sandpipers; eyes black; legs medium to short, black or blackish-green. **Vo** Prolonged harsh trilling whistles from ground or in low display flight on quivering wings. In flight and when flushed, commonly gives buzzing, whistled *skeeel* or slurred *screet*. In aggression or threat, a hoarse *gwrr-drr-drr-drr*.

CURLEW SANDPIPER

juv

ad win

juv

ad sum

ROCK SANDPIPER

tschuktschorum

ad win

juv

ad sum

STILT SANDPIPER

juv

ad sum

ad win

juv

ad sum
sakhalina

DUNLIN

ad win

juv

ad sum
arcticola

SPOON-BILLED SANDPIPER CTKJR
Eurynorhynchus pygmeus

Endangered. L 14–16cm; WT 29.5–34g. **SD** Rare, declining endemic; breeds only E Chukotka and Koryakia, winters India, SE Asia and S China. Rare coastal migrant Kamchatka to Taiwan, winters coastal Fujian, SE China. Monotypic. **HH** Coastal tundra with little vegetation; near lakes and marshes. On migration, coastal lagoons, estuaries and tidal mudflats; often with Red-necked Stint, but readily separated at distance – they favour wet sand/mud, on beaches they remain nearer waves, foraging in wet substrate with short jabbing action, back and forth and to sides. **ID** Small sandpiper with unique bill, though distinctive tip not always clear, especially in profile. Ad. breeding has brick red to orange head and neck, black feathers of upperparts fringed bright rufous; black streaks extend to sides, rest of underparts white. Non-breeding ad. plain grey above, white below, with rather prominent supercilium. Juv. brown-fringed back and wing-coverts, pale supercilium and very dark ear-coverts forming prominent patch. Closed wings reach just beyond tail. In flight, underwing largely white, upperwing dark with slight white wingbar, rump and tail mostly black with white sides. **BP** Short, black bill has wedge-shaped spatulate tip; eyes black; legs medium to short, black. **Vo** During aerial display gives insect-like buzzing trill, *zhree-zhree-zhree*. In flight a soft *puree, preep* or shrill *wheet*.

BROAD-BILLED SANDPIPER CTKJR
Limicola falcinellus

L 16–18cm; WS 34–37cm; WT 28–68g. **SD** Breeds NW Europe and Siberia. Winters Africa, India, SE Asia and Australia. *L. f. sibirica* breeds between the Lena and Kolyma, and around Kolyma mouth. Uncommon migrant SE Russia (Lake Baikal, Sakhalin, and Ussuriland), Kamchatka, Japan, Korea and China, with some wintering Taiwan. **HH** *L. f. sibirica* breeds in wet Arctic tundra. On migration often amongst Red-necked Stint and Dunlin on beaches, ricefields and mudflats. Forages sluggishly, recalling snipe rather than other sandpipers. **ID** Small sandpiper resembling large Long-toed Stint in breeding plumage, but small Dunlin in winter, with distinctive bill; often has conspicuous black carpal patch like Sanderling. Pale supercilium is double, splitting in front of eye. Ad. breeding is stint-like with heavily streaked cap, lightly streaked neck, upper breast and mantle (latter has prominent white lines), black scapulars, coverts and tertials broadly fringed orange-brown; breast streaks extend to sides, rest of underparts white. Non-breeding ad. plain grey above, the grey extending to upper chest, mantle feathers have dark centres, coverts white-fringed with black centres. Juv. like adult, with warm brown tones and warm buff/brown wash to lightly streaked breast. Closed wings reach just beyond tail, tertials almost to tail tip. In flight, underwing largely white, upperwing pale grey with slight white wingbar, central rump and tail mostly black with white rump-sides. **BP** Bill longish, straight, flattened slightly dorso-ventrally, but angles abruptly downwards at tip; eyes black; legs medium to short, greenish-grey. **Vo** During display flight gives dry buzzing Dunlin-like *suwir-suwir-suwir* with occasional faster trills and rattles. In flight and when flushed, dry trilling *ch-r-r-reep* or *tirr-tirr-terek* recalling Sand Martin.

BUFF-BREASTED SANDPIPER TKJR
Tryngites subruficollis

L 16–21cm; WS 43–47cm; WT ♂ 53–117g, ♀ 46–81g. **SD** Breeds Alaska and N Canada; winters southern S America. In E Asia, breeds on Ayon I. and Wrangel I., and coastal N Chukotka; accidental Kamchatka coast, N Kuril Is and Ussuriland, Japan, Korea, and Taiwan. Monotypic. **HH** Prefers drier habitats than most shore-birds, sometimes on sandy beaches, but also dry grasslands, including meadows, golf courses, airfields and short vegetation near wetlands, where resembles small golden plover. **ID** Elegant, rather erect sandpiper, resembling larger juv. Ruff, but has small rounded head, shorter bill and shorter tertials. Ad. has very plain orange-buff face, neck and underparts; crown darker, mantle and wing-coverts appear scaly (dark feather centres and pale buff/white fringes). Juv. has less pronounced buff wash to underparts and upperparts are browner. Closed wings long; tail long but primaries reach beyond tip, tertials almost to tail tip. In flight, upperwing plain brown, underwing largely white, with distinctive black crescent on primary-coverts; rump and long tail brown; tarsi hidden. **BP** Bill short, black; eyes prominent, black; legs long, yellow-ochre. **Vo** Generally silent, but occasionally gives low, dry rattling flight call, *krrrt* or *pr-r-r-reet* recalling Pectoral (p.192), and dry *chup* and *chitik* calls.

RUFF CTKJR
Philomachus pugnax

L ♂ 26–32cm, ♀ 20–26cm; WS 54–60cm, ♀ 46–52cm; WT ♂130–254g, ♀ 70–170g. **SD** Breeds N Europe across Russia; winters primarily in Africa and S Asia but also Australasia. In E Asia, N Yakutia and Chukotka to NE Sea of Okhotsk and Sakhalin. Migrates abundantly through Transbaikalia and uncommonly coastal S Russian Far East, China, Korea and Japan, some wintering Taiwan and coastal Fujian. Monotypic. **HH** Tundra to steppe; on migration, coastal and freshwater wetlands, and wet grasslands. **ID** Large, long-legged shorebird, with long neck, small head and short bill, typically erect, but back hunched. Ad. breeding is sexually dimorphic: ♂ typically 20% larger than ♀ and sports long, erectile ear-tufts and huge neck-ruff used in display; individually variable from white to black but may be orange, rusty or brown. Upperparts orange to black, with white fringes. Much smaller ♀ (or Reeve) dark brown; noticeably scaled on upperparts; dark-centred mantle, scapulars and coverts have pale brown fringes, and tertials boldly barred. Non-breeding ♂ quickly loses ruff and resembles ♀. In winter, both sexes paler than summer ♀; upperparts retain scaly appearance, but feather centres greyer brown. Juv. shares scaly upperparts with summer ♀, but has warm buff wash to head, neck and chest. Closed wings long; long tertials may completely cover primaries and tail tip (see Buff-breasted). In flight, both sexes have large white oval spots on sides of brown tail; underwing mostly white; upperwing dull brown with narrow white wingbar; toes extend beyond tail. **BP** Bill short, slightly curved, brown, pink or yellow, with yellow base (varying individually and with age and season); eyes black; legs long, yellow or green to orange-brown. **Vo** Typically silent, even at lek, but may give low croaking calls; single or double *kyuu*, *chuck-chuck* or *krit-krit* and shriller *hoo-ee* flight call.

SPOON-BILLED SANDPIPER

ad win

ad sum

juv

BROAD-BILLED SANDPIPER

juv

ad sum

sibirica

juv

ad win

BUFF-BREASTED SANDPIPER

ad win

juv

ad sum

ad ♂ win

ad ♀ sum

ad ♂ win

ad ♂ sum

juv

RUFF

display

WILSON'S PHALAROPE *Phalaropus tricolor* KJR

L 22–24cm; WS 35–43cm; WT ♂ 30–110g, ♀ 52–128g. **SD** Nearctic; winters in S America, but has strayed to Japan, and reported S Korea, N Yakutia and Wrangel I. Monotypic. **HH** Inland prairie pools and lakes, and coasts and estuaries on migration; wades more in characteristic crouching foraging gait and swims less than other phalaropes. **ID** Larger, more slender, longer-necked, longer-legged and longer-billed than other phalaropes; more like *Tringa*. ♀ breeding grey from forehead to mantle, with black band from lores and eye on neck-sides, white eyebrow, chin, throat and underparts, with orange wash on sides of neck/breast; upperparts grey and chestnut. ♂ duller, showing less contrast and less distinct pattern on upperparts. Non-breeding has pale grey head, neck and upperparts, with white supercilium, throat and underparts. Juv. dark blackish-brown above, with upperparts fringed rufous-brown, face and underparts white. At rest appears longer-tailed and longer-winged than other phalaropes. In flight, white rump contrasts with dark grey wings (no wingbar), mantle and mid-grey tail; underwing white. **BP** Bill needle-like, black; eyes black; legs longish, black (breeding) or pale yellow (non-breeding, juv.). **Vo** Essentially silent, but occasionally gives short, sharp *chew* or *puu*, and muffled nasal grunts in flight.

RED-NECKED PHALAROPE CTKJR
Phalaropus lobatus

L 16–20cm; WS 30–38cm; WT 20–48g. **SD** High latitudes of N America, Greenland, Iceland and Eurasia; winters at sea in Pacific and Indian Oceans. In region, N Russia to Bering Strait, Chukotka and Kamchatka, NE coasts of Sea of Okhotsk, N Sakhalin, and Commander and N Kuril Is. On migration occurs off all coasts, rare inland. Monotypic. **HH** Wet tundra pools and swamps. On migration at wet fields, freshwater, brackish and coastal wetlands, in bays, or very large numbers offshore. **ID** Small, delicate and elegant. Often very confiding. ♀ breeding brightly coloured, with dark grey head and upperparts, orange-brown bands on mantle and scapulars, and bright red band on neck-side and upper breast; chin, throat and eyebrow white. Lower breast and flanks grey, grading into white on belly and vent. ♂ duller, less contrasting grey and red on head and neck, less bright on upperparts. Non-breeding pale grey above (darker than Red Phalarope), white below, with blackish crown and black eye-patch reaching ear-coverts. Juv. back pattern of breeding ♂, but face pattern of non-breeding; more extensive grey on breast-sides. On water floats high, appears long-necked with small head; primaries reach tail-tip. In flight, black flight-feathers with narrow white wingbar, black rump and tail, with narrow white rump-sides. **BP** Bill fine, needle-like, black; eyes black; legs black, toes distinctly lobed. **Vo** Throaty, high-pitched twittering on breeding grounds, *kik-kik-kikik-tree-kik*; in flight a chirping *kip*; *chep* or longer *kerrek*.

RED PHALAROPE CTKJR
Phalaropus fulicarius

L 20–22cm; WS 37–43cm; WT 37–77g. **SD** High latitudes across America and N Russia from Urals to Bering Strait; winters off western S America and W & S Africa. On migration (rarely winter), abundant in Arctic seas of region and E Kamchatka, uncommon further south on E Russian coast, Kuril Is and off Japan; vagrant Sakhalin, Korea, NE China and Taiwan. Strong winds sometimes bring flocks inshore. Monotypic. **HH** Lakes in swampy tundra,
often near coast. On migration more likely at brackish and coastal wetlands, in bays or offshore. **ID** Slightly larger and thicker necked than Red-necked. ♀ breeding has black cap and face, white cheeks, and deep rufous neck and underparts. Upperparts black with broad orange-rufous fringes to scapulars, wing-coverts and tertials. ♂ breeding is less cleanly marked, rufous of underparts more broken with white. Non-breeding pale grey above, white below, with blackish crown and black eye patch extending to ear-coverts. Juv. has back pattern of breeding ♂, with some grey scapulars, and face pattern of non-breeding, but more extensive dusky-grey on breast-sides, and lacks mantle and scapular stripes of Red-necked. On water floats high; primaries reach tail tip. Slower, less erratic flight than Red-necked; black flight-feathers with narrow white wingbar, and grey rump and tail; in non-breeding plumage, very pale mantle contrasts with dark wings. **BP** Bill short, thicker than Red-necked, yellow with black tip (breeding), or black with pale yellow base; eyes black; short legs grey, toes distinctly lobed. **Vo** Trilling *tick-tick-tick* followed by murmuring *krree-kree* or *purii*, given by ♀ in display flight; rather explosive, metallic *pik* or *wit* in flight or when flushed, higher-pitched and clearer than Red-necked. **AN** Grey Phalarope.

ORIENTAL PRATINCOLE CTKJR
Glareola maldivarum

L 23–24cm; WS 59–64cm; WT 87g. **SD** S & E Asia; winters south to Australia. Summer visitor to S & E Chinese coasts and Taiwan north to Heilongjiang, W & SC Japan, also SE Transbaikalia and S Russian Far East. Migrant E China and Korea (may breed). Rarely overshoots beyond breeding range; accidental Lake Baikal, Hokkaido, S Kuril Is and Kamchatka. Monotypic. **HH** Wetlands and grasslands; breeds colonially, flocks on migration. Agile aerial feeder, but also runs after terrestrial insects. **ID** Medium-sized, slim and tern-like. Upperparts plain mid brown with warm tones, underparts pale brown with hint of peach. Flight-feathers and tail glossy black. Lores and line bordering throat black; chin and throat warm buff, bordered at rear with white, then black. Vent and undertail white, undertail-feathers with black tips. Young more uniform brown above and below, with narrow pale fringes affording scaly appearance. At rest, wings extend 2–3cm beyond tail tip (see Collared). Flight fast and agile, swallow-like, with long, narrow, dark wings (no white trailing edge), prominent white rump, short, shallow-forked black tail. Underwing-coverts rufous, contrasting with greyer underside to secondaries and darker primaries. Appears long-legged on ground, but toes barely reach tail-tip in flight. **BP** Bill short, black, red at base (ad.) or all black (juv.); eyes dark brown; legs grey. **Vo** Harsh, sharp and shrill *kuriri* and staccato *shtick*; grating *tar-rak* or *kr-d-d-ik* given only in flight.

COLLARED PRATINCOLE Extralimital
Glareola pratincola

L 24–28cm; WS 60–70cm; WT 60–95g. **ID** Resembles Oriental; has reached Hong Kong and may stray further east. In flight, tail longer with deeper fork and has white tips to secondaries forming trailing edge to innerwing. **Vo** *Kik*, *kirrik* and *stwick* flight calls are harsh and tern-like.

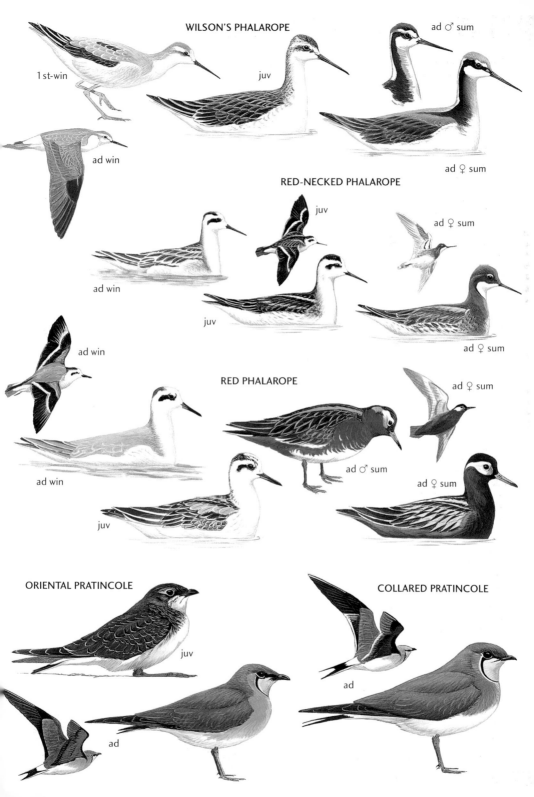

WILSON'S PHALAROPE

1st-win

juv

ad ♂ sum

ad ♀ sum

ad win

RED-NECKED PHALAROPE

juv

ad ♀ sum

ad win

juv

ad win

ad ♀ sum

RED PHALAROPE

ad ♀ sum

ad win

ad ♂ sum

juv

ad ♀ sum

ORIENTAL PRATINCOLE

COLLARED PRATINCOLE

juv

ad

ad

ad

BLACK-TAILED GULL　　　　　　　　CTKJR
Larus crassirostris

L 44–48cm; WS 126–128cm; WT 436–640g. **SD** E Asian endemic, from N Sakhalin, Kuril Is and coastal Russian Far East to Japanese coasts and offshore islands, offshore Korea, Shandong and Fujian, China; also Matsu (Taiwan); winters from Hokkaido to Taiwan and S China. Monotypic. **HH** Rocky islets, coasts and cliffs; coasts and at estuaries, rarely inland. **ID** Medium-sized, long, narrow-winged gull, with broad black tail-band. White head appears relatively flat-crowned, further accentuating long bill. Ad. breeding is white and grey, with mid to dark grey mantle and little contrast between mantle and primaries, white tertial crescent, black primaries with small white tips, and white tail with broad black subterminal band. White underparts may show pink flush in breeders. Winter ad. similar, but often dusky-grey on nape/rear crown, creating somewhat hooded appearance. Juv. extremely dark brown, with pale fringes to saddle, wing-coverts and tertials; could be confused with dark-morph skua. 1st-winter brown-winged, with grey wash to underparts, palest on head, face typically whitish, with clear white vent and rump, and all-black tail. 2nd-winter has grey mantle and less grey wash to underparts, mainly on neck-sides; tail white with broad black terminal band. In flight, ad. has all-black wingtips, white trailing edge to dark grey wings, and white outertail-feathers and broad black subterminal band. Flight graceful relative to other larger gulls. **BP** Bill long, yellow, black subterminal band and red tip in ad.; juv./1st-winter have pink bill with black tip, 2nd-winter has dull yellow or flesh-coloured bill with black tip; eyes yellow (ad.), brown/black until at least 2nd-winter; legs greenish-yellow in adult, pink until 1st-winter, thereafter becoming pale yellow. **Vo** Strong, though plaintive, nasal cat-like mewing: *aao* or *myaao*. Rather higher-pitched than larger gulls.

COMMON GULL　　　　　　　　　　CTKJR
Larus canus

L 40–46cm; WS 110–125cm; WT 394–586g. **SD** Breeds across Eurasia and northwest N America. In region, Russia east to Sakhalin, Yakutia, Chukotka and Kamchatka, also N China; winters (Oct–Apr) through Japan, Korea, E Chinese coast and occasionally Taiwan. Two races: *L. (c.) kamtschatschensis* (sometimes split as **Kamchatka Gull**) commonest throughout NE Asia; *L. c. heinei* breeds east to Lena River, and occurs in China and Japan in winter. **HH** Islands, coastal cliffs, beaches and marshes, from tundra to steppe and semi-desert. On migration/winter on lakes, rivers and coasts. **ID** Medium-sized gull resembling small, delicate Vega Gull (p.204), but has rounded head, plain face and rather gentle appearance. Ad. breeding has mantle and wings mid grey; prominent white tertial crescent separates grey of wings from black primaries. Head, neck, underparts and tail white, but in winter crown, ear-coverts, nape and neck-sides narrowly streaked or more broadly flecked brown; head streaking can be heavy. Juv. very dark, more coffee-coloured, with mantle and scapulars dark brown with broad pale fringes; dark carpal and secondary bars, and pale grey midwing bar; blackish outer primaries and greyer inner primaries; rump and tail mostly white with broad black terminal band. Kamchatka Gull matures more slowly than *heinei* and retains this plumage into mid or even late winter, only gradually replacing juv. feathers with grey. 2nd-winter frequently still has smudgy brownish-grey secondary-coverts, traces of secondary bar and underwing coverts can have dusky tips; more often has black in tail. Most *heinei* resemble ad. at this age; though some retain imm. lesser and median secondary-coverts, signs of immaturity usually restricted to primary-coverts and, in some cases, black in tail. In flight, ad. appears very white and grey, with prominent white mirrors in outermost black primaries, whilst 1st-winter appears darker, with much brown on upper surfaces and all-dark primaries. Flight more buoyant than larger gulls. **BP** Rather delicate bill varies from greenish- to bright yellow, lacks red spot of similar Vega Gull, generally paler in winter, subterminal markings range from complete black ring to plain, with blackish spot near tip of lower mandible when younger and dark smudge in winter adult; eyes dull yellow to brown/black, sometimes pale yellow; legs greenish to yellow. **Vo** Generally a nasal mewing *kya kya kyaa* or *gyu gyu*, a longer *glieeoo*, and in alarm *gleeu-gleeu-gleeu*.

MEW GULL　　　　　　　　　　　Extralimital
Larus (canus) brachyrhynchus

L 40–46cm; WS 110–125cm; WT 394–586g. **SD** N America, ranging to W Alaska. A likely stray to region. **ID** Separation difficult, but typical individuals differ from Kamchatka Gull in having paler grey upperparts, wing pattern, smaller size, shorter legs, 'gentler' appearance with larger eye and much shorter bill. In flight, less black in primaries with prominent white 'moons' separating black from grey feather bases extending to p8 or even p9, recalling Slaty-backed Gull's 'string of pearls' (p.206).

RING-BILLED GULL　　　　　　　　　KJ
Larus delawarensis

L 41–49cm; WS 112–127cm; WT 400–590g. **SD** N America, common in Pacific NW and Alaska; has reached Japan, also reported S Korea. Monotypic. **HH** Coasts, estuaries and rivers. **ID** Slightly larger and bulkier than Common Gull, with paler grey upperparts and heavier, distinctively patterned bill. Ad. has pale blue-grey mantle, scapulars and wing-coverts, black primaries with two white mirrors (2nd-winter usually one) and small white tips. Ad. breeding has white head, non-breeding has light brown flecking on crown, rear and sides of neck. In flight, wings somewhat narrower, more pointed than Vega (p.204). Ad. has pale grey inner primaries (paler than mantle), contrasting with black wingtip with small white mirrors above and below; tail white. 1st-winter has pale grey mantle and greater coverts, contrasting with dark brown, black-centred, lesser and median coverts, and in flight has grey panel in inner primaries, grey bar across middle of upperwing and pale underwing with broad dark brown border above and below; tail white with brown flecks and black subterminal band. 2nd-winter similar but lacks white primary tips and is duskier on head. Notched pattern on tertials and coverts can assist separation from *kamtschatschensis* Common Gull. **BP** Ad. bill rather thick, yellow with broad black ring near tip; eyes pale yellow with red orbital ring, may be blackish in winter; tarsi yellow. 1st-winter bill pink with black tip; eyes dark brown; tarsi pink. **Vo** Higher pitched, more nasal, than Vega Gull; also a soft *kowk*.

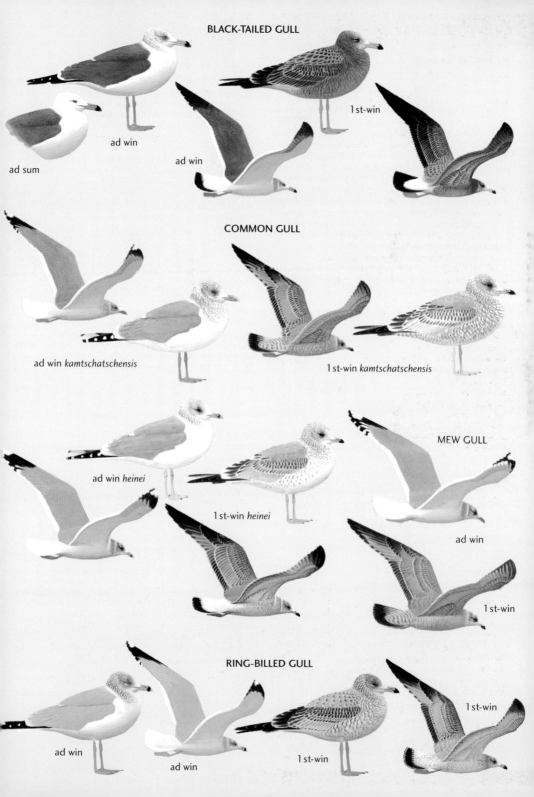

BLACK-TAILED GULL

ad win

ad sum

ad win

1 st-win

COMMON GULL

ad win *kamtschatschensis*

1 st-win *kamtschatschensis*

ad win *heinei*

MEW GULL

1 st-win *heinei*

ad win

1 st-win

RING-BILLED GULL

ad win

ad win

1 st-win

1 st-win

CALIFORNIA GULL   J
Larus californicus

L 51–58cm; WS 122–140cm; WT 0.432–1.045kg. **SD** N America, breeding inland and wintering on Pacific coast; has strayed to Japan in winter. Subspecies uncertain; either *L. c. albertaensis* or *L. c. californicus*. **HH** Coasts, estuaries and tidal mudflats. **ID** Medium-sized gull intermediate between Common (p.200) and Vega (p.204) gulls. Slightly more rounded head than Vega, but a relatively longer bill, and also rather long, narrow wings. Ad. has mid grey mantle and wings, white head, prominent white tertial crescent, with black in six outermost white-tipped primaries, outer two have large white mirrors. Non-breeding has variable amount of dusky streaking around eye and on hindneck. Juv. very dark, resembles juv. Black-tailed (p.200) but rump barred brown and tail dark brown, not black. Scapulars and wing-coverts of 1st-year have chocolate-brown centres with silvery-white edges, imparting overall scalloped appearance. By 2nd-winter develops grey mantle, legs bluish-grey, bill grey with blackish band near tip. In flight ad. appears very white and grey, with large black wingtips on both surfaces, clean-cut black broken only by prominent white mirrors in outermost black primaries. **BP** Bill appears quite long, yellow, with red gape line (spring/early summer), red gonys and, in winter, black subterminal band and white tip; black usually restricted to small area on gonys and tip yellow when breeding (see Ring-billed, p.200); narrow red orbital ring around usually dark eye (irides pale yellow in Ring-billed); legs yellow or greenish-yellow, brighter when breeding. **Vo** Deeper than Ring-billed, but harsher and higher than American Herring (p.206); a deep *gaaal*.

GLAUCOUS-WINGED GULL   CKJR
Larus glaucescens

L 61–68cm; WS 132–137cm; WT 0.9–1.25kg. **SD** Essentially Beringian species, ranging from Pacific NW and W Alaska to Commander Is, with most wintering on Pacific NW coast of America, but also on Kuril Is, S Sea of Okhotsk, Sakhalin and especially Hokkaido (mainly Nov–Mar); also reaches Korea (rare) and Fujian (accidental). Monotypic. **HH** Rocky offshore islets on Kamchatka coast and Commander Is. In winter, offshore, also rocky coasts, river mouths, occasionally beaches and especially harbours. **ID** Large, stocky, four-year gull, rather similar to Glaucous Gull; primaries never wholly white, always grey; generally weakly patterned and lacking contrast. Head white, rather flat-crowned, with especially large bill. Ad. mantle and wings mid grey, duller than other gulls; in winter crown, face and nape streaked or faintly barred brown. Juv. rather uniform pale or mid grey-brown lacking contrast in tail or wings; upperwing-coverts have pale fringes, giving scaled appearance in 1st-winter, but primaries same colour as body. By 2nd-winter mantle mid grey, rump white (often with heavy grey-brown wash), tail coffee-coloured, wing-coverts and primaries grey-brown, secondaries dark brown. At rest, ad. exhibits less contrast than most large gulls, as upperparts and primaries are similarly mid grey, wing-coverts largely framed by white trailing edge to secondaries; tertial crescent and smaller scapular crescent less conspicuous because mid grey wings very similar shade to grey of primaries. Like Glaucous and Slaty-backed (p.206), rather short-winged with wing projection beyond tail short, usually only three primaries. In flight, broad-winged and heavy; ad. shows little contrast; white trailing edge very narrow especially compared to Slaty-backed; outer primaries grey with narrow white tips and large white mirror near tip of outermost. Underwing shows little contrast between grey flight-feathers and wing-coverts. Juv./1st-winter in flight appear very uniform mid brown. **BP** Bill long and heavier than other large gulls in region, black in juv./1st-winter, develops small pink base by 2nd-winter (never as much as Glaucous Gull), ad. bill yellow with pinkish gape and oval red spot at prominent gonys; ad. has pinkish orbital ring and dark brown eyes; tarsi deep pink. **Vo** *Nyaao* or *kwa kwa*.

GLAUCOUS GULL   CTKJR
Larus hyperboreus

L 64–77cm; WS 150–165cm; WT 1.07–1.82kg. **SD** Arctic coasts of N America and Eurasia. In region *L. h. pallidissimus* breeds Russian Arctic east to E Chukotka and Bering Sea; winters (mostly Nov–Mar) on Kamchatkan coast, Sea of Okhotsk to Hokkaido and N Honshu (less common), Korea (uncommon) and coastal China (occasional). *L. h. barrovianus* of Alaska to W Canada also commonly reaches region in winter (e.g. E coast China and N Japan). **HH** Coastal tundra, cliffs; especially near seabird colonies. In winter offshore, also sandy coasts, river mouths and especially harbours. **ID** Large, stocky four-year gull, with white primaries; generally extremely pale. Head white, rather flat crown bulging slightly at forehead, neck thick with large bill. Ad. mantle and wings paler than all other large gulls; in winter nape streaked faintly brown. Juv. recalls Glaucous-winged, but paler, uniform tan or coffee-coloured with white-fringed wing-coverts, primaries always paler than rest, and no tail-band. Juv./1st-winter have distinctive black-tipped dull pink bill, gradually becoming pale at extreme tip by 2nd-winter, when entire plumage mostly white with some brown flecks on wing-coverts, mantle and tail, but some still quite brown; iris colour and bill pattern helpful in separation. At rest, tertial crescent inconspicuous because primaries also white; primary tips extend just beyond tail. In flight, broad wings, thick neck and large head give front-heavy appearance; flight-feathers tipped white in ad., buff-white in rather uniform juv./1st-winter, and translucent from below (dark in other large gulls). Underwing shows little contrast between pale grey flight-feathers and wing-coverts. *L. h. pallidissimus* large and heavily built, but *L. h. barrovianus* smaller, less bulky, longer winged and shorter-billed; however, features vary individually and between sexes, and overlap between races. *L. h. barrovianus* tends to have rounder head, slimmer bill and longer wing projection (may be mistaken for Iceland Gull (p.204), but wings not as long as Iceland and head larger, less rounded with proportionately smaller eye and larger bill). Adding to confusion, Glaucous hybridises with Glaucous-winged and American Herring (p.206) on W coast of N America; hybrids could easily wander to NE Asia. **BP** Bill long and heavy, in juv. basal two-thirds dull pink, with black tip, by 2nd-winter has dark band with pale tip, in ad. yellow with small oval red spot at gonys; juv./1st-winter have dark brown iris, 2nd-winter variably coloured (whitish and grey-green to brown), ad. has yellow, rarely red, orbital ring, and clear yellow iris; tarsi deep pink in ad., duller in young. **Vo** Similar to Vega Gull (p.204), but louder and coarser *myaaoo* or *kuwaa*.

CALIFORNIA GULL

ad win
albertaensis

ad win
californicus

1st-win

ad sum

GLAUCOUS-WINGED GULL

ad win

ad win

1st-win

1st-win

GLAUCOUS GULL

pallidissimus

ad win

ad sum

1st-win
(white)

ad win

juv/first-win

2nd/3rd-win

1st-win (white)

juv/first-win

ICELAND GULL · KJ
Larus glaucoides

L 52–60cm; WS 137–150cm; WT 557–863g. **SD** E Canadian and Greenland Arctic; winters American E coast, very rare W coast of USA, but has reached Japan and Korea. Two subspecies; *L. g. glaucoides* and *L. g. kumlieni* (Kumlien's is alternatively considered a hybrid *glaucoides* × Thayer's population). **HH** In winter, coasts and harbours. **ID** Medium-sized four-year gull. *L. g. glaucoides* resembles larger Glaucous at all ages (*L. g. kumlieni* very variable) and separated by head and bill proportions, and wing projection; but note *barrovianus* Glaucous (p.202) has fairly long primary projection, more rounded head and fairly thick-set bill. Overall Iceland Gull has more gentle appearance than Glaucous. Ad. saddle and wings pale grey, much like Glaucous; in winter, head and hindneck usually marked brown, but less heavily than Glaucous. Primary projection of ad. *glaucoides* long, entirely white and continuous with tertial crescent; *kumlieni* has primaries darker centred than upperparts, though extremes only lightly marked or almost black, so may resemble Thayer's at rest, but in flight dark coloration far more restricted. In flight, wings narrower than Glaucous and more narrowly pointed. Juv. similar to Glaucous though often greyer, primaries of *kumlieni* variable, from as pale as *glaucoides* to as dark as Thayer's. Whilst both can show subterminal arrowhead markings to primary tips, *glaucoides* is often uniformly white, and *kumlieni* shows dark centres to outer primaries, lacking in *glaucoides*, but with wear *kumlieni* becomes less distinctive as winter progresses. Base of bill becomes paler during 1st-winter in *glaucoides* but never as extensive as Glaucous and is more diffuse towards gape, with less 'dipped-in-paint' appearance. However, *kumlieni* more likely to retain mostly or all-black bill until late winter. By 2nd-winter *glaucoides* primaries tend to be paler than upperparts, whereas *kumlieni* darker. In flight both have pale undersides to primaries, but *kumlieni* has contrasting dark outer webs to upper surface of outermost feathers and often has dark subterminal markings on others; *glaucoides* more uniform, lacking dark markings. Rump and tail of *kumlieni* differ from more uniform *glaucoides* in having broad dark tail-band with contrasting white rump. **BP** Dark iris of juv. often becomes paler in 2nd-winter in *glaucoides*, less often in *kumlieni*. Ad. has pale yellow to yellowish-brown iris with purplish-red orbital ring, again *kumlieni* often darker; legs proportionately shorter than Glaucous and feet deep pink in ad., duller in young. **Vo** Similar to Glaucous but higher pitched.

THAYER'S GULL · KJR
Larus thayeri

L 56–64cm; WS 130–148cm; WT 0.846–1.152kg. **SD** High Canadian and Greenland Arctic; winters on American W coast, with small numbers in Japan (Nov–Mar) and Korea; accidental Kamchatka. Monotypic. A controversial taxon variously treated as race of Herring or Iceland Gull, with which it hybridises. **HH** In winter, on coasts and at harbours.

ID Large pale grey mantled (slightly darker grey than American Herring, p.206), four-year gull, with rounded head and rather small bill. Despite black primaries is closely related to Iceland and sometimes considered conspecific. Ad. mantle and wings pale grey, very similar to Iceland but typically slightly darker, with conspicuous tertial crescent; in winter, head, nape and neck extensively, but variably, streaked/smudged brown, as if wearing milk-coffee cowl. Underside of primary projection has limited black, though some Vega look very similar. Juv. rather plain, mottled tan-brown, with dark-centred tertials and pale-fringed dark brown primaries; by 2nd-winter mantle often grey but wings like younger birds and underparts paler. Usually exhibits contrast between pale rump and mostly blackish-brown tail. In flight, narrower winged than other large gulls; ad. has black in primaries largely restricted to outer webs of three outermost with the white inner webs typically continuous with mirrors in outer pair, plus limited black subterminally on 2–3 others. This results in strikingly white underwing, frequently only showing black trailing edge to outer primaries and (not always) narrow blackish leading edge. Beware of small juv./1st-winter Slaty-backed (p.206) which can show very similar primary pattern, though its wings look broader, more paddle-shaped and has narrow, slightly darker leading edge to primary tips, distinguishing it from *kumlieni* Iceland Gull. Tail-band variable, but always present, though vague in some. Underwing of ad. almost entirely pale, only narrow black primary tips visible. **BP** Bill long (averages longer than Iceland, but shorter than other large gulls), with gently curving culmen and weak gonydeal angle, black in juv./1st-winter, develops pink base by 2nd-winter, yellowish-green as ad. with small red spot; juv. to 2nd-winter has dark brown eyes, 3rd-winter to ad. has very dark yellowish-brown irides (some clear yellow), ad. also has purplish-red orbital ring; tarsi deep pink in ad., duller in young. **Vo** Calls resemble American Herring, but higher pitched. **TN** Formerly part of Iceland Gull.

VEGA GULL · CTKJR
Larus vegae

L 55–67cm; WS 135–150cm; WT 0.72–1.5kg. **SD** NE Asia, breeding from the Lena across Yakutia and Chukotka to Bering Sea. Winters on coasts of Kamchatka, Sea of Okhotsk, Japan, Korea, China and Taiwan. Monotypic. **HH** Breeds on grassy islands, cliffs and beaches. Winters at coastal wetlands and estuaries. **ID** Large, mid grey gull but commonly appears pale compared to much darker sympatric Slaty-backed (p.206). Vega is variable at all ages, but large head and prominent bill are consistent. Ad. has mid grey mantle and wings, white head, prominent white tertial crescent and white tips to black primaries, outermost primaries show one large and one small white mirror. Non-breeding has variable dusky streaking on head and often heavy blotching on hindneck and breast-sides. Juv./1st-winter can be very dark, with barred scapulars and chequered wing-coverts, broad dark tail-band and narrowly barred rump and uppertail-coverts; underparts dusky-brown. Upperwing has pale grey inner primaries and blackish-brown outers; secondaries dark brown. By 2nd-winter, mantle grey, wing-coverts paler brown almost grey, and underparts cleaner. In flight, ad. appears very white and grey, with contrasting black wingtips and prominent white mirrors on outer two primaries, visible above and below. Underside of flight-feathers show through as pale to mid grey (much darker in Slaty-backed). **BP** Bill deep, ad. yellow with large red spot at prominent gonys, juv./1st-winter black; thereafter pale pink with dark tip; ad. has orange-red orbital ring, brownish-yellow irides; tarsi strong flesh pink at all ages. **Vo** Rather vocal, giving long laughing display calls, and loud, guttural *guwaa*; *ao* or *gag-ag-ag* at other times. **TN** Formerly part of Herring Gull *L. argentatus*.

ICELAND GULL

ad win *glaucoides*

ad win *glaucoides*

ad win *kumlieni*

ad win *kumlieni*

juv/1st-win *glaucoides*

juv/1st-win *kumlieni*

THAYER'S GULL

ad win

ad win

juv/1st-win

2nd-win

juv/1st-win

VEGA GULL

ad win

ad win

2nd-win

1st-win

ad sum

juv

AMERICAN HERRING GULL CKJ
Larus smithsonianus

L 56–64cm; WS 135–147cm; WT 1.15kg. **SD** The common, large gull of northern N America, but vagrant (perhaps commoner than realised) to NE Asia (Japan, S Korea and E coastal China). **HH** Usually found amongst other large gulls at river mouths, estuaries and fishing harbours. **ID** Ad. is large mid grey gull; smaller with longer primary projection than Vega, saddle paler, closer to ad. Glaucous (p.202). Care needed to separate juv./1st-winter from dark Vega (p.204) and very variable Slaty-backed, both of which are potential pitfalls. From Vega combination of smaller size, slightly longer primary projection and solidly dark outer greater coverts bar distinctive at rest (but Vega can show dark outermost greater coverts, usually hidden by breast-feathers, though typically checkered). In flight, wings longer, thinner and darker, pale inner primary window obviously darker and less contrasting with surrounding dark feathers. Slaty-backed can be remarkably similar but is heavily built, large-headed and stout-billed with a much shorter wing projection; primaries often brownish with pale fringes at tips (always black in American Herring). In flight, primaries have contrasting pale inner and dark outer webs like Thayer's (p.204). Rather greyer western populations of American Herring can show contrasting warm brown collar to lower hindneck, separating paler head from mantle and merging with brownish underparts. Tail is black to base (or nearly so), but this is matched by some Vega and Slaty-backed, whilst latter can also share heavily barred rump/ uppertail-coverts. Hybrids with Glaucous or Glaucous-winged also possible. **BP** Bill can appear larger/longer than Vega, pale to deep yellow with red gonydeal spot (ad.) basally dull pink, but largely black in 1st-winter, increasingly pink with smaller black tip in 2nd-winter, dull yellow with black band near tip in 3rd-winter; eyes brown in juv., pale yellow by 3rd-winter with orange-yellow orbital ring in ad.; tarsi greyish-flesh at first becoming pale to mid pink from 2nd-winter. **Vo** A high, cracked *klaaw klaaw klaaw klaaw* and deeper *gyow gyow gyow gyow*. **TN** Formerly part of Herring Gull *L. argentatus*.

MONGOLIAN GULL CTKJR
Larus mongolicus

L 55–68cm; WS 140–155cm; WT 1.125kg. **SD** Breeding and winter ranges and migration routes poorly known; field identification requires caution and experience. Breeds east of Lake Baikal, Mongolia, NE China and Korea, wintering (Oct–Apr) mainly to Yellow Sea, Korea, Japan (uncommon), Taiwan (uncommon winter, commoner Feb/Mar migrant) and SE China coast. **HH** Usually found amongst other large gulls at river mouths, estuaries and harbours. **ID** Large four-year gull similar to Vega (p.204). Neck long, tapers from broad base to head when stretched (more parallel-sided in Vega), and has high, full-chested appearance when relaxed, this combined with longer primary projection often lends it a sleeker appearance. Ad. rounded head typically white, with any winter streaking usually faint and restricted to lower hindneck. Juv. dark brown upperparts, all feathers pale-fringed giving scaled appearance, head typically almost white, contrasting with black bill and brown-streaked neck and breast. Separation of 1st-/2nd-winter from Vega is beyond scope of this guide. At rest, broad white tertial and scapular crescents; black primaries have 4–5 (usually four) small white spots at tips and extend just beyond tail (longer winged than Vega). In

flight, ad. clean mid grey and white with extensive black above and below, appearing clear-cut across wingtips with large mirror on outermost primary (p10) and smaller, sometimes inconspicuous one on p9. Underside of inner primaries and secondaries pale grey, similar to Vega. Separation of Mongolian from Caspian Gull *L. cachinnans* and 'Steppe' Gull *L. (c.) barabensis* still unclear, but all three perhaps recorded Japan. **BP** Bill not normally as heavy-looking as Vega, deep orange-toned as spring approaches with large red gonys spot, whilst non-breeder paler, yellow (brighter in Vega) and often whitish at tip. Juv./1st-winter black, becoming pink-based with age. Ad. has red orbital ring, iris similar to Vega, yellowish but variably flecked darker, often creating dark-looking eye; legs pale flesh or greyish in winter, some becoming yellow in breeding condition, greyish to whitish in juv./1st-winter. **Vo** *Aa, aa* or *kwaa, kwaa*. **TN** Previously considered conspecific with Caspian Gull *L. cachinnans*.

SLATY-BACKED GULL CTKJR
Larus schistisagus

L 55–67cm; WS 132–148cm; WT 1.05– 1.695kg. **SD** East Asian endemic, restricted to coastal Koryakia, Kamchatka, Sea of Okhotsk, Shantar Is, Sakhalin, and Hokkaido and N Honshu. Winters in ice-free regions from Commander Is and Hokkaido down Sea of Japan to Korea and Bohai Gulf, and on Pacific coast to Nansei Shoto and Taiwan, occasionally SE China. Monotypic. **HH** Breeds colonially on isolated grassy islands, cliffs (locally on rooftops, e.g. Hokkaido), and common on rocky/sandy coasts. In winter, coasts, river mouths, beaches and harbours. **ID** Very large, very dark four-year gull. Head large, heavy and angular with large bill, body deep-keeled with stout legs, wings broad in flight, short at rest. Ad. has mantle and wings darker than all other gulls in region; at rest, broad, very conspicuous white tertial crescent, continuous with white trailing edge to secondaries, exposed below greater coverts. In flight, grey flight-feathers visible, creating contrasting underwing pattern. White trailing edge to secondaries and inner primaries broad and even. Continuing on outer primaries, a series of prominent white 'moons' on inner webs ('string of pearls') separate basal grey from subterminal black; eye-catching above and below. Outer primaries have 1–2 mirrors. Juv. very dark brown, gradually becoming paler, especially on head, in 1st-winter when mantle and scapulars have dark shaft-streaks. Greater coverts often more solidly dark, suggesting American Herring, as does all-dark tail and heavily barred rump/uppertail-coverts. Primary projection blackish to brownish, often with pale fringes, thus can recall Thayer's (p.204), but Slaty-backed much larger, bulkier with shorter wing projection than either Thayer's or American Herring. In flight suggests American Herring due to tail, rump/uppertail-coverts, greater coverts pattern, and dark underwing, but inner primary window paler, primaries often brownish not black, with obviously pale inner webs creating 'Venetian blind' pattern, like Thayer's; also structurally different, bulkier with broader, shorter wings. From late winter, increasingly very worn and bleached, with white coverts, brown primaries and irregular dark grey on mantle; similar-age Glaucous-winged (p.202) uniformly pale grey. By 2nd-winter saddle more ad.-like. **BP** Bill heavy, yellow with large oval red spot at gonys; juv./1st-winter black, thereafter pale pink with dark tip; ad. has reddish-pink orbital ring, and yellow eyes; tarsi deep pink. **Vo** *Kiwau* or *myaao*, but also longer series of notes *kwaau kwau kwau kwa a a a*.

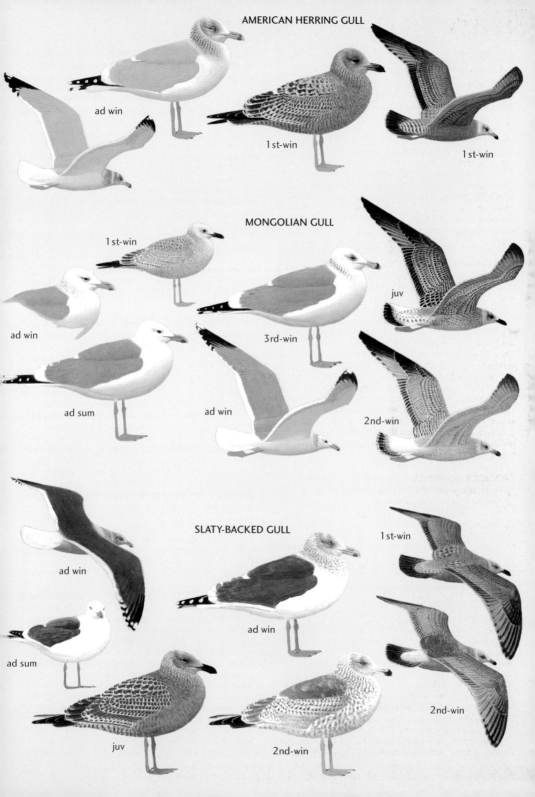

AMERICAN HERRING GULL

ad win

1st-win

1st-win

MONGOLIAN GULL

1st-win

ad win

3rd-win

juv

ad sum

ad win

2nd-win

SLATY-BACKED GULL

ad win

1st-win

ad win

ad sum

juv

2nd-win

2nd-win

HEUGLIN'S GULL CTKJR
Larus heuglini

L 51–65cm; WS 124–150cm; WT 0.55–1.2kg. **SD** Breeding and winter ranges and migration routes poorly known. Believed to breed outside region in N Russia west of Lena River; accidental to Sea of Okhotsk. *L. h. heuglini* seems rare in region. Small numbers of *L. h. 'taimyrensis'* winter on Chinese coast, Taiwan, Korea and Japan, but *'taimyrensis'* often considered an intergrade between Vega (p.204) and Heuglin's. **HH** Usually with other large gulls at river mouths, estuaries and harbours. **ID** Large, well-proportioned gull with rounded head and strong bill. Looks long-winged, tail extension beyond tertials about one-third of primary projection beyond tertials (see Vega, in which tail is equal to almost half primary projection). Following refers to *'taimyrensis'* which is distinctive at all ages: ad. darker grey upperparts than Vega, but paler than Slaty-backed (p.206). Non-breeding has fine streaking on hindneck, variable fine streaking on head and often has dusky surround to eye. In flight upperwing dark, outer primaries black, often with only one mirror, a second if present is small. Typically six, often seven feathers have black on both webs. Juv. saddle, lesser and median wing-coverts dark with narrow pale fringes, greater coverts mainly uniformly dark as American Herring (p.206). Juv. scapulars replaced by darkish grey 1st-winter feathers with darker anchors; grey fades to whitish during winter to give much paler overall appearance, with darker internal anchors prominent. Head becomes much whiter, and by late March may be moulting median wing-coverts, unlike Vega. Retained juv. tertials largely blackish and contrast with increasingly pale body. In flight, juv. very dark, inner primary window dull, less contrasting than Vega, and with dark greater coverts bar creates darker, more uniform upperwing, like American Herring. Shares largely black tail with latter, but rump and tail-coverts unbarred. In 2nd-winter dark ad. feathers visible in saddle, thus darker than Vega, more elegant than Slaty-backed. **BP** Bill deep and parallel-edged with distinct gonydeal angle, dark in juv., becoming pale-based with dark tip in winter, but in ad. washed-out yellow in winter, becoming brighter when breeding with large oval red spot at gonys, often spilling onto upper mandible at all seasons; ad. has red orbital ring and yellowish to whitish eyes, in winter variable flecked darker, thus can appear dark-eyed at distance; legs rather orange throughout winter, becoming yellow when breeding. **Vo** *Aa aa.* **TN** Previously considered conspecific with Herring Gull *L. argentatus* (and within Lesser Black-backed by some authorities).

PALLAS'S GULL CTKJ
Ichthyaetus ichthyaetus

L 57–72cm; WS 146–170cm; WT 0.9–2kg. **SD** C Asia to Mongolia, with most moving south and west to winter well outside region, though has strayed to Beidaihe (China), Korea, and even S Japan (almost annual Oct–Apr) and Taiwan. Monotypic. **HH** Breeds colonially at inland seas and saline lakes, but at other times, alone or in small groups at estuaries and tidal mudflats. **ID** Very large (larger than Vega Gull, p.204), four-year gull that develops more quickly than other large gulls. Structure eye-catching: crown peak at rear, combining with long bill to give long-headed appearance, tertials bulky, long primary projection and long legs. Ad. has mid grey mantle and wings averaging paler than Vega, with white tertial crescent and conspicuous apical spots

on primaries. Ad breeding has full black hood, but in winter this is reduced and may be just dusky around face and cheeks, but never entirely absent; white crescents above and below dark eye. In flight, front-heavy look due to large head and long, thick bill. Wings narrower than other large gulls, 'arm', inner primaries and primary-coverts mid to pale grey, outer primaries and coverts white creating broad wedge like Brown-headed (p.210) but with less black. Underwing white. Juv. has pale-fringed brown mantle, scapulars and wing-coverts, and fairly uniform greater coverts lack checkered pattern typical of most large gulls. Assumes ad.-like grey mantle and scapulars early in first winter, though lower hindneck and wing-coverts mainly brown, rump and tail white with broad black terminal band. In flight plain greyish-brown greater coverts form pale midwing panel. 2nd-winter more similar to 3rd-winter Vega but retains bold black tail-band. **BP** Bill long, thick, yellow at base, red at tip (supposedly only when breeding, but commonly also in winter), with variable black band forward of gonys; 1st-year has black-tipped pinkish-yellow bill; eyes black; legs greenish-yellow. **Vo** Generally silent away from colonies, but occasionally gives long deep, nasal crow-like *aaa, aagh* or *kra-ah* flight call. **TN** Formerly placed in *Larus*. **AN** Great Black-headed Gull.

RELICT GULL CKJ
Ichthyaetus relictus

L 39–45cm; WT 420–665g. **SD** C Asia, easternmost birds breeding in Mongolia and at Hulun Nur, NE China. Migrates (and winters), sometimes in large numbers, along W coast of Bohai Gulf, China. Rare winter visitor to Korea and vagrant to Japan. Monotypic. **HH** Inland at saline or semi-saline montane lakes, in winter coastal mudflats and estuaries, particularly sandy mudflats. **ID** Medium to large black-headed gull with distinctive wing pattern. Gait on ground almost plover-like. Ad. breeding is white with pale to mid grey mantle and wings, white tertial crescent and black primaries with white tips. Hood blackish-brown with white eye crescents, broad behind eye like Saunders's (p.212), unlike Black- and Brown-headed (p.210). Winter ad. lacks hood, has dusky-grey ear-coverts patch, some dark streaking on nape and sometimes the crown (cf. Brown- and Black-headed). In flight, black in outer primaries very variable, bases fade to white distally, with two conspicuous mirrors; upperwing resembles Franklin's (p.214) but pale innerwing far less contrasting, some can appear surprisingly white-winged. Underwing almost entirely grey/white except small black primary tips. 1st-winter white-headed with streaked hindneck, grey saddle and wings with brown lesser coverts and tips to grey median coverts; black primaries lack white tips. In flight, inner primaries and greater coverts grey, forming pale centre to wing, bordered above by dark brown carpal bar and below by prominent black subterminal spots on secondaries and inner primaries; primary-coverts and primaries black, forming bold wedge on leading edge. Can show one small mirror on outermost primary. Tail has narrow black subterminal band. **BP** Bill dark red, proportionately shorter with deeper gonys than Black- and Brown-headed, can be paler based with dark tip in winter, grey-brown in 1st-winter; eyes brown/black; legs dull, dark pink or red in ad., blackish-grey in young. **Vo** Laughing *ka-kak ka-ka kee-a*, downturned *kyeu* and prolonged *ke-arr*. **TN** Formerly placed in *Larus*.

ad win 'taimyrensis'

HEUGLIN'S GULL

1st-win

1st-win

ad sum 'taimyrensis'

ad sum heuglini

ad win heuglini

2nd-win heuglini

PALLAS'S GULL

ad win

ad sum

ad win

1st-win

3rd-win

2nd-win

RELICT GULL

ad sum

2nd-sum

1st-win

ad sum

2nd-win

1st-win

BROWN-HEADED GULL
Chroicocephalus brunnicephalus
CTJ

L 41–45cm; WT 450–714g. **SD** C Asia; generally winters in India and SE Asia, but vagrant to Bohai Gulf coast (China), Taiwan and Japan. Monotypic. **HH** Cold, high-altitude lakes; outside breeding season, coasts, estuaries and tidal mudflats. **ID** Medium-sized gull (slightly larger than Black-headed), with pale grey mantle and wings, dark chocolate-brown hood and black wingtips. Ad. breeding has dark chocolate-coloured hood extending to rear crown, though paler on face, with prominent broken white eye-ring like Black-headed. Non-breeding ad. lacks hood but has dark brown spot behind eye. In flight, wings slightly broader with rather more rounded 'hand' than Black-headed, and flight not as light. 'Arm' and inner primaries pale grey. Primary-coverts and bases to outer primaries white, otherwise black with 2–3 mirrors. Pattern resembles Pallas's but black more extensive. Underside of primaries largely black with mirrors, and underwing-coverts duskier grey than Black-headed. 1st-winter suggests Black-headed at rest, but underside of folded primaries lacks white. Differs in flight in that outer primaries and primary-coverts are black but bases of inner primaries white; almost reverse of Black-headed. **BP** Bill long, somewhat thicker than Black-headed, dark blood red when breeding, orange-red with black tip in non-breeder; orbital ring red; irides distinctly pale, yellowish, white or grey in ad., dark in juv./1st-winter; legs dull orange-red, or paler orange in winter/juv. **Vo** Resembles Black-headed, but lower, more guttural *krreeah-kreeah*, a harsh *gek-gek* and wailing *ka-yek ka-yek*. **TN** Formerly placed in *Larus*.

BLACK-HEADED GULL
Chroicocephalus ridibundus
CTKJR

L 37–43cm; WS 94–110cm; WT 195–325g. **SD** Primarily Palearctic. Breeds W Europe east to NE Yakutsk, SE Russia, Sakhalin, Kamchatka and Koryakia, and NE China. Common migrant across E China; common, even abundant migrant (Apr/May, Aug/Sep) in Japan. Winters locally in N Japan, fairly common around C & S Japanese coasts (Oct–Apr), Korea, Taiwan and E China coast south from the Chianjiang Estuary. Monotypic. **HH** Wetlands from coastal saltmarshes to inland lakes, rivers and swamps in temperate and taiga zones. On migration/in winter, coasts and inland rivers and lakes. May also frequent parks. **ID** Small grey and white gull with pale grey mantle and wings, dark chocolate-brown hood and distinctive white leading edge to primaries in flight, with only narrow black wingtip. Ad. breeding has dark brown hood extending to rear crown, with narrow broken white eye-ring. Non-breeding ad. lacks hood but has blackish spot behind eye, and grey-smudged crown. Juv./1st-winter resemble winter ad., but have brown tertials and wing-coverts, with grey midwing panel, black wingtips and narrow black terminal tail-band. In flight, wings narrow, somewhat tern-like, with pointed tip; mostly pale grey but outer primaries white, forming broad white leading edge, though all outer primaries have small black tips; underwing has largely black primaries, with only outermost 1–2 white; tail white. **BP** Bill long, slender, dark blood red, blackish-red when breeding, orange-red with black tip in non-breeder; eyes black/brown; legs dull red, paler in winter, orange in juv./1st-winter. **Vo** Highly vocal at colonies, fairly so on migration and less so in winter, though foraging flocks noisy at food frenzy: typically a long, strident *kyaar* and *krreearr*, with shorter, sharper *kek* notes. On territory also gives rasping 'long-call' like extended *krreearr*. **TN** Formerly placed in *Larus*.

SLENDER-BILLED GULL
Chroicocephalus genei
CKJ

L 42–44cm; WS 102–110cm; WT 220–350g. **SD** W Europe, N Africa and C Asia, but has strayed east to Beidaihe (China), Korea and S Japan (where has wintered). **HH** In non-breeding season, coastal wetlands including estuaries and lagoons. **ID** Elegant, medium-sized gull, marginally larger than Black-headed, but lacks hood, and has longer neck, forehead and bill. Ad. breeding essentially white with grey upperparts and wings, though underparts and even primaries may be suffused salmon-pink. Winter ad. has faint grey spot on ear-coverts. Longer-headed and longer-billed than Black-headed. Juv. has dusky-brown wing-coverts and flight-feathers, but more weakly patterned than equivalent-age Black-headed, and has slightly more prominent ear-coverts patch than ad. Graceful flight and upperwing pattern both similar to Black-headed; upperwing plain grey with black-tipped primaries, and white leading edge (outer four primaries); young similar to Black-headed but less heavily marked; also has white outer primaries in 1st-winter, and narrow black tail-band. **BP** Bill rather long, slender, blackish-red when breeding, red in non-breeding, but straw-yellow in 1st-year; eyes white or yellowish-white in breeder, pale brown in other plumages; legs somewhat longer than Black-headed, dull red in adult, pale orange in young. **Vo** Similar to Black-headed, but lower and drier *aaaa*, excited *ka-ka-ka*, drawn-out *kraaah…* and nasal *krerrr-krerrr*. **TN** Formerly placed in *Larus*.

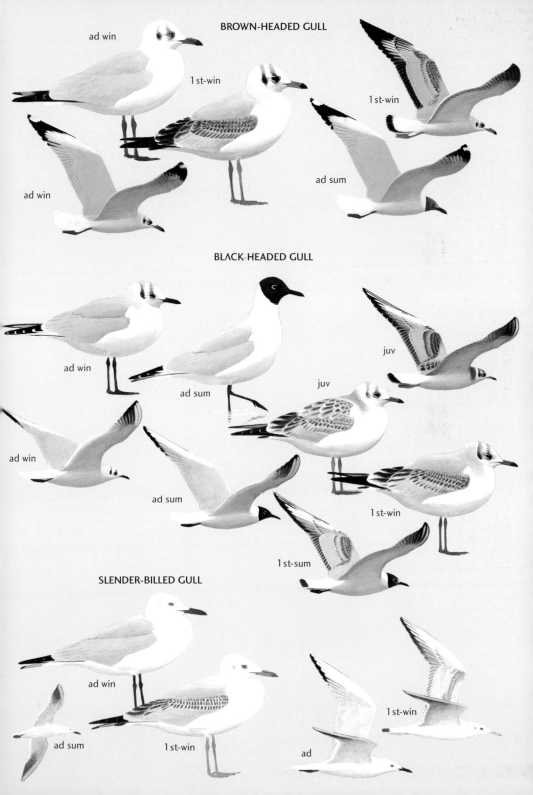

BROWN-HEADED GULL

ad win

1st-win

1st-win

ad win

ad sum

BLACK-HEADED GULL

ad win

ad sum

juv

juv

ad win

1st-win

ad sum

1st-win

1st-sum

SLENDER-BILLED GULL

ad win

ad sum

1st-win

ad

1st-win

BONAPARTE'S GULL   JR
Chroicocephalus philadelphia

L 28–34cm; WS 81–100cm; WT 170–230g. **SD** N America, ranging west into Alaska. Has reached Chukotka, C Honshu (Apr/May, Dec), and Taiwan. Monotypic. **HH** In non-breeding season, coastal estuaries. **ID** Small, tern-like gull, superficially similar to Black-headed (p.210), but smaller. Ad breeding has black (not brown) hood and narrow white crescents above and below eye. Mantle and wings pale to mid grey, wingtips at rest appear all black and extend well beyond tail. Winter ad. has dark ear-coverts spot and two faint grey bars on white crown. Juv. resembles juv. Black-headed, but has black ear-coverts spot and more pronounced, narrow, blacker M on upperwings. Wings narrow, flight light and graceful; upperwing like Black-headed but underwing much paler, the primaries being almost white, except for narrow black tips forming trailing edge. Tail white in ad., has narrow black terminal band in young. **BP** Bill fine, black; eyes black; tarsi dull pink. **Vo** Very different from Black-headed, a more rasping or grating, rather tern-like *gerrr reeek* or *tee-er*. **TN** Formerly placed in *Larus*.

SAUNDERS'S GULL   CTKJR
Saundersilarus saundersi

Vulnerable. L 29–33cm; WS 87–91cm. **SD** Range-restricted, rare E Asian species, breeding only in NE China on shores of Bohai Gulf, in Liaoning, Hebei and Jiangsu, and very locally on W coast of Korea; winters (Nov–Mar) in Korea, SW Japan (particularly Kyushu), Taiwan (mainly SW) and SE coastal China. Has strayed north to Vladivostok, Hokkaido, Kuril Is and Sakhalin. Monotypic. **HH** Nests on dry coastal flats with sparse vegetation above tideline. In winter, mudflats, estuaries and river mouths. **ID** Small, elegant gull, like small Black-headed, but has very distinctive wing pattern, tern-like call and stubby bill. Ad. breeding has black hood, broad white crescents surrounding all but front of eye, and pale mantle and upperwing. Non-breeding ad. has dusky-grey bar on crown, often eye to eye, and dark grey or blackish spot on ear-coverts. Juv. has dark brown on head where ad. is blackish, with brown wing-coverts, tertials and dark brown primary tips; tail, which is pure white in ad., has narrow terminal black band in juv. At rest, unlike Black-headed, black primaries show series of prominent white tips. Flight graceful and somewhat marsh tern-like, as

flies back and forth just a few metres above surface, dropping suddenly to snatch food, or settles briefly before continuing to quarter foraging grounds. Mantle and upperwing uniform pale grey, primaries have small black tips, and white outer primaries form white leading edge; in contrast underwing shows black wedge formed by black bases to primaries; juv. in flight shows black carpal spot especially when seen head-on. Prominent white trailing edge often useful in separation from Black-headed. **BP** Bill short, thick, black; eyes brown/black; legs dull red, dull orange in juv./1st-winter. **Vo** High-pitched, tern-like *eek-eek*, *teek-eek* or *kyi-kyi*. **TN** Formerly placed in *Larus*.

LAUGHING GULL   J
Leucophaeus atricilla

L 36–42cm; WS 98–110cm; WT 240–400g. **SD** Coasts of N America to northern S America, mainly on E coast and Gulf of California, but has strayed to Europe and Japan; subspecies presumably *L. a. megalopterus*. **HH** Beaches and harbours. **ID** Marginally larger than Black-headed and potentially confusable with Franklin's (p.214), but longer winged with much smaller white primary tips, and longer bill has more drooping tip. Ad. breeding resembles ad. Franklin's but eye-ring narrower and clearly broken behind eye. Non-breeding also resembles Franklin's, but less hooded, with only limited grey streaks or patches on rear crown. Flight graceful. Upper body and wings uniformly dark grey, wingtips largely black above and below, and lacks mirrors (compare Franklin's), underwing dusky-grey becoming black on primaries, unlike Franklin's. Juv. dusky-brown on head, neck and upperparts, lacking black hood of juv. Franklin's. 1st-winter distinguished from Franklin's by darker grey mantle, grey-brown hindneck and grey breast-sides to flanks (neck, breast and flanks clean white in Franklin's). Juv. brown upperwing-coverts largely retained by 1st-winter, which lacks white apical spots on closed wing, but Franklin's can lose spots due to wear. In flight, differs from Franklin's in having dark axillaries and grey-mottled underwing coverts (uniformly pale in Franklin's), as well as blackish inner primaries compared to latter's pale inner primary window. Terminal black tail-band broad and grey bases can give impression of black tail, band narrower in Franklin's with extensive white bases. 1st-summer may retain variable number of faded juv. coverts and primaries very worn (see Franklin's). 2nd-winter more closely resembles ad. **BP** Bill relatively long, thick, dull red with drooping tip, blackish with red tip in juv./1st-winter; eyes black; legs dull red, black in juv./1st-winter. **Vo** Common call a nasal, laughing disyllabic *kiiwa*, *kahwi* or *kee-agh*, and repetitive series of *ha ha hah* notes. **TN** Formerly placed in *Larus*.

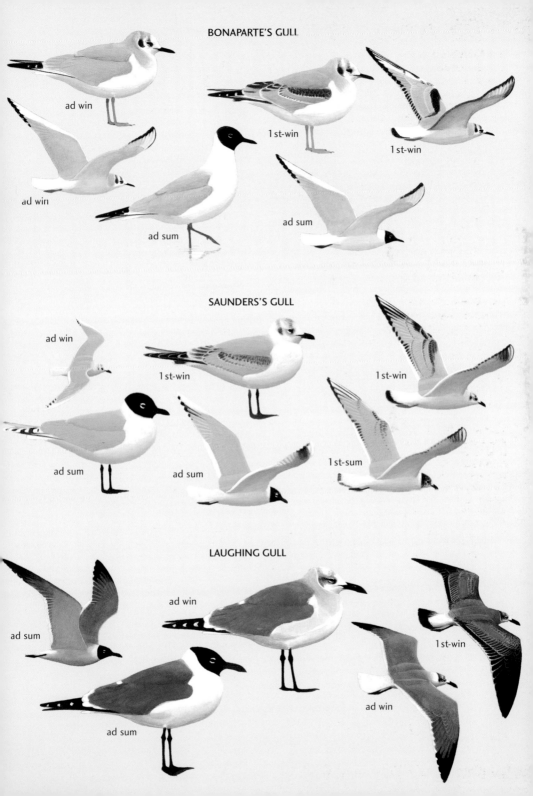

BONAPARTE'S GULL

ad win

1st-win

1st-win

ad win

ad sum

ad sum

SAUNDERS'S GULL

ad win

1st-win

1st-win

ad sum

ad sum

1st-sum

LAUGHING GULL

ad sum

ad win

1st-win

ad sum

ad win

FRANKLIN'S GULL   CTJ
Leucophaeus pipixcan

L 32–38cm; WS 85–95cm; WT 220–335g. **SD** Breeds mainly in C & W states of N America and migrates to western S America. Has strayed to Bering Sea, Japan, Bohai Gulf (Tianjin and Beidaihe) and Taiwan. Monotypic. **HH** Beaches and harbours. **ID** Small to mid-sized, rather dark-mantled gull, slightly smaller than (or overlaps) Black-headed, with distinctive head and wing patterns. Ad. breeding has full black hood, with broad white crescents above and below eye joining at rear, as in Saunders's (p.212) and Relict (p.208). Non-breeding ad. retains rather extensive black hood on rear crown and cheeks. At rest, primary extension long with prominent white tips (see Laughing, p.212). Wings slightly broader, shorter and blunter than Black-headed, dark grey, concolorous with back. Outer primaries subterminally black with white band separating black from basal grey. Juv. extensively brown on upperwing, mantle and neck, but head already has partly black hood, more clean-cut than greyer, more uniform young Laughing. 1st-winter has clearly demarcated partial blackish hood, white neck and underparts. Neat grey saddle contrasts with juv. wing-coverts. In flight, pale inner primary window with white tips extending trailing edge from secondaries. White tail has black band, broadest in centre. Underwing and breast clean white at all ages (compare young Laughing). Uniquely, undergoes complete moult into 1st-summer plumage, after which appears more adult-like but has more black in primary tips and some may show dark feathers in tail and secondaries until 2nd-winter. **BP** Bill short and thicker than Black-headed (p.210), bright red when breeding, black in non-breeding/juv.; eyes brown/black; legs bright red in breeding ad., dull blackish-red in winter, black in young. **Vo** Common calls include less penetrating hollow laughing *kowii*, *queel*, and softer *krruk* or *kaa* than other gulls, or shrill *guk*. **TN** Formerly placed in *Larus*.

LITTLE GULL   CTKJR
Hydrocoloeus minutus

L 24–30cm; WS 62–69cm; WT 88–162g. **SD** Widespread Palearctic species. In E Asia breeds in S Yakutia, from Lake Baikal to Sea of Okhotsk, and locally in NE China. Most winter well to west of region but has strayed to coastal Hebei and Jiangsu, Taiwan and several times to Japan, from Hokkaido to Izu Is and the Nansei Shoto; reported Korea. Monotypic. **HH** Well-vegetated freshwater lakes, wetlands and rivers in taiga and steppe. In non-breeding season, coasts, beaches and river mouths. **ID** Very small, elegant gull, like compact, diminutive Black-headed (p.210) in breeding plumage, but has unique dark grey underwing. Ad. breeding has more extensive black (not brown) hood than Black-headed, extending to nape; mantle/wings pale grey. Non-breeding ad. has black on rear crown and spot on ear-coverts. Juv. dark, with blackish wing-coverts and flight-feathers, and grey innerwing-panel, forming kittiwake-like

dark M on upperwing. Mantle dark brown/black, moulting to grey in 1st-winter, tail white with narrow black terminal band. 1st-winter has paler wing but retains prominent black carpal bar and primary tips, and has ad. winter-like head pattern. 1st-summer has poorly defined black hood, dark carpal bar and wingtips, but thereafter wings become steadily paler. At rest, primary tips rounded (though young have more pointed primaries), and protrude only just beyond tail. In flight, which is buoyant, wings slightly broader, shorter and more rounded than Black-headed. Wings (ad.) pale grey and uniform with mantle, but underwing dark sooty grey, almost black, with narrow white margin formed by white tips to flight-feathers. **BP** Bill very short, fine, black or reddish-black; eyes black; short tarsi bright red in ad., pale pink in juv./1st-winter. **Vo** Commonly, a rather hard, nasal, tern-like *keck*, often run together in rapid series *kek-kek-kek…*, particularly at colonies. **TN** Formerly placed in *Larus*.

ROSS'S GULL   CKJR
Hydrocoloeus rosea

L 29–34cm; WS 82–92cm; WT 120–250g. **SD** Breeds in region only in N Yakutia between Khroma and Kolyma rivers. Movements in E Asia unknown, but winters at sea near ice floes in Arctic Ocean, though individuals and even flocks have been recorded in N Hokkaido (Dec–Feb), with singles even south to Chiba, Honshu; also Liaoning (China) and reported from Korea. Monotypic. **HH** Open swampy tundra or pools and marshes in forest-tundra. In winter, occurs around sea ice and rarely along coasts, even harbours. **ID** Small, very distinctive gull with peaked head, long, narrow wings, wedge-shaped tail and unique plumage. Recalls Little Gull, but wings longer and more pointed, tail longer, also pointed. Ad. breeding has pale grey mantle and wings, narrow black necklace (broadest on nape), rosy-pink flush to head and underparts which may be retained in winter, and diamond-shaped tail. In winter necklace lost, but has small blackish ear-coverts spot. Juv. has brownish cap and nape, and dark-centred, buff-fringed scapulars; juv./1st-winter have grey upperparts, but dark primaries and carpal bar, forming prominent dark M on upperwing (Little typically has grey bar); long, diamond-shaped tail also has dark tip, and there is a dark smudge around eye and over ear-coverts, which may extend onto crown, nape and neck-sides. At rest, appears small-headed and long-winged, somewhat tern-like, with primaries extending well beyond tail; ad. extremely pale; 1st-winter has essentially black wings with white, midwing panel. In flight (light and buoyant, rather tern-like), upperwing grey with broad white trailing edge, and underwing darker grey, contrasting with paler belly, head and vent. **BP** Bill delicate, black; eyes black; legs red in ad., dull flesh-pink in young. **Vo** On breeding grounds, high melodious calls, a mellow barking *prrew* and tern-like *kik-kik-kik…*; generally silent at other times but occasionally gives *kuwa kuwa* flight call. **TN** Formerly *Rhodostethia rosea*.

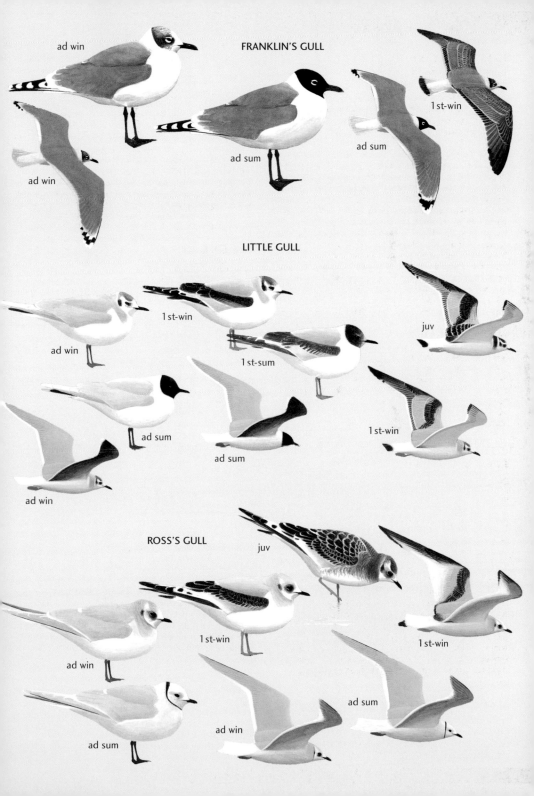

FRANKLIN'S GULL

ad win

ad win

ad sum

ad sum

1st-win

LITTLE GULL

ad win

1st-win

1st-sum

juv

ad sum

1st-win

ad win

ad sum

ROSS'S GULL

juv

ad win

1st-win

1st-win

ad sum

ad win

ad sum

IVORY GULL   KJR
Pagophila eburnea

L 40–47cm; WS 94–120cm; WT 520–700g. **SD** High Arctic, breeding in region only on islands north of New Siberia Is. Has occurred Wrangel I., and has strayed in winter to N & C Japan (Dec–Mar) and Korea. Monotypic. **HH** Crags or coastal cliffs, foraging offshore or amongst ice pack. In winter essentially pelagic; rarely visits coasts, including harbours; attracted to carcasses. Undergoing dramatic decline. **ID** Unique medium-sized, stocky gull, with large rounded head, short neck, similar in size to Common Gull (p.200), with pigeon-like proportions. Ad. all white. Juv./1st-winter also essentially white, but has dusky, blackish face and black spots on mantle, wing-coverts, flanks, and tail tips. At rest, primaries extend beyond tail. Flight buoyant and dove-like on rather broad wings. **BP** Bill short, greenish-grey or bluish-grey with yellow tip in ad., grey with black tip in young; eyes black, very prominent in ad.; very short tarsi black. **Vo** Grating, tern-like *kyuui* or *kree-kree*, or harsher, recalling Black-headed Gull (p.210), and high, mewing whistle *wheeew* or *preeo*.

SABINE'S GULL   KJR
Xema sabini

L 30–36cm; WS 80–91cm; WT 135–225g. **SD** Canadian, Alaskan and E Russian Arctic. Winters in eastern N Pacific, straying widely across North America and, in Atlantic, regularly to W Europe. In E Asia, Yakutia, E Chukotka and Wrangel I., with most presumed to winter off S America, hence its rarity in W Pacific. Accidental in Japan and Korea. Monotypic. **HH** Tundra, often at brackish wetlands near coasts. Away from breeding areas primarily pelagic. **ID** Small, elegant gull with forked tail and tricoloured upperwing, slightly smaller than Black-headed (p.210). Ad. breeding has dark grey hood, bordered at lower edge black; mantle and wing-coverts mid to dark grey. Non-breeding ad. has partial grey hood. At rest ad. has black primaries with white tips, extending well beyond tail. Juv. appears scalloped, with dark greyish-brown nape, breast-sides, back and wing-coverts, all with black subterminal margins; tail white with narrow black terminal band. 1st-winter has paler wings, but retains prominent brown carpal patch, primary tips lack white, with ad. winter-like head pattern. 1st-summer develops white primary tips and poorly defined grey hood. In flight, long-winged, graceful and buoyant; in ad. grey mantle and wing-coverts contrast strongly with white secondaries and inner primaries, and black outer primaries; uppertail-coverts/tail all white, tail shallowly forked; juv./1st-winter have browner grey mantle and wing-coverts, and narrow black tail-band. **BP** Bill short, black with yellow tip (ad.), or all black (juv.); eyes black; legs black, dull pink in juv./1st-winter. **Vo** Vagrants usually silent but occasionally give harsh, grating tern-like *kyeer* or *krrrree*.

BLACK-LEGGED KITTIWAKE   CTKJR
Rissa tridactyla

L 37–43cm; WS 91–105cm; WT 305–512g. **SD** Holarctic coasts, wintering at sea in northern oceans. In E Asia, *R. t. pollicaris* (**Pacific Kittiwake**) breeds on Arctic Ocean islands, coastal Yakutia,

Chukotka, Kamchatka, Commander Is, and Sea of Okhotsk to Sakhalin. Moves through Bering Sea, N Pacific and both coasts of Japan (common off N Japan, mainly Nov–Mar), uncommon in winter in Korea, and accidental to coastal E China and Taiwan. **HH** Precipitous cliffs or isolated rocky islands, otherwise coastal, near colonies. In winter, pelagic though occurs in sight of land, but less frequently along coasts or in harbours than other gulls, unless forced inshore by storms. **ID** Distinctive clean-looking, medium-sized gull, with black-tipped wings and shallow-forked tail. Ad. has pale grey mantle/wings, with neatly black-tipped wings, remaining plumage white with some grey smudging on rear crown in winter. Juv./1st-winter has grey upperparts, dark outer primaries and prominent black carpal bar forming prominent dark M on upperwing; the shallow-notched tail has narrow black terminal band; also blackish smudge behind eye and black band on nape. At rest at nest, does not adopt typical horizontal stance of other gulls, instead partly upright with body at 45°; appears round-headed, primaries extend just beyond tail; ad. pale; 1st-winter rather dark on closed wings. In flight (light and buoyant on rather narrow wings), upperwing grey with pure black tip, underwing is white. **BP** Bill short, pale yellow; eyes black; legs short, black in ad. (rarely orange-red), greyish in young (legs usually, but not always, visible in flight). **Vo** Highly vocal at colonies, with strident, high-pitched, nasal *kittee-wayek* calls; generally silent elsewhere.

RED-LEGGED KITTIWAKE   JR
Rissa brevirostris

Vulnerable. L 35–39cm; WS 84–90cm; 340–450g. **SD** Beringian endemic, locally common on islands of Bering Sea, Aleutians and on Commander Is. Winters more widely in Bering Sea and N Pacific south of Aleutians; accidental Pacific coast of Japan (mostly Jan–May, but also Aug and Nov). Monotypic. **HH** Found close to sea cliff colonies, but also off E Kamchatka; crepuscular, perhaps even nocturnal forager, over deep water. In winter pelagic, but accidentals in Japan have mostly been in harbours with other gulls. **ID** Superficially very similar to Black-legged, but range of characters separate them. Ad. slightly smaller than Black-legged, but head larger and more rounded, with steeper forehead, eye larger and more prominent. Shares grey mantle and wings with clean black tips with Black-legged; however, mantle and wings are darker grey, white trailing edge is much broader and underside of primaries much darker grey. In winter has dusky patch around eye and on ear-coverts. Juv./1st-winter like Black-legged, but nuchal bar dark grey, not black, as are ear-coverts and eye smudges; white trailing edge even broader, recalling Sabine's; white tail unmarked. Stance as Black-legged, but primaries extend well beyond tail. In flight, which differs from Black-legged in its stronger, 'rowing' action, the darker upperwing (and darker underside to primaries) and broader, white trailing edge are distinctive; tarsi often not visible in flight. **BP** Bill shorter, more arched than Black-legged, black in juv. and yellow in ad.; eyes black; legs very short, reddish in juv., bright red in ad. Leg colour alone unreliable, as Black-legged may occasionally have pink, yellow or orange legs. **Vo** Similar to Black-legged, but a much higher pitched squeal *suweeer*.

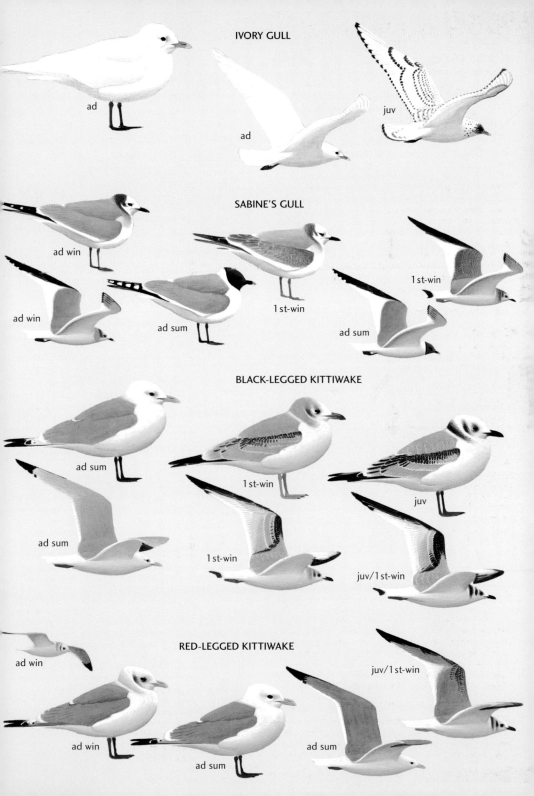

IVORY GULL

ad

ad

juv

SABINE'S GULL

ad win

ad win

ad sum

1st-win

ad sum

1st-win

BLACK-LEGGED KITTIWAKE

ad sum

ad sum

1st-win

1st-win

juv

juv/1st-win

RED-LEGGED KITTIWAKE

ad win

ad win

ad sum

ad sum

juv/1st-win

CASPIAN TERN
CTKJR

Hydroprogne caspia

L 48–56cm; WS 130–145cm; WT 574–782g. **SD** Uncommon but almost cosmopolitan. In E Asia, breeds in Transbaikalia and Russian Far East, and in coastal NE & E China around Bohai Gulf, wintering south along continental coast and presumably mixes with residents of SE Chinese coast; migrant Taiwan, Korea; scarce S Japan. Monotypic. **HH** Coastal wetlands, lagoons and lakes, on sand islands in rivers and offshore. Migration/winter at coasts, estuaries and rivers. **ID** Gull-sized, black-capped tern with large head, thick neck and enormous red bill; unmistakable. Ad. breeding white, with glossy flat black cap reaching to eye and nape, pale grey wings and mantle, and blackish primaries; tail short and forked. In winter, crown frosted with white tips and streaking. Juv. has less clean, but more extensive, black cap, and faint brown scaling on wing-coverts, more noticeable on mantle, becoming paler grey in 1st-winter. At rest, primaries extend well beyond tail. In flight, which is buoyant, has strong 'rowing' action with rather stiff, shallow beats; essentially pale grey upperwing and pale underwing with blackish primaries. **BP** Bill as long as head, bright red with small dusky tip in ad., dull orange in young; eyes black; legs short, black. **Vo** A rasping *kaa kyaaa* or *krrr-aaaack*, and more prolonged *kerrrrsch*. Juv. gives very different, whistled *wee-you*. **TN** Formerly placed in *Sterna*.

LESSER CRESTED TERN
J

Thalasseus bengalensis

L 35–43cm; WS 88–105cm; WT 185–242g. **SD** Breeds from Red Sea to NE Australia, including SE Asia, but has strayed to C Japan (presumably *T. b. torresii*). **HH** Warm coastlines, with other gulls and terns. **ID** Like diminutive Greater Crested Tern, but has sleeker, less shaggy crest and black forehead in breeding plumage. In flight, black of primaries more extensive on both surfaces. Mantle, rump, uppertail-coverts and tail concolorous grey (except white outer rectrices). **BP** Bill long, thin, orange-yellow to orange, paler at tip; eyes black; short tarsi black (ad.) or yellow (juv.). **Vo** Flight call a raucous *kirrrik* or *kurii*. **TN** Formerly placed in *Sterna*.

SANDWICH TERN
T

Thalasseus sandvicensis

L 36–46cm; WS 86–105cm; WT 130–285g. **SD** Wide range, Europe to Sri Lanka, eastern N America to Peru and Uruguay. Has strayed to N Taiwan (presumably nominate). **ID** Slightly larger than Lesser Crested and extremely white, with very long bill. Ad. breeding has crown and shaggy nuchal crest black to line of eye and forehead. Face and neck white, mantle, scapulars and wings pale grey. Non-breeding ad. develops white forehead, and grey-flecked black crown. Juv./1st-winter like winter ad., but has blackish-brown scaling on wing-coverts/tertials. Wings long and narrow, wingbeats strong with 'rowing' action on slightly flexed wings, showing darker grey outer primaries (forming indistinct dark wedge); rump and rather short tail both white, unlike other *Thalasseus*. **BP** Bill slender, black with yellow tip; eyes black; short tarsi, black. **Vo** Flight call a harsh *kerrick*. **TN** Formerly placed in *Sterna*.

CHINESE CRESTED TERN
CT

Thalasseus bernsteini

Critically Endangered. L 38–43cm. **SD** Extremely rare, known to breed only on small Taiwanese islands off Fujian coast within Greater Crested Tern colonies (e.g. Matsu). Has strayed in April to mouth of Ba Zhang River, on W Taiwan coast (presumably an extremely scarce passage migrant, perhaps associating with Greater Crested) and north to Bohai Gulf, so could appear elsewhere on Chinese coast, even Nansei Shoto or coastal Korea. Thought to winter in SE Asia. Monotypic. **HH** Subtropical coastlines, on offshore islets and pelagic. Returns to colonies in April. **ID** Large, slender tern resembling small Greater Crested. Ad. has extensive black cap reaching bill base, and paler grey mantle and wings, contrasting more strongly with black primaries than Greater Crested. Non-breeding ad. has white forehead and lores, and retains black cap from rear of eyes to nape, at all seasons has more white on crown than Greater Crested. Juv. browner than juv. Lesser Crested, with two dark bars on innerwing. In flight, black primaries contrast more noticeably with white underwing and extremely pale grey upperwing than in similar species; tail deeply forked. Like other *Thalasseus* (except Sandwich) has a grey tail, with white only on outer rectrices. In Chinese Crested rump, tail and uppertail-coverts are concolorous with pale silvery-grey mantle. **BP** Bill long, thin, pale orange with black near tip and extreme tip white; eyes black; legs short, black. **Vo** Harsh and higher pitched than Greater Crested. **TN** Formerly placed in *Sterna*.

GREATER CRESTED TERN
CTKJ

Thalasseus bergii

L 43–53cm; WS 125–130cm; WT 320–400g. **SD** Tropical and subtropical ocean tern that occurs in E Asia. *T. b. cristatus* breeds (May–Sep) north to islands off SE China, Taiwan and the Nansei Shoto and Ogasawara Is, with post-breeders reaching Honshu coast; vagrant Korea. **HH** Coasts and islands, also estuaries and river mouths. Forages in shallow inshore waters and at sea, resting on fishing buoys and floating platforms. **ID** Gull-sized, black-capped and crested tern. Ad. breeding white, with black crest from just in front of eye to shaggy nape; forehead and lores white, mantle and wings mid grey; primaries blackish; tail short and more deeply forked than Caspian. Rump, uppertail-coverts and tail darkish grey, concolorous with mantle. Outer rectrices white, but inner ones also grey. In winter has more extensive white on forecrown and around eye, retaining black only behind eye and on shaggy nape. Juv. has less clean-cut black cap, and dusky blackish-brown mantle/wing-coverts, becoming paler grey in 1st-winter. At rest, primaries extend well beyond tail. Flight like light Caspian, with similar wing pattern; strong with quite deep flaps, also glides on outstretched wings on migration. **BP** Bill long, thin, greenish-yellow or pale yellow in ad., dull orange in juv.; eyes black; legs short, black (ad.) or brown (juv.). **Vo** Calls alone or in duet. A coarse, purring *kr r ra a ar*, sharp rasping *kirrik* or *kurii kurii*, and hard rattling *skraach*. **TN** Formerly placed in *Sterna*. **AN** Swift Tern.

CASPIAN TERN

ad sum

ad win

juv

ad win

ad sum   *torresii*

SANDWICH TERN

sandvicensis

ad win

ad sum

LESSER CRESTED TERN

juv/1st-win

ad sum

CHINESE CRESTED TERN

ad win

ad sum

juv

cristatus

GREATER CRESTED TERN

ad win

ROSEATE TERN *Sterna dougallii* CTJ

L 33–43cm; WS 72–80cm; WT 90–110g. **SD** Wide-ranging but localised. *S. d. bangsi* breeds east into E Asia, where found (May–Sep) on coasts of Fujian, Taiwan, S Kyushu and the Nansei Shoto. Rarely strays north of breeding range, but sometimes typhoon-blown. **HH** Isolated, rocky offshore islets and atolls. Winter: moves south out of region. **ID** Medium-sized, elegant, black-capped tern with rounded head, long slender bill, and very deeply forked tail with greatly elongated outertail-feathers. Ad. extremely pale grey mantle and wings, but less white than Black-naped, and primaries have black shafts and outer vanes; underparts white with faint rosy tinge; rump and tail white but contrast little with mantle. Though fork is deep, tail usually held closed. Winter ad. loses black on forecrown. Juv./1st-winter mottled brown on crown and nape with dark forehead, whilst grey upperparts (mantle/scapulars) have broad blackish fringes affording heavy scalloped effect, and wing-coverts mid grey with dark fringes; tail shorter and unforked. At rest, despite long wings, outer rectrices of adult extend well beyond wingtips, which appear blackish. Wings narrow but shorter than Common; flight light with rapid, stiff shallow wingbeats, quite different from more graceful flight of Common or Arctic; hovers occasionally, more often seen power-diving into water for prey; underwing mainly white with some black near primary tips and in outermost primaries, but lacks dark trailing edge to underside of primaries typical of Common/Arctic. **BP** Bill black (non-breeding/juv.) or with deep red base (breeding); eyes black; legs short, orange (ad.), or black (juv.). **Vo** Flight call raucous and grating: *krraahk*, *kierrik* or *cherr-rrick*, a distinctive *gyuii* or *chu-vee*, higher *skivvik*, also *keeer* (in alarm), and in juv. *krrip*.

BLACK-NAPED TERN *Sterna sumatrana* CTJ

L 34–35cm; WS 61–64cm; WT 98–100g. **SD** Subtropical Indian Ocean and W Pacific. *S. s. sumatrana* breeds (May–Sep) off SE China, Taiwan and the Nansei Shoto. Rarely strays north of breeding range, but sometimes typhoon-blown. **HH** As Roseate. **ID** Small to medium-sized, slender, elegant, extremely pale greyish-white tern, lacking black cap, but has band from just forward of eye to nape. Wings and tail long, slender and long tail deeply forked. Ad. has such extremely pale grey mantle and wings as to appear virtually white. Juv./1st-winter mottled brown on crown and nape, whilst upperparts have buff to black subterminal marks, tail shorter and unforked. At rest, despite long wings, outer tail-streamers of ad. extend beyond wingtips. Flight very light, buoyant and often hovers. **BP** Bill long, thin and entirely black; eyes black; legs short, black. **Vo** Huskier, deeper *gui gui* call than Roseate, and various sharp notes.

COMMON TERN *Sterna hirundo* CTKJR

L 32–39cm; WS 72–83cm; WT 97–146g. **SD** Very wide N American and Eurasian range; winters in S Hemisphere. *S. h. longipennis* common across E Russian taiga south of Arctic Circle to Yakutia, Chukotka, Koryakia and Kamchatka, around Sea of Okhotsk, on Kuril Is, also NE China; *S. h. minussensis* Mongolia to Lake Baikal, scarce migrant E Asia. **HH** Coastal and inland colonies on sand bars, dunes and beaches, islets, beside

rivers, lakes. Migrates along coasts of Japan, Korea, SE China and Taiwan; also inland lakes, rivers and wet fields. **ID** Medium-sized, elegant, black-capped tern with long narrow wings and long, deep-forked tail; crown slightly flat and head somewhat elongated. Ad. *S. h. longipennis* differs markedly from European or N American subspecies in being greyer on mantle and wings, and on underparts when breeding. Non-breeding ad. has paler underparts, and whiter forehead; black carpal bar prominent. Juv./1st-winter has cap like winter ad., but grey upperparts and wings with blackish fringes to mantle, coverts and scapulars. Long, blackish wingtips level with outer tail-streamers; outermost webs of outertail-feathers dark-grey/black (white in Roseate). Flight light and buoyant, both hovers and plunge-dives; underwing mainly white with outer primary tips black, forming distinct dark panel and leaving only inner primaries translucent (see Arctic); juv. has dark carpal and secondary bars with paler midwing panel. *S. h. minussensis* has paler upperparts and upperwing than *S. h. longipennis*. **BP** Bill long, all black (*longipennis*) or crimson with blackish tip (*minussensis*); eyes black; legs short, usually black but sometimes blackish-red (*longipennis*), or noticeably red (*minussensis*), becoming black outside breeding season. **Vo** Particularly vocal when breeding, but also on migration. Hard chattering *kyi kyi kyi* and harsh, descending, oft-repeated long *kree-arr* or *keeeyurr* with strong emphasis on first part, and screeching *kzrrssh*. In aggression, a hard, rattling *k-k-k-k...*

ARCTIC TERN *Sterna paradisaea* JR

L 33–36cm; WS 76–85cm; WT 86–127g. **SD** Very wide N American and Eurasian range; winters in S Hemisphere. High-Arctic tundra coasts from Lena to Bering Sea, Koryakia, NE Sea of Okhotsk and coastal N & W Kamchatka. Seems astonishingly rare (overlooked?) further south in region, most move to E Pacific (where fairly common migrant in late summer), thence south for winter. Winters at sea, but migrates along coasts; accidental Japan mostly Jul–Aug. Monotypic. **HH** Breeds colonially along coastal shores, on offshore islands, also inland on tundra or damp meadows. **ID** Very similar to Common, but slightly smaller, with rounder head, smaller bill and shorter legs, and relatively longer, narrower wings; underparts clearly grey as *longipennis* Common. Ad. black cap extends lower on lores (nearly to gape) than Common. Tail streamers longer than Common and extend beyond wingtips. Non-breeding has white forehead. Juv./1st-winter cap like winter ad., but grey upperparts/ wings have dark grey fringes to mantle, coverts and scapulars, and dark carpal bar (far less prominent than Common); outermost webs of outertail-feathers dark grey/black (white in Roseate). Flight as Common; upperwing uniform grey, whilst underwing, which is mainly white, translucent and has only narrow black tips to outer primaries (forming narrow trailing edge); juv. has narrow, diffuse carpal bar, grey midwing panel and white secondaries, thus wing shades from dark at leading edge to pale. **BP** Bill short, fine, dark red (breeding), or black; eyes black; legs very short, red, duller in non-breeding/juv. **Vo** Very similar to Common, but higher pitched and squeakier, a buzzy *gyii-errr* and *ki-ki-ki-ki-ki*, a rapid *titkerri titkerri...* near nest, may press home attack on intruder with bill and claws while giving dry *raaaz* calls.

ROSEATE TERN

bangsi

ad sum

ad win

juv

BLACK-NAPED TERN

sumatrana

ad sum

juv

COMMON TERN

longipennis

ad sum

ad win

juv

ARCTIC TERN

ad sum

ad win

juv

LITTLE TERN CTKJR
Sternula albifrons

L 22–28cm; WS 47–55cm; WT 47–63g. **SD** Very widespread, from Europe to Africa and Asia to Australia. *S. a. sinensis* (may merit specific status) breeds (mid Apr–Sep) NE China, N Honshu and Korea, through Japan to Nansei Shoto, E coastal China and Taiwan. Has strayed north to Hokkaido and Sakhalin. Winters around Australia. **HH** Breeds on coasts and inland, on sandy shores and shingle-bedded rivers. At other times, coasts, coastal lagoons and tidal creeks. Highly susceptible to human disturbance and from ground predators. **ID** Small, rather delicate, pale grey tern. Ad. upperparts, including wings pale grey, cap black with small white forehead patch extending to just above eye, lores black. Non-breeding ad. has more extensive white on forehead and white lores. Juv. resembles winter ad., but mantle and wing-coverts have scaled appearance. Appears large-headed, wingtips fall just short of tail streamers in summer, extending beyond at other times. Wingbeats fast, wings narrow and set well forward, flies as if front-heavy, commonly hovers, dips and plunge-dives, repeating latter rapidly. Outer 2–3 primaries black with white shafts; rump grey, tail white and forked, with protruding tail-streamers. **BP** Long bill, pale yellow with small black tip in summer, blackish at other times; eyes black; legs orange, duller outside breeding season. **Vo** Rather noisy, sharp, high-pitched *kyi-kyi*, *ket* and rasping *kyik*. In display or alarm, more prolonged series of *kiri-kyik kiri-kyik...* calls. **TN** Formerly *Sterna albifrons*.

ALEUTIAN TERN CTKJR
Onychoprion aleuticus

L 32–34cm; WS 75–80cm; WT 83–140g. **SD** Beringian/Okhotsk Sea endemic, breeding locally only on Alaskan and Bering Sea coasts and Aleutians. In region breeds from Anadyr to S Kamchatka and Sakhalin. Might be expected to be common, at least in Hokkaido, but remarkably rare (overlooked?) in Japan (May/Jun, Aug–Oct); migration routes and wintering range poorly known, but includes coastal China, Taiwan (now known to be frequent migrant) and Korea (where increasingly reported) to Philippines, SE Asia (particularly Hong Kong, Sulawesi and Halmahera coasts). Monotypic. **HH** Breeds on open dry areas inland (but near sea), pools and lakes, or on dunes or beaches beside rivers, and shorelines; often with Common Tern (p.222) on Sakhalin and Kamchatkan coasts. Forages offshore and in shallow inshore waters and bays. **ID** Small dark tern. Ad. breeding is superficially similar to *longipennis* Common, but smaller, with black cap interrupted by white forehead extending to eye, black lores, dark grey mantle and wings, and long black primaries; underparts also grey as in *S. h. longipennis*, but call always distinctive. Non-breeding has more extensive white forehead, and white lores, but retains black nape; underparts whiter but upperparts remain dark grey. Juv./1st-winter very dark, cap all dark, blackish, reaching bill; upperparts dark grey, the mantle, scapulars and coverts fringed bright cinnamon, and neck and breast-sides also washed warm cinnamon. Ad. may appear dark grey with white only on forehead and in narrow band from chin to cheeks; dark primaries extend beyond white tail. Flight very light and agile; upperwing uniform dark grey, whilst underwing mainly pale grey with distinctive, prominent narrow black bar on trailing edge of secondaries and translucent inner primaries; outer primaries black from below; dark wings and mantle contrast clearly with white rump and white, forked tail; juv. in flight has very dark wing-coverts with brown tinge, and cinnamon wash visible on breast-sides. **BP** Bill short, fine and black in ad., black with slight yellowish cast to base in juv.; eyes black; legs short, orange in juv., black in adult. See Arctic and Common (p.220), for identification of potentially confusing juv. **Vo** Short, sharp, *eek eek*, recalling Saunders's Gull. **TN** Formerly *Sterna aleutica*.

GREY-BACKED TERN J
Onychoprion lunatus

L 36–38cm; WS 73–76cm; WT 115–177g. **SD** C & S Pacific; has reached the Iwo Is (Kita Iwo-jima and Minami Torishima), where perhaps formerly bred but status now uncertain. Monotypic. **HH** Breeds on oceanic islands, on beaches or low cliffs. Pelagic at other times. **ID** Medium-sized, grey-backed tern. Resembles Bridled Tern (p.224), but has grey, not sooty-black upperparts, more like Aleutian, but with dark rump and white underparts. Ad. breeding has cap and nape black, with white forehead extending beyond eye, eyestripe black; upperparts pale to mid grey on mantle and back; primaries, secondaries, rump and tail blackish-grey. Underparts pure white. Non-breeding ad. has white-streaked forecrown. Juv. has less distinct head pattern, buff-scaled upperparts and off-white breast-sides. **BP** Long slender bill, eyes and short tarsi all black. **Vo** Generally silent away from colonies, where gives distinctive high-pitched screeching calls recalling Sooty, but softer and less harsh. **TN** Formerly *Sterna lunata*. **AN** Spectacled Tern.

LITTLE TERN

sinensis

ad win

ad sum

ALEUTIAN TERN

juv

ad sum

ad sum

ad win

ad sum

GREY-BACKED TERN

juv

ad win

ad sum

BRIDLED TERN CTKJ
Onychoprion anaethetus

L 35–38cm; WS 76–81cm; WT 95–150g. **SD** Wide range in African, Middle Eastern, SE Asian and C American seas, also in tropical waters off Australia and north in Pacific. *O. a. anaethetus* occurs in southernmost part of region, on SE China coast to Fujian, off Taiwan, and in Yaeyama Is (Japan). Has appeared as far north as N Honshu and even Hokkaido; reported Korea. **HH** Breeds (May–Sep) on uninhabited rocky off-shore islets and forages inshore; at other times pelagic and departs region. **ID** Medium-sized dark-backed tern (size of Common Tern but longer-winged). Ad. has charcoal grey-brown mantle/wings, black cap and nape (contrasting with paler mantle) with narrow white forehead band extending above eye to a point; black loral line broad and long (see Sooty). Face, neck and underparts white. Non-breeding ad. has more extensive white forehead. Juv. has dusky-grey rear crown, ear-coverts and nape, mid to dark brown upperparts prominently scaled on mantle and coverts with buff fringes. Appears strongly pied, plain dark wingtips reach just beyond tail in non-breeder, just short of tail tip when breeding; closed tail appears white from sides. Flight graceful, buoyant and agile, dipping down to pick food from sea surface; blackish-tipped flight-feathers contrast with white underwing-coverts. Rump and tail dark charcoal grey in centre, tail deeply forked and outer sides of rump/tail white. Often perches on floating debris at sea. Ad. Bridled differs from Sooty in having dark grey, not black back, white forehead extending in narrow point over eye, and streamers of closed tail long and all white (shorter and darker in Sooty). Juv. Bridled is rather pale, whereas juv. Sooty is mainly sooty-black with white spots. **BP** Long bill, eyes and short tarsi all black. **Vo** Generally silent, except at colonies where gives low barking *kuu* or *kuraa*; yapping *wup wup*; a rising mellow whistled *weeeep* and a quavering *nyaauw*, and *ke-eeeee* in courtship. **TN** Formerly *Sterna anaethetus*.

SOOTY TERN CTKJ
Onychoprion fuscatus

L 36–45cm; WS 82–94cm; WT 147–240g. **SD** Ranges throughout subtropical and tropical oceans. Reaches just into southernmost part of region, where *O. f. nubilosa* breeds north to Yaeyama Is (southern Nansei Shoto) with *O. f. oahuensis* on Ogasawara Is. Small numbers migrate past Taiwan (May, Sep). Storm-blown birds occasionally reach SE China coast, Korea and Pacific coast of Honshu or even SE Hokkaido. **HH** Breeds (Apr–Sep) on rocky offshore, uninhabited islets and forages inshore; at other times pelagic and departs region.

ID Medium to large black-backed tern (nearly size of Greater Crested, p.218), superficially similar to smaller Bridled, but stockier and shorter tailed. Ad. blacker on mantle/wings, and black cap more extensive, extending forward of eyes and not contrasting with mantle. White forehead band broader, but does not extend as a superciliary line; black loral line narrows from eye to bill base. Face, neck and underparts entirely white, appearing particularly bright in contrast with the blacker upperparts. Non-breeding ad. has more white on forehead, sometimes giving appearance of superciliary to eye; hind-collar paler, more like Bridled. Juv. almost entirely sooty, blackish-brown, including head, neck and underparts, except white vent, with white tips to mantle/wing-coverts creating slightly spotted appearance; underwing-coverts pale grey, paler in juv. and darker in ad. (see Bridled for differences in upperwing, underwing and tail). Wings broader, beats stiffer than Bridled; rump and tail black, tail deeply forked and only outermost rectrices are white. Rarely if ever perches on floating debris at sea (compare Bridled). **BP** Long bill, eyes and tarsi all black. **Vo** Generally silent away from colony, where gives harsh, nasal *gii-ah* or *draaaaa* in threat, and nasal *wide-a-wake*, *ker-wacki-wack* or *ka weddy weddy* calls. **TN** Formerly *Sterna fuscata*.

GULL-BILLED TERN CTKJ
Gelochelidon nilotica

L 33–43cm; WS 76–108cm; WT 130–300g. **SD** Almost cosmopolitan, but local. *G. n. nilotica* breeds east to NE Inner Mongolia; *G. n. affinis* around Bohai Gulf and S & SE China; migrant Taiwan. Uncommon/rare in SW Japan where most are post-breeding or typhoon-blown (Jul–Oct); vagrant Korea. **HH** Breeds colonially at coastal wetlands, lagoons and lakes; on migration/winter at coasts, estuaries and rivers. **ID** Medium-sized, stocky tern with clean black cap, thick black bill and shallow-forked tail. Ad. breeding white, with glossy black cap to line of eye and onto nape, grey wings and mantle, darker grey primaries with blacker bar on trailing edge. Winter ad. paler above, lacks black cap but has black smudge through eye. Juv. pale brown on crown, hindneck and wing-coverts. Primaries extend well beyond tail. In flight, long-winged, buoyant with strong 'rowing' action; hovers and dives for fish, and hawks insects; upperwing plain grey but may show dark trailing edge to primaries; underwing whitish with distinct black trailing edge to long primaries; slightly forked tail and rump concolorous with back. **BP** Bill differs from other black-capped terns in having a reddish-base (juv.) or in being all black (ad.); eyes black; legs longer than other black-capped terns, black. **Vo** At colonies gives rattling *br-r-r-r*; but more commonly, in flight, a nasal yapping *ga-wik*, *kay-wek* or chivvying *kewick-kewick*.

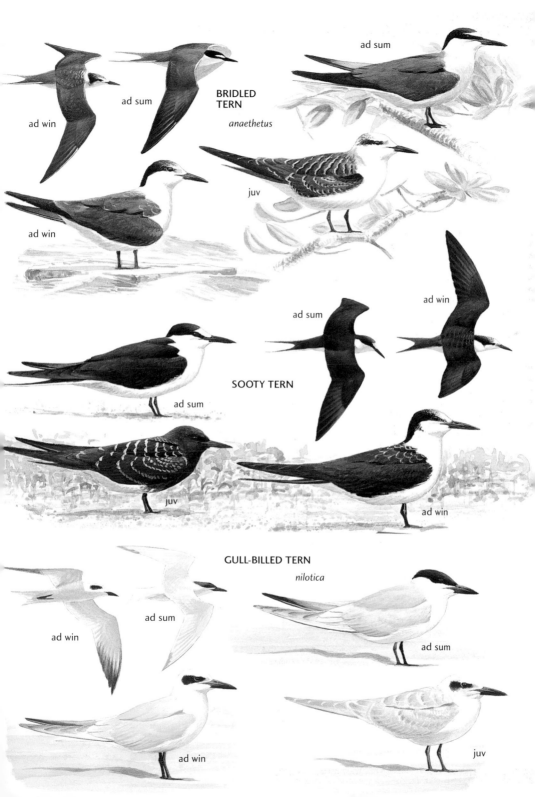

BRIDLED
TERN

anaethetus

ad win

ad sum

ad sum

ad win

juv

SOOTY TERN

ad sum

ad win

ad sum

juv

ad win

GULL-BILLED TERN

nilotica

ad win

ad sum

ad sum

ad win

juv

WHISKERED TERN
CTKJR

Chlidonias hybrida

L 23–29cm; WS 64–70cm; WT 60–101g. **SD** A widespread Palearctic marsh tern, but *C. h. hybrida* restricted to E Asia, breeding in extreme SE Russia (L Khanka region) and NE & E China. Scarce post-breeding migrant Japan, rare in spring, has occasionally wintered; occasional migrant in Sakhalin, Korea and Taiwan (where also winters). **HH** Breeds in loose colonies at swamps and lagoons on floating vegetation; on migration occurs at freshwater lagoons, lakes and pools, or flooded fields. **ID** Small, grey tern, with shallow-forked tail and rounded wings; typically dips and hawks for food from surface or above. Ad. breeding has full black cap, grey upperparts, pale grey wings, and dark grey to blackish-grey underparts, with only chin/cheeks ('whiskers') white. Non-breeding ad. dark with extensive white forehead and white lores, black ear-coverts, rear crown and nape; and all-white underparts. Juv. resembles winter ad., but mantle and wing-coverts have dark brown scalloping contrasting with paler grey rump and tail; flight-feathers blacker. At rest, appears slender, short-tailed, with wingtips extending beyond tail. Flight very light, changes direction frequently. Wingtips black, underwing pale grey or white, rump and short tail mid grey. **BP** Short bill deep blood red (breeding) or black; eyes black; tarsi (rather longer than other *Chlidonias*) blood red. **Vo** While foraging gives dry, metallic, rasping notes: *keh keh*; *ki-kitt*; *krche* or *kzzrt*.

WHITE-WINGED TERN
CTKJR

Chlidonias leucopterus

L 23–27cm; WS 58–67cm; WT 42–79g. **SD** Europe to E Asia, including SE Russia and adjacent NE China. Winters in Africa, S & SE Asia and Australasia. Migrates (May–Oct) through N and coastal China, Japan (scarce), Korea (scarce) and Taiwan. Some winter SE China and Taiwan. Monotypic. **HH** Inland freshwater habitats from taiga swamps to temperate lagoons with extensive floating vegetation, to river slacks and wet/flooded agricultural land. Migration/ winter on inland lakes, coasts, coastal wetlands and mangroves. **ID** Ad. breeding has sooty-black head, neck and underparts, except white vent/undertail-coverts; black mantle, dark grey back, white wing-coverts, grey flight-feathers, white rump and tail. Non-breeding ad. closely resembles Black, but has much paler upperwing, less extensive and less distinct black on crown (diffuse streaking rather than cap), and distinct black ear-coverts patch; lacks dark patches on grey breast-sides just forward of

wings. Juv. like winter ad., particularly in cap and lack of breast patches, but mantle and wing-coverts have dark brown scalloping and lack conspicuous pale tertial tips. At rest, slender, short-tailed, with very long primaries extending well beyond tail, legs longer than Black. Flight very light, buoyant, frequently banking and changing direction, wings broader than Black. Primary tips black, outermost primaries have black shafts creating black leading edge to primaries, darker than rest of upperwing, which is pure white on forewing; underwing-coverts black, contrasting strongly with grey flight-feathers; dark mantle contrasts strongly with white rump. Juv. appears very dark-backed in flight ('saddle' effect), dark-scalloped mantle contrasts with grey innerwing and, especially, with white rump and grey tail; juv./non-breeding ad. have dark carpal bar. **BP** Short bill, dark blood red (breeding) or black; eyes black; tarsi orange/red. **Vo** Calls include low, soft *kweek*, harder *kwek-kwek*, and Whiskered-like short *kesch* and *kek*, rasping *giri giri* or crackling *gzzrk-gzzrk gzzrk*. **AN** White-winged Black Tern.

BLACK TERN
CTKJ

Chlidonias niger

L 23–28cm; WS 57–65cm; WT 60–74g. **SD** Breeds Europe to C Asia as far as Lake Baikal; winters south to W African coast, where pelagic. Extremely rare in E Asia, where *C. n. niger* accidental to Japan (mainly Jul–Oct), from C Honshu to Okinawa, around Bohai Gulf and Yellow Sea in China, and Taiwan; reported Korea. **HH** Estuaries and marshes. **ID** Small, dark marsh tern, with shallow-forked tail, typically seen hawking low over wetlands. Ad. breeding is elegant, with sooty-black head, neck and underparts, except white vent/undertail-coverts; pale grey mantle, wings and tail. Non-breeding ad. has small neat black cap with teardrop extension behind eye; forehead, face and underparts otherwise white, but has dark grey breast-side patches just forward of wings (compare White-winged). Juv. resembles winter ad., particularly in black cap and breast patches, but mantle/ wing-coverts have dark brown scalloping with conspicuous pale tips, especially on tertials. At rest, slender, short-tailed, with very long primaries extending well beyond tail. Flight very buoyant, frequently banking and changing direction, with bill held down. Upperwing of breeder more uniform grey, whereas non-breeder has outer primaries black, darker than grey inner primaries and upperwing; underwing pale grey, rump and short tail mid grey (see White-winged); non-breeding ad. has dark forewing and secondary tips, but pale grey midwing panel. Juv. appears dark-backed in flight, the scalloped mantle contrasting with grey innerwing and grey rump/tail. **BP** Short bill is black (longer than White-winged); eyes black; short tarsi black (breeding) or dull orange/red. **Vo** Contact calls *kii kii*; *klit klit*; *kleep* and *kweeer* heard away from colonies, but nasal, shrill *kyeh* or *keef* only given at breeding sites.

WHISKERED TERN

hybrida

ad sum

ad win

ad win

juv

ad sum

WHITE-WINGED TERN

juv

ad win

ad win

ad sum

BLACK TERN

niger

ad win

juv

ad sum

ad sum

BROWN NODDY TJ
Anous stolidus

L 38–45cm; WS 75–86cm; WT 150–272g. **SD** Tropical and subtropical oceanic islands worldwide. In E Asia *A. s. pileatus* breeds as far north as Iwo and Ogasawara Is and the S Nansei Shoto (Apr–Sep), and off W (possibly N) Taiwan. Scarce further north in Nansei Shoto and storm-blown accidental along Pacific coast of Japan to Hokkaido. **HH** On and around isolated islets, where favours rocky crags with shady ledges. Its colonies, like those of Sooty and Bridled Terns, are prone to predation and disturbance by fishermen. **ID** Large, elegant, dark, long-tailed tern. Ad. rather uniform sooty-brown, except forehead to nape, which is whitish or very pale grey on forehead becoming pale grey on rear crown; face blackish-brown with narrow white crescents above and below eye (less prominent at long range than Black Noddy). Juv. resembles ad. but has sooty-brown crown. At rest, although wings are long and slender, tail is particularly long and extends beyond wingtips. Flight strong, on slightly arched wings with lazy 'rowing' action (somewhat gull-like); paler brown upperwing-coverts form noticeable panel and contrast with darker flight-feathers. Tail somewhat broad and long, parallel-sided with notched tip, but when closed may appear pointed. Appears rather long-bodied in flight, and dips to water to take food (does not plunge-dive). **BP** Long slender bill is black; eyes black; tarsi dark brownish-black. **Vo** Makes odd buzzing and grunting noises, a rattled, growling *kwuwaa* or *garrrh*, a hoarse *gee-aaa*, and a rising croaking *brraak*. In courtship flight a low *nek nek nek nek nek nekrrr*, and *geo geo*. Elsewhere rather quiet. **AN** Common Noddy.

BLACK NODDY J
Anous minutus

L 35–39cm; WS 66–72cm; WT 98–144g. **SD** Remote tropical and subtropical oceanic islands of Pacific, Caribbean and Atlantic. *A. m. marcusi* is a vagrant mainly to Iwo, Ogasawara and Nansei Shoto Is (Jul–Aug; perhaps formerly bred on Iwo Is). **HH** Oceanic islands, occasionally at Brown Noddy colonies; very rarely on coasts. **ID** Ad. smaller and blacker than Brown with much whiter forehead and crown, becoming grey on rear crown. Black face with prominent white crescent below eye. Juv. resembles ad. but has clean white cap. At rest, wings reach or extend just beyond tail tip. Flight resembles Brown, but upper- and underwing all dark, and action more buoyant with faster shallower wingbeats than Brown. Tail somewhat broad, greyish, parallel-sided with notched tip, but when closed may appear pointed. **BP** Black bill is longer (longer than head), thinner and straighter than Brown; eyes black; tarsi dark brownish-black. **Vo** Similar to Brown, but sharper, higher pitched, with a grating *keraa*, and *kuri kuri kuri* notes. **TN** Formerly considered race of Lesser Noddy *A. tenuirostris*. **AN** White-capped Noddy.

BLUE NODDY J
Procelsterna cerulea

L 25–28cm; WS 46–61cm; WT 41–69g. **SD** Breeds outside region in mid and S Pacific. *P. c. saxatilis* has reached Minami-Torishima and Kita-iwo-jima, Japan, and could conceivably occur elsewhere following typhoons. **HH** On and around remote oceanic islands, where forages storm-petrel-like by pattering at the sea surface. **ID** Small and pale. Ad. pale grey with whitish head, very pale blue-grey mantle and whitish-grey underparts; upperwing, (including coverts and flight-feathers), pale grey with darker trailing edge, underwing almost white with darker trailing edge; tail very shallowly forked. Juv. has mid brownish-grey wash to upperparts. **BP** Black bill is fine and sharply pointed; black eyes rather prominent in pale face; tarsi and toes black with yellow webs. **Vo** Sometimes gives a loud squealing call.

WHITE TERN J
Gygis alba

L 25–30cm; WS 76–80cm; WT 92–139g. **SD** Tropical and subtropical tern of Indian Ocean, Pacific and mid Atlantic. *G. a. candida* accidental in Japan, to Iwo and Ogasawara Is, Nansei Shoto and Pacific coast of main islands north to Hokkaido, though most records pre-1950s. Could conceivably appear almost anywhere in Pacific coastal region following typhoons. **HH** On and around vegetated coral atolls. **ID** Ad. small, elegant, all-white tern, with large dark eye particularly prominent on plain face, and sharply pointed bill that appears slightly angled upwards. Juv. has buff or pale brown barring on mantle, and spotting on crown, rump and upperwing-coverts. **BP** Bill largely black, but basally dull blue-grey; large eyes black; short tarsi blue-grey. **Vo** Calls include guttural *juku juku juku*; *heech heech* and descending nasal chatter. **AN** Fairy Tern.

BROWN NODDY

pileatus

ad

juv

ad

ad

BLACK NODDY

marcusi

ad

juv

BLUE NODDY

saxatilis

ad

juv

WHITE TERN

candida

ad

juv

PLATE 101 : SKUAS I

SOUTH POLAR SKUA TKJ
Stercorarius maccormicki

L 50–55cm; WS 127–140cm; WT 0.9–1.6kg. **SD** Breeds Antarctica but moves to N Atlantic and N Pacific in austral winter. In region, most frequent off Pacific coast of Japan (mostly May–Aug, but also at other times), also regular in Yellow Sea off S Korea, scarcer Taiwan. Should be looked for well offshore, primarily from coastal and inter-island ferries. Monotypic. Prior to early 1980s erroneously identified as Great Skua *Stercorarius skua*, and still appears in that species in older publications (Japan) or even some recent lists (China and Taiwan), but there appears to be no evidence of true vagrancy to the region, nor of the larger extralimital Brown Skua *S. antarcticus* (52–64cm) which has reached India and is a potential vagrant to region, though field identification poses many challenges. **HH** Pelagic during northern summer. **ID** Large skua resembling dark imm. gull in coloration with broad, rather blunt-tipped wings and heavy flight. Sexes similar, though ♀ larger than ♂. Ad. often considered dimorphic, but unlike morphs of smaller skuas, in South Polar there is complete gradation between extremes. Dark greyish-brown, with variable paler brown mottling on mantle, underparts and particularly head. Overall appearance a 'cool' brown bird, with pale, somewhat small head, lacking 'warmer', more rufous-brown tones, and dark-capped appearance of extralimital Great Skua or pale flecks of Brown. Dark ad. extremely dark, but commonly has pale 'frosting' on dark nape and pale crescent at base of bill; paler ad. may have sandier brown underparts with pale fringes to mantle. Juv. like ad., very similar to Great but typically has cold, grey-brown plumage. At rest (generally only on water, where floats rather high), primaries extend slightly beyond rounded tail. In flight, all ages show prominent white flashes at bases and along shafts of primaries, which are far more prominent than in smaller skuas; underwing-coverts black (dark brown in Great) contrasting with paler body; uniformly unbarred underwing-coverts, axillaries and upper- and undertail-coverts distinguish it from young Pomarine. 'Arm' broad, 'hand' short, with blunter wingtips than smaller species, tail broad, slightly wedge-shaped. **BP** Bill blackish-grey; large and heavy, but more slender than Brown; juv. has basal two-thirds blue-grey, tip black; eyes black; short tarsi black. **Vo** Occasional harsh gull-like *guaa* or *gwaa* calls, but generally silent in region. **TN** Formerly considered race of Great Skua. **AN** McCormick's Skua.

POMARINE SKUA CTKJR
Stercorarius pomarinus

L 42–58cm; WS 115–138cm; WT 550–850g. **SD** Breeds across northern N America and N Russia, including between the Lena and Bering Strait. Winters in equatorial latitudes. On passage, not uncommon off Kamchatka, the E coast of Japan and offshore in W Pacific. Recorded in all months off Japan, but most common Nov–Apr. Migrant Korea and E China; spring migrant in Taiwan Strait. Monotypic. **HH** Open tundra, in dry lichen areas and swampy areas near lakes/rivers; at other seasons coasts or well offshore; only very rarely inland (e.g. after typhoons). Should be looked for off headlands and occasionally larger river mouths or coastal wetlands. **ID** Large dark skua (about size of Black-tailed Gull, p.200), ad. and some subad. (not juv.) deep-chested, with prominent elongated, twisted, spatulate central rectrices in breeding plumage. Ad. sexes similar, and plumage similar year-round except central rectrices, with two colour morphs. Light morph commoner (>80%); dark cap reaching below gape, dark, scaly breast-band (sometimes reduced or lacking), barred breast-sides and flanks; vent all dark. Dark morph (5–20%) all dark except white wing flashes. Juv. most difficult to identify, ranging from very dark chocolate-brown to mid brown with pale buff-brown barring. Appears large-headed, with obviously heavier, more distinctly two-toned bill than Arctic, and prominently barred vent. Much thicker bill with more restricted dark tip (or more extensive and obvious pale base) affords it a distinctly imm. Glaucous Gull look (see p.202). Head may appear large and dark, or large and grey with fine pale grey spotting. Blunt-tipped central rectrices barely protrude. At rest, although long-winged, primaries of ad. do not reach tail tip; primaries of juv., however, extend beyond tail. In flight, which is steady and measured, appears rakish, but commonly heavy-chested; 'arm' appears evenly broad with comparatively slighter 'hand' (compare Arctic, p232). Small area of white in primary bases on upper- (4–6 primary shafts) and underwing, forming distinct 'flashes' (compare Arctic); ad. has dark underwing; juv. has pale-barred axillaries/underwing-coverts, and pale-barred rump/uppertail-coverts; central rectrices long (extending *c*.5cm), broad, with 'spoon'-like tips, giving lumpy appearance to tail tip. **BP** Bill pinkish with black tip in summer or grey with black tip; juv. bill prominently blue-grey with black tip even at long range; eyes black; legs short, blackish-grey. **Vo** Occasionally gives barking calls; on breeding grounds gives various harsh barking notes: *gyuwaa*; *gyu*; *gwee gwee* and yelping *vii vii*. Long-call a series of rising, nasal *weeek* notes, very different from other skuas. **AN** Pomarine Jaeger.

SOUTH POLAR SKUA

ad intermediate

ad dark

ad pale

juv intermediate

POMARINE SKUA

juv dark

juv pale

ad win pale

juv intermediate

ad ♀ sum pale

ad sum dark

ad ♂ sum pale

ad ♀ sum pale

ad sum dark

juv intermediate

ARCTIC SKUA CTKJR
Stercorarius parasiticus

L 37–51cm; WS 102–125cm; WT 330–610g. **SD** Very widespread and common across N American and Eurasian Arctic; winters in S Hemisphere. Breeds in N Russia from the Lena to Bering Strait, also south on Bering Sea coast to Kamchatka and NE Sea of Okhotsk. Not uncommon migrant off Bering and Okhotsk Sea and NW Pacific coasts, common off E Japan (commonest Apr–Jul, but records all months) and offshore in W Pacific. Regular spring migrant (Apr–May) Taiwan Strait; rare Korea. Monotypic. **HH** Open dry tundra and swampy areas near lakes/rivers; at other seasons, along coasts or well offshore; only rarely inland (e.g. after typhoons). More common inshore than other skuas where may pursue gulls and terns. Aerial engagements often more aerobatic and prolonged than Pomarine, which is more likely to tackle larger seabirds and often kills smaller victims. **ID** Medium-sized dark skua (about size of Common Gull, p.200), smaller and slighter than Pomarine, less deep-chested, with elongated pointed central rectrices in breeding plumage. Ad. sexes similar and plumage similar year-round except central rectrices, with two colour morphs (much like Pomarine). Light morph predominates in far north; dark cap does not enclose gape, usually with pale patch above bill base. Fewer have chest-band and this lacks scaling, also lacks barred flanks. Dark morph also common, all dark except white wing flashes, thus size, structure and flight pattern separate from Pomarine. Juv. appears small-headed; very dark chocolate-brown to mid cinnamon-brown with pale buff-brown head; upperparts have narrow rufous fringes and primaries narrow rufous tips; little or no barring on vent of darker juv., less prominently barred in paler juv. which tend to appear 'warmer' rusty-brown, rather than 'cooler' grey-brown of pale Pomarine juv. Central rectrices barely protrude (1–3cm), but are pointed, not blunt-tipped. At rest ad., though long-winged, primaries do not reach tail tip; in juv., however, primaries extend beyond tail. Flight fast, agile, rakish (even falcon-like) and lighter than Pomarine. Wings long, narrow (including 'arm', so tail looks longer than Pomarine and wing base narrower), 'hand' appears longer, more pointed than Pomarine. Small area of white in primary bases on upper- (3–5 primary shafts) and underwing form distinct 'flashes' (like Pomarine). ad. has very dark underwings; juv. has pale-barred axillaries/underwing-coverts and more uniform rump/uppertail-coverts, either dark or mostly pale with little barring; central rectrices long (extending c.5–8cm) and narrowly pointed. **BP** Bill grey with black tip, somewhat less heavy than Pomarine; gonydeal angle close to tip (mid-bill in Long-tailed); juv. bill blue-grey with black tip even at long range (pale base less extensive, thus lacks Glaucous-like appearance); eyes black; legs short, blackish-grey. **Vo** Usually silent, but on breeding grounds gives harsh gull- or tern-like, nasal *kewet kewet*, short barking *gek*, *gwaa* or *gii* notes, and long-call a crowing *feee-leerrrr*. **AN** Parasitic Jaeger.

LONG-TAILED SKUA CTKJR
Stercorarius longicaudus

L 35–53cm; WS 105–117cm; WT 230–350g. **SD** Breeds across northern N America and Eurasia; winters in Southern Ocean. *S. l. pallescens* occurs across northern E Siberia from the Lena to Bering Strait, also south along Bering Sea coast to Kamchatka and NE Sea of Okhotsk. Uncommon off Bering and Okhotsk Sea and NW Pacific coasts, scarce on passage (mostly Mar–Jun, but records in all months) off E Japan coast and offshore in W Pacific. Scarce migrant off E China, in Taiwan Strait (Apr–May) and Korea. **HH** Loose colonies on open dry tundra, often in hilly areas; at other seasons along coasts or well offshore; only rarely inland (e.g. after typhoons). **ID** Small, slim skua (about size of Black-headed Gull, see p.210), smaller and slighter than Arctic and flight more agile, buoyant and tern-like. May hover over breeding grounds, searching for rodents, and settles on sea more frequently than other skuas. Ad. breeding (both sexes) has brownish-grey mantle, upperwing-coverts and rump contrasting with black flight-feathers, black primary-coverts and tail. Finely pointed central rectrices are extremely long (extending c.12–24cm), at least as long as tail, typically longer. Wing 'flashes' narrower than Arctic or Pomarine, comprising two white primary shafts, appearing as stripes on wing, obvious even at distance but not present on underwing. Distinct cap is black. Cheeks and neck may have yellow wash like Pomarine or Arctic, but white lower neck/chest grade gradually into dusky-grey belly/vent – without sharp demarcation, although vent can appear quite dark. Non-breeding ad. has indistinct cap, dusky face- and neck-sides, and barred upper- and undertail-coverts; tail projections lost in winter. Juv., like Arctic, ranges from mid-brown to paler grey-brown, but typically pale-headed with pale upper belly. Distinctly barred vent, like Pomarine, but very much slighter with slimmer body and longer narrower wings even than Arctic. Tail projection of juv. short and bluntly rounded, not pointed. At rest ad., though long-winged, primaries do not reach tail tip; in juv., however, primaries extend beyond tail. Flight light and tern-like (lighter than Arctic), wings long, narrow and long-'handed', ad. shows strong contrast between brownish-grey upperwing-coverts and blackish flight-feathers forming distinct black trailing edge; white restricted to shafts of outer 2–3 primaries of upperwing only (distinct among skuas), whilst juv. has wing flashes above and below. Ad. has dark grey-brown underwing, juv. finely barred axillaries/underwing-coverts, and narrowly barred rump/uppertail-coverts. **BP** Bill short (about equal to distance from bill base to eye), blackish, shorter than Arctic, gonydeal angle not noticeable, nail covers about half of bill; juv. short, thick, basal half blue-grey, tip black; eyes black; legs short, black above tarsus, bluish-grey below with black, 'dipped in ink', webbed toes. **Vo** Usually silent, but on breeding grounds calls higher pitched than other skuas. Commonest calls a short, trilled *krrip*, nasal tern-like *kee-ur* and sharp *kl-dew*. Long-call comprises a rattle followed by a prolonged, plaintive *feeeeoo*. **AN** Long-tailed Jaeger.

ARCTIC SKUA

ad win pale

juv dark

juv pale

1st-sum pale

ad sum pale

juv intermediate

ad sum dark

juv dark

ad sum pale

ad sum dark

juv intermediate

LONG-TAILED SKUA

juv intermediate

juv pale

pallescens

2nd-sum

ad win

ad sum

juv dark

juv dark

juv pale

ad sum

juv intermediate

LITTLE AUK   JR
Alle alle

L 17–19cm; WS 40–48cm; WT 163g. **SD** Abundant Arctic Ocean species from Greenland east to NW Russian Arctic. Accidental or scarce breeder on islands of NE Russian Arctic, status unclear (presumably *A. a. polaris*), but pairs have also summered (bred?) in Least Auklet colony on Diomede Is. Regular in Alaska, mostly St Lawrence I., and has strayed south to Japan, even to Okinawa. **HH** Breeds on talus slopes on isolated Arctic islands; pelagic in winter. **ID** Small, dumpy auk, with a large head, thick neck and tiny bill. Ad. breeding has all-black upperparts, with 3–4 short white streaks on scapulars, and prominent white secondary band; wings and hood also black; underparts white. Lacks black hood outside breeding season, but black crown extends to eye, on nape and in broad bar on lower-neck-/breast-sides; white extends up behind eye. Juv. resembles winter ad. but less clean-cut and duskier. On water, floats high, when white on breast-sides prominent in breeding birds; often drags wings on water between dives. In flight, short wings slightly rounded at tip whirr and blur, all black except white tips to secondaries forming trailing edge to innerwing; underwing dark grey. Most prominent feature of non-breeder in flight is dark neck-band; tail all black, very short. **BP** Bill black; eyes black; tarsi blue-grey. **Vo** At colonies highly vocal chattering, and rising and falling screaming trill, but silent at sea. **AN** Dovekie.

BRÜNNICH'S GUILLEMOT   KJR
Uria lomvia

L 39–43cm; WS 65–73cm; WT 0.81–1.08kg. **SD** N Pacific and N Atlantic oceans, more northerly than Common Guillemot. In region, three subspecies recorded: *U. l. eleonorae* on New Siberia Is; *U. l. heckeri* at Wrangel I. and E Chukotka, and *U. l. arra* on Koryakia and Kamchatka coasts, Commander Is, Kuril Is and isolated islands of N Sea of Okhotsk. In non-breeding season evacuates freezing Arctic Ocean, Sea of Okhotsk and Bering Sea, but moves less far from breeding areas than Common Guillemot, though nevertheless fairly common on Pacific coast of Hokkaido and N Honshu (mainly Nov–Apr), less common in Sea of Japan to NW Honshu; reported E Korean coast. **HH** Breeds on ledges of cliffs, stacks and rocky islands. Flies some distance out to sea to feed; young tumble from nest prior to fledging and are then tended by ♂ alone. Largely pelagic in winter, but occasionally found inshore, in channels, bays and harbours, south from Hokkaido. **ID** Large, heavy-set black and white auk, similar to Common, but bill shape and overall coloration help separate them. Blacker in all plumages than Common. Ad. breeding black on head and neck, with clean border between strongly contrasting black and white underparts reaching point on mid neck; upperparts and wings blackish. May show fine indented line curving back from eye as Common. Non-breeding ad. paler but still blacker above than Common, from which separated by much darker head, black extending to lores, face and ear-coverts below level of eye and neck-sides. Underparts white, lacking dusky flank streaking of Common.

At rest, ad. has distinct white wingbar (tips of secondaries); wings short, pointed, fall short of tail tip. Flight like Common, but appears thicker necked and shorter beaked, thus squatter and heavier; darker upperwing relieved only by whitish trailing edge to secondaries; clean white underwing-coverts contrast with blackish flight-feathers, and white axillaries (dusky in Common); rump and tail narrower and blacker with white sides; toes project beyond tail. **BP** Bill, shorter, deeper, culmen more curved than Common, with gonydeal angle nearer midpoint rather than close to base, and variable white, grey or silver gape stripe extending back from base of black upper mandible; eyes black; tarsi blackish-grey, may stand on toes or rest on tarsi. **Vo** Ad. at colonies sounds aggressively angry, giving throaty roaring *aoorrr*, *quaaaaa* or *kuakuakuaaaa*, slightly lower pitched, harder and more crow-like than Common Guillemot. Juv. following ♂ at sea gives low plaintive whistle. **AN** Thick-billed Murre.

COMMON GUILLEMOT   TKJR
Uria aalge

L 38–43cm; WS 64–71cm; WT 0.945–1.044kg. **SD** N Pacific and N Atlantic. *U. a. inornata* breeds off Chukotkan and Kamchatkan coast, Commander Is, Kuril Is and isolated islands of Sea of Okhotsk, including those off Sakhalin; once also bred commonly around Hokkaido but extirpated due to over-fishing, the last few surviving on Teuri I. off W Hokkaido. Departs freezing Sea of Okhotsk and Bering Sea as winter advances, when common on Pacific coast of Japan, occasionally inshore, in channels, bays and harbours south from Hokkaido; less common in Sea of Japan to NW Honshu, near Korea, and accidental Taiwan. **HH** Breeds on ledges of high cliffs, stacks and rocky islands. Flies some distance out to sea to feed; young tumble from nest before fledging and tended by ♂ alone. Outside breeding season pelagic. **ID** Large, heavy-set, black-and-white alcid confusable only with Brünnich's. Ad. breeding has blackish-brown head and neck, with clean contrasting border between black and white underparts on lower neck; upperparts and wings blackish-brown. May show fine indented line curving back from eye, like crease in feathering. Non-breeding ad. much lower, crown to lores and nape blackish-brown, with fine black line curving back from eye on head-sides; white extends to level of eye on rear head-sides. Underparts white with dusky brownish or blackish streaks on flanks. Juv. like winter ad., blunter ended and fluffier. At rest, ad. has distinct white wingbar (tips of secondaries); wings short, pointed, fall short of tail tip. In flight, broad-based, pointed wings whirr rapidly; dark upperwing relieved only by whitish trailing edge to secondaries; dusky-white underwing-coverts contrast with blackish flight-feathers, and dusky axillaries; rump and tail appear broad and dark; toes project beyond tail. **BP** Bill, long, straight, black, narrowly pointed with gonydeal angle close to base; eyes black; tarsi blackish-grey, may stand on toes or rest on tarsi. **Vo** Ad. gives rolling, nasal *orrrr* or *aarrrr* and throaty *gwaaaaa* and *ururuuun* or staccato *ha ha ha ha* becoming louder; prolonged deep *ha-aaahr* at colonies. Juvenile at sea utters piping *pleeyou* begging call to accompanying ♂. **AN** Common Murre.

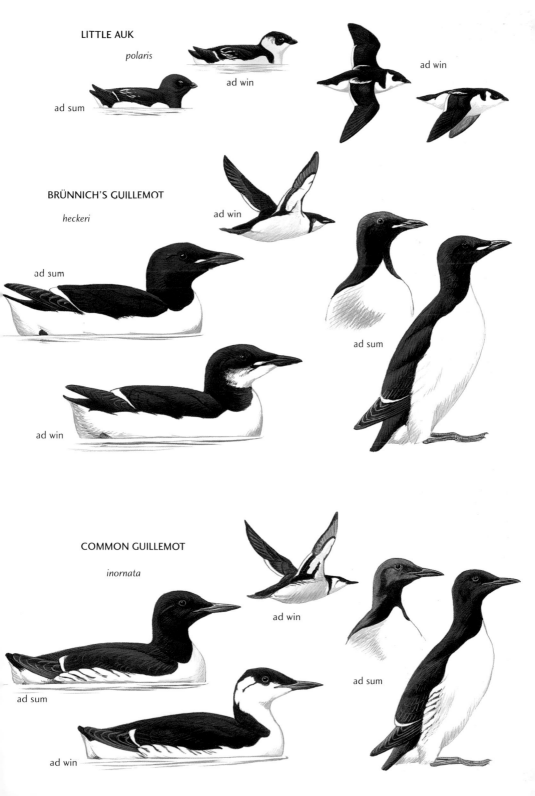

LITTLE AUK

polaris

ad win

ad sum

ad win

ad win

BRÜNNICH'S GUILLEMOT

heckeri

ad win

ad sum

ad win

ad sum

COMMON GUILLEMOT

inornata

ad win

ad sum

ad win

ad sum

BLACK GUILLEMOT *Cepphus grylle* R

L 30–32cm; WS 52–58cm; WT 450–550g. **SD** N American Arctic, Iceland and NW Europe, isolated populations in Russian Arctic, including Wrangel I., easternmost Chukotka and Bering Strait. E Asia possibly *C. g. mandtii*. **HH** Breeds on rocky coasts with scree and boulder talus slopes. May associate with Pigeon, or coincidentally select very similar nesting habitat. Winter: ice-free inshore waters, or polynas. **ID** Medium-sized, very dark auk. Ad. breeding is black with large round white patches on upperwing. Non-breeding retains wing pattern, but head, neck, mantle and rump white with variable black scaling on crown and mantle, streaking on neck, and white rump. Juv. has brown head- and neck-sides, barred black on neck and underparts, wing patch with dark spots. 1st-winter paler, wing patch more prominent, but still spotted dark. 1st-summer all black with white barring across mainly dark innerwing, rendering it perhaps indistinguishable from ad. *snowi* Pigeon Guillemot of C & S Kuril Is, except by underwing. At rest, floats high in water. Flight whirring; prominent wing patch obvious, but note mostly white underwing (all plumages); head appears small, and vent region heavy, short-tailed, but toes do not extend beyond tail. **BP** Bill black; eyes black; tarsi/feet bright red. **Vo** Very vocal, reveals bright red mouth lining while giving high, sibilant piping *seeeyou* or *swweeeeeer* and fine, pipit-like, *sit sit sit...* calls in quavering, rising and falling series; generally silent away from colonies.

PIGEON GUILLEMOT *Cepphus columba* KJR

L 30–37cm; WS 58cm; WT 450–550g. **SD** N Pacific and Bering Sea. Regular, scarce winter visitor N Japan, mostly E Hokkaido (Dec–Mar), rarely Honshu; reported Korea. *C. c. columba* NE Chukotka south to S Kamchatka; *C. c. kaiurka* Commander Is; *C. (c.) snowi* (probably distinct species, **Kuril Guillemot**) N & C Kuril Is, rare west to NW Sea of Okhotsk. **HH** Breeds along rocky coasts with scree and boulder talus slopes. Winters at sea or moves south to ice-free areas. **ID** Medium-sized, very dark auk, with distinct races. Ad. nominate and *C. c. kaiurka* like Black, but dark bar at base of greater coverts partially divides prominent white wing patch, visible at rest, more prominent in flight in breeding season. Dull to dark grey underwing distinguishes Pigeon from Black at all ages. Juv./1st-year/non-breeding ad. resemble same ages of Black, except grey underwing. Ad. *C. (c.) snowi* intermediate between Pigeon and Spectacled with all-black wings, variable white barring on coverts, in some broad enough to appear like a broken wing patch, but usually as series of narrow white bars (like dark 1st-summer Black); also shows small pale area around eye and resembles first summer Spectacled. *C. (c.) snowi* is most likely in Japan in winter. **BP** Bill black, slightly upturned, head held level or slightly uptilted; eyes black; tarsi/feet bright red. **Vo** As Black.

SPECTACLED GUILLEMOT KJR
Cepphus carbo

L 37–40cm; WS 65.5–69cm; WT 490g. **SD** Sea of Okhotsk (main stronghold Malminskiye and Shantar Is) south to Sakhalin and adjacent coasts, S Kamchatka, Hokkaido and N Honshu, but dramatic recent decline in Japan. Suspected breeding E Korean coast. Disperses south in winter; commoner Hokkaido, also Pacific shores of N Honshu; has strayed to continental Sea of Japan coasts, Tsushima and Korea. Monotypic. **HH** Rocky coasts, breeding amongst rock piles just above tideline. Winters in ice-free areas off rocky coasts, and headlands, occasionally bays and harbours.

ID Larger than Pigeon or Black. Sooty blackish-brown, except white patch around eye and curving back from it, and smaller white patches at bill base. White patches at tail-sides may reflect moult or individual variation. Non-breeding upperparts blackish-brown, underparts from chin, throat, neck-sides, breast to vent white. Face blackish, grading into white throat, reduced white ring around eye, head- and neck-sides dusky-grey, not black like forehead/crown. Forehead usually steep (see Long-billed Murrelet win). Wings all black; whirr in flight like other guillemots; tail all black, very short, toes often protrude beyond tip. **BP** Bill long, thin, black, reveals bright red mouth lining when calling; eyes black; tarsi/feet bright red (often visible). **Vo** At colonies gives high-pitched, rather tremulous whistles: *piii piii* or more rapid *pi pi pi* or *chi chi chi*.

LONG-BILLED MURRELET CKJR
Brachyramphus perdix

L 24–26cm; WS 43cm; WT 196–269g. **SD** Scarce regional endemic: Kamchatka, N Sea of Okhotsk, Sakhalin, also Kuril Is. Winter wanders to Hokkaido but as far as Kyushu and rarely N Nansei Shoto, Korea, Liaoning and Shandong. Monotypic. **HH** Almost exclusively at sea off forested regions (presumably tree-nester). **ID** Small, dumpy murrelet with rather hunched appearance, but elongated profile and slender bill. Ad. breeding is uniform dark brown, with distinctive pale brown throat and narrow white crescents above and below eye. Non-breeding resembles small, winter Spectacled Guillemot (size difficult to judge at sea), gently sloping profile, white crescents above and below eye, black of crown extends on nape and neck-sides, back blackish-grey, upperwings have prominent white scapular bar; pale grey underwing-coverts. Juv. resembles winter ad. but flanks duskier. At rest floats high, tail short, often cocked, wingtips extending beyond tail. In flight, wings narrow, pointed, beat rapidly; flight usually low, fast and direct; underwing-coverts pale. **BP** Bill rather long, appearing more so due to gently sloping profile, black; eyes black; tarsi black. **Vo** Thin, whistled *fii fii*. **TN** Formerly within Marbled Murrelet.

MARBLED MURRELET Extralimital
Brachyramphus marmoratus

Vulnerable. L 25cm; WS 43cm; WT 290g. **SD** E Bering Sea, Aleutians, NE Pacific. Occurs west to W Aleutians; very rare Gambell (Aug–Oct); may stray west to E Asia (e.g. Commander Is). **ID** Breeding ad. like Long-billed, but marginally smaller, has rustier and buffier markings on body, mantle and scapulars, and is less pale on throat. Non-breeding whitish scapular bar more prominent and has conspicuous white and dark collars, white extends high on head-sides almost to level of eye, and black extends to breast-sides (recalls winter Common Guillemot, see p.234); lacks pale ovals on nape-sides of Long-billed. Underwing always dark. Floats low in water.

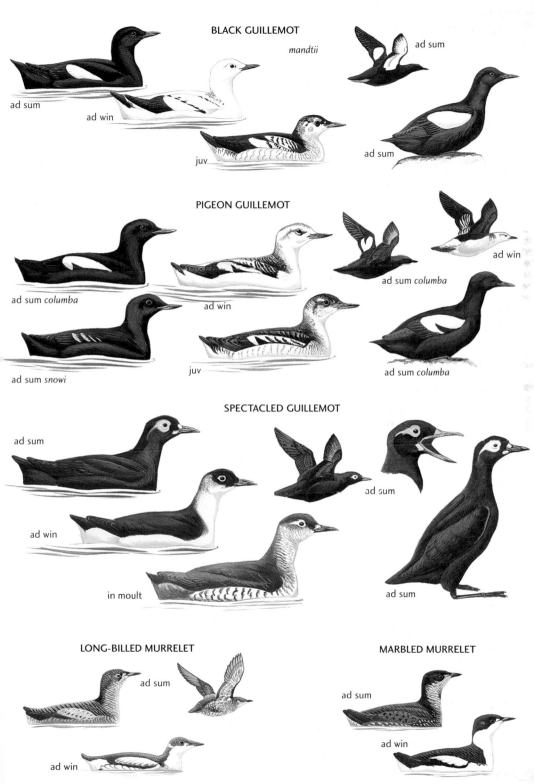

BLACK GUILLEMOT

mandtii

ad sum

ad sum

ad win

juv

ad sum

PIGEON GUILLEMOT

ad sum *columba*

ad win

ad sum *columba*

ad win

ad sum *snowi*

juv

ad sum *columba*

SPECTACLED GUILLEMOT

ad sum

ad sum

ad win

ad sum

in moult

ad sum

LONG-BILLED MURRELET

ad sum

ad win

MARBLED MURRELET

ad sum

ad win

KITTLITZ'S MURRELET JR
Brachyramphus brevirostris

L 22–23cm; WS 43cm; WT 224g. **SD** Breeds Pacific NW of Canada, Alaska (especially common Gulf of Alaska) and Aleutians, and in E Asia on Chukotski Peninsula, Koryakia and perhaps also NE Sea of Okhotsk. Believed to winter south on Kamchatka coast to Kuril Is; vagrant Japan. Monotypic. **HH** Alpine tundra, glacial moraines and scree slopes within 75km of coast. Winters offshore from breeding range south. **ID** Slightly smaller than Long-billed (p.236) but essentially similar in shape, except much shorter bill. Ad. breeding mid to pale brown, with speckled upperparts and flanks, and white belly and undertail-coverts. Suggestion of 'pale face', which is obvious in non-breeding plumage. Non-breeding has narrow, dark-grey crown, grey nape and upperparts, but all appear almost black in field. Face very white, eye therefore prominent, with white on lores, above eye, on rear head-sides and foreneck; grey patches extend onto lower-neck-/breast-sides forming near-complete collar, but underparts otherwise white; prominent white scapular bar. Juv. resembles winter ad. On water floats high, tail short, often cocked. In flight, dark wings narrow, pointed and contrast with somewhat golden-brown body; underwing dark in all plumages with more white in tail than Marbled. **BP** Bill very short, black; eyes black; tarsi black. **Vo** Low groaning *urrrhhn* and short quacking *urgh*.

ANCIENT MURRELET CTKJR
Synthliboramphus antiquus

L 24–27cm; WS 40–43cm; WT 177–249g. **SD** Pacific NW Canada, Alaska and Aleutians, and NE Asia south to Yellow Sea. *S. a. macrorhynchos* occurs off Kamchatka and Commander Is, and perhaps in NE Sea of Okhotsk; *S. a. antiquus* occurs Kuril Is, Sakhalin, off Hokkaido, N Honshu, Peter the Great Bay, Korea and Shandong (limits of subspecies ranges unclear). **HH** Nests in rock crevices and rock piles. In winter, Kamchatka and Sea of Okhotsk south through the Sea of Japan and Pacific coasts of Japan. Accidental Taiwan and E China mainland to Hong Kong. Common well out to sea, occasionally in bays and harbours. **ID** Slightly larger than Long-billed, similarly stocky but appears large- and rather flat-headed, with black, grey and white plumage and much shorter, stouter bill. Contrast between grey back and black crown an excellent field mark in all plumages. Ad. breeding plain mid grey above and on wings and rump, folded wing appears almost black, tail short and black, wingtips extend just beyond tail. Black of head reaches chin and throat, face-sides below eye and lower neck-sides, contrasting with grey mantle. Narrow white line of hoary feathers from behind eye to nape-sides, and also on black collar. Underparts white, including undertail-coverts, but flanks streaked grey. Non-breeding duskier on head, lacks hoary 'crest' and black on throat less extensive. Lacks white scapular bar of other murrelets (but see Japanese). Juv. resembles winter ad. On water floats low, tail short, often cocked, neck hunched, as if leaning forward. In flight, underwing white with almost vertical white patch extending onto neck-sides. On landing, simply stops flying and splashes into water breast-first from height of *c.*1m or more. Flocks often fly short distance then splash down together. **BP** Bill short, conical, pale horn, ivory or even yellow; eyes black; tarsi blue-grey. **Vo** At colonies various chips, trills, rasping sounds and a short *chirrup*, while at sea gives an abrupt high-pitched chipping *chi chi* like a young sparrow, or a whistled *teep*.

JAPANESE MURRELET CTKJR
Synthliboramphus wumizusume

Vulnerable. L 24–26cm; WT 183g. **SD** Isolated offshore islands around Japan, mainly from Izu Is southwest to Kyushu and off S Korea. Accidental to Taiwan, and Sakhalin and Kuril Is. Monotypic. **HH** Rocky islets and headlands, isolated stacks with cracks and crevices during early breeding season (Feb–May); usually close to shore in summer. Pelagic at other times, and may disperse away from breeding areas, inshore near headlands and bays in late winter/spring, year-round in Japanese waters. **ID** Small stocky auk, similar to Ancient, but has prominent white stripes on black crown-sides meeting on nape, slightly loose black crest and larger bill. General appearance is of horizontal black and white bands on head-sides, and broad black band from head to flanks, whereas Ancient has suggestion of vertical white and black bands on neck. Ad. breeding is plain mid-grey above and on wings and rump, folded wing appears almost black, tail short and black. Black of head extends only to chin and upper throat (further on throat in Ancient), but down sides in continuous band; face-sides below eye black. Underparts white but flanks blackish-grey. Non-breeding has less conspicuous crest. Like Ancient, lacks white scapular bar of other murrelets. Juv. resembles winter ad. On water floats low, tail short, often cocked, neck hunched. In flight, underwing white. **BP** Bill short, conical, pale blue-grey; eyes black; tarsi blue-grey to yellowish-grey. **Vo** Abrupt, bunting-like *chi chi chi chi* and, at colonies at night, *byubyubyu*. **AN** Crested Murrelet.

CASSIN'S AUKLET JR
Ptychoramphus aleuticus

L 23cm; WS 38cm; WT 150–200g. **SD** Aleutians and N American W coast to California. *P. a. aleuticus* breeds in Aleutians, SW & SE Alaska and British Columbia. Winters offshore, south to Baja California. Accidental Kamchatka, where rare offshore in winter; reported Japan. **ID** Stocky, large-headed grey auklet with short bill. Overall dark grey, darker on upperside, somewhat paler below. In flight, pale central stripe on otherwise dark underwing, paler belly and pale grey patch just behind base of wings. At long range, deceptively similar to larger, non-breeding Rhinoceros Auklet (p.242), but latter has larger dull yellow bill and uniform underwing. **BP** Very short bill is triangular, angled upwards, lower mandible has pale horn base, otherwise dark grey; eyes dull off-white or yellowish-white in ad., making eye very conspicuous, dark grey in juv. (Aug–Feb), all ages have white crescent above and in front of eye; tarsi blue-grey. **Vo** Silent away from colonies.

KITTLITZ'S MURRELET

ad sum

ad win

ad win

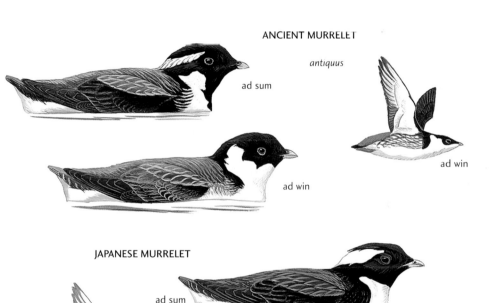

ANCIENT MURRELET

antiquus

ad sum

ad win

ad win

JAPANESE MURRELET

ad sum

ad win

ad win

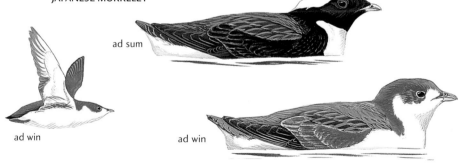

CASSIN'S AUKLET

aleuticus

ad

ad

PARAKEET AUKLET JR
Aethia psittacula

L 23–25cm; WS 44–48cm; WT 297g. **SD** Locally common Beringian endemic, breeding mainly on islands and coasts of Bering Sea and Sea of Okhotsk, the Aleutians and parts of Gulf of Alaska. In E Asia, offshore islets from E Chukotka to Kamchatka and Commander Is, Kuril Is and Sakhalin, also N Sea of Okhotsk (scarce). Post-breeding disperses to ice-free areas of Bering Sea, N Pacific and Sea of Okhotsk, straying occasionally to Hokkaido and N Honshu (Dec–Mar), with more records from Sea of Japan than Pacific. Monotypic. **HH** With other auklets at small mixed colonies, nesting in crevices in cliffs and rock piles. In winter pelagic, usually alone but may be among Crested or Least Auklet flocks. **ID** Dumpy, similar in size to Crested. Ad. breeding mostly dark grey above, but breast to vent bright white, upper breast and flanks smudged dark grey and white; single white plume extends behind eye. Non-breeding all dark above but white of underparts extends on throat and even chin is pale. Stands erect; on water floats quite high, tail short, often cocked, neck hunched. In flight, all-dark wings, dark head, white belly and prominent bill distinctive; head appears raised, wings whirr. Juv. resembles ad. in flight, but has narrow pale line on central underwing resembling Least, dusky flanks and neck, and lacks white scapular bar. **BP** Bill stubby, upturned and bright orange in ad., duller in non-breeding and all grey in juv.; eyes white; legs pale horn, webs dark blackish-grey. **Vo** At colonies gives rhythmic hoarse calls and descending squeals; silent at sea. **TN** Formerly *Cyclorrhynchus psittacula*.

LEAST AUKLET KJR
Aethia pusilla

L 12–14cm; WS 33–36cm; WT 85g. **SD** Abundant Beringian endemic, breeding mainly in Bering Sea. In E Asia, islands off E Chukotka, Diomede Is, Commander Is and perhaps N Kuril Is, as well as in N Sea of Okhotsk (largest colony in world, 6–10 million pairs, on Yamskiye Is). Monotypic. **HH** Nests under boulders on stable talus slopes. In non-breeding season typically pelagic, moving in large numbers through NW Pacific, particularly off E Hokkaido (mostly Oct–Apr) but also N Honshu; scarce in Sea of Japan and has strayed as far as Kyushu and Korea. Typically seen from ferries in winter, but also from shore and occasionally in harbours. **ID** Tiny, dumpy auklet: dark and scaly in summer, pale and pied in winter. Breeding ad. mostly slate-grey, variably mottled white on throat, chest and vent (some have very dark chests and flank barring, others only lightly barred below); white throat patch well defined and conspicuous, even in flight. Forehead has hoary white streaks and similar streaks extend behind eye. Non-breeding has dark upperparts and whitish scapular bar on closed wing; white underparts, including flanks, to chin. Face dark grey and white stripe behind eye less prominent. Stands erect; on water floats quite high, tail short, often cocked, neck hunched. In flight, upperwing uniformly dark and underwing mostly so, but has pale bar on centre of wing; flight whirring, very low over water, even flying through wave tops; flocks appear like smoke in distance. **BP** Bill tiny, black in breeding season with red tip, black with dull orange tip in non-breeding,

juv. all dark; eyes white; legs pale grey, webs dark blackish-grey. **Vo** At colonies highly vocal, but differs from other alcids in giving continuous chatter of high-pitched churrs and trills, including a soft buzzy *bii bii...* or *chi....*

WHISKERED AUKLET KJR
Aethia pygmaea

L 17–22cm; WS 36cm; WT 99–136g. **SD** Scarce and very local Beringian endemic, breeding only in Aleutians, Commander Is, Kuril Is, and N Sea of Okhotsk. Pelagic in winter when very rarely reported. Rare Japan (Feb–Apr), vagrant Korea; occasionally from shore. In N American seas prefers turbulent waters, particularly where tidal rips occur, and probably favours same in Asian range. Monotypic. **HH** Colonies in rock cracks and burrows, generally on uninhabited islets. Often in flocks at other times, typically pelagic, dispersing from breeding areas as seas freeze. **ID** Similar to Crested, but size closer to Least. Ad. breeding stunning. Seven fine, wispy facial plumes: single black forehead crest, with white plumes extending both back and up, and back and down, from bill base (forming white V on face), and back from behind eye. Undertail-coverts whitish. Non-breeding loses upper white facial plume and other plumes less prominent. Juv. has paler grey underparts and face with dark eyestripe and facial bar. Stands erect; on water usually floats quite high, tail short, often cocked, neck hunched or stretched up. In flight, wings dark above and below; undertail paler grey than Crested; whirrs low over water. **BP** Bill tiny, bright orange in breeding season, dull orange in non-breeding, and grey in juv; eyes white; tarsi blue-grey. Breeding birds and nest burrows have distinct citrus scent. **Vo** Noisy at and around colonies, giving buzzing, chattering notes in series: *beedeer beedeer beedeer beedeer bideer bideer bidi bidi bidi bidee*, and high kitten-like *meew*, but silent at sea.

CRESTED AUKLET CJR
Aethia cristatella

L 23–27cm; WS 40–50cm; WT 285g. **SD** Beringian endemic, breeding only in Bering Sea, Aleutians and Sea of Okhotsk. In E Asia, coasts of E Chukotka, on Commander Is, Kuril Is, N Sea of Okhotsk and off Sakhalin. Post-breeding disperses through W Pacific, past Hokkaido and N Honshu; rare in Sea of Japan; accidental China. Monotypic. **HH** As Whiskered Auklet. **ID** Size as Ancient and Japanese Murrelets but dumpier, with sooty grey-black plumage recalling smaller Whiskered Auklet. Breeding ad. has short shaggy black crest curling forwards, prominent beak and gape, and short white plume behind eye. Non-breeding has thinner (or absent) crest, white cheek plume less prominent, bill inconspicuous. Stands erect; on water floats low, tail short, often cocked, neck hunched. In flight, pot-bellied; wings dark above and below, rather long; whirrs low over water; undertail-coverts and vent dark grey (pale in Whiskered). **BP** Bill stubby, in breeding season bright orange with swollen upcurved gape, non-breeding has dull orange-black bill, juv. grey; eyes white; tarsi blue-grey. Birds and nest burrows have discernible citrus scent. **Vo** At colony, odd mix of hooting, honking and barking calls, including a nasal *kyow*; but silent at sea.

PARAKEET AUKLET

ad sum

ad win

LEAST AUKLET

juv

ad sum

ad win

ad sum

ad sum

WHISKERED AUKLET

ad sum

juv

ad win

ad sum

CRESTED AUKLET

ad sum

ad sum

ad sum

ad sum

ad win

juv

ad sum

RHINOCEROS AUKLET CKJR
Cerorhinca monocerata

L 32–38cm; WS 56–63cm; WT 520g. **SD** N Pacific. In E Asia, breeds on Kuril and Shantar Is, Sakhalin, N Japan (world's largest colony, *c*.350,000, on Teuri I., off W Hokkaido); also Sea of Japan coast perhaps as far as N Korea. Disperses at sea, though occurs Japanese waters year-round. Rare Kyushu, scarce winter visitor Korea; accidental Liaoning. Monotypic. **HH** Burrow nester on islands with deep clifftop soil, feeding relatively near colonies. Returns at dusk and scurries rapidly to burrow. Pelagic at other seasons. **ID** Large, size of Horned Puffin but predominantly blackish-grey upperparts, with only belly and vent whitish and not clear-cut from grey of flanks or breast. Head angular, with flat crown merging into long, heavy bill, white plumes behind eye and prominent white whiskers in breeding plumage. Long profile like Guillemot, but bill more like winter puffin. Non-breeding lacks or has reduced facial plumes, is duller overall and resembles juv. Tufted, which has deeper orange bill, more rounded head, all-dark belly and orange toes. Less upright than N Pacific auklets, more horizontal and neck more stretched. On water floats quite high, tail short, often cocked, but may swim with only head and shoulders above surface. In flight, all-dark wings, upperparts and head (appears raised), contrast with dull white belly/vent; flight heavy on whirring wings with bend at 'wrist'. **BP** Bill dull orange with grey culmen, sporting bizarre off-white horn only when breeding, duller and yellower in non-breeding season with grey culmen, and dark grey in juv.; eyes orange; legs pale grey, webs dark blackish-grey. **Vo** On return to colony gives deep throaty gurgling *gugu gugu* and moaning *woo woo* in rising and falling series.

TUFTED PUFFIN JR
Fratercula cirrhata

L 33–41cm; WS 64–66cm; WT 780g. **SD** N Pacific, Bering Sea and Sea of Okhotsk. In E Asia, common from Chukotka, Kamchatka, Kurils and N Sea of Okhotsk to Sakhalin, and in tiny numbers off SE Hokkaido. Scarce elsewhere, but has occurred off N Honshu, mainly off Pacific coast. Monotypic. **HH** Colonial, in deep burrows in thick soil. Feeds diurnally close to breeding sites; thereafter disperses at sea but nomadic rather than migratory. **ID** Largest puffin. All-black with large, clownish head. Ad. blackish-grey, offset by large, triangular white face patch, and long yellow plumes behind ear-coverts. Non-breeding has only faint outline of face patch and plumes absent or indistinct. Juv. browner with less conspicuous bill. Stands very upright at colonies. On water floats high, neck short, head very rounded. Flight heavy and direct, wings whirring, tail short, toes extend beyond tip. Longer wings with bend at 'wrist' and slightly rounded tips visible in flight (recalls Rhinoceros Auklet, but differs from most auklets and murrelets). **BP** Bill of breeding ad. very deep and short, largely red-orange with yellow-orange base, grey at base of lower mandible; non-breeder has smaller, grey-based orange bill (sheds outer sheath); juv. bill smaller and duller; eyes yellowish-white (brown in juv.), narrow eye-ring red; tarsi bright orange in ad., pale grey in juv. **Vo** Rather quiet, even at colonies, a low, groaning *kurrr*; silent at sea.

HORNED PUFFIN JR
Fratercula corniculata

L 32–41cm; WS 56–58cm; WT 620g. **SD** N Pacific, Bering Sea and Sea of Okhotsk. In E Asia, breeds on Wrangel I. and Chukotka, around Kamchatka, Kuril and Shantar Is, and islands and coasts of N Sea of Okhotsk. Remains near colonies in summer; post-breeding disperses short distances at sea, though has reached Japan (Dec–Mar), mainly off Hokkaido, but also to latitude of C Honshu. Monotypic. **HH** Breeds in crevices in cliffs and forages inshore. More pelagic post-breeding. **ID** Large, dumpy alcid with large, clownish head. Ad. upperparts blackish-grey, underparts and large face patch white. Crown, nape, throat and neck-sides blackish, forming complete collar. White face and enormous bill distinctive; eye prominent, appearing triangular due to fleshy horn above eye; fine black line extends behind eye. Non-breeding has dusky to dark grey face, bill less prominent. Juv. similar to non-breeding ad., but has less conspicuous bill. Stands very upright beside nests, forming groups before taking flight. On water floats quite high, neck appears short. In flight, all-dark wings and blackish upperparts contrast with white belly and face; flight heavy and direct, wings whirring. **BP** Bill of breeding ad. very deep and short, with high culmen, yellow at base and gape but red at tip, non-breeding has smaller, grey-based reddish-orange bill, and juv. has less deep and duller bill; eyes black with narrow red/orange crescents above and below, and small conical, blue-grey 'horn' above eye (lacking in juv., reduced in winter ad.); tarsi bright orange (usually visible in flight). **Vo** At colonies, low rhythmic groaning *aaaa* or *orr-orr*, but silent at sea.

PALLAS'S SANDGROUSE CKJ
Syrrhaptes paradoxus

L 27–41cm; WS 63–78cm; WT ♂ 250–300g, ♀ 200–260g. **SD** C Asian deserts and semi-deserts between 40° N and 50° N, from Caspian Sea to E Gobi. Breeds across dry regions to Heilongjiang, wintering south to Bohai Gulf. Vagrant Korea (old records only) and Japan. Monotypic. **HH** Open steppe with short vegetation. Outside breeding range, dry grasslands and agricultural land. **ID** Pigeon-sized ground-dweller. Head small, yellowish or orange-buff, neck and breast grey, upperparts barred buff, grey and black, shoulders and breast buff, belly patch black. Short legs, long, narrow grey primaries and long, pin-tail plumes all conspicuous. ♂ has grey extending to nape and crown, heavier black barring on mantle and rump, extending to lower neck and breast as narrow gorget, with limited spotting on wing-coverts. ♀ has dark spotting on crown/nape, narrow black necklace on throat, buffier barring on mantle/rump and more spotting on wing-coverts. Walks quickly, pecking like dove. In flight, black belly patch, pale underwing with dark trailing edge to inner primaries and secondaries forming dark wingbar, long pointed wings (with attenuated needle-like outer primaries) and tail (with elongated central feathers) all prominent. Flight fast, high, narrow outer primaries thrum, pin-tail noticeable. **BP** Bill very short, grey; eyes orange/brown; tarsi feathered, whitish-buff. **Vo** A characteristic *quat*, sharp *kiki kiki*, melodic *ten-ten*, murmuring *tryou-ryou* or *cu-ruu cu-ruu cu-ou-ruu* and, in flight, a rolling *por-r-r*.

RHINOCEROS AUKLET

ad sum

juv

ad win

TUFTED PUFFIN

ad sum

ad sum

juv

ad win

ad sum

HORNED PUFFIN

juv

ad win

PALLAS'S SANDGROUSE
(not to scale)

ad ♂

ad ♀

ad ♂

ROCK DOVE CTKJR
Columba livia

L 29–35cm; WS 60–71cm; WT 180–355g. **SD** Fairly common as a wild bird in N Africa, Palearctic, Middle East, India and N China, and may occur in region in E Inner Mongolia. Elsewhere, 'feral' form, derived from birds domesticated for food, now almost cosmopolitan, and widespread in NE Russia, and throughout Japan from Hokkaido south to Taiwan, also Korea. Usually overlooked by birdwatchers as being 'not wild', thus range poorly documented. **HH** Natural habitat comprises crags and cliffs on rocky coasts or in mountains, as feral species has adapted to urban and rural human environment, occurring around city buildings, villages and farms. Survives well even in regions with harsh winters, either in urban areas or farms with livestock or grain stores. **ID** Medium-sized pigeon. Head small; plumage extremely variable. Wild-type blue-grey with metallic green and purple sheen on neck, blackish-grey primaries and tail tip, and two black bars on wing-coverts. Tail grey, with white outer feathers, narrow black subterminal band and grey tip (see Hill Pigeon). Walks readily on ground with nodding head. In flight, appears long-winged, with more pointed wingtips than most doves; wild birds and wild-types reveal pale underwing with dark border (flight-feather tips dark), and pale grey to white lower back and rump contrasting with mid grey tail; whirls in flocks around feeding/roosting sites, usually very site-specific. Flight fast, beats rapid. Wings often held in steep dihedral as bird banks and turns. Feral flocks usually contain many colour forms, from brown and white to almost black, but four common: all dark, blackish; pied with bold black and white markings; chequered, mostly like wild form, but with fine white bars or spots; and brown type which is largely brown except flight-feathers and tail. **BP** Bill short, narrow, black, with raised grey cere at base of upper mandible; eyes orange-brown; tarsi orange-red. **Vo** Commonly gives burbling or moaning cooing *bru-u-oo-u*, or *oo-oo-oo* chorus of several birds together. In display, gives rattling clatter of wingbeats on take-off. **AN** Feral Pigeon.

HILL PIGEON CKR
Columba rupestris

L 33–35cm. **SD** High-altitude C Asia, Transbaikalia to S Russian Far East, and Himalayas to NE China, Korea (very local; coasts, offshore islands, towns and patchily in interior), and coastal Russian Far East. Could stray to Japan. Subspecies in region *C. r. rupestris*. May be overlooked amongst Rock Doves as such flocks often ignored by birdwatchers. **HH** Hills and low mountains with crags and caves, open country with rocky outcrops, and human-settled landscapes.

ID Medium-sized pigeon very similar to wild-type Rock Dove. Head small, plumage mostly pale grey, with two prominent, but narrower, black bars on wing-coverts, and underparts paler on lower breast/belly than Rock. In flight, resembles Rock but has bold white lower back/upper rump, contrasting with dark grey lower rump and tail, with broad white subterminal band and broad black terminal band. **BP** Bill short, narrow, black, with raised grey cere at base of upper mandible; eyes orange-brown; tarsi orange-red. **Vo** Cooing calls resemble Rock, but gurgling sounds higher pitched and more halting, *gut gut gut gut*.

STOCK DOVE CKJ
Columba oenas

L 32–34cm; WS 60–69cm; WT ♂ 303–365g, ♀ 286–290g. **SD** Essentially W Palearctic, from W Europe to Tibet, but has strayed to Beidaihe (China), Korea and S Japan. Subspecies probably *C. o. oenas* which is a long-distance migrant. **HH** Rural areas, woodland edge and farmland with old trees, mostly in lowlands. **ID** Medium-sized pigeon similar to wild Rock Dove and Hill Pigeon, but neater, more compact. Mostly pale blue-grey, with purplish, green and pink sheen on neck, vinous breast and otherwise grey underparts. Head, upperparts and wing-coverts largely plain grey, with two short black bars on closed wing, black flight-feathers, grey rump and broad black terminal tail-band. In flight like Rock, but flight-feather tips black, primary bases and wing-coverts paler, see also back/rump/tail pattern. **BP** Bill short, narrow yellowish tip, pink base; eyes dark reddish-brown; tarsi dark pink. **Vo** When breeding gives rhythmic series of muffled *uruu*, *wuruuu* or hoarse *hoo-hooo* calls, but migrants generally silent.

ASHY WOODPIGEON T
Columba pulchricollis

L 31–36cm; WT 330g. **SD** Himalayas to Yunnan and Thailand, with isolated population resident in E Asia, on Taiwan. Monotypic. **HH** On Taiwan confined to broadleaf hill forests (1,000–2,500m, some descending in winter). Typically in canopy, often in large flocks but sometimes descends to ground to take grit. **ID** Medium-sized woodpigeon; generally rather dark and long-tailed. Head pale grey, neck has pale creamy buff collar from chin to nape (where hatched with black) below, which darkens to glossy green on neck-sides, slate-grey with purple gloss on chest, grey on belly and buff on vent. The mantle and wing-coverts are dark slate, almost black, the flight-feathers, rump and tail are all greyish-black. Flight slow, heavy like other woodpigeons; appears very dark, relieved only by pale vent, throat and nape. **BP** Bill narrow, grey at tip, red at base; eyes pale yellow; tarsi deep red. **Vo** A deep *ooop* or *whoo*; song comprises slow series of short, deep, powerful *whoo whoo whoo* or *whoop* notes.

ROCK DOVE

ad

feral variants

HILL PIGEON

rupestris

ad

ad

ASHY WOODPIGEON

ad

STOCK DOVE

oenas

ad

ad

BLACK WOODPIGEON

CTKJR

Columba janthina

Endangered (*C. j. nitens*). L 37–43.5cm. **SD** E Asian endemic. *C. j. janthina* restricted to islands off C & S Honshu, Kyushu (Tsushima) and Ryukyu Is; *C. j. nitens* rare (30–40? birds), found only on Ogasawara and Iwo Is; *C. j. stejnegeri* occurs on Yaeyama Is, in southernmost Nansei Shoto. An unknown race occurs on Ulleung I., and off S Korea; also E Shandong, China (rare, perhaps extinct). Accidental in Taiwan and Russian Far East. Generally considered to be resident, yet mysteriously appears regularly in small numbers during passage periods on islands in Sea of Japan, and occasionally even mainland Japan. **HH** Favours mature broadleaf evergreen subtropical and temperate forest, where mostly feeds on berries in canopy, or seen in display flight over canopy, but also shows fondness for pines in Nansei Shoto. Silent whilst feeding, but moves heavily and audibly in canopy as it forages. **ID** Largest and darkest pigeon in region. Brownish-black with metallic green gloss on neck-sides and purple gloss on shoulders. *C. j. nitens* has reddish-violet or purplish-brown on head and neck, whilst *C. j. stejnegeri* appears smaller than nominate with shorter tail. In flight, uniformly dark, rather crow-like, with rather long neck, broad wings and somewhat long fanned tail; flight slow, heavy with deep beats. **BP** Bill somewhat longer than other pigeons, narrow, black, horn-coloured at tip; eyes dark reddish-brown; tarsi deep red in ad., paler in juv. **Vo** Prolonged continuous deep moaning or growling *u wuu…* or *u uu…* somewhat like lowing cow; also dry *gnerrr* like bleating goat. Calling birds give single, loud wing clap on take-off. **AN** Japanese Woodpigeon.

ORIENTAL TURTLE DOVE

CTKJR

Streptopelia orientalis

L 33–35cm; WS 53–62cm; WT 165–274g. **SD** Ranges from S Urals across Russia and from Himalayas and India to Japan. *S. o. orientalis* occurs throughout NE China, Korea and most of Japan, and as summer visitor to Russian Far East, Sakhalin, S Kuril Is and Hokkaido; *S. o. stimpsoni* resident in Ryukyu Is from Amami-Oshima south; *S. o. orii* resident in Taiwan. **HH** From open boreal forest to subtropical forest, in temperate zone particularly around woodland edge, farmland, parkland, and large suburban and even urban gardens; commonly forages on ground, moving slowly and deliberately. **ID** Medium-sized, rather dumpy, with prominent neck bars and rather rufous upperparts. Forehead blue-grey, head and face pale pinkish-brown, grading into blackish-brown upperparts. Mantle, scapulars, and upperwing-coverts dark with extensive blackish-brown centres (see European Turtle Dove), with rounded, rather broad rufous fringes, giving strongly scaled appearance; distinguished by black and white striped patch on rear neck-sides. Lower back, rump and uppertail-coverts grey; slightly rounded blackish tail has pale grey tips forming narrow terminal band above. Underparts pale pinkish-brown, whiter on belly and vent. Flight rather sluggish, appears dark with little contrast, except on undertail which is largely black with broad pale grey terminal band. *S. o. stimpsoni* rather darker, especially on neck and head, but forehead appears whiter. *S. o. orii*, primarily in montane Taiwan, only doubtfully distinct from nominate. **BP** Bill slender, dark grey; eyes orange; tarsi dull pink. **Vo** Soft deep *kuu*; *uu* or *pwuu*. In breeding season gives soft, repetitive hollow crooning: *or-doo doo-doo hoo-hoo hoo hoaw* or *der-der-pou-pou der-der-pou-pou*. **AN** Rufous Turtle Dove.

EUROPEAN TURTLE DOVE

Extralimital

Streptopelia turtur

L 27–29cm; WT 99–170g. **SD** Potential vagrant from its extensive, largely W Palearctic range, migrates east as far as NW China but winters in sub-Saharan Africa. **ID** Distinguished by slightly smaller size, more delicate appearance, but particularly by dark centres to mantle, scapulars and upperwing-coverts being narrower and more pointed.

RED TURTLE DOVE

CTKJR

Streptopelia tranquebarica

L 20.5–23cm; WT 104g. **SD** Widespread resident, from Indian subcontinent to SE Asia, and E China south of Changjiang. *S. t. humilis* occurs through E coastal China from Bohai Gulf south; northern birds migratory, those in south are resident, including those on Taiwan. Vagrant to Russian Far East, Korea and very rare winter visitor S Japan (mainly Kyushu and Nansei Shoto, Oct–May). **HH** Scrub, at woodland edge in open country and farmland with trees, hedges and isolated stands of trees. Moves quietly in trees when foraging; sometimes on ground. **ID** Small, somewhat dumpy or compact, grey-headed dove, with rather dark plumage and black hind-collar. ♂ has pale grey head, narrow all-black hind-collar, deep brownish-pink upperparts (darkest on wing-coverts) and pale grey rump; flight-feathers and tail blackish-grey with white tips to black outer tail-feathers. Underparts pale brownish-pink. ♀ generally duller, less pinkish-brown. Juv. has wing-coverts fringed buff, affording overall scaled appearance recalling Oriental Turtle Dove, but single broad black neck bar distinctive. In flight wing action is light and agile; appears somewhat dark with contrasting dark primaries, grey head and white corners to tail. **BP** Bill mid grey to blackish-grey; pale grey orbital skin, irides brown to blackish-brown; tarsi dull grey to dark purplish-red. **Vo** Soft *go goroo*, a *croo-croo-croo*, and purring *gurr gurr-gr-gurr*. **AN** Red Collared Dove.

BLACK WOODPIGEON

ad *nitens*

ad *janthina*

ad *janthina*

ORIENTAL TURTLE DOVE

ad *orientalis*

ad *orientalis*

ad *stimpsoni*

1st-win *orientalis*

EUROPEAN TURTLE DOVE

ad

1st-win

RED TURTLE DOVE

humilis

ad ♂

ad ♀

SPOTTED DOVE CTK
Streptopelia chinensis

L 27.5–30cm; WS 53cm; WT 128–160g. **SD** Wide range, from Himalayas and India across SE Asia to Indonesia and E China. *S. c. chinensis* is resident throughout E China north to Bohai Gulf; *S. c. formosana* resident on Taiwan. Accidental Korea. **HH** Lowlands, in farmland, villages, urban and suburban areas and parks, also moist deciduous forest. **ID** Medium-sized, rather long-tailed dove, with prominent dark grey neck patch covered with fine silvery spots. Head grey, upperparts, (including back, rump and uppertail-coverts) generally plain brown, but somewhat scalloped; flight-feathers and tail mostly blackish-brown. Long graduated tail appears narrow, mainly blackish-brown with white tips to outer feathers forming contrasting white corners. Underparts dark pinkish-grey, whitish on vent. Flight typically rather slow and low; appears plain brown except white corners to tail. **BP** Bill dark grey; eyes orange; tarsi dull dark pink. **Vo** Song a purring crooning.

EURASIAN COLLARED DOVE CTKJ
Streptopelia decaocto

L 28–33cm; WS 47–56cm; WT ♂ 150–196g, ♀ 125–196g. **SD** Resident from W Europe and S Asia to China. *S. d. xanthocycla* occurs E coast China around Bohai Gulf to Korea (localised). Accidental offshore Japan. Small introduced population established in C Honshu, is apparently of *S. d. decaocto*. **HH** Favours agricultural land and villages, where often relatively tame and approachable. Moves quietly amongst vegetation and on ground in search of food. **ID** Small- to medium-sized dove, rather delicate, elegant and small-headed. Mostly pale, tan or pinkish-grey, with browner wings, wing-coverts and tail, black flight-feathers; white-bordered black hind-collar is distinctive. Flight action light and agile, with pale mantle and shoulder contrasting with grey wing-coverts and secondaries, and blackish primaries. Uppertail buffish-brown with black bases and white tips to outer tail-feathers, underwing-coverts pale and flight-feathers grey, with broad black base to underside of tail which has broad white tip. **BP** Bill black; eyes dark reddish-brown; tarsi dull pink. Hybrids with domestic strains are typically paler overall, with paler undertail-coverts, and shorter, two- (rather than three-) syllabled calls. **Vo** ♂'s song is a soft, repetitive three-note crooning: *coo-coooo-coo*; *popoopou* or *hwa-hoooo huu*.

BARRED CUCKOO DOVE CT
Macropygia unchall

L 37–41cm; WT 153–182g. **SD** Rather uncommon, ranges from Himalayas to Indonesia. *M. u. minor* occurs in SE China, in N Fujian and Guangdong, and has strayed north to Shanghai, also Kinmen (Taiwan). **HH** Small flocks in dense subtropical forest on montane slopes. **ID** Large, long-tailed brown dove, heavily barred, with pale vent. Upperparts, wings and tail blackish-brown, with rufous-brown fringes to mantle, back, rump, scapulars and

wing-coverts, and blackish-brown tail heavily barred rufous-brown. ♂ head and underparts pinkish-grey, with iridescent green on hindneck, and narrow black scalloping from neck to breast-sides. ♀ darker, lacks iridescence, reddish-brown on head and underparts, barred on crown, throat and breast. In flight, appears all dark and long-tailed, flies fast through canopy; on ground may raise long tail. **BP** Bill short, black; orbital skin blue-grey, eyes blue with pink ring; tarsi dull purplish-brown. **Vo** Prolonged series of slightly rising coos, *huwuuuu*.

PHILIPPINE CUCKOO DOVE T
Macropygia tenuirostris

L 38.5cm; WT 157–191g. **SD** Essentially endemic to Philippines. *M. a. phaea* occurs in region only on Lanyu I. off SE Taiwan, and as a vagrant to adjacent main island coast. **HH** Small flocks in tropical forest. **ID** Large, long-tailed dove. Upperparts uniform dark brown (lacks barring of Barred Cuckoo Dove). ♂ head and neck pinkish-grey, with no iridescent hindneck patch or barring; underparts dark pinkish-brown to cinnamon on vent. ♀ darker, with black barring on hindneck, mantle and breast, but plain cinnamon crown and throat. In flight, appears all dark and long-tailed, flies fast through dense vegetation. **BP** Bill short, pinkish-brown; orbital ring crimson, eyes dull yellow with crimson outer ring; tarsi orange-pink. **Vo** A surprised, quite deep *waow*, rising in middle and a soft *hoo-hoo*, like Emerald Dove but higher-pitched. **AN** Slender-billed Cuckoo Dove.

EMERALD DOVE CTJ
Chalcophaps indica

L 23–27cm; WT 108–160g. **SD** Ranges from India to Australia. *C. i. indica* occurs only in S China, on Taiwan, and in Yaeyama Is, S Japan. **HH** Alone or in pairs, foraging on ground in dark tropical or subtropical broadleaf forest. **ID** Small, rather squat, short-tailed dove, smaller than most others in region. Ad. rather dark, with white or silver-grey forehead and band above eye, grey crown and nape (♂ extensive, ♀ restricted to forecrown); mantle dark greyish-pink, scapulars and wing-coverts iridescent green, with ♂ having white area on carpals. Underparts dark greyish-pink, paler on belly. Juv. darker and browner, lacking grey crown, with rufous wingbar formed by tips to greater coverts, and extensively barred belly. On ground moves slowly and hesitantly; clatters on take-off, flies fast and low through vegetation. In flight, narrow white band on forewing, two white bars across otherwise dark grey lower back; rump and tail dark blackish-grey, with pale grey sides. **BP** Bill bright red or orange-red; eyes large, black; tarsi red. **Vo** A very deep, rhythmical mellow cooing, almost humming, repeated several times, *whooo-whooo… uu uu uu…* or *tuc-cooo*, the *tuc* often inaudible. **AN** Green-winged Pigeon.

SPOTTED DOVE

chinensis

ad

EURASIAN COLLARED DOVE

xanthocycla

ad

BARRED CUCKOO DOVE

minor

ad ♂

ad ♀

PHILIPPINE CUCKOO DOVE

phaea

ad

ad ♀

ad ♂

indica

EMERALD DOVE

PLATE 111 : PIGEONS AND DOVES IV

ORANGE-BREASTED GREEN PIGEON T
Treron bicinctus

L 28–29cm. **SD** Ranges from S India and Himalayas to Indonesia; *T. b. domvilii* (rare resident on Hainan) has strayed to Taiwan. **HH** Fruiting trees in lowland forest. **ID** Medium-sized, rather colourful green pigeon. ♂ has pale yellowish-green face, grey nape and neck, lilac above orange breast-band, yellow underparts and dark chestnut vent. ♀ duller with green chest. Closed wing has yellow and black bars on greater coverts. In flight, almost black primaries, yellow innerwing with black bars on greater coverts and secondaries; uppertail-coverts cinnamon, tail grey, darker band at base, paler band at tip. **BP** Bill grey; whitish eye-ring, irides red; tarsi dull red. **Vo** Varied, but lacks modulated crooning notes of other doves; instead gives various guttural cackles and croaks

WHITE-BELLIED GREEN PIGEON CTKJR
Treron sieboldii

L 31–33cm. **SD** Indochina to S & SE China, Taiwan and Japan. *T. s. sieboldii* a locally fairly common summer visitor to Hokkaido and N Honshu, south of this resident (supplemented by winter visitors) through main Japanese islands; migrants also pass through Jiangsu and Fujian. Accidental Sakhalin and S Kuril Is, Hebei (China) and Ulleung I., Korea (where perhaps regular migrant). *T. s. sororius* (?= *T. s. sieboldii*) resident Taiwan. **HH** Favours broadleaf evergreen forest with fruiting trees; in north, deciduous forest with fruiting trees and vines; occasionally urban parks and secondary forest in Taiwan below *c*.2,000m. Bizarrely, flocks visit coasts in some areas to drink saltwater; presumably to aid digestion of some fruits. **ID** Medium-sized, rather colourful green pigeon, with creamy flanks, vent and undertail-coverts. ♂ head, neck and breast yellowish-green, mantle grey, wing-coverts and secondaries dull dark green, but broad scapular patch maroon. ♀ mostly plain green, except pale flanks and vent. Both sexes have dark patches or scaling on rear flanks and vent, but far less extensive than in Ryukyu Green Pigeon. Wings rounded at tip, flight lazy but fast; usually singly or pairs, but flocks outside breeding season. **BP** Bill blue-grey at base, horn-coloured at tip; orbital skin blue-grey, irides have blue and orange rings; tarsi dull reddish-pink. **Vo** Song a long drawn-out, strongly inflected, mournful *oh aooh* and slightly more fluty *ooaa aaoo*. **TN** Formerly *Sphenurus sieboldii*. **AN** Japanese Green Pigeon.

TAIWAN GREEN PIGEON TJ
Treron formosae

L 25–26cm. **SD** Restricted range in N Philippines (*T. f. filipinus*) and Taiwan (*T. f. formosae*). In Taiwan, rare on main island and scarce on Lanyu I. Accidental to Yaeyama Is. **HH** Tropical lowland evergreen forest, typically feeding on fruits in canopy, but shy and difficult to observe. **ID** Rather small, plain, green pigeon (smaller than White-bellied). Head yellowish-green, with golden cap from forehead to nape (unlike similar, larger Ryukyu Green Pigeon). Upperparts dark green, scapulars deep brownish-maroon (♂) or greenish-brown (♀), wings blackish-brown with very narrow yellowish-green fringes to greater coverts and secondaries. Underparts green,

vent pale yellowish-green, each feather with dark blackish-green centre, undertail-coverts long with dark blackish-green centres and yellow fringes. Tail broad, long and rounded. Wings somewhat rounded at tip, flight heavy but fast above canopy. **BP** Bill blue-grey, rather stout-based; irides dark orange-brown with dark blue ring; tarsi dull dark pink. **Vo** Song a long, fluty mournful note, suddenly rising and wavering near end, *po-aa-poaaoo* (like Japanese bamboo flute), deeper than but not as inflected as White-bellied. **TN** Formerly considered conspecific with Ryukyu Green Pigeon, with which it was known as Whistling Green Pigeon *Sphenurus formosae*.

RYUKYU GREEN PIGEON TJ
Treron permagnus

L 33–35cm. **SD** Restricted to Nansei Shoto Is: *T. p. permagnus* in N, from Tanegashima to Okinawa; *T. p. medioximus* elsewhere. Accidental in winter Taiwan. **HH** Not uncommon in subtropical forest, even suburban parks and gardens. Can be very tame and approachable. **ID** Rather large, plain, heavy green pigeon (larger and bulkier than White-bellied), with distinctive undertail pattern. Deep green head, upperparts and underparts; scapulars deep brownish-maroon, wings blackish-brown with very narrow greenish fringes to greater coverts and secondaries. Lesser wing-coverts of ♂ have slight reddish-violet tone, those of ♀ deep greenish-brown. Tail broad, long and rounded, flanks green, but vent feathers blackish-green with yellow fringes. Undertail-coverts long with dark blackish-green centres, narrowly fringed yellow, and extend almost to tail tip (slightly shorter with much broader fringes in Taiwan Green). Wings somewhat rounded at tip, flight heavy but fast over canopy. *T. p. permagnus* larger than *T. p. medioximus*, but no other differences. **BP** Bill blue-grey, rather stout-based; irides have deep scarlet and dark blue rings; tarsi dull pink. **Vo** Song a long, fluty mournful note, suddenly rising and wavering near end, *po-aa-poaaoo*, deeper, but not as inflected or as mournful as White-bellied. Also a staccato *puppupupupu*.

BLACK-CHINNED FRUIT DOVE TJ
Ptilinopus leclancheri

L 26–28cm; WT ♂ 174g, ♀ 153–159g. **SD** Essentially endemic to Philippines, with *P. l. longialis* of northern Philippines also on Lanyu I. and *P. l. taiwanus* on mainland Taiwan, where a rare resident, mainly in south; accidental S Japan. **HH** Lowland tropical forest, where shy. **ID** ♂ small with pale grey hood extending to nape and upper breast. Chin black, band bordering hood on lower breast purple, belly greyish-green and vent cinnamon. Mantle, scapulars, wing-coverts, tertials and rump bright green, flight-feathers and tail blackish-brown with green tone. ♀ has green head and neck, and dull purple pectoral band. Juv. lacks pectoral band. Both ♀ and juv. have face to vent considerably darker grey than ♂ (and Emerald Dove, p.248). In flight, flight-feathers appear black, contrasting with green upperparts and wing-coverts. Birds overhead can look very dark. **BP** Bill bright yellow with red base to lower mandible (♂), dull yellow (♀); eyes red (♂) or dark brown (♀); tarsi dark reddish-pink. **Vo** A deep, rather drawn-out *brrrrooooo*, easily lost in forest cacophony.

ORANGE-BREASTED GREEN PIGEON

domvilii

ad ♂

ad ♀

WHITE-BELLIED GREEN PIGEON

sieboldii

ad ♂

ad ♀

ad ♀

TAIWAN GREEN PIGEON

formosae

ad ♂

RYUKYU GREEN PIGEON

medioximus

ad ♂

ad ♂

ad ♀

BLACK-CHINNED FRUIT DOVE

longialis

PLATE 112: INTRODUCED PARROTS

TANIMBAR CORELLA *Cacatua goffini* T

L 30–32cm; WT 300g. **SD** Endemic to Tanimbar Is (Indonesia), but widely traded; introduced and breeding locally in Taiwan (and Singapore). **HH** Urban parks and gardens. **ID** A white cockatoo-like bird with a short rounded crest. Plumage almost entirely white, but has pink lores, pink feather bases (mostly concealed) and yellow undersides to wings and tail. **BP** Bill basally grey, pale yellow or horn-coloured at tip; orbital ring pale blue, eyes brown (♂) or reddish-brown (♀); tarsi grey. **Vo** Loud screeches.

SULPHUR-CRESTED COCKATOO T
Cacatua galerita

L 45–55cm; WT 815–975g. **SD** Natural range extends from New Guinea to Tasmania, but widely traded; introduced and breeding locally in Taiwan (subspecies uncertain). **HH** Lowland woodland and parkland. **ID** A very large, white cockatoo, with a long, loosely pointed, erectile forward-curving crest; much of crest bright sulphur yellow, pale yellow patch on cheeks. Underside of wings and tail pale yellow. Wings broad and rounded, flight uneven with stiff beats followed by short glides. **BP** Bill blackish-grey; orbital skin white, irides dark brown (♂) or reddish-brown (♀); tarsi blue-grey. **Vo** A distinctive raucous screeching, commonly given in flight and when perched.

WHITE COCKATOO *Cacatua alba* T

Vulnerable. L 46cm; WT 550g. **SD** Natural range restricted to N Moluccas, Indonesia. Threatened by habitat loss and high levels of trapping for cagebird trade. Introduced and breeding locally in Taiwan. **HH** Wide range of forest types including mangrove and hill forest, to 900m in natural range. Prefers areas with large mature trees for nesting and communal roosts. **ID** A large, all-white cockatoo with yellow wash to underside of wings and tail, and a long white loose nuchal crest affording very large-headed appearance. **BP** Bill large, greyish-black; eyes black (♂) or reddish-brown (♀), with broad bare, blue-white eye-ring; tarsi bluish-grey. **Vo** Loud, high, nasal screeches, particularly at roosts, and lower, rapid chattering in flight.

ALEXANDRINE PARAKEET J
Psittacula eupatria

L 50–62cm; WT 198–258g. **SD** Widespread from India to Vietnam. Established feral population in C Honshu. Subspecies uncertain, probably nominate. **HH** Wooded parks and suburbs. **ID** Large, pale green parakeet (like large Rose-ringed) with green crown and upper face, and grey-green nape and lower face; distinctive red shoulders and very long, largely blue tail with pale yellowish-white tip. ♂ has black chin and fore-collar, with turquoise and pink hind-collars. ♀ lacks chin patch and collar. **BP** Bill very large, deep red; eye-ring pink, irides pale grey; tarsi grey. **Vo** Squawking or screeching *tskraau* or *skyurt*.

ROSE-RINGED PARAKEET TJ
Psittacula krameri

L ♂ 39–42cm, ♀ 27–36cm; WS 42–48cm; WT 95–143g. **SD** Widespread in India and Africa, and common in cagebird trade. Populations of *P. k. manillensis* have become established in various parts of the world, including parts of Japan (Tokyo) and Taiwan. Beware wide range of other escapee Asian parakeets, which may also be encountered in warmer parts of Japan and Taiwan, but which are beyond the scope of this guide. **HH** Well-wooded suburban and urban parks, and tree-lined streets, where noisy and conspicuous. **ID** Somewhat large-headed, long-tailed green parakeet (lacks red shoulders of larger Alexandrine). Almost entirely pale yellowish-green. ♂ has black chin patch, narrow black collar, below which is a narrow rose-pink hind-collar. ♀ lacks chin patch and collars. In flight, very long narrow tail highly conspicuous, wings green, beats rapid. **BP** Bill large, upper mandible deep red, lower black; eye-ring pale green/grey, irides black; tarsi grey. **Vo** Noisy, commonly chattering loudly in trees and squawking in noisy groups in flight, a piercing *kyik kyik kyiek*, or noisy squealing *keeew* and grating *krech-krech-krech*. **AN** Ring-necked Parakeet.

RED-BREASTED PARAKEET J
Psittacula alexandri

L 33–38cm; WT 133–168g. **SD** Himalayas, SE Asia and S China to Indonesia. *P. a. fasciata* occurs only as an established feral bird in Tokyo. **HH** Well-wooded suburban and urban parks, and tree-lined streets, where noisy and conspicuous. **ID** A large, green and pink parakeet. Largely green, but head grey with narrow black band from just above bill to eyes, and black chin and sides to lower face. ♂ upperparts green, yellowish-green on wing-coverts, with elongated blue central rectrices with greenish-yellow tips. Deep vinous-pink from throat to lower breast, but belly, flanks and undertail-coverts green. ♀ has duller pink underparts. **BP** Bill large, upper mandible deep red with yellow tip, lower blackish-pink or black; eyes pale yellow; tarsi grey. **Vo** Honking *yaink yaink* and wailing *knyaouw* calls very different from other parakeets.

RED LORY *Eos rubra* T

L 31cm. **SD** Restricted to certain islands in the Moluccas, but widely traded; introduced and breeding locally in Taiwan. **HH** Lowland urban parks and gardens. **ID** Plumage almost entirely red, except black primaries, secondary-coverts and tips of secondaries; tertials and undertail-coverts dark blue. In flight, red speculum on largely black secondaries. **BP** Bill orange-yellow to orange; orbital skin dark grey, eyes red; tarsi blackish-grey. **Vo** Calls include considerable mimicry. **TN** Formerly *Eos bornea*.

BUDGERIGAR TKJ
Melopsittacus undulatus

L 18cm; WS 30cm; WT 26–29g. **SD** Australian endemic, widely traded and very common cagebird; escapes occur widely in warmer regions of Japan and Taiwan, often mixing with sparrows and finches; even recorded Pyongyang, N Korea. Monotypic. **HH** Rural and suburban lowlands with woodland. **ID** Small-headed, long-tailed small 'parrot'. Wild types largely green with yellow head and upperparts, black spots on throat, narrow black barring on crown and face, and black scallops on mantle and wing-coverts. Rump and underparts green; long, pointed tail deep blue. Domesticated varieties may be almost any colour or combination of colours, from yellow to blue. Wings pointed, with pale yellow wing-stripe above and below contrasting with blackish flight-feathers, long central rectrices entirely dark green. Flight rapid, whirring. **BP** Bill very short, face flat-looking, horn-coloured with blue cere (♂) or grey with rusty-orange cere (♀); eyes white; tarsi grey. **Vo** Dry chattering, soft warbling, and chirruping with scratchy notes, also a distinctive high-pitched screech in alarm.

TANIMBAR
CORELLA

SULPHUR-CRESTED
COCKATOO

SULPHUR-CRESTED
COCKATOO

TANIMBAR
CORELLA

WHITE
COCKATOO

ad

ad

ad

ad

ad

ad

WHITE
COCKATOO

ad

d ♀

2/5/19

O-BREASTED
RAKEET
fasciata

ad ♂

juv

TANIMBAR
CORELLA

ad

RED-BREASTED

ROSE-RINGED

ad ♂

ad ♂

ad ♂

ad ♂

ad ♂

ad ♀

ROSE-RINGED
PARAKEET
manillensis

ALEXANDRINE
PARAKEET

ad ♀

ad ♂

ad
'natural'

ad
'blue'

ad
'yellow'

ad

BUDGERIGAR

RED LORY

CHESTNUT-WINGED CUCKOO   CTKJ
Clamator coromandus

L 45–47cm; WT 77g. **SD** India to Indonesia north to E China. Summer visitor to E China, regular but scarce spring migrant (mostly Apr/May) to Taiwan, accidental Korea and Japan. Monotypic. **HH** Favours scrub, broadleaf forest and forest edge. Brood parasite of laughing-thrushes. **ID** Large, long-tailed cuckoo, proportions somewhat magpie-like. Ad. head black, with prominent erectile nuchal crest, sides of neck and collar white, mantle, rump and long graduated tail black, latter with white tips; underparts white with warm orange wash to chin, throat and upper breast; scapulars, upperwing-coverts and flight-feathers all chestnut. Vent and undertail-coverts black. Juv. has rufous-scaled upperparts, lacks orange wash on underparts. Flight rather slow and heavy like coucal. **BP** Bill arched, sharply pointed, black; eyes red; tarsi black. **Vo** A loud hoarse *kurii kurii*, harsh *creech-creech-creech*, series of double metallic whistles *breep breep* and a cackling rattle *ghee ghe-ghuh-ghuh-ghuh-ghuh*. **AN** Red-winged Cuckoo.

PLAINTIVE CUCKOO   CT
Cacomantis merulinus

L 18–23.5cm; WT 22–25g. **SD** E Himalayas to S China, south through SE Asia to Indonesia and Philippines. *C. m. querulus* breeds east to Fujian. **HH** Open woodland, second growth and cultivation. Brood parasite of cisticolas, prinias and tailorbirds. **ID** Small slim cuckoo. Ad. hood and mantle grey, back, wings and tail darker grey-green. Underparts orange-buff. Long, somewhat rounded tail black with white barring below. Hepatic ♀ dark rufous-brown, barred black, including on crown, nape and rump. Juv. dark brown above narrowly barred black, and off-white below washed orange-buff and narrowly barred black on chin, throat and breast. **BP** Bill arched, black; eyes reddish-brown; tarsi yellowish-brown. **Vo** Whistled series of accelerating calls falling in pitch and volume *pwee pwee pwee pee-pee-pee-pee* or *tee-tee-tee-tee-tita-tita-tita-tita-tee*, a mournful, speeding and rising *tay-ta-tee tay-ta-tee*, and a harsh screeching *tchree-tchree* (latter possibly by ♀).

LARGE HAWK CUCKOO   CTKJ
Cuculus sparverioides

L 38–40cm; WT 150g. **SD** From Himalayas through SE Asia to the Sundas. *C. s. sparverioides* an uncommon summer visitor to SE & E China and Taiwan, accidental Hebei, Korea and Japan. **HH** Open deciduous woodland in lowlands and at low to mid elevations in mountains (to at least 1,500m in Taiwan). **ID** Large, can appear confusingly like a sparrowhawk, best separated using jizz. Ad. crown and face dark grey, chin black; upperparts dark grey-brown, upper breast rufous, rest of underparts white with heavy streaking on upper breast and barring on lower, vent white. Tail, long, full, grey with 4–5 broad blackish bars and grey/white tip. Juv. finely barred mantle, more broadly barred back, rump and wings. Underparts white, with rufous wash on neck-sides,

vertical dark gular stripe, bold vertical flecks on breast, narrow horizontal dark bars on belly and flanks; vent white. **BP** Bill short, arched, blackish with yellow gape line; eye-ring yellow, irides yellow (ad.) or brown (juv.); tarsi yellow. **Vo** Shrill series of loud and piercing whistled notes, rising in pitch *pee-pee-ah pee-pee-ah pee-pee-ah...*, with hysterical crescendo. Calls at night, often in flight. ♀ gives short, more tuneful, purring whistles *turr durr durr* (rising in volume). **TN** Formerly *Hierococcyx sparverioides*.

NORTHERN HAWK CUCKOO   CTKJR
Cuculus hyperythrus

L 32–35cm. **SD** NE China and S Russian Far East, Sakhalin, and throughout Korea and main islands of Japan. Winters S & SE China. Accidental Taiwan. Monotypic. **HH** Coniferous, mixed and deciduous broadleaf forest on slopes. Not uncommon, but elusive and very secretive. Brood parasite of flycatchers. **ID** Medium-sized cuckoo that may appear confusingly like sparrowhawk. Ad. crown, cheeks and chin mid grey, throat, neck-sides and nuchal crescent white; upperparts plain grey with prominent white scapular crescent. Underparts washed warm rufous from neck-sides and breast to flanks. Tail, long, quite broad, grey with 3+ narrow blackish bars, a very broad blackish band near tip bordered above with rufous and tipped rufous. Juv. wings brown with grey bars, and blackish streaks on lower neck, breast and flanks. **BP** Bill short, arched, grey above, greenish-yellow at base and tip; eye-ring yellow, irides reddish-brown; tarsi yellow. **Vo** Highly vocal, calls by day and night from perch and in flight; a far-carrying frenetic, shrill *ju-ichi*, beginning weakly but accelerating and becoming stronger and higher pitched: *ju-ichi ju-ichi ju-ichi*. **TN** Formerly part of Hodgson's Hawk Cuckoo *C. (Hiercoccyx) fugax*.

INDIAN CUCKOO   CTKJR
Cuculus micropterus

L 32–33cm; WT 119g. **SD** S & SE Asia to Indonesia. *C. m. micropterus* a summer visitor to SE, E & NE China, Russian Far East along Amur River and through Korea. Accidental Taiwan (though common on Kinmen), and Japan (Apr–Jun). **HH** Common but elusive visitor to lowland deciduous and evergreen forest and forest edge. Brood parasite of Ashy Drongo (p.300). **ID** Medium-sized cuckoo closely resembling Eurasian (p.256), from which separated by smaller size, darker eyes, mantle colour and tail pattern. Ad. hood to lower neck mid grey (♂) or brownish-grey (♀); upperparts entirely dark grey (♂ slaty, ♀ dark brown), rump and tail darkest. Underparts white, with narrow to broad slate bands across breast, flanks and undertail-coverts. Wings dark brownish-grey. Tail long, dark grey above (concolorous with rump) with broad blackish terminal band, prominent narrow grey/white tips to all feathers and white marginal spots (often concealed when tail closed); below pale to dark grey with bands of white spots. Juv. has whitish or buff scaling on head and back. **BP** Bill short, arched, grey above, yellowish below and at gape; eye-ring yellow or grey, irides brown (see Eurasian); tarsi yellow. **Vo** Highly vocal, giving a repetitive, loud, four-syllable call rendered *one more bottle*; *orange-pekoe* or *kwer-kwah-kwah-kurh*, the last syllable lowest and longest.

CHESTNUT-WINGED
CUCKOO

ad

PLAINTIVE CUCKOO

ad ♂

querulus

ad ♀
hepatic

LARGE HAWK
CUCKOO

sparverioides

ad ♂

ad ♂

juv

NORTHERN
HAWK CUCKOO

juv

ad ♀

ad ♂

juv

micropterus

INDIAN CUCKOO

EURASIAN CUCKOO   CTKJR
Cuculus canorus

L 32–36cm; WS 54–60cm; WT 115g. **SD** Britain to Japan; winters Africa, S & SE Asia. *C. c. canorus* common visitor NE China east to Japan and north to Chukotka, Kamchatka and Commander Is. Accidental Taiwan. **HH** Often conspicuous in open grassland, farmland, reedbeds, marshes, also parkland, low woodland and taiga forest to limit of tundra. Brood parasite of *Acrocephalus* warblers, accentors and pipits. **ID** Medium-sized cuckoo best separated from similar Himalayan and Oriental by voice. Ad. has mid grey hood with little contrast from dark grey upperparts. Underparts white, greyish-black bars on breast and flanks narrower than Oriental; vent and undertail-coverts white (not buff). Long uppertail-coverts as back, but tail darker blackish-grey than upperparts, with large white tips, undertail dark with white spots on outer feathers and large white tips to all. ♀ as ♂, though may have rufous wash to upper breast, or hepatic (rufous-brown with narrow black bars, palest on breast and belly, but rump unbarred). Juv. finely barred brown and grey across upperparts, rump and tail, with white nape patch. In flight, underwing rather uniform, coverts whitish, lightly barred; broader based silvery-white underwing panel, so appears much whiter than Oriental, with less contrasting, shorter central stripe, reaching onto to p5–6. Flight heavy with shallow beats. **BP** Bill short, arched, grey above, yellowish below and at gape (bill usually longer than Oriental, but much overlap); eye-ring yellow, irides yellow/orange (ad.) or dark (juv.); tarsi yellow. **Vo** Highly vocal, calling from post, bush or treetop; ♂ cocks tail and droops wings when calling. Distinctive, loud repetitive *cuk-kooo* call usually disyllabic. In flight, gives typical call or guttural chattering; ♀ may give descending series of bubbling or cackling *squip quip quipurrrr...* calls in flight.

HIMALAYAN CUCKOO   CT
Cuculus saturatus

L 29cm; WT 90.6g. **SD** Summers S Asia from Himalayas to SE China, resident parts of SE Asia, Indonesia, winters Philippines, New Guinea to E Australia. *C. s. saturatus* in S China and Taiwan. **HH** Secretive in mixed forest and orchards in hilly country; favours canopy. Brood parasite of *Phylloscopus* and *Seicercus* warblers. **ID** Slightly darker than similar Eurasian, but with more contrast between paler grey head and darker grey upperparts; has broader, blacker bars more widely spaced on white lower breast, flanks and belly; wings fall short of tail tip. Many have distinctly yellowish-buff undertail-coverts, lacking bars, but vague dark splodges on vent. ♀ either as ♂, with some brown on neck-sides, or hepatic (with barred rump). Juv. like Eurasian. Himalayan and Oriental essentially identical, but Oriental's wings longer (some overlap) and may average larger. In flight, Himalayan's striking underwing pattern recalls Common Snipe (p.170), the white centre (axillaries, greater coverts and primary bases) contrasts strongly with dark secondaries, lesser and median coverts, with 3–6 white bars on underside of dull grey primaries. Flight more rapid and direct than Eurasian. **BP** Bill slightly shorter than Eurasian; eye-ring yellow,

irides orange/brown, or dark (juv.); tarsi yellow. **Vo** A series of rather Hoopoe-like four-noted phrases *huk-hoop-hoop-hoop*, the first syllable slightly higher pitched and the phrases well separated. **TN** Formerly considered a race of Oriental Cuckoo.

ORIENTAL CUCKOO   CTKJR
Cuculus optatus

L 30–34cm; WT 99g. **SD** N Asia from W Russia to Chukotka. Common NE Russia in Yakutia, Chukotka, SE Russian Far East, Sakhalin, Kuril Is, Kamchatka and Commander Is, also Korea and Japan, and on migration Taiwan. Winters SE Asia to E Australia. **HH** Secretive visitor to taiga and mature mixed forest further south; favours canopy. Brood parasite of *Phylloscopus* warblers. **ID** Closely resembles Eurasian; virtually identical to Himalayan. Many Oriental have distinctly yellowish-buff undertail-coverts, lacking bars, but sometimes with distinct spots. ♀ when hepatic has barred rump. Juv. like Eurasian and Himalayan. In flight, underwing shows bold white barring on inner primaries and outer secondaries, and has barred underwing-coverts. Flight like Himalayan. Much less often in open than Eurasian. **BP** Same as Himalayan. **Vo** ♂'s call/song, given from high canopy branch, far-carrying, deeply resonant and rhythmic; even, double notes in series: *po-po...po-po...po-po...* or *boop-boop...boop-boop...boop-boop*, sometimes preceded by a hoarse intake *gua*. Commonly gives a very rapid multisyllabic *po-po-po-po-po-po-po* before settling into typical monotonous pattern of 10–20 bisyllabic notes. **TN** Formerly *C. saturatus*. **AN** Horsfield's Cuckoo.

LESSER CUCKOO   CTKJR
Cuculus poliocephalus

L 22–27cm; WT 52g. **SD** Himalayas to Korea, Japan, winters mainly S Asia, E Africa, some SE Asia. Summer visitor NE China, Korea, Russian Far East, Japan (Nansei Shoto to SW Hokkaido). Monotypic. **HH** Common, though elusive, in deciduous and evergreen broadleaf forest in lowlands and hills. Brood parasite of *Cettia* warblers. **ID** Smaller, slimmer and more compact than preceding cuckoos, hood variable but sometimes less extensive, only to upper neck. Blackish barring on white underparts wider spaced even than Oriental, so appears whiter, enhanced by unbarred vent and buffy-grey undertail-coverts; undertail blackish with white spots, uppertail grey with narrow white tip. Hepatic ♀ bright rufous with indistinct barring or blotches on crown and nape. Narrow white panel on underwing, the white coverts quite strongly barred dark, and rump/uppertail-coverts relatively darker than Oriental, contrasting less with tail. **BP** Bill short, arched, grey above, yellowish below and at gape; eye-ring yellow, irides brown (compare Common/Oriental); tarsi yellow. **Vo** Highly vocal, particularly early morning and late afternoon, but also at night, and in flight over canopy. Loud six-note whistle *tep-pen-kaketaka* or *that's your cho-ky pepper* rising over first four syllables, with emphasis on *choky* and falling on final slurred double syllable (pattern strongly recalls Gray's Warbler). ♀ gives high-pitched piping *pipipipipipipi* in flight.

EURASIAN
CUCKOO

ad

ad ♀
hepatic

ORIENTAL
CUCKOO

saturatus

ad ♂

HIMALAYAN
CUCKOO

ad ♀
hepatic

juv

ad ♂

juv

LESSER CUCKOO

ASIAN DRONGO CUCKOO CTKJ
Surniculus lugubris

L 25cm; WT 35g. **SD** India to SE China and Indonesia. *S. (l.) dicruroides* occurs Fujian and, as accidental, on Kinmen (Taiwan), in Korea and Japan (May). Relationships within *S. lugubris* complex currently under investigation, but *S. (l.) dicruroides* is probably a separate species, **Fork-tailed Drongo Cuckoo**. **HH** Uncommon and secretive, at forest edge and in scrub. **ID** Small to medium, slim, black cuckoo (drongo-like but behaves like a cuckoo). Ad. glossy blue-black (some have white thighs and nuchal patch), with long, deeply notched (drongo-like) tail. Vent/undertail-coverts black with narrow white bars which continue on underside of outertail-feathers. Juv. has white-spangled wing-coverts, rump and tail tip. **BP** Bill arched, black; eyes brown; tarsi blue-grey. **Vo** Ascending series of 4–7 piercing rising whistles, *pii pii pii...*, by day or night.

ASIAN KOEL CTKJ
Eudynamys scolopaceus

L 43cm. **SD** India to Indonesia and Australia. *E. s. chinensis* is a summer visitor to much of E China south of Chiangjiang River, rare migrant Taiwan (but scarce breeder Kinmen), accidental S Japan (mainly Yonaguni, Yaeyama Is, but also Kyushu), and reported Korea. **HH** Skulking but vocal bird of dense primary and secondary forest, open forest, plantations, orchards, scrub and gardens. Brood parasite of Large-billed and House Crows. **ID** Large, heavy-set cuckoo, with somewhat long, broad tail. ♂ glossy greenish-black. ♀ blackish-brown with numerous fine buff spots on wings, back and rump, buff-barred tail, buff streaks on throat and barring across entire underparts. **BP** Bill heavy, arched and slightly hooked, dark grey-green or dull-green; eye-ring dull blue, irides red; tarsi blue-grey. **Vo** Name derives from voice; shrill, fast, oft-repeated slurred whistles *ko-el* rising in pitch and frequency; a loud repetitive *kow-wow*, by day or at night. Also a wide range of other sweet to harsh notes.

GREATER COUCAL CT
Centropus sinensis

L 47–52cm; WT ♂ 236g, ♀ 268g. **SD** India to SE Asia, Indonesia and Philippines, and S & SE China. *C. s. sinensis* ranges east to Fujian and Zhejiang. Common resident Kinmen (Taiwan), scarcer Matsu (Taiwan), absent Taiwan mainland. **HH** Forest edge, dense scrub and reedy areas. **ID** Very large, heavy-set cuckoo, with short rounded wings and long graduated tail. Ad. a large version of Lesser Coucal, but separated by glossier black plumage, brighter, cleaner chestnut mantle, wings and wing-coverts. Juv. dark brownish-black with brown mantle, back and wings boldly barred black. Flight typically low, laboured, with slow flaps. **BP** Bill very thick, arched, slightly hooked, black in ad., greyish-black in young; eye-ring dull blue, irides reddish-brown; tarsi blackish-grey. Often walks on ground (hence old name, crow-pheasant). **Vo** Various *hoop* and *tok* notes and harsh *skaah*; song comprises series of hooting *boop boop boop* notes falling in pitch but increasing in speed.

LESSER COUCAL CTKJ
Centropus bengalensis

L 31–42cm; WT 88–152g. **SD** India to Indonesia. *C. b. lignator* a common resident over much of SE & E China from Fujian South, and Taiwan. Not a strong flier, but has strayed to Korea and the Nansei Shoto. **HH** Scrub, farmland, grassland and marsh edges. **ID** Large, heavy-set cuckoo, with rather short, rounded wings and long graduated tail. Very similar to Greater Coucal, but smaller, bill and tail shorter, and underwing-coverts chestnut (not black). Ad. breeding dull or dirty black, with dull brown back and warmer rufous-brown wings. Non-breeding ad. has pale straw-coloured streaking on crown, mantle, face and upper breast, underparts otherwise rufous-white barred dusky. Juv. resembles non-breeding ad. but has dark barring on brown wings and tail, tawny head and neck with dark streaks, and tawny underparts. Flight as Greater. **BP** Bill thick, arched, slightly hooked, black in ad. greyish-horn in young; eye-ring dull blue, irides reddish-brown; tarsi blackish. **Vo** Song a series of double, hollow notes followed by staccato phrases, *huup huup huup-uup tokalok tokalok*.

EASTERN GRASS OWL CTKJ
Tyto longimembris

L ♂ 32–36cm, ♀ 35–38cm; WT ♂ 265–375g, ♀ 320–450g. **SD** Indian subcontinent to Philippines and Australia. *T. l. chinensis* occurs in E China, including Fujian, Zhejiang, Shandong and Hebei; *T. l. pithecops* resident in Taiwan to Penghu (Taiwan) and S Japan and Korea. **HH** Open areas with tall grassland in which it roosts by day. Crepuscular. In Taiwan, low grasslands of W foothills. **ID** Medium-sized owl with prominent heart-shaped facial disc and long narrow wings, recalling extralimital Barn Owl *T. alba*. Upperparts dark brown and tawny, with some fine silvery spots on crown, mantle and wings (lacks heavy streaking on chest of Short-eared, which may occur in similar habitat, see p.268), flight-feathers barred tawny, grey and smudgy blackish-brown; underparts creamy buff with fine black spots (not streaks like Short-eared), belly and vent whitish. Facial disc grey-buff with tawny rim, blackish-brown vertical line on forehead and smudges between eyes and bill. Broad wings have golden-buff patch at base of primaries, pale tail short with dark bars. **BP** Bill whitish; eyes appear relatively small, black; tarsi white to whitish-pink. **Vo** Various shrieking and screaming sounds in flight, like Barn Owl. In alarm gives bill clicks, tongue snaps and snoring sounds. **TN** Formerly part of Grass Owl *T. capensis*.

ASIAN DRONGO CUCKOO

dicruroides

ad

juv

chinensis

ASIAN KOEL

ad ♂

GREATER COUCAL

sinensis

juv

ad ♀

ad

ad

EASTERN GRASS OWL

chinensis

imm

ad

lignator

LESSER COUCAL

MOUNTAIN SCOPS OWL CT
Otus spilocephalus

L 17–21cm; WT 55–120g. **SD** Himalayas to Borneo. *O. s. latouchi* occurs in SE China and *O. s. hambroecki* in Taiwan. **HH** On Taiwan, widespread in dense montane broadleaf forest at 100–2,500m. **ID** Small, squat, dark brown scops owl, with short ear-tufts, poorly defined facial disc, distinctive scapular bar of white spots and pale central belly stripe. *O. s. latouchi* has upperparts from crown to tail finely mottled dark and pale brown with rufous tones, a faint hind-collar, facial disc pale brown rimmed blackish-brown, and underparts pale brown with silver spots and dark brown flecks. Smaller headed and slimmer than Collared Scops. *O. s. hambroecki* differs in having darker brown upperparts, prominent pale hind-collar, pale facial disc and longer, two-tone (pale and dark brown) ear-tufts. **BP** Bill pale horn to creamy white; eyes yellow; tarsi feathered brown, toes grey. **Vo** Calls frequently, a level fluty two-note whistle, *toot-too*, *plew-plew* or *poop-poop*, repeated monotonously at 5–10-second intervals for several minutes. **AN** Spotted Scops Owl.

COLLARED SCOPS OWL CTKJR
Otus lettia/semitorques

L 23–25cm; WS 55–59cm; WT 100–170g. **SD** Taxonomy unclear. Some authors retain *O. bakkamoena* as extralimital **Indian Scops Owl**, leaving *O. lettia* as **Collared Scops Owl** and *O. semitorques* as **Japanese Scops Owl**. Taking only *O. lettia/semitorques* into account, ranges E Nepal and Indochina to Russian Far East, Korea and Japan. Race *ussuriensis* occurs SE Siberia, Korea and NE China to the Chiangjiang (and has strayed to Japan). South of the Chiangjiang *erythrocampe* occurs, whilst in Taiwan *glabripes* is resident, and on Sakhalin, Kuril Is and Japan *semitorques* is a widespread resident, though perhaps only summer visitor/migrant in Hokkaido, but *pryeri* (smaller and redder than *semitorques*, lacking toe feathers, with redder eyes than *semitorques*) is resident on Izu Is and in Nansei Shoto (Okinawa to Iriomote) and may be a separate species. **HH** From lowlands to low mountains: farmland, tree-lined streets, parks, woodland edge and forest (deciduous/mixed in N, subtropical evergreen in S), where hunts mainly terrestrial insects, reptiles and small mammals, occasionally birds. Crepuscular/nocturnal. **ID** Rather large scops owl with particularly large head and conspicuous ear-tufts. Generally greyish-brown or dull brown mottled black and buff, with pale brown nuchal collar; upperparts browner, face and underparts greyer, face rimmed blackish. **BP** Bill short, hooked, grey; eyes yellowish-brown to orange (*lettia* group) or reddish-brown (*semitorques* group, but orange-yellow in *pryeri*), pupil generally appearing large; tarsi feathered brown in N subspecies, toes unfeathered in S subspecies (e.g. *pryeri*), toes yellowish. **Vo** Very variable, not only between races/species, but also regionally and sexually. Includes a frog-like *whuk* and whistled *pew*, also *kwooo, kwaa, kuui* and *wowowo* calls. Dog-like barking calls in breeding season and cat-like *myau-myau* year-round described from Japan, but on Okinawa *pew pew pew*; *kyo kyo kyo* or *woffu-woffu-woffu-woffu*; elsewhere *kwee-kwee* and *pew-u pew-u*. In Taiwan, a surprised *wuh!* repeated every 4–10 seconds. Considerable variation renders subspecific, even specific, confirmation difficult.

ORIENTAL SCOPS OWL CTKJR
Otus sunia

L 18–21cm; WS 50.5–52.6cm; WT 75–95g. **SD** Himalayas and S Asia to Indochina, north to Sakhalin. Migratory in NE Asia. *O. s. stictonotus* in E China (resident south of Changjiang) and NE China, Taiwan and Korea; *O. s. japonicus* occurs throughout Japan (Apr–Sep; scarce in N and absent S half of Kyushu), and perhaps same subspecies S Russian Far East and on Sakhalin. **HH** Parks, large gardens, woodland edge and montane forest (deciduous and mixed in N and evergreen broadleaf in S), also coniferous taiga edge. Crepuscular/nocturnal. **ID** Small scops owl with two colour forms, grey (common) and rufous (scarce), with prominent ear-tufts. Grey morph overall mid to dark greyish-brown, with black spotting on crown, fine black streaking on mantle, prominent white spots on scapulars (forming bar) and white spots in primaries. Facial disc grey with whiter brows and blacker margin; grey-brown underparts finely vermiculated with strong dark streaks. Rufous morph is reddish-brown, with a more prominent scapular bar. Wings rather short, only 4–5 primaries extending beyond tertials. Best distinguished from similar species by voice. **BP** Bill pale grey-horn; eyes yellow to pale orange; lower tarsi unfeathered, feet grey to greyish-flesh. **Vo** Soft, monotonous, three-note phrase, *bu-po-so*, or *buk-kyok-koo* (first syllable sometimes too soft to hear), often repeated at night, occasionally calls during day, especially on spring migration.

EURASIAN SCOPS OWL Extralimital
Otus scops

L 16–20cm; WS 53–64cm; WT 60–135g. **SD** Summer visitor east to Baikal and may stray to region. **ID** Very similar to Oriental Scops perhaps distinguishable only by longer primary projection (6–7 primaries visible beyond tertials), its plaintive repetitive *tu tu tu* call and feathering on lower tarsi.

ELEGANT SCOPS OWL TJ
Otus elegans

L 19–22cm; WS 60cm; WT 100–107g. **SD** Endemic resident of E Asia, with *O. e. elegans* widespread through the Nansei Shoto; *O. e. interpositus* on Daito Is; *O. (e.) botelensis* on Lanyu I. off SE Taiwan, which may be a separate species. **HH** Tropical and subtropical evergreen broadleaf forest from sea level to at least 550m in hills. Crepuscular/nocturnal.

ID Small, dark rufous-brown scops owl with short ear-tufts, lacks collar of Collared Scops, and is more spotted than Oriental. Dark spotting on crown, fine vermiculations and dark streaks on mantle, and distinct white scapular bar. Prominent facial disc with white brows and chin/lower face. Underparts finely barred and strongly streaked. Readily distinguished by range and voice. **BP** Bill dark grey; eyes yellow; tarsi feathered, feet rather large, toes grey. **Vo** Highly vocal. ♂ gives very repetitive and rather resonant coughing whistle, *ko-ho ko-ho*; or *u-hu* or *pu-wu* call, and is easily attracted by imitations. ♀ sometimes responds with a nasal *nnya* or *niea*, also a high-pitched chittering recalling fledgling kestrels. **AN** Ryukyu Scops Owl.

ad *latouchi*

MOUNTAIN
SCOPS OWL

COLLARED
SCOPS OWL

ad *lettia*

ad *hambroecki*

ad *pryeri*

ad *semitorques*

EURASIAN
SCOPS OWL

ad *stictonotus*

ORIENTAL
SCOPS OWL

ad *japonicus*

ad

ELEGANT
SCOPS OWL

elegans

ad

SNOWY OWL   CKJR
Bubo scandiacus

L ♂ 55–64cm, ♀ 60–70cm; WS 125–166cm; WT ♂ 0.7–2.5kg, ♀ 0.78–2.95kg. **SD** Widespread in N Holarctic, across Russian Arctic; in region east to Chukotski Peninsula, also Wrangel and other Arctic islands, and on Commander Is. Wanders south in winter as far as Korea (one record), Hokkaido (very rare Nov–Mar) and NE China (rare). Monotypic. **HH** Mainly open tundra, but wanders to taiga and temperate regions in winter (particularly following crashes in prey populations), where favours forest edge, coastal grasslands or alpine regions. Commonly diurnal. **ID** Very large white owl, with rather small head lacking ear-tufts. ♂ almost entirely white, with some black spotting on wing-coverts and sparse black bars on tertials. ♀ generally larger, from distance appears greyer because of extent of black streaking on crown and nape, and narrow black barring on mantle, wings, underparts and tail; face, neck and upper breast white; at close range retains overall appearance of being white. In flight, ♂ ghostly; ♀ appears white with heavy grey barring on tail and finer barring on wings. **BP** Bill greyish-black and mostly concealed by facial feathering; eyes large, black-rimmed lids contrast with bright golden-yellow irides; tarsi feathered grey-white, toes large, claws black. **Vo** Generally silent except on breeding ground when gives far-carrying, surprisingly duck-like barking *guwa guwa* or *krek-krek-krek...* in alarm, and a high drawn-out scream and muffled hooting or booming *goo goo* given by ♂. **TN** Formerly *Nyctea scandiaca*.

EURASIAN EAGLE OWL   CKJR
Bubo bubo

L 56–75cm; WS 138–170cm; WT ♂ 1.5–2.8kg, ♀ 1.75–4.2kg. **SD** Widespread but scarce, from SW Europe and Scandinavia to NE Russia. Some winter nomadism possible. *B. b. ussuriensis* ranges in E Asia as far as NE China, Russian Far East east to Magadan, perhaps Chukotka; *B. b. borrisowi* in Sakhalin and S Kuril Is, and an extremely rare resident N Hokkaido; *B. b. kiautschensis* is scarce resident in Korea and also occurs in E & SE China. Subspecific taxonomy is debated. **HH** Forested mountains with rocky outcrops and cliffs for nesting. Crepuscular/nocturnal. **ID** Very large owl, dark brown and tawny on upperparts with black and grey mottling, creamy brown to tawny underparts with heavy blackish-brown spotting on breast, and fine barring and more elongated streaking on belly and flanks; large facial disc has black rim and long ear-tufts also black-edged. In flight, wings extremely broad and appears rather plain, rusty brown. **BP** Bill dark grey; eyes deep orange; legs feathered, toes yellow. **Vo** On territory gives very deep, hoarse coughing *uh-hoo* or *whoohuu* repeated infrequently, far-carrying; clicks beak if disturbed or gives hard barking *ka ka-kau*.

BLAKISTON'S FISH OWL   CJR
Bubo blakistoni

Endangered. L 60–72cm; WS 180–190cm; WT ♂ 3.15–3.45kg, ♀ 3.36–4.6kg. **SD** Scarce NE Asian endemic, threatened by habitat loss and disturbance. *B. b. doerriesi* in extreme NE China (very rare: Heilongjiang, Jilin, Inner Mongolia), SE Russia (Primorye, Khabarovsk, Magadan), whilst *B. b. blakistoni* occurs in S Kuril Is, Sakhalin and Hokkaido. **HH** Mature, mixed boreal riparian forest in lowlands and mountains: requires rivers, lakes or springs that do not freeze in winter, and enormous cavities for nesting. Crepuscular/nocturnal. **ID** Enormous. More uniformly brown than Eurasian Eagle Owl, lacks black outline to dusky brown face and ear-tufts are broader, looser. *B. b. blakistoni* has chin/throat white, especially noticeable when throat bulges during calling. Upperparts cool, mid brown with blackish-brown shaft-streaking to mantle and back, scapulars more broadly streaked blackish-brown, closed wings heavily barred dark brown and pale sandy brown; tawny-buff underparts have many narrow elongated shaft-streaks and fine dark crossbars. *B. b. doerriesi* differs in having white nuchal patch. **BP** Bill dark greyish-horn; eyes yellowish-orange; tarsi feathered, toes have only very short feathering or none, pinkish-grey. **Vo** Very deep, echoing duet: ♂ gives deep *boo-boo*, immediately followed by ♀'s even more sonorous single *bu*; this three-note duet is repeated in short or long series at *c.*1-minute intervals. **TN** Formerly *Ketupa blakistoni*.

TAWNY FISH OWL   CT
Bubo flavipes

L 48–55cm. **SD** Very scarce resident from Himalayas to Indochina and SE China. Occurs E China (Jiangsu; Zhejiang) and on Taiwan (very rare and local). Monotypic. **HH** Dense riparian forest in foothills to *c.*1,500m, but tolerates bamboo and degraded riparian forest with plentiful shade. Crepuscular/nocturnal. **ID** Large, rufous owl with large, laterally spreading ear-tufts. Smaller than Blakiston's or Eurasian Eagle; much more rufous-brown than Blakiston's above and below; upperparts rich rufous orange-brown heavily streaked black (resembling Eurasian), underparts dark tawny or mid rufous-brown, with bold blackish-brown shaft-streaking lacking fine crossbars (unlike Blakiston's), face darker, more orange rufous-brown than Blakiston's or Eurasian Eagle, but shares loose broad ear-tufts of former, albeit much more horizontal. Closed wings heavily barred black and rufous; chin/throat may appear white, especially when calling. In flight, wings and tail appear strongly banded above and below. **BP** Bill horn or black with grey cere; eyes yellow; tarsi feathered, toes grey. **Vo** Variable, including cat-like mewing notes and deep hooting *whoo-hoo* or *buh-huh wooo*; like Blakiston's, pairs duet near nest around dusk and dawn. **TN** Formerly *Ketupa flavipes*.

SNOWY OWL

ad ♂

ad ♀

EURASIAN
EAGLE OWL

kiautschensis

ad

TAWNY FISH OWL

BLAKISTON'S
FISH OWL

doerriesi

ad

ad

BROWN WOOD OWL *Strix leptogramma* CT

L 40–53cm; WT 500–700g. **SD** S Asia to Indonesia; *S. l. ticehursti* in SE China; *S. l. caligata* Taiwan. Rare and secretive. **HH** Subtropical montane broadleaf forest (Taiwan *c.*200 to >2,300m); prefers dense, undisturbed areas. **ID** Large, somewhat elongated. Small-headed; unusual facial disc, appears spectacled with dark shadows around eyes, pale buff bordered blackish-brown; black chin bar above whitish throat. Overall dark chocolate-brown, especially crown and neck; fine cinnamon barring on mantle; very fine grey bars on scapulars and wings, and pale scapulars. Underparts buff with narrow dark brown breast-band and more wide-spaced barring on belly and flanks. **BP** Bill pale bluish-grey; eyes dark brown; tarsi feathered brown. **Vo** Eerie screams, short series of abrupt barking hoots, similar to opening of Ural Owl's call, but more explosive at close range; a deep *goke-goke-ga-looo*.

CHINESE TAWNY OWL *Strix nivicola* CTK

L 45–47cm. **SD** Himalayas to E Asia. *S. n. nivicola* SE China; *S. n. yamadae* Taiwan; *S. n. ma* NE China (Hebei, Shandong) and Korea; not common. **HH** Deciduous, mixed and coniferous woodland/forest in lowlands and mountains. Usually >2,000m in Taiwan. **ID** Medium-sized, resembles small, dark brown Ural Owl. Large rounded head, broad wings and short tail. Overall warm dark reddish-brown, with pale striping on dark crown, darker mottling on upperparts, whitish scapular bar and dark barring on wings and tail; grey morph rare. Underparts buff with extensive blackish-brown shaft-streaking and associated fine dark barring. Wings extend just beyond tail (see Ural). Facial disc pale brown with white eyebrows and narrow dark rim, and pale X on face. **BP** Bill horn; eyes black; tarsi feathered brown. **Vo** Loud, repetitive, *hu-hu*, repeated slowly. **AN** Himalayan Wood Owl.

URAL OWL *Strix uralensis* CKJR

L 50–62cm; WS 124–134cm; WT ♂ 500–950g, ♀ 0.57–1.3kg. **SD** Scandinavia to E Asia. In region, NE China, Korea (rare S, commoner N), Sakhalin to Okhotsk, and main Japanese islands. Debate over range of subspecies: *S. u. daurica* Baikal to Amurland; *S. u. nikolskii* Transbaikalia, E Amurland, Sakhalin, NE China and Korea; *S. u. japonica* S Kurils and Hokkaido; *S. u. hondoensis* N Honshu; *S. u. momiyamae* C Honshu; *S. u. fuscescens* S Honshu, Kyushu and Shikoku. **HH** Mature forest. Often roosts in open. **ID** Large, greyish-brown with large head; pale facial disc radially streaked brown, rimmed narrowly white and black. Upperparts cool grey-brown, heavy blackish-brown streaking on head and mantle; wings and tail greyish-brown with paler and darker bars and prominent off-white scapular bar. Underparts pale buff with dark blackish-brown shaft-streaks, heaviest on neck and breast, more widely spaced on belly. Tail long, somewhat wedge-shaped, barred, extending well beyond wings (see Chinese Tawny). *S. u. japonica* overall much paler grey-brown than southern races; face very pale, white spots bordering facial disc, largely white underparts and pale grey upperparts streaked dark. *S. u. hondoensis*, *momiyamae* and *fuscescens* doubtfully separable, but much smaller than northern races, much darker dusky-brown, with heavier streaking on back and breast and adjoining facial disc, more closely resembling Chinese Tawny. Flight slow and *Buteo*-like on broad wings. **BP** Bill dull yellow; eyes small, irides black; tarsi feathered whitish. **Vo** Deep, gruff *guhu ... hoo huhu hoo-hoooh* or *gouhou ... guroske go ho*, with clear pause between introductory call and sequence of hoots that dies away at end; mainly autumn and late winter. Also *goh goh goh*, a harsh *gyaa gyaa* or *schrank* (♀ only), similar to Grey Heron, and loud barking given by nesting ♀.

GREAT GREY OWL *Strix nebulosa* CR

L 59–70cm; WS 128–148cm; WT ♂ 0.8–1.175kg, ♀ 0.925–1.70kg. **SD** Northern Holarctic. *S. n. lapponica* Scandinavia to E Chukotka south to Amur and Sakhalin, very rare NE China. **HH** Forest-tundra and mature taiga, often by clearings or meadows. Wanders south in winter. Partly diurnal. **ID** Very large. Imposing presence heightened by massive head and extremely large facial disc (narrow concentric circles of grey and blackish, outer border black), with grey crescents between eyes forming large pale X above bill, and black and white 'bow-tie' pattern on chin. Overall mid grey, with dark grey shaft-streaking and finely vermiculated upperparts, with broken, pale scapular bar; underparts grey, paler on belly, with dark streaking (heaviest on neck and upper breast, looser on belly and flanks). Wings broad, grey with buff-brown at base of outer primaries, tail long and barred with broad dark band at tip. Flight light and slow, stalls to drop onto prey. **BP** Bill dull yellow; eyes close-set, small and 'staring', irides bright yellow; tarsi feathered grey-brown. **Vo** Territorial ♂ gives series of deep booming hoots: *bvoo bvoo bvoo ...* becoming quieter and lower; ♀ responds with emphatic low whistled *iihwew*. ♀ also gives higher-pitched hoots. Snaps bill loudly when alarmed or aggressive.

NORTHERN HAWK OWL *Surnia ulula* CKR

L 35–43cm; WS 69–82cm; WT ♂ 270–314g, ♀ 320–345g. **SD** Northern Holarctic; *S. u. ulula* Chukotka, Koryakia and Kamchatka, south through Russian Far East and Sakhalin, winters to NE China. Irruptive or nomadic. **HH** Taiga and forest edge, often diurnal, frequently conspicuous on treetops. **ID** Distinctive, medium-sized owl, with long tail and relatively small head, lacking ear-tufts. Forehead grey speckled black, face greyish-white with prominent white brows, black line above eye curves back to form blackish-grey rim of facial disc and splits to form band on sides of nape. Pale grey nape patches create false face. Upperparts dark grey-brown with white spots, wings mostly dark grey-brown with white-spotted scapulars and coverts, and barred flight-feathers; prominent pale scapular bars; tail dark grey-brown narrowly barred white. Underparts white with narrow grey barring on breast to undertail-coverts. Wings narrow with pointed tips; *Accipiter*-like fast flaps interspersed by glides; swoops up to land. **BP** Bill yellowish-horn; eyes large, irides yellow; toes feathered, whitish. **Vo** In early spring gives long bubbling trill, *popopopopo...* or *lülülülülü...* (higher, longer than Tengmalm's); shrill chattering alarm resembles that of small falcon *kwikwikwikwikwi*.

BROWN WOOD OWL

ticehursti

ad

ad *yamadae*

CHINESE TAWNY OWL

ad *ma*

ad *nivicola*

URAL OWL

ad *japonica*

ad *hondoensis*

GREAT GREY OWL

lapponica

ad

NORTHERN HAWK OWL

ulula

ad

EURASIAN PYGMY OWL CR
Glaucidium passerinum

L 16–17cm; WS 34–36cm; WT ♂ 50–65g, ♀ 67–77g. **SD** Scandinavia across Russia to Sakhalin. *G. p. orientale* breeds through Russian taiga below Arctic Circle, from C Siberia to W shore of Sea of Okhotsk, Amurland and Sakhalin; also adjacent NE China. **HH** Mature, open coniferous or mixed boreal forest, where nests in woodpecker holes. Commonly crepuscular. **ID** Tiny plump owl, large-headed and short-tailed. Upperparts uniform dark greyish-brown with small spots on crown, head-sides and wing-coverts, and narrowly barred tail. Facial disc indistinct, with grey-brown and white concentric rings, and short white eyebrows giving austere look. Pale nape patches create false face. Darkly barred grey-brown on throat and flanks, but breast and belly whitish with dark shaft-streaks. Flight fast and bounding, like woodpecker. **BP** Bill dull horn; eyes small, close-set, irides yellow; toes feathered, whitish, claws black. **Vo** Song a series of 10–15 rising sharp bullfinch-like fluty whistles, *pjuu pjuu pjuu...* or *deu deu deu*, also rapid series of mellow whistles. ♀ gives thin thrush-like *tseeeh*. Autumn 'song' differs: *chuuk-chüük-cheek-chiik*.

COLLARED OWLET CT
Glaucidium brodiei

L 15–17cm; WT ♂ c.53g, ♀ c.63g. **SD** Resident from Himalayas to Borneo and across China. *G. b. brodiei* reaches S & E China; *G. b. pardalotum* occurs in Taiwan. **HH** Wide range of forest types from subtropical to temperate, in foothills to high mountains, preferring taller trees. Descends to lower elevations (to c.100m) in winter. Active day and night. Frequently mobbed by small birds. **ID** Tiny plump owl; large-headed (lacks ear-tufts), short-winged and rather long-tailed, resembling Eurasian Pygmy Owl, but has prominent pale spotting on dark crown, barred breast-sides and flanks, and broken streaking on underparts which appear to have white bands on chest, belly and flanks. Facial disc indistinct, but has short white eyebrows and white 'choker', and austere look. Two prominent dark nape spots ringed buff, creating prominent false face. Upperparts reddish- or greyish-brown; underparts buff or rufous-brown; tail narrowly banded. Flight fast. **BP** Bill horn; eyes close-set, irides yellow; toes yellow. **Vo** Soft series of clear, level four-note hoots, *toot-toot-toot-toot*, repeated for several minutes day or night.

ASIAN BARRED OWLET C
Glaucidium cuculoides

L 22–25cm; WT 150–175g. **SD** Himalayas to SE Asia. *G. c. whitelyi* resident in SE China and has strayed northeast to Shandong. **HH** Gardens, parks, secondary and primary forest, temperate to tropical, from lowlands to mountains. Mainly nocturnal. **ID** Small, plump, heavily barred owl. Head appears rather small and rounded; short, narrow white eyebrows and white whisker lines break facial disc into two crescentic halves. Wings short, falling well short of tail, white-edged scapulars form distinct but narrow bar. Overall dark rufous-brown with narrow buff-barred upperparts. Underparts

barred rufous-brown and pale rufous on breast and flanks, with white collar, central breast stripe and broken stripes on flanks merging in white lower belly. Flight fast. **BP** Bill horn; eyes close-set, orange-yellow; legs grey, toes yellow. **Vo** Prolonged bubbling series of barbet-like whistles, *hoop hoop hoop wup-wup-wup pupupupupu*, and a quavering trill increasing steadily in volume, *wu u u u u u u u*; also various squawks, churrs and barks. **AN** Cuckoo Owlet.

LITTLE OWL *Athene noctua* CTKJR

L 21–23cm; WS 54–58cm; WT ♂ 167–177g, ♀ 166–206g. **SD** Widespread, from W Europe to Transbaikalia and NE China. *A. n. plumipes* is a common resident NE China, rare resident N Korea and Ussuriland, and a rare visitor further south in Korea. Japanese and Taiwanese records may refer to escapes.

HH Semi-desert, steppe, farmland and open woodland, isolated tree rows and windbreaks. Commonly diurnal or crepuscular. **ID** Medium-sized, compact owl with long legs. Broad-headed, lacking ear-tufts but has prominent dark-rimmed facial disc with pale eyebrows, whisker lines and chin bar. Flat crown is mid-brown finely spotted white; back, wings and tail sandy to dark brown with paler spots on mantle and scapulars, tail has broad pale bars. Underparts buff, broadly streaked mid-brown, but lower belly/vent whitish. Undulating flight on very rounded wings; appears short-tailed. May perch prominently on posts, bobbing when excited. **BP** Bill yellow-horn; eyes large, staring, irides lemon-yellow; legs and toes feathered pale buff or white. **Vo** Calls by day or night somewhat wigeon-like, plaintive high-pitched whistled *quew, kee-you* or *kaaaooh*; also a low, mellow hooting in slow series, and high-pitched staccato *kir-rik-kir-ik....*

TENGMALM'S OWL *Aegolius funereus* CJR

L ♂ 21–25cm, ♀ 25–28cm; WS 50–62cm; WT ♂ 90–115g, ♀ 120–195g. **SD** Wide range in Holarctic. *A. f. magnus* occurs NE Siberia to Yakutia, Chukotka, Anadyr and Kamchatka; *A. f. sibiricus* ranges across taiga to NE China, Russian Far East, Sakhalin and western shores of Sea of Okhotsk. Scarce, but possibly widespread resident in Hokkaido. **HH** Mature boreal forest, using old woodpecker holes for nests, particularly those of Black Woodpecker. Nocturnal. **ID** Small but very large-headed spotted owl. Prominent white and grey facial disc, with almost complete black border and raised black eyebrows affording 'surprised' look; chin and lower disc rim brownish-black. Crown spotted grey, nape, mantle and wings dark brown with large white spots, especially on scapulars; underparts whitish with large brown spots especially heavy in band across upper breast. Wings and tail both broad, rounded, brown, spotted with grey in five rows across primaries, three on tail. **BP** Bill yellowish-horn; eyes large, irides yellow; tarsi feathered buff. **Vo** On territory, ♂ gives far-carrying repetitive series of short, rapid hoots, in bouts lasting many minutes: *po-po-po-po-po...* rising slightly in pitch and becoming clearer, but ending rather suddenly; also a sharp whistled *skiew*. **AN** Boreal Owl.

EURASIAN PYGMY OWL

orientale

ad

COLLARED OWLET

brodiei

ad

grey morph

rufous morph

ASIAN BARRED OWLET

whitelyi

ad

LITTLE OWL

plumipes

ad

TENGMALM'S OWL

magnus

ad

juv

BROWN HAWK OWL CTKJR
Ninox scutulata

L 27–33cm; WS 66–70cm; WT 172–227g. **SD** Resident from Himalayas and Indian subcontinent to Indonesia, and in S China, Taiwan and Ryukyu Is. *N. s. florensis* a summer visitor to Russian Far East, N Korea, C & NE China; *N. (s.) japonica* (possibly a full species, **Japanese Hawk Owl**) is a summer visitor to E China, S Korea and Japan (migrants in Taiwan presumably this subspecies); *N. s. totogo* (possible split as **Ryukyu Hawk Owl**) resident in Taiwan, Lanyu I.) and Ryukyu Is. **HH** Gardens, parks, secondary and primary forest, temperate to tropical, from lowlands to *c.*1,500m. Mainly nocturnal, hunting large insects in flight; often drawn to streetlights. **ID** Medium-sized, 'elongated' owl with no facial disc or ear-tufts. Small, grey-brown, rounded head and long barred tail give distinctive shape. *N. s. florensis* and *N. s. japonica* largely dark chocolate-brown on head and upperparts, with heavy, brown streaking on whitish underparts; white spot above bill, between eyes; chin and vent white. *N. s. totogo* even darker, with darker face and more heavily streaked underparts. **BP** Bill greyish-black; eyes oversized, irides yellow-orange; legs brown, toes yellow. **Vo** Call given frequently on territory consists of deep, paired hoots repeated for many minutes at a time: *hoho hoho hoho* …. *N. s. totogo*, however, gives more evenly spaced *ho ho ho ho* notes, trailing away. **AN** Northern Boobook.

NORTHERN LONG-EARED OWL CTKJR
Asio otus

L ♂ 35–38cm, ♀ 37–40cm; WS ♂ 90–100cm; WT ♂ 220–305g, ♀ 260–435g. **SD** Wide range in Holarctic. *A. o. otus* occurs across boreal Russia to SE Russian Far East and NE China, uncommon winter visitor Korea (may breed), winters E China, and resident in N Japan, winter visitor south from C Japan. **HH** Coniferous and mixed forests, windbreaks and woodland with clearings or adjacent open areas, from lowlands to foothills. Northern birds migratory/irruptive. Nocturnal/crepuscular, but day-roosts (often communal in winter) sometimes conspicuous. **ID** Medium-sized, 'elongated', slender and long-winged owl. Facial disc orange-buff, with white X formed by eyebrows and whiskers, rimmed narrowly black above and broadly white below. Alert birds erect long black and buff ear-tufts, though these may be barely visible when relaxed. Upperparts dark greyish-brown and tawny, with heavy blackish-brown streaking on mantle and wings, white scapular spots and strong grey barring on flight-feathers; extensive tawny-buff panel at base of primaries. Underparts buff with long, narrow dark shaft-streaks, with some side bars, pattern continues on belly (see Short-eared). Adopts upright camouflaged posture against tree trunk if disturbed. In flight, very similar to Short-eared, particularly when seen in low light and any subtle plumage distinctions difficult or impossible to assess; wings shorter and broader than Short-eared, orange-buff with dark bars; buff-brown tail is narrowly barred grey, upperwing dark with dark carpal patch, orange-buff primary base panel and 4–5 narrow blackish bands on grey flight-feathers (see Short-eared); underwing pale brown, with broad black carpal crescent, and 3–5 blackish bands on grey primary tips; undertail brown narrowly barred grey. Flight slow, wavering, even erratic, on long narrow wings, 'rowing' wingbeats interspersed with glides. **BP** Bill dark grey; eyes orange; tarsi feathered buff. **Vo** Territorial song comprises long, slow series of low moaning *boo boo boo*; *oo oo oo oo* or *hooom hooom* notes by ♂, occasionally in duet; ♀ gives higher, softer *sheoof* and sighing *hnyauu*. Alarm a nasal, barking *wrack wrack-wrack*. Juv. gives high, metallic squeaking *kee-kee*.

SHORT-EARED OWL CTKJR
Asio flammeus

L 35–41cm; WS 95–110cm; WT ♂ 200–450g, ♀ 280–500g. **SD** Very wide range in Holarctic and Neotropics. *A. f. flammeus* occurs across Russia to Chukotka, Kamchatka and Commander Is, to SE Russian Far East and Sakhalin; also NE China. Winters in much of E China; uncommon winter visitor (Oct–Apr) to Korea and Japan. **HH** Open, generally lowland, habitats from tundra and taiga to steppe and desert. Favours dry country or grasslands and reedbeds bordering swamps; in winter also in open farmland. Northern birds migratory/irruptive. Crepuscular, sometimes diurnal, especially in winter when commonly found at riversides and marshes, roosts on ground, or may perch in open on post. An owl of this type, hunting or disturbed in open country in winter in daylight, is probably Short-eared, whereas one roosting in a bush or tree and reluctant to be disturbed is likely to be Long-eared. **ID** Medium-sized owl, with large head and prominent facial disc with variable dusky patches around eyes. Face whitish-buff to dull brown with darker radial streaking, narrowly bordered white and has white chin; eyebrows and whisker lines pale grey to pale brown, and not particularly contrasting; ear-tufts rarely visible. Ground colour whitish-grey to warm brown and tawny: upperparts heavily mottled black and grey, some have fairly prominent scapular bar; mantle and wings small; buff panel at base of primaries. Darker on breast than belly, creamy to buff or pale brown with heavy streaks concentrated on upper breast, with long narrow streaks on lower breast (lacking side bars), unpatterned belly/vent. In flight, upperwing has white or grey trailing edge, black carpal, pale buff-brown flight-feathers with 2–3 black bands on primary tips (see Long-eared) and 3–4 narrow black bands on secondaries. Underwing very pale (white when seen against dark terrain), with broad blackish bands on primary tips, broad black carpal crescent, and black line to rear of axillaries formed by dark tertials; buff tail broadly barred black above, undertail narrowly barred black. Flight much like Long-eared, typically low, quartering like harrier, may hover or stall then drop onto prey. **BP** Bill dark greyish-black; eyes yellow (not orange); tarsi feathered, buff. **Vo** Typically silent, but ♂ gives series of deep, rather quiet hoots, *uh uh uh*…, or mellow *oop-oop* likened to Hoopoe, in brief series. Loud wing-clapping noise in rapid series during display flight. Also on territory gives loud *giyau* and softer *ge ge ge*…. Alarm a wheezy *chef-chef-chef*.

BROWN HAWK OWL

ad *japonica*

ad *japonica*

ad *totogo*

ad

**NORTHERN
LONG-EARED OWL**

otus

ad

SHORT-EARED OWL

flammeus

GREY NIGHTJAR *Caprimulgus jotaka*   CTKJR

L 27–32cm; WT 60–108g. **SD** Summer visitor Transbaikalia, S Russian Far East, E & NE China, Korea and Japan (May–Nov); accidental Taiwan; winters SE Asia, Indonesia and Philippines. **HH** Heavily forested areas with clearings, or at treeline in mountains, where crepuscular/nocturnal. **ID** The common nightjar of E Asia. Medium-sized, dark grey and grey-brown, heavily barred and streaked blackish-brown and grey, with few if any rufous tones except spots on wing-coverts and bars on primaries. Crown flat, face dark with greyish-black chin/throat and neck-sides, white malar curves below dark brownish-grey ear-coverts, and has broad white chin. Flight erratic; when flushed flies rather lazily. ♂ has white bases to outer primaries (much reduced and buffish-brown in ♀) and broad, white, subterminal band on undertail (absent in ♀). **BP** Bill (with very broad gape) grey; eyes black; tarsi grey. **Vo** Monotonous, repetitive *chuckchuckchuckchuck* or *kyokyokyokyokyo* at dusk, dawn and by night, from high perch or ground, commonly in prolonged series lasting several minutes. Sharp wing-clapping in flight. **TN** Formerly part of Jungle Nightjar *C. indicus.*

EUROPEAN NIGHTJAR   Extralimital
Caprimulgus europaeus

L 24.5–28cm; WS 52–59cm; WT 51–101g. **SD** Potential vagrant, as *C. e. dementievi* migrates from Africa to temperate breeding grounds as far east as S Transbaikalia, N China and NE Mongolia. **ID** Distinguished by slightly smaller size, voice (a continuous churring from perch and soft *quoit quoit* in flight), narrow, black lanceolate streaking on grey crown, nape and mantle, browner underparts and broad white tips to outertail-feathers of ♂.

SAVANNA NIGHTJAR *Caprimulgus affinis*   CT

L 20–26cm; WT ♂ 54–86g, ♀ 75–110g. **SD** India to Indonesia. *C. a. amoyensis* resident SE China; *C. a. stictomus* on Taiwan. **HH** Urban areas in tropics (roosts on roofs) or in dry open areas, grasslands near rivers and gravel riverbeds, from lowlands to 1,500m. Crepuscular/nocturnal. **ID** Small, rather pale buff-brown nightjar. Upperparts pale to mid-brown with blacker spotting, some rufous spotting on wing-coverts and extensively on flight-feathers; underparts dark buff-brown, barred darker, pale on vent. Extensive rufous-barred secondaries and inner primaries; ♂ has long white patch at base of four outer primaries and distinctive white outertail-feathers; ♀ has buff at base of primaries and lacks white outer tail. **BP** Bill greyish-black; eyes dark brown; tarsi grey-brown. **Vo** Hard, strident *dheet*, *chweep* or *chuck* when flushed. Song a slightly dry, high-pitched, twangy *twee-ik* that is slightly inflected, often repeated for several minutes in high flight or from rock perch. **AN** Allied Nightjar.

HIMALAYAN SWIFTLET   TKJ
Aerodramus brevirostris

L 13–14cm, WT 12.5–13g. **SD** Himalayas to C China and SE Asia to Java; resident and partial migrant. *A. b. innominatus* breeds close to region in EC China and has strayed to Taiwan, Korea and S Japan in spring. **HH** Typically occurs in high forested mountains, nesting in rock crevices, but ranges over various habitats including lowlands and cultivation. **ID** Small, blackish-brown swiftlet, with long, blunt-tipped wings, white outer-primary shafts, and short tail with shallow fork (albeit deep for swiftlet). Generally uniform, but broad rump band varies from greyish- to brownish-black, contrasting with darker brown mantle and blackish-brown tail; underparts generally paler greyish-brown, particularly on throat, scaled paler on undertail-coverts, and blackish-brown underwing-coverts contrast with paler flight-feathers. **BP** Bill black; eyes large, irides brown; toes black. **Vo** Various squeaking, buzzing, clicking and twittering notes, but vagrants usually silent. **TN** Often placed in *Collocalia.*

WHITE-THROATED NEEDLETAIL   CTKJR
Hirundapus caudacutus

L 19–20cm; WS 50–53cm; WT 101–140g. **SD** Himalayas and C Siberia east, wintering as far as Australia and New Zealand. *H. c. caudacutus* occurs NE China, E Russia from Lake Baikal to Amur River mouth, Sakhalin and Kuril Is; Korea, Japan (N Honshu and Hokkaido) as a summer visitor (Apr–Sep), and on migration through S Japan and Chinese coast. **HH** Not uncommon in montane regions with mature forest. Generally high overhead, but sometimes descends to lakes to drink, or over flooded fields or agricultural land to feed. **ID** The largest and longest-winged swift in region. Long scythe-shaped wings, bulky, rather long body, and short blunt tail, with fine needle-like points visible only at close range. Ad. generally rather blackish-brown, but has sharply defined white throat and large U-shaped vent patch which extends forwards on flanks to rear of wings; forehead and supraloral also narrowly white. Upperparts dark brown with conspicuous pale silvery-grey 'saddle', wings and tail glossed metallic green, white on inner tertial tips, and has metallic blue-green patch on innerwing-coverts and outer tertials. Juv. duller with less contrasting mantle. Similar Silver-backed Needletail has dusky greyish-brown throat, (not clearly defined white), more diffuse but brighter 'saddle', and lacks white on tertials. **BP** Bill short, black; eyes brown; toes black. **Vo** A high-pitched, shrill 'screaming' *tsuiririri juriri* similar to Pacific Swift; wings audible.

SILVER-BACKED NEEDLETAIL   T
Hirundapus cochinchinensis

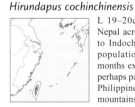

L 19–20cm; WT 76–86g. **SD** From Nepal across S China, wintering south to Indochina and Indonesia; isolated population on Taiwan, recorded all months except Dec (mostly Mar–Nov), perhaps partly resident or may winter in Philippines. Monotypic. **HH** Forested mountains of Taiwan, at low and mid elevations (200–2,500m). **ID** Very large swift, very similar to White-throated, but appears smaller, shorter bodied and slimmer. Weakly defined greyish-white throat with diffuse edge (can look darkish in field), more diffuse but brighter (almost white) 'saddle' visible at long range, and lacks white on tertials. Wings appear shorter and more cigar-shaped, with broader secondary bulge giving pinched-in appearance at base. **BP** Bill short, black; eyes brown; toes black. **Vo** Generally silent, but wings make audible 'swoosh' when low overhead (as does White-throated) also gives short, thin descending trills.

GREY NIGHTJAR

ad ♀

ad ♂

ad ♀
stictomus

SAVANNA
NIGHTJAR

ad ♂ *amoyensis*

ad ♂ *stictomus*

ad ♂
amoyensis

ad ♂
stictomus

ad

innominatus

HIMALAYAN
SWIFTLET

ad

caudacutus

WHITE-THROATED NEEDLETAIL

ad

SILVER-BACKED
NEEDLETAIL

(swifts not to scale)

COMMON SWIFT
Apus apus

CKJR

L 16–19cm; WS 40–44cm; WT 36–52g. **SD** From NW Europe across Russia to Lake Baikal and NE China; winters in Africa. *A. a. pekinensis* breeds NE China and Transbaikalia; winters in Namibia and Botswana. Likely on spring migration anywhere in NE Asia; reported from both Japan (e.g. Yaeyama and Hegura) and Korea. **HH** Lowlands to mountains, in habitats ranging from desert and arid steppe to temperate and boreal forested zones, also urban centres. Nests colonially on cliffs or under roofs of buildings; frequently forages in flocks and often highly vocal. **ID** Medium-sized, uniform dark blackish-brown swift with all-dark rump and only slightly forked tail, and lacks contrast on upperwing. Throat white, forehead pale grey. **BP** Bill black; eyes brown; very short tarsi black. **Vo** Gives shrill buzzing screams during high-speed display flight, sometimes in screaming parties: *vzz-vzz vzzzz*, or *sriiirr*.

PACIFIC SWIFT
Apus pacificus

CTKJR

L 17–18cm; WS 43–54cm; WT ♂ 42.5g, ♀ 44.5g. **SD** NE Asia from upper Ob River northeast to Kamchatka, also China; winters south to Indonesia and Australia. *A. p. pacificus* is a summer visitor (Apr–Oct) NE Russia to Yakutia, Chukotka and Kamchatka, Kuril Is, Sakhalin, S Russian Far East, Japan, Korea and NE China; *A. p. kanoi* resident in SE China, Taiwan and Lanyu I. **HH** Tropics to Arctic, lowlands to high mountains, where nests colonially in cliff crevices, but also under roofs of buildings; often forages in flocks and usually highly vocal. Common in breeding range and on migration. **ID** Medium-sized rather long and slender-bodied swift, with long, narrow sickle-shaped wings. Plain blackish-brown (usually appears black) with narrow, but very conspicuous white rump band and deeply forked tail; tail and primaries blackest, mantle and wing-coverts slightly browner, tertials and secondaries paler still. Throat whitish, with some paler, browner scaling on flanks, belly and vent, most marked in young birds. *A. p. kanoi* blacker with greenish sheen to upperparts, smaller rump and throat patches, and less deeply forked tail. **BP** Bill black; eyes brown; very short tarsi black. **Vo** Small parties and flocks close to breeding sites commonly give shrill, high-pitched, trilling screams: *tsiririri* or *juriri* and harsh *spee-err*; calls are more sibilant, less buzzy than Common Swift. **AN** Asian White-rumped Swift.

HOUSE SWIFT
Apus nipalensis

CTKJ

L 12–15cm; WS 28–35cm; WT 20–35g. **SD** Nepal to SE Asia, Philippines, SE China and Japan. *A. n. nipalensis* presumed resident in SC Japan (mostly coastal, though some evidence of local migration) and common resident of coastal SE China; *A. n. kuntzi* is common resident of Taiwan. **HH** Colonial; favours montane and coastal cliffs and caves, but also nests regularly on buildings in coastal towns (and urban areas in Taiwan). Locally common

in breeding range and on migration. **ID** Small, rather stocky, blackish-brown swift (usually appears black) with broad white rump band wrapping onto flanks, and short, slightly notched tail, the notch disappearing when fanned in rather fluttering flight. Wings shorter, broader and broader tipped, less scythe-like than other swifts, and trailing edge to wing often appears straight. Throat greyish or whitish. *A. n. kuntzi* has heavily streaked rump. **BP** Bill black; eyes brown; toes black. **Vo** High-pitched twittering trill, *chiiririri* or *jurirri*, and soft screaming *vzz-vzz* like weak Common Swift. **TN** Formerly within Little Swift *A. affinis*.

RUFOUS HUMMINGBIRD
Selasphorus rufus

R

L 9.5cm; WS 11cm; WT 2.9–3.9g. **SD** Winters in Mexico and breeds in Pacific NW of USA and Canada, as far as SC Alaska, with spring overshoots reaching as far afield as Chukotski Peninsula, Russia. Monotypic. **ID** Compact, rather short-winged hummingbird with mainly orange tail. ♂ largely orange-rufous from crown to rump/uppertail-coverts, scapulars and wing-coverts deep green, flight-feathers black (a few ♂ have splotchy green crown and mantle); tail extends beyond wings. Metallic rufous-orange chin and throat may appear dark wine red or black, depending on light; face orange; upper breast white, rest of breast and belly orange-rufous, vent white. ♀/juv. have green crown, nape, mantle, back and rump, but extensive orange-rufous at base of tail, which has black subterminal band and white tips to three outer feathers. Chin and throat largely white, with orange-red central spot. **BP** Long slender black bill is longer than head; irides black, with white teardrop behind eye; tarsi extremely short, black. **Vo** A hard chipping *tyuk*.

RED-HEADED TROGON
Harpactes erythrocephalus

C

L 31–35cm; WT 75–110g. **SD** Himalayas to Sumatra, north to SE China. In E Asia *H. e. yamakenensis* in Guangdong and Fujian. **HH** Uncommon resident of dense tropical and subtropical forest (600–1,500m) including areas with bamboo. Perches still and very upright in lower and middle canopy, sallying out for food. **ID** Large, long-tailed and pigeon-sized trogon. ♂ has deep red hood with black lores, divided from red underparts by narrow white and broader rufous-brown pectoral bands. Upperparts and uppertail rufous-brown, wings black with narrow white barring on wing-coverts and prominent white primary shafts. ♀ has cinnamon-brown hood to mid chest, rest of underparts pale red. Uppertail of ♂ and ♀ has black outer border and black tip; undertail graduated and strongly pied, black with white edges and large white tips. Wings short, flight undulating. **BP** Bill short, broad, dark blue-grey with prominent black rictal bristles; eye-ring deep blue, irides dark red; tarsi grey. **Vo** Series of short, hollow downslurred *tyaup* notes repeated 5+ times, a more melancholy *pluu-du* and grating or croaking alarm, *tewirrr*.

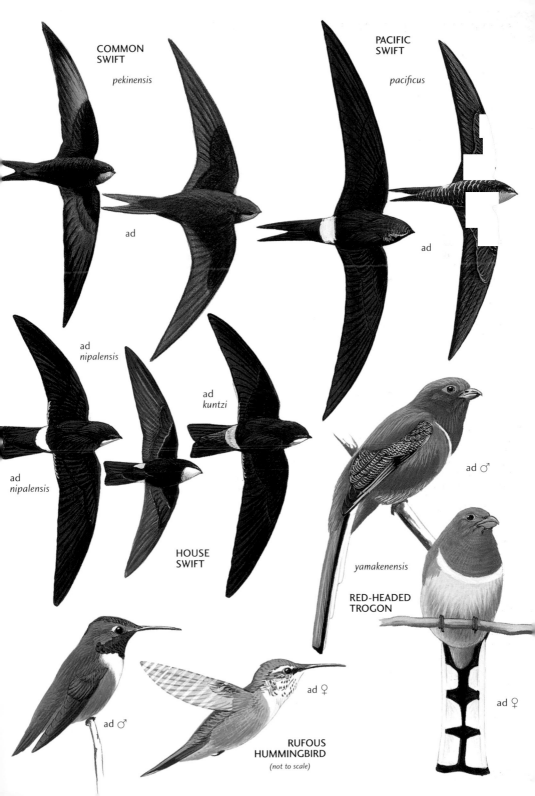

COMMON SWIFT

pekinensis

ad

PACIFIC SWIFT

pacificus

ad

ad
nipalensis

ad
nipalensis

ad
kuntzi

HOUSE SWIFT

RED-HEADED TROGON

yamakenensis

ad ♂

ad ♀

RUFOUS HUMMINGBIRD

(not to scale)

ad ♂

ad ♀

DOLLARBIRD *Eurystomus orientalis* CTKJR

L 27–32cm; WT 109–186g. **SD** India to Australia and E Asia. *E. o. calonyx* is a widespread, uncommon to locally common summer visitor (May–Sep) to S Russian Far East, Japan south of Hokkaido, Korea, NE & E China, and migrant (May and Sep) through Taiwan (has attempted to breed). **HH** Forest and woodland edge, especially in river valleys, typically dependent on natural nesting cavities. Sits upright on prominent perch, frequently on treetops or wires; sallies for large insects. **ID** Medium-sized, stocky and large-headed bird; rather long-winged and short-tailed. Ad. head, flight-feathers and tail blackish-brown, mantle, wings and rump dark green tinged blue, with bright purplish-blue carpal and edge of closed wing; underparts bluish-green with brighter cobalt-blue throat. Juv. darker with poorly defined wing patch. In flight, rather long, broadly rounded wings with whitish or pale blue patches (the size of old dollar coins, hence name) near base of primaries. Flight loose, with somewhat floppy beats. **BP** Bill short, broad-based, red (black above, reddish-orange below in young); narrow eye-ring red, irides brown; toes red. **Vo** Harsh, guttural *khya khya-a*, a hoarse, noisy *shraak* or *grek* and in flight a cackling *ge ge gegegeegegege* or shorter *cher-cher*. **AN** Broad-billed Roller.

RUDDY KINGFISHER CTKJR
Halcyon coromanda

L 25–27cm; WT 60–92g. **SD** Largely resident, locally migratory, from NE India to Indonesia. Summer visitor to NE Asia, wintering south to Philippines: *H. c. major* in Japan north to W Hokkaido (May–Aug), Korea (scarce) and adjacent Jilin, a migrant through coastal E China, vagrant Sakhalin; *H. c. bangsi* resident in Nansei Shoto, Taiwan and Lanyu I. **HH** Scarce, solitary and rather secretive, but vocal; favours mature deciduous or dense broadleaf evergreen forest, with lakes, rivers and streams. **ID** Large, rather uniform ruddy-orange kingfisher. Upperparts generally orange or rufous-brown, with distinct violet sheen on upper back and bright, narrow, turquoise patch on lower back/rump. Underparts paler rufous-orange, chin/throat almost white and vent pale. *H. c. bangsi* overall darker with more extensive violet or purplish sheen to upperparts, neck and chest than *H. c. major*, with a larger pale blue rump patch. **BP** Bill very large, especially deep at base, orange-red; eyes brown; toes orange-red. **Vo** ♂ sings mainly early morning from concealed perch in canopy but also during day when raining. Song a high-pitched but descending, rolling whistled trill: *pyorrrr* or *kyorororo*; also an explosive rattle, but silent outside breeding season.

WHITE-THROATED KINGFISHER CTJ
Halcyon smyrnensis

L 27–28cm; WT 75–108g. **SD** E Mediterranean and Middle East across S Asia to S China, SE Asia and Philippines. *H. s. fokiensis* (perhaps = *H. s. fusca*) of S China is common resident Fujian, and has strayed to Taiwan (also occurs on Kinmen and Matsu) and Yaeyama Is (Apr and May). **HH** Typically associated with freshwater and coastal wetlands: marshes, lakes, ponds, rivers and wet ricefields. **ID** Large, boldly patterned, brown, blue and white kingfisher. Ad. deep chocolate-brown head and under-

parts, with extensive white patch from chin on centre of breast; mantle, back, wings, rump and tail bright blue. Juv. duller with paler bill and dusky breast markings. In flight, contrasting dark brown wing-coverts, blue secondaries, and white primaries with black tips. **BP** Bill very large, especially deep at base, bright red; eyes brown; toes red. **Vo** A sharp, woodpecker-like *kyo, kya, chik* or *kit*, a thin, short *pi* and angry rattling *krich-krich...*. Song a descending tittering trill recalling Ruddy Kingfisher, but higher pitched, *tiy-dy-y-y-y-y-y-y*, and loud cackling *chake ake ake ake ake*. **AN** Smyrna Kingfisher.

BLACK-CAPPED KINGFISHER CTKJR
Halcyon pileata

L 28–30cm; WT 67–91g. **SD** Largely resident, from India to Indonesia. In region, resident S China but summer visitor to E China north to Liaoning, and Korea, rare migrant and winter visitor to Taiwan, rare migrant (Apr–Jun) Japan and S Russian Far East. Monotypic. **HH** Temperate regions, where typically associated with deciduous woods and wetlands. In subtropics and tropics, more varied lowland habitats from coastal mangroves, wetlands and woods to inland forest clearings and cultivated areas with water; commonest below 600m in China. On migration, wide range of wooded and wetland habitats. **ID** Large, boldly patterned, blue, black, white and rufous kingfisher. Ad. head black with white chin, collar and upper breast; mantle, rump and tail deep blue, as are secondaries and primary-coverts (mantle and upperwing have purple-violet sheen); remaining wing-coverts black; primaries white with black tips, but hidden at rest. Lower breast, belly, flanks and vent rufous-orange. Juv. duller (including bill) with buff collar and dusky scaling on breast. In flight, contrasting black wing-coverts, blue secondaries, white primaries with black tips. **BP** Bill very large, especially deep at base, dull red; eyes brown; toes dark red. **Vo** A short *ki ki*, and in breeding season a louder, fast, hard rattle, *kyoro kyoro...* or *kikikikikiki*, recalling White-throated but less grating.

COLLARED KINGFISHER CTJ
Todiramphus chloris

L 23–25cm; WT 51–100g. **SD** Ranges from NE Africa through SE Asia to Indonesia, Philippines and N Australia. Generally resident, but has strayed to coastal Fujian and Jiangsu, Taiwan, even SW Japan in winter (Oct–Mar). Of 50 subspecies, unclear which has reached region, but *T. c. collaris* of Philippines perhaps most likely. **HH** Typically associated with coastal wetlands, especially mangroves, beaches and wooded shores, also cultivated lowlands. **ID** Medium-sized, bright turquoise-blue and white kingfisher, with distinctive white collar. Upperparts, from crown to tail, iridescent greenish-blue, whilst underparts are white. Broad black band from base of bill to nape separates white underparts from green-blue crown; also has white patch between eye and base of bill. **BP** Bill very large, especially deep at base, upper mandible and tip black, lower mandible whitish/horn; eyes brown; toes dark grey/black. **Vo** During breeding season gives a chattering, rattling *kii kii kii kii*; calls include strident *krek-krerk*. **TN** Formerly *Halcyon chloris*. **AN** Mangrove Kingfisher.

DOLLARBIRD

calonyx

ad

RUDDY KINGFISHER

ad *major*

ad *bangsi*

WHITE-THROATED KINGFISHER

fokiensis

ad

ad *bangsi*

BLACK-CAPPED KINGFISHER

ad

COLLARED KINGFISHER

ad

ad

collaris

ORIENTAL DWARF KINGFISHER · T
Ceyx erithaca

L 14cm; WT 14–21.5g. **SD** India to Indonesia; in region has strayed to S Taiwan. *C. e. erithaca* from India and SE China to Sumatra likeliest subspecies. **HH** Occurs along streams or at pools within various forest types. **ID** A tiny, bright rufous and black kingfisher, with a large head and very short tail. Like miniature Ruddy Kingfisher (p.274). ♂ largely rufous-orange, but with a black forehead, blue-black mantle, black wings and violet sheen to crown and to rump (although rufous-backed morph also occurs). ♀ similar but lacks violet sheen to crown. Juv. is a duller version of ad. **BP** Bill long, red; eye dark brown surrounded by black eye-patch; toes red. **Vo** Call consists of very high-pitched, shrill *tsriet-tsriet* or *tseet* and softer *tjie-tjie-tjie* notes given when perched and in flight. **AN** Black-backed Dwarf Kingfisher.

COMMON KINGFISHER · CTKJR
Alcedo atthis

L 16–20cm; WS 24–25cm; WT 19–40g. **SD** Ranges from W Europe to E Asia. *A. a. bengalensis* a fairly common resident across SE Russia from Transbaikalia to Amur Estuary, Sakhalin and Kuril Is, NE China, Korea, Japan and Taiwan, northern birds migratory. **HH** Wide range of wetlands, lakes and ponds, wooded streams and rivers, where flow nil or gentle. On migration and in winter also on coasts, at harbours and estuaries. Typically plunge-dives for small fish, either from secluded perch or from hovering flight. **ID** Small, blue, green and orange kingfisher. Ad. has bright metallic green head and wings, scaled dark green and pale blue on crown and spotted turquoise on upperwing-coverts; upper back, rump and tail shining blue. Orange spot on lores and band across ear-coverts, white neck-sides and chin; otherwise entire underparts and underwing-coverts bright orange. Juv. duller with dark breast-band and black toes. Flight low, fast and direct. **BP** Bill long, black in ♂, but orange-based in ♀; eyes brown; toes red. **Vo** Often calls in flight, a strongly whistled *tiii, tjii* or *peeet*; repeats calls rapidly or gives series of hard *chit-it* or *tee titi titi titi* notes when agitated. **AN** River Kingfisher.

BLYTH'S KINGFISHER · C
Alcedo hercules

L 22-23cm. **SD** Ranges from E Nepal to S China; scarce in region, locally in NW Fujian; found recently during breeding season. Monotypic. **HH** Occurs along shady streams in broadleaved evergreen hill forest (to *c.*900m). **ID** Like a large version of Common Kingfisher; bright blue above and rufous below. Green on face, neck and wings, with fine pale blue spangles across crown and nape. Lores orange, white patch on rear ear-coverts, blue spots on wing coverts and brilliant blue back, rump and short tail. Creamy white on chin/throat, bright rufous-orange from neck to undertail-coverts. **BP** Bill long, heavy, black (base red in ♀); eyes dark brown; legs and feet orange. **Vo** A loud, sharp whistled *chhee*, similar to Common, but deeper and more hoarse.

CRESTED KINGFISHER · CKJR
Megaceryle lugubris

L 41–43cm; WT 230–280g. **SD** Himalayan foothills to E China, Japan and Kuril Is. *M. l. guttulata* throughout E China, northeast to Jilin, and formerly regular but now rare N Korea; *M. l. lugubris* through most of S & C Japan, from Kyushu and Shikoku to N Honshu; *M. l. pallida* local resident in Hokkaido that has strayed to Kunashir (Russia); those in Liaoning (NE China) either *M. l. lugubris* or *M. l. guttulata*. **HH** Scarce resident, favouring cold, fast-flowing rivers and streams in forested mountains, moving downriver, rarely to coasts, in winter. **ID** Very large grey and white kingfisher, rather stout with an overly large, crested head and erectile plumes from forehead to nape. Overall, black above and white below; upperparts narrowly barred white on shaggy crest, back and wings, and more broadly barred white on rather long tail. Largely white underparts have band of grey streaks from malar region to breast-sides, and broader band, similarly grey with white, on upper breast. ♂ has some rusty-orange in malar stripe and chest band, and white underwing-coverts. ♀ lacks orange but has rusty underwing-coverts. Wings broad, flight steady, direct. *M. l. pallida* larger and paler grey overall than other races. **BP** Bill long, largely black but blue-grey at base and horn-coloured at tip; eyes dark brown; toes grey. **Vo** Loud, grating, chattering *kyura kyura* or *kyara kyara* given by both sexes, a powerful *chek chek* or *ket ket* and wader-like *wick* or *pik pik-wik* in flight. **TN** Formerly *Ceryle lugubris*. **AN** Greater Pied Kingfisher.

PIED KINGFISHER · CT
Ceryle rudis

L 25–31cm; WT ♂ 68–100g, ♀ 71–110g. **SD** Wide-ranging Afro-Asian species of warm or tropical regions. *C. r. insignis* occurs in E China south of Changjiang River, also Kinmen (Taiwan), but not mainland Taiwan. **HH** Sociable, in pairs or noisy groups at wetlands including wet agricultural fields, rivers, estuaries and mangroves. Commonly hovers above water when foraging. **ID** Medium-sized black and white kingfisher with loose crest restricted to hindcrown. Smaller, more slender, with smaller head and longer, thinner bill than Crested Kingfisher. Ad. upperparts black, heavily spotted white, supercilium and neck-sides white, both sexes with black bar on side of neck. ♂ has two black chest-bands, one broad, the other narrow. ♀ has single breast-band, often broken, thus appearing as black breast-side patches. Juv. like ♀, with brown fringes to feathers of face, throat and breast. **BP** Bill long, black; eyes brown; toes black. **Vo** Various loudly whistled chattering notes, including *tik twikik; cheek* and *chit-it-it*, either in flight or from perch; and a low-pitched *trrr* in alarm. **AN** Lesser Pied Kingfisher.

ORIENTAL DWARF KINGFISHER

erithaca

ad ♂

COMMON KINGFISHER

ad ♂

ad ♀

bengalensis

ad ♀

ad ♀

ad diving for fish

BLYTH'S KINGFISHER

ad ♂

CRESTED KINGFISHER

ad ♀ *guttulata*

PIED KINGFISHER

insignis

ad ♀

ad ♂

ad ♂

BLUE-TAILED BEE-EATER   CT
Merops philippinus

L 29–36cm; WT 29–43g. **SD** Indian subcontinent to SE Asia and New Guinea. *M. p. javanicus* occurs north to SE China, in S Fujian and Kinmen I. (Taiwan), but not mainland Taiwan. **HH** Wide range of habitats from open country to woodland edge, farmland and even suburban gardens. **ID** Largely green with blue lower back, tail and elongated (up to 7cm) central rectrices. Ad. has crown, nape and mantle green, black eyestripe bordered below with blue; chin and cheeks yellow with brown half-collar on foreneck. Green breast to belly, but undertail-coverts blue, like lower back, rump and tail (see Blue-throated Bee-eater). Juv. duller and bluer than ad., with white fringes to feathers of head and body. **BP** Bill long, slender, curved, black; eyes bright red (ad.) or brown (juv.); short tarsi greyish-black. **Vo** Gives *cheer-it* call singly or in trilled or rattled series in flight.

RAINBOW BEE-EATER   TJ
Merops ornatus

L 19–28cm; WT 20–33g. **SD** Migrant that ranges from S Australia to New Guinea, the Lesser Sundas and Sulawesi, with southern breeders wintering in north of range; has strayed north to Taiwan and Japan. Monotypic. **ID** Medium-sized blue, green and brown bee-eater with elongated central tail-feathers (7cm). Ad. similar to Blue-throated Bee-eater, but crown suffused rusty-brown, chin and throat grade from yellow to brown and black, rump and vent azure-blue, tail black. Stripe from bill through eye black and very broad on ear-coverts, where bordered below by blue; underparts pale green. ♂ has long, narrow tail-streamers with spatulate tips. ♀ has shorter, broader central rectrices. Juv. recalls ♀; upperparts olive with narrow brown nuchal collar; chin and throat pale yellow, lacks black gorget of ad., lower throat dull rufous-brown merging into pale olive breast. **BP** Bill long, slender, curved, black; eyes dark red (♂) or reddish-brown (♀); toes black. **Vo** A melodious, fluty *piru*. **AN** Australian Bee-eater.

BLUE-THROATED BEE-EATER   CT
Merops viridis

L 21–30cm; WT 34–41g. **SD** Indonesia, Borneo and Philippines to S China. *M. v. viridis* is a summer visitor north to coastal Zhejiang, and accidental on Kinmen I. (Taiwan). **HH** Sociable, colonial, lowland species of open country with woods and along coasts. **ID** Medium-sized blue, green and brown bee-eater with greatly elongated (up to 9cm) central tail-feathers. Ad. has black stripe from bill through eye to ear-coverts, barely contrasting with dark chocolate-brown crown to mantle; blue lower back, rump and tail, dark green wings with blacker, green-fringed primaries. Chin, throat and face-sides blue; breast and belly pale green. Wings long, pointed and, in flight, have black band on trailing edge. Juv. lacks chocolate-brown, instead crown and mantle dark green; chin and cheeks range from straw-coloured to pale blue, breast and belly blue. **BP** Bill long, slender, curved, black; eyes reddish-brown; toes grey. **Vo** Series of 3+ loud, clear *phil-ip phil-ip phil-ip*; *pit pit pit* or *kerik-kerik-kerik* notes given in flight.

HOOPOE   CTKJR
Upupa epops

L 26–32cm; WS 44–48cm; WT 47–89g. **SD** Widespread Afro-Palearctic and Oriental species; a summer visitor from Scandinavia to Russian Far East, wintering in Africa, S & SE Asia, China south of the Chiangjiang River. *U. e. saturata* (perhaps = *U. e. epops*) summers from SC Russia to Transbaikalia and S Russian Far East, C, E & NE China, and Korea. Year-round in SE China. A scarce, early spring overshoot migrant to Japan (Feb–Jun; has bred), and scarce in Taiwan. **HH** Favours open woodland, forest edge, groves and thickets, especially in river valleys, and in parks and gardens. Commonly forages for invertebrates in dry soil or sandy ground. Largely dependent on natural nest cavities, but sometimes utilises holes in buildings. **ID** Unmistakable medium-sized bird, with small head, large erectile crest and long decurved bill, giving strange hammer-headed appearance. Ad. overall pinkish-buff, mantle darker buff-brown; wings barred black and white; crest longer than head, usually held flat, comprises long loose graduated feathers each tipped black, forming barring across crown. Juv. duller with shorter crest and bill. In flight, which is slow and undulating, reveals broad, barred, square-ended wings, white lower back, and black tail with one broad curving white band across middle; erects fan-like crest on alighting (and in aggression). **BP** Bill long (5–6 cm), thin, decurved, black, greyish-pink at base; eyes brown; toes black. **Vo** Harsh *guwaai* call, but territorial song a muted series of hollow trisyllabic hoots, *oop-oop-oop oop-oop-oop*....

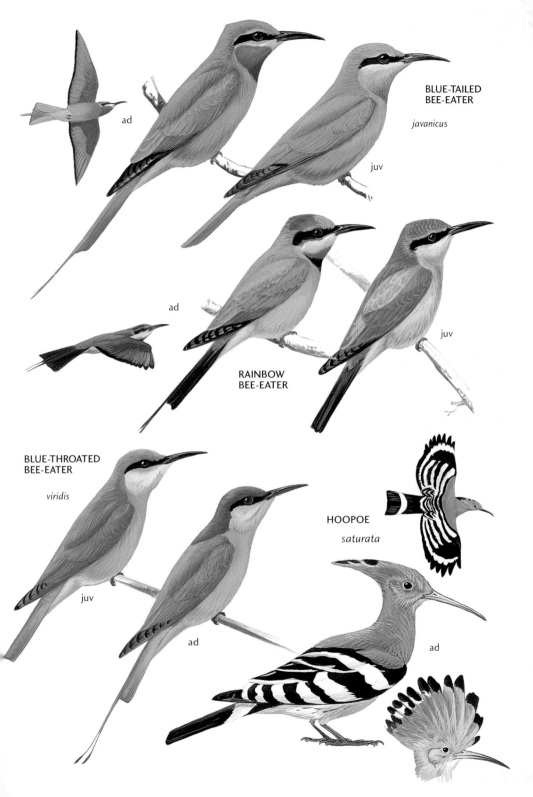

ad

**BLUE-TAILED
BEE-EATER**

javanicus

juv

ad

juv

**RAINBOW
BEE-EATER**

**BLUE-THROATED
BEE-EATER**

viridis

juv

ad

HOOPOE

saturata

ad

GREAT BARBET C
Megalaima virens

L 32–35cm; WT 164–295g. **SD** NW Himalayas to SE Asia. *M. v. virens* occurs in S & E China south of Changjiang Valley. **HH** Deciduous and evergreen forests, from sea level to *c.*2,400m. **ID** Very large, bull-headed, green and brown barbet with very large pale bill. Head deep steel-blue with black forehead, lores and chin, mantle brown, wings, rump and tail green. Underparts yellow streaked brown, green on breast and flanks, with red vent. Flight deeply undulating like large woodpecker. May gather in crowns of fruiting trees to forage or call. **BP** Bill large, stout, pale horn or yellow with grey tip to upper mandible; eyes dark brown; legs short, tarsi grey. **Vo** Various calls include a churring, screechy *keeah*, and loud territorial *kay-oh* (repeated regularly at one-second intervals), and a faster, mournful *peeao- peeao-peeao-peeao*....

TAIWAN BARBET T
Megalaima nuchalis

L 20–22cm; WT 62–123g. **SD** Endemic to Taiwan. Monotypic. **HH** Common and widespread in all forest types from lowlands to mountains, but also in sub-urban and even urban areas (0–3,000m). Sometimes forms flocks. Nests in tree holes. **ID** Bright green with a rather large blue head and large bill, broad rounded wings and rather broad, short tail. Head mainly pale to mid blue, with very narrow black forehead, pale yellow forecrown (shading to white), pale blue crown (shading to green), mustard-yellow throat and red loral spot, with black band curving back from eye to nape-sides, and narrower black lateral throat-stripe. Also has red spot on mantle. Underparts largely green, somewhat yellowish-green on belly and vent, with narrow turquoise-blue upper-breast-band and narrow red crescent forming lower-breast-band. Often perches motionless in foliage, but flight rapid, audibly whirring and woodpecker-like. **BP** Bill large, stout, black with prominent rictal bristles; eyes brown; legs short, tarsi dark grey. **Vo** Both sexes give somewhat muffled, whinnying, hollow whistle, *pyorrrrrr*, sounding rather frog-like and often repeated. More frequently gives repetitive series of popping notes, *ko ko ko...*, *pup pup pup... or tokaratok tokaratok*. **TN** Formerly placed within Black-browed Barbet *M. oorti*. **AN** Mueller's Barbet.

EURASIAN WRYNECK CTKJR
Jynx torquilla

L 16–17cm; WT 30–50g. **SD** Summer visitor from W Europe across Russia to E Asia, wintering in Africa and S & SE Asia. In region summer visitor to N & NE China, northern Korea, and E Russia from Baikal to Sakhalin and S Russian Far East, northeast to Yakutia and NE Sea of Okhotsk; also Hokkaido (May–Sep), scarce (overlooked?) migrant through Korea and Japan, where occasionally winters (Oct–Apr), from central Honshu to Kyushu, and in SE China; rare Taiwan. Monotypic. **HH** Open deciduous woodland, riparian thickets, forest edges, hedgerows and clumps of trees in cultivated land,

and in winter also in reedbeds. **ID** Small, highly cryptic, rather atypical long-tailed woodpecker. Overall grey-brown, upperparts greyer with broad blackish-brown band on crown and back, also from lores to ear-coverts, neck and scapulars, and malar region. Underparts warmer buffy-grey with narrow dark brown bars from chin to flanks. Wings rusty-brown with blackish spotting, and blackish barred flight-feathers; tail long, grey and rounded at tip, with 3–4 narrow blackish-brown bars. Does not drum. Usually feeds on ground. **BP** Bill conical, short, pointed, horn-coloured; eyes brown; toes grey-brown. **Vo** On territory, rather high-pitched, repetitive ringing *quee-quee-quee* or *kyii kyi kyi kyi* reminiscent of Lesser Spotted Woodpecker, but lower and hoarser. When disturbed at nest gives a hissing *shuu shuu* while writhing its neck.

SPECKLED PICULET C
Picumnus innominatus

L 10cm; WT 9–13g. **SD** Resident from Himalayas to S & E China, ranging as far as Sumatra and Borneo. *P. i. chinensis* in coastal China north to Changjiang River. **HH** Low hills, in thickets in mixed deciduous forest. Forages on bark of branches and twigs. **ID** Tiny, very active woodpecker, with tiny bill and short tail. Plain olive-green upperparts become somewhat browner on crown and mantle, with masked appearance created by dark dull olive-grey to black ear-coverts and submalar stripe contrasting with long pale supercilium (to nape), malar stripe and chin. Face superficially appears barred black and white; forehead orange streaked black; underparts buff/grey with prominent black scaling from lores to flanks. ♀ lacks orange forehead. In flight, wings plain (unbarred), short black tail with white outer and central feathers distinctive. **BP** Bill short, fine, conical, grey; eyes brown, surrounded by pale blue-grey orbital skin; tarsi blue-grey. **Vo** A sharp *tsit*, high-pitched twittering *chi-tititititi*, squeaky *sik-sik-sik*, but also a loud, persistent tapping that sometimes becomes a loud 'drum roll'.

RUFOUS-BELLIED WOODPECKER CKJR
Hypopicus hyperythrus

L 20–25cm; WT 53–74g. **SD** Largely resident, from NW Himalayas to SE Asia, with migrant population (*H. h. sub-rufinus*) in Heilongjiang (NE China) and SE Russian Far East (possibly N Korea), passing through NE & E China, and wintering SE China; rarely C & S Korea and accidental Japan. **HH** Mixed broadleaf and coniferous forests. **ID** Rufous, black and white woodpecker. ♂ has bright red cap and nape, white face, black back, rump and tail (mantle barred white), wing-coverts tipped white and flight-feathers barred white (no white shoulder patch). ♀ has white-spotted black crown. Underparts rufous-orange, pale red on vent. Juv. black above spotted white, with dark buff underparts heavily mottled black. **BP** Bill long, dark grey upper mandible, yellow below; eyes reddish-brown; tarsi grey. **Vo** A rapid *ptikitititit* and *tik-tik-tik-tik*, whilst territorial ♂ gives rapid, sputtering rattle, *ki-i-i-i-i-i*; *chit-chit-chit-r-r-r-h* or *kirritrick*. Also a more nasal *twicca-twicca-wicca-wicc-wicca*. Drumming short and accelerating, fading at end. **TN** Formerly placed in *Dendrocopus*.

GREAT BARBET
(not to scale)

virens

ad

TAIWAN BARBET

ad

EURASIAN WRYNECK

ad

RUFOUS-BELLIED WOODPECKER

subrufinus

ad ♀

ad ♂

SPECKLED PICULET

chinensis

ad

JAPANESE PYGMY WOODPECKER CKJR
Yungipicus kizuki

L 13–15cm; WT 18–26g. **SD** S Russian Far East, Sakhalin, S Kuril Is, Japan, Korea and NE China. Several races (some controversial): *Y. k. permutatus* Liaoning, N Korea, SE Siberia; *Y. k. seebohmi* Sakhalin, Hokkaido, S Kurils; *Y. k. nippon* Shandong, Hebei, S Korea, Cheju-do, Honshu; *Y. k. shikokuensis* SW Honshu, Shikoku; *Y. k. kizuki* Kyushu; *Y. k. matsudairae* Yaku-shima, Izu Is; *Y. k. kotataki* Tsushima, Oki Is; *Y. k. amamii* Amami-Oshima; *Y. k. nigrescens* Okinawa; *Y. k. orii* Iriomote. **HH** Lowland to subalpine (to 2,100m) woodland, various forest types, also riparian thickets, scrub, urban parks and gardens; especially forages on thinner branches; often with mixed species flocks. **ID** Small, dusky, ladder-backed woodpecker. Upperparts dusky-brown, greyer on crown, mantle plain, but back narrowly barred blackish-brown and white, tail black. Face dusky-brown with pale supercilium merging into whitish patch on sides of hindneck; malar region, chin/throat and upper breast also whitish, lateral throat-stripe dusky-brown merging into dark neck-sides and dusky underparts, which have darker spotting. Underparts off-white streaked grey-brown in northern races, buff with heavy dark brown streaking elsewhere. Becomes smaller and darker in N–S cline, southern populations also having heavier streaked underparts. ♂ has tiny red spot on sides of hindcrown, visible only when wind ruffles feathers or in territorial disputes. **BP** Bill short, sharply pointed; eyes brown to reddish-brown (age-related); short tarsi blackish-grey. **Vo** Calls frequently, sharp *khit* or *khit-khit-khit* notes, more frequently a highly distinctive buzzy *kzzz kzzz* or an agitated *kikikiki*. Drumming extremely faint; feeding taps can be confused with foraging Varied Tit battering nuts. **TN** Formerly placed in *Dendrocopos*.

GREY-CAPPED PYGMY WOODPECKER
Yungipicus canicapillus CTKR

L 14–16cm; WT 20–27g. **SD** Across southern C Asia, S & SE Asia to Sulawesi and Borneo, and across China, Taiwan, Korea to S Russian Far East. *Y. c. doerriesi* SE Siberia, E Manchuria, Korea; *Y. c. scintilliceps* E China, Zhejiang to Hebei; *Y. c. nagamichii* S China east to Fujian; *Y. c. kaleensis* Taiwan. **HH** Uncommon resident in mixed deciduous lowland and montane woodland (c.100–2,500m in Taiwan), scrub and gardens. Sometimes with mixed-species flocks. **ID** Small, robust, rather dusky woodpecker; more pied than Japanese. Upperparts contrasting, with plain grey crown, black nape and mantle, but white back; wings black with extensively white-barred coverts and flight-feathers; tail black. Face off-white with dusky-buff ear-coverts and lateral throat-stripe; chin/throat whiter, grading into dusky-buff breast/belly with black streaking on chest and flanks; buff or white vent. In flight shows square, grey-white back patch. *Y. c. doerriesi* is largest, most distinctive race; black mantle/upper back contrast strongly with white lower back/rump, and white wing-covert patches recall Great Spotted (p.284). *Y. c. scintilliceps* slightly smaller, with white-barred upper back and less white in wings. *Y. c. kaleensis* recalls *scintilliceps*, but has little white on wing-coverts and heavily streaked buff underparts. **BP** Bill blackish-grey, short and sharply pointed; eyes reddish-brown; short tarsi black. **Vo** Drumming comprises weak taps. Otherwise, a strong *wick* (like weak Great Spotted or White-backed), a softer *cheep*, a rattled *tit-tit-erh-r-r-r-r-h* and squeaky *kweek-kweek*. **TN** Formerly placed in *Dendrocopos*. **AN** Grey-headed Pygmy Woodpecker.

LESSER SPOTTED WOODPECKER CKJR
Dryobates minor

L 14–16cm; WS 24–29cm; WT 16–26g. **SD** NW Europe to NE China and E Russia. *D. m. kamtschatkensis* Urals to N Sea of Okhotsk, Kamchatka, Koryakia, Anadyr basin; *D. m. amurensis* NE China, northernmost Korea, SE Siberia, Sakhalin, Hokkaido. **HH** Deciduous and mixed lowland woodland (often oaks or alders), near rivers, wetlands, or at forest edge. Quiet, unobtrusive; commonly forages at mid-level in trees, on underside of larger branches and thin snags. **ID** Smallest pied woodpecker. Small rounded head and fine bill contribute to gentle appearance. Superficially like Great Spotted, but smaller and more obviously pied (*kamtschatkensis*) or less sharply pied (*amurensis*). ♂ forehead off-white, crown red, with buff cheeks and black T-bar on face-sides reaching neither nape nor bill. ♀ has more extensive off-white forehead and black rear crown; black T-bar reduced. Upperparts resemble White-backed (p.284), with broad white transverse bars on folded wing; black back barred white (back white in *kamtschatkensis*), but rump and tail black. Underparts white to buffish, streaked finely with black, lacks red undertail-coverts. *D. m. kamtschatkensis* very white and black, with clean white underparts; *D. m. amurensis* less contrasting with greyer, well-streaked underparts. In flight, white-barred black wings and back distinctive, flight deeply undulating. **BP** Bill fine, grey; eyes brown; tarsi grey. **Vo** Very vocal when breeding, otherwise rather silent: high-pitched *gee-geegeegee*, or *ki ki* notes, weaker than Great Spotted, and a Wryneck-like, but less querulous, or a falcon-like *kee-kee-kee-kee*. Drumming rarely heard, but comprises brief, weak high-pitched rattling, often in two parts. **TN** Formerly placed in *Dendrocopos*.

RUFOUS WOODPECKER C
Micropternus brachyurus

L 21–25cm; WT 55–114g. **SD** India to SE Asia. *M. b. fokiensis* S China east to Fujian. **HH** Gardens and forest edge to secondary forest and mature open deciduous and evergreen forest, from lowlands to c.1,500m. **ID** Medium-sized, cream and rufous-brown. Head cream with black crown streaking, appears large, with loose rounded crest on rear crown/nape; ♂ has red patch from eye to ear-coverts. Mantle, back, wings and tail warm rufous-brown narrowly barred black. Chin and throat heavily streaked black, breast brown, flanks, belly and vent dark grey with some darker streaks on flanks and vent. **BP** Bill short, dark grey; eyes black; tarsi dark grey. **Vo** Loud, nasal *keenk-keenk-keenk*, harsh *tu-wic tu-wic t-wic tu-wicca* and shrill rattling notes. Drumming brief, up to five seconds, starts rapidly but soon slows, likened to stalling motorbike *bdddd-d-d—dt*, repeated every 2–3 minutes. **TN** Formerly *Celeus brachyurus*.

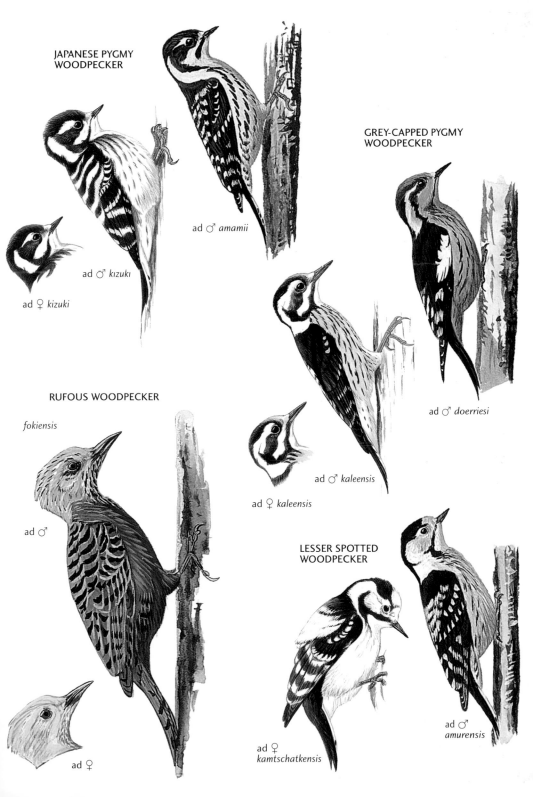

JAPANESE PYGMY
WOODPECKER

ad ♂ amamii

ad ♂ kızukı

ad ♀ kizuki

GREY-CAPPED PYGMY
WOODPECKER

ad ♂ doerriesi

RUFOUS WOODPECKER

fokiensis

ad ♂

ad ♀

ad ♂ kaleensis

ad ♀ kaleensis

LESSER SPOTTED
WOODPECKER

ad ♀
kamtschatkensis

ad ♂
amurensis

PRYER'S WOODPECKER J
Dendrocopos noguchii

Critically Endangered. L 31–35cm. **SD** Endemic to Yanbaru, N Okinawa. An aberrantly dark *Dendrocopos*; perhaps continuing the trend from Owston's Woodpecker. Monotypic. **HH** Mature, subtropical broadleaf evergreen forest near hilltops; feeds on ground and in canopy. **ID** Large, dark, with deep red belly and vent, and some red on breast and mantle. ♂ brighter than ♀, with deep red crown and nape; ♀ has brown crown contrasting with paler brown head-sides. Wings blackish-brown with several white spots in outer primaries; tail blackish-brown. **BP** Bill pale greyish-horn; eyes dark reddish-brown; strong tarsi blackish-grey. **Vo** Highly vocal year-round; a sharp *kwe kwe kwe*, whiplash-like *pwip* or *whit*, hard *kyo* or *kyu-kyu-kup* in contact and more rattling *kyararara*. 'Drum rolls' (three per minute) vary in length, but accelerate. **TN** Formerly *Sapheopipo noguchii*. **AN** Okinawa Woodpecker.

WHITE-BACKED WOODPECKER CTKJR
Dendrocopos leucotos

L 25–30cm. **SD** *D. l. leucotos* resident NW Europe to Sea of Okhotsk, S Russian Far East and NE China, Korea, Sakhalin and Kamchatka. *D. l. subcirris* S Kuril Is, and Hokkaido; *D. l. stejnegeri* N & C Honshu and Sado I.; *D. l. namiyei* W Honshu, Shikoku, Kyushu, Oki Is; *D. (l.) owstoni* **Owston's Woodpecker** from Amami-Oshima, perhaps specifically distinct; *D. l. takahashii* Ulleung-do; *D. l. quelpartensis* Cheju-do; *D. l. fohkiensis* NW Fujian; *D. l. insularis* montane Taiwan. **HH** Mature deciduous woodland, often in damp areas, near lakes, with many dead and dying trees. Like Black Woodpecker often feeds at base of trunks or on ground. **ID** The largest pied woodpecker. ♂ has a red crown, ♀ a black crown. Superficially like Great Spotted. Black face bar reaches bill base, connects to neck bar, but not to nape. Upperparts black with broad white transverse bars on folded wing; lower back/rump white (especially obvious in flight). Forehead, cheeks, neck- and throat-sides whitish, but underparts buffier, with narrow black streaks from chest to flanks; rear underparts suffused buffish-pink and vent pale red. Tail black with large white spots at sides. In flight, resembles Great Spotted but lacks scapular patch; white lower back/rump distinctive. Southern island races usually darker, with more extensive black on chest, heavier black streaking on deeper red underparts. Owston's markedly different from other races in region, being larger, very much darker, with broad black stripe on neck-sides to upper breast and back, almost entirely black upperparts, so heavily streaked on underparts as to appear almost all black, and lower belly/vent dark red. *D. l. takahashii* also distinctive, but closely resembles *stejnegeri* and *namiyei*, though wings and bill shorter, whiter above with paler face and underparts. **BP** Bill long, grey; eyes brown; tarsi grey. **Vo** Hard contact *kyo kyo* and a weak *wick*. ♀ with young, or pairs, sometimes give cackling series of harsh *sketchekekeke* notes. Calls perhaps inseparable from those of Great Spotted. In late winter/early spring ♂ drums powerfully for up to two seconds; the normally resonant roll accelerates from middle onwards, fades at end, rather like a dropped table tennis ball bouncing to a halt.

GREAT SPOTTED WOODPECKER CKJR
Dendrocopos major

L 20–24cm; WS 38–44cm; WT 66–98g. **SD** Resident Scandinavia, W Europe and N Africa across N Russia to China, Japan and Kamchatka. *D. m. brevirostris* NE China, W Sea of Okhotsk; *D. m. kamtschaticus* N shore of Sea of Okhotsk, Kamchatka; *D. m. japonicus* N & E Manchuria, Russian Far East, Sakhalin to S Kuril Is, N & C Japan, Korea; *D. m. cabanisi* S Manchuria to N Jiangsu. **HH** Widest range of woodland types of any woodpecker in region, including gardens and riparian scrub. Agile, forages on trunks, limbs, even amongst fruits on outer twigs, occasionally on ground. Largely resident, though some birds wander south for winter. **ID** The commonest pied woodpecker; ♂ has black crown with a small, bright red nape patch, ♀ lacks red on nape. Boldly black and white with large white shoulder patches, black bar from nape to base of bill, and large, deep red vent patch. Forehead, cheeks and underparts white. Tail black with large white spots at sides. Juv., confusable with White-backed, has less distinct white shoulder, more diffuse red vent and all-red crown (but black rim). In flight, wings black with white patches on innerwing, four narrow bands of spots on flight-feathers, and white bars on sides of otherwise black tail. **BP** Bill dark grey, chisel-shaped; eyes grey (nestlings) becoming dark brown then reddish-brown; short tarsi blue-grey. *D. m. brevirostris* has shorter, thicker bill than *japonicus*, larger white scapular patches, and whiter face and underparts. **Vo** Abrupt *kick* or *chick*, rather loud, sometimes given in rapid series *ke-ke-ke-ke* in alarm. Drumming brief (< 1 second), very rapid, ending abruptly, but repeated frequently, especially by ♂, in early spring.

EURASIAN THREE-TOED WOODPECKER
Picoides tridactylus CKJR

L 20–24cm; WS 32–38cm; WT 46–76g. **SD** Scarce. Resident Scandinavia, European Alps, across taiga to Kamchatka. *P. t. tridactylus* southern taiga to NE China, Ussuriland, Sakhalin; *P. t. crissoleucus* northern taiga to Sea of Okhotsk, Yakutia, Anadyr; *P. t. albidior* Kamchatka; *P. t. kurodai* SE Manchuria and N Korea; *P. t. inouyei* (near extinction?) Hokkaido. **HH** Mainly coniferous boreal/alpine or subalpine forest, but also mixed forest, typically above 650m; favours dead and dying forest. Flakes away bark to expose grubs or drills holes for sap. Quiet, shy; perhaps overlooked rather than scarce. **ID** Mid-sized, rather dark, large-headed woodpecker. Head blackish-grey with ochre-yellow (♂) or grey-streaked crown (♀), with white streaks extending from eye to nape and bill base to neck-sides. Upperparts largely slate-grey, but back white. Chin to vent white with black malar extending into heavy streaking on breast-sides, and barring on belly/vent (*crissoleucus* whiter below with almost no streaking). At rest, narrow, angled white bars on flight-feathers, tertials white-tipped. In flight, wings largely dark, with narrow rows of white spots on flight-feathers and white back distinctive. **BP** Bill grey; eyes black; tarsi grey. **Vo** Abrupt *bick*, *pwick* or *kip* in contact, softer than Great Spotted, a short shrill rattled alarm *kri-kri-kri-kri* and short powerful drumming (>1 second), accelerating and trailing off at end; given in paired bouts. **TN** Formerly part of Three-toed Woodpecker with American *P. dorsalis*.

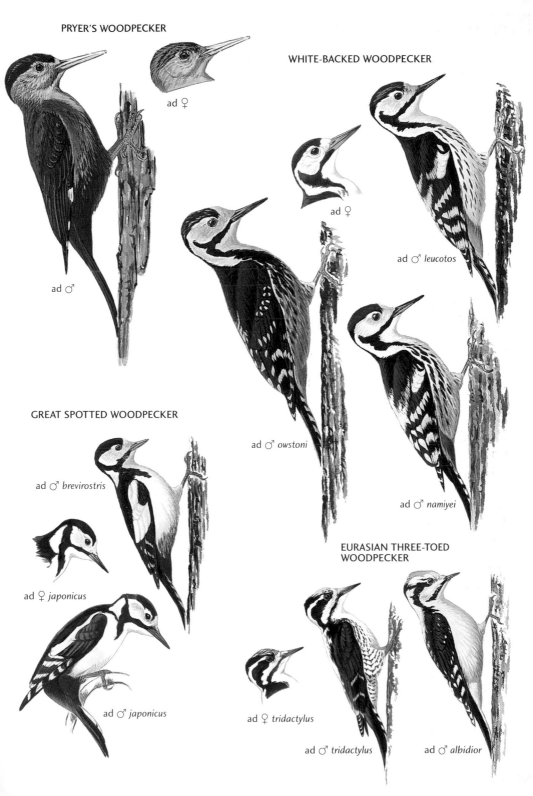

PRYER'S WOODPECKER

ad ♀

ad ♂

WHITE-BACKED WOODPECKER

ad ♀

ad ♂ leucotos

ad ♂ owstoni

ad ♂ namiyei

GREAT SPOTTED WOODPECKER

ad ♂ brevirostris

ad ♀ japonicus

ad ♂ japonicus

EURASIAN THREE-TOED WOODPECKER

ad ♀ tridactylus

ad ♂ tridactylus

ad ♂ albidior

WHITE-BELLIED WOODPECKER K
Dryocopus javensis

L 40–48cm; WT 197–347g. **SD** Disjunct range, from India to SE Asia and Philippines, and Korea. *D. j. richardsi* now extinct on Tsushima, S Japan, and in S Korea, but still extant (probably very rare) in C Korea north to NKorean border. **HH** Wide range of forest types, from lowland to montane regions. Most recent records of *D. j. richardsi* suggest it favours much-threatened lowland forests and that hill forest may represent suboptimal habitat, explaining its drawn-out decline. **ID** Very large pied woodpecker. Mostly black, with black upperparts, breast, lower belly and vent, but rump, primary bases, lower breast, flanks and belly all white. ♂ has crown and narrow malar region bright red, face otherwise black. ♀ has all-black head. In flight, white rump, belly, underwing-coverts and primary bases distinctive, contrasting with otherwise black plumage. **BP** Bill long, blackish-grey; eyes white or pale yellow; large tarsi dark grey. **Vo** Loud *kyaa kyaa* or *kiyaar* and rattling *kyekyekyekekekekekeke*; like Black Woodpecker, calls when perched and in flight. When feeding gives three loud pecks (apparently not in advertisement) on lower trunks, considerably muffled when feeding on sodden trunks. Drumming loud, brief (<two seconds) and accelerates, but slower than Black; up to three rolls per minute. **AN** White-bellied Black Woodpecker.

BLACK WOODPECKER CKJR
Dryocopus martius

L 45–55cm; WS 67–73cm; WT 250–370g. **SD** Wide-ranging Palearctic species: *D. m. martius* occurs from W Europe across taiga to Yakutia, Chukotka, Kamchatka, Russian Far East, Sakhalin, S Kuril Is, Hokkaido and N Honshu (rare) to Korea (scarce and local) and NE China. **HH** Extensive mature mixed woodland from lowlands to foothills, sometimes in natural pure conifer stands (if not dense), and in non-breeding season wanders to more marginal wooded habitat. Commonly forages low on larger trees, often near ground, but calls from high in trees. Excavates distinctive, deep vertical oval cavities. Often shy and unobtrusive, and surprisingly easy to overlook despite its size. **ID** Largest woodpecker in region, crow-sized, unmistakably black with dark blue gloss especially on wing-coverts. Large head, with angled nape, on thin neck. ♂ has all-red crown, ♀ a black forehead and red hindcrown. Juv. sooty-black, lacks gloss, with dull red crown and may have grey throat. Flight (somewhat crow-like) direct, but rather slow and appears loose, even clumsy, with head raised, until last few metres before landing, when more typically undulating. **BP** Bill large, ivory-coloured with dark grey tip; staring white irides make eyes appear large in black face; strong tarsi dark grey. **Vo** Loud, vibrant *kyoon kyoon*, far-carrying ringing *kweeoo kweeoo* or *kree-a kree-a kree-a* territorial call, and in flight a rapid *kyorokyorokyoro* or *krry-krry-krry*. Song comprises strident series of loud whistles, *kwee kvi-kvi-kvi-kvi*…or *kree-kree-kree.* Responds readily to imitation. When foraging gives powerful irregular tapping on dry wood, but true drumming sounds like distant salvo of machine-gun fire, very loud and powerful when close by and rather long (up to three seconds), up to four rolls per minute.

LESSER YELLOW-NAPED WOODPECKER C
Chrysophlegma chlorolophus

L 25–28cm; WT 57–83g. **SD** India to Sumatra, and SE China. *C. c. citrinocristatus* from N Vietnam to NE Fujian (now very rare in region). **HH** Subtropical evergreen forest (800–2,000m), moist deciduous, second growth and plantations, sometimes with mixed-species flocks. **ID** Dark green and grey woodpecker, with fluffy bright lemon-yellow nuchal crest. Head largely green, with narrow white streak from lores to lower ear-coverts, mantle, back and wings green tinged yellow; flight-feathers brown and rufous-brown with white spots; uppertail blackish-green. Underparts grey to sooty greyish-green, with pale grey bars on flanks. ♂ has red border to crown-sides. ♀ has red only on rear-crown-sides. Both yellow-napes show rusty-brown on folded wing, but this is broken by dark bars on Greater Yellow-naped. **BP** Bill rather short, blackish-grey above, grey below with yellow base; eye surrounded by narrow, grey orbital skin, irides reddish-brown; tarsi dark grey. **Vo** Highly vocal: a loud, shrill *kwee-kwee-kwee* in alarm, short *chak*, mournful *pee-ui; squieeer* and *kyee-kur-kur*; crest raised when calling. Drums only occasionally. **TN** Formerly placed in *Picus*.

GREATER YELLOW-NAPED WOODPECKER
Chrysophlegma flavinucha C

L 33–34cm; WT 153–198g. **SD** Himalayas to Sumatra, also S China east to NE Fujian. Subspecies in region, *C. f. ricketti*, now very rare. **HH** Various types of open forest from second growth to mixed mature subtropical and pine forest (800–2,000m). **ID** Large grey and green woodpecker, with rather long, pointed, pale yellow nuchal crest, the yellow extending on to the hindneck. Head and face grey, paler on ear-coverts, with broad white malar stripe; mantle, back, rump, scapulars and wing-coverts uniform bright green, flight-feathers barred black and brown, tail black. Only ♂ has dull reddish-brown-tipped crown-feathers. Chin and throat blackish-grey with white flecking; breast and flanks grey, vent pale grey. **BP** Bill rather long, dark grey; eye surrounded by narrow, blue-grey orbital ring, irides dark reddish-brown; tarsi grey. **Vo** Less vocal than Lesser Yellow-naped, but gives slow *chup* or *chup-chup*, downturned *kyaarr* in series, rich laughing *keeyuu*, and explosive *kchaer*. Pairs maintain contact using long accelerating series of notes: *kwee-kwee-kwee-kwee-kwee-kwee-kwee-kwi-kwi-kwi-kwi-wi-wi-wi-wik*. Occasionally drums weakly, loud at first, but wavering and weakening. **TN** Formerly placed in *Picus*.

WHITE-BELLIED WOODPECKER

richardsi

ad ♂

ad ♀

BLACK WOODPECKER

martius

ad ♂

ad ♀

LESSER YELLOW-NAPED
WOODPECKER

citrinocristatus

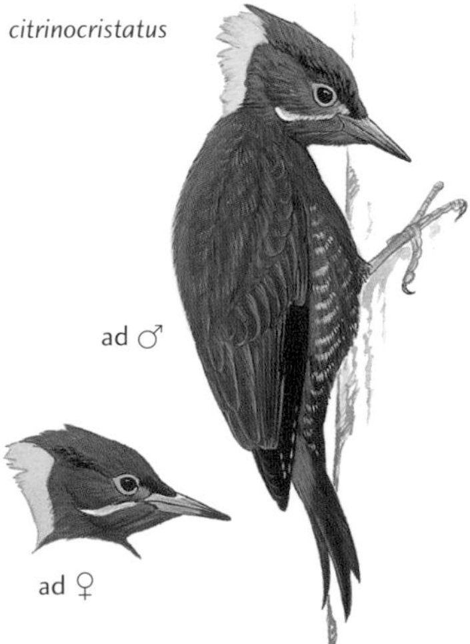

ad ♂

ad ♀

GREATER YELLOW-NAPED
WOODPECKER

ricketti

ad ♀

ad ♂

JAPANESE WOODPECKER   J
Picus awokera

L 29–30cm; WT 120–138g. **SD** Endemic to Japan. *P. a. awokera* Honshu (and associated offshore islets, Tobishima, Awashima, Sado; Shikoku, Kyushu and Tsushima; *P. a. horii* Tanega-shima; *P. a. takatsukasae* Yaku-shima. **HH** Open mixed deciduous and broadleaf evergreen forest in hills and mountains; commonest 300–1,400m. **ID** Distinctive green and grey woodpecker with heavily marked underparts. Head and face, breast and mantle essentially grey, lores black. ♂ has bright red crown, nape and malar patch. ♀ has grey crown with red on nape and centre of malar, rest of which is black. Mantle, scapulars, wing-coverts secondaries and tertials bright green, primaries blackish-brown with white spots; rump bright green, tail darker green with faint dark grey barring on outer feathers. Plain grey breast, off-white belly, flanks and undertail-coverts, all covered with rows of black chevrons. Populations decrease in size and become increasingly dark N–S. **BP** Bill has grey upper mandible with black tip, yellow below; eyes black; tarsi grey. **Vo** A strong, whistled *peoo peoo* or *piyoo piyoo piyoo* mostly in spring; whiplash-like *pwip pwip* and a frog-like *kere kerere*. Contact call a harsh chattering *jerrrrerrerr* in flight, downslurred and indignant in tone. Also a hard, abrupt *ket ket*. Drum roll is fast and rather long. **AN** Japanese Green Woodpecker.

GREY-HEADED WOODPECKER   CTKJR
Picus canus

L 26–33cm; WS 38–40cm; WT 110–206g. **SD** Resident from W Europe across mid latitudes of Russia to E Asia, south to Sumatra. *P. c. jessoensis* occurs from Transbaikalia to S Russian Far East, NE China, Korea, Sakhalin and Hokkaido; *P. c. sobrinus* in SE China (Shandong to Zhejiang); *P. c. tancolo* in montane Taiwan. In Japan breeds only in Hokkaido; replaced by Japanese Woodpecker on main islands further south. **HH** Mature open deciduous and mixed broadleaf forest, hill forest, and swamp forest, from lowlands to foothills; in Taiwan *P. c. tancolo* generally occurs in montane forest above 1,000m. **ID** Large green woodpecker, with distinctive ringing, whistled call. Overall appearance (*P. c. jessoensis*) dull greyish-green. Head small, rounded, grey with black loral stripe and narrow black malar; ♂ has small, bright red forehead patch, ♀ a black-streaked forehead. Greyish-green mantle and back, somewhat stronger olive-green on wing-coverts, flight-feathers blackish-brown with white spots; rump bright green (prominent in flight), tail dull blackish-green and unbarred. Whitish chin/throat, and dull pale grey throat to vent, latter lightly scaled darker. Closed wings have black and white bands on primaries. In flight, primaries and outer secondaries largely black with transverse rows of white spots. Forages on trunks and larger limbs, commonly resting conspicuously high in trees. *P. c. sobrinus* golden-green above and green below. *P. c. tancolo* smaller and darker than other subspecies with stronger grey face, deeper green upperparts and green underparts. **BP** Bill mid-length, dark grey, blacker at tip and base of upper mandible, pale horn at base of lower; eyes brown become blood red in older birds; tarsi grey. **Vo** Single *kik* and strongly whistled descending series of up to 20 *pyoo pyopyopyopyo* notes, slowing after fast start; responds readily to imitations. Drums quite regularly; can be loud, but mostly a short, quiet roll lasting 1–1.5 seconds, rather even-pitched, neither fading nor slowing.

PALE-HEADED WOODPECKER   C
Gecinulus grantia

L 25–27cm; WT 68–85g. **SD** E Himalayas to Indochina. *G. g. viridanus* local in S China east to Fujian (now very rare). **HH** Favours bamboo in evergreen and semi-deciduous forest, and secondary growth to 1,000m. **ID** Medium-sized olive and brown woodpecker with pale head. Face pale, crown yellowish-olive, ♂ has pinkish patch on central crown; head otherwise greyish, pale olive-green. Mantle pale olive-brown, back, rump and wings rufous-brown, flight-feathers barred black and buff, tail barred chestnut and black. Underparts dark olive-green. **BP** Bill short, stubby, pale yellowish-horn with grey base; eyes dark reddish-brown; tarsi grey (has only three toes). **Vo** Harsh, nasal *chaik-chaik-chaik-chaik* or *grrrit-grrrit-grrrit* and *grridit grrit-grrit*; and series of *kwee-kwee* notes. When alarmed/agitated gives loud rattling *kereki kereki kereki*. Also drums in short, fast even bursts. **TN** Sometimes considered conspecific with Bamboo Woodpecker *G. viridis*.

BAY WOODPECKER   C
Blythipicus pyrrhotis

L 27–30cm; WT 100–170g. **SD** Himalayas to SE Asia, and S China east to Fujian. Subspecies in region *B. p. sinensis*. **HH** Evergreen and mixed deciduous forest above 500m, typically in dense growth, often with bamboo. **ID** Large brown and black woodpecker with long pale bill. Head, face and neck-sides brownish-buff, rear crown-feathers extend in short, dark grey-brown crest; nape patch dull red (♂); mantle, back, rump and wings blackish-brown barred chestnut or cinnamon, bars narrow on mantle and scapulars, broader on wings, and broadest on tail. Chin/throat pale buff-brown, breast, flanks, belly and vent dark buff-brown to black. **BP** Bill long, slender, pale greenish-yellow; eyes reddish-brown; tarsi dark grey. **Vo** Calls include a squirrel-like chattering *kecker-rak-kecker-rak*, dry cackling *dit-d-d-di-di-di-di-dit-d-d-di-di*, and loud laughing *keek keek-keek-keek-keek-keek*, descending and accelerating. Not known to drum.

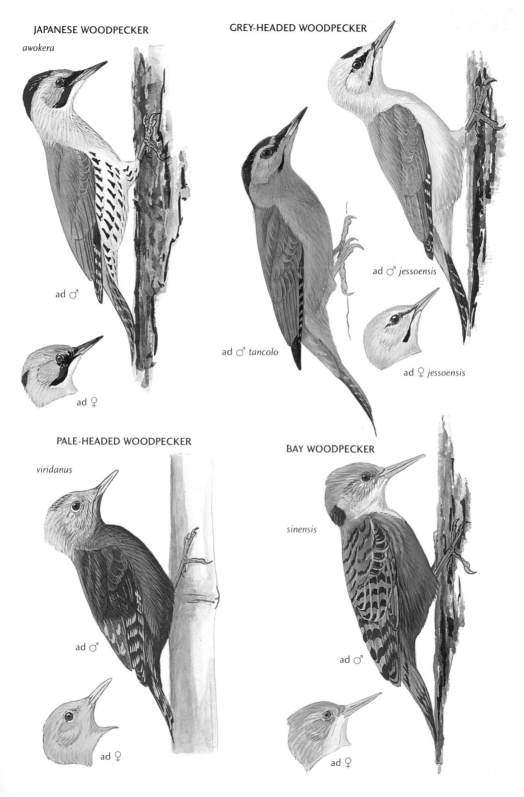

JAPANESE WOODPECKER

awokera

ad ♂

ad ♀

GREY-HEADED WOODPECKER

ad ♂ *jessoensis*

ad ♂ *tancolo*

ad ♀ *jessoensis*

PALE-HEADED WOODPECKER

viridanus

ad ♂

ad ♀

BAY WOODPECKER

sinensis

ad ♂

ad ♀

HOODED PITTA   J
Pitta sordida

L 16–19 cm; WT 42–72g. **SD** Ranges from Himalayan foothills to SE Asia, Philippines, Indonesia and New Guinea. Migrant *P. s. cucullata* is accidental in S Nansei Shoto, Japan. **HH** Forest, where forages for invertebrates on forest floor. **ID** Deep green with a black hood and brown crown. Upperparts deep green, underparts paler apple-green with deep red lower belly, vent and undertail-coverts. Lower back and rump pale blue. Wings blue with white patches, sky blue lesser and median coverts, primaries and tail black. **BP** Bill short, black; eyes dark brown; tarsi flesh-pink. **Vo** An explosive bisyllabic whistled *wieuw-wieuw*.

FAIRY PITTA   CTKJ
Pitta nympha

Vulnerable. L 16–20cm; WT 67–155g. **SD** Scarce and local summer visitor to Kyushu (and offshore islands), Shikoku and Honshu; Korea (e.g. Cheju-do), S China and Taiwan; winters primarily in Borneo. Monotypic. **HH** Inhabits dark broadleaf evergreen, subtropical and tropical forests, including secondary forest, mainly in lowlands to 1,200m, foraging on ground, but singing high in trees. **ID** Brilliantly coloured, with large head, long broad wings and very short tail (primaries extend beyond tip); rather erect. Ad: upperparts and wings bright apple-green; crown chestnut with black median stripe, and broad black mask from bill through eye and around nape, bordered above by broad yellow supercilium. Pale yellow chin, throat, neck and breast, bright red on belly, vent and undertail-coverts, bright blue on scapulars and rump/tail. Juv. duller, with large white spots on median coverts. In flight, black outer coverts and primaries, with white panel at base of primaries; blue tail has black subterminal band. **BP** Bill large, black (ad.) or dark brown with scarlet-orange tip (juv.); eyes large, irides black; long tarsi dull pink. **Vo** Quadrisyllabic, whistled song delivered as two paired notes, variously transliterated as: *pee-yu pee-yu, kwah-he-wwa-wu, hoo-hee hoo-hee* or *pwo-heh wo-heh* and even *shiropen-kuropen*; also a cat-like scolding call.

BLUE-WINGED PITTA   T
Pitta moluccensis

L 18–20cm; WT 54–146g. **SD** Summer visitor to northern SE Asia (S China to N peninsular Malaysia); winters to Sumatra and Borneo. Accidental in winter Taiwan, but records involve birds captured or injured, and perhaps relate to escapees. Monotypic. **HH** Wide range of forest habitats, in lowlands to 1,800m. **ID** Brightly coloured pitta recalling Fairy, but has broader black band across face, deeper green upperparts, extensive deep blue wing panel (on scapulars to secondaries) and very broad white panel on primaries. Underparts deep orange-buff. **BP** Bill large, black; eyes large, irides red-brown to dark brown; long tarsi dull pink to pale brown. **Vo** Silent away from breeding grounds.

LARGE WOODSHRIKE   C
Tephrodornis virgatus

L 18.5–23cm; WT 28–46g. **SD** India to Indonesia. *T. v. latouchei* occurs SE China north to E Fujian. **HH** Uncommon, in lowland broadleaf evergreen and mixed deciduous forest, particularly at edges and clearings; from lowlands to 1,500m. **ID** Medium-sized, rather heavy-set shrike-like bird with short tail. ♂ has grey crown and nape, and black mask from bill to ear-coverts; earth-brown mantle, back and wing-coverts, white rump, with dark earth-brown wings (relieved by rufous fringes to coverts and tertials) and tail. Chin and throat off-white, breast peach, belly and vent white. Tail rounded, plain brown. ♀ is duller with brown crown, and blackish-brown mask and duller, more buff, underparts. **BP** Bill stout, shrike-like but longer, with prominent rictal bristles; eyes yellow to pale brown; tarsi black. **Vo** Varied, harsh and noisy notes, including *wit wit wit* or *chew-chew*, a ringing *ki-ki-ki-ki* and harsh *chreek-chreek chee-ree*.

WHITE-BREASTED WOODSWALLOW   KJ
Artamus leucorynchus

L 17.5–19cm. **SD** Essentially resident in Philippines, Malay Peninsula, Borneo, Indonesia and Australia, but *A. l. leucorynchus* has reached Korea and Japan as a vagrant. **HH** Open areas including agricultural land, with isolated trees. **ID** Starling-like in proportions and flight, with rather straight, triangular wings and short broad tail. Ad.: hood and upperparts blackish-grey, blackest on face and chin; underparts, underwing, lower rump and uppertail-coverts white. Juv. lacks black hood, has buff scaling on brown crown, neck, mantle and back, and orange-buff wash on chest. **BP** Bill deep at base, sharply pointed, dull blue (brownish in juv.); eyes black; short legs and small feet dark grey/black. **Vo** Call, often given in flight, a harsh, rather hoarse *geet geet* or *pert pert*; also chatters and gives softer *ku ku ku* calls.

LARGE CUCKOO-SHRIKE   CT
Coracina macei

L 23–30cm. **SD** Widespread resident in S & SE Asia. *C. m. rexpineti* occurs in S & SE coastal China and Taiwan. **HH** Favours lowland to mid-elevation (c.200–1,500m in Taiwan) open woodland, clearings and edges; also savanna and cultivated areas. Often uses large leafless trees as lookouts. Characteristically flies with wings held below the horizontal, as do other large *Coracina*. **ID** Large, bulky grey bird, with pale grey wing-coverts and tertials contrasting with black flight-feathers. *C. m. rexpineti* is a particularly dark race. ♂ generally mid-grey with black forehead, face, chin and throat; black tail has white corners. Underparts paler grey than upperparts, fading to white, particularly on vent. ♀ lacks black mask, but has blackish-grey lores and patch behind eye, and fine grey barring on breast, flanks and rump. **BP** Bill stout, black; eyes dark reddish-brown; tarsi grey. **Vo** Noisy, calls include a loud, whistled *pee-eeo-pee-eeo, tweer*, and sharp, piercing *tweet-weet* in flight; song (rarely heard) is a melodious warble.

HOODED PITTA

cucullata

ad

FAIRY PITTA

ad

LARGE WOODSHRIKE

latouchei

BLUE-WINGED PITTA

ad

ad ♀

ad ♂

WHITE-BREASTED WOODSWALLOW

ad

leucorynchus

ad ♂

rexpineti

LARGE CUCKOO-SHRIKE

BLACK-WINGED CUCKOO-SHRIKE CTKJ
Coracina melaschistos

L 19.5–24cm; WT 35–42g. **SD** Migrant to much of China, including E, in summer; winters in India and SE Asia. *C. m. intermedia* occurs in coastal China almost to Bohai Gulf, an annual migrant in Taiwan, and accidental in offshore Korea and offshore Japan. **HH** Open montane forest (300–1,500m), forest edge, woodland and bamboo groves. On migration in Taiwan usually in coastal and lowland forests. **ID** Medium-sized with a long, graduated tail. Overall grey with black wings (no contrast between coverts and flight-feathers) and tail, latter with white tips to outermost feathers above and broad white tips below forming bars. ♂ dark ash-grey, blacker on face, paler near vent. ♀ paler with less black on face, white crescents above and below eye, and underparts grey to mid-chest, below that white with grey-barred undertail-coverts. In flight from below shows white patch at base of primaries. **BP** Bill thick, black with slight hook; eyes reddish-brown; tarsi dark grey. **Vo** A magpie-like chattering and raptor-like squealing. Song comprises a descending series of 3–4 high-pitched whistles: *wii wii jeeow jeeow.*

SWINHOE'S MINIVET CT
Pericrocotus cantonensis

L 18–19cm. **SD** Breeds across C,S & E China, mostly south of Changjiang River, east to coast. Accidental Taiwan. Monotypic. **HH** Deciduous, broadleaf evergreen and pine forests, to 1,500m. **ID** Small grey, black and white minivet. ♂ has extensive white forehead to rear of eye, black lores, charcoal-grey midcrown and ear-coverts to nape and upper back, then buffy-grey lower back and pale buffy-grey rump. Cheeks, chin, neck-sides and throat white, vinous-brown to vinous-grey breast and flanks, belly and vent white. Wings black with buff fringes to greater coverts and tertials, and yellowish-buff to white spots on inner webs of primaries and secondaries, typically concealed when perched, but shows as wingbar in flight. Tail long, black with white outer rectrices. ♀ has grey forehead and crown, off-white underparts and more extensive yellow wing patch visible on closed wing. Swinhoe's has more extensive white forehead and grey, not black, rear crown, with brown-tinged upperparts, vinous-brown/grey underparts, and wingbar less conspicuous than in similar Ashy Minivet. **BP** Bill thick, black, slightly hooked; eyes black; short tarsi black. **Vo** Call a metallic trill; flight call like that of Ashy Minivet. **TN** Sometimes considered conspecific with Rosy Minivet *P. roseus.*

ASHY MINIVET CTKJR
Pericrocotus divaricatus

L 18–21cm; WT 21g. **SD** Summer visitor (Apr–Oct) to Russian Far East (has reached Sakhalin), NE China, N Korea and Japan (Kyushu to N Honshu; though has declined in recent decades). Migrant throughout region, including through range of Ryukyu Minivet in Nansei Shoto. Winters in SE Asia and S India, with small numbers in Taiwan. **HH** Locally common, typically in mixed broadleaf evergreen or broadleaf deciduous forest where

frequents canopy. On migration may occur in almost any kind of woodland. **ID** Grey and white, rather wagtail-proportioned bird, but with distinctly upright posture. ♂ has white forehead reaching to eye, black mid and hindcrown and thick black eyeline merging on headsides with patch behind ear-coverts and black nape. Upperparts grey except black flight-feathers with pale fringes to primaries, and long black tail with white outer rectrices. Underparts white. ♀ has black lores, soft grey forehead, crown, ear-coverts and nape, upperparts slightly paler grey than ♂ and underparts off-white. Juv. similar to ♀ but brownish-grey; broad white tips to wing-coverts and paler brownish-grey crown and lores, mantle has white fringes. Tertials also fringed white, the last feature to wear in 1st-winter plumage. In flight, white underparts, white underwing-coverts, broad wingbar and call diagnostic. **BP** Bill thick, black, slightly hooked; eyes black; short tarsi black. **Vo** A high-pitched, shimmering, metallic, somewhat cicada-like trill: *hirihirihirin* or *dee-dee-dee-dee-de* rapidly ascending, with hesitant, tinkling quality, *tsure re re... reee*; or *hii-rii-rii*, especially on taking flight, drier and more quavering than Ryukyu Minivet.

RYUKYU MINIVET TKJ
Pericrocotus tegimae

L 18–21cm; WT 21g. **SD** Resident throughout Nansei Shoto and expanding north through S Kyushu, even to W Honshu and Shikoku, where breeding reported recently. Also recorded N Taiwan and Korea. Monotypic. **HH** Evergreen forest to 600m in Nansei Shoto (also suburban gardens); in Kyushu, resident to 1,700m in evergreen and mixed deciduous forest, sometimes adjoining cedar plantations. Some southward migration noted in Oct in S Kyushu, where increasing numbers winter. **ID** Structurally similar to Ashy Minivet, but has distinctly shorter wings and appears slightly smaller and shorter tailed. Ad.: dark ashy-black crown and ear-coverts concolorous with upperparts (♂), or a shade paler (♀), but still darker than ♀ Ashy. Black loral line broader, white forehead and supercilium much narrower, with split white eye-ring (a feature usually lacking on Ashy). Underparts less bright white than Ashy, with dark grey wash from neck-sides over breast. Juv. differs markedly from that of Ashy, having darker grey upperparts (crown pattern is 'ghost' of ad.) with slight pale mottling, and more white-spotted appearance due to neat brace of white spots on scapulars. Underparts dirty white, with some fine dark streaking on breast and flanks. In all plumages, tail has less extensive white on outer rectrices than Ashy. Fringes of closed primaries appear dark in all plumages, whereas they are white in Ashy. Primary projection slightly shorter than length of longest tertial (longer in Ashy). **BP** Bill thick, black, slightly hooked; eyes black; short tarsi dark greyish-flesh to black. **Vo** Call, given especially in flight, similar to Ashy but shriller, deeper and harder, a rather strained *schree... schree... schree...* the syllables running into one another, either ascending or on same pitch, and often given as two consecutive trills; also soft, whispery *tseet* in contact. Song comprises a call note followed by a high tinkling *schreee... ti ti ti titititititi!* **TN** Formerly part of Ashy Minivet.

BLACK WINGED CUCKOO-SHRIKE

intermedia

ad ♀

ad ♂

SWINHOE'S MINIVET

ad ♀

ad ♂

flock of Ashy Minivets

ASHY MINIVET

ad ♀

ad ♂

ad ♀

ad ♂

RYUKYU MINIVET

GREY-CHINNED MINIVET   CT
Pericrocotus solaris

L 17–19cm; WT 11–17g. **SD** Resident across much of S & SE Asia from E Himalayas to Sumatra and Borneo; *P. s. griseogularis* occurs S & SE China and Taiwan. **HH** Broadleaf evergreen and deciduous forest, sometimes pine forest, at 1,000–1,800m. In Taiwan occurs year-round at *c.*150–2,300m. Forms large flocks (several dozens to over 100) in winter. **ID** Small colourful minivet. Head and mantle dark grey, face including ear-coverts, chin and throat paler grey, wings and long tail greyish-black. ♂ has dull to bright orange underparts, lower back, rump and outer tail-feathers, and orange-tipped greater coverts, a large single orange panel at base of primaries and across secondaries. ♀ somewhat paler grey, with bright lemon-yellow underparts, and in wings and tail where ♂ is orange, but more olive-green on back and rump. In flight, shows colourful underwing-coverts and dark mid-wing bar. **BP** Bill short, black; eyes black; tarsi black. **Vo** Calls have more whispery quality than Ashy Minivet, including a soft, rasping *tsee-sip-sip*, repeated *tswee-seet*, slurred *swirrrririt* and shorter, softer *trip*. **AN** Yellow-throated Minivet; Mountain Minivet.

LONG-TAILED MINIVET   CT
Pericrocotus ethologus

L 17.5–20.5cm; WT 18g. **SD** Ranges from Afghanistan, Himalayas and NE India to SE Asia. *P. e. ethologus* occurs across C to NE China reaching Hebei and Beijing; accidental Beidaihe, and in Taiwan (Lanyu I.). **HH** Lowlands to mountains, in well-wooded areas of broadleaf and pine forest, and cultivation with groves of trees. **ID** ♂ has black hood and glossy blue-black upperparts except scarlet rump; black chin and throat, but rest of underparts crimson. Black wings have broad red patch on inner greater coverts and secondaries, with two extensions on primaries and outer tertials. Black tail has red outer rectrices. ♀ has narrow yellow forehead and short yellow supercilium just reaching eye, grey forecrown and ear-coverts to nape; mantle grey tinged green, back and rump yellowish-green; pale yellow throat, deeper yellow, with slight orange tone on breast and flanks. Resembles Scarlet Minivet, but smaller, more slender with proportionately longer tail, and lacks red (♂) or yellow (♀) spots at tips of secondaries and tertials. **BP** Bill short, black; eyes black; tarsi black. **Vo** A distinctive whistled *pi-ru*, or *prrr'wi* and a sibilant *swii-swii-swii-swii*.

SCARLET MINIVET   C
Pericrocotus flammeus

L 17–22cm; WT 19–24.5g. **SD** India to Indonesia, with *P. f. fohkiensis* in SE China to Fujian. **ID** ♂ like large, heavy-billed Grey-chinned Minivet, but more strongly contrasting. Hood, mantle, wings and tail glossy blue-black. Underparts, lower back, rump, extensive wing patch (including greater coverts, primary and secondary bases, and tips of secondaries and tertials) and sides of rather broad tail all bright vermillion. ♀ bright yellow where ♂ is vermillion; forehead, face and ear-coverts are yellow not black. Crown dark grey, nape, mantle and shoulders grey. **BP** Bill short, robust, black; eyes black; tarsi black. **Vo** Ranges from a loudly whistled *sweep-sweep-sweep-sweep* to a high-pitched, wagtail-like *sigit sigit sigit*, a trilling *hurr* and a soft *kroo-oo-oo-tup tu-turr*.

TIGER SHRIKE   CTKJR
Lanius tigrinus

L 18–19cm. **SD** Summer visitor (May–Aug) to E & NE China, S Russian Far East, Korea and C & N Honshu, Japan. Winters in SE Asia from the Philippines to Indonesia. Seemingly uncommon throughout range. Accidental Taiwan. Monotypic. **HH** Lowland forests, woodland edge, town parks, open country with bushes; typically perches inconspicuously in trees/bushes. **ID** Typical shrike with large head and long tail, but slightly oversized bill and uniquely barred back. ♂ has plain blue-grey crown and nape to upper mantle, broad black mask from forehead to ear-coverts; mantle, wings, rump and tail rufous-brown /chestnut, mostly with distinct black scaling, though tail less marked. Underparts white from chin to vent (lacks white in wing or abutting mask). ♀ resembles dull ♂, but is barred from cheeks to flanks, upperparts as scaled but ground colour less rufous, crown and nape duller, lores whitish, thus mask more restricted. Blue-grey limited to crown and nape – does not extend to upper mantle (compare ♂). 1st-winter recalls ad. winter, but has yellowish-brown upperparts, even in ♂, with more black scaling on wing-coverts; confusable with juv. Red-backed and Brown (nominate and *superciliosus*, see p.296), but note large head and eyes, large flesh-pink bill with only small dark tip, and characteristic vermiculations as well as pale spotting on upperparts. **BP** Bill stout, strongly hooked, black (pink in juv.); eyes large, black; tarsi black. **Vo** A harsh chattering similar to Brown Shrike: *gyun gyun gichigichigichigichi*. **AN** Thick-billed Shrike.

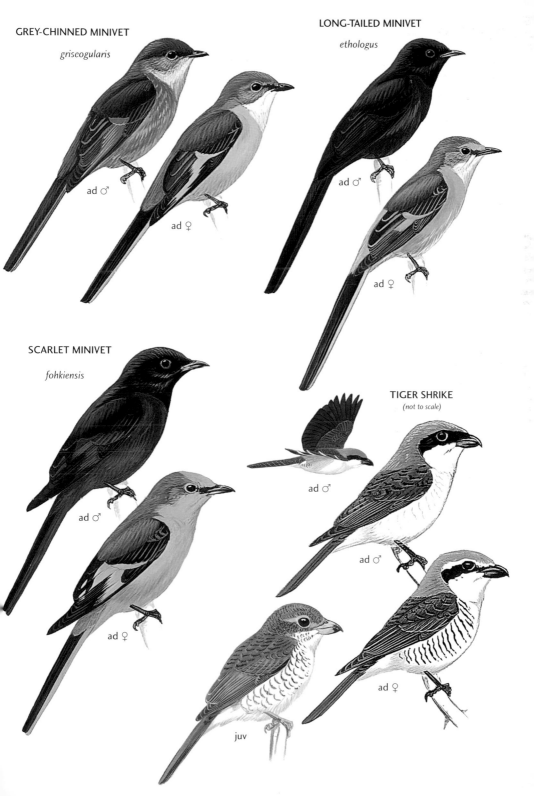

GREY-CHINNED MINIVET

griscogularis

ad ♂

ad ♀

LONG-TAILED MINIVET

ethologus

ad ♂

ad ♀

SCARLET MINIVET

fohkiensis

ad ♂

ad ♀

TIGER SHRIKE
(not to scale)

ad ♂

ad ♂

ad ♀

juv

BULL-HEADED SHRIKE CTKJR
Lanius bucephalus

L 19–20cm; WT 35–54g. **SD** North central China to Sakhalin and Japan. *L. b. bucephalus*: summer visitor NE China, S Russian Far East, Sakhalin, S Kuril Is, Hokkaido, year-round C & S Japan, and Korea, winter visitor Nansei Shoto, E China; accidental Taiwan. **HH** Clumps of bushes, trees, thickets and brush in open country, including farmland, with isolated trees, also forest edge. Can breed very early – some young fledge in May, allowing confusion with rarer spring migrant shrikes. **ID** Brown and grey with thick neck. Compared with Tiger (p.294), Brown, and Red-backed, has shorter, more rounded wings; wing-length is shorter than tail-length (Brown and Tiger have longer wings). ♂ crown, nape and flanks dark rufous-brown, with grey wing-coverts; back, rump, mask, wings and tail black, with narrow white border above mask and prominent white patch at base of primaries; underparts white, sometimes scaled. Black mask usually slightly broader at rear (usually narrower in Tiger/Brown). In winter upperparts become worn and greyer. ♀ browner; wing-coverts, back and tail grey washed brown; lacks white primary patch; underparts brown with narrow dark scales from cheeks to lower belly; face 'gentler' than other similar shrikes, lacks black mask, being brown behind eye, lores pale (see Brown). 1st-winter very like ad. ♀ (1st-winter ♂ already has white patch at base of primaries), but retains some juv. feathers, thus tips of primary-coverts buff. **BP** Bill thick, short, well-hooked, black; large eyes black; tarsi black. **Vo** Harsh chattering *ju ju ju* typical, but also *chi-chi-tyo-tyo*. Song rasping and coarse *kyiikyiikyiikyii chikichikichiki gyun gyun*, heard commonly until early autumn; also mimics other passerines.

BROWN SHRIKE CTKJR
Lanius cristatus

L 17–20cm; WT 30–38g. **SD** C Siberia to E Chukotka (except N tundra), N Sea of Okhotsk and Kamchatka; winters south to S & SE Asia. Summer visitor to most of region. Abundant migrant Taiwan. Winters from S Nansei Shoto and Taiwan throughout S China and SE Asia. Several subspecies migrate through E Asia: *L. c. cristatus* E Siberia from Baikal to Kamchatka, migration mainly continental, but vagrant to Sea of Japan islands and Nansei Shoto; *L. c. confusus* Heilongjiang, Russian Far East and Sakhalin, vagrant to Sea of Japan islands and Nansei Shoto; *L. c. lucionensis* Jilin, Liaoning and elsewhere in N & E China, Korea, Kyushu and Nansei Shoto; *L. c. superciliosus* (sometimes split as **Japanese Shrike**) coastal Russian Far East, S Sakhalin and Hokkaido to C Honshu, and regularly on migration in Honshu, Kyushu and Nansei Shoto. **HH** Thickets, forest edge, plantations, parks and gardens, and open country, including farmland, with isolated trees. **ID** Rather plain brown and somewhat slender shrike; sexes very similar, though ♀ usually duller, with less distinct mask and dark-scaled underparts (see Bull-headed); usually little or no contrast between nape and mantle, but stronger between back and rump/tail (see Isabelline). Wings dark brown, with no (or slight) wing patch at base of primaries. ♂ *superciliosus* has warm rufous-brown upperparts, broad black mask from forehead to ear-coverts

bordered by broad white supercilia joining on white forehead; chin and cheeks white, underparts have rich orange-buff wash to breast-sides and flanks. ♀ less bright, supercilia do not meet, greater wing-coverts fringed buff, and has dark brown scalloping on flanks. *L. c. lucionensis* duller, ash-grey on forehead, with variable white (sometimes none) above black mask, crown and nape grey, mantle greyish-brown; chin/throat/cheeks white, but underparts extensively buff-washed. *L. c. cristatus* is like pale *superciliosus* with colder, sandy-brown crown, mantle and rump, tail largely blackish-brown, white supercilia less contrasting, forehead narrower. *L. c. confusus* resembles *lucionensis*, but white forehead broader and head browner. *L. c. superciliosus* has most rufous upperparts and broadest supercilium. White supercilium and forehead usually broader in ♂ than ♀. Separated from Red-backed by short primary projection, from Isabelline by darker tail (♀ red-brown not orange-brown) with more graduated tip, slightly heavier bill and yellowish-buff underparts. **BP** Bill thick, short, well hooked, black; eyes large, irides dark brown; tarsi black. **Vo** Harsh chattering *che che che che* or *gichigichigichi* in alarm.

RED-BACKED SHRIKE *Lanius collurio* KJ

L 17–19cm; WT 25–35g. **SD** W Europe to C Asia. *L. c. collurio* accidental to Japan, has overwintered (Kyushu); reported Korea. **HH** Open country with low bushes. **ID** ♂ has pale grey crown and nape contrast with bold black mask bordered above by narrow white supercilium, and narrow black forehead. Warm reddish-brown mantle/scapulars and fringes to wing-coverts, secondaries and tertials. Tail long, round-tipped and almost pure black, but white sides broadest from base to mid tail. Underparts white or off-white; cheeks, chin, flanks and breast may be suffused warm peach. Flight-feathers black fringed pale brown with small white patch at base of primaries. ♀ has less boldly marked face, brown forecrown, grey hindcrown, off-white lores, broad brown patch behind eye, and distinct dark scaling formed by narrow tips to feathers of breast and flanks; tail dark brown with narrow white edges. Juv./1st-winter like ♀, but crown brown to dark grey with fine black stripes, upper back boldly barred and scaled black on brown. Underparts essentially off-white, but breast and flanks finely scalloped black. **BP** Bill grey; eyes dark brown; legs/feet black. **Vo** Harsh grating *schak-schak* and a tongue-clicking *tschek*.

ISABELLINE SHRIKE *Lanius isabellinus* J

L 16–18cm; WT 25–34g. **SD** Iran to Inner Mongolia; winters E Africa and India. Accidental S Japan. **ID** Recalls Red-backed, but paler, more sandy-brown with longer rufous tail recalling Brown. ♂ sandy with reddish-brown forehead and narrow black mask broadening from eye to ear-coverts, with narrow white border above lores and eye. Pale grey-brown nape, mantle, scapulars, and fringes to lesser/greater coverts, secondaries and tertials. Primary-coverts and primaries black, with small white patch at base of latter. Long tail with rounded tip is warm rusty-red, with darker, browner tips especially to central rectrices. Underparts white, with creamy cast to cheeks, breast-sides and flanks. ♀ face mask less bold, and often faintly vermiculated on breast and flanks. Juv./1st-winter like greyish-buff young Red-backed, but little or no scaling on much plainer grey-buff mantle and back; pale buff fringes to blackish-brown wing-coverts and flight-feathers. Underparts essentially off-white, with fine grey scaling on breast and flanks. **BP** Bill grey with paler base to lower mandible (ad.) or pinkish-grey with pink base (juv.); eyes dark brown; legs/feet dark grey. **Vo** As Red-backed.

BULL-HEADED SHRIKE

bucephalus

ad ♂ worn

juv

ad ♀

ad ♂

ad ♂

1st-win *cristatus*

BROWN SHRIKE

Taiwan
02/09/19

ad ♂ *cristatus*

ad ♀ *cristatus*

ad ♂ *lucionensis*

ad ♂ *superciliosus*

ad ♂ *cristatus*

RED-BACKED SHRIKE

collurio

ad ♂

ISABELLINE SHRIKE

isabellinus

ad ♀

juv

juv

ad ♂

LONG-TAILED SHRIKE CTKJ
Lanius schach

L 21–25cm; WT 33–42g. **SD** Generally resident in C & SE Asia, to New Guinea and north to S China: *L. s. schach* north to Zhejiang and recently to Hebei; questionably distinct *L. s. formosae* resident in lowland Taiwan; northern birds move south with some wintering in S Asia. Almost annual in recent winters Korea and Japan. *L. s. schach* (including *formosae*) sometimes split as **Rufous-backed Shrike** *L. schach*. **HH** Clumps and thickets in arid steppe and semi-desert, at forest edge and in open country, with isolated trees, and tea and other plantations. **ID** Rather large, black, grey and rufous shrike with long, graduated tail (lacking white) and rounded tip. Black mask extends high onto forehead (more so in *L. s. formosae*), crown and mantle grey, wing-coverts, lower back and rump rufous, wings black with white patch at base of primaries (conspicuous in flight), rufous fringes to tertials and outer rectrices. White chin, face-sides and breast, washed rufous on flanks, darker rufous on belly/vent. **BP** Bill short, well hooked, black; eyes large, irides black; tarsi black. **Vo** Harsh screeching and chattering calls: *gidigidi, giiitt, gijigiji..., gyiip*, and a warbling song mimicking other birds. **AN** Rufous-backed Shrike.

GREAT GREY SHRIKE CKJR
Lanius excubitor

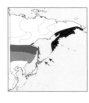

L 24–27cm; WT 48–81g. **SD** Summer visitor to boreal Eurasia and N America. Some resident (in S), northern birds migratory. In E Asia, *L. e. sibiricus* (C & E Siberia) resident from Baikal to Sakhalin, and scarce summer visitor to N Yakutia, E Chukotka, N Sea of Okhotsk and Kamchatka. Winters (Oct–Mar) S Russian Far East, NE China and rarely Hokkaido. *L. e. bianchii* of S Kuril Is and Sakhalin winters to N Japan. Rare in winter further south in Japan and Korea, where care should be taken in separation from Chinese Grey Shrike and potential vagrant extralimital Southern Grey Shrike *L. meridionalis* of the 'Steppe' race *pallidirostris* (reported Korea). N American *L. e. borealis* may occur NE Russia. Populations resident, migratory and irruptive. **HH** Forest-tundra in summer, also forest edges, wetland margins and desolate open country, in winter also agricultural land with isolated trees. **ID** Large, grey, black and white shrike. Ad. pale grey from crown to lower back, rump paler or white, particularly noticeable in flight (see Chinese Grey). Black mask from lores (narrow) across ear-coverts (broad); wings black with broad white band at base of primaries (prominent on closed wing), secondaries and tertials black, latter with white tips; tail black with white outer feathers, and white tips to all but central feathers. Underparts white or pale grey with clear barring. Juv./1st-winter like washed-out ad., with less clearly marked black mask, weak to strong brown tones to face and underparts, and strongly vermiculated flanks and breast, latter generally more diffuse. In flight, white wing patch restricted to primary bases on both surfaces. *L. e. sibiricus* has slightly brown- or ochre-toned dark grey upperparts, contrasting with white rump; underparts of young and some ad. coarsely vermiculated. *L. e. bianchii* is smaller, paler, greyer and lacks brown tones to upperparts. *L. e. borealis* has black lores and distinct pale supercilium, and very pale grey or, more frequently,

pure white rump. **BP** Bill slender, short, well hooked, black, pale pink base in young; eyes large, irides black; tarsi black. **Vo** Song rather quiet, rhythmically repeated squeaky notes and hoarse trills interspersed with shrill chattering; typically silent outside breeding season, but occasionally gives coarse chattering *gijigijigiji* or harsh *check-check*. **AN** Northern Grey Shrike.

CHINESE GREY SHRIKE CTKJR
Lanius sphenocercus

L 28–31cm; WT 80–100g. **SD** Restricted to E Asia, mostly China. *L. s. sphenocercus* a summer visitor to N & NE China, S Russian Far East and N Korea; winter visitor (Oct–Mar) and spring migrant to Korea, E & SE China, very rare winter visitor/accidental S Japan and Taiwan. *L. (s.) giganteus* (probably separate species, **Tibetan Grey Shrike**) of E Tibetan plateau, to Sichuan and Gansu, migrates to E China and could stray to region. **HH** Steppe, semi-desert, scrub, around thickets and bushy clumps in river valleys and meadows, wetlands, marshes, and wet agricultural land. **ID** Very large, grey, black and white shrike; larger and longer-tailed than Great Grey. Upperparts pale to mid-grey from crown to lower back, rump darker grey than Great Grey, less contrasting with tail, which is dark greyish-black. Black mask somewhat narrower on ear-coverts than Great Grey, with a narrow white supercilium from bill to ear-coverts; wings black with large white patch at base of primaries, narrower band across secondaries, connecting quite broad white outer fringes to tertials, which are also prominently tipped white. Appears short-winged, tips falling near tail base, but primaries extend well beyond tertial tips. In flight, very prominent white bar on primaries, narrower white bar on secondaries, narrow white bar at edge of wing-coverts and white trailing edge to tertials, thus much whiter in wing than Great Grey. Underparts white. Tail graduated. with longer central and shorter outer feathers; tail-length longer than wing-length (Great Grey has shorter tail, with tail-length rather shorter than wing-length). **BP** Bill thicker than Great Grey, short, well hooked, black (base pale in 1st-winter); eyes large, irides black; tarsi black. **Vo** Harsh, chattering *ga-ga-ga* and harsh *check-cherr* resembling Great Grey.

MAROON ORIOLE T
Oriolus traillii

L 25–27cm. **SD** Himalayas to Indochina. *O. (t.) ardens* is a scarce local resident of Taiwan; may well be specifically distinct (with Indochinese populations). **HH** Low forested mountains. **ID** Unmistakable. Bright black and crimson oriole. ♂ has black hood to breast and black wings; otherwise entirely deep crimson, though white bases to body-feathers often visible when plumage ruffled. ♀ hood slightly less black, upperparts darker reddish-brown and underparts grade from reddish to off-white on belly, with broad blackish streaking on lower belly/flanks. Juv. has paler underparts with heavy black streaking. **BP** Bill slender, pale grey; eyes grey-brown (juv.) to pale straw yellow or white (ad.); tarsi pale grey. **Vo** A squawking *kee-ah*, cat-like mewing, and *ga-ga-ga-ga-ga...* in alarm. Song a rich fluty two-note whistle, *pi-lo i-lo* or *wu-wu wu-wu*, to which ♀ responds with a whistle.

LONG-TAILED SHRIKE

ad ♂
schach

juv

ad ♂
formosae

GREAT GREY SHRIKE

juv *sibiricus*

ad ♂ *sibiricus*

ad ♂ *sibiricus*

SOUTHERN GREY SHRIKE

ad ♂
pallidirostris

ad ♂
pallidirostris

CHINESE GREY SHRIKE

ad ♂
sphenocerus

ad ♂
giganteus

ad ♂
sphenocerus

MAROON ORIOLE

ardens

ad ♂

ad ♂

BLACK-NAPED ORIOLE   CTKJR
Oriolus chinensis

L 24–28cm. **SD** Resident across much of S & SE Asia, from India to Sulawesi, with *O. c. diffusus* migrating to China, Korea and S Russian Far East for summer. Scarce resident breeder S & E Taiwan, common Korea, scarce (spring) to rare (autumn) migrant offshore Japan (has bred Honshu), scarce migrant Taiwan; has reached Sakhalin. Non-*diffusus* races observed on Taiwan presumably cagebird escapes. **HH** Lowland deciduous woodland, plantations and parkland, where secretive and best found by voice. **ID** Unmistakable, the only large yellow bird in region. ♂ bright golden-yellow with broad black 'bandana' from lores to nape, black wing-coverts, and yellow-fringed black flight-feathers; tail black with yellow corners. ♀ more lemon-yellow, greener on wings and has narrower black 'headscarf'. Juv./1st-winter greenish-yellow, lacks head scarf, but has blackish-streaked white belly/breast, black areas are grey; by spring has brownish nape-band and prominent black streaking on underparts. **BP** Bill large, bright flesh-pink (greyish above, pink below in juv.); eye appears large, with narrow pink eye-ring, iris pale brown; tarsi dark grey/black. **Vo** Call harsh, nasal and somewhat jay- or cat-like, *niiie, myaa* or *gyaa*. Song a clear fluty whistling, *lwee wee wee-leeow*, recalling Japanese Grosbeak but more powerful.

EUROPEAN GOLDEN ORIOLE   Extralimital
Oriolus oriolus

L 22–25cm; WT 56–79g. **SD** *O. o. kundoo* (**Indian Golden Oriole**) ranges almost to Lake Baikal and could stray to region. **ID** Recalls Black-naped, but smaller. ♂ has black lores, lacks black nape-band, and has brighter, yellower upperparts and more extensive black on wing. ♀ usually olive-green above with black wings, and yellow only on flanks and vent – underparts off-white with prominent streaking from throat to lower belly, though some more yellow on upperparts. **Vo** Calls higher pitched than Black-naped.

BLACK DRONGO   CTKJR
Dicrurus macrocercus

L 27–30cm. **SD** Resident in S & SE Asia to SE Russia. *D. m. cathoecus* is a summer visitor across China to Jilin and Heilongjiang, with spring migrants reaching Korea, offshore Japan and maritime S Russian Far East; *D. m. harterti* resident throughout lowland Taiwan. **HH** Open country with trees, agricultural fields and urban parks, frequently perching on poles, wires and treetops. **ID** Large all-black bird with a long tail. Entire plumage black, but gloss variable, having a blue cast on breast and back. Tiny white spot may be visible at corner of gape. Long tail is deeply forked and tips curve upwards and outwards. Juv./1st-winter has duller wings and tail, rather less forked tail (uncurved), and whitish-scaled uppertail-coverts, breast and undertail-coverts. **BP** Bill heavy, black with prominent rictal bristles; eyes reddish-brown to black; tarsi black. **Vo** Harsh, rasping calls: *zyee; shyaa*.

ASHY DRONGO   CTKJ
Dicrurus leucophaeus

L 24–29cm. **SD** Resident from S Asia to Indonesia, with *D. l. leucogenis* a summer visitor across C & E China northeast to Bohai Gulf, with migrants (and some in winter) reaching Taiwan, and overshooting to offshore Korea and offshore Japan, with late autumn records in Nansei Shoto. **HH** Forest edge and open woodland, but also in parks. **ID** A pale, washed-out version of Black Drongo. Entire plumage ash-grey, darker on wings and tail, and black at base of bill, especially lores, with dusky ear-coverts, though some have pale grey lores and white facial patch (unclear whether individual, age-related or subspecific variation). Underparts paler ash-grey becoming even paler on belly. Long tail more deeply forked than Black, but also has outward-curved tips. **BP** Bill heavy, black, with short rictal bristles; eyes reddish-brown to black; tarsi grey. **Vo** Chattering calls recall Bull-headed Shrike and sometimes imitates other passerines. Song a simple phrase: *chochobyuui*.

BRONZED DRONGO   T
Dicrurus aeneus

L 23–25cm. **SD** Ranges from India to Borneo. *D. a. braunianus* resident in montane Taiwan. **HH** Primary and secondary and forest edge, from low to mid-elevations (*c*.100–2,000m), never in open lowland fields (see Black). Frequently flocks with Grey-chinned Minivet (p.294). **ID** Resembles Black Drongo, but smaller, glossier with stronger metallic blue or greenish sheen to head-, neck- and breast-sides and wing-coverts, and tail less deeply forked and blunter tipped. **BP** Bill slender, black; eyes dark reddish-brown; tarsi black. **Vo** Wide range of loud, harsh calls to clearer whistled notes including *tsweep-tswee-tsweep*. A great mimic, often including perfect imitation (*kyik-kirrrr*) of Japanese Sparrowhawk; dawn song a clearer, Japanese Thrush-like whistling.

HAIR-CRESTED DRONGO   CTKJR
Dicrurus hottentottus

L 31–32cm. **SD** Resident in SE Asia with *D. h. brevirostris* a summer visitor to C & E China north to Hebei; spring migrants overshoot as far as offshore Korea, offshore Japan and maritime S Russian Far East. **HH** Favours forests, and open areas with trees. **ID** A large, velvety black version of Black Drongo. Entire plumage black, with dark green gloss on head, breast, back and wing-coverts, strongest on wings and tail; some feathers of crown and breast have iridescent blue sheen (spangles). Long, loose crest of filament feathers rising from forehead. Long tail is broader at tip, barely forked and tips curl upwards and outwards. **BP** Bill heavy, grey, strongly arched to fine tip, with short bristles on forehead; eyes reddish-brown to black; tarsi black. **Vo** Various harsh screeching calls and melodious notes.

BLACK-NAPED ORIOLE

diffusus

ad ♀

ad ♂

EUROPEAN GOLDEN ORIOLE

kundoo

ad ♀

ad ♂

BLACK DRONGO

cathoecus

imm

ad

ASHY DRONGO

leucogenis

ad

BRONZED DRONGO

ad

braunianus

HAIR-CRESTED DRONGO

ad

brevirostris

BLACK-NAPED MONARCH   T
Hypothymis azurea

L 15–17cm; WT 8.9–13.4g. **SD** India to Indonesia north to S China. *H. a. oberholseri* resident on Taiwan. **HH** Favours middle and lower levels of broadleaf forest and woodland thickets, except in high montane forest; also secondary forest and urban parks. **ID** Slender, small-headed and long-tailed flycatcher superficially resembling slightly larger Blue-and-white Flycatcher. ♂ almost entirely deep cobalt blue, except sooty black tufts at base of upper mandible and on rear crown, and narrow black half-collar on throat; breast grades from blue to bluish-grey on lower breast and belly, to white on vent. Wings blue with black flight-feathers; tail blackish-blue with blue fringes to outer rectrices, tip rounded. ♀ has blue head and chest, but grey underparts, becoming whiter below tail. Mantle, wings, back and tail dark grey-brown with blue fringes to outer tail-feathers. Juv. like ♀. **BP** Bill small, slender, dark blue-grey; eye appears small, irides black; tarsi black. **Vo** Contact call a harsh *chee chweet*, also a series of 3+ whistles: *treet-treet-treet*. Song a clear, ringing *pwee-pwee-pwee-pwee*.

ASIAN PARADISE FLYCATCHER   CTKJR
Terpsiphone paradisi

L ♂ 47–48cm, ♀ 20–21cm; WT 20–22g. **SD** Essentially resident from Afghanistan to Indonesia, but *T. p. incei* is a summer visitor to E & NE China and S Russian Far East, wintering in India and SE Asia. Accidental to annual migrant Taiwan, rare Korea, though probably is (or was) local breeder in NW, and suspected Japan. **HH** Shady mature mixed deciduous broadleaf forest in N and broadleaf evergreen forest in S, and in coastal forest on migration in Taiwan. **ID** Distinctive, long-tailed flycatcher. ♂ has two colour morphs, brown and white. Brown morph has glossy black hood and nuchal crest (hood bluer than Japanese Paradise Flycatcher), with dark grey hind-collar and breast-band; upperparts (mantle, wings and tail) mainly plain rufous-brown, belly and vent white. Tail long, but extremely elongated central feathers of ad. ♂ (young ♂ lacks full tail) project up to 30cm, and may be more than twice tail-length.

Apart from black hood ♂ white morph is entirely white, though whether white morph exists in *T. p. incei* is unclear. ♀ resembles ♂, but is duller with smaller crest, black of head tinged blue and lacks elongated tail-feathers (very closely resembles ♀ Japanese, but has warmer, brighter chestnut-brown mantle and wings, and grey of breast, all more contrasting with black hood). **BP** Bill blue; eye large, with blue eye-ring, irides black; tarsi black. **Vo** Harsh, nasal and abrupt: *chet-trkh-chettr* or *dzh-zee dzh-zee*; also a loud *chee-tew* in contact. Song a low fluty descending phrase: *ee-tym-lyou-lyou-goo*.

JAPANESE PARADISE FLYCATCHER   CTKJR
Terpsiphone atrocaudata

L ♂ 35–45 cm, ♀ 17–18cm; WT 18.7g. **SD** Range-restricted and uncommon breeding bird of Japan (absent Hokkaido), S & C Korea and Taiwan (Lanyu I.); winters from Philippines to Sumatra. Accidental Ussuriland and Hokkaido. Some in Nansei Shoto perhaps resident, remainder migratory, moving along Chinese E coast, also Taiwan, and wintering in peninsular Malaysia and Sumatra. *T. a. atrocaudata* breeds through most of Japanese/Korean range, whilst *T. a. illex* is resident (perhaps migrant) on Nansei Shoto, and *T. a. periophthalmica* is considered resident on Lanyu off SE Taiwan, but presence in winter uncertain. **HH** Shady mature deciduous or evergreen broadleaf forest in temperate areas, and dark subtropical evergreen forest further south. **ID** Like Asian Paradise Flycatcher but slightly smaller. ♂ has black hood with purplish-blue gloss, becoming blackish-grey on chest, off-white to white on belly and vent. Mantle, wings and rump dark purplish-chestnut, tail (including extremely long central feathers) black. No white morph. Imm. ♂ like ad., but has shorter central tail-feathers. ♀ closely recalls ♀ Asian, but is duller, darker brown on mantle, wings and tail, generally lacking rufous tones, though some are confusingly rufous. *T. a. illex* is smaller and darker, whilst *T. s. periophthalmica* is similar, but ♂ has purplish-black mantle and grey-black underparts except belly, which is white; ♀ resembles other races but is darker. **BP** Bill blue, short, broad-based; eye large, with broad blue eye-ring, irides black; tarsi black. **Vo** A coarse *gii* or *bii* and querulous *jouey*. Song a somewhat rasping, whistled *tski-hi-hoshi hoi-hoi-hoi*; or *fi-chii hoihoihoi*. **AN** Black Paradise Flycatcher.

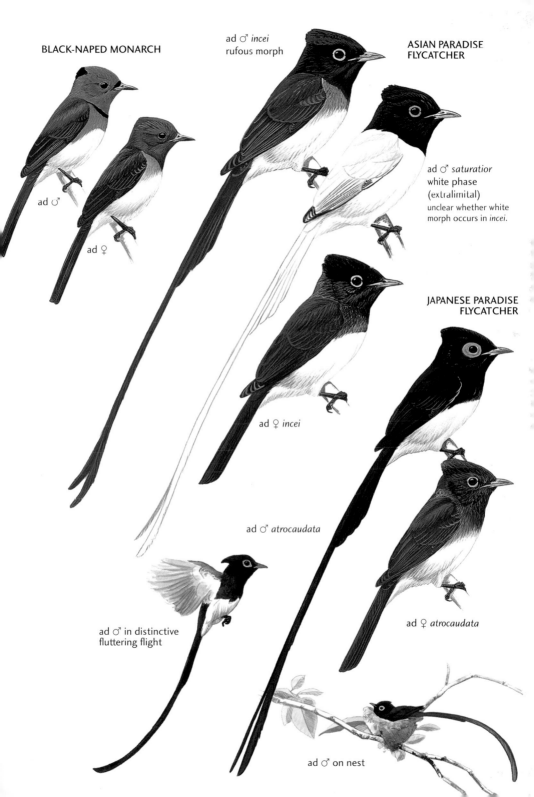

BLACK-NAPED MONARCH

ad ♂

ad ♀

ad ♂ *incei*
rufous morph

ASIAN PARADISE FLYCATCHER

ad ♂ *saturatior*
white phase
(extralimital)
unclear whether white
morph occurs in *incei*.

ad ♀ *incei*

JAPANESE PARADISE FLYCATCHER

ad ♂ *atrocaudata*

ad ♂ in distinctive
fluttering flight

ad ♀ *atrocaudata*

ad ♂ on nest

SIBERIAN JAY   CR
Perisoreus infaustus

L 26–31cm; WS 40–46cm; WT 73–101g. **SD** Norway to Russian Far East and NE China: *P. i. yakutensis* from lower Lena to Anadyr and Sea of Okhotsk; *P. i. tkatchenkoi* from middle Lena to W Amurland; *P. i. maritimus* from E Amurland and NE China (Heilongjiang) to Russian Far East and Sakhalin. **HH** Fairly common resident throughout range, favouring coniferous taiga forest. Secretive in breeding season, but typically encountered in pairs or in small flocks outside breeding season. **ID** Rather plain, small, long-tailed jay, with large rounded head. Dark brown head, particularly dark on crown, grey-brown upperparts and underparts, becoming rather buffy orange-brown on flanks; wings and tail dark grey-brown, wing-coverts, rump and tail-sides also rusty-orange. **BP** Bill strong, black, very short, with pale-feathered base to upper mandible; eyes black; tarsi blackish. **Vo** Generally rather silent, but sometimes gives mewing *geay* and upslurred, whistled *kui* notes. Song is a very quiet mix of twittering, whistling and mewing sounds.

EURASIAN JAY   CTKJR
Garrulus glandarius

L 32-37cm; WS 52–58cm; WT 140–190g. **SD** W Europe to E Asia. Common, mostly sedentary, but with clear evidence of strong migratory movements through Hokkaido (perhaps to continent), and elsewhere irregularly or irruptively. Various subspecies, sometimes split as **Brandt's Jay** *G. brandtii* (including *pekingensis*); **Japanese Jay** *G. japonicus* (including *tokugawae, hiugaensis, orii*) and **Himalayan Jay** *G. bispecularis* (including *taivanus*). **HH** Wide range of forest types, including deciduous and evergreen broadleaf woodlands, mixed forests with conifers, and open country with trees outside breeding season. Often solitary (when skulks) or in family parties, but flocks on migration and sometimes in winter. Taiwanese birds generally at 1,000–2,500m, descending to c.100m in winter. **ID** Large, big-headed, colourful corvid. Northern *G. g. brandtii* (Urals to NE China, Russian Far East, Korea, Sakhalin, S Kuril Is and Hokkaido) has dark cinnamon head and wash to breast, face generally plain with broad black lateral throat-stripes; mantle and scapulars pale grey-brown, rump white, tail black; wings largely black, but has white fringes to primaries and broad white patch at base of secondaries, wing-coverts turquoise-blue barred black and white. Underparts grey with cinnamon wash to breast, white on vent and undertail-coverts. *G. g. japonicus* of Honshu, Shikoku, Tsushima and Oshima, *G. g. tokugawae* of Sado I., *G. g. hiugaensis* of Kyushu and *G. g. orii* of Yaku-shima, Japan (inseparable in field) have pale-grey forehead streaked black, a black face, head uniform with back. *G. g. sinensis* of C, S & E China and *G. g. pekingensis* of Shanxi, Hebei, Beijing and S Liaoning (also reported Korea) are paler creamy buff above and below. *G. g. taivanus* of Taiwanese mountains, has plain cinnamon-grey head, black above base of bill and broad black malar; wings show little white except on outer

fringes of primaries, but cobalt-blue barred black very extensive on coverts and outer secondaries. All subspecies in flight have broad, rounded wings, broad tail and characteristic somewhat laboured, erratic flight, often at some height over forest. **BP** Bill strong, blunt, black; eyes reddish-brown in *G. g. brandtii* and *G. g. taivanus*, white in other subspecies; tarsi grey-brown. **Vo** Generally silent, but can be very vocal, responding quickly and loudly to intrusion; mostly gives rasping, harsh *gsharrr* but also soft sweet whistles. Commonly imitates other birds.

LIDTH'S JAY   J
Garrulus lidthi

Vulnerable. L 38cm. **SD** Endemic resident on Amami Is, S Japan. Monotypic. **HH** Subtropical broadleaf evergreen and coniferous forest, forest edges and gardens, but prefers mature forest. Forages largely on ground; forms social roosts outside breeding season. **ID** Large, bright, cobalt and brown jay. Beautiful combination of deep cobalt-blue hood, wings and tail, and deep vinaceous-brown mantle, back, rump, breast, belly and vent. Face (forehead, lores and chin) black, the chin with fine white streaking; wings largely deep blue with fine black barring on coverts; primaries, secondaries and tertials largely black with cobalt outer fringes and fine white spots at tips; tail broadly tipped white, particularly on underside. **BP** Bill blunt, thick, very pale straw-yellow, bluish-grey at base of lower mandible; eyes black with pale orbital ring; tarsi grey. **Vo** Calls varied, harsh, but drier and slightly thinner than Eurasian Jay, less commonly imitates other species: deep *gaa* or *gyaa, kyui, pyuru, pyuui,* also *skerr skerr* or *cher cher.* No true song, but sometimes gives a soft murmuring *kyuh, kyukyu, kuku.*

AZURE-WINGED MAGPIE   CTKJR
Cyanopica cyanus

L 33–37cm; WS 38–40cm; WT 65–79g. **SD** Restricted to E Asia. *C. cyanus* occurs from Lake Baikal to Lesser Hinggan Mountains, Heilongjiang (N Manchuria) and Russian Far East; *C. c. japonica* in N & C Honshu, Japan; *C. c. stegmanni* in Greater Hinggan Mountains and Changbai Mountains, China; *C. c. koreensis* in Korea; *C. c. interposita* northeast to Shandong and Hebei; *C. c. swinhoei* east to Jiangxi and Zhejiang. **HH** Resident and locally common, sociable and often in noisy groups. Deciduous hill forests, thickets in river valleys, orchards and large urban parks. **ID** Slender, pale corvid, with long graduated tail. Upperparts cleanly marked, with black cap, grey mantle and rump, pale to deep powder-blue wings and tail; primaries black with blue outer fringes, tail has broad white tip; underparts and collar clean white. In flight, short rounded wings and long tail, and rapid flapping interspersed by long glides distinctive. **BP** Bill slender, short, black; eyes black; tarsi dark-grey/black. **Vo** Various calls, a screeching *ray-it wit-wit-wit,* harsh, repetitive *zhreee, gyuuui, geh, quehi, kyururururu* and *zweep zweep zweep.* **TN** Formerly considered conspecific with Iberian Azure-winged Magpie *C. cooki.*

SIBERIAN JAY

maritimus

ad

ad *japonicus*

EURASIAN JAY

ad *brandtii*

ad *taivanus*

LIDTH'S JAY

ad

AZURE-WINGED MAGPIE

ad *koreensis*

ad *koreensis*

juv *koreensis*

TAIWAN BLUE MAGPIE T
Urocissa caerulea

L 64–69cm. **SD** Endemic to Taiwan, where locally common. Monotypic. **HH** Typically encountered in small parties in lower montane (100–1,200m) broadleaf and secondary forests, and forest edge. **ID** A large, extremely long-tailed colourful magpie. Almost entirely deep cobalt-blue; hood, breast and underwing-coverts black, wings blue with white tips to tertials. Tail long and graduated, with black and white to each feather; underparts somewhat paler blue, paler still on belly and white on vent. **BP** Bill strong, arched and chilli red; eyes conspicuous, pale yellow; tarsi bright red. **Vo** A cackling (somewhat Common Magpie-like) metallic *kyak-kyak-kyak-kyak*, and various softer and harsher calls and whistles.

RED-BILLED BLUE MAGPIE CT
Urocissa erythrorhyncha

L 65–68cm. **SD** Himalayas to SE Asia, northeast across China to Bohai Gulf. *U. e. erythrorhyncha* resident in SE China and *U. e. brevivexilla* in NE China to S Liaoning; a small introduced population breeds locally on Taiwan. **HH** Forest-edge, also scrub and villages, where sociable and highly vocal. **ID** Large, extremely long-tailed, colourful magpie. Plumage largely bright blue; hood and breast black; crown to nape grey-white; wings blue with white tips to tertials, and graduated tail with black and white tips to each feather; lower breast to vent white. **BP** Bill strong, arched and chilli red; eyes brown; tarsi bright red. **Vo** A range of harsh screeching calls and whistles.

GREY TREEPIE CT
Dendrocitta formosae

L 34–40cm. **SD** Himalayas to SE Asia and E China north to Changjiang. *D. f. formosae* resident in lower montane forest of Taiwan; *D. f. sinica* in SE China. **HH** Broadleaf and secondary forests, urban parks and gardens, from 400–1,200m (0–2,000m in Taiwan), sometimes in noisy groups or mixed flocks. **ID** Resembles Common Magpie in proportions, though long graduated tail is fuller and more rounded. Face and chest sooty

blackish-brown, upperparts dark ash-grey on crown, browner on mantle and grey on rump. Uppertail-coverts grey-brown and tail black; wings black with small white patch at base of primaries. Underparts: dark blackish-brown on chest, becoming paler buff-brown on belly and grey on lower belly; vent chestnut. *D. f. sinica* has white rump. **BP** Bill strong, slightly arched and pointed, grey-black; eyes reddish-brown; tarsi grey-black. **Vo** Extremely varied, including harsh chattering and squawking *klok-kli-klok-kli-kli* along with more musical notes. In Taiwan, commonest call a harsh, chattering *pJER j-j-jerjerjerr!*, also commonly gives an abrupt squawking *kerLINK!* **AN** Himalayan Treepie.

COMMON MAGPIE CTKJR
Pica pica

L 40–51cm; WS 52–60cm; WT 182–272g. **SD** Widespread from Europe across Palearctic. In E Asia occurs from S China to S Russian Far East, with isolated population in Kamchatka north to S Chukotka, also Korea, Japan (essentially Kyushu, where probably introduced, but has recently colonised SW Hokkaido) and Taiwan. Recent evidence indicates that both taxa in E Asia, *P. (p.) camtschatica* (**Kamchatka Magpie**) of Anadyr to Kamchatka, and *P. (p.) sericea* (**Oriental Magpie**) of Amurland, Korea, Kyushu, NE & E China and Taiwan, differ sufficiently in DNA from western forms to be considered separate species; however, their exact ranges and any morphological/vocal differences are still unclear. **HH** Open wooded habitats (avoids dense forest), thickets, agricultural land with scattered trees, suburban and urban parks and gardens; often places messy domed nest atop poles, pylons and trees. **ID** Large, black and white, long-tailed corvid. Hood and upperparts largely black with variable purple, blue and green iridescence; scapulars and inner webs of primaries white (visible in flight), primaries have black fringes and tips; black chest, white belly, black vent. Long tail is graduated, broadest at mid point. Flight laboured, with frequent wingbeats. **BP** Bill strong, pointed, black; eyes black; tarsi black. **Vo** Various harsh chattering *kasha kasha* and cackling calls that are higher and more 'tinny' than European birds. **AN** Eurasian Magpie.

TAIWAN BLUE MAGPIE

RED-BILLED BLUE MAGPIE

erythrorhyncha

ad

ad

GREY TREEPIE

formosae

ad

COMMON MAGPIE

ad *camtschatica*

ad *camtschatica*

Taiwon
02/09/19

ad *sericea*

SPOTTED NUTCRACKER
CTKJR
Nucifraga caryocatactes

L 32–38cm; WS 52–58cm; WT 140–190g. **SD** From Scandinavia to Kamchatka south through S Russian Far East, NE China, Korea and Japan; also high montane Taiwan. *N. c. macrorhynchos* across Russia to Yakutia, Chukotka and Kamchatka, Sakhalin, NE China and Korea; *N. c. japonica* in S Kurils and Hokkaido, Honshu and Kyushu; *N. c. owstoni* on Taiwan; *N. c. interdicta* in NE China to Hebei, Shandong and Liaoning. **HH** Fairly common in pure coniferous taiga and lowland mixed forests, and alpine conifers, including stone pine forest in south; some (numbers vary greatly between winters) descend lower in winter, when may also occur in small flocks. Actively caches seeds during autumn as winter food store. **ID** Dark brown, broad-winged, short-tailed corvid. Entirely mid-brown to dark chocolate-brown, heavily spotted white on face, neck, mantle and underparts (less spotted on *owstoni*). Lores whitish, eye-ring white, cap and nape blackish-brown, wings and tail also dark blackish-brown, tail has white corners. In flight, broad wings, white vent and short tail very noticeable; flight undulating. **BP** Bill slender and rather long, sharply pointed, black; eyes black; tarsi black. **Vo** Recalls Eurasian Jay, but deeper more growling *graarr* or *zhrrerr*, also given in brief series, *zhree-zhree-zhree*, sometimes higher pitched; also a range of whistles and mimicry.

RED-BILLED CHOUGH
CK
Pyrrhocorax pyrrhocorax

L 35–41cm; WS 73–90cm; WT 260–350g. **SD** Resident from European Alps to montane continental E Asia, including S Siberia, Mongolia, with *P. p. brachypus* in NE China to Shandong and Hebei. Vagrant Korea. **HH** Typically in mountainous grasslands, where forages on ground in open, but at somewhat lower altitudes and more steppe-like habitats in NE China; usually found in flocks. **ID** All-black corvid, with broad wings and distinctly 'fingered' primaries and short broad tail; thigh feathers long, appearing like 'baggy trousers'. Plumage black with dark bluish gloss. Flight aerobatic and agile, gliding and soaring on broad wings. **BP** Bill long, slender, decurved and sharply pointed, bright red (dusky-yellow in young); eyes black; tarsi bright red. **Vo** Very vocal, commonly gives abrupt, harsh *chee-ow*, typically in flight but also when perched.

WESTERN JACKDAW
J
Coloeus monedula

L 33–34cm; WS 67–74cm; WT 180–260g. **SD** Common and widespread from W Europe, with *C. m. soemmerringii* to extreme NW China and SC Siberia. Accidental Hokkaido, and potentially to rest of region. **HH** Open wooded areas with crags, cliffs or settlements. Typically sociable, highly vocal and gregarious, but solitary as vagrant. **ID** Small black corvid with rather rounded head and short wings. Ad. generally black with metallic blue sheen to scapulars, but distinctly pale grey nape and head-sides. Juv. lacks silver-streaked ear-coverts (see Daurian Jackdaw). Wings rather short and pointed, with fast pigeon-like wingbeats and on ground gait pigeon-like. **BP** Bill short, slender, black; eyes white; tarsi black. **Vo** Highly excitable and vocal in flocks, giving abrupt, loud but quite pleasant *chyak*. **TN** Formerly placed in *Corvus*.

DAURIAN JACKDAW
CTKJR
Coloeus dauuricus

L 33–34cm; WS 67–74cm. **SD** Across E Asia, from Tibetan plateau and Mongolia to continental coast. Resident over much of range, and summer visitor to northern regions from Lake Baikal east, including northern Korea. Locally common, or uncommon, winter visitor (Nov–Mar) to E China, Korea, and Kyushu; rare winter visitor elsewhere in Japan and on Taiwan. Monotypic. **HH** In winter, very sociable and typically found with Rooks (see p.310) on agricultural land, typically ricefields with wooded areas nearby for roosting. **ID** Small, black and white (or pale grey) corvid. Ad. somewhat variable; extremely pale form has black hood and upperparts (including wings, tail and vent) and white nape, collar, breast and belly. Darkest forms have more grey, less white. Juv./1st-winter largely black, closely resembling Western Jackdaw, but has variable grey or silver on sides of head and nape, and dark eye. **BP** Bill short, slender, black; eyes black; tarsi black. **Vo** Highly excitable and vocal, giving abrupt, loud *chyak* calls, probably indistinguishable from Western Jackdaw. **TN** Formerly placed in *Corvus*.

HOUSE CROW
TJ
Corvus splendens

L 41–43cm; WS 76–85; WT 252–362g. **SD** Indian subcontinent, Burma and SW China, but also introduced or beneficiary of ship-assisted arrival to peninsular Malaysia, Singapore, S Africa, Middle East. Has occurred more widely as a vagrant (also probably ship-assisted) including to Taiwan and Japan (subspecies unclear, but perhaps nominate of India). **HH** Human habitation, agricultural land and rubbish dumps. As a vagrant, generally in coastal areas. **ID** Medium-sized grey and black crow, slightly smaller than Carrion (p.310), with strongly peaked crown. Forecrown, face and throat black; nape, neck, breast and flanks grey, shading to black on back, wings, tail and lower belly. **BP** Bill slightly arched, quite thick, black; eyes black; tarsi black. **Vo** Highly social and vocal giving range of calls, including a dry, flat *kaaa-kaao*.

SPOTTED NUTCRACKER

ad *japonica*

ad *interdicta*

ad *japonica*

RED-BILLED CHOUGH

brachypus

ad

WESTERN JACKDAW

soemmerringii

ad

DAURIAN JACKDAW

ad

ad

juv

HOUSE CROW

ad

flock feeding

ROOK *Corvus frugilegus* CTKJR

L 44–46cm; WS 81–99cm; WT 280–340g. **SD** W Europe across Russia, with *C. f. pastinator* (**Eastern Rook**) in Mongolia, C & E China to S Yakutia, Russian Far East, Korea and NE China, and in winter (Nov–Mar) E China, Korea and S Japan, particularly Kyushu and W Honshu; increasing evidence (range expansion?) of marked migration through W Japan; accidental Taiwan. **HH** Nests colonially in woods, forest edges, copses, foraging on open land; in winter, typically forages on agricultural land, often in large flocks, commonly with other crows. **ID** Large corvid with rather small head, slender neck and broad tail. Plumage black with bluish-purple iridescence; thigh feathers shaggy, like baggy trousers. In display, broad, blunt-ended tail is fanned. In flight, head protrudes more than Carrion Crow, primaries very distinctly 'fingered', wing bases narrow, tail full. **BP** Bill slender, very sharply pointed, black with grey-white base to both mandibles; eyes black; tarsi black. *C. f. pastinator* lacks bare chalky white facial skin of extralimital western race *C. f. frugilegus* (**Western Rook**) and has more sharply pointed bill. **Vo** Dry, flat, nasal *kaak*, somewhat akin to Carrion but weaker and harsher.

CARRION CROW *Corvus corone* CTKJR

L 45–47cm; WS 93–104cm; WT 370–570g. **SD** W Europe to E Asia. *C. (c.) orientalis* (some inconclusive evidence for specific status as **Oriental Crow**) from Afghanistan across C Asia and N Russia from Yenisei to Yakutia, Chukotka, Kamchatka, Russian Far East, Sakhalin, south to N China, Korea and Japan, but absent most of Nansei Shoto; accidental Taiwan. **HH** Common in settled countryside, mountains and coasts, also cities (especially winter), flocks in winter, foraging over agricultural land. Commonly walks, whereas Large-billed often hops. **ID** Large, all-black crow with slight metallic blue sheen. Similar size to some small races of Large-billed, but separated by low profile – continuous slope from bill tip to slightly rounded crown. In flight, primaries 'fingered', but each primary broader than Rook's. **BP** Bill slender, pointed, slightly arched (not as slender or pointed as Rook, nor as blunt and deep as Large-billed), black; eyes black; tarsi black. **Vo** Harsh rolling *caw* repeated 3–4 times. Calling posture usually differs from Large-billed – tail often depressed, body hunched, head thrust forward.

COLLARED CROW *Corvus pectoralis* CT

L 48–54cm. **SD** Vietnam to C, E & NE China. Monotypic. Resident over much of China, including coast north to Hebei, also Kinmen (Taiwan); has strayed to mainland Taiwan. **HH** Agricultural land and rural habitation, replaces Carrion in such habitats in E China, but now rare. **ID** Resembles glossy Carrion, but has prominent grey-white band on hindneck, mantle and neck-sides to breast. **BP** Bill slender, pointed, black; eyes black; tarsi grey-black. **Vo** Usually a higher pitched *kaaarr* than Large-billed.

LARGE-BILLED CROW CTKJR
Corvus macrorhynchos

L 46–59cm; WS 100–130cm; WT 450–650g. **SD** Afghanistan to Philippines. Largest races in north, smallest in south: *C. m. mandschuricus* NE China, Russian Far East and Korea; *C. m. japonensis* C & S Sakhalin, S Kuril Is and most of Japan; *C. m. connectens* Amami-Oshima and N Ryukyu Is; *C. m. osai* Yaeyama Is; *C. m. colonorum* C, S & E China and Taiwan. Split has been proposed, giving **Japanese Crow** *C. japonensis* (including *connectens* and *osai*) and **Large-billed Crow** *C. macrorhynchos* (including *mandschuricus* and *colonorum*). **HH** Mostly common resident (or present year-round), though some northern birds move south in winter. In Japan and Korea common in urban and rural areas; elsewhere more rural, associated with woodlands, margins of cultivated land, even montane forest. Gathers at large roosts (often with Carrion) during much of year; in winter, typically forages in large mixed flocks. Aggressive, will harass raptors and even pedestrians near nests. Commonly moves on ground using bouncing hop; also walks. **ID** Very large all-black crow, with noticeable but variable purple/blue sheen, especially on scapulars and wings. Head large with flatter crown than Carrion, more vertical forehead emphasised when forehead-feathers erect. Chin-feathers also sometimes erected, giving shaggy-throated appearance resembling Northern Raven (for which sometimes mistaken). Beware small *C. m. osai* of Nansei Shoto; close to Carrion in size, with typical bill shape but less prominent forehead and distinctive Rook-like call, in grassland and agricultural land. In flight, primaries distinctly 'fingered', slightly more widely than Carrion, far more so than Northern Raven. **BP** Bill, black, deep, bluntly curved to tip, so upper mandible distinctly arched (with steep forehead, has very different profile from Carrion); eyes black; tarsi black. **Vo** Almost laughing *awa-awa-awa* to harsh, clear: *kaaw*, *gwarr* or *kaa kaa*, hoarser than Carrion, sometimes intermixed with throaty rattling sounds. When calling, tail usually raised and body often horizontal. **AN** Thick-billed Crow; Jungle Crow.

NORTHERN RAVEN *Corvus corax* CKJR

L 54–69cm; WS 120–150cm; WT 0.8–1.56kg. **SD** N America and Eurasia. In E Asia *C. c. kamtschaticus* in N & NE China, N Korea, Yakutia, Chukotka, Kamchatka, Russian Far East, Sakhalin and S Kuril Is. Scarce winter visitor to Hokkaido. **HH** Favours rugged terrain: forested mountains, river valleys, coasts with rocky capes; in Kamchatka and Hokkaido also volcanoes and calderas. Almost invariably in pairs, or small family parties in winter; larger groups may gather at food sources. Powerful and aerobatic flier. Highly vocal, even playful, engaging in impressive aerobatics. **ID** Large corvid with long, narrow-tipped wings, long, wedge-shaped tail and distinctive vocalisations. All black with metallic purplish-blue gloss; throat-feathers shaggy, giving bearded appearance. Profile similar to Carrion, lacks steep forehead of Large-billed. Wings long, primaries less noticeably 'fingered' than other large black corvids. **BP** Bill large, deep, black; eyes black; tarsi black. **Vo** Distinctive honking and croaking calls, e.g. *prrok-prrok-prrok*; sometimes with high-pitched fluty notes and almost musical *kapon kapon*.

ROOK

pastinator

ad

ad calling

juv

flock

CARRION CROW

orientalis

ad

ad calling

COLLARED CROW

ad

LARGE-BILLED CROW

ad *japonensis*

ad *connectens*

ad *japonensis*

ad

NORTHERN RAVEN

kamtschaticus

BOHEMIAN WAXWING   CTKJR
Bombycilla garrulus

L 19–23cm; WT 50–75g. **SD** Common, widespread Holarctic waxwing. E Palearctic *B. g. centralasiae* ranges across Russia south of treeline from Urals to Sea of Okhotsk, Yakutia, Koryakia and Kamchatka, wintering to NE China, Korea and Japan. Irruptive and nomadic with vagrants reaching Taiwan and Fujian. **HH** Breeds in fairly open coniferous and deciduous boreal forest, favouring birch and pine. In winter, wide range of forest types from uplands to parks, gardens and roadside trees, wherever there are berry-bearing trees. **ID** Largest waxwing; may appear sleek or plump. Overall buffy, grey-brown, somewhat pinkish on face and head, browner on mantle and greyer on lower back and rump. Narrow black mask from forehead to nape (below crest), chin also black (less extensive in ♀). Erectile swept-back crest orange- to pinkish-brown. Flight-feathers and primary-coverts black with broad white tips (forming bars) to primary-coverts and secondaries, white outer tips and yellow inner tips to primaries; secondaries have elongated red, wax-like tips. Tail black with yellow terminal band (broader in ♂). Underparts paler pinkish-buff, dull chestnut/orange on vent. In flight, pale pinkish-buff upperparts contrast with black primaries/secondaries; flight starling-like – fast, erratic. **BP** Bill short, black with yellow-horn to blue-grey base to lower mandible; eyes dark brown; tarsi dark grey/black. **Vo** Highly vocal in winter when utters pleasant high-pitched sibilant trills: *ssirrreee ssirrreee*, or *chirrri chirrri*, with bell-like quality, often given by whole flock before take-off. Song a slow, halting mix of pleasant call notes and harsher sounds.

JAPANESE WAXWING   CTKJR
Bombycilla japonica

L 15–18cm; WT 54–64g. **SD** Breeds only in Russian Far East, around W Sea of Okhotsk, lower Amur River, and NE China. Winters throughout Japan (generally far rarer than Bohemian), and irregularly in NE China south to Shandong, and Korea; accidental Taiwan. Autumn and spring migrations rather late (Nov/Dec, May). Monotypic. **HH** Breeds in mixed deciduous and coniferous taiga with berry-bearing understorey. Winters in same habitats as Bohemian and small numbers often join flocks of latter, though also forms monospecific groups. **ID** Smaller, slightly plainer and greyer than Bohemian. Grey-brown of mantle/back lacks warmer tones, with more extensive grey on wings and tail. Black mask continues to rear of crest, even to tip (does not extend to crest in Bohemian). Erectile crest otherwise orange- to pinkish-brown. Largely grey secondaries and outer webs of primaries, with broad red scapular band and narrow red line at tips of secondaries; lacks yellow tips to primaries and waxen tips to coverts of Bohemian, but has broader white tips to both webs of primaries. Tail has narrower black band and dull red (not yellow) terminal band. Underparts pinkish-buff, with distinctive yellow patch on central belly. Undertail-coverts dull orange/red. In flight, look for narrow red wingbar; yellow belly patch and red tail-band. **BP** Bill short, black; eyes dark reddish-brown; tarsi grey/black. **Vo** Much like Bohemian, but trills higher pitched, more silvery and shorter, *chiri chiri chiri or hiiii hiiii*; also high-pitched whistles.

NORTHERN GREAT TIT *Parus major*   CR

L 14–17cm; WT 14–22g. **SD** Europe across S Siberia to mid Amur Valley. Subspecies *P. m. major*. Some hybridisation between Northern and Eastern in E Amur Valley. **HH** Deciduous and mixed, woodland, scrub, parks and gardens; readily joins mixed flocks in winter. **ID** Large, noisy, with black stripe down pale yellow belly. Head black with extensive white cheeks and small white nape patch; mantle, wing-coverts and rump yellowish-green, tail quite long, blue-grey to black with white outer feathers. Both sexes have belly stripe (broader in ♂, does not reach base of legs in ♀); in juv. does not reach lower belly, and latter also has yellowish underparts and undertail-coverts. Wings blue-grey with prominent white greater covert bar and white fringes to blackish tertials. **BP** Bill black; eyes black; legs rather strong, tarsi dark grey/black. **Vo** Calls including strong *tee-cher tee-cher*.

EASTERN GREAT TIT *Parus minor*   CKJR

L 15cm; WT 11–20g. **SD** Common resident NC China to Sakhalin and Japan. *P. m. minor* NC & NE China, Russian Far East, Sakhalin, Kuril Is, Korea and Japan south to N Kyushu; *P. m. kagoshimae* S Kyushu and Goto Is; *P. m. dageletensis* Ulleung-do, Korea; *P. m. amamiensis* Amami-Oshima and Tokuno-shima; and *P. m. okinawae* Okinawa. **HH** Temperate and subtropical forests, and wooded habitats from mixed deciduous to broadleaf evergreen, scrub, suburban gardens and parks. **ID** Distinctive black, grey and green tit with stripe on belly. Differs from Northern in having mantle alone yellowish-green; back, wing-coverts and rump blue-grey, tail black (with white outer feathers), lacking any yellow tones (broad black stripe from chin to belly is broadest in ♂ and reaches undertail-coverts; narrow in ♀). Wings have white fringes to blackish tertials. Tail largely white below, black at base and narrowly down centre (see Green-backed, p.314). **BP** As Northern. **Vo** Typically demonstrative and repetitive *bee-tsu bee-tsu bee-tsu....*; *tea-cher-tea-cher*, also *tsupi tsupi*. Song contains repeated call notes: *tsutsupii tsutsupii, tsupii tsupii, tsupi tsupi tsupi*. **TN** Formerly within Great Tit. **AN** Japanese Tit, East Asian Tit.

SOUTHERN GREAT TIT *Parus cinereus*   CTJ

L 13–17cm. **SD** S & SE Asia. *P. c. commixtus* S & E China; *P. c. nigriloris* S Nansei Shoto. **HH** Various forest types, but prefers broadleaf evergreen and mangroves; also gardens. **ID** Very dark, grey tit. Head black, cap extends below eye and chin/throat patch to malar region, nape and neck-sides black with small white cheeks. Mantle, back and scapulars mid grey, wing-coverts and wing-feathers black with narrow grey fringes and faint greater covert bar, tertials have white fringes. Tail long, greyish-black with white outer feathers. Black of throat and neck-sides reaches upper chest and central belly as broad black stripe (broader in ♂); underparts dull mid grey, dark grey on flanks. **BP** As Northern. **Vo** Very different from Northern and Eastern, having an electric jarring quality: *zerr zerr zerr*. Song a jaunty, repetitive two-note whistle. **TN** Formerly within Great Tit. **AN** Cinereous Tit.

BOHEMIAN WAXWING

centralasiae

ad

juv

ad

NORTHERN GREAT TIT

major

ad ♂

ad

juv

**JAPANESE
WAXWING**

ad

**EASTERN
GREAT TIT**

minor

ad ♂

SOUTHERN GREAT TIT

ad ♂
nigriloris

ad ♂
commixtus

GREEN-BACKED TIT   T
Parus monticolus

L 12–13cm; WT 12–17g. **SD** Himalayas to S China and Indochina. *P. m. insperatus* in Taiwan. **HH** Broadleaved mid-level montane forests of Taiwan (800–2,800m), though some descend to 200m or even lower in winter. **ID** Small, bright tit recalling Northern Great Tit (p.312) but separated by green mantle and white wingbars. Head, throat and broad belly stripe black, cheek patch white, rounded at rear (smaller, more rounded than in Northern Great), nape white; underparts bright yellow. Wings blue-black with white fringes to coverts forming double white wingbars (yellow in juv.), white tips to black tertials and blue-grey fringes to flight-feathers; tail blue-grey above, black below with white spots near tip (see Northern Great). ♀ has shorter belly stripe, ending in centre of belly (almost to vent in ♂). **BP** Bill short, black; eyes black; tarsi bluish-black. **Vo** Not very vocal; call is a hard, dry *chichichi* and song consists of a simple phrase of 3–4 melancholy whistles, *tsew tsew tsew (tsew)*.

YELLOW TIT   T
Parus holsti

L 13cm. **SD** Endemic resident of mid-elevation montane forests of Taiwan. Monotypic. **HH** Favours canopy of broadleaf forest, but also in conifers amongst or adjacent to broadleaf forest, though seldom visits conifer plantations. Generally at 1,000–2,500m, but small numbers descend lower (to 200m or less) in winter. **ID** Unmistakable bright yellow crested tit. ♂ forehead to nape and prominent spiky crest on rear crown are blackish blue-grey, with white edge to rear of crest; eye patch and line to bill black; mantle and back very dark blue-grey; wings and tail black with blue-grey fringes, white tips to greater coverts and tertials. From bill to just above eye, cheeks and underparts bright lemon-yellow, deepest on face, becoming white on vent with black spot between legs. ♀ slightly duller, less blue, more greenish-grey on mantle, face more lemon-yellow, lacks black belly spot. **BP** Bill black rather stout (recalls Varied Tit, p.316); eyes black; tarsi dark grey. **Vo** Range of calls in alarm, including rather deep *jerrr* and Northern Great Tit-like *pitchou*. Also commonly gives an indignant, abrupt *tsih chichi ch ch cheh!* Song a repetitive, rather high-pitched whistled phrase *pit pit tsee pit pit tsee, tsu-wee-wee tsu-wee-wee-tsu*, a deeper slower *tsu-weee tsu-weee…* or a combination. **AN** Taiwan Tit.

YELLOW-BELLIED TIT   CK
Periparus venustulus

L 10cm; WT 9–12.5g. **SD** Endemic to wooded areas of SE China (to Zhejiang) & NE China (to Hebei), experiencing periodic irruptions, when has reached Korea. Monotypic. **ID** Resembles a small, colourful Coal Tit, with more rounded head and shorter, thicker bill. Cap, chin and throat black, contrasting strongly with broad white nape and narrow white cheeks; mantle dark grey with black streaking, wings blackish with double row of white spots (tips of coverts) on dark wing; tail blackish with white outer tail-feathers. Underparts yellow. **BP** Bill short, thick, black; eyes black; tarsi dark grey. **Vo** Call a high-pitched, nasal *si-si-si-si*. **TN** Formerly *Parus venustulus*.

YELLOW-CHEEKED TIT   CT
Periparus spilonotus

L 14cm; WT 18–23g. **SD** Himalayas to Indochina and S China. In region *P. s. rex* occurs east to Fujian, and a frequent escapee on Taiwan. **HH** Open, broadleaf evergreen forest (800–2,745m). **ID** Large, yellow, black, grey and white, crested tit. ♂ has yellow forehead, yellow lores, face, rear crown and nape, whilst crown, crest, supercilium from eye to ear-coverts, and stripe from chin to vent are black. Mantle dark grey heavily streaked black, wings black with white tips to coverts forming double wingbar, white bases of primaries forming patch, and white tips to tertials. Tail rather long, black with white outer webs to outer feathers. Breast-sides to vent-sides off white. ♀ greenish-yellow with yellow wingbars. **BP** Bill black; eyes black; tarsi blue-grey. **Vo** Alarm call a loud, deep *cherrrr*, deeper than Eastern Great Tit. **TN** Formerly *Parus spilonotus*.

COAL TIT   CTKJR
Periparus ater

L 10–12cm; WT 8–10g. **SD** Common and wide-ranging resident, with *P. a. ater* from NW Europe to E Russia, S Kamchatka, Kuril Is, Sakhalin, south through Russian Far East, NE China, Korea and Hokkaido; *P. a. pekinensis* Shanxi to Hebei and Shandong; *P. a. insularis* S Kuril Is and Japan; and *P. a. ptilosus* high mountains of Taiwan. **HH** Typically found in mature coniferous and mixed forests, in montane and lowland regions, less often in parks and gardens; readily joins mixed-tit flocks in winter, and is often first to respond to 'pishing'. Favours needle-leaved trees more than other tits. **ID** Very small, rather slim black and grey tit, with rather large head and short tail. Black head has distinct, fine crest (very pronounced in Taiwanese *P. a. ptilosus*), white nape and white cheeks. Mantle, rump and tail dark bluish-grey, wings blacker with pale fringes to coverts forming two narrow wingbars, and white tips to tertials. Black bib broadens widely across upper chest, diffusing into grey of neck and buff or off-white of breast and belly. **BP** Bill slender, finely pointed, black; eyes large, irides black; tarsi dark grey. **Vo** Call thin, Goldcrest-like *see-see-see* and *tsuu tsu-tsu-tsu-chi-ririri*. Song is rather repetitive high and fast, *tse tse peen tse tse peen…* and a ringing *s'pee, s'pee, s'pee*, clearer and higher than Eastern Great Tit. **TN** Formerly *Parus ater*.

GREEN-BACKED TIT

insperatus

ad

YELLOW TIT

ad

YELLOW-BELLIED TIT

ad ♂

ad ♀

YELLOW-CHEEKED TIT

rex

ad ♂

COAL TIT

ater

ad

AZURE TIT
Cyanistes cyanus
CKJR

L 12–14cm; WT 10–16g. **SD** Ranges across rather narrow latitudinal belt from Belarus to Baikal and Mongolia, with *C. c. yenisseensis* in S Russian Far East and NE China. Vagrant Hokkaido and reported Korea. **HH** Range of woodland types from pure deciduous to mixed, including riparian willows and thickets around wetlands. Flocks in non-breeding season. **ID** Small, pallid blue-grey and white tit with somewhat long tail. Pale whitish-grey crown and nape, with dark blue-grey eyestripe which joins broader, blacker hind-collar; mantle, scapulars and rump pale blue-grey, tail deeper blue with white outertail-feathers and rounded white tip; wings largely deep blue with broad white bar on greater coverts and prominent white tips to tertials. Underparts white. **BP** Bill very stubby, black; eyes black; legs black. **Vo** Call a thin repetitive *chi chwee chi chwee chi chwee...*, *jii jii* or *tsee-tsee-tserrr de-de-de*. Song variable, but includes chattering and churring notes and short trills. **TN** Formerly *Parus cyanus*.

VARIED TIT
Poecile varius
CTKJ

L 11–15cm; WT 16–18g. **SD** E Asian endemic; resident Japan (and S Kuril Is) from Hokkaido to Nansei Shoto; also NE China, Korea, Taiwan. Three species may be involved. **Japanese Varied Tit**: *P. v. varius* in NE China, Korea, S Kurils and Japan; *P. v. sunsunpi* on Tanega-shima and Yaku-shima; *P. v. namiyei* on N Izu Is; *P. v. amamii* in N Ryukyus from Amami to Okinawa; and *P. v. olivaceus* on Iriomote I. **Owston's Varied Tit** *P. (v.) owstoni* on S Izu Is (Miyake, Mikura and Hachijo). **Taiwan Varied Tit** *P. (v.) castaneoventris* in Taiwan. **HH** Deciduous and evergreen broadleaf forests, also woodland, scrub, parks and gardens in winter when readily joins mixed-tit flocks. **ID** Large, noisy, blue-grey and pale chestnut tit. Forehead, lores, cheeks and small nape patch off-white or creamy yellowish-white; rest of head, crown, face-sides to eye level, chin and throat black. Hind-collar and most of underparts deep rufous or pale to mid chestnut, with buff upper breast. Mantle, wings, rump and tail plain blue-grey. 'Japanese Varied Tit' races largely inseparable in field, but become progressively darker N–S. Owston's Varied Tit, however, is considerably larger and darker, with rusty-orange forehead, small nape patch and cheeks, more extensive black bib, and chestnut of cheeks is continuous with all-chestnut underparts. Upperparts darker brownish blue-grey. Differs also in breeding ecology and vocalisations. 'Taiwan Varied Tit' much smaller than 'Japanese' with pure white face and nape patch, lacking any buff or orange; upperparts dark blue-grey and underparts deep chestnut. **BP** Bill rather strong, black (particularly deep in *P. v. owstoni*); eyes black; tarsi rather strong, black. **Vo** Calls include a nasal *nii nii* or *tsuee tsuee*, rasping *vay vay vay* and weak *tsu tsuu tsu*, also a thin, high *see see see*, recalling Goldcrest, in alarm a loud *tsutsu bee bee bee*. Song a slow thin *tsuu tsuu pee tsuu tsuu pee* and repetitive *tsi-turrr*, with much regional and local variation. In Taiwan, calls higher pitched, and also gives a jaunty *chay chikyechay cherr* and high, rapid *tsititititititi*, which may function as a song. **TN** Formerly *Parus varius*.

MARSH TIT
Poecile palustris
CKJR

L 11–13cm; WT 10–13g. **SD** Common resident of Europe, with disjunct population in E Asia. *P. p. brevirostris* (sometimes split as **Asian Marsh Tit** *P. brevirostris*) occurs across E Siberia, N Manchuria and extreme N Korea; *P. p. ernsti* on Sakhalin; *P. p. hensoni* S Kurils and Hokkaido; *P. p. jeholicus* in N Hebei, Liaoning and N Korea; and *P. p. hellmayri* in NE China and S Korea. **HH** Wide range of forest types from mature deciduous to woodland edge, riverine thickets, scrub, urban parks and gardens, commonly forming mixed flocks with other tits, nuthatches and woodpeckers in winter. **ID** Small and pale with structure of Eastern Great, head rather small, tail rather long and square-tipped. Crown and nape glossy black, small rather neat black bib, mantle, wings, rump and tail uniform dull ash-grey brown; underparts, cheeks and head-sides white, breast and belly off-white, with grey/buff wash on flanks. Wings often have pale fringes to secondaries, thus very similar to Willow (criteria for separation, such as call and wing patterns, of western subspecies less helpful in E Asia). **BP** Bill short, stubby (thicker and blunter than Willow), black with paler bluish-grey cutting edges. Eyes large, black; tarsi dark grey. **Vo** Various calls include thin *tseet*, and hard churring *chichi jeejee*. Strongly whistled spring song *pew-pew-pew* or *cho cho cho* is reminiscent of Eurasian Nuthatch and very like Willow. Song is also similar to Willow, but stronger. **TN** Formerly *Parus palustris*.

WILLOW TIT
Poecile montanus
CKJR

L 11–14cm; WT 8–14g. **SD** Common resident from NW Europe to Russian Far East, NE China and Japan: *P. m. baicalensis* Yenisei to Russian Far East, NE China and N Korea; *P. m. anadyrensis* Koryakia and Anadyr; *P. m. kamtschatkensis* Kamchatka to Kuril Is; *P. m. sachalinensis* Sakhalin; *P. m. restrictus* throughout Japan. **HH** Wide range of forest types from coniferous and mixed to mature deciduous and woodland edge, scrub and urban parks and gardens, but favours areas with dead trees more than Marsh Tit. **ID** Very similar to Marsh Tit, but differs slightly in structure, appearing larger headed and thicker necked; shorter tail has more rounded tip. Crown and nape duller black than Marsh, lacking gloss; bib less neat, generally more extensive than Marsh; and pale fringes to tertials and secondaries form more prominent wing panel. **BP** Bill short, stubby (finer, more pointed than Marsh), black; eyes large, black; tarsi dark grey. **Vo** Marsh and Willow Tits of W Europe readily distinguished by voice, but eastern Marsh calls more closely resemble those of Willow. Call a harsh, nasal *chichi jeejee*. Song a high, strident *tsupii tsupii pipii pipii* or *cho cho cho*, slightly more metallic then Marsh. **TN** Formerly *Parus montana*.

AZURE TIT

yenisseensis

ad

VARIED TIT

ad *castaneoventris*

ad *varius*

ad *owstoni*

MARSH TIT

ad *brevirostris*

Marsh

ad *hellmayri*

Willow

WILLOW TIT

ad *restrictus*

Songar
see plate 145

ad *baicalensis*

SONGAR TIT
Poecile songara
C

L 12–14cm; WT 8–12g. **SD** Tien Shan Mountains through N & SC China to NE China. *P. s. stoetzneri* just reaches region in Hebei. **ID** Resembles Siberian and Willow (p.316) Tits, but mantle, wings and tail dark greyish-brown, and seemingly lacks wing panel. Cap, from forehead to nape dull brownish-black, black chin and throat patch extensive and broad, leaving cheek patch broad on face-sides but narrowing to a bar below eye to bill base. Underparts dusky grey-brown with rusty wash on flanks; thus much darker than Willow and lacking contrast between grey underparts and rusty flanks of Siberian. **BP** Bill blunt-tipped, black; eyes black; tarsi blue-grey. **Vo** Calls include harsh *bjee bjee bjee* and *tsujee tsujee* notes. **TN** Until recently considered conspecific with Willow Tit. Formerly known as *Parus songara*.

SIBERIAN TIT
Poecile cincta
R

L 12–14cm; WT 11–14g. **SD** Northern N America, Scandinavia to Yakutia and inland Chukotka. Subspecies in region *P. c. cincta*. **HH** Prefers coniferous taiga forest, but occasionally in deciduous. Essentially resident, but may wander in winter, when also joins mixed-species flocks. **ID** Resembles Marsh or Willow Tits (p.316), but mantle and back tawny-brown, wings and tail dark grey, and cap, chin and throat blackish-brown contrasting with extensive, dirty white cheeks. Underparts off-white, dusky, with rusty-brown wash on flanks. **BP** Bill rather blunt-tipped, black; eyes black; tarsi dull grey-brown. **Vo** Song a harsh, buzzy series of notes, *chiyurr chiyurr chiyurr…*, whilst contact call is thin, a fine *tsee* or loud *tsin-tsin*. **TN** Formerly *Parus cincta*. **AN** Grey-headed Chickadee.

SULTAN TIT
Melanochlora sultanea
C

L 20cm; WT 34–49g. **SD** Himalayas to SE Asia and southern China. In region *M. s. seorsa* occurs S & SE China east to Fujian. **HH** Canopy of lowland evergreen forest. **ID** Very large black and yellow crested tit. ♂ largely glossy blue-black with bright yellow crown and erect crest, and bright yellow underparts from chest to vent. ♀ similarly crested, but is olive to dark brown where ♂ is blue-black, and has scalloped

upperparts. Juv. as ♀, but dark parts duller and crest shorter. **BP** Bill somewhat stout, black; eyes dark brown; tarsi dark grey. **Vo** Various calls include a hard *tji-jup*, squeaky *whit* and shrill *squear-squear-squear*. Song a mellow *piu-piu-piu-piu…*.

YELLOW-BROWED TIT
Sylviparus modestus
C

L 9–10cm; WT 5–9g. **SD** Kashmir to Laos and Vietnam; *S. m. modestus* in SE China to Wuyishan and mountains of N Fujian. **HH** Montane broadleaf evergreen forest, often with mixed-species feeding flocks. **ID** Small, atypical plain, compact and short-tailed tit. Upperparts entirely greyish-olive-green, with short, pale yellow-olive supercilium, short, rounded crest on rear crown, and pale yellow-olive bar on tips of greater coverts. Underparts paler yellowish-olive. **BP** Bill short, black; eye appears large on rather plain face, irides black; tarsi dark grey. **Vo** Various calls include a thin *psit*, harder *chip* and rapid trilling *sisisisisisi*. Song a series of up to 15 mellow notes: *pli-pli-pli-pli-pli…* or *piu-piu-piu-piu…*.

CHINESE PENDULINE TIT
Remiz consobrinus
CTKJR

L 9–12cm; WT 7.5–12g. **SD** N & NE China to Transbaikalia and Russian Far East, migrating through or wintering locally (Oct–May) in E China, Korea and C Honshu south to Kyushu; accidental in winter to Taiwan. Monotypic. Division of the penduline tits into as many as four species remains debatable, and is a subject ripe for additional study. **HH** Generally in or around reedbeds, marshes with tall grasses, and thickets and open woodland near wetlands, to 2,000m; forms small flocks in winter, but numbers and movements at this season erratic. **ID** A small reedbed specialist. Rather pale; ♂ has grey crown and nape contrasting strongly with black mask from forehead to ear-coverts, bordered above with white; upperparts warm rufous-brown, especially on neck-sides and mantle, to grey-brown back and rump. Wings and tail black, former have chestnut wing-coverts forming wingbar in flight. Creamy white chin/throat, grading to off-white and buff on belly. Tail short, notched. ♀ duller, with brown mask and lacks warm rufous tones. **BP** Bill finely pointed, conical, grey; eyes black; tarsi black. **Vo** Call a very thin, high-pitched, but descending, drawn-out *tseeooo*; also a ringing *tsi* or *chii-chii-chii* (similar to Japanese White-eye, but thinner), given by foraging flocks. Song *chiiichuri chuichuii*. **TN** Formerly part of Eurasian Penduline Tit *R. pendulinus*.

SONGAR TIT

stoetzneri

ad

SIBERIAN TIT

cincta

ad

SULTAN TIT
(not to scale)

seorsa

ad

CHINESE PENDULINE TIT

ad ♀

ad ♂

YELLOW-BROWED TIT

modestus

ad

GREY-THROATED MARTIN TJ
Riparia chinensis

L 10–13cm; WT 9.5–17g. **SD** Resident of S Asia, SW China, the Philippines, and in region *R. c. chinensis* is resident in Taiwan; accidental S Japan. **HH** Close to wetlands and over various open grassland habitats. **ID** Closely resembles Sand Martin, but underparts duskier, off-white to buff on chin/throat, lacks contrast between throat and ear-coverts, and has less clear demarcation between upperparts and underparts; lacks distinctive brown breast-band of Sand Martin, instead has grey-brown wash on chest. In flight, underwing rather dark. **BP** Bill tiny, black; eyes black; tarsi brown/black. **Vo** Calls include a distinctive Long-tailed Tit-like rasping *chitrr* and ringing *chit-it*, in addition to quiet twittering calls very similar to Sand Martin. **TN** Formerly within Brown-throated Martin *R. paludicola*.

SAND MARTIN CTKJR
Riparia riparia

L 12–13cm; WS 26.5–29cm; WT 11–19.5g. **SD** Widespread Holarctic martin: *R. r. taczanowskii* ranges across Russia from Lake Baikal to S Russian Far East, across Yakutia to E Chukotka and Kamchatka; *R. r. ijimae* in Sakhalin, Kuril Is and N Japan. Summer visitor (May–Aug); both E Asian races winter from E China south to SE Asia. Migrant through Korea and Taiwan. **HH** Almost exclusively near water; along rivers, at lakes and wetlands, nesting in colonies in vertical sand cliffs or banks. **ID** Small, compact, rather slender hirundine. Ad. grey-brown with pale underparts and distinctive, clearly defined brown breast-band. Head small, 'face' dark, with white crescent on neck curving up behind auriculars. Upperparts uniform mid-brown; underparts white or pale buff from chin to vent, with broad, mid brown breast-band. Young have paler fringes to mantle and wing-covert feathers, grey-buff face and throat, and less clearly marked breast-band. At rest, wingtips reach tail tip; in flight, wings appear rather broad in 'arm'; dark on upper surface, dusky on underwing; tail short, clearly notched. **BP** Bill tiny, black; eyes black; tarsi brown/black. **Vo** Common flight calls dry and scratchy: *chirr-chirr*, also *ju ju ju*, or *juku juku juku ju ju*; around colonies, flocks give excited harsh chattering including typical *trrrsh*. **AN** Bank Swallow.

PALE MARTIN CTR
Riparia diluta

L 12cm. **SD** Occurs from Pakistan, Kazakhstan and C Siberia east to Lena River, Transbaikalia and C, S & SE China: *R. d. gavrilovi* just reaches extreme NW of region at Lena River; *R. d. transbaykalica* in Transbaikalia; *R. d. fohkienensis* in C & SE China; accidental Taiwan. **HH** Large bodies of water. **ID** Difficult to separate from Sand and Grey-throated Martins, and field identification criteria remain uncertain, but upperparts particularly pale grey-brown, chin and throat white with very little contrast on face; underparts have poorly defined breast-band (though *R. d. gavrilovi* has more clear-cut breast-band). In flight, paler than related species with extremely shallow tail notch and very pale underwing. **BP** Bill tiny, black; eyes black; tarsi brown/black. **Vo** A grating twittering and hard *ret* or *brrit* around colonies. **TN** Formerly within Sand Martin. **AN** Eastern Sand Martin.

TREE SWALLOW R
Tachycineta bicolor

L 12–13cm; WS 34–37cm; WT 17–25.5g. **SD** Widespread N American species migrating northwest as far as W Alaska, which has strayed to St Lawrence (USA), and to NE Russia (e.g. Wrangel I.) an old Japanese record was probably an individual brought from the Aleutian Is aboard a fishing vessel. Monotypic. **HH** Around wetlands and open areas. **ID** Smart, very contrasting, rather stocky swallow between Barn Swallow (p.147) and Sand Martin in size, with broad wings and short, barely notched tail. Upperparts very dark. Underparts from chin to vent clean white. ♂ has distinctive blue-green sheen on head, back, wing-coverts and rump. ♀ has less distinctive green cast to otherwise blackish plumage, and white tertial tips. Young uniform grey-brown above with white tertial tips and may show incomplete dusky breast-band leading to confusion with Sand Martin. At rest, wingtips reach or extend just beyond tail. In flight, from above appears uniformly dark, almost blackish, with white crescents visible at rump-sides at all ages; from below dark underwing contrasts with white body. **BP** Bill tiny, black; eyes black; tarsi pinkish-grey. **Vo** Twittering and soft chirping notes, *quuii, tsuwi* and scratchy *tzeev*.

GREY-THROATED MARTIN

chinensis

ad

SAND MARTIN

ijimae

ad

PALE MARTIN

gavrilovi

ad

ad

TREE SWALLOW

1st-win

ad

ad

PURPLE MARTIN   R
Progne subis

L 19–21cm; WT 48–64g. **SD** Long-distance migrant between S and N America. Polytypic. Accidental NE Russia, subspecies uncertain. **HH** Open areas, forest edge and waterbodies. **ID** Large, stocky martin. ♂ has glossy blue-black body with black wings and tail (moderately forked) and sooty-grey underwing-coverts. ♀ duller, forehead, neck and throat grey, throat finely streaked, underparts greyish-white with dusky streaks. Juv. has upperparts and throat grey-brown, underparts grey-white; 1st-year ♂ like ad. ♀ but has blue feathers on head and underparts; 1st-year ♀ has paler underparts and browner upperparts than ad. **BP** Bill short, broad-based, black; eyes dark brown; tarsi dark brown. **Vo** Song comprises guttural gurgling; calls include a *tchew-wew*, *zwrack* and *zweet*.

BARN SWALLOW   CTKJR
Hirundo rustica

L 17–19cm; WS 32–34.5cm; WT 16–22g. **SD** Almost global. Common summer visitor (Mar–Sep) over much of region almost to Arctic Circle (scarcer and later arriving in N). *H. r. gutturalis* across China, Taiwan, Japan, Korea, Kuril Is and lower Amur River; *H. r. tytleri* breeds mainly SC Siberia and Mongolia, also occurs in NE China, so may occur elsewhere on migration. *H. r. saturata* NE China from Hebei to Russian Far East, Sea of Okhotsk, Chukotka, Kamchatka and Commander Is. E Asian subspecies winter from N Australia to S Asia occasionally northeast to Taiwan. *H. r. erythrogaster* (a possible split as **American Barn Swallow** *H. erythrogaster*) migrates northwest to Alaska. As yet unrecorded, it is a likely vagrant to NE Russia. **HH** Generally in lowland habitats from urban to rural areas, often near water, particularly in areas with domestic livestock. Nests under eaves of buildings. Congregates in roosting flocks, especially on or prior to migration, sometimes in trees or on wires. **ID** Highly active aerial insectivore. Mid-sized swallow. Ad. has glossy steely-blue upperparts and largely white underparts. Forehead, chin and throat deep brick red. *H. r. gutturalis* has narrow blue-black band on upper chest bordering chestnut throat (see Pacific Swallow), then white to vent, but some have buff cast to underparts; underwing-coverts also white. *H. r. tytleri* is dusky brick orange over entire underparts. *H. r. erythrogaster* has rufous throat that reaches chest, breast-band is reduced to sides; throat usually darker than rest of underparts. Buff-orange underparts and incomplete blue upper-breast-band provide ready separation from most Eurasian forms, but ♀ *erythrogaster* can be almost white-bellied, though usually show some faded peach colour; rarely if ever appear as clean white as Eurasian forms. Separation from *H. r. tytleri* difficult though tends to be noticeable contrast between rather orange throat of *erythrogaster* and more washed-out, almost peach-coloured underparts; *tytleri* tends to be more evenly rufous-orange over entire underparts. However, underparts of many ♂ *erythrogaster* can appear quite consistently rufous-orange from chin to vent, with no obvious contrast between throat and rest of underparts. Asian races tend to have narrower, but more distinct breast-band and are either white below or have more uniform rufous colouring to throat and underparts than *erythrogaster*. Tail long (streamers

2–7cm) in all races, deeply forked, outer feathers longest in ♂ and very short in juv. Juv. duller, dusky below, with dull orange throat and indistinct chest-band. Flight-feathers blue-black, tail has subterminal band of white spots revealed when spread. Flight fast and aerobatic, either high, or low over ground or water. In flight, underwing-coverts clean white (except *H. r. tytleri*; also see Pacific). **BP** Bill short (gape wide) black; eyes black; tarsi black. **Vo** Short, hard twittering *chubi*, or *veet-veet*. Song a rapid, rambling squeaky or scratchy twittering *pichi kuchu chiriri....* Alarm call a sharp and agitated *vitveet*, *siflitt* or *flitt*!

PACIFIC SWALLOW   TJ
Hirundo tahitica

L 13cm; WT 11–16g. **SD** Locally common resident of S India, SE Asia and Philippines to New Guinea and Pacific islands. *H. t. namiyei* Taiwan, Lanyu and Ryukyu Is. Possible split, with *H. t. namiyei* becoming **Small House Swallow** within *H. javanica*. **HH** Various lowland habitats from suburban (even urban in Taiwan) to rural areas, often near water. Nests under eaves or bridges. **ID** Small to mid-sized swallow with glossy steely blue upperparts, and largely grey or buff underparts. Smaller than Barn Swallow, which it resembles. Ad. has duskier, greyer underparts, lacks blue-black chest-band (but see *H. r. gutturalis*), and has short tail. Forehead, chin and throat patch deep brick red; underparts dusky grey-buff, including underwing-coverts; undertail-coverts dark with pale fringes creating distinctive hatched pattern. Tail short, lacks outer 'streamers,' slightly forked with subterminal row of white spots. Juv. closely resembles juv. Barn Swallow, but duller, duskier below and lacks chest-band. Flight-feathers blue-black. Flight fast and aerobatic, often low over ground or water. In flight, underwing-coverts dusky-grey. **BP** Bill short (gape wide) black; eyes black; tarsi black. **Vo** Similar to Barn, but shriller. Calls frequently in flight, giving *je je je* or *ju ju ju* and a starling-like *skreet* or *vitt*. Song a rapid rambling series also resembling Barn Swallow, *juku juku tsiriri....*

EURASIAN CRAG MARTIN   CK
Ptyonoprogne rupestris

L 14–15cm; WS 32–34.5cm; WT 17–30g. **SD** NW Africa and S Europe to C Asia and NE China, wintering in Mediterranean, W, N & NE Africa, Middle East, India and S China. Ranges across N China to Bohai Gulf. Reported Korea. Monotypic. **HH** Around crags, ravines, cliffs and gorges; also sometimes around buildings. **ID** Medium-sized, dusky martin, resembling Sand Martin (p.320). Upperparts plain dusky-brown, darker on wings and tail, but latter has subterminal row of large white spots, tail slightly notched. Underparts plain dusky-brown, paler on throat, darker on belly, very dark on vent, with pale buff fringes. Flight more sluggish than other hirundines; underwing-coverts very dark blackish-brown. **BP** Bill short, broad, black; eyes dark brown; very short tarsi yellowish-brown. **Vo** Song a quiet, fast series of twittering notes; also *prrrt*, a hard rather high-pitched clicking *pli*, and Northern House Martin-like *tshir* and *trit*.

PURPLE MARTIN
(not to scale)

ad ♂

ad ♀

BARN SWALLOW

ad *gutturalis*

juv *gutturalis*

ad *tytleri*

ad

PACIFIC SWALLOW

ad

namiyei

EURASIAN CRAG MARTIN

NORTHERN HOUSE MARTIN   CKJR
Delichon urbicum

L 13–14cm; WS 26–29cm; WT 15–23g. **SD** Palearctic and northern Africa. *D. u. lagopodum* widespread from Yenisei to NE China, NE Russia, Yakutia, Chukotka, N Koryakia coast and possibly Kamchatka; winters south to Africa and SE Asia. Rare migrant to Japan and Korea, some winter SE China. **ID** Very clean black and white, typically larger, longer tailed with deeper fork, and more glossy upperparts than Asian House Martin. More sharply defined black cap does not extend to cheeks or chin. Ad. upperparts steel blue-black, often glossy, strong contrast with large white rump, uppertail-coverts and lower back; underparts, from chin to vent, very white. Juv. duskier, more closely resembles Asian. Wings broader based and shorter than swallow; underwing-coverts whitish-grey; tail short, sharply forked (more deeply than in Asian), undertail-coverts white. **BP** Bill short, black; eyes black; legs feathered (white), feet grey-pink. **Vo** Noisy at colonies, and in flight away from colonies; a steady pleasing twittering *juriri juriri*, a stronger *prrit* or *brit*, very similar to Asian, but lower pitched.

ASIAN HOUSE MARTIN   CTKJR
Delichon dasypus

L 13cm; WT 18g. **SD** Restricted to Himalayas, SE & E Asia. *D. d. dasypus* SC China to Japan, Korea and NE China, Russian Far East and Kuril Is, migrating to SE Asia, Indonesia and Philippines; *D. d. nigrimentale* S & E China and Taiwan, migrating to SE Asia. **HH** In montane or coastal regions with cliffs, nesting on crags, under bridges or in tunnels. In Taiwan summers at high and mid elevations, in foothills in winter, on plains on migration. **ID** Small, compact, dusky martin. *D. d. dasypus* much like Northern but less neat. Black cap extends onto face and ear-coverts, and just below bill on chin. Upperparts dull steel blue-black, contrasting with rather small, grey-streaked, white rump (smaller than Northern). Throat and neck-sides grey-white, breast, flanks and belly dusky grey-white, even buff, often with narrow streaking; underwing-coverts, undertail-coverts usually dusky off-white. *D. d. nigrimentale* cleaner, whiter on underparts except wing-coverts. In flight, uppertail-coverts glossy bluish-black, tail fork shallower than Northern; when tail fanned appears square-ended. **BP** Bill short black; eyes black; legs feathered (white), toes grey-pink. **Vo** *Juriri juri* or *ju ju piriri*.

RED-RUMPED SWALLOW   CTKJR
Cecropis daurica

L 16–17cm; WS 32–34cm; WT 19–29g. **SD** Widespread Europe, Africa and E Asia. *C. d. japonica* a fairly common, local summer visitor S, E & NE China to E Amurland, also Sakhalin, Korea and Japan (scarce Hokkaido), wintering south to SE Asia and Australasia; *C. d. daurica* from N Mongolia, Inner Mongolia to W Amurland, winters in S & SE Asia. **HH** Lowland habitats typically close to water. Nest a mud bottle attached under eaves, bridges or cliffs. **ID** Slightly larger than Barn (p.322), which it resembles. ♂ upperparts glossy blue-black, rump patch brick red; tail

blue-black with very long outer tail-feathers (5–6cm). Supercilium, head-sides to rear of ear-coverts and nape-sides dusky brick red; many may have rufous hind-collar. Face and underparts narrowly streaked dark grey (more heavily streaked in *C. d. japonica* making field separation from Striated difficult, though *C. d. japonica* clearly has red extending quite far around neck like a collar; in Striated red confined to neck-sides). Vent white, undertail-coverts black. ♀ has shorter tail. Juv. also has shorter tail, duller browner upperparts, paler rufous areas and less distinct streaking. Flight-feathers blue-black; pale rump and white vent contrast with black undertail-feathers giving tail oddly detached look. In flight, underwing-coverts creamy buff in *C. d. japonica* (see Striated), and not heavily streaked. Flight fast and aerobatic. **BP** Bill short black; eyes black; tarsi black. **Vo** Barn Swallow-like *jubi chibi* notes, but also a nasal *tveyk*. Song a rapid complex series of twittering notes: *jubitt-juru-juri-churujuri*, slower, lower and harsher than Barn; in alarm a sharp *kiir!* **TN** Formerly *Hirundo daurica*.

STRIATED SWALLOW   T
Cecropis striolata

L 19cm; WT 22g. **SD** Mainly resident, SE Asia to Philippines north to Taiwan. *C. s. striolata* a common resident C & S Taiwan. **HH** Lowlands to *c*.1,500m, usually near cultivation. **ID** Closely resembles Red-rumped; criteria for field separation unclear, though generally larger. Orange-red on head-sides perhaps less extensive, not extending to near centre of nape, thus lacks 'collar', and blue-black crown less cap-like. Face and underparts very heavily streaked, with broader black streaks (but note *japonica* Red-rumped is intermediate between *daurica* Red-rumped and Striated in extent of streaking). Vent white, undertail-coverts black. Flight somewhat lazier, slower than other swallows, often soars. In flight, underwing-coverts pale off-white with buffy wash (see Red-rumped). **BP** Bill short black; eyes black; tarsi black. **Vo** Flight calls differ from Red-rumped: a harsh *pin* and *quitsch*, or a gravelly *kvertch*. Song consists of soft twittering. **TN** Formerly *Hirundo striolata*.

CLIFF SWALLOW   R
Petrochelidon pyrrhonota

L 13–15cm; WS 34cm; WT 17–27g. **SD** Widespread N America; winters in S America. *P. p. pyrrhonota* reaches as far north as W & C Alaska; has strayed to Aleutians and NE Russia (e.g. Wrangel I.). **ID** Rather stocky, slightly larger than Sand Martin (p.320), with broad, rounded wings and short, square tail. Recalls Red-rumped, but tail square-ended, and lacks black vent. Ad. has steel-blue cap and mantle, rufous face and collar, pale rufous rump, white forehead and black chin, blackish-grey wings and tail, buff underwing-coverts, greyish-buff underparts with black patch on upper breast and dark spotting on undertail-coverts. Juv. duller with dark ear-coverts, pale nape and rump, and spotted undertail-coverts. In flight, note pale buff collar, pale tawny rump, white streaking on mantle, 'braces' on scapulars (not in juv.), and broad wings. **BP** Bill, very short, black; eyes black. **Vo** A low, husky *verr* and dry rolled *vrrrt*.

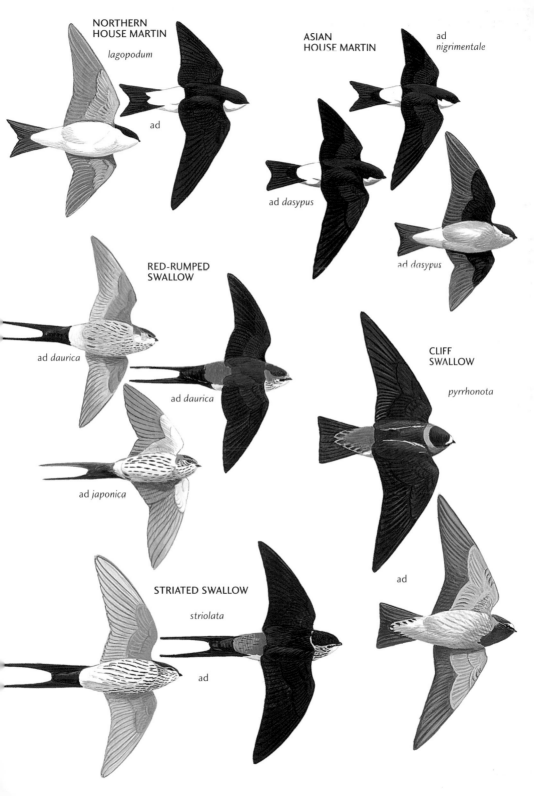

NORTHERN
HOUSE MARTIN

lagopodum

ad

ASIAN
HOUSE MARTIN

ad *nigrimentale*

ad *dasypus*

ad *dasypus*

RED-RUMPED
SWALLOW

ad *daurica*

ad *daurica*

ad *japonica*

CLIFF
SWALLOW

pyrrhonota

ad

STRIATED SWALLOW

striolata

ad

LONG-TAILED TIT
Aegithalos caudatus
CKJR

L 13–17cm; WT 7–10g. **SD** Common Palearctic resident. *A. c. caudatus* ranges from N Europe across Russia to N Sea of Okhotsk and Kamchatka, south through S Russian Far East, Sakhalin, S Kuril Is, Hokkaido, N Korea and NE China; *A. c. vinaceus* in NE China to Hebei; *A. c. glaucogularis* in E China north to Zhejiang; *A. c. trivirgatus* in Honshu; *A. c. kiusiuensis* in Shikoku and Kyushu; and *A. c. magnus* in C & S Korea, and on Tsushima (Japan). Some consider splitting this taxon into two: **Northern Long-tailed Tit** *A. caduatus* (including *caudatus, trivirgatus, kiusiuensis* and *magnus*) and **Silver-throated Tit** *A. glaucogularis* (including *glaucogularis* and *vinaceus*). **HH** Mixed and deciduous forest, woodland edge, and in winter also scrub and reedbeds, often in small flocks, sometimes with other species. **ID** Small, pale, long-tailed bird, with short rounded wings; tail of up to 9cm. Ad. (*A. c. vinaceus*) largely white with broad black band from lores to upper back; wings, rump and tail also black, latter with white outer feathers. Band of dull pink extends across scapulars to lower back. Underparts whitish, sometimes with pink wash on flanks, brighter on vent. Northern *A. c. caudatus* has entirely white face and head-sides, whilst southern *A. c. glaucogularis* of E China and other races in Japan and Korea have broad dark bands on head-sides and are generally darker. Juv. shorter tailed, duskier and lacks pink. **BP** Bill short, stubby, black; eyes small, black; tarsi black. **Vo** Range of thin, high-pitched trisyllabic *sree-sree-sree* calls and deeper, stronger churring *cherrrr cherrrr* or softer *prrrr prrrr* notes. Song a complex, thin soft twittering *chii-chii-chii-tsuriri-juriri*.

BLACK-THROATED TIT
Aegithalos concinnus
CT

L 10–11cm; WT 4–9g. **SD** Resident from Himalayas to Indochina, with *A. c. concinnus* in S, C & E China, north to Changjiang River, and Taiwan, though some evidence of migration (Matsu, Taiwan). **HH** Open mixed forest in mountains, from 700–3,200m, often in family parties or flocks, and mixing with other species. Locally abundant Taiwan. Small numbers descend in winter to 500m or less. Spring and autumn migrants on Matsu found below 150m. **ID** Small active, often sociable bird. Forehead to nape bright chestnut, as is breast-band and flanks. Upperparts, including wings and tail, dull blue-grey. Black mask extends from bill to sides of nape; underparts white with triangular black spot on throat, which is less defined in 1st-year. **BP** Bill tiny, black; eye appears large, irides bright lemon-yellow; tarsi pale pink. **Vo** Foraging groups give high-pitched *tsip* and *trr* contact calls. Song a short high-pitched phrase. **AN** Red-headed Tit.

BIMACULATED LARK
Melanocorypha bimaculata
KJ

L 16–17cm; WT 47–62g. **SD** W & C Asian species, migrating to NW China. Accidental in winter to Ryukyu Is (Dec–Apr); reported Korea. **HH** Semi-desert, rocky and sandy steppe with sparse vegetation, and on mountainsides. **ID** Large, heavy-set, heavy-billed, short-tailed lark, only likely to be confused with extralimital Calandra Lark *M. calandra* (which breeds closer to region than Bimaculated, in SW Mongolia and SC Siberia). Upperparts sandy-brown with dark centres to most feathers, crown finely spotted dark, wing-feathers dark brown with sandy fringes; most prominent feature is black patch forming narrow collar on foreneck and streaking on sides of upper breast. Head large, strongly patterned, with prominent white supercilium, darker crown, dark ear-coverts, dark line through eye, white crescent below eye forward to bill, and white chin with narrow malar stripe. Tail short, blackish-brown, including sides, but has broad white band at tip. In flight, upperwing lacks white trailing edge of Calandra, and little contrast between upperwing-coverts and remiges; tail short (often fanned), white-tipped (white-sided in Calandra), and underwing brownish-grey (blackish in Calandra). **BP** Bill large, heavy, rather blunt, grey culmen, otherwise yellow; eyes black; strong legs dull orange. **Vo** Calls generally harsh and drawn-out, also recalling Eurasian Skylark – a chirruping *biru*, but with drier, rasping *tchur* and *ju ju* notes recalling Asian Short-toed.

MONGOLIAN LARK
Melanocorypha mongolica
CJ

L 18–22cm. **SD** Occurs in Tuva, Mongolia, Transbaikalia, and N China east to Inner Mongolia and Beijing. Rare winter visitor to Beijing and Hebei; accidental Hokkaido. Monotypic. **HH** Damp grassy areas in steppe. **ID** Large, pale lark with very large bill. Face plain, pale sandy-brown, with broad eye-ring and stripe extending back from eye. Upperparts warm brown on crown and nape, somewhat greyish on hindneck; mantle grey-brown with black streaking, blackish-brown rump and tail, with white outer tail-feathers. Broad black band across neck-sides to throat and breast. Wings blackish-brown, with buff fringes to coverts and tertials, and broad white or pale fringes to flight-feathers. Underparts: chin white, breast to vent off-white with warmer rusty wash on flanks. Juv. lacks rufous and black neck-band. **BP** Bill pale pink or yellow with grey culmen and tip; eye appears small, irides dark brown; tarsi pale pinkish-grey. **Vo** Various harsh and high-pitched calls.

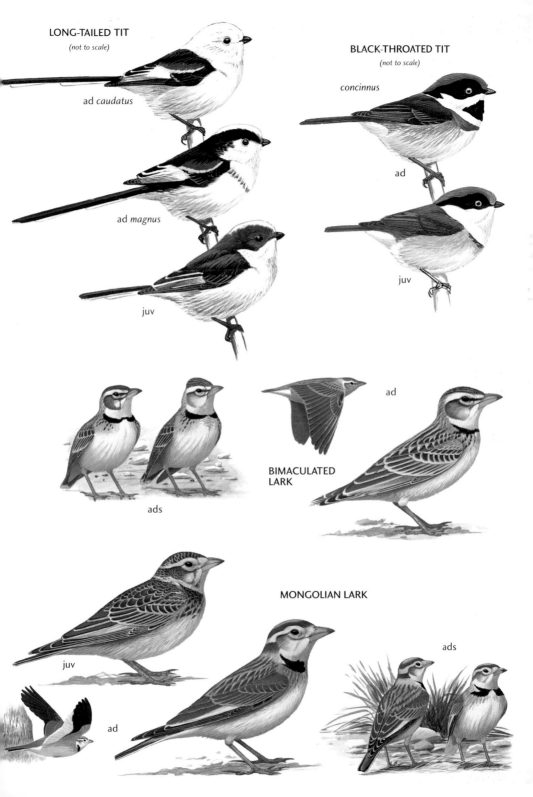

LONG-TAILED TIT
(not to scale)

ad *caudatus*

ad *magnus*

juv

BLACK-THROATED TIT
(not to scale)

concinnus

ad

juv

ads

BIMACULATED LARK

ad

ad

MONGOLIAN LARK

juv

ad

ads

GREATER SHORT-TOED LARK CTKJ
Calandrella brachydactyla

L 14–15cm; WT 20–26g. **SD** Black Sea to N Mongolia, Transbaikalia and W & N China to Inner Mongolia; winters east to Bohai Gulf, perhaps further south. NE birds either *C. b. longipennis* or *C. b. dukhunensis*. Rare migrant and winter visitor Korea and Japan. **HH** Open steppe with sparse vegetation; visits agricultural fields. **ID** Small, plain lark with unstreaked breast and variable black patches on upper-breast/neck-sides. Upperparts sandy-brown, crown somewhat darker, appearing slightly capped, with dark streaking; broad pale supercilium and eye-ring, mantle and back sandy-brown with darker streaks, wings blackish-brown with broad pale fringes to coverts, and tertials (very long, reaching primary tips). Underparts generally white, with warm rufous-buff wash to breast-sides and black patch. **BP** Bill stout, rather round-tipped, grey above, horn below; eyes black; tarsi flesh-pink. **Vo** Often calls on take-off. In flight, a dry sparrow-like *tjrip* or *dreet*, but also has *prrt* similar to next species. Other calls skylark-like *jiru jiru*, or sparrow-like *chun chun* or *pyu pyu*. Song a dry, faltering chirruping and twittering.

ASIAN SHORT-TOED LARK CTKJ
Calandrella cheleensis

L 13–14cm; WT 20–27g. **SD** Mongolia, Transbaikalia, N & NE China to Bohai Gulf, Jilin and Heilongjiang, south to Shandong. Nominate *cheleensis* ranges into region. Rare migrant Korea; rare winter visitor Japan. **HH** Dry steppe or dry agricultural land. **ID** Small, compact, rather dark sandy-brown lark, closely recalling Greater Short-toed, but lacks black neck patches. Instead has necklace of narrow streaking on upper breast. Upperparts sandy-brown heavily streaked dark brown; face rather plain with narrow buff supercilium, eye-ring and lores, darker narrow malar on otherwise off-white chin; wings blackish-brown with paler fringes to coverts and tertials; tertials short, *not* covering primary tips; tail blackish-brown with white outer feathers. Underparts off-white with buff lower neck and flanks, and dark necklace of streaks. **BP** Bill stubby, greyish-horn; eyes black; tarsi brownish-pink. **Vo** Dry, buzzy *drrrrd* recalling Sand Martin, and *chirrick*, *puri*, *chui* or *pichu*. In flight, a purring *prrrt* or *prrr-rrr-rrr*. Song, given in flight, richer, more varied than Greater Short-toed, more prolonged, including mimicry of other species, including larks, but intermixing of own dry calls is characteristic. **TN** Sometimes considered within Lesser Short-toed Lark *C. rufescens.*

CRESTED LARK Galerida cristata CK

L 17–19cm; WT 35–50g. **SD** More than 35, mostly resident, subspecies from Mediterranean to Mongolia, NE China and Korea. *G. c. leautungensis* S Manchuria and NE China; *G. c. coreensis* Korea (major decline and range contraction due to habitat loss and pesticide use). **HH** Semi-deserts, dry agricultural land, and arid industrial/urban wasteland. **ID** Slightly larger than Eurasian Skylark (ES), with far more prominent, spiky crest. Overall greyish-brown and heavily streaked; upperparts dull grey-brown, lacking warmer buff or even orange tones of ES. Underparts whitish, with variable fine streaking on breast and flanks. Most prominent feature is long crest, which extends well past nape, even when flattened. Face plainer than ES, with pale crescents above and below eye, but lacks pale supercilium. Primary projection very short (long in ES), tail dark with buff/brown outer tail-feathers. In flight, wings broad and rounded, upperwing lacks white or buff trailing edge, underwing rusty-brown, and tail has brown, not white, outer feathers. **BP** Bill longer, slightly more arched than ES, upper mandible grey, lower pink; eyes black; tarsi dull grey-pink, hindclaw same length as toe. **Vo** Typically a melancholic whistled *treeleepeeu* or *twee-tee-too*; also a gentle flight call *dvuuee* or *too-ee*. Song, from ground or in display flight, rich and varied, including whistled call notes.

EURASIAN SKYLARK Alauda arvensis CTKJR

L 16–19cm; WT 26–50g. **SD** Common W Europe to NE Siberia; 13 subspecies, five in region. Taxonomy and range limits of eastern taxa disputed; criteria for separation unclear. *A. a. kiborti*: Transbaikalia, Mongolia and NW Manchuria, winters further south in China; *A. a. intermedia* (including *nigrescens*) Lena to Kolyma basin, south to N China, Korea and Russian Far East, winters to E & SE Asia; *A. a. pekinensis* NE Russia from Magadan to Koryakia, Kamchatka and N Kurils, perhaps Hokkaido, winters south to Japan and E China; *A. a. lonnbergi* N Sakhalin and lower Amur, moves south in winter. *A. a. japonica* S Sakhalin, S Kurils, Japan south to Ryukyus, also moves south in winter. Although *A. a. japonica* sometimes elevated to species as **Japanese Skylark** (*A. a. japonica* and *A. a. lonnbergi* reportedly sympatric in parts of Sakhalin, and *A. a. japonica* and *A. a. pekinensis* sympatric in S Kurils), also considered within Oriental Skylark; no convincing criteria separating *A. a. japonica* from all other races of Eurasian/Oriental have been published. **HH** Agricultural land and short grasslands; mostly present year-round, but northern birds migratory and wintering ranges unclear. **ID** Medium-sized, buffy-brown lark with distinct, broad, rounded crest and obvious display flight. Overall brownish-grey with buff, even orange, tones. Crown finely streaked blackish, crown-feathers may be raised in distinct crest; face appears gentle, with broad whitish supercilium and brown ear-coverts patch; mantle and coverts have dark-centred grey-brown feathers, and dark flight-feathers broadly fringed rusty-brown (paler when worn). Slightly smaller *A. japonica* has somewhat darker upperparts, more rufous-brown lesser covert patch. Underparts have band of short, fine black streaks across upper breast and neck-sides, contrasting with white belly with hardly any streaking on flanks, whereas *A. a. pekinensis* has less contrasting underparts pattern and streaks extending well onto flanks. At rest, primary projection beyond tertials long, but that of *A. a. japonica* shorter than *A. a. lonnbergi*. Tail short, dark brown with paler central and white outer feathers. Upperwing has variably distinct or indistinct white or buff trailing edge, underwing grey-brown. **BP** Bill short, upper mandible grey, lower horn; eyes black; tarsi dull grey-pink; hindclaw longer than toe. **Vo** Call, typically given on take-off, a repeated, rolling *chirrup*, *chrr-ik*, *byuru* or *prreet*. May sing from perch, but typically in high display flight (wings fluttering while hovering, tail fanned), when difficult to locate; a very prolonged torrent of rapid chirrups, whistles and complex phrases interspersed with calls.

GREATER SHORT-TOED LARK

juv

ad *longipennis*

ad

ad *dukhunensis*

ASIAN SHORT-TOED LARK

cheleensis

ad

juv

CRESTED LARK

coreensis

ad

juv

EURASIAN SKYLARK

ad

ad *intermedia*

ad

juv

ad *lonnbergi*

ad *japonica*

ORIENTAL SKYLARK CTJ
Alauda gulgula

L 15.5–18cm; WT 24–30g. **SD** Fairly common S & C Asian species, which occurs in region in E China north to Jiangsu, and lowland Taiwan (including Penghu), and has strayed to S Japan. Largely resident, but northern birds perhaps move south in winter. Races in region are *A. g. wattersi* (Taiwan); *A. g. coelivox* (SE China); and *A. g. weigoldi* (E China including Changjiang Valley). Identification criteria unclear and taxonomic position of skylark taxa in NE Asia uncertain, with some considering *japonica* Eurasian Skylark to be a race of this species. **HH** Dry agricultural land and steppe. **ID** Small to medium-sized, very similar to Eurasian Skylark (p.328), from which best distinguished by slightly smaller size, paler upperparts, short primary projection, short tail with buff, not white, edges; buff, not white, trailing edge to wing (but see *A. a. japonica*), more prominently white behind eye (fresh plumage), and slightly longer, finer bill. Indistinct rufous wing-panel formed by fringes to primaries. May also raise small crest, and shares similar song-flight with Eurasian. **BP** Bill short, fine, upper mandible grey, lower pink/horn; eyes black; tarsi dull grey-pink; hindclaw longer than toe. **Vo** Call, typically given at take-off, *swit switswit* or a dry buzzing *drzz*, or *bazz bazz*, with Eurasian-like notes. Display flight song very similar to Eurasian Skylark but less varied, more repetitious, higher in pitch.

SHORE LARK CJR
Eremophila alpestris

L 14–17cm; WT 26–46g. **SD** Widespread Northern Hemisphere species, with more than 40 subspecies. Races in region likely to involve: *E. a. flava* across N Palearctic to Yakutia and Chukotka, south to Lake Baikal, east to Amurland; *E. a. brandti* east to Mongolia, Inner Mongolia and W Manchuria; *E. a. arcticola* a possible vagrant to region from Alaska and NW Canada. Winters in region in S Russia, NE China and Japan (scarce Sep–Mar). **HH** Breeds both in mountains (above treeline), and on steppe and tundra, generally in arid areas. In winter also in short-grass areas, fallow cultivation, coasts and on beaches. **ID** Distinctive, slender, medium-sized lark, with prominent black mask, black breast-band and fine, black horns. *E. a. brandti* has broad pale supercilium extending across forehead to base of bill. Upperparts rather plain mid to sandy brown, more rufous on nape and rump, with fine streaking on mantle. Underparts white to buff (*E. a. flava* has some yellow on face), with broad buff streaking on flanks. ♂ smartly attired, ♀ similar but less boldly patterned. *E. a. arcticola* may also have variable pale yellow on face and throat, but always less than *E. a. flava*, with none on lores or eyebrow. In flight long-winged, upperwing plain, lacks pale trailing edge, underwing off-white, tail long, blackish with pale grey or brown central, and white, outer feathers. **BP** Bill stubby, grey at base, blackish at tip; eyes black; tarsi black. **Vo** Calls high, weak and thin, *seeh*; *see-tu* or *chit chit-see*, sometimes a harsher *prsh* or *tssr*, and liquid *tur-reep*. Song, given in flight or from rock perch includes rapidly repeated short tinkling phrases and rippling trills followed by short chatter. **AN** Horned Lark.

ZITTING CISTICOLA CTKJR
Cisticola juncidis

L 10–14cm; WT 8–12g. **SD** Common Old World species, from Mediterranean Europe to southern Africa and across S & SE Asia to China and Japan: *C. j. brunniceps* is a summer visitor to Korea (predominantly W) and N & C Honshu, resident in W & S Japan to Ryukyu Is, but northern birds winter to Philippines; *C. j. tinnabulans* ranges north to NE China and south to Taiwan and Philippines. **HH** Locally common in grasslands and reedbeds, typically at wetland margins, but also in rice and sugarcane fields. Performs distinctive display flight. **ID** Small, generally brown bird, noticeably darker on crown, more rufous on back and rump, with off-white underparts, browner on flanks; supercilium broad and pale in front of eye, becomes buff on head-sides behind eye, otherwise face rather plain. Mantle, tertials and coverts black fringed buff. Tail graduated, typically fanned in flight, with black subterminal band and white tip; from below tail appears broadly banded black and grey. In winter, dark crown more streaked, supercilium broader, paler more prominent (see Golden-headed). **BP** Bill sharp, arched, grey with blackish tip; eyes black; tarsi flesh-pink. **Vo** Loud, hard *zit zit zit*, or in flight a high, weak *tsiek*, slightly broken in tone. During courtship, ♂ calls while rising, hovering and circling in undulating display flight over territory; when rising it gives metallic *chin chin chin* or *dzip dzip dzip* notes, while descending it gives harder *chat chat chat* notes. **AN** Fan-tailed Warbler.

GOLDEN-HEADED CISTICOLA CT
Cisticola exilis

L 9–11cm; WT 6–10g. **SD** Resident from Himalayas and S India to SE Asia and Australia. *C. e. courtoisi* occurs in S & E China north to Zhejiang; *C. e. volitans* endemic to Taiwan (mainly in lowlands). **HH** Dense grasslands often associated with water. **ID** Small *Cisticola*, very similar in most plumages to Zitting. ♂ breeding distinctive with pale unstreaked sandy cap, face and nape (almost white in *C. e. courtoisi*, golden-yellow in *C. e. volitans*), back brown with broad blackish-brown streaks, rump rufous-brown, tail blackish-brown with pale fringes and buff tips, wings blackish-brown with broad pale brown fringes; underparts white on chin/throat, yellow-buff on breast, belly and flanks, off-white on lower belly and vent. Winter ♂ resembles ♀, with plain grey-buff face, faint, poorly marked grey supercilium, streaked crown, but unstreaked rufous nape. ♀ very similar to Zitting summer ad. (note bill colour), but crown brown with dark streaks, face very pale, ear-coverts unstreaked buff-brown, underparts creamy with some brown on breast-sides, flanks pale; tail shorter. From below, tail appears black with narrow grey bands at tips of graduated feathers. **BP** Bill dark grey-brown above, pink below (all year); eyes brown; tarsi dull pink. **Vo** A high, wheezy drawn-out scolding *ch(w)eeesh* or *keeesh*, not particularly loud; *bzee bzee* in alarm; song from perch or in display flight comprises several bleating calls, followed by scratchy, buzzy notes and a more liquid note: *bzee tooli bzee tooli*.

ORIENTAL SKYLARK

coelivox

ad

ad

SHORE LARK

juv

ad ♂ *arcticola*

ad ♂ *brandti*

ad ♀ *flava*

ad ♂ *flava*

ZITTING CISTICOLA

tinnabulans

ad

juv

ad ♂ sum *courtoisi*

GOLDEN-HEADED CISTICOLA

ad ♂ sum *volitans*

CHINESE HILL WARBLER CK
Rhopophilus pekinensis

L 17cm. **SD** Scarce and seemingly declining. *R. p. pekinensis* occurs in N & NE China (Liaoning) and N Korea (predominantly W). **HH** Rough scrub and pines with tall grasses. **ID** Rather dull grey-brown and long-tailed. Supercilium (from eye) grey, cheeks grey-brown, lores and malar black, and throat white. Upperparts mid ash-brown heavily streaked blackish-brown on crown, mantle and wings; rump plain brown, long graduated tail finely barred darker on central feathers, with pale grey tips to outer feathers. Underparts white with chestnut/rufous-brown streaking on breast-sides and flanks; vent rufous-brown. **BP** Bill, heavy, pink; eyes white; tarsi dull greyish-pink. **Vo** Calls/song includes duetting of mellow *chee-anh* notes, and series of short sweet syllables, each starting high and falling away: *dear dear dear....* **AN** White-browed Chinese Warbler.

STRIATED PRINIA CT
Prinia crinigera

L 16cm; WT 14–17g. **SD** Resident from Afghanistan to E China. *P. c. parumstriata* in E China; *P. c. striata* on Taiwan. **HH** Widespread, except in lowlands and highest montane regions. Grasses, scrub, second growth; mainly modified habitats and never in dense forest. **ID** Rather large, slender, brown, very long, loose-tailed *Prinia*. Generally mid brown, with fine blackish streaking on crown, broad blackish streaking on mantle, scapulars and rump; tail dark brown, long, graduated. Wings short, blackish-brown with mid brown fringes to feathers. Underparts grey-buff, chin/throat plain, but breast streaked and spotted dark grey, belly plain yellowish-buff. *P. c. parumstriata* greyer, *P. c. striata* paler. **BP** Bill black (brown in winter); eyes red-brown; long tarsi brownish-pink. **Vo** Call is a dry *tk tk tk tk*. Song is a rather dry, rhythmical ringing *krri krri krri krri*; *tsuritsuritsuritsuri* or jaunty, repetitive *chitzereet chitzereet...* **AN** Brown Hill Warbler.

HILL PRINIA C
Prinia atrogularis

L 16–20cm; WT 8–16g. **SD** E Himalayas to W Sumatra. *P. a. superciliaris* in S & SE China east to Fujian. **HH** Grasses, scrub, forest edge and forested hillsides above 600m. **ID** *P. a. superciliaris* is a rather distinctive prinia, with black-streaked neck-sides in breeding plumage. Head grey; narrow white supercilium reaches rear of ear-coverts and contrasts with black lores, eyestripe and dark grey ear-coverts; chin and throat white, contrasting with bold black neck streaking.

Upperparts unstreaked olive-brown, but long, graduated tail and short, rounded wings somewhat darker brown. Flanks to vent buff to orange-buff, belly white. Sexes similar, but ♂ larger and longer tailed. **BP** Bill fine, black above yellow below; eyes pale brown; tarsi pinkish-brown. **Vo** Calls includes a soft *prr-prr-prr* and scolding *chrrr-chrrr-chrrr*. Song a series of mechanical, buzzy notes: *zee-szelik zee-szelik zee-szelik*.

YELLOW-BELLIED PRINIA CT
Prinia flaviventris

L 12–14cm; WT 6–9g. **SD** From Himalayas across S Asia to Borneo. *P. f. sonitans* in S & SE China to Fujian; also Taiwan. **HH** Rather shy bird of tall grasses, reeds and scrub, from lowlands to *c*.900m. **ID** Slender, olive-green, very long, loose-tailed *Prinia*. Breeding ad. quite colourful, with grey head, white throat and breast, and yellow belly. Upperparts brownish to olive-green; crown dark ash-grey, head-sides and ear-coverts pale ash-grey, lores black, narrow white supercilium reaches just beyond eye. Underparts: off-white chin/throat to neck-sides, breast pale whitish-lemon, lower breast to vent lemon-yellow. Wings short, dark olive-green, rump plain, tail long and graduated, dark olive-green or brown. **BP** Bill slender, blackish-grey above paler below; eyes reddish-brown; long tarsi yellowish-brown. **Vo** A distinctive nasal mewing *pzeeeu* and repetitive *chink-chink*. Song: *tee dee dew dew*, sometimes repeated rapidly, with slightly protesting quality, owing to rising first note and falling second, whilst the last two notes are on the same pitch; also a chirping, trilling: *chirp didli-idli-u didli-idli-u didli-idli-u.*

PLAIN PRINIA CTJ
Prinia inornata

L 11cm; WT 6–9g. **SD** Resident from Pakistan to SE Asia. *P. i. extensicauda* ranges over S & SE China north to Changjiang; *P. i. flavirostris* endemic to Taiwan. Accidental Okinawa. **HH** Tall grasses, reeds, wetlands and cropfields, and adjacent scrub and isolated trees, where small parties often noisy and conspicuous. Shuns highest montane regions (otherwise locally abundant). **ID** Rather dull, unstreaked prinia with plain greyish earth brown upperparts, short, pale buff supercilium, buff lores and buffish-brown ear-coverts, affording pale-faced appearance. Tail long, graduated, with pale tips; wings short, earth brown, with greenish tone to outer webs of primaries. Buffy-brown throat and chest grades to pale yellowish-buff belly and vent. **BP** Bill black (with pale tip and pink base in *P. i. flavirostris*); eyes pale brown; tarsi pink. **Vo** Calls varied, including a nasal *beep*, plaintive *tee-tee-tee* (see previous species) and buzzing *bzzp* and *zzpink*. Song a rapid trill: *tlick tlick tlick*, slightly more buzzing than, but similar to, Zitting Cisticola song.

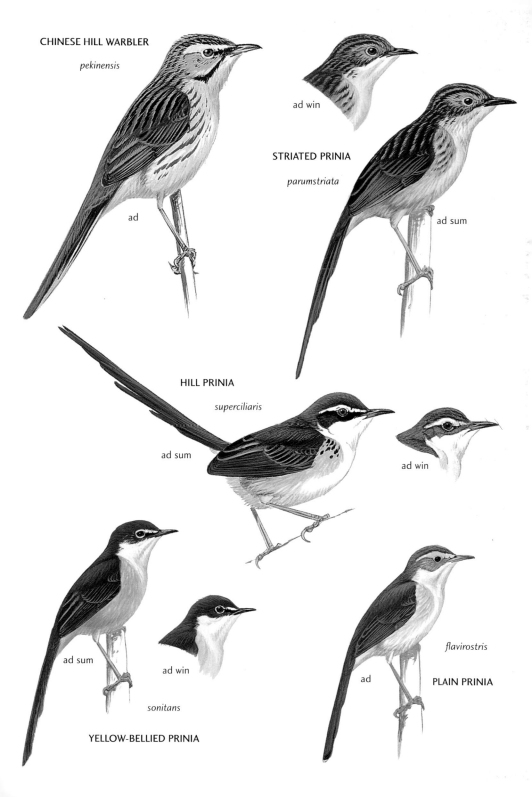

CHINESE HILL WARBLER

pekinensis

ad

ad win

STRIATED PRINIA

parumstriata

ad sum

HILL PRINIA

superciliaris

ad sum

ad win

ad sum

ad win

sonitans

YELLOW-BELLIED PRINIA

flavirostris

ad

PLAIN PRINIA

COMMON TAILORBIRD C
Orthotomus sutorius

L 10–14cm; WT 6–10g. **SD** Widespread in S & SE Asia, with *O. s. longicauda* in SE China east to Fujian. **HH** Keeps to lower vegetation and thick brush in gardens and secondary forest, to 1,500m; active and vocal. **ID** Slender, long-tailed bird that typically holds tail erect. Bright rufous forehead and crown, with pale grey eyestripe, slightly darker grey cheeks, dark grey nape, bright olive-green mantle, scapulars and rump, short, blackish-brown wings fringed olive-green, and long, graduated dark brown tail. Underparts whitest on chin/throat and central belly, with buff-grey wash on breast-sides and flanks, undertail-coverts also white. ♀ shorter-tailed, with darker streaking on cheeks and more extensive buff-grey wash on chest. **BP** Bill long (exceeds eye to bill base distance), dark above, pinkish below; narrow red orbital ring and black eyes; tarsi pink. **Vo** A repetitive *cheep cheep cheep cheep* contact note, and loud, rapid *pit-pit-pit* when agitated. Song loud and abrupt: *chubit chubit chubit chubit* or *pitchik pitchik pitchik pitchik*. **AN** Long-tailed Tailorbird.

COLLARED FINCHBILL CT
Spizixos semitorques

L 21–23cm. **SD** E China to N Vietnam. *S. s. semitorques* E China largely south of Changjiang River; *S. s. cinereicapillus* on Taiwan, though very rare in N. **HH** Favours scrub, bamboo, woodland edge, second growth and fruit orchards. In Taiwan occurs in foothills (mainly below 1,500m). **ID** Differs from other bulbuls by unique combination of blunt, pale bill, grey head, and green plumage. Upperparts dark green, paler, somewhat yellower green on underparts; darker on scapulars and blacker at tip of rather broad-tipped green tail. Hood dark blackish-grey, throat black, with white-streaked ear-coverts and neck-sides; white half-collar separates grey hood from green underparts, and is broadest in centre. *S. s. cinereicapillus* has forehead, crown, nape and tail tip paler grey. **BP** Blunt, rather deep bill is pale yellow/horn at tip, blue-grey at base; eyes dark reddish-brown; tarsi pale greyish-pink. **Vo** An upslurred *chrup chrup* or *wirrrrp* throatier than Himalayan Black or Chinese Bulbuls; also a *whit whit weet a weet*. Song a series of mellow chirrups and chortles, or series of trisyllabic notes, *chuwichu-chuwichu-chuwichi-chuwi*, somewhat reminiscent of Red-billed Leiothrix song, but less melodious.

RED-WHISKERED BULBUL CTJ
Pycnonotus jocosus

L 18–20.5cm; WT 25–31g. **SD** Natural range is from India to SE Asia northeast into S China (Guangdong, S Jiangxi and S Fujian). Common cagebird and established in many other regions. Escapees/feral population in C Japan, and probably established Taiwan; all may involve nominate. **HH** Often in wooded habitat near habitation, in villages, urban parks and gardens, also scrub and secondary forest. Typically sociable and noisy. **ID** Dark, medium-sized, rather slender bulbul with distinctive head pattern and crest. Ad. has black head with erect spiky crest; ear-coverts red above white, separated from white chin/throat by narrow black malar. Upperparts largely dusky grey-brown, but nape and neck-sides black. White of throat/face separated from rest of underparts by black neck bar, below which grades to dusky grey-brown on flanks and breast-sides; centre of breast and belly white, vent contrastingly red. Juv. has shorter, browner crest, lacks red postocular patch, and vent duller orange or buffish-pink. **BP** Bill slender, black; eyes dark brown; tarsi black. **Vo** Calls include *queee-kwut* and a musical *per'r'p*. Song rich and complex: *wit-ti-waet queep kwil-ya queek-kay*.

BROWN-BREASTED BULBUL C
Pycnonotus xanthorrhous

L 20cm; WT 24–31g. **SD** China, Burma and N Indochina; *P. x. andersoni* in E & S China north to Changjiang River. **HH** Typically in scrub, tangles and second growth on hillsides above 800m, where noisy and conspicuous. **ID** Rather dumpy, dull-coloured bulbul. Upperparts dull grey-brown, darkest on wings and tail, with black cap from forehead and lores to nape (and slight nuchal crest); short malar also black with red spot at gape; ear-coverts brown. Underparts, white on chin and throat, with faint dusky grey-brown wash on breast, paler on flanks and whiter on belly, vent buff-ochre to yellow. **BP** Bill heavier and blunter than Red-whiskered, black; eyes dark brown; tarsi greyish-black. **Vo** A thin *ti-whi* and harsher *chi* and *brzzp*. Song a simple and repetitive *chirriwu'l whi'chu whirri'ui*.

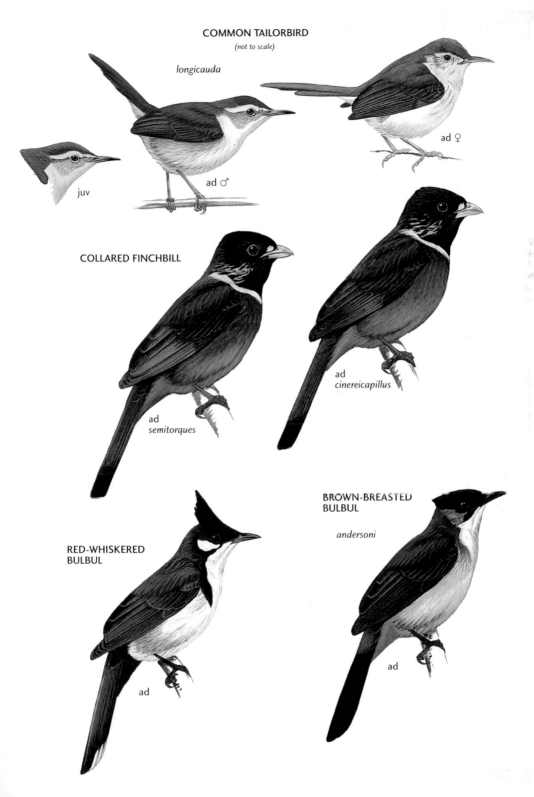

COMMON TAILORBIRD
(not to scale)

longicauda

ad ♀

juv

ad ♂

COLLARED FINCHBILL

ad
cinereicapillus

ad
semitorques

BROWN-BREASTED BULBUL

andersoni

RED-WHISKERED BULBUL

ad

ad

CHINESE BULBUL   CTKJ
Pycnonotus sinensis

L 18–19cm. **SD** C & E China to Vietnam. *P. s. sinensis* resident across S & E China north to Changjiang, but expanding north with regular records in Hebei and Shandong, and on migration rare offshore Korea (may breed); *P. s. formosae* N & W Taiwan; *P. s. orii* S Nansei Shoto, north to Okinawa; accidental Honshu (race uncertain). **HH** Often in lightly wooded habitat or open woodland and cultivated areas near habitation, in villages, urban parks and gardens, also scrub and secondary forest. Sociable and noisy, often perching conspicuously on treetops, poles or wires. **ID** Pale bulbul with a black head and broad white 'bandana' from eye to nape (where broadest), bordered below with black, also has tiny white spot on lores and larger spot on rear ear-coverts; chin and throat white bordered by black malar. Mantle and back greyish-green, rump plain grey, tail dark grey with green outer fringes to feathers; wings greyish-brown with yellowish-green outer fringes to flight-feathers; greyish-buff breast and flanks, off-white belly, vent white. **BP** Bill black; eyes dark brown; tarsi black. **Vo** Generally rather coarse and shrill, but song includes more melodious tones. Calls: *ju ju* or *byu byu*, sometimes a continuous *piyopiyopiyopiyopiyo* and somewhat nasal *vyer*. Song a strident, whistle-warble: *chip-chop-chop-twee*; also *plit prilyor trilor* often repeated; *kyo kyan pyuu, pikkyo pik-kyo pyuu* or *bukikyopyuu*. In Nansei Shoto, vocalisations more subdued with 'thicker' calls. **AN** Light-vented Bulbul.

TAIWAN BULBUL   T
Pycnonotus taivanus

Vulnerable. L 18–19cm. **SD** Endemic to lowlands and slopes below *c*.1,600m of E & S Taiwan. Declining, mainly due to hybridisation with Chinese Bulbul in natural overlap zone, and following ceremonial mass-releases of Chinese Bulbuls within its range. Monotypic. **HH** Wooded habitat near habitation, cultivated areas in villages, urban parks and gardens, also scrub and secondary forest. **ID** Closely resembles Chinese Bulbul, but has unbroken black cap to nape (no white 'bandana'), white lores and ear-coverts separated from white chin and throat by prominent, quite long broad black malar. Mid to dark grey mantle, scapulars, rump and tail, with olive-green tone to primaries and outer tail-feathers. Throat and ear-coverts white, becoming pale grey on neck which is concolorous with breast; flanks also grey, belly centre and vent white. Hybrid Taiwan/Chinese resembles Taiwan, but has varying amount of white on crown and nape (though less than Chinese), may have dusky cheeks, less

clear-cut malar stripe (often broader and more diffuse) and a yellow rather than orange spot at bill base. **BP** Bill heavier and blunter than Chinese, black; eyes black; tarsi black. **Vo** Calls include short choppy and more continuous chattering notes. Song slightly higher and sweeter than Chinese, a spaced, rhythmic *pri tü rrilit!* **AN** Styan's Bulbul.

SOOTY-HEADED BULBUL   C
Pycnonotus aurigaster

L 19–21cm; WT 40–50g. **SD** Resident in SE Asia, S & SE China. *P. a. chrysor-rhoides* in SE China east to Fujian. **HH** Around habitation, in gardens and parks, also scrub, second growth and forest edge, mainly in lowlands. Often in noisy groups, sometimes with other bulbuls. **ID** Rather dumpy, dull-coloured bulbul.
Ad. cap, including crown to below eye, lores and chin, and small nuchal crest all dull black; cheeks pale grey-brown, rest of upperparts mid grey-brown, except pale grey rump, white uppertail coverts and grey-white tip to otherwise blackish-brown tail. Underparts, including chin and throat, very pale grey-brown; vent deep dull red. Juv. resembles ad. but has browner crown and duller, pale yellow vent. Closely recalls Brown-breasted Bulbul (p.334), but separated by lack of prominent malar, greyer, more patterned upperparts, and rump/vent coloration. **BP** Bill rather thick and blunt, black; eyes dark brown; tarsi grey-black. **Vo** Varied, including sharp single or double notes and chattering and gurgling calls. Song *whi-wi-wiwi-wiwi*. **AN** Black-capped Bulbul.

MOUNTAIN BULBUL   C
Ixos mcclellandii

L 21–24cm; WT 27–41g. **SD** Himalayas to SE Asia and S China. *I. m. holtii* in SE China east to Fujian. **HH** Scrub, forest edge and forests, generally above 1,000m. **ID** Large olive-green and grey-brown bulbul. Ad. has rufous head, face and nape with a loose, spiky, brown-streaked crest that is often raised; mantle grey, back, wings and tail olive-green. Tail rather long and square-tipped. Underparts: ear-coverts rufous, chin and throat grey-white (often fluffed out), chest orange-buff, then yellowish-white on belly and flanks, vent pale yellow. Juv. duller, browner with shorter crest, and cinnamon tinge to vent. **BP** Bill, long slim, dark brown above, pale grey below; eyes reddish-brown; tarsi pinkish-brown. **Vo** Noisy (flocks call almost constantly); metallic, high-pitched chirps, *tsiuc* or *chewp*. Song of comprises 4–5 loud descending notes.

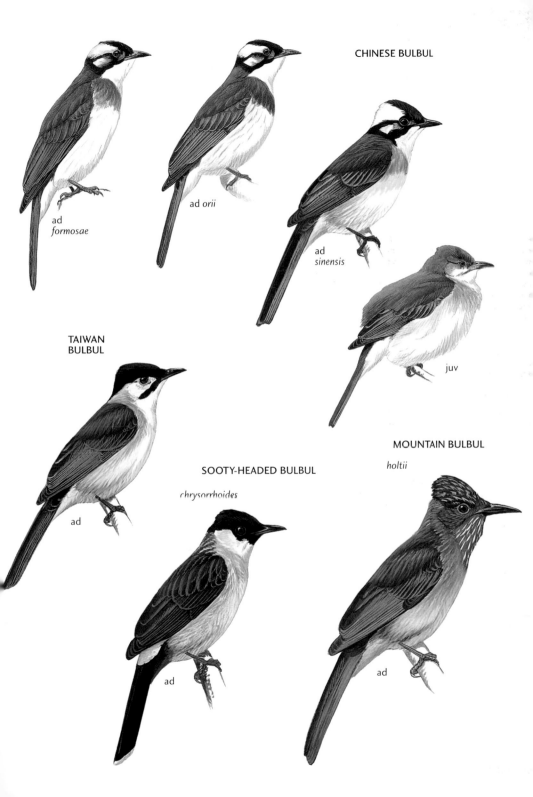

CHINESE BULBUL

ad
formosae

ad orii

ad
sinensis

juv

TAIWAN
BULBUL

MOUNTAIN BULBUL

holtii

SOOTY-HEADED BULBUL

chrysorrhoides

ad

ad

ad

BROWN-EARED BULBUL　　　　　　　CTKJR
Microscelis amaurotis

L 27–29cm. **SD** Northernmost bulbul and endemic to region. Mainly resident, but marked local movements by *M. a. amaurotis*, which ranges across S Sakhalin, Hokkaido, Honshu, Kyushu and N Izu Is, and C & S Korea, wintering to NE China and Ryukyu Is, and reaches Jiangsu and Zhejiang, China, and Taiwan. *M. a. matchiae* occurs in S Izu Is, Tanega-shima and Yaku-shima; *M. a. ogawae* in N Ryukyu Is; *M. a. pryeri* in C Ryukyu Islands; *M. a. stejnegeri* in S Ryukyu Is; *M. a. squamiceps* in Ogasawara Shoto; *M. a. magnirostris* on Iwo Is; *M. a. borodinonis* on Borodino Is; *M. a. harterti* common resident of Lanyu, Green and Turtle Is (off E & SE Taiwan). Accidental (race uncertain) mainland Taiwan. **HH** Wide range of forest types (deciduous, mixed and evergreen broadleaf), from mtn foothills (to c.1,600m) to lowlands. In winter also in rural/agricultural areas with scattered trees and in suburban and urban gardens and parks. **ID** Large, rather drab bulbul, conspicuous by its harsh calls, strongly undulating flight and long tail. Ad. *M. a. amaurotis* is overall dull, dark ash-grey with fine paler 'frosting' on head, slightly shaggy feathering on rear crown (sometimes raised as crest) and larger pale grey spots on chest, with diagnostic dark chestnut crescent on ear-coverts. Upperparts plain grey, browner on wings and tail. Underparts have brown wash on flanks and undertail-coverts black with white fringes. Juv. duller, browner, and lacks silvery wash. Flight typically undulating; underwing-coverts dark brown; long tail broadens towards square tip. Southern races smaller, darker, with dark brown (not grey) mantle and more extensive chestnut ear-patch joining dark chestnut-brown of neck, breast and underparts. **BP** Bill sharply pointed, black (larger in southern races); eyes dark reddish-brown; tarsi dark grey. **Vo** Noisy; highly social and vociferous. Calls varied but almost invariably loud, shrill and drawn-out, including *wheesp*; *whee-eesp*; *shreep shreeeep*; *piiyo piiyo*, or *piiyy piitt* and *pii-hyara, piyopiyopiyo* and *hiihii*. **TN** Formerly in *Hypsipetes*.

CHESTNUT BULBUL　　　　　　　　CT
Hemixos castanonotus

L 21.5cm. **SD** Restricted to S & SE China and Hainan. In region, *H. c. canipennis* in Fujian; accidental Kinmen I (Taiwan). **HH** Lowland broadleaf evergreen forest and second growth, to 1,000m. **ID** Medium-sized, smart, two-tone bulbul. Ad. upperparts warm chestnut-brown, with black mid and hindcrown and crest; forehead, lores, ear-coverts and neck-sides chestnut, malar black; wings blackish-brown with pale fringes to coverts, secondaries and tertials; tail dark grey-brown, black at broad square tip. Chin and throat white (throat feathers may be puffed-out), contrasting with black malar and chestnut cheeks, and grey breast and flanks; belly and vent white. Juv. has pale brown crown, dull brown face and upperparts, and brownish-grey breast-band. **BP** Bill long, slightly decurved, black; eyes dark reddish-brown; tarsi black. **Vo** Noisy, particularly in flocks; various calls include harsh single notes and churrs. Clearly whistled, distinctive three-part song is a taunting *whi-wi-wu* (last note lowest). **TN** Formerly within Ashy Bulbul *H. flavala*.

HIMALAYAN BLACK BULBUL　　　　　CTJ
Hypsipetes leucocephalus

L 23.5–26.5cm. **SD** Himalayas to SE Asia and S China. *H. l. leucocephalus* is summer visitor across SE China north to Changjiang River, accidental offshore Japan; *H. l. nigerrimus* resident in Taiwan. **HH** Broadleaf evergreen and mixed deciduous forests. *H. l. nigerrimus* mainly in lowlands and urban parks to montane forest of Taiwan, less common at high altitudes. Social, flocks in winter. **ID** Medium to large black bulbul with rather loose, ragged 'crest' on rear crown and long, square-ended or slightly notched tail. *H. l. leucocephalus* has all-white hood and white tips to grey belly and vent feathers. *H. l. nigerrimus* all black (including head), with conspicuous pale-grey outer fringes to primaries and secondaries. **BP** Bill prominently orange (*H. l. leucocephalus*) or bright red (*H. l. nigerrimus*); eyes dark brown; tarsi bright orange or red. **Vo** Noisy and conspicuous, calling almost continuously in flocks, perched or in flight. A whining, plaintive *keer* and cat-like *meow* or *nyeeer*; flight call a strident *tsit* or *tseesp*. Song a repeated series of 3–4 clear, alternating whistles, often uttered by 2–3 birds simultaneously in an alternating jumble of notes with a jaunty rhythm.

MARSH GRASSBIRD　　　　　　　　CKJR
Megalurus pryeri

Vulnerable. L 12–14cm; WT 12–17g. **SD** Endemic to region; local and restricted in range and habitat. *M. p. pryeri* very local in N, C & W Honshu; *M. p. sinensis* coastal Liaoning and Hebei and S Russian Far East, wintering in SE China (along middle Changjiang River), and Korea (old, pre-1980 records). **HH** Occurs year-round in reedbeds and rank grassland near rivers or swamps. Like Zitting Cisticola (p.330), can balance while grasping separate reed stems with each foot. Sings from atop tall reed or in distinctive curving display flight up and over territory. **ID** Medium-sized warbler, resembling rather plain Zitting Cisticola, but with longer tail. Generally mid to rufous-brown on upperparts, including wings and tail, with fine black streaking on crown, unstreaked nape and bold black scaling on back; tertials also blackish. Underparts whitish with brown wash on breast-sides and flanks. Lores and ear-coverts pale grey-brown, supercilium broad but extends only from eye, as does indistinct eyestripe. Tail long, rounded, typically fanned in flight. *M. p. sinensis* is paler, cooler buff-brown and less rufous-brown on upperparts, with streaked nape. **BP** Bill short, somewhat thick, black above with pink base to lower mandible; eyes black; tarsi dull brownish-pink. **Vo** Alarm a hard *jutt jutt* or *chut chut*, lower than Japanese Bush Warbler and recalling Siberian Rubythroat's low *jut* call. Song, given during display flight: *churuchuruchuru chochiri chohiri* and *jukukuku kyururuuru*. In autumn and winter gives hard *chak chak chak* and fast repetitive *chur-chur-chur-chur* in contact. Also a dry, slightly rising *trrik* or *trrrrett*, occasionally louder and even thrush-like in tone, *churrrrek!* **TN** Often placed in *Locustella*. **AN** Japanese Marsh Warbler; Japanese Swamp Warbler.

BROWN-EARED BULBUL

ad
amaurotis

ad
pryeri

ad
harterti

CHESTNUT BULBUL

canipennis

ad

ad
nigerrimus

**HIMALAYAN
BLACK BULBUL**

ad
leucocephalus

ad
sinensis

ad
pryeri

MARSH GRASSBIRD

ASIAN STUBTAIL CTKJR
Urosphena squameiceps

L 9.5–10.5cm; WT 8–10g. **SD** E Asian endemic. *U. s. squameiceps* S Sakhalin, S Kuril Is, Japan, winters to S Japan (S Kyushu to Nansei Shoto), E China and SE Asia. *U. s. ussurianus* (perhaps doubtfully distinct) Russian Far East, Korea, NE China (Heilongjiang to Hebei), winters through E China to SE Asia. Migrant Taiwan. **HH** Coniferous taiga to mixed temperate and broadleaf evergreen forests of semitropical areas; often skulking in lower, dense vegetation, including dwarf bamboo, or on forest floor. **ID** Tiny, dark, warm brown, crown dark brown (visibly scaled darker at close range) with long off-white supercilium wrapping round onto nape, and long blackish eye stripe. Upperparts warm, mid brown. Very short tail, sometimes cocked (compare similar-sized but much darker wren) barely extends beyond wingtips. Underparts off-white, washed buff on undertail-coverts. **BP** Bill fine, black, with pinkish-yellow base to lower mandible; eyes black; tarsi pale pink. **Vo** Common call a high-pitched *tchick*, a wren-like hard *tyutt*, *chap* or *kyip*, and *tchak* like Japanese Bush Warbler but slightly 'wetter'. Song a rapid, repetitive insect-like 'white noise', low at first, rising slightly and steadily in pitch and volume, but pulsating: *tsitsutsatsitsutsatsitsutsa...*; *see-see-see-see-see-see-see* or *shiri-shiri-shiri-shiriri*. On migration a lower, though still high-pitched and penetrating *sti-titit!*

JAPANESE BUSH WARBLER CTKJR
Cettia diphone

L 15–18cm; WT 10.5–21g. **SD** E Asian endemic. Sakhalin and throughout Japan. Northern birds migratory, some reaching E coast China (Shandong to Zhejiang), Taiwan, and Nansei Shoto. Generally common, but very secretive. Much variation in size between sexes and subspecies in *Cettia*, making sexing (and even specific identification) difficult. *C. d. sakhalinensis* Sakhalin and S Kuril Is migrates to SE China, presumably via Japan; *C. d. cantans* main Japanese islands (and associated offshore islands, with northern birds moving south to winter, presumably in Honshu), S Korea, Taiwan; *C. d. riukiuensis* winters in Nansei Shoto (probably = *sakhalinensis*, i.e. a single race that breeds further north and winters south; however, this has yet to be confirmed); *C. d. restricta* was considered an extinct endemic of Daito Is (birds rediscovered there in 2003 were not *restricta*), now known to be resident on Okinawa; seemingly more similar to *cantans*; *C. (d.) diphone* endemic S Izu, Ogasawara and Iwo Is (possibly merits specific status). Vocally distinct Tokara Is form requires further study. Full taxonomic review may confirm presence of more than one species. **HH** Dense woodland undergrowth, forest, riparian shrubbery, bushy scrub and wooded parks; typically skulks in low vegetation, dwarf bamboo and other ground cover, though sometimes sings in open. **ID** Greyish-brown to warm brown, darker on upperparts (some races have olive-green fringes to remiges), slightly browner on forehead, pale grey-brown below, greyer on face, whiter on chin and throat, greyish-buff on flanks; tail long and prominent, rounded at tip. Could be confused with Dusky Warbler or *Locustella*, but larger

and longer tailed than former with heavier bill, and supercilium dull, buffy grey-brown and eyestripe weak, distinguishing it from latter. Tail rather wide, especially at tip. Rather variable; island races generally smaller, darker and greyer. *C. d. cantans* (commonest bush warbler in Japan) upperparts yellowish-brown or olive-tinged, with warm chestnut-brown fringes to flight-feathers, supercilium pale buffish-brown, eyestripe dark brown; underparts dirty white, tail appears somewhat long, bill thin and brown, legs rather long. *C. d. sakhalinensis/riukiuensis* generally plain greyish-brown, with darker brown crown than *cantans*. *C. (d.) diphone* very similar to *cantans* but smaller bodied, with a more elongated profile and much longer bill; forehead and crown rufous-brown, supercilium yellowish-white, flight-feathers have rufous fringes, with buffish wash to breast, and rest of underparts yellowish-white. *C. d. restricta* is smaller, overall darker and browner than other races, with strong red-brown tone to forehead, wings, and tail. **BP** Bill somewhat blunt (though notably long in *C. (d.) diphone*), blackish above, yellowish-pink below; eyes black; tarsi brownish-pink. **Vo** Extremely vocal. A distinctive, single, hard dry *tchak*, *chek* or *chatt*. Song varied, rich and liquid: long whistle followed by explosive burst of three notes *pheeuw hou-ke-kyo*, also *hoo-hokekyo hii-hikekyo* or continuously *pipipipi kekyo kekyo kekyo....* A quieter song given later in day is a furtive low whistle, *ho—hohoho hoit*. Also descending staccato series of trisyllabic notes, with a rippling precursor in alarm: *tirrrrrrr chepi chepi chepi che-pichew che-pichew che-pichew.* ♀ utters descending *hee-hee-hee* when breeding (until chicks fledge).

KOREAN BUSH WARBLER CTKJR
Cettia canturians

L 14–17cm. **SD** E & NE China, Korea and Russian Far East. Two races (perhaps species). *C. c. borealis* is a widespread summer visitor to E China north of Changjiang R, NE China, also N Korea and S Russian Far East, winters south to S Nansei Shoto (Yaeyama), Taiwan (common), SE China and SE Asia; range and overlap with following taxon unclear. *C. c. canturians* **Manchurian Bush Warbler** occurs C & E China, migrates to S China, Taiwan (scarce) and SE Asia. Generally common, but very secretive. **HH** As Japanese. **ID** Compared with Japanese, has a more earth brown or even chestnut tone to upperparts/wings, distinctive, rather bright orange-rust forehead, more prominent pale grey supercilium, paler throat and strongly buff-washed underparts (Korean, *C. c. borealis*). The largest and most rufous-brown birds, with warmer chestnut-brown crown somewhat less distinct from rest of upperparts, and stronger rusty-brown fringes to flight-feathers, are seemingly Manchurian. ♀ have less distinctly coloured crowns, but show cleaner contrast with pale lores and supercilium than Japanese. Beware considerable size difference between sexes. **BP** Bill heavier and more stub-tipped than Japanese, blackish above, yellowish below; eyes black; tarsi (sturdier than Japanese) orange-brown to pink. **Vo** Extremely vocal. Calls similar to Oriental Reed's *trek*; most commonly a rolled *chrek* or stronger *trrrek!*, also a vaguely Radde's-like *trrt* or *prrrt* (very different from *tchak* of Japanese). Song recalls Japanese but is weaker, less rich, with a shorter opening whistle, a chuntering *ho hokeryon* or *pu-hu-hu* followed by a short *chirweeu* (*borealis*) or rapid, piping *tul-tul-tul-tul-tul-tul-tu* (*canturians*). **TN** Formerly within Japanese Bush Warbler.

ASIAN STUBTAIL

squameiceps

ad

JAPANESE BUSH WARBLER

ad *cantans*

ad *riukiuensis*

ad ♂ *canturians*

ad ♀ *canturians*

KOREAN BUSH WARBLER

ad ♂ *borealis*

BROWN-FLANKED BUSH WARBLER CT
Cettia fortipes

L 11–12.5cm; WT 8–11.5g. **SD** Himalayas across China to E coast north to Changjiang River. *C. f. davidiana* in E China; *C. (f.) robustipes* (sometimes considered specifically as **Strong-footed Bush Warbler**) endemic resident of mid-elevation Taiwan. **HH** Favours dense undergrowth, clearings with rush-like grasses and grassy margins of mid-level forests, also edges of cultivation and bushy hillsides. **ID** Rather uniform, dusky-brown *Cettia* lacking much contrast between upper- and underparts. Head pattern indistinct: pale grey-brown or buff supercilium most noticeable between bill and eye, very faint behind eye; indistinct blackish-brown eyestripe is narrow, does not reach bill base and hardly extends behind eye. Upperparts dusky greyish earth brown. Underparts paler greyish earth brown on face, throat, breast-sides, and even paler on belly, flanks and vent. Separated from Japanese Bush Warbler (p.340) with difficulty, mostly by smaller size; Brown-flanked is more heavily washed buffish-brown on throat and breast, and is browner still on flanks, also has less prominent supercilium and slightly smaller bill. **BP** Bill long (exceeds length of lores), greyish-brown; eyes dark brown; legs brownish-pink, feet large, brown. **Vo** Song is a loudly whistled high-pitched *weeee*, followed by an explosive *wichyou*. On Taiwan, *C. (f.) robustipes* gives dry clicks similar to *borealis* Korean Bush Warbler, but higher pitched: *trrk trrk*, occasionally rolled to a thicker *trrrek*. *C. f. davidiana* probably gives hard *chuk* or *tchuk tchuk* (like other races). Yellow-bellied has similar calls but even higher pitched.

YELLOW-BELLIED BUSH WARBLER CT
Cettia acanthizoides

L 9.5–11cm; WT 6g. **SD** From E Himalayas across China south of Changjiang River. *C. a. acanthizoides* in E China; *C. a. concolor* endemic to higher elevations of Taiwan (split by some as **Taiwan Yellow-bellied Bush Warbler**). **HH** Scrub, bamboo thickets, roadside vegetation and forest edge above 1,000m (1,500m in summer) in China; above 2,440m in Taiwan. **ID** Tiny, compact, very active; the smallest *Cettia* in region. Crown slightly rufous earth brown, pale buff-brown supercilium broad but rather short, short black eyestripe somewhat indistinct; ear-coverts dusky grey-brown. Upperparts entirely warm mid-brown, including wings and tail, rufous fringes to flight-feathers and tertials form panel on closed wing, may also show some white (from underparts) at bend of wing; tail previously falsely depicted as extremely short, but proportions actually typical of most *Cettia* (tail length equal to or exceeds length of primaries), tip slightly notched. Chin to breast buffish-yellow, neck-sides yellowish-brown, flanks, belly and vent distinctly dull yellow. **BP** Bill short, grey-brown, paler on lower mandible; eyes black; tarsi orange-brown. **Vo** Very high-pitched slightly 'wet' sounding *tsik*, most like Asian Stubtail call; also a rasping *brrrr*, and short rattled *chrrt chrrt....* Song extraordinary: a long build-up of high-pitched (squeaky gate-hinge) whistled *seee* notes, reaching an impossibly high-pitched climax, pausing, then breaking into a lower pitched trilling cadence that does not fall as far as the ascent; *chirrrrrrrrrrrrrrr.* **AN** Verreaux's Bush Warbler.

SIBERIAN BUSH WARBLER CKJ
Bradypterus davidi

L 12cm; WT 10g. **SD** From SC Siberia to E Asia. *B. d. davidi* breeds SE Russia (Transbaikalia to W Amurland) and NE China (south to N Hubei); rare (or overlooked?) summer visitor to N Korea, accidental offshore Korea and may have strayed to Japan. Probably winters SE Asia. **HH** Extremely skulking poorly known species of taiga, in forest glades with damp grasses, thickets beside streams and open grassy areas on scrubby hillsides. Winters in reedbeds, tall grasslands and scrub. **ID** *Bradypterus* resemble *Cettia* in many respects, but have longer and graduated tails. Generally dark brown with short broad wings. Pale, buffy supercilium, grey ear-coverts, grey-white chin, throat and breast, with necklace of fine black streaks (more prominent in ♂). Underparts rufous-brown, long undertail-coverts dark brown with white crescentic tips; tail rounded. Darker, more reddish-brown back than Chinese Bush Warbler. In non-breeding plumage throat spotting less distinct (even completely obscured) and underparts and supercilium yellow-toned. **BP** Bill dark grey with black tip (lower mandible pale in non-breeding season); eyes dark brown; tarsi pinkish-brown. **Vo** A rasped *tschuk* and low *tuk.* Song an insect-like series of rising, short rasping notes: *dzeeep dzeeep dzeeep dzeeep....* **TN** Formerly within Spotted Bush Warbler *B. thoracicus*. **AN** Père David's Bush Warbler; Baikal Bush Warbler.

CHINESE BUSH WARBLER CR
Bradypterus tacsanowskius

L 13cm; WT 10g. **SD** Ranges across S Siberia from Yenisei through S Transbaikalia to extreme S Russian Far East (Khabarovsk region), also NE, C & S China. Winters in S China and SE Asia. Monotypic. **HH** Rare and poorly known, extremely skulking bird of grassy meadows, thickets and woods in river valleys, and montane areas with lush grasses, especially larch forest in summer and grasses and reedbeds in winter. **ID** Upperparts, crown, ear-coverts, wings and tail olive-brown, but buffy-white lores, eye-ring and slight supercilium lend a pale-faced appearance. Lower ear-coverts and sides of neck/breast olive-brown with off-white streaking. Off-white throat and belly centre; breast and flanks greyish-brown to brownish-buff, sometimes with faint brown streaking on lower neck; long undertail-coverts have some pale scaling. Closely resembles Siberian Bush Warbler but paler above, with less neck-streaking and darker flanks. Identification criteria (other than voice) of Asian *Bradypterus* still poorly known. **BP** Bill thin, black when breeding, otherwise blackish-brown above, pink below; eyes dark brown; tarsi pinkish-brown. **Vo** Call uncertain, but perhaps a low *chir-chirr.* Rasping song distinctive, a scraping insect-like low *tze-tze-tze* or electronic buzzing *dzzzeep-dzzzeep-dzzzeep.*

BROWN-FLANKED BUSH WARBLER

ad fresh *davidiana*

ad *robustipes*

YELLOW-BELLIED BUSH WARBLER

ad *concolor*

ad fresh *acanthizoides*

ad worn *acanthizoides*

SIBERIAN BUSH WARBLER

davidi

juv

ad

CHINESE BUSH WARBLER

ad plain

ad streaked

BROWN BUSH WARBLER   C
Bradypterus luteoventris

L 13cm; WT 10g. **SD** E Himalayas to N Vietnam and C & S China to E coast south of Changjiang River. Monotypic. **HH** Extremely skulking, in dense grasses, low scrub and forest clearings on slopes (2,000–3,000m); descends in winter. **ID** Rather plain, generally unmarked *Bradypterus* with faint buff supercilium. Upperparts plain, mid brown from forehead to rump, also wings and tail. Wings short and broad, tail long, full and rounded. White on chin and throat to upper breast and central belly; face, breast-sides, flanks and belly warm buff-brown, undertail-coverts orange-brown, essentially unmarked. Very similar to Russet Bush Warbler. **BP** Bill fine, blackish-brown above, yellowish-pink below; eyes dark brown; tarsi brownish- or yellowish-pink. **Vo** Call a hard *tak*. Song a quiet, monotonous rapid dry reeling *tutututututu…* also likened to clicking sewing machine: *tk tk tk tk tk tk tk tk tk tk….*

TAIWAN BUSH WARBLER   T
Bradypterus alishanensis

L 13cm; WT 10g. **SD** Endemic to Taiwan. Monotypic. **HH** Confined to central mountains, common *c*.1,200–3,800m in cold-temperate bamboo thickets and temperate coniferous forest edges, where keeps low in dense scrub, weeds and thick grass. Descends in winter. Skulking and very elusive, but can be very approachable. **ID** Medium-sized, generally dark brown bush warbler with narrow, indistinct buffy supercilium, short rounded wings and longish broad, graduated tail. Upperparts (including head) olive-brown with rufous tinge; tail more olive. Flight-feathers and tail marginally darker than rest. Chin and throat whitish-buff, with small brown speckles below in spotted form, otherwise dingy white, washed grey on neck-sides and olive-grey or brown on flanks and undertail-coverts, buff tips to latter give indistinct scaly appearance. **BP** Bill thin, distinctively black; eyes dark brown; tarsi brownish- to greyish-pink, feet appear oversized. **Vo** Call a sharp, dry *tick* or *stip* and rapid, scratchy *ksh ksh ksh*. Song (clearer, sweeter, more piercing than other Asian *Bradypterus*) given from well before sunrise, a frequently repeated series of pulsing whistles interspersed by a ticking noise: *ti-ti-teer ti-ti-teer ti-ti-teer….* or *hwee tiki hwee e tiki hwee e tiki hwee tiki*, recalling a cricket and tree frog together. **TN** Formerly within *Bradypterus seebohmi* before this species was split; see Russet Bush Warbler.

RUSSET BUSH WARBLER   C
Bradypterus mandelli

L 13cm; WT 10g. **SD** E Himalayas to SE Asia and S China. In region *B. m. melanorhynchus* occurs in E China (Hubei to Fujian), wintering in SE China and N SE Aisa. **HH** Skulks in dense scrub, thickets and margins of cultivation and forest on slopes. **ID** Upperparts from crown to tail including wings dark chestnut-brown, face slightly paler, with pale grey-brown supercilium extending only weakly behind eye. Chin/throat pale grey with fine dark grey streaks; underparts grey-white, sometimes very grey on breast, with orange-brown to dark brown sides, flanks, belly and vent; long undertail-coverts have pale tips forming well-patterned crescents. Closely resembles Siberian Bush Warbler (see p.342), but generally darker with longer tail and larger bill. **BP** Bill blackish-grey, pale horn at base of lower mandible; eyes brown; tarsi brownish-pink. **Vo** An emphatic *shtuk*. Song comprises a series of paired notes, first a nasal buzz followed by a sharp click: *cre-ut cre-ut cre-ut cre-ut…* **TN** Russet Bush Warbler was *B. seebohmi* before the species was split; *seebohmi* now refers to Javan Bush Warbler.

LANCEOLATED WARBLER   CTKJR
Locustella lanceolata

L 11–13cm; WT 9–13g. **SD** Wideranging summer visitor (May–Sep) from W Russia to W Chukotka, Kamchatka, Sakhalin, Hokkaido, NE China and N Korea, migrates through coastal E Asia, including Taiwan; winters in SE Asia and Philippines. *L. l. lanceolata* occurs east to Sea of Okhotsk and Kamchatka; *L. l. hendersonii* occurs from Sakhalin and S Kuril Is to Hokkaido (formerly to Honshu). **HH** Around wetlands, in reedbeds, flooded scrub and damp grasslands with scattered bushes, also dry grassy areas, fallow or abandoned fields, where best located by unmistakable song. Moves mouse-like through short vegetation; difficult to flush. **ID** The smallest and most distinctive *Locustella*. Mid olive-brown plumage very finely streaked black on crown, heavily on mantle and finely on underparts. Supercilium long, but faint. Scapulars, coverts and tertials all broadly centred blackish-brown, with pale olive-brown fringes. Underparts whitish on throat, buff/brown on breast and sides, with prominent narrow streaking on breast-sides and flanks, and fine teardrop spots on vent and long undertail-coverts. Dark tail is rather short and rounded at tip, lacking contrast. **BP** Bill sharp, dark grey culmen and tip, with yellowish-pink lower mandible; eyes dark brown; tarsi pale brownish-pink. **Vo** A short, hard *chu chu*, fainter *tack*, metallic *pit* or *chit* (similar to Pallas's Grasshopper Warbler, but quieter and slightly drier and squeakier), and harsh, scolding series of *cheek-cheek* notes. Song by day comprises short bursts of very fast, rather high, insect-like churring or metallic reeling, *chirir-iririririri* or *chichichichi*; nocturnal song similar but continues for many minutes at a stretch.

BROWN BUSH WARBLER

TAIWAN BUSH WARBLER

ad plain

ad spotted

RUSSET BUSH WARBLER

melanorhynchus

ad

LANCEOLATED WARBLER

ad

lanceolata

1st-win

PALLAS'S GRASSHOPPER WARBLER   CTKJR
Locustella certhiola

L 13–14cm; WT 13–22g. **SD** N Asia, in summer from Ob to Yakutia and Sea of Okhotsk, also S Russian Far East and Sakhalin, south to NE China. Winters SE & S Asia, migrating through east of region, some winter S Nansei Shoto. *L. c. certhiola* (including *minor*) E Transbaikalia, NE China to S Sea of Okhotsk, migrates through E China; *L. c. rubescens* Irtysh east to Kolyma River and N Sea of Okhotsk, recorded on migration Hebei and Japan. **HH** Wetlands with reedbeds and wet meadows with bushes, in taiga and steppe. Extremely skulking (except when singing); flies reluctantly, forages close to ground. **ID** Small, distinctive *Locustella*. Generally mid-brown; almost black crown with fine grey-brown streaking, usually heavy dark streaking on brown mantle, scapulars/wing-coverts and tertials blackish-brown with pale fringes; rump contrastingly more rufous- or rusty-brown, with dark centres to uppertail-coverts. Long, broad supercilium grey-white. Underparts grey-buff with browner wash to breast-sides, flanks and vent. Dark rusty-brown tail has rounded tip, with blackish subterminal band and greyish-white tips (best seen from below); vent and undertail-coverts unstreaked. Juv. has yellowish wash to chest and flanks, fine streaking across throat extends to flanks, clear white tips to tertials. *L. c. certhiola* has heavy black streaking on mantle and secondaries. *L. c. rubescens* darker overall, darker brown upperparts (less contrastingly streaked); underparts also darker with browner flanks and undertail-coverts. **BP** Bill blackish-grey above with dark-tipped yellowish-pink lower mandible; eyes dark reddish-brown; tarsi pink. **Vo** Calls range from ticking *pit*, dry clicking *chat* and dry rolling rattle *trrrrrrrt*; also abrupt, explosive *dt dt dt* in alarm. Song, from exposed perch or short song-flight, lacks typical *Locustella* reeling; instead a more *Acrocephalus*-like quickly repeated series of whistled, trilled or rattled phrases, often culminating in loud *sivih-sivih-sivih* or *see-wee-seewee-seewee*. Song recalls Middendorff's but more complex: *jiriiri… chirichiri chuichuichui*. **AN** Rusty-rumped Warbler.

MIDDENDORFF'S WARBLER   CTKJR
Locustella ochotensis

L 13.5–14.5cm; WT 19–23g. **SD** Locally common breeding endemic around Sea of Okhotsk from Amur Delta, Sakhalin and Hokkaido, throughout Kamchatka to Kuril Is. Migrates through Korea, Taiwan and E China; winters Philippines and Indonesia. Monotypic. **HH** Wetland margins with bushes, woodland fringes and open areas e.g. headlands with dwarf bamboo. **ID** Large, plain, with typical *Locustella* elongated shape, low-sloping head/bill profile and longish tail. Ad. brown overall, greyer on crown, ear-coverts and nape; pale creamy white or pale buff supercilium, weak before eye but extends well back to ear-coverts; mantle more olive with rather darker, reddish-brown centres, forming diffuse streaks; rump/uppertail-coverts more yellowish- or rufous-brown. Tail long, rounded, often fanned slightly in flight, showing subterminal black spots and white tips. Juv. has yellow-tinged face and underparts, dark-streaked breast, and warmer olive-brown flanks than ad. Differs from Pallas's in having weakly streaked upperparts, tail typically has less contrasting dark subterminal band, and from Pleske's by shorter bill, browner upperparts, paler underparts and darker more prominent eyestripe. **BP** Bill sharp, short, dark grey upper mandible, yellowish- or pinkish-grey base to dark-tipped lower mandible; eyes dark brown; tarsi brownish- or flesh-pink. **Vo** An abrupt *tit tit tit…*. Song from bush or in song-flight a dry rattling: *che-tit che-tit che-tit-chewee-chewee-chewee-chewee*.

PLESKE'S WARBLER *Locustella pleskei*   CTKJR

Vulnerable. L 15–17cm; WT 16–24g. **SD** Breeds only on small islands off extreme S Russian Far East, S Korea, nearby Chinese coast, and W & C Japan east to Izu Is. Migrates through Korea, SE China and Taiwan. Winters mainly Philippines and Indonesia, scarce Taiwan. Monotypic. **HH** Breeds in grasslands/dwarf bamboo; winters in scrub, reedbeds and mangroves. Creeps and hops stealthily like a pipit. **ID** Very like Middendorff's; separated primarily on distribution, larger bill and subtle colour differences. Pleske's has shorter, more greyish-buff supercilium, dark lores (darker than crown) and weak eyestripe (thin or absent behind eye). Upperparts uniform greyish olive-brown (lacks richer coloration and contrast of Middendorff's). Only mantle shows subtle broad dark streaks; wing-coverts and tertials olive with pale buff fringes, flight-feathers have silver-brown fringes. Underparts dusky pale grey (slightly darker than Middendorff's), darkest on breast-sides; tail like Middendorff's, but dark brown with narrow off-white tips to outer 3–4 feathers, undertail-coverts buff. **BP** Bill long (longer than distance from rear of eye to bill base), dark grey with pale tip and greyish- or yellowish-pink lower mandible; eye-ring, irides dark brown; tarsi sturdy, brownish-pink, strong short hindclaw. **Vo** An abrupt, hard *stit it it* or *tschup-tschuptschup*. Song from exposed perch breezy, slightly electric 3–4-part wavering phrase *swee swee swee swee* (slower and harder than Ijima's Warbler's song, which sounds similar at distance). Also a dry chirping: *tski tski tski…*; *chitti chuichuichui* or *chiririri-chui-chui-chui*. **AN** Styan's Grasshopper Warbler.

GRAY'S WARBLER *Locustella fasciolata*   CTKJR

L 16.5–18cm; WT 24–32g. **SD** Siberia to Sakhalin and N Japan. *L. f. fasciolata* SC Siberia, Transbaikalia, NE China to Ussuriland, winters SE Asia to New Guinea. *L. f. amnicola* (sometimes considered distinct as **Stepanyan's Warbler**) Sakhalin, Hokkaido, S Kuril Is; winters Indonesia and Philippines; migrates via Taiwan. **HH** Extremely skulking in undergrowth of lowland forest, taiga, bushes near streams. **ID** Largest *Locustella*; entirely unstreaked. Ad. upperparts dark olive-brown. Face grey or greyish-white, with dark eyestripe and long greyish-white supercilium extending to nape. Greyish-white chin/throat and face with grey-brown wash to breast-sides and dusky-brown wash on flanks more extensive than in other *Locustella*. Tail rusty-brown, long, rounded, with rusty-brown rump and long, rusty undertail-coverts. Juv. warmer brown upperparts and yellower or olive, rather than grey, tones to face and underparts. *L. f. amnicola* slightly warmer above and more strongly buff below. **BP** Bill long, slightly decurved, dark grey, base of lower mandible yellowish or pinkish; eyes dark brown; strong tarsi pinkish-brown. **Vo** Vociferous: call a guttural *gu gu gu*; song loud and repetitive, from deep cover day and night, hurried and accelerating: *chot-pin chot-pin-kake-taka*; also *choppin chipicho*.

PALLAS'S GRASSHOPPER WARBLER

certhiola

1 st-win

ad

MIDDENDORFF'S WARBLER

ad

1 st-win

PLESKE'S WARBLER

ad

1 st-win

1 st-win *fasciolata*

ad *fasciolata*

GRAY'S WARBLER

1 st-win *amnicola*

ad *amnicola*

THICK-BILLED WARBLER CTKJR
Acrocephalus aedon

L 18–19cm; WT 22–31g. **SD** SC Siberia to S Russian Far East and NE China, wintering from S China to SE Asia and India. *A. a. stegmanni* from upper Ob east via Transbaikalia to S Russian Far East and NE China, migrating through E China to winter to south. Regular migrant to offshore Korea (rarer on mainland) and accidental offshore Japan. **HH** Very different habitat from other *Acrocephalus*, being found in woodland/forest edge, dense thickets, scrub and bushy areas. **ID** Large, plain, short-winged and long-tailed *Acrocephalus* superficially resembling Oriental Reed, but more shrike-like proportions and plainer face appears 'gentler' with larger eye, and lacks supercilium and eyestripe of latter, with distinctive, pale lores. Upperparts rusty-brown, slightly greyer on face and crown, warmer on lower back and rump. Underparts off-white, whitest on chin/throat, with buff wash to flanks and vent. Tail long and graduated. Primary projection short. **BP** Bill large, thick (shorter, thicker, blunter than Oriental Reed, with more curved upper mandible), dark greyish-brown above, paler yellowish- or pinkish-horn below; large eyes, dark reddish-brown; strong tarsi bluish-grey. **Vo** Low muffled *tuc*, sometimes in rolled series when agitated, *tuc tuc tuc trruc trruc trrrc* (recalling Dusky Warbler), a wheezy *wep* or *jep*, strong, hard *chack*, *chock* or *tack*, often repeated; and occasionally *skeesh* sounding like air escaping from a pump. Alarm call is a loud *jah jah* or *bzee bzee*. Song a chattering warble combining extensive mimicry with twittering call notes and loud whistles in repeated phrases. **TN** Formerly *Phragmaticola aedon rufescens*.

ORIENTAL REED WARBLER CTKJR
Acrocephalus orientalis

L 17–19cm; WT 22–29g. **SD** Summer visitor to E Asia, from Mongolia and Transbaikalia to S Russian Far East, Sakhalin, much of Japan, Korea and E China south to Fujian. Winters on Taiwan and in Philippines, S & SE Asia and Indonesia. Monotypic. **HH** Very noisy bird of large reedbeds, or small reedbeds adjacent to ricefields, and marshland vegetation with bushes, from sea level to *c.*1,000m. **ID** The common large reed warbler in region, readily distinguished by size and voice. Face pattern somewhat weak; supercilium broad and whitish from bill to eye and indistinct pale brown to rear, bordered only below by weak, narrow blackish-brown eyestripe. Overall olive-brown with grey cast to crown, back, and paler, greyer rump; tail rather long and somewhat square-ended. Underparts off-white on throat and belly, with indistinctly dark-streaked breast, warm buff wash on flanks and vent. Primary extension shorter than length of visible tertials. When singing, reveals bright orange-red mouth lining. In flight thrush-like, with plain wings and long tail rounded at tip, with whitish tips to outer feathers. **BP** Bill long and rather thick, dark grey-brown above and at tip, yellowish- or pinkish-brown on most of lower mandible; eyes brown; strong tarsi pinkish-grey, toes grey. **Vo** A deep thick *turrr* or *chichikarr*, loud arresting *tack!*, sharp 'tick' and slurred *trek*, *krak*, *kirr* or *ge*. Song, often delivered from conspicuous perch atop reeds/bushes or wires, a series of loud, ratcheting gravelly calls and dry chuckles, repeated somewhat cyclically: *kawa-kawa-kawa-kawa-gurk-gurk-eek-eek-kawa-gurk*; *kiruk kiruk kiruk jee jee* or *gyo-shi-gyogyoshi ke-ke-ka-ka-shi-shi-shi-shi...* and *gyogyogyogyogyoshi gyoshigyogishi*. **TN** Formerly within Great Reed Warbler *A. arundinaceus.*

BLACK-BROWED REED WARBLER CTKJR
Acrocephalus bistrigiceps

L 13.5cm; WT 7–11g. **SD** Summer visitor (Apr–Sep) to E Asia, from CE China to S Russian Far East, Sakhalin, Hokkaido south to C Honshu, locally Kyushu; surprisingly scarce in Korea. Winters uncommonly in Taiwan and, outside region, in S China and SE Asia. Monotypic. **HH** Reedbeds, scrubby grassland, woodland fringes near wetlands and rivers, particularly with willows; streams and wetlands to 1,500m. Usually perches conspicuously atop reeds or bushes when singing, revealing bright yellowish mouth lining. **ID** The common small reed warbler in region, readily distinguished by size and face pattern from Oriental Reed with which it often shares habitat. Ad. has prominent white supercilium that broadens behind eye, ending rather squarely and bordered above by long, narrow black brow and below by fine black eyestripe. Generally mid-brown above, warmer and more rufous on rump, and whitish below with white chin/throat contrasting with brown ear-coverts, and warm buff- or pale brown wash to sides and flanks. Juv./1st-year has distinctive yellowish-buff wash to upper- and underparts, and broader less defined brow. **BP** Bill fine, short, dark blackish-grey above, paler at base of lower mandible; eyes dark brown; tarsi dark brownish-pink. **Vo** Harsh churring *kurr*, hard *jat jat* and low *trruk*. Song a prolonged jumbled mixture of notes: *chi chi chi chur jee jee jee jurr chi-ur chi-ur chi-ur chi-ur* interspersed with dry rattles, harsh trills and mimicry; also *kirikiri-pi gyoshi kyoriri-piririri....*

STREAKED REED WARBLER CTJ
Acrocephalus sorghophilus

Vulnerable. L 12–13cm. **SD** Very rare and poorly known E Asian endemic, presumed to breed in very limited area of Liaoning and Hebei, migrating through E coastal China to winter in Philippines. Accidental S Japan (Yonaguni-jima) and Taiwan. Monotypic. **HH** Presumed to breed only in reedbeds, but on migration found in agricultural fields. **ID** Rather similar to Black-browed Reed Warbler in size and structure, also has black lateral crown-stripes, but distinguished by fine black streaking on crown, narrower black margins to crown, broader pale creamy-buff supercilium behind eye, paler brown nape and mantle, with narrow black streaking on mantle; wings show broad pale fringes to dark greyish-brown coverts and tertials, and narrow pale fringes to flight-feathers; rump uniform mid brown, tail dark brown with buff fringes and tips; white chin/throat with buff wash on flanks. **BP** Bill fine (somewhat stronger than Black-browed), culmen dark blackish-brown, with pale ochre edges, lower mandible entirely pale ochre; eyes dark brown; tarsi grey. **Vo** Calls undescribed; song is rasping and churring, likened to a weaker version of Oriental Reed Warbler. **AN** Speckled Reed Warbler.

ORIENTAL REED WARBLER

1st-win

THICK-BILLED WARBLER

stegmanni

ad

ad

BLACK-BROWED REED WARBLER

1st-win

STREAKED REED WARBLER

autumn/1st-win

ad spring

ad

BLUNT-WINGED WARBLER   C
Acrocephalus concinens

L 13–14cm; WT 7.7–8.9g. **SD** Disjunct range includes parts of C Asia, NE India and adjacent areas, with *A. c. concinens* breeding only in E China north to Hebei. Winters to S & SE Asia. **HH** Grasslands, reedbeds and scrub associated with wetlands. **ID** Rather drab, olive-grey or olive-brown, unstreaked *Acrocephalus*, with very short wings not reaching base of tail (others in genus do so), and short primary projection. Upperparts brownish-olive, somewhat more rufous-brown on rump/uppertail-coverts. Underparts off-white, with yellowish-buff tinge to breast and on flanks to vent. Resembles Manchurian Reed, but distinguished by short white supercilium extending only from bill to eye, not bordered above by black, and blacker tail with brown outer edges. **BP** Bill dark greyish-brown above, pale pinkish-brown below (longer, thicker than Paddyfield); eyes dark brown; tarsi brownish-grey. **Vo** A quiet *tcheck* and soft *churr*. Song varied, short phrases comprising a mix of slurred whistles and buzzing notes.

MANCHURIAN REED WARBLER   CTK
Acrocephalus tangorum

Vulnerable. L 13–14.5cm. **SD** Breeds NE China and extreme SE Russia, wintering in SE Asia. Reported Taiwan and Korea. Monotypic. **HH** Emergent marshland vegetation. **ID** Small, rather drab reed warbler, appearing short-winged and long-tailed. Recalls Black-browed Reed (p.348), Paddyfield and Blunt-winged, but generally darker with more prominent black margin above longer white supercilium (which narrows behind eye), and narrow black eyestripe. Upperparts mid to dark brown, somewhat warmer buffy-brown on rump. Chin/throat white, warm buff-brown wash to flanks and undertail-coverts, central belly paler. Appears much more rufous-brown in fresh plumage (autumn/winter). **BP** Bill rather thick and long, giving elongated impression to head, greyish upper mandible, pink lower mandible; eyes dark brown; tarsi pinkish-brown. **Vo** Various sharp, hard notes: *chik-chik, chr-chuck* and *zack-zack*. Song a series of warbled phrases mixed with higher pitched notes.

PADDYFIELD WARBLER   TKJ
Acrocephalus agricola

L 12–14cm; WT 8–11g. **SD** From Black and Azov seas to C Asia east to Tuva. Winters mainly S Asia. *A. a. agricola* accidental Taiwan, Japan and reported Korea. **HH** Reedbeds, tall grasses, willows and birches fringing wetlands. **ID** Unstreaked *Acrocephalus* closely resembling Eurasian Reed and Black-browed Reed, but has more prominent white supercilium, broadest behind eye and bordered above by narrow, blacker margin to crown and below by shorter darker eyestripe than Eurasian Reed, though has less clearly defined black lateral crown-stripes than Black-browed Reed; appears to have rounded head due to erectile crown feathers; neck-sides pale. Upperparts warm sandy to mid brown, darker on crown, and contrastingly rusty-brown on rump and uppertail-coverts; whitish from chin to vent with buff wash on sides and lower belly; tertials typically dark with paler fringes. Tail rather long and distinctly rounded. **BP** Bill rather thick, sharply pointed, greyish-brown upper mandible, pale yellowish- or pinkish-buff lower mandible

with dark tip; eyes dark brown; tarsi pinkish-brown. **Vo** A hard *check* or *tack*, rolling *cherrr* or *trrr*, and harsh *cheeer*. Song a long series of unbroken melodious chattering phrases, mostly consisting of mimicry.

EURASIAN REED WARBLER   C
Acrocephalus scirpaceus

L 12–14cm; WT 8–19.7g. **SD** N Africa, Europe and Baltic States to Afghanistan; winters in sub-Saharan Africa. Eastern race *A.* (*s.*) *fuscus* ('**Caspian Reed Warbler**') accidental to Jiangsu (E China). **HH** Reedbeds and willow thickets in or near wetlands. **ID** A plain grey-brown reed warbler resembling Paddyfield, with a long profile and prominent bill. Upperparts pale greyish-brown, more rufous-brown on rump. Underparts whitish with rusty wash on flanks. Crown grey-brown, with noticeable rounded peak at rear; eyestripe dark brown, supercilium white, prominent only from bill to eye, with no dark margins. Primary projection very long (see Manchurian Reed and Paddyfield). *A.* (*s.*) *fuscus* is darker, greyer brown than nominate. **BP** Bill long, rather thick, with broad blunt tip, upper mandible dark brown, lower mandible yellowish- or pinkish-brown; eyes dark reddish-brown; tarsi grey-brown. **Vo** A short hard *che, chk* or rolling *trr-rr*.

BLYTH'S REED WARBLER   J
Acrocephalus dumetorum

L 12–14cm; WT 8–16g. **SD** Breeds from Baltic to C Russia and Afghanistan; winters to India and Burma. Accidental Hokkaido and Okinawa. **HH** More arboreal and less associated with wetlands than most *Acrocephalus*; typically occurs in lightly wooded country, in riparian and flooded deciduous forests, also forest edge and overgrown orchards. Occupies similar habitats in winter and on migration. **ID** A plain, 'cool' even drab greyish-brown unstreaked reed warbler; rather slim, short-winged and long-billed. From crown to tail rather uniform cool olive-brown, short, rounded wings somewhat browner; face pattern distinct, with short white supercilium prominent only between bill and eye, eyestripe dusky dark brown. Underparts plain, off-white from chin to vent, with buff flanks. **BP** Bill rather long, thick, greyish-brown upper mandible, with pale pinkish-brown base to lower mandible; eyes dark brown; legs/feet pinkish-brown. **Vo** Hard, jarring and scraping sounds: *thik, chak* and *cherr* interspersed with sweet whistles.

BOOTED WARBLER
Hippolais caligata   J

L 11–12.5cm; WT 7–11g. **SD** From W Russia to SC Siberia and NW Mongolia, wintering in S Asia. Accidental Japan. Monotypic. **HH** Dry scrub in steppe. **ID** Shape and size resemble *Phylloscopus*, but plumage recalls *Acrocephalus*. Small, very plain, greyish-brown, with distinctly rounded crown, rather prominent supercilium, and somewhat short, square-ended tail. White supercilium, long and broad, but diffuse behind eye, contrasting with rather dark forecrown, narrow black eyestripe and narrow, darker border to crown. Upperparts pale, sandy, grey-brown from crown to rump, including scapulars; wings and tail blackish-brown with pale, grey-brown fringes to coverts and secondaries, and whitish fringes to tertials; may suggest a pale wing-panel. Wingtips just reach tail base, with short primary projection about half of longest tertial; tail greyish-brown, rather square-tipped, with off-white fringes to outer tertials, undertail-coverts short. **BP** Bill very short, slender, blunt, dark upper mandible, pale lower mandible with dark tip; eyes black; tarsi pinkish-grey. **Vo** A harsh, hard *chet, chek* or *chat*, and short, dry trills: *tr'r'rk*.

BLUNT-WINGED WARBLER

concinens

ad

1st-win

ad

MANCHURIAN REED WARBLER

1st-win

EURASIAN REED WARBLER

fuscus

ad

ad

PADDYFIELD WARBLER

agricola

1st-win

BLYTH'S REED WARBLER

ad

1st-win

BOOTED WARBLER

WILLOW WARBLER *Phylloscopus trochilus*　　KJR

L 11–12.5cm; WT 6.3–14.6g. **SD** From British Isles to E Chukotka. *P. t. yakutensis* breeds from Yenisei east into region at high latitudes of E Siberia to Anadyr; winters in Africa; rare, probably annual migrant (Sep–Nov) offshore Japan; accidental Korea. **HH** Breeds in wooded habitats or thickets, from tundra and taiga to desert regions, but usually where there are birches. On migration, woodland and scrub. *P. t. yakutensis* regularly wags tail downward, recalling Siberian Chiffchaff. **ID** Rather plain *Phylloscopus* lacking wingbars or other prominent features, somewhat resembling grey-brown Arctic Warbler (p.356). Has prominent off-white supercilium generally shorter than in Arctic; dark eyestripe is prominent. Crown to rump greyish-brown with only hint of green on rump; wings and tail darker brown with faint green outer fringes, and long primary projection. Underparts vary from dusky-olive to greyish-white, with some streaking on breast-sides, and yellow underwing-coverts. Pale supercilium above dark eyestripe, lack of wingbars and leg colour distinguish it from very similar Siberian Chiffchaff. **BP** Bill fine, blackish-brown with pink-based lower mandible; eyes black; tarsi pale to dark pinkish-brown to warm brown. **Vo** A sweet, slightly hesitant and weak upslurred *phwee*, sharply pitched *hewit* or *hoowee*. Song a jerky but gradually descending warble *pt'i wee p'tew wee-itew.*

SIBERIAN CHIFFCHAFF *Phylloscopus tristis* CKJR

L 11–12cm; WT 6–10.9g. **SD** Urals across Yakutia to NW Chukotka; in E Asia breeds at high latitudes mostly between Lena and Kolyma, wintering mainly in India. Accidental migrant coastal China, offshore and mainland Korea, and offshore Japan; rare winterer Japan. Monotypic as treated here (the only taxon of the group likely to occur), but possibly a race of widespread Common Chiffchaff *P. collybita*. **HH** Breeds in thickets and coniferous forest in taiga; winters in similar (or, more usually, drier) habitats to Dusky Warbler. **ID** A plain brown, buff and white *Phylloscopus*. Pale greyish-brown upperparts, weak olive on back and rump, and flight-feathers fringed olive. Yellow restricted to small area at bend of wing. Pale buff supercilium and strong dark eyestripe; ear-coverts, neck- and breast-sides, and flanks cool buff, lacking any hint of yellow. White chin and throat, off-white to cream below. Primary projection short (about half of tertial length). In worn plumage may show short, pale wingbar (see Greenish Warbler, p.360). **BP** Bill short, fine, predominantly black; eyes dark brown; legs black or blackish-brown. **Vo** A mournful, high-pitched, monosyllabic *heet, eet, iiihp* or *viip*; similar to Daurian Redstart, but slightly longer and softer. Song a repetitive *chui-picho chi-picho picho picho*..., faster and more hesitant, with notes run together and more musical than Common Chiffchaff, recalling Greenish.

WOOD WARBLER *Phylloscopus sibilatrix*　　J

L 11–13cm; WT 6.4–15g. **SD** Breeds Europe to SC Siberia, and winters in Africa. Accidental offshore Japan (Sep–Oct). Monotypic. **HH** On migration/in winter, bushes, low trees and woodland. **ID** Rather cleanly marked *Phylloscopus*, with bright yellowish-green upperparts; wings and tail blackish-brown with broad pale green fringes. Supercilium, chin/throat and upper breast bright lemon-yellow, whilst belly and vent are pure white. Wings long, with long primary projection, but has fairly short tail. **BP** Bill dark brown above, pale yellowish-pink below; eyes dark brown; legs brownish- or yellowish-pink. **Vo** A powerful *jii*, sharp *zip* or soft, sad *hwui, chui* or *tyouyou*. Song a series of brief notes followed by a shimmering trill: *zip.. zip.. zip.. zip.. zip zip zip zip-zip-zip-zipzipzipfurrrurrrr.*

DUSKY WARBLER *Phylloscopus fuscatus*　　CTKJR

L 10–13cm; WT 8.5–13.5g. **SD** Occurs east from C Siberia. *P. f. fuscatus* ranges across Russia from Ob River northeast to S Chukotka, Koryakia and Kamchatka, also S Russian Far East, Ussuriland and Sakhalin, NE China and N & C Korea; winters from coastal SE China and Taiwan to S & SE Asia. Scarce migrant Japan and S Korea, scarce winter visitor Japan. **HH** In summer favours low, sparse taiga, scrub or bushes near swamps or edges in lowlands, or in thickets on slopes, where sings from prominent perch. On migration/in winter skulks in dense vegetation, where very active and vocal, often flicking tail and wings. Favours reeds or tall grasses with some low trees near streams/ditches, rivers or ponds. **ID** Medium-sized warbler, recalling Siberian Chiffchaff but darker upperparts lack any green tones and has rather short tail. Also resembles Radde's (p.354), but slightly slimmer, smaller headed, shorter tailed and thinner-billed. Upperparts dark greyish-brown. Eyestripe clear, dark, especially from bill to eye. Supercilium narrow, white between bill and eye, off-white or buff behind; white arc below eye. Wings rather short, primary projection short, tail quite long. Off-white chin/throat, breast and flanks grey-white, rear flanks and vent have warmer buff-brown wash. **BP** Bill short, fine, blackish above, paler horn at base and sides of lower mandible; eyes dark brown; thin legs and small feet brownish-pink to dark orange-brown. **Vo** Call (rather similar to Japanese Bush or Winter Wren) a hard, dry, but somewhat muffled *tak tak tak, tchak, chett* or *chack*, or slightly wetter *chett-chett*, often repeated rapidly; sometimes a rolled *trrac*. Song a rapid series of monotonous notes delivered from a bush top: *chubichubichubi chochochocho* and including a distinctive, repetitive *chewee chewee chewee* phrase. Higher pitched and less varied than Radde's.

TICKELL'S LEAF WARBLER *Phylloscopus affinis* CKJ

L 10–11cm; WT 5.5–7.75g. **SD** Breeds Himalayas and W China; accidental to E China, offshore Japan and Korea. Monotypic. **HH** Skulks in low scrub. **ID** Small to medium-sized compact warbler resembling Siberian Chiffchaff in form, but has brighter plumage. Mid to dark smoky-olive from crown to tail, with long, pale yellow supercilium, broad and distinct eyestripe (weak on lores but stronger behind eye and extending well onto ear-coverts). Face yellowish-olive. Chin, throat and breast rather bright lemon-yellow, fading to buff on breast-sides, grey/olive on flanks and whiter on belly. No wingbars. Outer tail-feathers have inner vanes and narrow tips white. **BP** Bill short, dark greyish-brown above, pale pinkish-brown below; eyes dark brown; tarsi quite bright brownish-pink. **Vo** A sharp, husky *chup* or *chep* and repetitive *tak-tak*. Song is repetitive, *chuchuchu*, and *churuchuruchuru*, or a loud *chep-chi-chi-chi-chi-chi.*

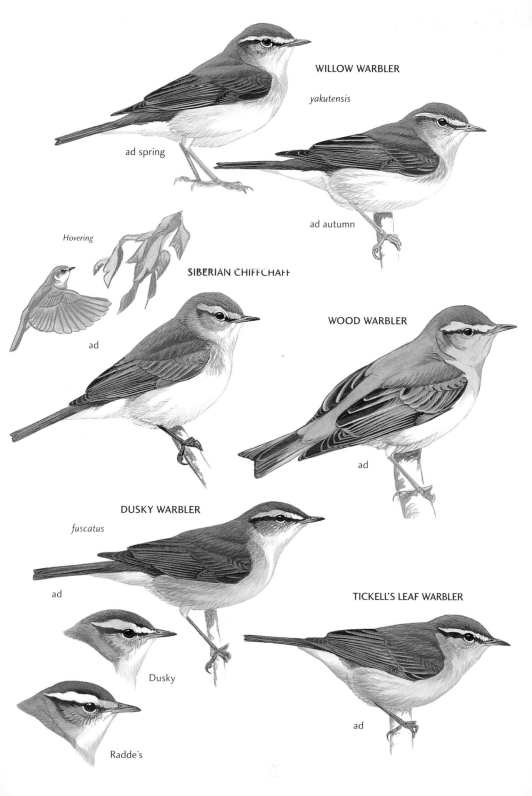

WILLOW WARBLER

yakutensis

ad spring

ad autumn

Hovering

SIBERIAN CHIFFCHAFF

ad

WOOD WARBLER

ad

DUSKY WARBLER

fuscatus

ad

Dusky

Radde's

TICKELL'S LEAF WARBLER

ad

BUFF-THROATED WARBLER   C
Phylloscopus subaffinis

L 10.5–11cm; WT 6.25–7.5g. **SD** Breeds across C,S & SE China including Fujian and Zhejiang. Winters S China to SE Asia. Monotypic. **HH** Breeds in montane scrub and forest up to 3,600m, but winters to lowlands. Usually skulks in dense low vegetation. **ID** Medium-sized, rather undistinguished warbler, resembling Dusky (p.352) but has rather darker brown upperparts, though wings and tail have olive fringes. Supercilium broad, yellowish-buff, eyestripe grades into dark cheeks and dusky ear-coverts. Yellowish-buff chin to belly, more strongly buff on vent. Like Tickell's Leaf (p.352) has white inner webs to outer tail-feathers and narrow white tips, but difficult to see in field. **BP** Bill short, fine, dark culmen and tip, yellow sides to upper, and base to lower, mandible (which has extensive dark tip); eyes and tarsi dark brown. **Vo** Rasping *chrrup* or *chrrip* somewhat different from Tickell's Leaf, more insect-like. Song a series of soft notes resembling Tickell's Leaf, but lacks first *chep* note: *chi-chi-chi-chi-chi.* **AN** Buff-bellied Leaf Warbler.

YELLOW-STREAKED WARBLER   C
Phylloscopus armandii

L 12cm; WT 8–10.5g. **SD** WC & N China, with *P. a. armandii* east to Liaoning and Jilin; migrant Shandong (hence likely offshore Korea). Winters to Indochina. **HH** Groves of deciduous trees (willow and poplar) in subalpine regions in summer, but low scrub at lower elevations in winter. **ID** Rather plain, plump, medium-sized warbler closely resembling Radde's (most readily separated by call, structural differences and leg colour). Upperparts mid to dark brown with olive fringes to wings and tail. Pale buff supercilium longer and broader than Dusky (p.352), and more uniform, lacking stronger contrast between fore and rear parts of Radde's; and lacks contrast with dark eyestripe which blends into dusky ear-coverts. Pale cream chin/throat, darker olive on chest and flanks, heavily buff on vent; yellow throat streaks may be faint or bright, and can extend to breast or even belly. **BP** Bill short, sharp, upper mandible and tip dark brown, paler yellowish-brown at base of lower mandible; eyes dark brown; tarsi strong, pinkish-brown to yellowish-horn. **Vo** A sharp, bunting-like *tzic.* Song recalls Radde's, but weaker: *zschetteretterettetterette twittwitttittititti djielidjieli djitt djitt djitt djehl djehl djuid djuid djuid dwitt dwitti dwitti.*

RADDE'S WARBLER   CTKJR
Phylloscopus schwarzi

L 12.5–13.5cm; WT 8–15g. **SD** From Ob River via Transbaikalia to S Russian Far East, Sakhalin, NE China and N Korea, wintering in S China and SE Asia. Probably regular migrant (and in winter) Taiwan, rare offshore Japan, scarce migrant Korea. Monotypic. **HH** Favours low mixed and deciduous thickets, scrub, edge and taiga with dense undergrowth near water; usually active

near forest floor, though ♂ sings from higher perch. **ID** Large-headed, dark *Phylloscopus* with rather stout bill and tarsi. Long, prominent supercilium, buff and diffuse between bill and eye, usually broader in front of eye, narrower, cleaner and creamy-white behind, and extending to nape-sides (in Dusky supercilium is diffuse and broad behind eye, see p.352); dark brown eyestripe also prominent. Upperparts dark olive-brown, slightly greener on rump. Underparts dusky, off-white on throat, with buff wash to chest and flanks, and distinctive cinnamon-buff undertail-coverts. Belly may be yellow-tinged in autumn. Larger, generally more dark olive-green than Dusky, with longer broader eyestripe and supercilium, heavier bill, stouter legs and larger feet. Lacks throat streaks of very similar Yellow-streaked, which has more uniform supercilium and thinner bill and legs than Radde's. In fresh plumage Radde's has greener cast to upperparts and yellowish cast to supercilium and underparts, differing from overall browner appearance of Dusky. **BP** Bill rather short and thick, with dark brown upper mandible and yellowish or pink at base of lower mandible; eyes dark brown; legs vary from dark orange-pink to pale brown. **Vo** Call a strong, bush warbler-like *check check*, *chep*, *chrep*, *tep* or *pwek* with softer, more throaty quality than Dusky, often with stuttering delivery; scolding alarm is *trrr-trick-trr.* Song loud, melodious with repetitve phrases, erratic and somewhat similar to Siberian Blue Robin (but lacks high-pitched *hi hi hi hi* introductory phrase): *chuchuchuchu chop-cho-cho-cho-cho* (slightly lower than Dusky).

CHINESE LEAF WARBLER   C
Phylloscopus yunnanensis

L 9–10cm. **SD** Range poorly known, but mostly C & NE China, Hebei and Beijing to Liaoning. Winters in N Laos and N Thailand, has also reached coastal Hebei and potential vagrant to Shandong and offshore Korea. Monotypic. **HH** Favours montane mixed, low deciduous, broadleaf and conifer forests, mainly at 1,000–2,800m but to *c.*200m in NE China. Usually active in upper forest levels. **ID** Small, active warbler, closely resembling Pallas's Leaf Warbler (p.356), but paler, more washed-out grey-green, with long supercilium, dingy or buff in front of eye and off-white behind, above dark greyish-brown eyestripe, pale greyish-olive crown-stripe (which may be reduced to rear-crown spot, even nearly absent), whiter wingbar(s) contrasting with dark bases to greater coverts, and pale whitish-yellow rump. Underparts mostly whitish or pale yellow with olive-grey tone to breast. **BP** Bill short, fine, dark blackish-brown above and at tip, pale yellowish-brown at base of lower mandible; eyes dark brown; tarsi pinkish-brown. **Vo** Calls, given in series and very frequently during breeding season, are strongly whistled *tueet-tueet-tueet...*; in non-breeding season gives single *swit* or *tueet* notes. Song, from treetops, is a dry, repetitive and prolonged *tsiridi-tsiridi-tsiridi-tsiridi-tsiridi.* **TN** Formerly *Phylloscopus sichuanensis.*

BUFF-THROATED WARBLER

ad

YELLOW-STREAKED WARBLER

armandii

ad

ad fresh

RADDE'S WARBLER

ad worn

CHINESE LEAF WARBLER

ad

PALLAS'S LEAF WARBLER
CTKJR

Phylloscopus proregulus

L 9–10cm; WT 4.5–7.5g. **SD** C & S Siberia; Yenisei and Transbaikalia to S Yakutia and N Sea of Okhotsk, Sakhalin, S Russian Far East, NE China and N Korea; migrates via Korea, E China to winter south of Changjiang, including coastal SE China and SE Asia. Rare but annual migrant offshore Japan, scarce migrant Korea; scarce migrant, rare winter visitor Taiwan. Monotypic. **HH** Coniferous and mixed taiga <1,700m; migration/winter wider range of scrub/woodland at lower altitudes. **ID** Very small, bright, often hovers, revealing distinctive pale yellow coronal stripe, lemon-yellow rump and short tail. Prominent lemon-yellow supercilia contrast with long black eyestripes and dark crown sides. Wings brownish-grey with two deep yellow wingbars; tertials blackish broadly fringed white (contrast variable). Underparts greyish-white, to pale yellow on vent. **BP** Bill fine, blackish-brown above, paler horn at base below; eyes black; legs dark brownish-pink or brownish-grey. **Vo** Calls infrequently, a strong, nasal *chuii*, *dju-ee*, or *hueet*, softer, quieter and deeper than Yellow-browed, and less rising. Song surprisingly loud, rich and varied; very prolonged series of clear high whistles and wren-like trills, from concealed perch near top of tall tree.

YELLOW-BROWED WARBLER
CTKJR

Phylloscopus inornatus

L 10–11cm; WT 4.3–6.5g. **SD** Urals to Yakutia, Chukotka, Sea of Okhotsk, Amur Delta, Sakhalin, NE China, perhaps N Korea. Winters mainly coastal SE China to S & SE Asia, also Taiwan. Autumn migrants rare/annual offshore Japan, common Korea. Monotypic. **HH** In broadleaf and edges of coniferous taiga (1,000–2,440m). Migration/winter in drier deciduous habitats. Appears agitated, frequently flicks wings. **ID** Small, pale olive-green but variable. Well-marked birds tend to greener tones, others much paler and greyer; wingbars either prominent or partly obscured. Poorly defined pale rear crown stripe, prominent long, pale (yellowish/whitish) supercilium, and weak black eyestripe. Two creamy wingbars on median and greater coverts, the latter usually broad with prominent dark border; primaries fringed green; tertials black broadly fringed cream or whitish (though contrast often lacking), secondaries finely tipped white. Underparts creamy white. **BP** Bill fine, blackish-grey above, paler yellowish-horn at base below; eyes dark brown; tarsi pinkish-brown. **Vo** Calls frequently, a rather loud, distinctive upslurred *tsweeoo*, *chuii* or *chiiii* (recalling Japanese White-eye), rising terminally; or penetrating *sweeet*. Song: a short, ringing *veet-veet-tzhee*, or high and thin recalling Goldcrest: *chiweest chiweest chiweest west-weest*.

HUME'S WARBLER *Phylloscopus humei*
CKJ

L 10–11cm; WT 5–8.8g. **SD** Afghanistan to Himalayas, north to EC China and SC Russia, *P. h. mandellii* (**Mandelli's Leaf Warbler**) SC Russia to C China, east to N Hebei; *P. h. humei* WC Asian mountains east to S Baikal, also C Himalayas. Winters S & SE Asia. Accidental (most likely *P. h. mandellii*) offshore Korea and Japan. **HH** Conifer forests (1,200–2,500m) in N China, and wider range of woodlands, dry deciduous forest, riparian areas

and gardens during migration/winter. **ID** Small, dull grey-green, very like Yellow-browed (YB); plumage-based ID not always possible. In autumn (fresh), greyer, more 'washed-out' than YB; lesser coverts wingbar less distinct, but greater coverts wingbar broad and white. Cheeks appear paler, supercilium more buff or dull yellowish-white, crown lacks green tone; tertials less contrasting, with greyer (less black) centres; wingbar slightly buff; underparts mostly dull off-white with yellowish tone to flanks. **BP** Bill darker than YB's, mostly black; eyes dark brown; tarsi dark blackish- or orange-brown. **Vo** Call forceful, whistled *dsweet*, *tsui* or *weesoo*, drier sparrow-like *chirp* and clearly disyllabic, rasping *juwheet* (YB call more slurred and whispery). Simple drawn-out insect-like buzzing song fades after several seconds: *bzzzzzzzzzzzzzzeeeeeeeeeeeeeeeo*. **TN** Formerly within Yellow-browed Warbler.

ARCTIC WARBLER *Phylloscopus borealis*
CTKJR

L 11–13cm; WT 7.5–15g. **SD** Scandinavia to NE Russia and Alaska. Common migrant through Japan, Korea, scarce Taiwan. *P. b. borealis* Siberia east to Kolyma River (possibly further), south through Russian Far East, winters SE Asia and Indonesia; *P. b. xanthodryas* E Chukotka, Kamchatka, Kuril Is, Sakhalin, Hokkaido, N Honshu, winters SE Asia, Philippines, Indonesia; *P. b. kennicotti* W & C Alaska, migrates through E Asia including Japan, Taiwan, winters Philippines, Indonesia, Nansei Shoto. All three taxa move through Japan, possibly E China, Taiwan and presumably Korea. Possibly two or more species. **HH** Common from tundra edge and taiga to temperate zone, in deciduous, mixed and, to lesser extent, coniferous forests and thickets <2,500m. Migration/winter in wide range of wooded habitats <1,800m. **ID** Large, dark green, with long head, broad dark eyestripe from bill base to nape-sides, long white supercilium reaches only to lores, not bill. Upperparts rather dark olive-green (greener in juv., duller in ad.); underparts off-white, breast yellow tinged (obvious in many juv., lacking in most ad.), vent white. Two very narrow, whitish wingbars; upper bar sometimes obscure; worn spring individuals may show almost no wingbars. Primary projection long. *P. b. borealis* has browner upperparts, pale brown underparts, faintly yellow-washed flanks, and long, slender yellow-white supercilium; slightly finer than *P. b. xanthodryas*. *P. b. xanthodryas* brighter green above, has broader, yellower wingbars and the yellowest underparts (especially flanks), whiter on belly. *P. b. kennicotti* as nominate but bill finer, shorter and less broad-based, slightly brighter green upperparts and brighter yellow underparts. Kamchatka and N Kuril birds (sometimes considered as *P. b. examinandus*) slightly larger than *P. b. xanthodryas* with heavier bill and less yellowish underparts. Those in NE Hokkaido and Sakhalin are slightly smaller than in Kamchatka. **BP** Bill rather heavy, dark brown above, orange/pink at base of lower mandible, typically with dark tip; eyes dark brown; tarsi dull orange-brown to yellowish-pink, toes paler and brighter in juv.. **Vo** Hard, dipper-like *dzit*, *dzik* or *bjjt*, a softer *vit*, and whistled *ryu ryu*. Song: *borealis* and *kennicotti* a repetitive, mechanical, insect-like *zirriri...*, *drree-ree-ree-ree*,or *chirrit-chirrit chirrit-chirrit*, *xanthodryas* same rhythm, but softer *jup chorichori chorichori*, *chi-chirra chi-chirra chi-chirra*, *ji-ji-ro ji-ji-ro* or *chichori chichori*, becoming louder.

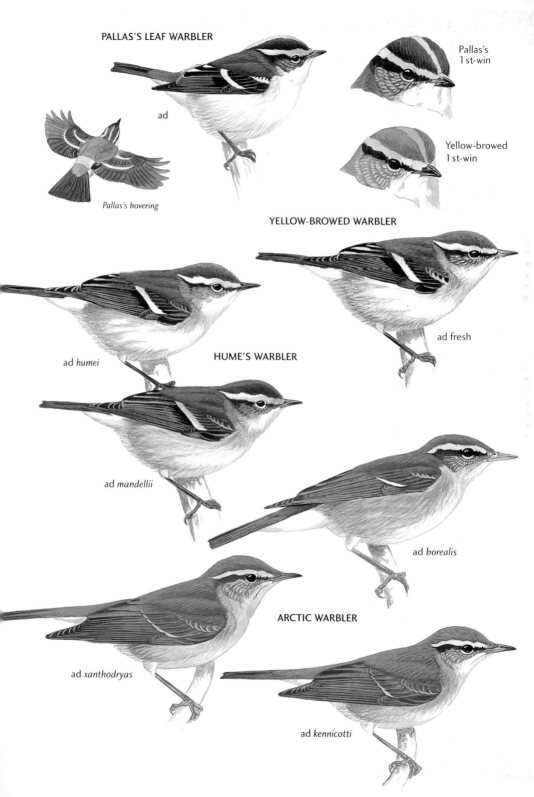

PALLAS'S LEAF WARBLER

ad

Pallas's
1st-win

Yellow-browed
1st-win

Pallas's hovering

YELLOW-BROWED WARBLER

ad *humei*

HUME'S WARBLER

ad fresh

ad *mandellii*

ad *borealis*

ARCTIC WARBLER

ad *xanthodryas*

ad *kennicotti*

PALE-LEGGED LEAF WARBLER   CTKJR
Phylloscopus tenellipes

L 10–11cm; WT 13.5g. **SD** Breeds only in continental E Asia, from Amur River southeast to NE China and N & C Korea. Fairly common migrant Korea, rare Japan. Winters in SE Asia. Monotypic. **HH** Breeds in thickets, riparian woods and sparse forest to 1,800m; on migration in lowland wooded habitats. Often wags tail downwards (unlike Arctic, Eastern Crowned or Ijima's Warblers). **ID** Small, dark, olive-brown warbler with distinctly pale legs, and very long creamy-white supercilium, narrow (and sometimes buff) between bill and eye, broader behind, above broad, dark brown eyestripe. Closely resembles Sakhalin Leaf Warbler and their field separation is uncertain (though song diagnostic). Rather plain, dark grey-brown crown usually contrasts little with generally dark nape/mantle, with olive-brown cast especially on rump (beware light conditions); underparts off-white. Wings brown, with green fringes to flight-feathers and may show two faint wingbars. Primaries may appear shorter, with only 5–6 primaries extending beyond tertials. Commonly pumps tail, which is brown with olive-brown fringes. **BP** Bill moderately fine, with dark brown culmen, pale pink edges to upper mandible, and pale pinkish-horn below; eyes dark brown; tarsi typically whitish-pink, though sometimes pale blue-grey. **Vo** Hard *tit tit* or metallic *piit* is similar to Red-flanked Bluetail's *heet*, but probably indistinguishable from Sakhalin Leaf, though harsh, shivering, cricket-like song, in short rapid phrases, is distinctive: *chi chi chi chi chi chi chi...* or *see see see si si si si si...* with tones somewhat akin to 'songs' of Lanceolated Warbler, Asian Stubtail, or weaker Ashy Minivet, with greater emphasis at beginning and slight upward inflection.

SAKHALIN LEAF WARBLER   CTKJR
Phylloscopus borealoides

L 11.5cm; WT 10.7g. **SD** Breeds only on Sakhalin, Kunashir, Hokkaido and in mountains of C & N Honshu (Apr–Oct). Autumn migration period shorter than Arctic Warbler. Winters in SE Asia. Scarce migrant offshore Korea; rare E China coast and Taiwan. Monotypic. **HH** Typically in mixed montane taiga forest with strong conifer element, to treeline. On migration in lowland wooded habitats. **ID** Small, dark olive-brown warbler with very long whitish supercilium, narrow between bill and eye, broader from eye to nape. Closely resembles Pale-legged. Strong contrast between greyish-toned crown/nape, and greenish (or brownish) mantle, brown back, and somewhat rufous-brown or rusty rump, brown wings and tail; flight-feathers and tertials fringed greenish-olive. May show faint wingbars. Primaries appear longer than Pale-legged, with 7–8 primaries extending beyond tertials. Underparts off-white, greyish-buff on breast-sides and flanks. Tail has narrow whitish border to inner webs of outermost three feathers; appears rather short and square-cut compared with Arctic (p.356), like Pale-legged it is brown with olive-brown fringes, and is often wagged downwards. **BP** Bill moderately fine, dark brown above, pale horn or pink at base of lower mandible, extreme tip of pale horn; eyes dark brown; tarsi dull pinkish-brown. **Vo** Call seemingly inseparable from Pale-legged, though on breeding grounds gives louder, more emphatic *tsit tsit*. Song is very different from Pale-legged: a thin, repetitive very high-pitched, three-note metallic whistle – *hee-tsoo-kee hee-tsoo-kee hee-tsoo-kee*, sometimes broken in midst of series, then recommenced after brief interval. **TN** Formerly included in *P. tenellipes*.

LARGE-BILLED LEAF WARBLER   C
Phylloscopus magnirostris

L 12.5–13cm; WT 11.5–13g. **SD** NE Afghanistan to C China, wintering in S India, Sri Lanka and C Burma. Recently discovered in the region, in Wulingshan reserve, Hebei. Monotypic. **HH** Breeds in glades in broadleaf and evergreen forest, at 1,500–1,800m in region (elsewhere to 4,000m), often near streams or rivers, descending in winter, when also occurs in thickets, near cultivation, and isolated wooded areas. **ID** Large, rather dark greenish-olive *Phylloscopus*, with a conspicuously large bill. Recalls Greenish (p.360), but larger, with longer yellowish-white supercilium (whiter at rear) and broad dark eyestripe and distinctive vocalisations. Crown dark greenish-olive, contrasting markedly with supercilium; yellowish-white and olive-mottled cheeks and ear-coverts; rest of upperparts olive-green, the wings and tail slightly brighter, the wing-coverts with yellowish-white wingbars; outer tail-feathers brown, narrowly tipped white, with white inner webs. Underparts greyish or yellowish-white, with olive-grey wash on sides/flanks. Juv. has browner or more olive-brown upperparts and greyer breast streaked yellow. **BP** Bill large, brown with dark tip and paler (yellowish, orange or pale pink) base, upper mandible may show small hook at tip; eyes black; tarsi blue-grey or grey-brown to pink. **Vo** Typical bisyllabic tit-like call, often accompanied by tail-flicking, is rising *pe-pe*, *chee-wee* or *dir-tee*, but also gives longer, ascending *yaw-wee-ee*. Song a clear, descending, whistled five-note phrase *tee-ti-tii-tu-tu* or *si si-si su-su* the last two notes drawn-out; sometimes commences with more call-like *zi zi zu zu*.

PALE-LEGGED LEAF WARBLER

ad fresh

ad worn

SAKHALIN LEAF WARBLER

ad

Pale-legged

Sakhalin

Large-billed

LARGE-BILLED LEAF WARBLER

ad

GREENISH WARBLER   TKJ
Phylloscopus trochiloides

L 10–11.5cm; WT 6.5–10.5g. **SD** Breeds from NE Europe to C Siberia and W & C China, wintering in S & SE Asia. Accidental offshore Japan, Korea and Taiwan. Subspecies involved possibly *P. t. viridanus* (ranges east to Baikal and C Mongolia) or perhaps *P. t. obscuratus* (of C and possibly NE China). Greenish and Two-barred represent a rare example of an avian ring species; treated separately here because their populations in E Asia appear to behave as distinct species. **HH** Mixed taiga woodland and thickets in hills and mountains to 4,500m. On migration/in winter, scrub, thickets and woodland/forest in lowlands. **ID** Small, dark green warbler recalling Siberian Chiffchaff, but has very long whitish supercilium to nape, broader behind eye, above long dark olive eyestripe, with off-white crescent below eye; head appears rather large and rounded. Upperparts plain dark olive-green, underparts off-white. Primaries and secondaries dark brown with narrow pale green fringes and a very narrow, yellowish-white greater coverts wingbar; but in fresh autumn plumage may have pale yellow tips to median coverts, suggesting a second bar (more common in *P. t. obscuratus*); primary projection of Greenish/Two-barred shorter than in Arctic (p.356). **BP** Bill fine, dark brown above, pale horn or yellow to pinkish-orange below; eyes dark brown; tarsi dull dark brown to brownish-pink. **Vo** A loud, slightly disyllabic *psueee, tisli* or slurred *chli-wee*; in alarm a sharp *si-chiwee* or persistent *tsit tsit*. Song commences with *chis-wee* call note and becomes a rapid, lively jumbled warble with descending phrases, terminating in a short trill or rattle.

TWO-BARRED GREENISH WARBLER   CTKJR
Phylloscopus (trochiloides) plumbeitarsus

L 11.5–12cm; WT 9g. **SD** E subspecies (or possibly full species) of Greenish Warbler, of E Siberia; in region from Lake Baikal to W & N Sea of Okhotsk, Sakhalin and Sikhote-Alin mountains of S Russian Far East, moving through E China to winter in SE Asia. Status of birds in NE China uncertain. Accidental Taiwan (though more numerous Matsu), scarce migrant Korea and offshore Japan. **HH** Breeds in montane taiga, in mixed deciduous and conifer forests and thickets, to 4,000m, groves of broadleaf trees, and on migration occurs in scrub and second growth. **ID** Mid-sized *Phylloscopus* with double wingbars and long, yellowish-white supercilium. Resembles Greenish, Arctic (p.356) and Yellow-browed (p.356), but is cleaner, darker green above, and whiter below than Greenish, with double wingbars (greater coverts bar broader and longer

than Greenish, that on lesser coverts narrower, yellow-white, but distinct). Supercilium almost reaches bill base, unlike in Arctic. Primary projection of Two-barred/Greenish shorter than Arctic. **BP** Bill fine, dark brown above, pink or yellow below; eyes dark brown; tarsi blackish-grey to dull reddish-brown. **Vo** A flat, trisyllabic *chi-wi-ri* or *chururi*, similar in tone to White Wagtail and Eurasian Tree Sparrow, but can also recall disyllabic call of Greenish Warbler, making separation on voice extremely difficult and potentially unreliable. Song a rapid series of whistles, warbles and chattering notes containing slurred and jumbled phrases. **TN** Sometimes split as Two-barred Warbler *P. plumbeitarsus*.

EASTERN CROWNED WARBLER   CTKJR
Phylloscopus coronatus

L 11–12cm; WT 8–10.5g. **SD** Breeds only from C & NE China to Amurland and Ussuriland, Sakhalin, Japan (Hokkaido to Kyushu) and Korea. Common migrant through region (autumn passage earlier and less protracted than Arctic Warbler), though rare Taiwan; winters NE India, southern SE Asia to Java. Monotypic. **HH** Breeds in mixed broadleaf forest in lowlands and foothills. On migration, found in all types of wooded habitats. **ID** Large olive-green warbler, resembling Arctic (p.356), but has distinct yellow/green fringes to flight-feathers, and darker rear crown. Crown dark greyish-green with prominent paler grey-green coronal stripe (does not reach forehead), with dark olive-green lores; long, narrow whitish supercilium, from bill to nape (usually thin and yellowish in front of eye, broader behind and almost joining at nape), narrow dark olive-green eyestripe, blackish on lores; crown-sides dark. Crown slightly higher than Arctic, Sakhalin (p.358) or Ijima's (p.362). Rest of upperparts have strong green cast, with yellowish-green fringes to flight-feathers. Commonly has single, narrow, yellowish-white greater covert wingbar. Underparts clean, rather bright off-white, with distinct pale yellow wash to vent/undertail-coverts. Juv. has browner upperparts, duller yellow undertail-coverts, and duller, less well-defined crown-stripe. **BP** Bill rather thick, broad-based, quite long, dark grey or brown above, pale orange or bright yellow below, lacks dark tip; eyes dark brown; tarsi dark pink to grey or greyish-brown. **Vo** Unlike other *Phylloscopus* of region rarely calls, but occasionally gives strong *chi* or *chiu* (confusingly similar to Ijima's downslurred *chiu*), but more strident and cheery, a soft *phit phit*, or harsher, nasal *dwee*. Sings frequently, song consists of a simple, strong, rather harsh 3–4-note phrase, *pichew pichew bwee* or *chiiyo chiiyo chiiyo biiii* (the final note very distinctively nasal, although sometimes lacking).

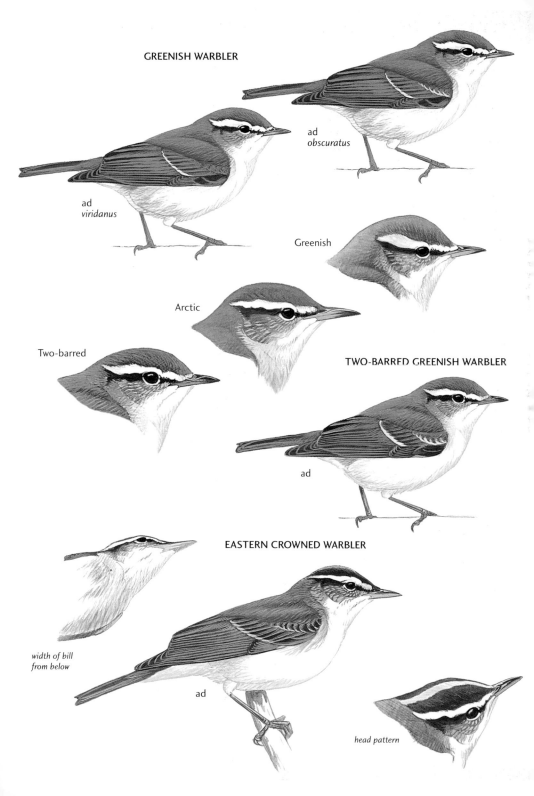

GREENISH WARBLER

ad
obscuratus

ad
viridanus

Greenish

Arctic

Two-barred

TWO-BARRED GREENISH WARBLER

ad

EASTERN CROWNED WARBLER

*width of bill
from below*

ad

head pattern

IJIMA'S WARBLER TJ
Phylloscopus ijimae

Vulnerable. L 11–12cm. **SD** Endemic to Izu and Tokara Is, Japan (Apr–Sep); migrants occur on Honshu and Kyushu; migrates via Taiwan to Philippines, but rarely encountered outside breeding areas. Monotypic. **HH** Mature deciduous, mixed and evergreen broadleaf subtropical forest, where mostly rather active in canopy. On migration also mixed woodlands, pine forest, bamboo, scrub and alder thickets. Migration/ winter seems to prefer coastal evergreen forest (<1,000m) as when breeding. **ID** Like a small Eastern Crowned Warbler (p.360), with plain greyish-green crown (no coronal stripe), ear-coverts and nape. Supercilium long, buffish-white above dark olive-brown lores and eyestripe; appears whitest above and below eye, as if 'spectacled'; eyestripe is narrower than eye (same width as eye in Arctic and Eastern Crowned). Mantle and back greyish-green or grey-brown with green tinge; flight-feathers grey-brown with bright green outer fringes, greater coverts have narrow white tips forming indistinct wingbar, outer tail-feathers also show green tone, otherwise grey-brown. Off-white chin and sides of throat, pale grey breast and flanks with variable yellowish-green wash on vent. **BP** Bill fine, upper mandible dark brown to tip, lower mandible dull yellow, orange or pinkish-orange, less broad-based than Eastern Crowned; eyes dark brown; tarsi brownish-pink. **Vo** Calls frequently, a soft *pee or hee* with falling intonation, or *hu-eet*, a downslurred *se-chui, chiu, tiu* or *twee* reminiscent of Coal Tit (and very similar to Eastern Crowned), but with a melancholy ring to it. Song very variable: *chubi chubi chubi chui chui chui pii chobi chobi*, also Coal Tit-like, and a very sibilant *shiri-shiri-shiri fisisisisi*, or silvery wavering *swisswisswisswisswiss*, occasionally slowing into more enunciated *tsu wiss tsu wiss tsu wiss tsu wiss*. Occasionally a repeated *tseeoo tseeo tseeo* between song bursts, similar to some parts of Tree Pipit song.

CLAUDIA'S LEAF WARBLER C
Phylloscopus claudiae

L 10.5–12cm; WT 7.5–10g. **SD** Core breeding range Sichuan, with population in N Hebei and Beijing, and perhaps others between. Perhaps accidental on continental coast. Winters from Himalayan foothills to Bangladesh, and S China to SE Asia. Monotypic. **HH** Forages in canopy, sometimes hanging tit-like below branches, when characteristically flicks wings alternately, slowly. **ID** Closely resembles Eastern Crowned (p.360), but smaller, brighter; also similar to Kloss's. Crown-sides blackish-green, centre green with whiter nuchal stripe; supercilium long, pale yellow in front of eye, whiter at rear, eyestripe narrow, black, wraps up at rear, so lateral crown-, supercilium and eyestripe form three prominent stripes on nape-sides. Upperparts green with two short, pale whitish-yellow wingbars. Underparts white (lacks Eastern Crowned's obvious yellow undertail-coverts). Tail has narrow white edges to outer two feathers. **BP** Bill dark above, pale orange-yellow or pink below; eyes black; tarsi greenish- or bluish-green to yellow or yellowish-pink. **Vo** A loud *pit-cha* or *pit-chew-a*. Song a prolonged or extended trilling comprised of call notes. **TN** Formerly within Blyth's Leaf Warbler *P. reguloides*.

HARTERT'S WARBLER C
Phylloscopus goodsoni

L 10.5–12cm. **SD** Restricted to E & SE China; race breeding in Fujian is *P. g. fokiensis*. Winters further south in China. Regular migrant on coast, likely to stray to S Japan and Korea. **HH** As Claudia's. **ID** Closely resembles Claudia's (and vocal distinctions unknown), but has broader, darker lateral crown-stripes, pale yellow median crown-stripe, short pale yellow supercilium and short, black eyestripe. Upperparts green with two short, pale whitish-yellow wingbars; tail green with small amounts of white on inner webs of outer two feathers. Underparts white. **BP** Bill dark above, pale orange-yellow or pink below; eyes black; tarsi greenish- or bluish-green to yellow or yellowish-pink. **Vo** Distinctions from previous species uncertain, probably very similar disyllabic *pit-cha* or trisyllabic *pit-chew-a*. **TN** Formerly within Blyth's Leaf Warbler *P. reguloides*.

KLOSS'S LEAF WARBLER C
Phylloscopus ogilviegranti

L 10–11cm; WT 6.4g. **SD** Vietnam, southern China northeast to Fujian. *P. o. ogilviegranti* of N Guangdong occurs to NW Fujian. **HH** Mixed forest. Characteristic quick simultaneous flicks of wings while foraging. **ID** Rather bright green, with broad yellow supercilium to nape-sides, dark olive-green eyestripe merging into neck-sides, yellow stripe mainly on rear crown to nape, and broad blackish-green crown-sides. Slightly smaller than similar Claudia's and Hartert's, but with more prominent, yellower coronal stripe, pinker legs and white underside to tail. Upperparts bright green, including wing-coverts, with two whitish-yellow wingbars. Underparts off-white with some yellow streaking; much of outer tail-feathers white, thus closed tail has visibly white underside. **BP** Bill dark brown above with pink base to lower mandible; eyes black; tarsi pale pinkish-brown. **Vo** Call *pitsui* or *pitsitsui*. Song commences with single high-pitched call, followed by *tit-sui-titsui-titsui* or *see-chee-wee see-chee-wee see-chee-wee*. **TN** Formerly within White-tailed Leaf Warbler *P. davisoni*.

SULPHUR-BREASTED LEAF WARBLER CT
Phylloscopus ricketti

L 10–11cm. **SD** C & SE China to Fujian, accidental Taiwan, wintering N SE Asia. Monotypic. **HH** Mixed forests in low hills/mountains to 1,500m. **ID** Very bright; upperparts bright green with broad bright yellow supercilium contrasting with broad blackish-green lateral crown-stripes, and narrow eyestripes, yellowish coronal stripe extends to nape; two, variably distinct, yellow wingbars, that on greater coverts broader and far more distinct. Underparts from chin to vent entirely bright yellow. **BP** Bill short, fine, dark brown above, yellowish/pinkish below; eyes dark brown; tarsi pale, from yellowish to brownish-pink. **Vo** Call bisyllabic: *pee-chew pee-chew*. Song a series of high-pitched accelerating phrases: *sit siri sii-sii see-chew sit sweety sweety sweety swee-chew*.

IJIMA'S WARBLER

ad

CLAUDIA'S LEAF WARBLER

ad

HARTERT'S WARBLER

fokiensis

ad

KLOSS'S LEAF WARBLER

ogilviegranti

ad

SULPHUR-BREASTED LEAF WARBLER

ad

WHITE-SPECTACLED WARBLER   C
Seicercus affinis

L 11–12cm. **SD** From Himalayas to N SE Asia and S China east to coast, and in Fujian, where isolated population of *S. a. intermedius* perhaps deserves specific status. **HH** Montane bamboo thickets in damp broadleaf evergreen forests at 1,000–1,750m, lower in winter. **ID** *S. a. intermedius* a distinctively grey-headed, green and yellow warbler, very similar to larger Bianchi's. Crown and supercilium grey (sometimes green), with broad, black lateral crown-stripes, face green with yellow eye-ring. Upperparts generally green, including wings, with narrow yellow tips to greater coverts forming wingbar. **BP** Bill blackish-brown above, pale orange below; eyes large, black with pale yellow eye-ring; tarsi brownish-yellow to greyish-pink. **Vo** A soft clear whistled *ti-tiuu-tit*, various rolling whistles and a drawn-out *churrrruwedichi*. Song is complex, involving clear whistled notes, commencing hesitantly then accelerating.

ALSTRÖM'S WARBLER   C
Seicercus soror

L 11–12cm. **SD** Restricted to EC & SE China; in region occurs locally in Zhejiang and Fujian; winters to SE Asia. Monotypic. **HH** Undergrowth, bushes and low trees in warm temperate evergreen broadleaf forest, at lower elevations than Bianchi's Warbler. **ID** Rather large-headed with distinctive grey crown, black lateral crown-stripes, yellow underparts and white in outer tail-feathers. Upperparts green, face olive-grey, wings dark grey with narrow green fringes. Chin to vent lemon-yellow. Tail rather short. From very similar Bianchi's by larger bill, shorter tail with less white in outer pairs of rectrices and no wingbar. **BP** Bill rather large, blackish-brown above, pale orange below; eyes large, black with yellow eye-ring; tarsi pale greyish-pink to pinkish-brown. **Vo** Distinctive, short, high-pitched *tsi-pit* or *tsrit*. Song a simple, repetitive series of whistled phrases, lacking trills, so structurally similar to Bianchi's, but higher pitched and weaker: *chip chu-se-sis-chu-se-sis... chip chu-se-sis-chu-se-sis... chip ple-di-tsi-ple-di-tsi... chip chu-se-si-sis-chu-se... chip chu-wee-tsi-chu-wee.* **AN** Plain-tailed Warbler.

BIANCHI'S WARBLER   C
Seicercus valentini

L 11–12cm. **SD** Breeds C & SE China, winters south to N Vietnam. *S. v. latouchei* restricted to S & E China north to Fujian, though has strayed to Hebei coast. **HH** Lower storey of cool temperate forest above *c.*1,800m. **ID** Rather large-headed with distinctive grey crown, black lateral crown-stripes, yellow underparts, white underside to tail and white in outer pairs of rectrices. Upperparts green, face olive-grey, wings dark grey with narrow green fringes and pale yellow tips to greater coverts forming single wingbar. Underparts, from chin to vent, lemon-yellow. **BP** Bill blackish-brown above,

pink or orange below; eyes large, black with yellow eye-ring; tarsi greyish-pink or pale pinkish-brown. **Vo** A short, forceful, whistled, robin-like *diu*. Song comprises simple whistled phrases, repetitive but with long pauses, and lacks trills: *chu wee-chu-wee-chu... chu chu-wee-chu-wee... chu chu-will-yu-chu-will... chu chu-see-chu-see... chu chu-see-chu-see.* **TN** Formerly within Golden-spectacled Warbler *S. burkii.*

CHESTNUT-CROWNED WARBLER   C
Seicercus castaniceps

L 9.5cm; WT 4–6g. **SD** E Himalayas to SE Asia, across S China, wintering in SE China northeast to Fujian; likely to reach Taiwan. Race in SE China is *S. c. sinensis.* **HH** Montane forest where favours canopy. **ID** Small, very active grey, green and yellow warbler. Resembles White-spectacled Warbler in structure, but foreparts grey from face to breast, crown mid chestnut-brown and underparts yellow. Crown and supercilium pale to medium chestnut-brown, with black lateral crown-stripes from above eye to nape; lores, face, neck-sides and mantle grey, scapulars and back dull olive-green, rump bright lemon-yellow, tail dull green with white inner webs to outer rectrices; wings blackish-green with two broad yellow wingbars and narrow yellowish-green fringes to all but primaries. Grey-white face, chin, throat and upper breast, lemon-yellow belly, flanks and vent. **BP** Bill dark brown above, pale yellowish-pink below; eyes large, black with narrow white eye-ring; tarsi pale yellowish-pink, underside of toes almost white. **Vo** A wren-like *tsik* and hard bisyllabic *chee-cheee.* Song a series of very high, thin notes: *see see see-see-see-see-see.*

LESSER WHITETHROAT   CTKJR
Sylvia curruca

L 12.5–14cm; WT 9.5–18g. **SD** Europe and W Asia east to the Lena, wintering in Africa, Middle East and S Asia. *S. c. halimodendri* from Kazakhstan to Mongolia and *S. c. curruca* (including '*blythi*') from Europe to just west of region in Transbaikalia. In region, accidentals to all countries presumably '*blythi*'. **HH** Woodland edges, thickets and scrub in open country, from lowlands to over 2,500m. In north also in taiga-forest. Usually skulks in dense vegetation. **ID** Small and compact; grey-brown upperparts with darker grey crown, somewhat blacker wings and tail. Lores and ear-coverts black contrasting with crown and, particularly, white chin and throat. Underparts white with grey-brown wash to breast-sides/flanks. Tail rather long with prominent off-white inner webs to outer rectrices, recalling buntings. **BP** Bill rather short, black; narrow pale grey eye-ring, irides dark reddish-brown with black ring; tarsi black. **Vo** Short, hard and abrupt *tek*, *chett*, *tac* or *cha.* Two-part song given from dense cover, a short scratchy chattering warble followed by a rattling trill.

DESERT LESSER WHITETHROAT   C
Sylvia minula

L 12cm; WT 8–13g. **SD** From Caspian Sea to Gobi Desert, wintering in deserts of Middle East. Reported in coastal NE China (Beidaihe/Happy I.). **ID** Similar to Lesser Whitethroat, but smaller, paler (especially mantle, rump and scapulars), with less pronounced mask, shorter bill and shorter primary projection.

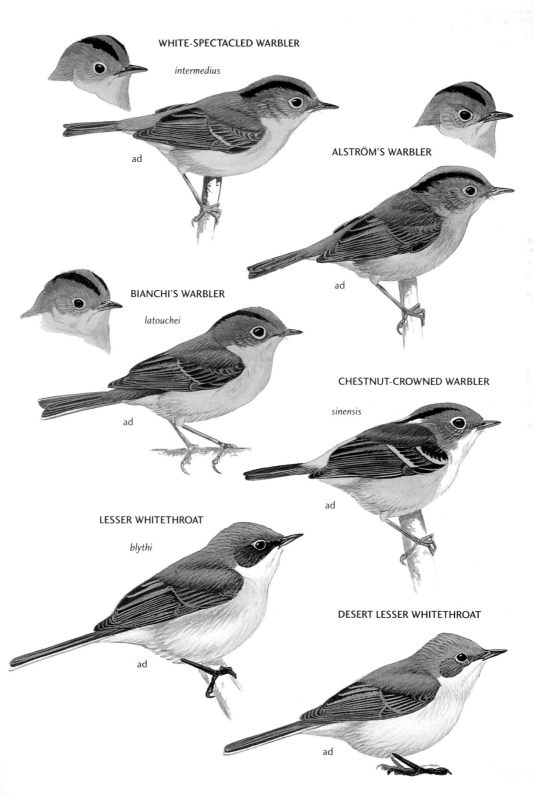

WHITE-SPECTACLED WARBLER

intermedius

ad

ALSTRÖM'S WARBLER

ad

BIANCHI'S WARBLER

latouchei

ad

CHESTNUT-CROWNED WARBLER

sinensis

ad

LESSER WHITETHROAT

blythi

ad

DESERT LESSER WHITETHROAT

ad

RUFOUS-FACED WARBLER   CT
Abroscopus albogularis

L 8cm; WT 4g. **SD** E Himalayas to SE Asia, S & E China. *A. a. fulvifacies* in S China east to Fujian and Zhejiang, and Taiwan. **HH** Undergrowth and bamboo thickets of secondary and evergreen forests (on Taiwan *c.*200–2,500m). Sociable outside breeding season, sometimes in large single-species flocks or mixed bird 'waves'. **ID** Tiny and very active. Forehead, supercilium and head-sides bright rufous-brown, lateral crown-stripes black broadening at nape, crown olive; upperparts, wings and tail bright olive-green with lemon-yellow and white lower back and rump; underparts bright white, but throat black with narrow white streaks or flecks, broad white malar and yellow wash on upper-breast-sides. **BP** Bill short but rather broad, arched, dark grey culmen and tip, cutting edges and entire lower mandible dull pink; eyes large with grey orbital, irides black; tarsi dull pinkish- or yellowish-brown. **Vo** In flocks utters whimpering, twittering notes, a common, high-pitched drawn-out 'chirp' *chipchee*, or a rising cricket-like *heeee*. Song includes 'chirps' and a short scratchy jumble of tinny very high-pitched whistles *tititiriiii tititiriiii tititiriiii*. **AN** White-throated Flycatcher-warbler.

BLACK-NECKLACED SCIMITAR BABBLER   T
Pomatorhinus erythrocnemis

L 25cm. **SD** Endemic to Taiwan. Monotypic. **HH** Skulking resident of montane forests (100–2,200m), in undergrowth, forest edge, scrub and tall grass. **ID** Large, rusty chocolate-brown scimitar babbler. Head rather plain, dull greyish-earth-brown, with rufous-brown forehead. Black lateral throat-stripe extends as deep gorget of heavy black stripes on neck-sides and upper breast. Upperparts dark earth- to dark rufous-brown. Underparts white (no dark shaft-streaks) on chin/throat and breast (with dense dark streaks on latter); whitish-grey belly, duskier flanks and rufous/chestnut vent. Tail long, graduated, wings short, somewhat rounded. **BP** Bill, long, de-curved, dark blackish-brown, paler at tip; eyes yellowish-orange; tarsi brownish-grey. **Vo** Alarm a grating, insistent *kvrrit-vit vit vit vrrit vrrit*. Also, a querulous whistled *weeowu* by ♂ in spaced series or followed almost instantaneously by piped *poh poh* from ♀. ♂ call can sound similar to distant Taiwan Hill Partridge. **TN** Formerly known as Spot-breasted Scimitar Babbler and placed within Rusty-cheeked Scimitar Babbler *P. erythrogenys*.

GREY-SIDED SCIMITAR BABBLER   C
Pomatorhinus swinhoei

L 25cm. **SD** SE China from Guangxi to Fujian. *P. s. swinhoei* occurs east to NW & C Fujian. **HH** Skulking resident of montane forests (300–2,200m) in undergrowth, forest edge and scrub. **ID** Large chocolate-brown scimitar babbler. Lacks black moustachial of closely related Black-necklaced, from which now separated. Head and neck largely dull grey-brown, forehead and lower ear-coverts with rufous tinge; scapulars and back fox-red, wings and

tail dark earth brown. White chin/throat and breast, with breast streaking forming neat gorget not extending to upper belly; belly and flanks suffused grey, vent fox-red. Tail long, graduated, wings short, somewhat rounded. **BP** Bill long, decurved, blackish-brown; eyes orange; tarsi brownish-grey. **Vo** Differences from previous species unclear. **TN** Formerly within Rusty-cheeked Scimitar Babbler *P. erythrogenys*.

TAIWAN SCIMITAR BABBLER   T
Pomatorhinus musicus

L 19–21cm. **SD** Common endemic resident of Taiwan. Monotypic. **HH** Lower and mid-level (below 2,500m) mixed and evergreen forests, secondary growth, scrub, tall grass and bamboo thickets; also suburban parks and gardens. Not shy, may approach observer closely. **ID** Boldly patterned face and a rather flat crown. Head slate-grey with broad white supercilium from above bill onto nape-sides; dark grey crown contrasts with rich chestnut nape, mantle and neck-sides, forming broad hind-collar; back, short wings, rump and long tail dark earth brown to blackish-brown on central rectrices. Face, chin/throat and breast bright white, with deep gorget of slate-grey or black oval streaks from neck-sides to breast; lower belly, flanks and vent chestnut. **BP** Large bill sharply pointed, strongly decurved, black above, pale horn below; eyes lemon-yellow; tarsi grey. **Vo** Foraging pairs or groups near ground commonly give repeated, upslurred, querulous *vway* amongst series of low churring *jerr jerr* notes, also a low, whistled *tu-ti-tu-tu* and whooping *kuo-kuei*. Increases in volume when agitated. Song seems to mimic other birds (e.g. Rufous-capped Babbler, even duet of Black-necklaced Scimitar Babbler). **TN** Formerly within Streak-breasted Scimitar Babbler.

STREAK-BREASTED SCIMITAR BABBLER   C
Pomatorhinus ruficollis

L 19–21cm. **SD** Himalayas across S & E China east to Fujian; disjunct population in SE China is *P. r. stridulus*. Common resident. **HH** Lower and mid-level mixed and evergreen forest, secondary growth, scrub, tall grasses, bamboo thickets, parks and gardens. **ID** Has boldly patterned face and is rather flat-crowned. Head is dark brown with broad white supercilia and some silvery flecks on crown-sides; crown olive, nape and neck-sides bright rufous, back, short wings, rump and long tail all dark brown. Face, chin/throat and breast bright white, with deep gorget of dark reddish-brown streaks from neck-sides across breast, streaks longer, broader and browner on belly; lower belly, flanks and vent dark brown. **BP** Bill (smaller than Taiwan Scimitar Babbler) is sharply pointed, strongly decurved, black above, pale horn below; eyes lemon-yellow; tarsi grey. **Vo** Differences from Taiwan Scimitar Babbler unclear. **AN** Rufous-necked Scimitar Babbler.

RUFOUS-FACED WARBLER
(not to scale)

fulvifacies

ad

**BLACK-NECKLACED
SCIMITAR BABBLER**

ad

**GREY-SIDED
SCIMITAR BABBLER**

swinhoei

ad

**TAIWAN
SCIMITAR BABBLER**

ad

**STREAK-BREASTED
SCIMITAR BABBLER**

stridulus

ad

TAIWAN WREN-BABBLER   T
Pnoepyga formosana

L 9cm. **SD** Endemic resident of Taiwan. Monotypic. **HH** High montane forests (*c*.1,000–3,000m), where highly skulking in low, dense undergrowth, keeps to ground, often feeding beneath fallen leaves. At rest and while singing, the wings are flicked open and shivered repeatedly. **ID** Tiny, virtually tail-less bird. Ad. upperparts dark warm brown, with some buff (even rufous) scaling to crown and mantle, wings dark brown and reach tail tip. Underparts: breast and belly rather dark, as black feather centres are prominent and pale buff fringes narrow, affording scaled effect from chin to vent. Scaling less distinct and white fringes broader at chin, but scaling still present. Juv. dark brown with dark rufous fringes to flight-feathers. Spotting and scaling appears in 1st-winter. **BP** Bill rather long, slender, dark (blackish) above and at tip, pale horn at base of lower mandible; eyes large, irides dark brown; short tarsi dull brownish-pink. **Vo** A surprisingly loud, rather jarring repetitive *ger*, wheezy *feesh!* and quite piercing *kit, kit-it!* Song consists of a strident whistled phrase, the first note separated from the rest by a distinct pause: *wheet... weeyu wuluit*. **TN** Formerly has been placed in both Pygmy Wren-babbler and Scaly-breasted Wren-babbler *P. albiventer.*

PYGMY WREN-BABBLER   C
Pnoepyga pusilla

L 9cm. **SD** Himalayas to SE Asia, also S, C & E China north to Zhejiang. **HH** Secretive, in low, dense undergrowth, where keeps to ground. **ID** Tiny, virtually tail-less bird. Upperparts dark brown, plain (unspotted) on crown, ear-coverts and neck-sides, but speckled pale brown on mantle, back and rump; wings dark brown and reach tail tip. Chin/throat whitish-brown, rest of underparts the pale-fringed feathers each having a dark, blackish-brown centre, appearing as rather intense scaling over entire underparts to vent. **BP** Bill rather long, slender, black above and at tip, pale horn at base of lower mandible; eyes large, irides dark brown; tarsi dull brownish-pink. **Vo** A sharp *tchit* or *tsick!* Song consists of 2–3 piercing whistles, *ti-ti-tu* (repeated monotonously) or *tseet tsuut*. **AN** Brown Wren-babbler.

RUFOUS-CAPPED BABBLER   CT
Stachyris ruficeps

L 11–12cm. **SD** E Himalayas across southern China to N SE Asia: *S. r. davidi* S, C & E China north to Zhejiang; *S. r. praecognita* throughout Taiwan. **HH** Dense undergrowth, scrub, tall grass, bamboo thickets and forest. **ID** Small, warbler-like bird with bright rufous cap. Upperparts dark brown, including wings yellowish-buff; underparts grey-brown with yellowish wash to chin/throat, belly and vent, and fine dark spotting on chin. Juv. lacks rufous cap. **BP** Bill black above and at tip, base of lower mandible pink; eyes dark brown; tarsi brown. **Vo** Calls consist of sharp churrs. Song, a descending series of rather bullfinch-like whistles: *pu-pu-pu-pu-pu-pu*, sometimes disarmingly similar to human whistles and will respond to imitation. **AN** Red-headed Babbler.

CHINESE BABAX   C
Babax lanceolatus

L 26–28cm. **SD** NE India across southern China to Fujian: *B. l. latouchei* in SE China. **HH** Generally skulking, lives close to ground in dense undergrowth, scrub and thickets of hill and montane forests. **ID** Resembles medium-sized streaky laughingthrush. Dark brown crown, mantle, neck, scapulars, back and rump, streaked grey and brown; tail long, graduated with rounded tip; wings short, dark earth brown. Face distinctive with white supercilium, black eye patch and broad black malar stripe, ear-coverts streaked grey and brown. Chin, throat and breast cream with heavy brown streaking on neck- and breast-sides and flanks. **BP** Bill arched, sharply pointed, black; eyes large, irides yellow; strong tarsi brownish-pink. **Vo** Loud, creaky, wailing *ou-phee-ou-phee*.

TAIWAN WREN-BABBLER

ad

ad dark

PYGMY WREN-BABBLER

juv

ad pale

ad
davidii

ad
praecognita

RUFOUS-CAPPED
BABBLER

CHINESE BABAX
(not to scale)

latouchei

ad

PLATE 171 : LAUGHINGTHRUSHES I

MASKED LAUGHINGTHRUSH   C
Garrulax perspicillatus

L 30cm. **SD** Resident of Vietnam and southern China north to Jiangsu, accidental Shandong. Monotypic. **HH** Forages on ground in range of habitats, from town parks to rural areas, with scrub, dense grass and bamboo thickets, commonly in small noisy parties. Laughingthrushes typically have short wings and long tails, flight consists of series of rapid wingbeats followed by long glide. **ID** A large, plain laughingthrush. Upperparts drab ashen grey-brown from crown to tail, though latter blacker toward tips of outer feathers. Forehead, face, earcoverts and chin black. Underparts ash-brown, paler on belly and grading to rufous on vent. **BP** Bill strong, black; eyes black; tarsi strong, brownish-pink. **Vo** Calls include harsh chattering and loud *jhew* or *jhow* notes.

RUFOUS-CROWNED LAUGHINGTHRUSH   T
Garrulax ruficeps

L 28cm. **SD** Endemic to mountains of Taiwan. Monotypic. **HH** Mid-level and high montane forests of Taiwan. Descends in winter. Moves through all levels of forest in large, vocal flocks. **ID** Dark, earth brown laughingthrush with prominent white chin and throat. Upperparts mostly mid to dark brown, but forehead crown and nape rich chestnut; wings and tail plain mid brown, tail is graduated thus appears full and has broad white tips to outer feathers. Lores, subloral patch, narrow forehead band and very narrow chin band black, eye patch also black, blending into blackish-brown ear coverts. Chin, lower-face-sides, throat and upper breast bright creamy white, forming broadly oval throat/breast patch when seen face on; sides of breast and narrow breast-band dull greyish-brown, lower breast and belly creamy or golden-buff, but rear underparts and undertail-coverts off-white, with grey flanks, and two-thirds of underside to rectrices on folded tail white. **BP** Bill black; eyes dark brown; tarsi dark grey. **Vo** Rather quiet chattering *trrreee…* from foraging flocks, occasionally a louder *chuk-wi*, but groups probably chatter noisily if sufficiently disturbed, like other laughingthrushes. **TN** Formerly within White-throated Laughingthrush *G. albogularis.*

LESSER NECKLACED LAUGHINGTHRUSH   C
Garrulax monileger

L 27–28cm. **SD** E Himalayas to SE Asia and across S China, with *G. m. melli* east to Fujian. **HH** Sociable, forages on ground in montane forests. **ID** Strongly marked, with prominent narrow black necklace. Grey-brown crown, back, wings and tail, but rufous mantle and neck-sides. Face boldly marked, with white supercilium, broad black loral patch, black eye-ring and eyestripe to ear-coverts, merging with broad black necklace on neck-sides; ear-coverts white, malar streak black, also merging with necklace. Tail long and graduated, grey-brown with black subterminal bands and white or buff tips to outer feathers. Wings short, brown with brown primary-coverts and white outer fringes to primaries. Chin and throat white bordered below with rufous, then black of necklace, lower breast and belly white, but flanks, thighs and vent rufous-orange. **BP** Bill strong, black; yellow eyes conspicuous; strong tarsi brownish-pink. **Vo** Calls consist of strange piping notes.

GREATER NECKLACED LAUGHINGTHRUSH
Garrulax pectoralis   C

L 27–34.5 cm. **SD** E Himalayas to SE Asia and across S & E China; *G. p. picticollis* ranges north to Changjiang River and east to Fujian and Zhejiang. **HH** Forages on ground in noisy groups in forested hills. **ID** Larger and stockier than Lesser Necklaced, with broader black necklace. Grey-brown crown, back, wings and tail, but rufous mantle and neck-sides, with white supercilium and lores, black eye-ring and eyestripe (extends to ear-coverts); black lines on white face, from malar to ear-coverts, contrast with broad grey gorget from rear of ear-coverts down neck-sides to breast. Tail long and graduated, grey-brown with black subterminal bands and white tips to outer feathers. Wings short, brown with black primary-coverts and white outer fringes to primaries. Chin and throat white bordered below by grey gorget; lower breast and belly white, but flanks, thighs and vent dark rufous-orange. **BP** Bill strong, black; large-looking dark brown eyes with black eye-ring; strong tarsi dull brownish-pink. **Vo** Soft contact calls to louder, mournful 'laughter' and short whistles.

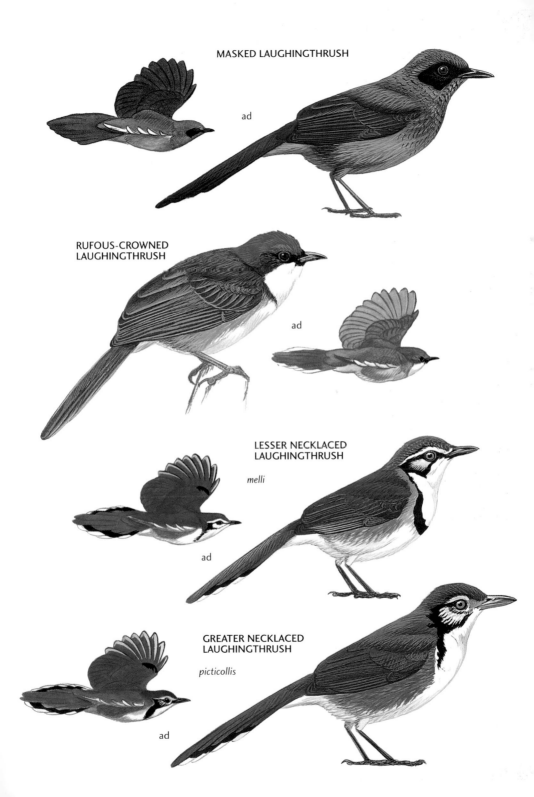

MASKED LAUGHINGTHRUSH

ad

RUFOUS-CROWNED LAUGHINGTHRUSH

ad

LESSER NECKLACED LAUGHINGTHRUSH

melli

ad

GREATER NECKLACED LAUGHINGTHRUSH

picticollis

ad

BLACK-THROATED LAUGHINGTHRUSH   T
Garrulax chinensis

L 23cm. **SD** Natural range SE Asia to S China, but established locally in Taiwan (subspecies uncertain), where apparently increasing. **HH** In Taiwan, lowland and hill forest. **ID** Rather dull-coloured grey-brown laughingthrush with prominent white face patch. Forehead, eye surround, chin, throat and upper breast black, with narrow black line extending from eye as border to white ear-coverts. Crown, nape, neck-sides and breast grey, rest of upperparts including wings and underparts, grey-brown. Brown tail rather long, graduated, with black outer feathers and broad black tip. **BP** Bill blackish-grey; eyes reddish-brown; tarsi brownish-pink. **Vo** Calls include a cackling chorus. Song rich and thrush-like, but an excellent mimic combining notes from many other species including laughingthrushes, pittas, Taiwan Hill Partridge, scimitar babblers and serpent eagle.

PLAIN LAUGHINGTHRUSH   C
Garrulax davidi

L 29cm. **SD** Restricted to C & N China northeast to W Heilongjiang: *G. d. davidi* S Hebei; *G. d. chinganicus* in N Hebei to NW Manchuria. **HH** Scrub and thickets at 400–1,600m. **ID** Rather drab grey-brown laughingthrush. Upperparts grey-brown from crown to tail, supercilium paler grey, chin and rictal bristles black. Underparts pale grey-brown. **BP** Bill strong, arched, bright yellow with green tip; eyes black; strong tarsi brownish-pink. **Vo** Series of *wiau* notes in alarm or for contact. Song a loud series of short notes: *wiau wa-wikwikwik woitwoitwoitwoit*. **AN** Père David's Laughingthrush.

MOUSTACHED LAUGHINGTHRUSH   CJ
Garrulax cineraceus

L 22cm. **SD** Occurs from NE India, with *G. c. cinereiceps* in C, S & E China from Fujian north to Jiangsu. Introduced and now widespread on Shikoku, Japan. **HH** Scrub, bamboo thickets and broadleaf forests and plantations above 200m, sometimes near habitation. **ID** Rather small, boldly marked brown laughingthrush. Black from forehead to nape, but otherwise warm earth brown from back to tail. Face white, with narrow black line behind eye, black malar and black-streaked neck-sides. Wings brown with black primary-coverts, pale blue-grey primaries, tertials and secondaries terminally black with narrow white tips; tail long, brown with broad black subterminal band and white tip. White chin and throat with narrow grey streaks merging with black-streaked face- to neck-sides; breast, flanks and vent orange-brown. **BP** Bill strong, grey above, yellow below; eyes yellow with narrow black ring; strong tarsi brownish-pink. **Vo** Musical notes and thrush-like chattering; song a loud *diu-diuuid*.

RUSTY LAUGHINGTHRUSH   T
Garrulax poecilorhynchus

L 28cm. **SD** Endemic resident of Taiwan. **HH** Mature, dense forest or woodland, where usually keeps to undergrowth in mid montane regions. Descends lower in winter. Typically in flocks of 20–30 which noisily rustle leaves. **ID** Large, overall rather rufous laughingthrush with distinctive two-tone bill and blue eye-ring. Largely dark earth- or rufous-brown from crown to rump and tail, with some black around eye and on chin. Underparts: rufous-brown throat and upper breast, whilst lower breast, belly and flanks are dark smoky-grey, contrasting with white undertail-coverts. Wings short and rounded, greyish-brown tinged rufous, with grey outer fringes to primaries. Tail long, somewhat rounded, rufous-brown with white tips to 1–2 outermost feathers. **BP** Bill large, horn-coloured at tip, blue-grey at base; eyes appear very large and black, with broad blue orbital ring; strong tarsi pale grey. **Vo** Very vocal, uttering an impressive array of clunks and whirrs while foraging. Some calls very loud; include a characteristic double whistle *tyu wee* (second note rising) that is easily imitated and which birds respond to. **TN** Formerly within Grey-sided Laughingthrush *G. caerulatus*.

BUFFY LAUGHINGTHRUSH   C
Garrulax berthemyi

L 28cm. **SD** Resident across S & SE China to NW Fujian and Zhejiang. **HH** Favours undergrowth of mature forest in mid-montane areas, descending in winter. **ID** Large, somewhat dark rufous laughingthrush with distinctive two-tone bill and blue eye-ring; distinctly paler than Rusty Laughingthrush being largely pale olive-rufous from crown to rump. Lores, lower forehead and chin black; underparts creamy buff-brown on throat and upper breast, and pale silvery grey on lower breast, belly and flanks. Wings and tail rich chestnut, contrasting rather strongly with paler body, the tail with large white tips to three outermost feathers. **BP** Bill large, horn-coloured at tip, blue at base; large black eyes with broad blue ring; strong tarsi grey. **Vo** As Rusty Laughingthrush. **TN** Buffy was formerly considered within Rusty, and previously both were lumped within Grey-sided Laughingthrush *G. caerulatus*.

BLACK-THROATED LAUGHINGTHRUSH

ad

PLAIN LAUGHINGTHRUSH

ad

davidi

MOUSTACHED LAUGHINGTHRUSH

cinereiceps

ad

RUSTY LAUGHINGTHRUSH

ad

BUFFY LAUGHINGTHRUSH

ad

CHINESE HWAMEI CTJ
Garrulax canorus

L 24–25cm. **SD** S & EC China with feral population on Taiwan; introduced and now well established in C & S Japan. **HH** Favours low montane forest, dwarf bamboo or dense vegetation on hillsides, and is especially fond of narrow stream valleys. Often located by noisy habit of turning or hopping amongst leaf litter on forest floor. Very active when feeding, but often hard to see in dense foliage. **ID** Earth brown, thrush-like bird with short wings and long, full tail. Distinctive grey-white 'spectacles' comprising prominent white eye-ring and brow extending across head-sides. Crown and nape reddish-brown, ear-coverts and chin earth brown, neck slightly paler, rest of upperparts dark brown with narrow black streaking on crown, nape and mantle; wings dark brown, as tail but latter narrowly and indistinctly barred blackish-brown. Underparts rufous-brown with fine streaking from chin to lower breast. **BP** Bill pale horn; eyes pale grey-brown, with bluish-grey orbital skin; tarsi yellowish-brown. **Vo** Very soft contact call, a slurred thrush-like *trrrrrr*, which can be difficult to hear. Song comprises four loud melodic whistles introduced by a mimicked phrase (e. g. call of Japanese Sparrowhawk). Highly prized as a cagebird for its rich song. **TN** Melodious Laughingthrush.

TAIWAN HWAMEI T
Garrulax taewanus

L 24–25cm. **SD** Endemic to Taiwan, where at risk from increasing hybridisation with introduced Chinese Hwamei. Monotypic. **HH** Secretive resident of scrub, tall grass, forest edge and dense undergrowth from lowlands to mid montane regions. **ID** Earth brown, thrush-like bird with long, full tail. Lacks prominent 'spectacles' of Chinese Hwamei. Upperparts dark brown with narrow black streaking on crown, nape, mantle and back, dark brown tail narrowly barred blackish-brown. Underparts paler brown with very fine black streaks on face, neck-sides and from chin to lower breast; belly unstreaked grey. Rather plain brown face with paler buff eye-ring which extends very slightly behind eye but does not become a brow. Characteristic beautiful hood of golden-buff scaling to mid back. Hybrids (Chinese x Taiwan Hwamei) resemble Chinese, but have reduced off-white or very pale powder blue eyebrow. **BP** Bill pale horn; eyes pale grey-brown, with bare yellowish orbital skin; tarsi yellowish-brown or brownish-orange. **Vo** Similar to Chinese, but less melodic and lacks mimicry. **TN** Formerly within Chinese Hwamei.

WHITE-BROWED LAUGHINGTHRUSH C
Garrulax sannio

L 25cm. **SD** Resident in SE Asia, S China. *G. s. sannio* in SE China east to Fujian. **HH** Favours scrub, forest edge, clearings and bamboo thickets at mid elevations. **ID** Mid-sized, brown laughingthrush with white face pattern. Dark earth brown forehead to

nape; mantle, back and rump mid grey-brown, short wings warm brown, as is full, graduated tail. Face largely creamy white with a broad creamy supercilium from lores to nape-sides and creamy white ear-coverts, rest of face including dark postocular stripe warm earth brown. Underparts warm brown to grey brown from throat to belly, cinnamon on vent. **BP** Bill blunt, dark grey; eyes black; strong tarsi brownish-pink. **Vo** Cackling, buzzing and harsh ringing notes.

WHITE-WHISKERED LAUGHINGTHRUSH T
Garrulax morrisonianus

L 26–28cm. **SD** Endemic to upper montane forests of Taiwan. Monotypic. **HH** Tall grass, dense undergrowth, thickets and forest edge scrub, from 2,000m, but also comes into open at roadsides and clearings. Sociable, commonly encountered in noisy parties. Often confiding. **ID** Large, dark brown laughingthrush with distinctive face pattern. Head dark brown, scaled greyer on forehead and crown, with yellow forehead spot, broad white supercilium from lores to nape-sides (broadest at rear) and broad white malar; chin, face, mantle, scapulars and back dark earth brown, rump ash-brown, long tail is blue-grey or ash-brown, blacker at tip and orange at sides towards base. Wings blue-grey with orange outer fringes to primaries and green fringes to secondaries. Underparts dark brown, with faint pale scaling, darker grey on flanks, chestnut to blackish-brown on lower belly and undertail-coverts. **BP** Bill thrush-like, pale yellowish-horn; eyes black; strong tarsi brownish-pink. **Vo** While foraging utters a thin, quiet *swip*, also various indistinct pips and squeaks, quickly becoming more staccato when alarmed, finally bursting into loud churring *skit-pyorrrrrrrr*. Song a strong, melodious bell-like whistle: *di di di di di...* and *tsip pee pe wee*.

RED-TAILED LAUGHINGTHRUSH C
Garrulax milnei

L 25cm. **SD** Restricted to northern SE Asia and S China, with *G. m. milnei* only in NW Fujian. **HH** Shady undergrowth and bamboo thickets in montane evergreen forest. **ID** Large, sooty-brown and orange laughingthrush. Bright orange-brown forehead to nape, grey-brown with dark scaling on mantle, neck-sides, scapulars and rump; wings and tail bright rufous-brown to crimson. Lores, narrow band at base of bill, short supercilium, chin and throat all black, with grey-brown ear-coverts. Underparts from neck to vent dull grey-brown, possibly violet in good light, with narrow chest scaling. **BP** Bill blunt, dark grey; eyes black, with narrow blue-grey eye-ring extending back from top of eye; strong tarsi dark grey. **Vo** Various calls include strong chattering. Song comprises loud whistled phrases: *uuu-weeoo eeoo-wee* or *uuuu-hiu-hiu*.

CHINESE HWAMEI

ad

ad

TAIWAN HWAMEI

WHITE-BROWED
LAUGHINGTHRUSH

sannio

ad

ad

WHITE-WHISKERED
LAUGHINGTHRUSH

RED-TAILED
LAUGHINGTHRUSH

milnei

ad

STEERE'S LIOCICHLA　　　　　　　　　T
Liocichla steerii

L 17–18cm. **SD** Endemic to Taiwan. Monotypic. **HH** Mid- and upper-level forests, c.1,000–2,500m, but some descend as low as c.200m in winter. Favours lower storeys of forest, forest edge scrub, bamboo groves, tall grass and dense vegetation, sometimes at roadsides. Very social and vocal, often on ground. **ID** Colourful, with a long graduated tail and overall dark olive-green plumage, contrasting with narrow yellow forehead, yellow-orange loral and subloral streak. Crown grey, with broad, blackish-grey supercilium bordered below by yellow streaking from eye around rear ear-coverts to neck-sides. Dark olive scapulars, yellow outer fringes to primaries, chestnut secondary-coverts and tertials, and dark blue-grey secondaries; rump dark blue-grey, tail dark olive-green with broad black subterminal band and narrow white tips to outer feathers. Underparts mid-grey with olive breast extending to scapulars/mantle. **BP** Bill short, slender; eyes black; tarsi dark grey. **Vo** Foraging pairs and groups give a low, buzzy *jerr jerr jerr*, increasing in volume when alarmed. Calls deeper than those of Taiwan Scimitar Babbler, more 'swallowed' also a *wee chuwee*. Song consists of several evenly spaced strident whistles with a lilting rhythm: *swiih seeoo seeseesee.*

RED-BILLED LEIOTHRIX　　　　　　　CTJ
Leiothrix lutea

L 13–16cm. **SD** Himalayas across southern China to Zhejiang and south to N Vietnam; *L. l. lutea* in SE China. Introduced, now naturalised in S Japan (Kyushu, Shikoku and central Honshu). **HH** Highly social, in groups in dense undergrowth of dark secondary forest, *Cryptomeria* plantations and montane bamboo scrub to above 1,000m (Japan). **ID** A grey-brown babbler with olive-green crown and bright orange breast. Face plain, 'gentle', with large pale yellow eye patch, dark blackish-grey malar streak and grey ear-coverts. Upperparts rather plain olive-grey (except greener olive crown and nape), darker on wings and tail, grey wing-coverts contrast with yellow/orange outer fringes to base of primaries and orange outer fringes to secondaries; tail blackish-brown, broad and square-tipped, slightly notched. Chin and throat yellow, breast deeper yellow/orange, belly to vent pale yellow, flanks grey. Juv./1st-year has

slightly duller markings and reduced orange-red fringes on folded wing. **BP** Bill short, red, black at base; black eyes appear large; tarsi orange-pink. **Vo** Call similar to Japanese Bush Warbler, but slightly thicker: *chwet*, also a buzzy rattling *zye zye* or *fii-fii-fii* in alarm. Song a rich, mellow, rather rapidly warbled phrase, recalling Japanese Grey Thrush. **AN** Pekin Robin.

WHITE-BROWED SHRIKE-BABBLER　　　C
Pteruthius flaviscapis

L 14–16cm. **SD** Himalayas to SE Asia. *P. f. ricketti* in S, C & E China north to Fujian. **HH** Forages in lower and upper canopy of broadleaf forests, searching for insects on branches and twigs. **ID** Sexually dimorphic. ♂ has black head and mask, with bold white band from above eye to nape-sides; mantle, back and rump grey, wings black with orange-brown tertials and white tips to primaries, short, square-tipped tail is black; underparts white with orange-brown wash to flanks. ♀ has grey head with white band from eye to nape, olive-grey mantle and scapulars, green wings and tail; white chin/throat to buffy-orange flanks, belly and vent. **BP** Bill short, broad, hook-tipped, dark grey; eyes appear small, pale blue-white; tarsi pink. **Vo** Various grating and churring sounds in alarm, and a short *pink*. Song a monotonous *chip-chip-chap-chip-chap*. **AN** Red-winged Shrike-babbler.

GREEN SHRIKE-BABBLER　　　　　　　C
Pteruthias xanchochlorus

L 12-13cm. **SD** Ranges from N Pakistan to SE China; an isolated population, *P. x. obscurus*, occurs locally in region only in NW Fujian. **HH** Found in mixed montane broadleaved evergreen and subalpine forest, descending somewhat in winter. **ID** A small, robust olive-green forest bird. Crown, nape and sides of face all grey; mantle, rump and wing coverts all olive-green, has narrow white wingbar, primaries black fringed green; short, broad tail also black and has narrow white tip. Throat, sides of neck and breast all white; flanks, belly, underwing and undertail-coverts lemon-yellow. **BP** Bill short, stubby, blue-grey with black tip; eyes mid-brown with white eye-ring; legs and feet grey. **Vo** A tit-like *jerr jerr-jerr jer-ri* call and a monotonous song consisting of a series of *whitu-whitu-wheet* notes.

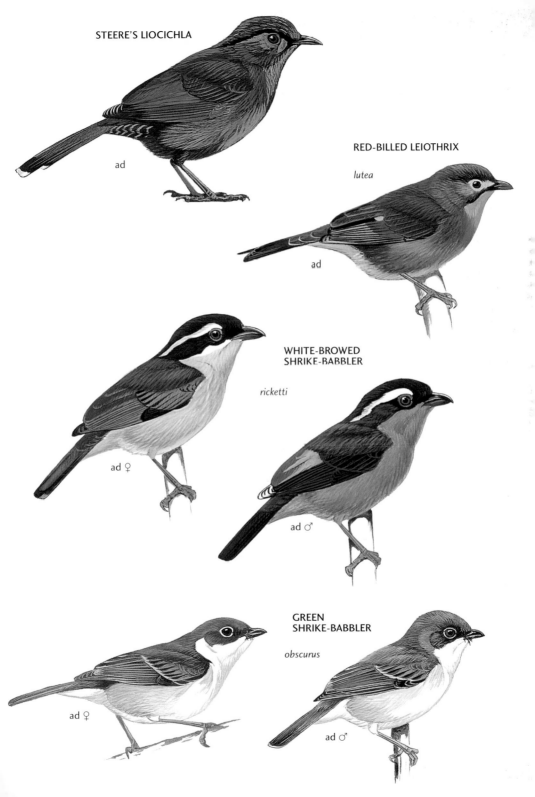

STEERE'S LIOCICHLA

ad

RED-BILLED LEIOTHRIX

lutea

ad

WHITE-BROWED SHRIKE-BABBLER

ricketti

ad ♀

ad ♂

GREEN SHRIKE-BABBLER

obscurus

ad ♀

ad ♂

TAIWAN BARWING   T
Actinodura morrisoniana

L 18cm. **SD** Endemic to montane forests of Taiwan. Monotypic. **HH** Favours lower and mid levels, in shrubs and low canopy of deciduous and coniferous forests, from *c*.1,500 to over 3,000m. Forms small, noisy, active groups. **ID** Thrush-like, with a rounded head, broad rounded wings, and broad tail with rounded tip. Largely dark brown, but warm dark reddish-brown on head, olive-brown on mantle, scapulars and rump; wings and tail greyish-brown narrowly barred black across tertials and rectrices and orange-brown across primaries and secondaries. Band of mid-grey, streaked pale grey and black, extends from mantle over neck-sides to breast; belly and flanks warmer brown with dark grey streaks. Graduated tail has broad black subterminal band and narrow white tips to outer feathers. **BP** Bill slender, arched, black; eye has narrow orange crescents above and below, irides dark brown; tarsi pink. **Vo** A soft *jiao-jiao* or more hurried *jia-jia-jia* in alarm; also a prolonged, repetitive *jip jip jip trrrrr trrrr trrrr…* and various purring and trilling notes, but generally rather quiet.

TAIWAN FULVETTA   T
Alcippe formosana

L 11–12cm. **SD** Endemic to Taiwan. Monotypic. **HH** Upper montane areas, where found in dense scrub and bamboo. **ID** Small but distinctly marked fulvetta with prominent pale eyes, extensive throat streaking and rufous wing panel. Upperparts dull, dark ash-brown on head, face, mantle and back, lores black, ear-coverts streaked, crown mouse-brown with pale greyish-brown lateral crown-stripes extending to nape-sides. Wings warm orange-brown contrasting with pale grey-fringed black primaries; rump and outer rectrices plain orange-brown, central tail dark blackish-brown; tail rather long and round-tipped. Underparts grey-white, strongly streaked brown on breast. **BP** Bill short, pinkish-grey; eye appears large in plain dark face, with white eye-ring and irides; tarsi dark brownish-pink. **Vo** Contact call a high squeaky *tee-trrr*; the most common call is a song-like *see dee* (second note higher). **TN** Formerly within Streak-throated Fulvetta *A. cinereiceps* (see Grey-hooded Fulvetta).

GREY-HOODED FULVETTA   C
Alcippe cinereiceps

L 11–12cm. **SD** Ranges across EC China, with *A. c. guttaticollis* east to Fujian. **HH** Undergrowth of evergreen forest, thickets, dense scrub and bamboo. **ID** Small but distinctly marked fulvetta with prominent pale eye. Upperparts dull, dark ash-brown on head, face, mantle and back, with indistinct dark lateral crown-stripes extending to nape-sides. Wings warm orange-brown with pale grey-fringed black primaries, rump and outer tail-feathers orange-brown, central tail dark blackish-brown. Tail rather long

and round-tipped. Underparts grey-white, with faint throat streaking. *A. b. guttaticollis* has vinous-brown crown bordered by grey-brown lateral stripes. **BP** Bill short, black; eye appears large in plain dark face, with white eye-ring and irides; tarsi dark brownish-pink. **Vo** Contact call a tit-like *cheep*, also a high, thin *see-see*. Song a three- or four-note rattle. **TN** Formerly part of Streak-throated Fulvetta *A. cinereiceps*, which has been split.

DUSKY FULVETTA   CT
Alcippe brunnea

L 13cm. **SD** C, S & E China: *A. b. superciliaris* E & SE China north to Zhejiang; *A. b. brunnea* Taiwan. **HH** In small flocks in low shrub layer of evergreen and deciduous forests, above 400m, common in lower and mid montane forests of Taiwan (100–2,500m). Often terrestrial, foraging close to or on ground amongst leaf litter, where hard to observe, but not shy. **ID** Two-tone, grey and dark brown fulvetta with large head and short, rounded tail. Forehead to tail dark brown, lateral crown-stripes black extending from forehead to neck-sides. Face plain grey, with indistinct buff orbital ring, short, narrow black moustachial, broad grey malar, off-white chin, and neck-sides and rest of underparts plain dull grey. **BP** Bill strong, black; eyes orange-brown; tarsi dark brownish-pink. **Vo** A repetitive low *jurr* or purring *churrk*, given from ground, a high *wheet churrk*, and louder, rising *brueet* in alarm. Song a short, rather gloomy warble. **AN** Gould's Fulvetta.

GREY-CHEEKED FULVETTA   CT
Alcippe morrisonia

L 12–14cm. **SD** From northern SE Asia across C, S & E China to Zhejiang: *A. m. hueti* SE China; *A. m. morrisonia* on Taiwan. **HH** Forest and forest edge at low to mid elevations, *c*.100–2,500m (generally replaced in coniferous forests by Dusky in Taiwan), usually in noisy parties, often in mixed-species flocks. One of the commonest species in Taiwan, and a standard component species of bird 'waves', but becomes quieter when breeding (late Mar–Apr). **ID** Rather plain and drab with very large head. Head, face and mantle plain, dull mid grey, with indistinct black lateral crown-stripe extending to nape, and broad white (*A. m. morrisonia*) and buff (*A. m. hueti*) eyering. Wings, back and tail dark brown. Grey of face extends as lateral throat-stripe, chin/throat dusky, underparts buff, slightly more orange-brown on flanks and undertail-coverts. Tail rather long. **BP** Bill short, black; eyes orange-brown; tarsi yellowish-brown. **Vo** While foraging utters curt *jrr* sounds, but in alarm a noisy, rasping *zee zee zee*. Song a high sweet warble followed by a drawn-out, slightly wheezy note.

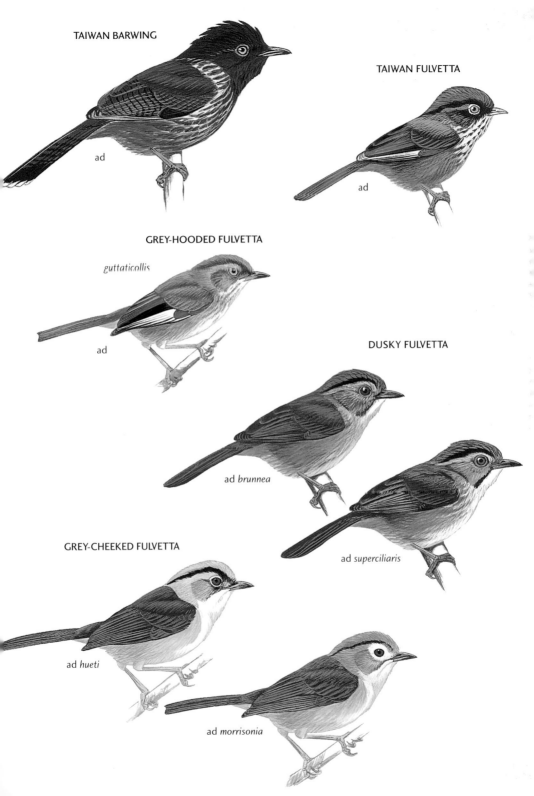

TAIWAN BARWING

ad

TAIWAN FULVETTA

ad

GREY-HOODED FULVETTA

guttaticollis

ad

DUSKY FULVETTA

ad *brunnea*

ad *superciliaris*

GREY-CHEEKED FULVETTA

ad *hueti*

ad *morrisonia*

TAIWAN SIBIA   T
Heterophasia auricularis

L 23–24cm. **SD** Endemic to Taiwan. Monotypic. **HH** Montane deciduous forests mostly above 1,000m during breeding season, but descends to lowlands (locally even breeds in coastal forests) in winter. Travels in flocks in lower and upper canopy. May congregate at flowering or fruiting trees. **ID** Large, grey and orange thrush-like bird, with rather flat crown. Foreparts largely dark grey, blackest on crown and nape, bold white stripe from bill to nape-sides extends as fine plumes. Wings and long tail black with white outer fringes to primary bases and tips to graduated rectrices. Chin, foreneck and breast dark grey, becoming deep foxy orange on sides and belly, darker foxy orange on undertail-coverts, back and rump. When singing, exposes fine white plumes on underwing. In flight, flashes white in primaries, orange rump contrasts with black tail with white lateral tips. **BP** Bill strong, black; eyes dark brown; tarsi dull orange. **Vo** The song consists of an ecstatic whistled phrase, the last part loudest and sounds like distant wolf whistle: *switswitswit Sweeeoo!* In alarm, and contact, the call is a hard, tinny, rattling *kirrrrr* recalling Japanese Pygmy Woodpecker, but more drawn-out and much louder *tik tik trrck trrrck.* **AN** White-eared Sibia.

INDOCHINESE YUHINA   C
Yuhina torqueola

L 13cm. **SD** Thailand to SE China, north to Fujian and Zhejiang. Monotypic. **HH** Lower forest canopy, above 400m. **ID** Rather plain yuhina, with prominently rounded crest. Grey head and crown, narrow supercilium white flecked grey, ear-coverts rich chestnut extending as broad, paler chestnut nuchal collar boldly streaked white, malar streaked black, mantle and scapulars grey-brown, wings and tail blackish-brown, tail has white outer fringes. Underparts, white from chin to vent. **BP** Bill rather heavy, tit-like, blunt, dull pink with grey tip; eyes black; tarsi bright pink. **Vo** Call a repetitive *ser-weet ser-weet ser-weet* or *di-duit di-duit;* flock members maintain a continuous chattering. Song a series of high-pitched *tchi-chi* notes. **TN** Formerly within Striated Yuhina *Y. castaniceps.*

TAIWAN YUHINA   T
Yuhina brunneiceps

L 12–13cm. **SD** Endemic to hill forest on Taiwan, where abundant. Monotypic. **HH** Regular 1,000–3,300m, but commonest bird at 1,500–2,500m, and descends as low as 200m in winter. Shrub layer, lower and upper canopy. **ID** Smart, crested, tit-like yuhina. Crown and top of spiky crest chocolate-brown bordered narrowly black; forehead, supercilium, hindcrown and rear of crest, pale grey; nape grey, rest of upperparts dark ash-brown, darkest on wings and tail. Face attractively striped, with very narrow black stripe behind eye, black malar connects to narrow black line marking rear edge of ear-coverts which wraps to eyestripe. Chin/throat white with gorget of indistinct black streaks, breast pale grey, belly and vent off-white; flanks have broad rusty-orange streaks. **BP** Bill slender, black; eyes dark brown; tarsi dull ochre. **Vo** Foraging flocks produce steady chatter of soft *pit pit* contact calls. Song a clear three-part whistle *tswee sit seeyou!* each note evenly spaced, the first rising, second and third level. Beware rather similar-toned songs of Taiwan Sibia, Steere's Liocichla and Taiwan Wren-Babbler.

BLACK-CHINNED YUHINA   C
Yuhina nigrimenta

L 11cm. **SD** Ranges from Himalayas and northern SE Asia to SE China: *Y. n. pallida* in SE China east to Fujian. **HH** Canopy of scrub, secondary forest, around clearings and in montane forest. **ID** Small, rather plain, crested yuhina. Forehead, crown and pointed crest grey with black streaking, lores and chin black, face, head-sides and nape grey; upperparts dull greyish-brown, darkest on wings and tail. Underparts white with grey-brown wash to flanks and vent. **BP** Bill short, black above, pink on lower mandible; eyes black; tarsi pink. **Vo** A constant nervous buzzy chattering *wh'rr'rr.* Song a soft *whee-to-whee-de-der-n-whee-yer* or high-pitched ringing *uu ii uui ii uui uu ii uui ii uui uu ii uui uui ii uui….*

TAIWAN SIBIA

ad

ad

INDOCHINESE
YUHINA

TAIWAN YUHINA

ad

BLACK-CHINNED
YUHINA

pallida

ad

WHITE-BELLIED ERPORNIS CT
Erpornis zantholeuca

L 11–13cm. **SD** Himalayas to SE Asia. *E. z. griseiloris* in SE China to Fujian, and Taiwan. **HH** Mid and upper storeys of hill and montane broadleaf forest, particularly at higher elevations (but only low and mid elevations, *c.*100–2,500m, on Taiwan). **ID** Somewhat tit-like, active bird with rounded crest. Upperparts, including wings and square-tipped tail, bright olive-green. Face (including lores and crescent above eye), and underparts off-white, throat prominently white, vent yellow. **BP** Bill short, somewhat stout, grey above and at tip, pink below; eyes black; tarsi pale pink. **Vo** Call is a loud, nasal, scolding *vray vray vray* or *nher-nher* rising in pitch, is similar to call of Varied Tit, and a metallic *chit*. Song is a high-pitched descending trill: *si-i-i-i-i....* **TN** Formerly White-bellied Yuhina *Yuhina zantholeuca*.

BEARDED TIT CKJ
Panurus biarmicus

L 14–15.5cm; WT 12–18g. **SD** W Europe to NE China. Birds from E Europe to E Asia attributed to *P. b. russicus*. In region, from S Transbaikalia to N & NE China, with some wintering south to Bohai Gulf. Vagrant Korea and Japan in winter. **HH** Restricted to reed and cattail beds fringing rivers, lakes and marshes; forages at bases but climbs to tops of reeds in calm weather. **ID** Rather plain cinnamon-brown bird with long (*c.*7cm) graduated tail with a rounded tip. ♂ has plain pale blue-grey head with prominent black lores and drooping black 'moustaches'; mantle, back, tail and flanks warm tawny-brown; underparts off-white suffused pink, except undertail-coverts which are black; wing-coverts, tertials and primary tips black, folded primaries appear largely white. ♀ lacks grey hood, black 'moustaches' and black undertail-coverts. Flight rather weak on rapid whirring wingbeats. **BP** Bill short, finely pointed, yellow-orange; eyes small, white in ♂, orange in ♀; tarsi dark-grey to black. **Vo** A metallic *ching ching* or *chveen chveen* given while bird moves up and down reeds, and commonly in flight. Song a three-part phrase: *tschin-dschik-tschrää*. **AN** Bearded Reedling.

SPOT-BREASTED PARROTBILL C
Paradoxornis guttaticollis

L 18cm. **SD** Occurs in NE India, northern SE Asia, and S,W & SE China east to Fujian. Monotypic. **HH** Tall grasses, scrub and secondary forest in mountains. **ID** Rather drab parrotbill with black and white face. Cinnamon-brown forehead, crown and nape, rest dull mid brown; face, lores and chin largely black, with white scaling behind eye and on cheeks. Underparts mainly buff, with slight gorget of streaks on throat. **BP** Bill large, deep, short, hook-tipped and yellow-orange; eyes large, irides dark brown; tarsi dark-grey. **Vo** Calls including social chattering, various coarse and metallic notes, a sibilant *chut-chut-chut*, strident *du-du-du* and rapid series of 3–7 mellow whistled *whit-whit-whit...* notes.

VINOUS-THROATED PARROTBILL CTKJR
Paradoxornis webbianus

L 12–13cm. **SD** Common resident from N Vietnam to NE China; also adjacent S Russian Far East, Korea and Taiwan. Accidental offshore Japan. *P. w. suffusus* in SE China north to Fujian; *P. w. webbianus* in E China including S Jiangsu and N Zhejiang; *P. w. bulomachus* on Taiwan, *P. w. fulvicauda* in NE Hebei, China, and Korea; *P. w. mantschuricus* in NE China and adjacent Russia. **HH** Generally in scrub, riparian thickets, woodland edge and fringes of reedbeds; often encountered in noisy, wandering flocks. **ID** Small plain pinkish-brown long-tailed bird. Overall mid sandy- or pinkish-brown, darker and more chestnut on crown and wings, with paler buffy-brown underparts. Long tail is darker, with grey-brown outer feathers and squarish tip. **BP** Bill very short and stubby, grey with pale/horn tip; eyes small, black, very prominent on open face; tarsi grey. **Vo** Flock members give thin chattering *chii chii chii chii*. Song a high-pitched repetitive *rit rit piwee-you wee-ee-ee*. **AN** Webb's Parrotbill.

WHITE-BELLIED
ERPORNIS

griseiloris

ad

BEARDED TIT

russicus

ad ♂

ad ♀

juv ♂

VINOUS-THROATED
PARROTBILL

SPOT-BREASTED
PARROTBILL

ad

ad
mantschuricus

ad
webbianus

ad
bulomachus

GOLDEN PARROTBILL   CT
Paradoxornis verreauxi

L 10–12cm. **SD** Resident in inland northern SE Asia and SE China: *P. v. pallidus* in NW Fujian; *P. v. morrisonianus*: high mountains of Taiwan. **HH** Forms flocks in bamboo thickets and scrub around montane evergreen forests. **ID** Small, rather bright parrotbill. Upperparts warm orange-brown, brightest on crown, tail-sides, greater coverts and secondaries, browner on back and scapulars. Face dusky, blackish-grey around eye; short supercilium, lores and ear-coverts dusky greyish-white, whilst short, very broad malar is whiter. Wings blackish with pale fringes to flight-feathers; tail long, slightly graduated, blackish with orange-brown outer fringes to feathers. Chin and throat patch black, breast and belly greyish-white, flanks warm orange-brown. *P. v. morrisonianus* (described here) is generally duller and greyer than other races. **BP** Bill very short, stubby, dull pink; eyes black; tarsi pale pink. **Vo** Contact calls while foraging *pit* and *tup*, higher and quieter than Streak-throated Fulvetta, also a purring *prrp prrp*, a rattling *trrr'it* and various chattering and churring calls. Song a thin, high-pitched *ssii-ssii-ssu-ssii*.

SHORT-TAILED PARROTBILL   C
Paradoxornis davidianus

L 10 cm. **SD** Ranges from S China to Thailand. In region, *P. d. davidianus* is an uncommon resident in NW Fujian. **HH** Found in bamboo stands in broad-leaved evergreen forest from lowlands into hills. **ID** A small brown parrotbill. Head and upper mantle warm rufous-brown, back and wings grey-brown with pale fringes to black primaries, rump and tail warm brown. Chin and throat black, rest of underparts plain grey-brown. **BP** Bill short, stubby greyish-pink; eyes dark brown appearing rather large in plain face; legs and feet dull greyish-pink. **Vo** Calls consist of subdued twittering; song is high-pitched rapidly rising *ih'ih'ih'ih'ih'ih'*.

GREY-HEADED PARROTBILL   C
Paradoxornis gularis

L 18cm. **SD** SE Asia across southern China north to S Jiangsu. *P. g. fokiensis* in SE China. **HH** Sociable; occurs in scrub, bamboo thickets and forest, from undergrowth to canopy, in hills and low mountains. **ID** Clean, large-headed parrotbill, with large bright bill. Head largely pale grey, with black border to crown from forehead to nape, lores grey, but broad malar white, broad white brow above and narrow white crescent below eye; mantle, back, wings and tail uniform mid-brown. Neat black chin contrasts with clean white breast to vent, with some cinnamon-brown on flanks. **BP** Bill large, deep, short, hook-tipped and bright orange; eyes large, irides dark brown; tarsi blue-grey. **Vo** Calls including low twittering, soft *chip*, rattled *chrrrat* and harsh slurred *jieu-jieu-jieu*.

NORTHERN PARROTBILL   RC
Paradoxornis polivanovi

L 18cm. **SD** Restricted to NE Mongolia and NW Manchuria (*P. p. mongolicus*), NE China (Liaoning and Heilongjiang) and Lake Khanka region of extreme southern Russian Far East (*P. p. polivanovi*). **HH** Dense reedbeds and associated marshland vegetation. **ID** Long-tailed marsh bird with rather swollen bill. Head large, with pale grey crown and broad brown band extending from sides of forecrown to nape, bordered below by black line from eye to nape; lores black, with a white crescent above eye. Face and neck-sides grey with long, dark-centred 'mane' on neck. Upperparts rusty-brown, the mantle partially obscured by 'mane'. Wings very short, largely dark brown to black, but has paler panel in secondaries and white fringes to tertials. Off-white or pale grey chin and throat, contrasting strongly with rusty-brown upper breast to flanks and belly. Long graduated tail is grey-brown with black outer feathers and white tips to outermost. **BP** Bill short, deep-based, yellow; eyes dark brown; legs brown. **Vo** Short trills and nasal whistles. **TN** Formerly within Reed Parrotbill. **AN** Polivanov's Parrotbill.

REED PARROTBILL   C
Paradoxornis heudei

L 18cm. **SD** Restricted to E China, including Jiangsu and perhaps Hebei. Monotypic. **HH** Dense reedbeds and associated marshy vegetation. **ID** Long-tailed marsh bird with rather swollen bill. Differs from Northern in head/face pattern. Head large, with dark grey crown and broad brownish-black band extending from sides of forecrown to nape; lores and from eye to lower mandible black, but has short white brow. Face and neck-sides grey with long, dark-centred 'mane' on neck. Upperparts rusty-brown, the mantle partly obscured by 'mane'. Wings very short, largely dark brown to black, but with paler panel in secondaries and white-fringed tertials. Off-white or pale grey chin and throat, upper breast to flanks and belly dull orange-brown, brightest on breast. Long graduated tail grey-brown with black outer feathers and white tips to three outermost. **BP** Bill short, deep-based, yellow; eyes dark brown; legs brown. **Vo** Vocal distinctions from Northern unclear. **AN** Chinese Parrotbill.

GOLDEN PARROTBILL

ad
pallidus

ad
morrisonianus

**SHORT-TAILED
PARROTBILL**

davidianus

**GREY-HEADED
PARROTBILL**

fokiensis

ad

ad

REED PARROTBILL

ad

**NORTHERN
PARROTBILL**

polivanovi

ad

CHESTNUT-FLANKED WHITE-EYE   CKJR
Zosterops erythropleurus

L 10–12cm. **SD** Breeds only in NE China and S Russian Far East, possibly N Korea, but migrates through E & S China to winter in SE Asia. Uncommon to common migrant Korea, and very rare offshore Japan. Monotypic. **HH** Favours mature mixed deciduous and coniferous forests, and riparian woodland, but on migration occurs in any type of woodland. **ID** Small, bright green rather warbler-like bird, very similar to Japanese White-eye. Hood and upperparts bright yellow-green. Yellow chin/throat and vent, and clear white belly, but usually has distinct broad chestnut patch at sides (fainter in young, making separation from *japonicus* Japanese White-eye tricky). Primary projection 75–100% length of exposed tertials (longer than Japanese). **BP** Bill finely pointed, grey above and at tip, pink below and at base; eyes appear large, surrounded by prominent broad white eye-ring, broken at lores, irides black; tarsi grey. **Vo** *Chii chii* or *tsee-plee*, similar to Japanese but less powerful and less clear.

JAPANESE WHITE-EYE   CTKJR
Zosterops japonicus

L 10–11.5cm. **SD** Throughout Japan, though only summer visitor to N Honshu and Hokkaido; resident C & S Japan, S Korea, Taiwan, S China and N Indochina; summer visitor to much of E China; winters in SE Asia. Little range overlap with Chestnut-flanked. *Z. j. yesoensis* on S Sakhalin and Hokkaido; *Z. j. japonicus* Honshu, Kyushu, Shikoku and Tsushima; *Z. j. stejnegeri* Izu Is and introduced to Ogasawara Is; *Z. j. alani* Ogasawara and Iwo Is; *Z. j. insularis* Tanega-shima and Yaku-shima; *Z. j. loochooensis* Ryukyus; *Z. j. daitoensis* Borodino Is; *Z. j. simplex* E & SE China and Taiwan. **HH** Deciduous and evergreen broadleaf forests, especially with flowers and fruits, but on migration in any type of woodland, gardens and parks. In Taiwan, abundant in secondary growth and built-up areas with trees and scrub. **ID** Small, bright green rather warbler-like bird. Hood and upperparts bright yellow-green, with prominent broad white eye-ring; yellow breast. Underparts off-white to buffy-grey, darker on flanks, but lacks clear contrast between white belly and chestnut of Chestnut-flanked; breast and flanks have reddish-brown wash, undertail-coverts yellowish. *Z. j. japonicus*, in particular, often has rather brown, even maroon, tone to flanks, making separation from Chestnut-flanked more difficult. *Z. j. simplex* rather small with pale yellowish-green upperparts, yellow throat and vent, and clean white to off-white underparts. Strong black loral stripe extends below eye. *Z. j. loochooensis* and *Z. j. daitoensis* smaller bodied than *Z. j. japonicus*, with greyish-white breast and flanks. Small race, migrant through the region with yellower upperparts than *Z. j. japonicus* and greyish-white breast and flanks, may refer to *Z. j. simplex*; other subspecies probably inseparable in field. **BP** Bill slightly arched and fine-pointed, dark grey above and at tip, paler blue-grey below and at base; eyes orange to orange-brown; tarsi grey/black. **Vo** A high, thin twittering: *tsee tsee, chii chii puu chii chii, chu*. Song is rapid and complex: *chuichui chochopiichui*.

LOWLAND WHITE-EYE   T
Zosterops meyeni

L 10–12cm. **SD** Essentially Philippine species, but *Z. m. batanis* occurs on Lanyu I., off SE Taiwan. **HH** Tropical evergreen broadleaf forest, especially with flowers and fruits, also secondary growth and gardens. **ID** Small green white-eye. Hood and upperparts dull olive-green, with prominent broad white eye-ring. Underparts off-white to buffy-grey, darker grey or dusky-brown on flanks, bright yellow on chin, throat, lower belly and vent. **BP** Bill slightly heavier than Japanese, dark grey above and at tip, paler grey below and at base; eyes brown; tarsi grey/black. **Vo** The call is a hard chattering, high-pitched *chii chii chii*. *Z. m. meyeni* of Philippines also gives a Yellow Wagtail-like *spiz* in flight.

BONIN HONEYEATER   J
Apalopteron familiare

Vulnerable. L 12–14cm. **SD** Endemic to Ogasawara Is, Japan; *A. f. familiare* of Mukojima and Chichijima is presumed extinct; *A. f. hahasima* survives only on Haha I. **HH** From gardens with tall bushes, to plantations, woodland edge and open subtropical forest. **ID** Superficially resembles large white-eye (often treated as congeneric), being yellowish-green and grey overall, but has unique black face pattern. Forecrown, central forehead stripe reaching bill, small patch above rear of eye and elongated triangle pointing down face below eye, are all black; sides of forehead, rear ear-coverts, chin and throat bright yellow, neck-sides and flanks greyish-yellow. Crown, mantle and back olive-green, wings brighter with yellow outer fringes to flight-feathers; tail dark grey-olive. **BP** Bill rather long, arched, black; eyes appear large, with white crescents above and below, irides dark reddish-brown; tarsi black. **Vo** Soft *pee-yu, chui, weet* and *pit* notes, a loud explosive *tit-tit*, and when mobbing gives harsher *weet-weet* and *zhree-zhree…* calls. Song, given from high perch and occasionally in flight, is a melodious warbling *tu-ti-ti ti-titu-tuoo*; also *chui churiripyuuyo* and *fiyo chui chuchee feeyo*. **TN** Sometimes placed in *Zosterops*. **AN** Bonin White-eye.

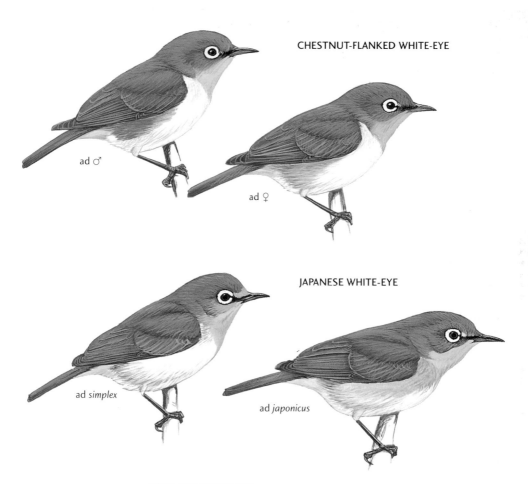

CHESTNUT-FLANKED WHITE-EYE

ad ♂

ad ♀

JAPANESE WHITE-EYE

ad *simplex*

ad *japonicus*

LOWLAND WHITE-EYE

batanis

ad

BONIN HONEYEATER

hahasima

ad

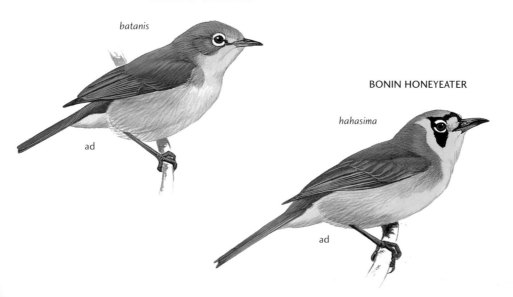

FLAMECREST T
Regulus goodfellowi

L 9cm; WT 7g. **SD** Common Taiwanese endemic, where restricted to central mountains. Monotypic. **HH** Upper montane coniferous forests (c.2,000m to scrub just above treeline at c.3,600m), where often the most abundant bird. Often performs sallies to glean insects from upper branches. **ID** Boldly marked, colourful firecrest, with particularly prominent face pattern. Crown black with narrow yellow (♀) or orange (♂) median stripe (usually well hidden). When ♂ is excited, flame-orange feathers are exposed in a dome, completely altering head shape. Very broad supercilium, lores, narrow forehead band and face white, with broad black patch around eye, and short, narrow, black malar stripe. Chin white, throat buff-grey, ear-coverts, neck-sides and nape grey. Mantle, scapulars and short tail bright green; rump, flanks, belly and vent bright lemon-yellow, central belly off-white. Wings short, black with broad white covert tips forming single distinct wingbar, white tips to black tertials and green fringes to primaries and secondaries. **BP** Bill fine, black; eyes appear very large because of black eye-ring set in white face, irides black; tarsi brownish-pink. **Vo** High pitched call is an abrupt *tsi tsi tsi*; sometimes gives a thin *see see see* identical to Goldcrest. Song consists of a regular series of high-pitched call notes. **AN** Taiwan Firecrest.

GOLDCREST CTKJR
Regulus regulus

L 9cm; WT 5–7g. **SD** British Isles to Japan. In region, *R. r. japonensis* is summer visitor to NE China, Amurland, Sakhalin, S Kuril Is; resident Hokkaido, N & C Honshu, breeds sparingly in N & C Korea, winter visitor to S Japan, Korea, E China; rare winter visitor Taiwan. Some evidence exists to separate this race as **Asian Goldcrest**. **HH** Typically in coniferous and, to lesser extent, mixed forests, but on migration found in any type of woodland or even scrub. **ID** Tiny; one of smallest birds in region; large-headed and short-tailed. Resembles *Phylloscopus* warblers, but smaller, neater, plain olive-green upperparts, duller, more olive below. Crown yellow, bordered by broad black lateral crown-stripes that do not merge on forehead (both sexes), ♂ alone has orange central crown-stripe with yellow margins; short, narrow black malar, white lores and spectacles, grey head-sides, ear-coverts and nape; wings largely black with two broad white bars, yellowish fringes to primaries, and white-fringes to tertials. **BP** Bill fine, short black; eyes appear especially large in broad pale eye-ring, irides black; tarsi dull brownish-pink or brighter orange-yellow (breeding ♂). **Vo** Thin, high-pitched insistent *sree-sree-sree*, similar to Eurasian Treecreeper but slightly less insistent and shorter; also a quiet *seeeh* and *zick*. Song regionally variable, either a complex series of thin, sibilant notes in a rapid trill *chiichiichii chiriri tsutsutsutsu-tjii-tsitsi-chocho*, or alternating high and low notes *tee-de-dee tee-de-dee*.

RUBY-CROWNED KINGLET R
Regulus calendula

L 9–11cm; WT 5–8g. **SD** Widespread summer visitor across northern N America from Newfoundland to Alaska. Vagrant to NE Russia. Subspecies probably *R. c. grinnelli*. **HH** Favours coniferous forests when breeding, but also deciduous forest, woodland and scrub at other times. **ID** Tiny grey-green 'goldcrest', with rather plain ash-grey head, face and mantle, broken white eye-ring, black wings with two narrow white bars, white outer fringes to tertials, and yellowish-green fringes to primaries contrasting with neat black patch at base of secondaries; tail black with yellowish-green fringes. ♂ has narrow red median crown-stripe (often concealed), lacking in ♀. Creamy white chin/throat, buff to olive on flanks. **BP** Bill fine, black with fine bristles at base; irides dark brown, appearing black in contrast with eye-ring; tarsi dark brown, toes slightly paler. **Vo** Fine *tsee-tsee* notes, a low *tak* and loud *zerr*.

WINTER WREN CTKJR
Troglodytes troglodytes

L 9–10cm; WT 7–12g. **SD** Very widespread across Holarctic with more than 40 subspecies recognised, and probably three species involved, with Old World clade deserving separation from two Nearctic clades. In region, winter visitor to Chinese coast north to Shandong and resident in high mountains of Taiwan, also resident NE China, Korea and throughout Japan and Kuril Is, and summer visitor to Russian Far East and Sakhalin. *T. t. idius* NE China to Hebei; *T. t. dauricus* SE Siberia, extreme NE China, Korea and Tsushima, Japan; *T. t. pallescens* Commander Is and possibly Kamchatka; *T. t. kurilensis* N Kuril Is; *T. t. fumigatus* S Kuril Is, Sakhalin, Japan and N Izu Is; *T. t. mosukei* S Izu Is and Borodino Is; *T. t. ogawai* Tanega-shima and Yaku-shima; *T. t. orii* Daito Is (extinct); *T. t. taivanus* resident on Taiwan (but occasional records in lowlands and on offshore islands presumed migrants or vagrants, possibly of other subspecies). **HH** Locally common, typically in rather dark woodland, particularly favouring stream sides, rocky areas, and well-vegetated damp gullies, also damp thickets; in winter in almost any riparian habitat including stream-side willows and reeds. **ID** Tiny, cock-tailed bird renowned for active, energetic behaviour and extraordinarily loud song, easily heard over rushing water. Appears generally dark brown, but finely barred black and grey on wings, tail, and underparts, and has long, narrow brownish-white supercilium (varies between subspecies). Wings short, rounded, tail short and habitually cocked. Flight very fast, whirring from one patch of vegetation to next (compare Asian Stubtail, p.340). Distinctive *T. t. fumigatus* is very dark brown with a longer, darker bill and longer tail than other races. *T. t. pallescens* is rather duller and greyer than others. *T. t. taivanus* somewhat less rufous than most others. **BP** Bill slender, sharply pointed, blackish-brown with grey/horn base to lower mandible; eyes dark brown; tarsi pinkish-brown. **Vo** An indignant chatting *tet tet-tet, tji tji* or *chet chet chet*, similar to Japanese Bush Warbler, but higher. Song astonishingly loud, penetrating and far-carrying for such a diminutive bird, rapid and complex: *pipipi chui chui chiyo chiyo chuririri*. **TN** Proposed change to *Nannus troglodytes*.

FLAMECREST

ad ♀

ad ♂

GOLDCREST

juponensis

ad ♂

ad ♀

**RUBY-CROWNED
KINGLET**

grinnelli

ad ♀

ad ♂

ad
taivanus

ad
fumigatus

WINTER WREN

ad
pallescens

EURASIAN NUTHATCH CTKJR
Sitta europaea

L 12–17cm; WT 21–26g. **SD** W Europe to E Asia. *S. (e.) arctica* (**Siberian Nuthatch**) Lena River basin east to Yakutia, Chukotka and Koryakia; *S. e. albifrons* Kamchatka; *S. e. asiatica* S Siberia, NE China to Sea of Okhotsk and Hokkaido; *S. e. amurensis* N Hebei to S Russian Far East, Korea, Honshu, Shikoku and N Kyushu; *S. e. roseilia* S Kyushu; *S. e. bedfordi* Cheju-do (Korea); *S. e. sinensis* NC & E China north to Hebei, also SE & S China and Taiwan. **HH** Common resident; occurs in a wide range of wooded habitats, but mainly deciduous and evergreen broadleaf forests, sometimes in conifers, and also woodland, scrub, parks and gardens in winter when readily joins mixed tit flocks. Acrobatic, feeding on trunks and larger branches. **ID** Large-headed, stub-tailed and rather vociferous. Upperparts rather dull leaden blue-grey (paler blue-grey in some races). Very long thick black eyestripe extending to nape-sides, bordered above from forehead to just behind eye by narrow supercilium; wings blackish-grey, tail grey in centre, black at sides with white patches near tips. Underparts largely white or off-white, with variable chestnut, from lower belly and flanks to vent in *S. e. amurensis/S. e. roseilia*, but *S. e. asiatica* is much paler blue-grey above, all-white below except for limited chestnut scalloping on undertail-coverts; *S. e. sinensis* has entirely buffy-chestnut underparts, darkest on flanks and vent. From below, wings show black and white carpal patches, and tail is largely black with white corners. **BP** Bill very strong, wedge-shaped, black; eyes black; legs strong, feet large, black. **Vo** Calls include a loud, liquid *plewp plewp plewp* or *pyo pyo pyo*, drawn-out *ziit* and forceful *twett*. Another common call is a high-pitched mouse-like *spee tee tee tee*. Song, also loud and penetrating, comprises whistled series of loud, high, clear trilled or rippling notes: *pipipipipipi* or *fififififi….* and *jujujuju….*

CHINESE NUTHATCH CK
Sitta villosa

L 11–12cm. **SD** China north to E Manchuria, N & C Korea and Ussuriland. Race in region is *S. v. villosa*. Rare and irruptive winter visitor to S Korea. **HH** Locally common resident of coniferous forests and town parks; in Korea strongly associated with *Pinus densiflora*. **ID** Resembles Eurasian but smaller, duller and darker. ♂ has black crown, broad black eyestripe and supercilium from forehead to nape-sides; broad black eyestripe and rest of upperparts dark bluish-grey; chin, cheeks and throat white, rest of underparts dull greyish-fulvous. ♀ duller, with grey, not black, crown and eyestripe, and paler upperparts. Wings and stubby tail plain. **BP** Bill finely wedge-shaped, black; eyes black; tarsi and toes large, black. **Vo** Utters distinctive harsh, scolding *schraa*. Also a nasal *quir quir* and piping *wip wip wip* in either short or prolonged series. Easily overlooked until voice known. Song a series of rising whistles. **TN** Snowy-browed Nuthatch.

WALLCREEPER *Tichodroma muraria* C

L 15–17cm; WT 15–19.6g. **SD** Highly specialised species of high mountains, from S Europe to N China. *T. m. nepalensis* Afghanistan to N & C China and S Mongolia. Nomadic rather

than migratory, wintering to foothills in much of E China northeast to Bohai Gulf, and occasionally further afield, so could appear anywhere. **HH** Montane areas with crags, gorges and cliffs, but winters at lower elevations; also at stone ruins and rocky river beds. Forages on rock faces. Uncommon. **ID** Attractive, as much for its unusual habits as its plumage, but highly cryptic except in flight. Ad. mostly ash-grey, with black rump, belly, wings and tail, but summer ♂ has black face, throat and breast. In winter, and ♀, chin/throat white, breast pale grey. Wings broad and rounded, flight buoyant and butterfly-like. Scapulars deep crimson, and coverts and outer fringes of primaries and secondary bases also crimson, whilst primaries have large round white spots. Tail short, mostly black with white outer feathers and corners. **BP** Bill long, slender and slightly arched, black; eyes brown; legs short, toes very long, blackish. **Vo** Generally silent, except in breeding season, occasionally gives low, whistled *tseeoo*.

EURASIAN TREECREEPER CKJR
Certhia familiaris

L 12–15cm; WT 8–11g. **SD** W Europe to NE China, Korea, Russian Far East and Japan. *C. f. daurica* in S Siberia (Urals to Sakhalin), NE China, N & C Korea, S Kuril Is and Hokkaido; *C. f. japonica* in Honshu, Kyushu and Shikoku. **HH** Fairly common in temperate mixed deciduous broadleaf and coniferous forests. Generally alone or in pairs, but joins mixed-species flocks in winter, when may also disperse lower, even visiting urban parks. Creeps mouse-like up tree trunks and larger limbs, then flies to base of another tree. **ID** Slender, dark brown bird, the upperparts streaked dark grey and buff; wings mottled grey-brown and orange-brown, rump and tail plain mid brown. Supercilium and underparts creamy white, except buff-washed flanks and vent. Long tail, serves as support when climbing upwards, much like woodpecker. In flight, distinct pale wingbar on mid underwing. **BP** Bill long, fine, arched, blackish above, pink below; eyes black; tarsi large, pinkish-brown, with long hindclaw. **Vo** A very high-pitched sibilant and slightly buzzy *tsee tsee* or *tsuu tsuririri* recalling Goldcrest, but more prolonged. Song recalls Winter Wren, but weaker and shorter: *pichi pii pii chii chii chiririri* falling in pitch to final flourish.

ASIAN GLOSSY STARLING TJ
Aplonis panayensis

L 17–20cm. **SD** Disjunctly from Assam to Indonesia. Accidental early spring (Mar/Apr) S Japan and Lanyu I., Taiwan; race unknown. **HH** Woodland and forest edges; introduced Taiwan, where local but fairly numerous in urban areas and parks. **ID** Mid-sized, plain, black starling. Ad. entirely deep glossy black with metallic green-glossed face, neck and breast, and purple gloss to remaining plumage. Juv. blackish-brown above with some green gloss, buff below with heavy dark streaking. **BP** Bill blunt, more crow-like than starling-like, black; eyes large, irides bright red (ad.) or dull orange (juv.); tarsi blackish-grey. **Vo** A Common Starling-like *chank* and shrill *sreep*.

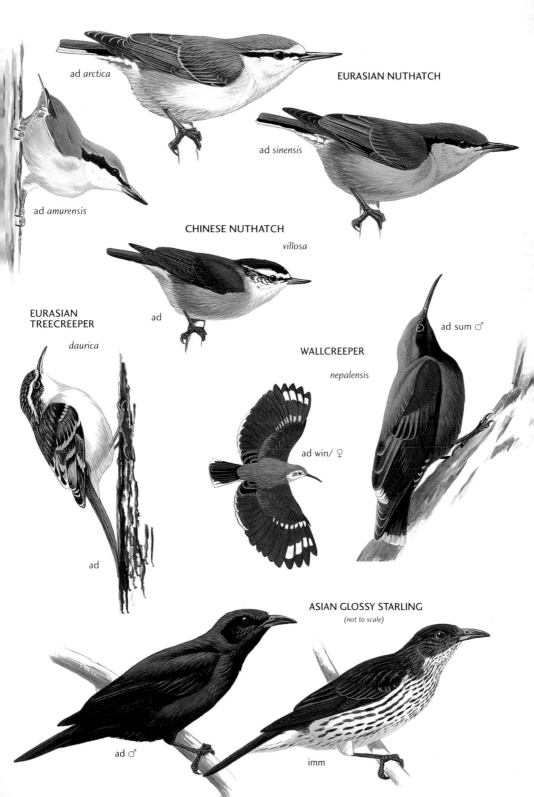

ad *arctica*

EURASIAN NUTHATCH

ad *sinensis*

ad *amurensis*

CHINESE NUTHATCH

villosa

ad

EURASIAN TREECREEPER

daurica

WALLCREEPER

nepalensis

ad sum ♂

ad win/ ♀

ad

ASIAN GLOSSY STARLING
(not to scale)

ad ♂

imm

GREAT MYNA   T
Acridotheres grandis

L 23–25cm. **SD** Assam to Indochina. Introduced and breeding locally on Taiwan. **HH** Open areas including parks and gardens. **ID** Like large Jungle Myna, with erectile forecrown crest and variable bare skin around eye; upperparts uniform blackish-grey; wings black with prominent white patch at base of primaries (visible on upper- and underwing in flight); tail black with broad white tip; underparts uniform dark grey, contrasting strongly with white vent, undertail-coverts and undertail. **BP** Bill bright yellow; eyes dark reddish-brown; strong tarsi bright yellow. **Vo** An explosive, shrill *skree'e'ch*, high-pitched *chuur-chuur*, harsh *kaar* and soft *piu*. Song a varied jumble of repetitive phrases comprising harsh, chortling and shrill notes. **AN** White-vented Myna; source of much confusion with Javan Myna *A. javanicus*

CRESTED MYNA   CTJ
Acridotheres cristatellus

L 25.5–27.5cm; WT 113g. **SD** Southern China to N SE Asia. *A. c. cristatellus* across E Indochina and S & SE China north to Changjiang River; *A. c. formosanus* a rare local resident of lowland Taiwan, but more numerous on Kinmen (introduced Javan Myna much commoner on Taiwan mainland and commonly mistaken for Crested); nominate established locally elsewhere, e.g. in Tokyo and Beijing. **HH** Open areas with scrub, around cultivation and suburban parks; forages in small (Taiwan) or large (Kinmen) flocks, largely on agricultural land and often alongside introduced species. **ID** Large, mostly black, starling-like bird with unusual short bushy crest at base of bill. Entirely black except white panel at base of primaries, white scalloping on vent (albeit rather indistinct and difficult to see), white corners to rounded tail and narrow tips to rectrices. In flight shows short, broad white primary bar. **BP** Bill pale, ivory-coloured or even pale greenish-straw, with orange base to lower mandible; eyes prominent in black face, irides bright orange; legs dull yellow-orange. **Vo** Hard *kyuru kyuru* or *kyutt kyutt* notes.

JAVAN MYNA   TJ
Acridotheres javanicus

L 23–25cm. **SD** Endemic to Java, but introduced to C Honshu, Japan, and Taiwan, where feral populations now established; common in Taiwan. Monotypic. **HH** Agricultural land, parks and gardens. **ID** Mid-sized rather dark-grey myna, with bright orange bill and legs, and short, curly black crest above bill. Head, neck and wings black; underparts, mantle, back, scapulars, rump and tail dark grey; white ventral area small. Wings have broad white bases to primaries, and broad white tips to outer tail-feathers. In flight, white on wings, vent and tail diagnostic. Differs from Crested in being slightly smaller with shorter crest, greyer body, larger white tail patches, narrower white wing spots and yellower bill. **BP** Bill bright orange; eyes white to pale yellow (pale blue in young); tarsi yellowish-orange (pale yellowish-flesh in young). **Vo** A guttural chattering *kyuru kyuru*. Song complex *gyugyu kiru-rikirurukyororii*. **AN** White-vented Myna (see Great Myna).

JUNGLE MYNA   T
Acridotheres fuscus

L 24.5–25cm. **SD** Ranges across Indian subcontinent, from Pakistan to Malay Peninsula. Introduced and breeding locally on Taiwan. **HH** Occurs in open grassy areas, including farmland, riversides and roadsides. **ID** A grey myna, recalling Javan, but less black and has orange/yellow bill and eyes, and very short crest. Overall plumage grey with black lores, brownish-black wings, black flight-feathers with white patch at base of outer primaries. Tail black with white tips to outer feathers. Underparts mid to pale grey, dull white vent, white undertail-coverts and undertail. In flight, shows only small white wing patch. **BP** Bill orange, blue-grey at base; eyes yellowish-white to pale orange; tarsi dull yellow-orange. **Vo** A repetitive *tiuck-tiuck-tiuck* and high-pitched *tchieu-tchieu*.

BANK MYNA   TJ
Acridotheres ginginianus

L 23–25cm. **SD** Native of S Asia, found locally in Japan and Taiwan (also Guangdong and Hong Kong), though records presumably relate to escaped or released birds, or feral populations. Monotypic. **HH** Villages and towns with trees, parks and gardens. **ID** A blue-grey myna with black cap and cheeks. Resembles Common Myna, but black of head extends only to face, and has short crest on forehead. Chin, throat, breast, mantle, scapulars, back and rump all dark grey, whilst belly and vent are buff-brown. Wings and tail black and outer tail tips yellowish-buff. In flight, shows orange-buff patch at base of primaries, yellowish-buff tail tips and border, and orange-buff underwing. **BP** Bill dull yellow; bare skin around eyes reddish-orange, eyes brown; tarsi dull dark yellow. **Vo** Croaks, clucks and screeching notes as well as whistles, warbles and mimicry.

COMMON MYNA   TJ
Acridotheres tristis

L 23–25cm; WT 106g. **SD** *A. t. tristis* widespread in S & SE Asia, ranging north to S China, but also a frequent cagebird that has escaped and become established in some areas, e.g., in E China, C Japan and Taiwan. **HH** Agricultural areas and villages, also gardens and parks in towns, where forages on ground. **ID** Fairly large brown myna, with blackish-brown hood, wings and tail. Underparts grey-brown, except white lower belly, vent and undertail-coverts. In flight, shows prominent broad white patch at base of primaries, white tail tips and white underwing. **BP** Bill yellow; bare skin around eyes yellow, irides orange-brown; tarsi orange-yellow. **Vo** Both weak and harsh scolding notes (*chake chake*), whilst song combines skilled mimicry with tuneless chattering, gurgling and whistling notes: *hee hee chirk-a chirk-a chirk-a....* **AN** Indian Myna.

GREAT MYNA

ad

CRESTED MYNA

cristatellus

ad

JAVAN MYNA

ad

ad

JUNGLE MYNA

ad

BANK MYNA

ad

COMMON MYNA

tristis

ad

BLACK-COLLARED STARLING   CT
Sturnus nigricollis

L 28cm. **SD** SE Asia and S China east to Fujian. Established exotic in low-lying areas of Taiwan, and common resident on Kinmen and Matsu (Taiwan), off Fujian coast. Monotypic. **HH** Forages on open ground, typically in farmland, often around domestic animals. **ID** A large, boldly pied, myna-like starling. Mainly white, it has a white head with yellowish eye patches and ear-coverts, a broad black collar and a black mantle, scapulars and wings, the latter with white fringes to most feathers. Rump white, tail black with sides and rounded tip white; underparts white. **BP** Bill black; bare skin around eyes yellow, irides dark brown; tarsi very pale grey. **Vo** Very vocal. Discordant and harsh, e.g. a loud *werr* and bee-eater-like purred whistle *prrü*; also soft, piping *pü pü-pü-pü* recalling a hesitant Rufous-capped Babbler. Song a short flourish of frizzling notes *tcheeuw-tchew-trieuw*.

ASIAN PIED STARLING   TJ
Sturnus contra

L 23cm. **SD** Resident in S Asia from India to Indonesia. Introduced (probably various races) with established feral populations in Taiwan and C Honshu. **HH** Agricultural areas with trees, damp grasslands and around habitation. Also parks and gardens in residential areas. **ID** Mid-sized black and pale buff starling. Ad. upperparts glossy black from crown to lower back, also tail and wings (with white band across scapulars), rump and uppertail-coverts pale greyish-buff; lores and ear-coverts dusky off-white; chin, throat and upper breast/neck-sides glossy black, with green-glossed head; breast, flanks and undertail-coverts pale greyish-buff. Juv. dull brown where ad. black, with paler bill. **BP** Bill long, sharply pointed, yellow with red at base of lower mandible; large bare red orbital patch, irides off-white (ad.) or pale brown (young); tarsi yellow. **Vo** Various high-pitched musical notes, a chattering *cheek-cheurk* and descending *treek-treek-treek*.

DAURIAN STARLING   CTKJR
Sturnus sturninus

L 16–19cm. **SD** Fairly common, E Asian starling, breeding across NE China, N & C Korea and S Russian Far East to Ussuriland; winters in S China and southern SE Asia. Migrant through E China, a scarce migrant offshore Japan and Korea; has reached Sakhalin and accidental Taiwan. Monotypic. **HH** Typically in groves of trees, woodland and forest edges, also rural and semi-rural areas, around cultivation and villages. **ID** Small, rather neat, grey, black and buff starling. ♂ has plain face with head and entire underparts mid grey; rear crown patch and mantle black with metallic violet/purple iridescence, wings and tail black with green gloss; at rest, broad white bar on scapulars, white tips to median coverts and orange-brown bases to primaries; rump, vent and undertail-coverts cinnamon-buff; belly greyish-buff. ♀ has less distinct wing pattern, is browner where

♂ is black, and lacks gloss. Juv. Daurian and Chestnut-cheeked very similar, but former usually has white scapular bar and pale tips to tertials. **BP** Bill fine-pointed, black; eyes prominent in plain face, irides black; tarsi dull greyish-brown. **Vo** A guttural *kyuru kyuru* similar to Chestnut-cheeked Starling; song comprises various trills and whistles interrupted by mimicry of other species. **AN** Purple-backed Starling.

CHESTNUT-CHEEKED STARLING   CTKJR
Sturnus philippensis

L 17–19cm. **SD** Range-restricted E Asian migrant, breeding only from C Honshu to Hokkaido, S Sakhalin and S Kuril Is (Apr–Oct); winters in SE Asia, Philippines. Scarce to rare migrant Korea, Taiwan, E China. Monotypic. **HH** Typically around groves of trees in rural and semi-rural areas, cultivation and villages. **ID** Small, rather dusky, grey, black and buff starling. ♂ has head creamy white, with plain face, but reddish-chestnut patch on ear-coverts and neck-sides; nape grey, mantle and scapulars black, with violet- or purple-glossed scapulars, mantle and upper back; wings strongly green glossed; grey lower back and buff rump contrast with black tail. Underparts creamy on throat and upper breast, dark to pale grey from breast to vent. Grey-white scapular bar and grey-white panel near base of primaries. ♀ much plainer grey-buff, head/face plain, mantle brown, wings brown with indistinct pattern, lower back and rump buff. **BP** Bill fine-pointed, black; eyes prominent in plain face, irides black; legs blackish-grey. **Vo** A harsh *jee* and throaty *gyuru gyuru*. Song a rapid and complex series of notes: *pyuu-kirurip-kyuru-kyuru-gyiip*. **AN** Violet-backed Starling; Red-cheeked Starling.

WHITE-SHOULDERED STARLING   CTKJR
Sturnus sinensis

L 18–20cm. **SD** Range-restricted Asian migrant, breeding only in Vietnam, S & SE China and wintering in Taiwan, SE Asia and Philippines. Accidental offshore Korea, Ussuriland, Honshu, Shikoku, scarce annual winter visitor Nansei Shoto, and migrant or winter visitor to S Kyushu. Monotypic. **HH** Mostly in dry farmland with isolated stands of trees; often perches on wires. **ID** A pale, grey, black and white starling. ♂ almost white-headed, becoming pale grey from nape and ear-coverts to darker grey on back; lower back pale grey, rump pale buff; chin/throat white grading to grey on underparts; wings and tail black, with prominent white shoulder patch and band on tail tip. ♀ has darker brownish-grey head and upperparts, wings and tail browner with reduced white shoulder and only white corners to tail. White shoulders and tail pattern distinctive in flight. Juv. like ♀, but lacks white shoulder. **BP** Bill fine-pointed, whitish-grey; eyes appear small and indistinct in plain face, irides white or bluish-white; legs blue-grey. **Vo** A throaty chattering *kyuru kyururii*, *kaar* or *gi gi* similar to Chestnut-cheeked Starling, and soft *preep* when flushed. Calls generally higher pitched than other starlings of region. **AN** Chinese Starling; Grey-backed Starling.

BLACK-COLLARED
STARLING

juv

ad

ASIAN PIED
STARLING

ad

ad ♀

ad ♂

DAURIAN
STARLING

ad ♂

WHITE-SHOULDERED
STARLING

ad ♀

CHESTNUT-CHEEKED
STARLING

ad ♀

ad ♂

ROSY STARLING CKJ
Pastor roseus

L 19–24cm; WT 67–88g. **SD** E Europe to C & W Asia, wintering S & SW Asia, but has strayed to E China (Shanghai), S Japan and Korea. Monotypic. **HH** Rather arid agricultural or steppe areas, often near cliffs or rocky areas, also settled environments, foraging over dry grasslands and agricultural land. **ID** ♂'s unique shaggy crest and glossy black and pink breeding plumage make it unmistakable. Crest, hood, wings, vent, rump and tail all glossy black (purplish on head, blue-green on wings), remaining plumage pale buffy-pink. Non-breeding and 1st-winter duller, with much of pink obscured by buff fringes. ♀ similar to ♂ but duller. Juv. in autumn (more likely to wander out of range), plain sandy-brown, paler than juv. Common Starling, with broad pale fringes to wing-feathers, and stout bill. **BP** Bill rather thick, pink with black base (breeding ad.) or yellow-horn with grey culmen (juv./1st-winter); eyes black; tarsi pink (ad.) or dull yellowish-orange (juv.). **Vo** Rather variable calls, but generally short, harsh and rasping, some similar to Common Starling: *kyururi, chit,* or *baaht.* **TN** Formerly placed in *Sturnus.*

CHESTNUT-TAILED STARLING T
Sturnus malabaricus

L 21cm. **SD** Ranges across Indian subcontinent to mainland SE Asia and S China. Introduced (race uncertain) and established locally on Taiwan (e.g. Kaohsiung), but genuine vagrancy by *S. m. nemoricola* (S China to Kinmen I.) also likely. **HH** Open wooded areas. **ID** Ad. has pale grey to silver-grey head with loose, pointed plumes on crown and neck (darker shaft-streaks of other subspecies absent in *S. m. nemoricola*); upperparts darker plain grey, blackish-grey on wings and tail, latter with chestnut outer feathers. Underparts warm rufous-brown to brick-orange. Juv. plainer, duller grey-brown with rufous-brown sides and tips to tail. **BP** Bill yellow at tip, blue at base; eyes white; tarsi dull pink. **Vo** Sharp disyllabic metallic notes, tremulous whistles and harsh rasping notes. Song comprises rambling warbles, chortles and squeaky churrs.

RED-BILLED STARLING CTKJ
Sturnus sericeus

L 24cm. **SD** Resident or local migrant across much of S & E China, some moving south to Vietnam, occasionally Laos, Thailand and Philippines, for winter. Common in China in winter, with some large roosts reported. Scarce Taiwan and scarce to rare migrant/winter visitor (Nov–Apr) in Nansei Shoto and Kyushu, Japan (sometimes in small flocks); rare migrant Korea. Monotypic. **HH** Lowland agricultural areas and open areas with scrub. **ID** A large starling. ♂ has distinctive pale hood, yellowish on crown, face and head-sides, and white on chin, throat and upper breast; upperparts mid slate-grey, blackest at edge of pale hood, rump pale grey, wings black with white panel at base of primaries, black tail slightly rounded at tip. ♀ pale buffy-grey, white on chin/throat, darker on mantle, very pale on rump; wings and tail as ♂. In flight note pale head, white wing patch and contrast between pale rump and black tail. **BP** Bill long, slender and sharply pointed, red with grey culmen and black tip; eyes black; legs bright red-orange (♂) or dull orange (♀). **Vo** A chattering *jree-eep or zhree-eep.* **AN** Silky Starling.

WHITE-CHEEKED STARLING CTKJR
Sturnus cineraceus

L 22–24cm. **SD** Widespread in E Asia. Resident (or local migrant) throughout much of Japan, Korea, and Transbaikalia to Shandong; summer visitor to NE China, S Russian Far East, Hokkaido, Sakhalin, C & S Kuril Is, and winter visitor to Nansei Shoto and most of E & SE China, also Taiwan south to SE Asia and Philippines. Monotypic. **HH** Favours deciduous forest fringes, groves of trees and habitation, often in urban areas, also agricultural land. Common; highly gregarious pre- and post-breeding, and in winter. **ID** A large, dark grey and brown starling. Hood blackish, but forehead, chin and cheeks white with some dark streaking; upper- and underparts, wings and tail mid brown, with white rump, white fringes to secondaries and tips to tail-feathers. ♀ slightly duller, young plainer still and greyer but has pale off-white cheeks. In flight, distinctive starling structure combined with white rump, tail tips and underwing-coverts, and pale panel in secondaries diagnostic. Some ads. have much whiter heads. **BP** Orange bill black-tipped in ad., all dull orange in young; eyes distinct on white face, irides black; legs deep orange, dull brownish-orange in young. **Vo** A highly social and vocal bird, with families and flocks maintaining contact using various throaty chattering, even creaking sounds: *chir-chir-chay-cheet-cheet, gyuru, gyee* and *chi chi,* but commonest call a harsh *jah.* **AN** Grey Starling.

COMMON STARLING CTKJR
Sturnus vulgaris

L 20–23cm; WT 82g. **SD** Widespread resident in Europe (and N America, where introduced) and summer visitor around Lake Baikal, mid-Lena River and W China. In region, *S. v. poltaratskyi* strays east to coastal China (as far as Hebei), a scarce/rare winter visitor (Nov–Apr) to offshore, S & W Japan, offshore and coastal China, and Taiwan; has also reached Sakhalin. **HH** Habitation, often urban areas, also agricultural land; in region, many records involve birds consorting with White-cheeked Starlings. **ID** A generally dark starling, sexes similar but breeding and non-breeding plumages differ. Ad. breeding black with strong oily or metallic purple/green iridescence, especially on neck/breast; wings and tail black with brown fringes to feathers; some fine spotting, buff or grey-brown tips to scapulars, and some spots on nape, mantle and flanks. Non-breeding moults into fresh plumage, largely with pale buff tips, so in winter appears heavily spotted, forehead almost entirely pale buff, head with tiny spots, lores black, underparts finely spotted white, upperparts more coarsely so with buff; mantle in winter has green sheen in sunlight. Juv. drab, mid brown with pale throat; differs from Rosy Starling in less prominently fringed wing-feathers and duller, grey bill. In flight, broad-based rather triangular wings and shortish blunt tail give distinctive shape. **BP** Bill, straight, fine-pointed, yellow in breeding season, dull grey-black at other times, brownish in young; black eyes indistinct in dark face; tarsi dark orange in summer, dull brown-orange in winter. **Vo** Varied guttural chattering notes *gyaa, gyuru* and *gyeee.* Often gives high-pitched 'radio-tuning' whistles as subsong during winter. **AN** European Starling.

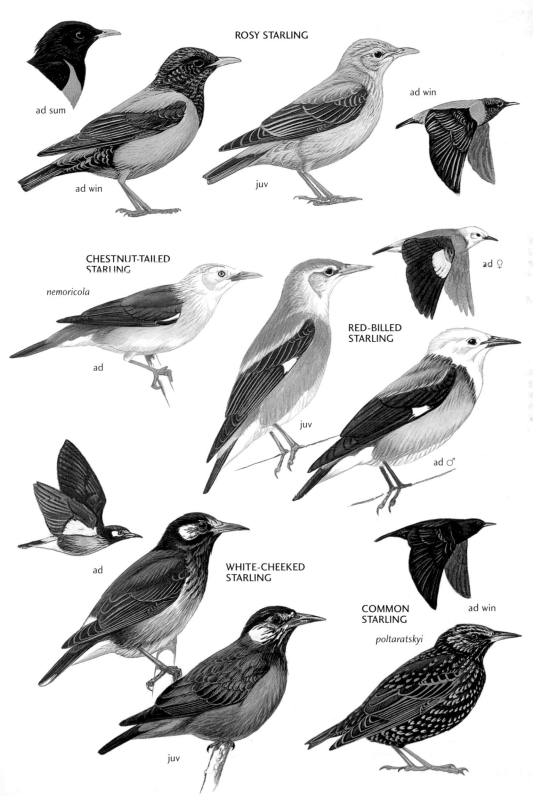

ROSY STARLING

ad sum

ad win

ad win

juv

ad win

CHESTNUT-TAILED STARLING

nemoricola

ad

ad ♀

RED-BILLED STARLING

juv

ad ♂

ad

WHITE-CHEEKED STARLING

ad win

COMMON STARLING

poltaratskyi

juv

BLUE WHISTLING THRUSH CT
Myophonus caeruleus

L 29–35cm; WT 136–231g. **SD** C & SE Asia, across China to E coast and northeast to Bohai Gulf. Accidental to Taiwan and resident on Matsu. Subspecies in region *M. c. caeruleus*. **HH** Hill forests near streams and rivers, at moderate to high altitudes, foraging on ground and at edges of fast-flowing rivers. **ID** Very large, entirely blue-black thrush (largest in region), with violet- and silvery-spotted upperparts from crown to rump, royal-blue lesser wing-coverts, silver-grey tips to median coverts and purplish-blue gloss to wings and tail; underparts black with extensive metallic blue tips forming scaled pattern. **BP** Bill rather stout, black; eyes reddish-orange to dark brown; long tarsi black. **Vo** A sharp piercing whistle, harsh *dzhee* and in alarm a loud screeching *eer-ee-ee*, *kreee* or *scree*. Song, given year-round, from high perch or, occasionally, in flight, is a rather disjointed series of loud, beautiful whistles interspersed by harsh or grating notes, terminating in a rising *tzeeet-tzuit-tzuit-zuit-zuit*.

TAIWAN WHISTLING THRUSH T
Myophonus insularis

L 28–30cm. **SD** Endemic to low and mid-level forests (0–2,700m) on Taiwan. Monotypic. **HH** Mainly in damp areas, typically near streams and rivers amid dense broadleaf evergreen forest, where forages on ground. Also near human structures, where often nests. Flicks tail up, then slowly lowers and fans it, especially while calling. **ID** Large but rather slim, glossy, blue-black thrush. Mostly black, strongly glossed metallic blue on forehead, lesser coverts and tail; throat, neck-sides, breast and flanks heavily scaled metallic blue. **BP** Bill strong, black; eyes bright deep red or reddish-brown; legs rather long, tarsi black. **Vo** A screeched, rising *zhreeeeeh, tssseeoo* or *sui yi*, often uttered at first light. Song a short set of loud, lushly melodious rising and falling whistles, with high-pitched piping notes and short trills.

ORANGE-HEADED THRUSH CTK
Zoothera citrina

L 20–23cm; WT 47–67g. **SD** Across S & SE Asia and SE China to Indonesia. *Z. c. melli* in S China east to Guangdong and Fujian, reaching northeast to Hebei coast on migration; escapees Taiwan; reported Korea. Uncommon resident and partial migrant; migratory habits make it likely to appear beyond normal breeding range. **HH** A shy bird of dense ground cover in moist deciduous and evergreen forests, also thickets, bamboo groves and plantations. Crepuscular. **ID** Brightly coloured ground thrush. ♂ (*Z. c. melli*) has head and chest warm orange, face paler orange-buff with two vertical dark blackish-brown subocular and auricular bars. Mantle, wings and tail bluish-grey. White median coverts form wing patch; belly and vent also white. ♀ duller with upperparts browner or more olive-brown, and wings warm brown. **BP** Bill dark grey; eyes large, irides black; tarsi brownish-pink. **Vo** A thin *tzzeet*, subdued *tjuck*, and screeching *teer-teer-teerr* in alarm. Attractive whistled song comprises series of loud, sweet, lilting phrases, recalling Eurasian Blackbird or Song Thrush, with some phrases repeated and incorporating rich trills and mimicry. Differences of *Z. c. melli* song from nominate (of NE India) unclear; nominate described as: *wheeur-tee-lee wheeper-pree-pree-pelee wheeper pree-pree-pelee rhi-ti-li-tili-wheeper-wo-wheeper-wo kwir-loo-twi-lili kwir-loo-twi-lili*.

SIBERIAN THRUSH CTKJR
Zoothera sibirica

L 20.5–23cm; WT 60–72g. **SD** Breeds from C Siberia to Russian Far East, wintering in SE Asia. *Z. s. sibirica* occurs from Yenisei and Baikal region to N Sea of Okhotsk, S Russian Far East and NE China; *Z. s. davisoni* restricted to Sakhalin, S Kuril Is, and N & C Japan (Apr–Sep). Scarce migrant Korea, accidental Taiwan. Both subspecies reported E China and Korea, but only *Z. s. davisoni* in Japan. **HH** Mixed deciduous/evergreen broadleaf and coniferous montane forest at lower latitudes, in lower taiga forest further north. **ID** Mid-sized thrush with very dark plumage. Ad. ♂ deep dark grey, blackest on head and throat, with long, broad silvery-white supercilium. Some narrow white scaling on flanks, lower belly white, vent/undertail-coverts slate scaled white. 1st-year ♂ less black above head, has less crisp yellow-buff supercilium, some buff on ear-coverts and throat, and retains juv. body-feathers, hence duller grey overall; pale brown spots often on crown, throat, breast and belly, and brownish feathers retained in greater wing-coverts and remiges. ♀ warm olive-brown with pale, yellowish-buff supercilium, submoustachial and chin. Upperparts plain brown, underparts pale buff but so heavily scaled dark brown as to appear dark overall. On take-off reveals white corners to tail. In flight, double white bar on slaty-black underwing diagnostic (like White's Thrush but more contrasting, see p.400). *Z. s. davisoni* larger and darker than nominate, ♂ almost entirely blue-black with more contrasting silver-white supercilium and only narrow white tips to undertail-coverts; ♀ darker olive-brown. **BP** Bill black in ♂, grey-tipped with yellow base in ♀; eyes dark brown; tarsi bright yellow. **Vo** Typical call a high, thin *tsi*, rather similar to Elegant Bunting, or sharper *zit*, but in flight/on migration a wavering sibilant quite drawn-out *siiiiiih*. Sometimes a loud *chirr* or rattling *chrssss* in alarm, and more typical thrush-like *jack*. Song, given from high perch, very simple, slow, fluty two-note whistled phrases, given rather hesitantly and commonly followed by a fainter more sibilant and higher pitched note: *kiron-tsee, kyoro-tsuii* or *kyoron chii*, also described as *hueee-tirrrr... heeeuu-tirrr*. Also *tvee-tring tvee-tryu tvee-kvee tvee-kwi-tring yui'i-tss*. Occasionally includes other wavering notes: *tyui'i – tss yui'i-tss* or *tss sss ss*.

ad foraging
in stream

**BLUE WHISTLING
THRUSH**

ad

caeruleus

ad
in sunlight

ad ♀

ad foraging
in stream

ad

**ORANGE-
HEADED THRUSH**

TAIWAN WHISTLING THRUST

melli

ad ♂

ad ♀ foraging
in leaflitter

juv

1 st-year ♂
sibirica

ad ♂
sibirica

**SIBERIAN
THRUSH**

ad ♀

ad ♂ *sibirica*

ad ♂ *davisoni*

ad ♂ *sibirica*

ad ♂ *davisoni*

WHITE'S THRUSH *Zoothera dauma* CTKJR

L 24–30cm; WT 88–103g. **SD** Breeds from Urals to Sakhalin, Korea and Japan, and from Himalayas to Indonesia and Philippines; winters NE India, S China and SE Asia. Common migrant Chinese coast, summer visitor N China and Korea, but many winter in C & S Japan, and resident Taiwan. *Z. d. aurea* E European Russia to N Mongolia, NE China to Sea of Okhotsk, wintering to S China, Taiwan and SE Asia. *Z. d. toratugumi* (questionable taxon) Amurland, Korea to Sakhalin, S Kuril Is and Japan, wintering to E China and Taiwan. Two indeterminate or undescribed taxa (mistakenly ascribed to *Z. d. horsfieldi*) resident on Iriomote-jima (Yaeyama Is) and Taiwan (rare); these may be referrable to *Z. d. dauma*. Frequently treated as two species, **White's Thrush** *Z. aurea* (including *aurea* and *toratugumi*) in N, and **Scaly Thrush** *Z. dauma* (various subspecies), in S, but validity of subspecies limits uncertain. **HH** Dark taiga-forest, wooded steppe and mature mixed and montane forests (500–1,600m). Winters in range of forest types, wooded parks and gardens. Frequents damp shady areas, where feeds quietly on forest floor; walks rather than hops like other thrushes; has curious double-bobbing creeping gait. **ID** Very large thrush, with proportionately short tail. Sexes alike. Unmistakable, golden olive-brown covered with black crescentic scales. Upperparts darker, underparts paler, whiter on belly and vent; wings and tail plain brown. Face varies from pale and 'open' to distinctly barred, with pale lores and eye-ring, and black malar, subocular 'teardrop' and ear-coverts crescent. Primary-coverts black-tipped, median/greater coverts buff-tipped, affording pattern to otherwise plain wings. On take-off reveals white corners to tail (like Mistle Thrush, p.410), and has white axillaries and boldly barred black and white underwing. White's (*aurea*/*toratugumi*) slightly larger on average, with stronger, more rounded scaling, longer bill (with white to flesh-pink base to lower mandible), stronger eye-ring and auricular crescent than Scaly, but not always separable in field. May show paler face and yellowish centres (to black-tipped crescents) of mantle, back and inner scapulars; in hand, tail has 14 feathers. S Japanese and Taiwanese taxa smaller, darker and seemingly resident in restricted ranges; voice undescribed. **BP** Bill rather long compared with other thrushes, stout, dark grey at tip, horn at base; eyes large with narrow white orbital, irides black; tarsi flesh-pink. **Vo** A hard *ga ga*, sometimes *gyororururu* or raspy *rraattchh* in alarm, and low chuckling note on flushing. Song (*aurea*/*toratugumi*), typically given at night or dawn, a slow, soft, mournful and drawn-out whistle, repeated after several seconds, and may solo or duet, two whistled notes, one significantly higher pitched: *hyeeee jeweeee* or *twooo-chuooo*; also *huuwiiiii wiiiuuuu piii-yuuu*.

AMAMI THRUSH J
Zoothera major

Critically Endangered. L 29–30cm. **SD** Amami I., Japan. Monotypic. **HH** Undisturbed, dark, broadleaf evergreen primary forest at 100–400m; active dawn and dusk. Lacks curious double-bobbing creeping gait of White's Thrush, instead moves like *Pluvialis* plover, with fast runs punctuated by abrupt pauses to listen for worms. **ID** Closely resembles White's Thrush, slightly larger than Scaly race *dauma* but not White's *aurea*/*toratugumi*; face paler with subtly whiter malar, darker narrow throat-stripe, more prominent black bar below eye and at rear of ear-coverts, gold spots on crown larger than in Scaly race. Has fewer (12,

not 14) tail-feathers, and different song. In flight, reveals white corners to tail and bold white bar on blackish underwing as in White's. **BP** Bill long, stout, deeper than White's, dark grey at tip, yellow at base of lower mandible; eyes large, surrounded by bare pink skin, particularly at rear, irides black; tarsi brownish-orange or brownish-pink. **Vo** A thin flat *tsih*, lower and more piercing, less wavering than Pale Thrush, also whispery *sih* and very high, thin, descending *tseeooo* like White's in alarm. Song, given mainly pre-dawn, especially in late Feb–Mar, more closely resembles Japanese Thrush and is pleasantly musical, quite unlike White's: a 3–4-part series of melodic whistles usually ending in a flourish, *chieli... cheelü... tridüli-tsi!* or *tsuii chee kyoronchee*. Terminal flourish sounds like speeded-up Siberian's song. Occasionally a breezy, high *treer* note is inserted. **TN** Formerly within White's Thrush.

VARIED THRUSH *Ixoreus naevius* R

L 21.5–24cm; WT 65–100g. **SD** Migratory, ranging from NW California to Alaska and NW Canada; accidental Wrangel I. (probably *I. n. meruloides*). **HH** In native range, dark, damp conifer forests. **ID** Largely orange, black and grey. ♂ has slate-grey crown, orange postocular supercilium, black bar from bill to neck-sides, merging with breast-band; mantle, scapulars, rump and tail dark grey; double orange wingbar and orange bases to primaries and outer fringes of secondaries and tertials, wings otherwise largely black. Orange of chin/throat separated from that of lower breast/belly by broad black breast-band; flanks scaled grey, vent white. ♀ has orange areas paler, brown not grey upperparts, and brown not black breast-band. **BP** Bill black with yellow base to lower mandible; eyes black; tarsi pink. **Vo** A dry *chup*, soft *tiup* and short humming whistle in flight.

GREY-CHEEKED THRUSH JR
Catharus minimus

L 17–18.5cm; WT 26–50g. **SD** Primarily breeds across N America, wintering in S America, but *C. m. aliciae* also nests in Chukotka west to Kolyma R and south to Penzhina River. Vagrant in autumn in W Aleutians and has reached offshore Japan (Hegura-jima). **HH** In summer favours low trees and bushes in forest tundra and at upper limits of montane taiga. On migration visits range of wooded areas, but usually those with a heavy shrub layer. **ID** Small, greyish-olive-brown thrush with very plain, cold grey face, broken pale grey eye-crescents and pale throat bordered below by loose necklace of dark spots extending to breast. Underparts off-white, with extensive olive-grey on flanks. At rest, long primary projection beyond tertials. Tail, often cocked, long, dull brown. In flight, typical *Catharus* underwing, with two white wingbars contrasting with blackish central bar and dark flight-feathers, recalling Siberian Thrush (p.398). **BP** Bill yellowish-brown below with dark grey tip and culmen; black eyes appear large in plain face, with indistinct pale grey eye-ring; long tarsi dull pinkish-brown. **Vo** A shrill, drawn-out *tsiiew* or thin *seee*; flight call high-pitched, a penetrating, nasal *jee-er*. Song a high-pitched series of descending notes, commencing with a slowly repeated *wee-oh* or *chook-chook*, then a staccato *chee-chee*, before ending with *wee-oh wee-oh* or *wheee-e*.

WHITE'S THRUSH

ad *aurea*

ad *aurea*

1st-win
aurea

ad *dauma*
(to which Taiwan and Yaeyama
birds may be referrable)

ad *aurea*
foraging

ad *toratugumi*

AMAMI THRUSH

ad

ad Amami

ad ♂

GREY-CHEEKED
THRUSH

aliciae

1st-year

VARIED
THRUSH

meruloides

1st year ♂

ad

ad ♂

ad ♂

ad ♀

ad ♂

SWAINSON'S THRUSH R
Catharus ustulatus

L 16–20cm; WT 25–45g. **SD** Breeds across N America, from Alaska to Newfoundland, and winters C & W S America. Vagrant Aug–Sep St Lawrence I. Accidental to Ussuriland. **HH** Mainly coniferous forests. **ID** Small greyish-olive-brown thrush with very plain grey face. Pale buff throat bordered below by loose necklace of dark brown spots on neck and breast. Underparts greyish-buff, white on vent. Closely resembles Grey-cheeked (p.400), but has buffy-cream face, browner (less grey) cheeks and darker up-perparts. **BP** Bill black with pale pink base to lower mandible; eyes dark brown, with buffy-cream eye-ring; tarsi dull pink. **Vo** A mellow *whit*, emphatic *fweep* and double *whit-burrr*.

HERMIT THRUSH R
Catharus guttatus

L 16–18cm; WT 18–37g. **SD** Breeds across North America, from Alaska to Newfoundland, wintering south to C America. Vagrant Aug–Sep St Lawrence I. (USA). Accidental to Chukotka (presumably *C. g. guttatus*). **HH** Range of forest types from taiga to arctic-alpine, occurring particularly at clearings and edges. **ID** A small *Catharus* with rusty-red rump, uppertail-coverts and tail. Face rather plain, with complete white eye-ring, upperparts including wings cool, grey-brown tinged olive, but warmer buff fringes to greater coverts and rusty fringes to closed primaries make it appear brighter than slightly larger Grey-cheeked Thrush. Short primary projection. Underparts cleaner and brighter than Grey-cheeked, whiter from chin to belly, with large dark but sparse spots; flanks unmarked plain grey; undertail-coverts white. Tail often raised upwards then slowly lowered; wings often flicked. In flight, typical *Catharus* underwing pattern (recalling larger *Zoothera* thrushes), with two buff bars. **BP** Bill dark grey on culmen and tip with pale yellow base to lower mandible; black eyes in white facial area; long tarsi dull pinkish-brown. **Vo** A soft, dry single or double *chuck* or *chup*, also a rising whiney *weee* or *zhweeee*, and in flight a whistled *peew*.

GREY-BACKED THRUSH CTKJR
Turdus hortulorum

L 20–23cm; WT 61–69g. **SD** Endemic to continental E Asia. Breeds from W Sea of Okhotsk south through S Russian Far East and NE China to Korea; winters SE China, N Vietnam and Laos. Common migrant Korea and E China, rare Japan and Taiwan. Monotypic. **HH** Thickets and open mixed deciduous forest, broadleaf evergreen forest and riparian woodland, in hills and low mountains to 1,100m. **ID** Rather pale. Ad. ♂ has blue-grey

hood, including head, neck and chest; mantle, back and rump also grey; wings and tail blackish-grey. Orange-brown breast and flanks, white central belly to vent. Imm. ♂ has streaked upper breast. ♀ has pale to mid olive-brown head and upperparts, and dark blackish-brown wings and tail; underparts creamy white, spotted dark grey or black on throat and breast-sides, with orange-brown wash on breast and flanks. On take-off reveals orange-brown underwing-coverts. **BP** Bill yellow, duller or browner in ♀ especially at tip; narrow yellow eye-ring, irides black; tarsi pinkish-yellow. **Vo** Various *duiitt*, *zwiip* and *shiririip* notes. Chuckling *chuck-chuck* and *tsee* in alarm recall Eurasian Blackbird, but less ringing. Song, from high perch, melodious and similar to Japanese Thrush: *kyorohii-kyorori*. Also series of short, quickly whistled and repeated phrases, followed by higher trilled notes: *tvet-tvet-tvet...qui-qui-qui...tviu-tviu-tviu-tve...tevetii-tevetii-tevetii...kyuu-qwo...tvii-tvii-tvii...trrryuuu trevtii-trevtii-trevtii*.

JAPANESE THRUSH CTKJR
Turdus cardis

L 21–22cm. **SD** Breeds only in Japan (Hokkaido to Kyushu) and outside region in EC China. Winters SE China. Rare Taiwan; accidental Sakhalin, Moneron I., and scarce migrant Korea. Monotypic. **HH** Mature deciduous and evergreen broadleaf forest, from sea level to 1,200m. In winter, forest, woodland and urban parks. **ID** Small, compact thrush. Ad. ♂ almost entirely glossy black, with all-black upperparts, head and breast, but white belly and flanks covered with black arrowhead spots, vent clear white. 1st-winter ♂ similar to ad., but dark grey rather than black, has brownish-black head and breast, and white flecking on breast. 1st-summer is dimorphic, recalling ad. ♂ or like ♀. ♀ plain brown above, creamy below with buffy-orange wash from neck-sides to flanks (and on underwing-coverts), and narrow black malar stripe merging with small black spots on breast to flanks. **BP** Bill yellow; ♂ has prominent yellow eye-ring, irides black; tarsi dull orange or pale pinkish-yellow. **Vo** A thin *tsweee* or *zwii*, a *tsuuu*, hollow *chuk* and chuckling *kyokyokyo* recalling Eurasian Blackbird. Song, from high perch, very attractive comprising varied mix of rich fluting whistles and warbles: *kyoroi kyorokyoro kyokokyoko kokiiko kiiko* or *kyoroon-kyoroon kyoko-kyoko* with phrases often repeated several times in quick succession.

SWAINSON'S THRUSH

1st-year

HERMIT THRUSH
guttatus

1st-year

GREY-BACKED THRUSH

1st-year ♂

ad ♂

ad ♂

ad ♀

ad ♂
Grey-backed

1st-win ♂

ad ♀

1st-sum ♂

JAPANESE THRUSH

1st-sum ♂

1st-sum ♂

ad ♂

ad ♀ / 1st year

ad ♂

PLATE 188: THRUSHES III

EURASIAN BLACKBIRD *Turdus merula* CTKJ

L 23.5–29cm; WT 85–105g. **SD** From British Isles to SE China, wintering to S Europe, Middle East and NW India. Isolated eastern population, including *T. m. mandarinus* of S & SE China, is seemingly colonising NE China and N Korea. This taxon (along with others from Asia) sometimes considered as **Eastern Blackbird** *T. mandarinus*) though position of related taxa (e.g. 'Tibetan Blackbird' *T. maximus*) makes situation unclear. Migratory, with northern birds moving to S China in winter. Scarce winter visitor (Nov–May) to N Korea (where occasional in summer), Taiwan, and rare migrant Korea; accidental Japan. **HH** Wide range of forest types, to 4,000m, but prefers deciduous areas, open woodland and parks, often foraging in open near edges. **ID** Ad. ♂ largely sooty-black. 1st-year ♂'s differ little between races, with dark blackish-brown upperparts, somewhat browner wings and brownish-black underparts, but may show rufous tone to throat-sides. ♀ *mandarinus* dark sooty blackish-brown, with paler underparts, narrow buffish-white throat and rust-brown tinge to breast and flanks; generally lacks spots on lower throat and breast. Juv. dark rufous-brown with some buff streaking and mottling. All the Eastern Palearctic races have longer primary projection and broader-based wings than western forms. Flight heavy and undulating, with head and neck prominent. **BP** Bill yellow (♂) or has dull dark grey-brown tip and culmen and yellow base to lower mandible (♀); narrow eye-ring pale yellow (♂ only), irides black; tarsi black. **Vo** Alarm a loud chattering *plik plik plik*, when excited gives persistent *twink twink twink*. Also chuckling *pyuck pyuck* or *kyott kyott* while feeding, and buzzy *dzeeb* and high thin *sri* flight calls. Song of Eastern is fluty and melodious, with limited repertoire (shorter with less mellow musical whistling than western birds); a series of rather repetitive forced whistles (*piew-piew* or *tieuw-quoit*), often interrupted by almost myna-like chuckles.

ISLAND THRUSH *Turdus poliocephalus* T

L 21–22cm; WT 40–86g. **SD** From Taiwan to Philippines, Borneo and SW Pacific islands with *c*.50 subspecies recognised; **Taiwan Thrush** *T. (p.) niveiceps* highly distinctive, probably merits specific status. **HH** Rare resident of mid-level montane forest, 1,000–2,500m, on Taiwan. In summer (Apr–Jul) often sings from conspicuous perch in canopy; at other seasons shy and generally elusive, but congregates at fruiting trees; occasionally forages on ground and in winter often in small flocks with other thrushes. **ID** Rather small thrush with very pale head. *T. p. niveiceps* shows pronounced sexual dimorphism. Ad. ♂ has white or pale creamy yellow hood, black or dark blackish-brown mantle to rump, tail and wings black. Underparts blackish-brown below white upper breast, with strong rufous-orange wash to flanks, and black-streaked white vent. 1st-year ♂ like ad. but has browner flight-feathers and duller head and underparts. ♀ lacks hood, but has prominent white postocular supercilium, dusky face with narrow, dark lateral throat-stripe and off-white chin and throat; upperparts dark blackish-brown, and underparts more uniform dull orange-brown than ♂. **BP** Bill grey-tipped, yellow-horn; eyes black; tarsi dull yellow. **Vo** Calls very similar to Eurasian Blackbird; a short *tchook* or *tchack* and

higher pitched scolding or chattering. In flight a high *seeee*. Alarm comprises quite loud, staccato chattering notes, *chyuk chyuk chyuk*; *chik-chik-chik* or *chak-chak-chak*. Song a short but rich series of melodious high-pitched whistles and lower fluty notes, increasing in speed from slow start, not as clear or far-carrying as Japanese Thrush.

GREY-SIDED THRUSH *Turdus feae* CT

Vulnerable. L 22–23.5cm. **SD** Scarce NE Chinese species, breeding in region in Hebei and Beijing. Winters SE Asia. Scarce migrant Chinese coast; accidental Taiwan. Monotypic. **HH** Mixed deciduous broadleaf and coniferous forests (1,000–1,900m, but mainly above 1,500m). Winters also at high altitude (mainly 1,500–2,600m), preferring broadleaf evergreen forest. **ID** Rather plain, recalling Eyebrowed and Pale (p.406) Thrushes. Upperparts rather dull earth brown, but white supercilium, black lores, white crescent below eye and white chin recall Eyebrowed. Underparts lack warmth of Eyebrowed. ♂ has white chin, grey throat, dark grey-brown breast, grey belly and flanks, white centre to belly and vent. ♀ has more extensive white on throat, brown breast, belly and flanks. **BP** Bill tip and upper mandible dark grey, lower basally yellow; eyes dark brown; tarsi brownish-pink. **Vo** Flight call a thin *sieee* or *zeeee*, thinner than Eyebrowed; also a crisp rattle. Song a hesitant or jerky series of short, repeated phrases with short trills and richer notes often interspersed by long pauses, e.g. *sit tewau...wet too...chit-to-too*.

EYEBROWED THRUSH *Turdus obscurus* CTKJR

L 21–23cm; WT 61–117g. **SD** Breeds across C Siberia to Kamchatka, Sakhalin; winters from NE India to Indonesia and S Japan to S China. Common migrant Japan, Korea and Taiwan. Monotypic. **HH** Secretive in dark taiga in lowland and montane areas. On migration/winter in mature deciduous and evergreen forest, open woodland and parkland. Shy and flighty. **ID** Resembles slim Brown-headed Thrush (p.406), but has distinctive head pattern. ♂ has blue-grey hood, with prominent white supercilium above black lores, narrow white streak from bill to below eye, short white malar and chin streak. Upperparts mid-brown, darkest on wings; underparts orange-brown, particularly on breast and flanks, and white from central belly to vent. ♀ lacks blue-grey hood; crown, ear-coverts and neck grey-brown, with blackish-grey malar and more extensive white on chin/throat; orange-brown on flanks and upper-breast-sides not as bright. 1st-year ♂ resembles ♀, but has more grey on head and less white on chin/throat. In flight, very pale brownish-grey axillaries and underwing-coverts; also white spots at tail corners. **BP** Bill has blackish tip, grey upper and yellow lower mandible; narrow yellow eye-ring, irides black; tarsi dull brownish-yellow. **Vo** Contact and flight call a thin *zieeh, seee, tsuii* or *seep*, also thin *sip-sip* or *zip-zip*, and occasional deeper chuckling *tuck-tuck* or *kyott-kyott* notes similar to Brown-headed but thinner, sharper and stronger; also a rattled *turrr*. Song simple and resembles Siberian's, but usually comprises three not two notes, but also disjointed simple phrases, *teveteu...trrryutetyuté...trrryutetyutyu* followed by high trills or buzzes.

EURASIAN BLACKBIRD

ad ♂
mandarinus

ad ♂
mandarinus

ad ♂
mandarinus

ad ♀
mandarinus

ISLAND THRUSH

ad ♂
niveiceps

ad ♀
niveiceps

ad ♂

1st-year ♂

ad ♂
niveiceps

GREY-SIDED THRUSH

ad ♂

ad ♂
niveiceps

juv *niveiceps*

EYEBROWED THRUSH

ad ♂

ad ♂

ad ♀

1st-year

ad ♂

PALE THRUSH CTKJR
Turdus pallidus

L 22–23cm; WT 64–90g. **SD** Wide range from Yenisei to Sea of Okhotsk and S Kamchatka. In region breeds from W Sea of Okhotsk coast through Russian Far East, NE China to Korea; winters and migrant (Oct–May) in C & S Japan, S Korea and much of S & E China. Visitor Taiwan. Monotypic. **HH** Breeds in open woodland, both deciduous and coniferous forest. Winters at lower elevations, on wooded hillsides, in forest, wooded areas including parks and gardens, where forages on ground. **ID** Rather drab grey and brown thrush, like dull Brown-headed. ♂ has grey hood, upperparts including tail mid-brown, darkest on wings, almost black on flight-feathers; underparts pale brown below grey hood, paler brownish-buff on belly and sides, central belly and vent white. ♀ has paler upperparts, with white of submoustachial separated from throat by dark malar. In flight, rather plain, except for diagnostic white corners to tail. **BP** Bill has dark grey tip and culmen, dull yellow base and lower mandible; narrow yellow eye-ring, irides black; tarsi pale brownish-yellow. **Vo** Alarm- and contact calls include a discordant thin *tzeee* or *tsuii* in flight, stronger *shi-ri-riip* and deep chuckling *kyott kyott* or *gyo-kyo-kyo* when disturbed. Song similar to Brown-headed: *kiichororii kyoron kyoron* or a simple monotonous whistled *tuvee-tulee tulee-tuvee*, but often sings a four-note *kyoron-kyoron-kyoron-tree*, instead of Brown-headed's three-note *kyoron-kyoron-tree*.

BROWN-HEADED THRUSH CTKJR
Turdus chrysolaus

L 23–24cm; WT 64–90g. **SD** Endemic to E Asia, main breeding range Sakhalin to C Honshu: *T. c. orii* on Kuril Is and winters south to Japan and Ryukyus; *T. c. chrysolaus* on Hokkaido and N & C Honshu, wintering from C Honshu through S Japan and SE China, also Taiwan, to Philippines. Uncommon migrant offshore Korea. **HH** Mature deciduous broadleaf forest in lowlands (riverine plains in Sakhalin), from sea level to 1,200m in Hokkaido (2,400m in Honshu). In winter and on migration in mature broadleaf forest, also parks, gardens and agricultural areas. **ID** ♂ has mid olive-brown upperparts including wings and tail, with blackish-brown hood. Underparts orange-brown, white on central belly and vent. ♀ has head concolorous with upperparts, throat pale and lightly streaked. Juv. has indistinct supercilium, dark spotting on neck, throat and chest, orange-brown wash to flanks. *T. c. orii* separable by slightly larger size and darker (sootier) crown, not concolorous with dark olive-brown back; face and throat blacker (more closely resembling Izu Thrush, but lacks clear contrast between head, breast and hindneck, and tail is not black). On take-off reveals pale grey underwing-coverts and orange-brown axillaries. **BP** Bill dull yellow with dusky-grey tip and culmen (all grey in juv.; thicker and longer in *T. c. orii*); narrow yellow eye-ring, irides black; tarsi pinkish-yellow. **Vo** Alarm call a harsh *chuck-chuck*, deep chuckling or bubbling notes, and thin *tsurii* or *zeeee* flight call. Song a pleasant, simple three-part: *kiron...kiron.. tsee*; or *kyoron...kyoron...tsureep*; less varied than Japanese but similar to Siberian. Can recall whistle of Ryukyu Robin, but slower and more subdued.

IZU THRUSH J
Turdus celaenops

Vulnerable. L 23cm. **SD** Endemic to Izu Is south of Tokyo, the Tokara Is, and Yaku-shima south of Kyushu; also winter visitor to Oshima and rare winter visitor to S Honshu. Monotypic. **HH** Mainly shady deciduous and evergreen broadleaf forest, and nearby agricultural land and gardens in Izu Is, but also mixed juniper/rhododendron forest to treeline on islands in SW of range. **ID** Very dark-headed, brown thrush. Ad. ♂ recalls Taiwan race of Island Thrush, but has black (not white) hood reaching upper breast; mantle, back, rump and scapulars warm, dark rufous-brown, wings blackish-brown, tail black. Breast-sides, flanks and belly-sides warm orange-chestnut; central underparts and vent white or largely white with 4–5 deep rufous spots (as in confusingly similar *orii* Brown-headed Thrush which occurs, for example, on Tokara Is on migration). 1st-year can look confusingly similar to nominate Brown-headed. 1st-winter ♂ resembles ad., but has greyer head and breast with some streaking on throat. ♀ like ♂, but grey-brown on head, not black, and has off-white chin narrowly streaked grey and extensive streaking on undertail-coverts (lacking in Brown-headed). **BP** Bill bright orange-yellow (♂), with dark tip and culmen (♀) (grey with yellow base in young); eye-ring bright yellow, irides black; tarsi brownish-orange. **Vo** Throaty *kyop kyop* or *quwatt quwatt* notes, and thinner, more sibilant *tsuii* or *tyii*; bubbling alarm is deeper, more guttural than Pale or Brown-headed Thrushes. Song, from low perch, a brief *gyororott jiitt* or *kyoro-ruru jiip*, alternatively described as like a deeper, more thrush-like version of Japanese Robin's 3–5 buzzing trills *tsurrrrrr...turrrrrr...tzurrrrrrr...tsizi*.

BLACK-THROATED THRUSH CTKJ
Turdus atrogularis

L 24–27cm; WT 54–110g. **SD** Breeds from W Urals to Lake Baikal and NW Mongolia, wintering to SW, S & SE Asia. Monotypic. Accidental in Korea, Japan and Zhejiang, China; reported Taiwan. Closely related to Red-throated Thrush (p.408). Monotypic. **HH** Many forest types and thickets, but favours evergreen forest below 2,200m. In winter may be found higher, in subalpine areas, also scrub-covered hillsides and gardens. **ID** Grey, black and white thrush. Ad. ♂ has dark ash-grey upperparts from forehead to rump, somewhat darker wings and tail, with face to breast black and contrasting strongly with mainly white underparts. Winter/1st-winter ♂ has greyer face, and white fringes to black feathers of throat/breast. ♀ recalls 1st-winter ♂, but has largely white chin/throat and more white streaking in malar region. 1st-winter ♀ very like Red-throated, but lacks red in tail, and like Naumann's and Dusky (p.408), but greyer and lacks rufous in wings, tail or flanks. **BP** Bill tip and culmen black with yellow base to lower mandible; eyes black; tarsi dark brownish-pink. **Vo** Alarm and flight calls closely resemble Naumann's and Dusky, a thin *tseep*, harsh *chack-chack* and soft chuckling. Song comprises slow, fluty phrases interspersed by chattering and low husky notes: *chip-chip-chip...chi-chi-chi-chi...cherr-vóó, cherr-véé, chet-chet-chet...*. **TN** Formerly within Dark-throated Thrush *T. ruficollis.*

PALE THRUSH

BROWN-HEADED THRUSH

ad ♂

ad ♂

ad ♂

ad ♀

1st-year

chrysolaus

ad ♂

ad ♀

2/13/19
Japan

IZU THRUSH

ad ♂

ad ♂

ad ♂

ad ♀

ad ♂
Brown-headed

1st-win ♂

BLACK-THROATED
THRUSH

ad ♂

1st-year ♀

ad ♂

RED-THROATED THRUSH *Turdus ruficollis* CTKJ

L 24–27cm; WT 63–103g. **SD** Breeds SC Siberia, from E Altai and Baikal to N Mongolia and Transbaikalia, wintering to N India, N Indochina and NW China. Accidental Taiwan, Korea and Japan. Monotypic. **HH** Favours evergreen, montane forest and upland riverine forests. Winters in scrub, orchards and juniper woodland. **ID** Grey, red and white thrush. Ashy-grey from crown to tail, paler on back and blackish-grey on wings, with black lores and dull red outer tail. **Ad.** ♂ has deep red supercilium, neck-sides, chin/throat and breast, and white underparts; in winter red of throat narrowly fringed white. 1st-winter ♂ less bright; pale to deep orange-red supercilium and, to some extent, lower cheeks, malar, throat and breast, with variably streaked submalar and upper breast. ♀ has paler supercilium, white chin, black-streaked throat-sides, and pale rufous breast; 1st-winter ♀ very like Black-throated but has rufous outer tail, also recalls 1st-winter Naumann's/ Dusky but lacks rufous in wings and flanks. Underwing-coverts pale orange-brown. Occasional birds in Japan and Korea appear intermediate between Red-throated and Naumann's, suggesting hybrids, but age/sexual differences unclear in this complex. **BP** Bill has blackish tip and upper mandible with yellow base and lower mandible; eyes black; tarsi dull brownish-pink. **Vo** Alarm and flight calls as Black-throated, and song also similar but lacks chattering quality, being a series of simple, slow fluty phrases, *chulee chooee chewee, chisya chivya chuya.* **TN** Formerly within Dark-throated Thrush *T. ruficollis.*

NAUMANN'S THRUSH *Turdus naumanni* CTKJR

L 23–25cm; WT 63–81g. **SD** SC Siberia from Yenisei to the Lena in summer, wintering in E Asia, SE Russian Far East, N Korea, E & S China; small numbers S Korea and very small numbers Japan. Monotypic. Naumann's breeds further S than Dusky, and generally winters W of region. **HH** In summer, woodland from taiga to edges of lowland tundra. In winter, flocks in mid-elevation to lowland mixed forest, parks, agricultural land and gardens, often foraging in open. On migration often in large flocks mixed with Dusky. **ID** Variable, but ad. generally mid grey-brown from forehead to back, with rusty-red or brick-orange lower back, base and outer tail (always lacking in Dusky). ♂ has face buff or orange, with brick orange supercilium, neck-sides and malar, throat bordered by grey-brown lateral throat-stripes. Underparts variably rufous-orange, with white scaling from lower chest to vent, and white central belly. ♀ has more dark streaking on throat-sides, little or no rufous-orange on face and less red on underparts. 1st-winter has throat streaking extending to neck- and breast-sides. In flight, outer tail-feathers distinctly rufous-orange, wings show only limited (or no) rufous-brown panel on secondaries. Age and sex-related plumage variation unclear and identification further confused by hybridisation. **BP** Bill has blackish tip and culmen with yellow lower mandible; eyes black; tarsi dull brownish-yellow. **Vo** A shrill, nasal *cheeh*, a harsh *shak* or *chack* in alarm which is commonly repeated, sometimes in series. Song (rather fluting and melodious) comprises series of unhurried whistled phrases concluding in soft twittering: *tvee-tryuuu-tee... tvee tryuuuu-tvee* (middle phrase longest and accentuated), *tsee-tsee-tsee.* **TN** See Dusky. **AN** Rufous-tailed Thrush.

DUSKY THRUSH *Turdus eunomus* CTKJR

L 23–25cm; WT 55–106g. **SD** Breeds from NC to NE Russia east to Yakutia, Chukotka, Kamchatka and Sakhalin. Generally north of Naumann's, with range extending further W and E, and also winters further E; migrates (Oct–May) in large numbers through Japan, S Korea, E & S China; scarce Taiwan. Monotypic. **HH** In summer, occupies similar habitats to Naumann's, but generally at higher latitudes and altitudes. In winter/on migration as Naumann's. **ID** Extremely variable but generally dark with much black on face and underparts, extensive rusty-red in wings (coverts, primary bases, tertials) and rump, but none in tail. Much more contrasting than Naumann's, though can be confusingly similar. ♂ has prominent creamy white supercilium, chin, throat, breast and neck bar, with blackish-brown lores and ear-coverts, narrow dark brown lateral throat-stripe, black breast-band and heavy black scaling on breast and flanks. Crown to back olive brown, tail dark brown; wings appear mostly bright rufous-brown due to extensive rufous fringing to scapulars, coverts, secondaries and tertials. ♀ is poorly marked and less contrasting than ♂, but also has dark brown tail and distinct rufous wing-panel, albeit less conspicuous than in ♂, and some lack rufous in wings (see ♀ Black-throated, p.406). In flight, tail lacks rufous but much of wing appears bright rufous-brown; also has rusty-red axillaries. **BP** Bill has blackish tip and culmen with yellow lower mandible (brighter than Naumann's); eyes black; tarsi dull brownish-yellow. **Vo** Varied sibilant and chattering calls: *kii kii, chirii, tsuii* or *shrree* and *quwatt quwatt*, also strident *chek-chek-chek-chek* and staccato alarm *chuck*. Song similar to Naumann's, but first phrase more strongly emphasised and the whole interspersed by harsher notes or short trills, terminating in series of hard, dry notes: *tryuuuu-tvee-tryu... tyuu-trrryu-uute tryuute tryuute frrrrr ket-ket-ket.* Song in NE Russia described as: *veet tyulir-tyulir fru-fru fir-fee veet-veet tyulir-tyulir che-che-che-che veet-veet-veet fru-fru pryupee-pryupee*, perhaps indicating regional song variation. **TN** Formerly considered race of *T. naumanni* (then referred to specifically as Dusky Thrush).

FIELDFARE *Turdus pilaris* CJR

L 22–28cm; WT 81–141g. **SD** Iceland to Lake Baikal and Lena, wintering as far as SW Europe and N Africa. Monotypic. Accidental NE China (Liaoning/Jilin border) and several winter records in Japan. **HH** Deciduous and mixed forests of taiga zone, but in winter visits open agricultural land with isolated trees and wooded countryside. Flocks in winter, either alone or with other thrushes. **ID** Large, grey and brown thrush, almost as large as White's (p.400), with distinctive blue-grey head, rump and uppertail-coverts. Ad. head, face, nape, lower back, rump and uppertail-coverts pale blue-grey; mantle, scapulars and wing-coverts warm rufous-brown, tail and wings blackish-brown. Underparts off-white with warm buff-orange wash to breast, narrow dark streaks on throat-sides becoming larger darker spotting and scalloping on breast-sides and flanks. Rather upright stance; appears deep-chested. 1st-winter less clearly and brightly marked. Flight strong, undulating, flashes pure white axillaries and underwing-coverts. **BP** Bill pale, brownish-yellow; eyes dark brown; long tarsi blue-grey. **Vo** Alarm and flight calls generally loud and hard, 2–3 notes, *tchack-tchack* or *chachacha*, but also a thin, nasal *zreep* or squeaky *gih*.

RED-THROATED THRUSH

ad ♂

1st-win ♂

ad ♂

ad ♂

1st-win ♀

ad ♂

1st-win ♀
Naumann's

ad ♂
Naumann's

NAUMANN'S THRUSH

1st-win ♂

ad ♂

ad ♂

Naumann's x
Dusky hybrid

ad ♂

ad ♂

DUSKY THRUSH

1st-win ♂

ad

ad ♂

ad ♂

1st-win ♀

ad Fieldfare

ad

FIELDFARE

REDWING   KJR
Turdus iliacus

L 20–24cm; WT 46–80g. **SD** Widespread, with *T. i. iliacus* from NW Europe to E Siberia and Kolyma R; wintering well W of region in Europe and NW Africa. Accidental in winter to Japan and Korea. **HH** Riparian thickets and mixed woodland; winters, with other thrushes, in agricultural and suburban areas with berry- or fruit-bearing trees/shrubs. **ID** Small, slender thrush, resembling Naumann's (p.408) or an unstreaked Eyebrowed Thrush (p.406), but has prominent supercilia and deep rusty-red flanks and underwings. Upperparts drab earth brown, darker on wings and tail. Ear-coverts dark brown, bordered by long creamy supercilium; chin to belly off-white with narrow dark streaks forming malar and merging into heavy streaks on breast-sides and somewhat weaker streaks on belly; flanks and underwing distinctively rusty-red. **BP** Bill brown with lower mandible mostly pale yellow with dark tip; eyes black; tarsi yellowish- or pinkish-brown. **Vo** Flight call a long, thin high-pitched buzzing *tsee* or *shirii*; an abrupt *chup* or *chittick*; alarm a harder, rattling *trrt-trrt-trrt* or *chet-chet-chet*.

SONG THRUSH   J
Turdus philomelos

L 20–23cm; WT 50–107g. **SD** W Europe to L Baikal; winters in Europe, Africa and SW Asia. *T. p. nataliae* occurs closest to region in W & C Siberia. Accidental in winter to Japan. **HH** Various types of open woodland in lowlands and hills, foraging in open on ground. **ID** Medium-sized dark olive-brown thrush. Upperparts olive-brown, face plain with faint malar of black spots (lacks prominent supercilium of Redwing, and black crescent on rear ear-coverts and black subocular stripe of otherwise very similar Chinese Thrush); underparts off-white, with warm buff wash to breast, fine dark spotting on throat-sides, larger spots on breast and fainter browner spots on flanks. In flight, yellowish-buff underwing-coverts (reddish in Redwing, buffish-brown in Chinese). **BP** Bill dark blackish-grey above and at tip, paler at base of lower mandible; eyes dark brown; tarsi pinkish-brown. **Vo** Call, commonly given on take-off or when flushed, a sharp, anxious *tsui* or *tseeu* and soft but sharp *tip* or *tsipp* or *zit* similar to, but softer than, Rustic Bunting. Alarm call is a series of sharp, scolding *tix-ix-ix-ix* or *stuk-stuk-stuk* sounds, or an explosive chatter *tikikikikikik*.

CHINESE THRUSH   CK
Turdus mupinensis

L 23cm. **SD** Occurs only in S & C China to NE China, including Hebei. Winters in S & SE China. Reported as accidental to S Korea. Monotypic. **HH** Mixed broad-leaf and conifer forests (1,300–3,200m). **ID** Medium-sized dark olive-brown thrush that closely resembles more widespread but westerly Song Thrush. Upperparts mid brown, face grey-brown with black crescent on rear ear-coverts, vertical black stripe below eye and black malar; wings dark brown with more prominent white spots at tips of median and greater coverts than Song Thrush; underparts off-white, with warm buff wash to breast and sides, and fine dark spotting on throat-sides, larger spots on breast and fainter browner spots on flanks. In flight, reddish or cinnamon-buff underwing-coverts (like Redwing). **BP** Bill dark grey above, paler at base of lower mandible; eyes dark brown; tarsi yellowish- or pinkish-brown. **Vo** Alarm and flight calls and song all very similar to Song Thrush. Song comprises slow series of phrases, some repeated, others slurred, and interspersed with pauses lasting several seconds: *drip-dii-du dudu-du-twi dju-wi-wi chu-wii-wr'up chu-wi'i-wu-wrrh dju-dju-wiii'u*. **AN** Naumann's Song Thrush.

MISTLE THRUSH   J
Turdus viscivorus

L 27–28cm; WT 93–167g. **SD** W Europe to Lake Baikal, wintering in Europe, Africa and SW & S Asia. Several winter records in Japan. *T. v. bonapartei* occurs closest to region in SC Siberia. **HH** Open woodland, primarily deciduous, including in montane areas, descending in winter to lowland forest and parkland. Rather vigilant and wary. **ID** Large, greyish olive-brown thrush. Upperparts and face plain grey-brown, paler on rump; tail long with white tips to outer three feathers. Underparts off-white with buff wash to breast and heavy dark spotting, small and concentrated on throat-sides and upper breast, larger on belly, blackest and largest on flanks and lower belly. Lacks golden tones of White's Thrush (see p.400). In flight, note pale rump, whitish underwings and white tail-corners (compare smaller, plainer Song Thrush). **BP** Bill large, blackish-brown with pale base to lower mandible; eyes dark brown; tarsi yellowish-brown. **Vo** Flight- and alarm calls a dry rattling *trrrrk* or *zer'r'r'r'r*; also an abrupt *tuc* or *kewk*.

REDWING

iliacus

ad

1st-year

ad

CHINESE THRUSH

ad

juv

ad

SONG THRUSH

nataliae

ad

ad

ad Song

ad feeding by smashing snail shell

MISTLE THRUSH

bonapartei

ad

ad

ad

juv

GREEN COCHOA
Cochoa viridis

C

L 27-29cm. **SD** From Himalayas to Thailand and SE China. In region, rare resident in Fujian. Monotypic. **HH** Moist broadleaved evergreen hill-forest canopy; lethargic, takes fruit and insects. **ID** A bright blue, black and green thrush-like bird. Crown, nape and face bright blue; mantle, back and rump green; base of primaries and much of secondaries bright silvery-blue, wings otherwise black; tail bright blue with broad black tip. Underparts dull green. **BP** Bill thrush-like, black; eyes dark brown with orange eye-ring; legs and feet greyish-pink. **Vo** A series of monotone whistles.

LESSER SHORTWING
Brachypteryx leucophrys

C

L 11–13cm; WT 12.5g. **SD** Himalayas to SE Asia. *B. (l.) carolinae* (may merit species status as **Caroline's Shortwing**) ranges north to S & SE China east to Fujian. **HH** Favours damp undergrowth, often near streams, of moist montane broadleaf forest (1,000–3,200m). Descends in winter. **ID** A small, upright, short-winged and short-tailed bird of forest floor. ♂ *B. l. carolinae* (unlike extralimital nominate which is slaty-blue) resembles ♀. Ad. upperparts uniform earth brown, with slight rufous tones; weak white supraloral reaches to just above eye. Chin and throat white with brown flecking, breast and flanks browner, belly and vent off-white. Juv. is spotted version of ad. **BP** Bill fine, sharply pointed, black with pale base; eyes black; long tarsi greyish-pink. **Vo** A short, downslurred sibilant *psueeet* and subdued, hard *tack*. Song a sweet high-pitched warble commencing with several short sibilant notes followed by a jumbled flourish. **AN** Brown Shortwing.

WHITE-BROWED SHORTWING
Brachypteryx montana

CJ

L 12–13cm; WT 27.5g. **SD** Himalayas to SE Asia and Indonesia. *B. m. sinensis*, restricted to SE China north to Fujian. Record from Hegura-jima, Japan, possibly an escapee. **HH** Rather secretive inhabitant of dense, shady undergrowth in mature forest where prefers damp overgrown areas close to streams or fallen trees, forages near ground. Altitudinal range in region uncertain, but probably above 1,000m in summer, perhaps descending in winter. **ID** ♂ *B. m. sinensis* overall dark, slaty blue-grey, with prominent, short white supercilium (sometimes concealed), narrowly white-fringed carpal, and paler blue-grey underparts, the central belly and vent being almost white. ♀ dull brown above with rufous-brown wings and tail; face plain dull brown, lacks ♂'s supercilium but has pale eye-ring; underparts, pale brown on face, throat, breast and flanks, whiter on central belly. **BP** Bill fine, black; eyes black; tarsi blackish-brown. **Vo** A hard *tack* and rattled *tt-tt-tt* in alarm accompanied by wing-flicking. Song a series of single silvery notes quickening to a plaintive, formless babbling before terminating abruptly. **AN** Blue Shortwing.

TAIWAN SHORTWING
Brachypteryx goodfellowi

T

L 13cm. **SD** Endemic to Taiwan. Monotypic **HH** Mid- and upper-elevation montane forest (*c.*1,000–3,000m). Not shy but cryptic and very secretive, preferring dense, shady undergrowth in mature forest, especially damp areas near streams, foraging close to ground, often in leaf litter. **ID** Small, all-dark shortwing, with a short tail and wings. ♂ largely olive-brown except prominent white brow starting just above lores and broadening on crown-sides (sometimes concealed); wings and tail dark olive-brown; face and underparts (to breast/flanks) paler, more rusty-brown, greyer on centre of belly and vent. ♀ (and 1st-year) slightly paler dull brown above with dark rufous-brown wings and tail, underparts pale brown with pale fringes to feathers of throat and breast, and whiter on centre of belly; eyebrow reduced/concealed. **BP** Bill fine, quite long, black; eyes black; tarsi brown. **Vo** Generally rather quiet, but when agitated gives low *sher*. ♂ restless when singing, scrambling from perch to perch and flaring white brows. Song a series of loud single notes quickening to a brief plaintive babbling, an explosive jumble of descending notes with a gravelly quality that stop abruptly: *sheeee shegli-shegli-gli-gli-gli.* **TN** Formerly within White-browed Shortwing.

BLUETHROAT *Luscinia svecica*

CTKJR

L 13–15cm; WT 12–25g. **SD** Very widespread in taiga zone, from Scandinavia to extreme E Chukotka south through Koryakia to N & C Kamchatka, S Russian Far East and NE China; also N Alaska and Yukon. Winters SE China, S & SE Asia and N Africa. Migrates through E China; scarce migrant or winter visitor Japan, Korea and Taiwan. Subspecies breeding in NE Asia *L. s. svecica* (**Red-spotted** or **Arctic Bluethroat**), but *L. s. przewalskii* winters to E China. **HH** Locally common on breeding grounds, but scarce/overlooked on migration/in winter when rather secretive. Favours brushy tundra, forest edge or clearings; in winter often in scrub and grassy areas near water, but skulks in bushes or on ground. **ID** Face, chest and tail pattern distinctive. Ad. ♂ breeding has prominent white supercilium, blue chin and throat, with central red spot, and unmistakable chest pattern – banded orange, blue, black, white and orange. Upperparts plain olive-brown, with some dark streaking on mantle. Underparts creamy white with buff flanks. Imm. and winter ad. ♂ have more washed-out face and chest, and have blue restricted to lower throat/upper breast. ♀ lacks blue and orange of chest, instead has white supercilium and throat, contrasting with black malar and chest band of black streaks. Tail of both sexes short but distinctive even in brief flight views: brown centre and black tip with bright rufous patches at sides of base. Stance rather upright, when droops wings and cocks tail. *L. s. przewalskii* probably indistinguishable from *L. s. svecica* in winter. **BP** Bill fine, black; eyes black; long tarsi black. **Vo** A hard dry *tatt* or *chack*, subdued *turrc* and *shtik-shtik*, also sharp *trac*, and whistled *heett* or *hueet* (rather *Phylloscopus*-like) when surprised, a hoarse *bzrew* and in winter *skwink.* Song, from perch or in flight, powerful and varied, a speeding cascade of mixed notes (some musical, others harsh), e.g. *djip-djip-djip, trr-trr-trr* and *zrew churuchuru churichuri*, but also mimics many other birds and insects.

GREEN COCHOA
(not to scale)

ad

LESSER SHORTWING

carolinae

ad ♀

ad ♂

WHITE-BROWED SHORTWING

sinensis

ad ♂

ad ♀

TAIWAN SHORTWING

ad ♂

ad ♀

ad ♀

BLUETHROAT

svecica

ad ♂ sum

ad ♂ win

ad ♀/1st-win

EUROPEAN ROBIN
CKJ
Erithacus rubecula

L 14cm; WT 14–25g. **SD** W Palearctic as far east as the Ob in W Siberia; *E. r. tataricus* occurs closest to E Asia. Northern birds move south, some as far as S Europe and N Africa. Accidental coastal China, Japan (May/Nov) and Korea (Mar). **HH** Shady forest, cool woodlands, parks and gardens, where often forages on ground. **ID** Resembles Japanese Robin, but darker earth-brown from crown to tail, and orange-red of face and breast bordered by grey from forehead and ear-coverts to flanks; belly off-white. Upright stance and appears rather plump. **BP** Bill fine, sharply pointed, blackish-brown; eyes large, irides black; tarsi blackish-pink. **Vo** Call a sharp *tic*, often repeated in excited rapid-fire series, *tic-tic-tic-tit*, even *tic tic tikerititit*. In alarm a high-pitched thin *tsiih*.

JAPANESE ROBIN
CTKJR
Luscinia akahige

L 14–15cm. **SD** E Asian endemic. *L. a. akahige* summers (Apr–Oct) in Japan, Sakhalin and S Kurils, wintering to Ryukyus and SE China. *L. a. tanensis* Izu Is (and perhaps islands off S Kyushu, but those on Yaku-shima resemble nominate) may be resident. Rare migrant and winter visitor Taiwan, vagrant Korea. **HH** Mature montane (600–2,500m) broadleaf (deciduous and evergreen) forest, with fir/spruce, often in shadier areas, where difficult to see. Perky and upright, foraging on ground. **ID** ♂ largely warm rufous-brown from crown to rump and dark grey below, but lores, face, neck and breast bright reddish- or orange-brown, and tail bright rufous-brown. Grey of underparts darkest where meets orange of breast, and may appear as blackish border. Paler grey on belly becoming yellowish-white in centre and on vent. ♀ less bright, with less contrasting breast markings and lacks black border to orange breast. ♂ *L. a. tanensis* differs in having no black border to orange breast, with paler grey belly and flanks. **BP** Bill black (*L. a. akahige*) or black with yellowish base (*L. a. tanensis*); eyes black; tarsi yellowish-pink. **Vo** A hard, tacking *tun tun* or *tsu*, or high-pitched straight *hiii* and thin metallic *tsip*. Song a loud, distinctive and quite far-carrying drawn-out trill, commencing with a single high note followed by a somewhat rattling trill: *peen-karakararara*. Repetitions occur at different pitches in series: *tyurrrrrr…. pyorrrrrrr… tyurrrrrr* etc. *L. a. tanensis* gives slightly slower trills and notes are harder.

RYUKYU ROBIN
TJ
Luscinia komadori

L 14–15cm. **SD** Endemic to Nansei Shoto of Japan: *L. k. komadori* occurs on islands off SW Kyushu and Tanega-shima to Tokuno-shima in N Ryukyus, wintering to S Ryukyus; *L. k. namiyei* probably resident Okinawa, but birds from C Nansei Shoto apparently migrate in winter, some reaching as far north as S Kyushu, and others south to Yaeyama; accidental Taiwan. **HH** Understorey of shady evergreen forest, often in gullies or near streams, below 600m. Forages on ground; often at forest edge or on tracks in early morning. **ID** Perky, black-breasted orange robin. ♂ *L. k. komadori* has bright orange-rufous upperparts (including wings), black forehead, face and chest, and white or greyish-white from belly to vent, with blackish flank patches. *L. k. namiyei* differs in having a rufous-orange forehead and more uniform dark grey belly and flanks. ♀s very similar, though *L. k. komadori* often has some black spotting on breast, whereas *L. k. namiyei* is pale greyish-buff on breast, and grey-white below. **BP** Bill black; eyes black; long tarsi yellowish-brown. **Vo** A high-pitched, rising whine *swiiii* or *hii-hii*, hard warning *gu gu*, and *kirrick* alarm-note accompanied by tail-flicking and wing-quivering. Song (nominate) a series of melodious yodelling whistles, changing in pitch with each repetition, typically given around dawn: *pyolololo…. trelulululu…. tyululululu….* tending to stop after three repetitions, then starts again; similar to Japanese Robin but more musical.

SIBERIAN RUBYTHROAT
CTKJR
Luscinia calliope

L 14–16cm; WT 16–29g. **SD** Widespread summer visitor (May–Aug) to taiga zone south of Arctic Circle, from Urals to NE China and Sea of Okhotsk. Winters south to S China, Taiwan and S Asia. *L. c. calliope* breeds from C Urals to Anadyr, N & C Korea, NE China, and winters in S & SE Asia; *L. c. camtschatkensis* breeds in Kamchatka, Commander and Kuril Is Sakhalin and Hokkaido, wintering to Nansei Shoto, SE Asia and Philippines. Migrates through Korea and E China. **HH** Locally common and conspicuous on breeding grounds; scarce (overlooked?) on migration/in winter when rather secretive. Favours treeline/alpine zone in south of breeding range, lower altitude taiga-forest edge in north and on some offshore islands. In winter skulks in grassy areas with bushes and near wetlands with reeds. **ID** Initially recalls Bluethroat (p.412), but face pattern distinctive and ♂ simply stunning. Upperparts largely plain mid olive-brown, with wings and tail, with paler brown chest (greyer with wear), greyish-brown on flanks and whitish on belly and vent. ♂ unmistakable; short, clear white supercilium separated from broad white malar by bold black lores, with black lateral throat-stripe and blackish-grey face, but outstanding feature is the brilliant metallic ruby chin/throat, which pulsates when singing. Young ♂ less boldly marked. ♀ has less boldly marked face, and white chin/throat, though some have pale pink wash to chin. When flushed, appears chunky, short-tailed and plain brown. **BP** Bill fine, black; eyes black; long tarsi yellowish-pink. **Vo** A deep, gruff, muffled *vehp* or *jütt*, and Fieldfare-like *schak* in alarm, or a combined *huitt-tak-tak* (recalling Red-flanked Bluetail). ♂ gives whistled *cue-ee*, *ti-lui* or *feeyoou(eet)* on territory, on migration and in winter. Song sweet, complex, penetrating but slightly melancholic warbling, mixing clear whistles and harsh notes: *kyoro-kiri hyogori kii-kyorochirii*; and *choichoi chorori chuichui chiichirichirii*.

EUROPEAN ROBIN

tataricus

ad

JAPANESE ROBIN

akahige

ad ♂

ad ♀

RYUKYU ROBIN

ad ♂
namiyei

ad ♂
komadori

ad ♀
komadori

ad ♂

SIBERIAN RUBYTHROAT

calliope

ad ♀/1st-win

ad ♂

ad ♀ variant

SIBERIAN BLUE ROBIN   CTKJR
Luscinia cyane

L 13–14cm; WT 11–18g. **SD** From upper Ob to Sea of Okhotsk east to Magadan, Sakhalin, Kuril Is, N & C Japan, NE China and Korea (Apr–Sep). Subspecies in region is *L. c. bochaiensis*. Accidental Taiwan; in winter SE China and SE Asia. **HH** In breeding season favours mixed taiga-forest, forest edge or clearings; often around gullies and streams or fallen trees. Generally skulking, close to or on forest floor, often in dwarf bamboo or dense ground cover, sometimes singing at mid levels. On migration, found in scrub and grassy areas near water, whereas typical winter habitat is lush tropical forest and secondary growth. Stance rather horizontal, pumps tail when on ground. **ID** ♂ *L. c. bochaiensis* unmistakably deep blue, black and white. Upperparts uniform deep blue, wings and tail somewhat blue-black, with bright white underparts. Lores and line from bill base to breast-sides jet black. Young ♂ browner on face and wings (lacking black), but has variable blue on rump, uppertail-coverts and tail. ♀ very pale to mid olive-brown with very plain face, but often has dull, dark blue uppertail-coverts/ tail base, with few distinguishing marks except tarsi. Some lack dark blue uppertail-coverts and have slight scaling on chest, recalling Swinhoe's Robin (p.418), but stance of latter usually rather erect, whilst Siberian Blue is usually horizontal and ducks its head. **BP** Bill black, sometimes showing pink gape; eyes appear large (especially ♀), irides black, with very narrow white eye-ring; long tarsi very pallid pink. **Vo** A thin *chip chip*, a *se-ic*, low *chuck-chuck-chuck* and, in alarm, hard *tak tak* or *dack*. Song similar to Japanese Robin's but more varied, and starts with series of hard call-like notes: *zit zit zit* or *chitt chitt chitt* followed by rapid, explosive trilling rattle *lololololololo* or *hichohichohicho chochocho*, or *tsutsutsutsu hin-kararara*.

WHITE-BROWED ROBIN   T
Luscinia indica

L 13–15cm; WT 16g. **SD** Nepal to Vietnam. *L. (i.) formosana* on Taiwan; may be specifically distinct (as **Taiwan Robin**). **HH** Dense undergrowth in montane mixed forest (2,000–3,400m), foraging on or near ground. **ID** Unusually drab chat with little sexual dimorphism. Generally dark ash-brown above and buff-brown below, somewhat yellowish on undertail-coverts. ♂ has prominent narrow white supercilium from bill to rear ear-coverts, with black lores and blackish-grey crown and ear-coverts. ♀ has narrower, less distinct supercilium, with face, chin and underparts more uniform dull yellowish-grey, more yellowish-buff or pale brown on vent (see Johnstone's). **BP** Bill dark grey; eyes large, irides black; tarsi black. **Vo** A low *chac*, frog-like *kerrt* in alarm, and louder, more urgent, shrike-like *shek-shek-shek*. Song a short series of quiet staccato wolf whistles, *whit wrreoo*, from low perch. **TN** Formerly *Tarsiger indicus*.

JOHNSTONE'S ROBIN   T
Luscinia johnstoniae

L 12–13cm. **SD** Endemic to Taiwan. Monotypic. **HH** Forest edge and undergrowth in montane areas (1,700–3,800m). Often above treeline. Prefers less dense vegetation than White-browed Robin, and habitat selection more flexible. **ID** ♂ has mostly grey and black upperparts broken by broad white stripe from above eye, narrowing on nape-sides, and bright red breast-band extending as narrow collar on hindneck and scapulars. Rarely entire mantle deep crimson-red. Mantle/ back usually dark grey, wings, rump and tail black, with narrow reddish and white bands on uppertail-coverts. Face, chin and throat black, then below chest-band, breast washed orange-buff and belly, flanks and vent off-white. ♀ ash-grey on head, mantle, back and rump, blacker on tail, with faint white supercilium from bill to just behind eye, and narrow pale crescent on uppertail-coverts. Chin/throat ash-grey with fine white flecks, underparts dull greenish-olive or greyish-green, becoming off-white on lower belly/vent (see White-browed). **BP** Bill greyish-black; eyes large, irides black; tarsi black. **Vo** Low grating *tuc tuc* notes interspersed with high *pi pi pi* notes, and low, burped *grruit* in alarm, sometimes combining high piping with harsh grating calls: *sipsipsip grrgrrgrr sipsip grrgrr*. Song a series of jaunty, loud and clear, two-part whistles. **TN** Formerly *Tarsiger johnstoniae*. **AN** Collared Bush Robin.

RED-FLANKED BLUETAIL   CTKJR
Luscinia cyanura

L 13–15cm; WT 10–18g. **SD** Finland and N Russia to Sakhalin, N Sea of Okhotsk, Yakutia, Kamchatka, Kuril Is, S Russian Far East, NE China, N & C Korea, N & C Japan. Common migrant through E Asia. In winter S Japan, S Korea, E China and Taiwan to SE Asia. Monotypic. **HH** Subalpine evergreen or mixed forests at lower latitudes, sometimes to sea level, and typically in lowland taiga at higher latitudes. In winter, various woodland types, even suburban parks. Quite confiding, often sings from obvious perch at mid-levels. Forages on ground. Restless, often flicks tail and wings. **ID** Plump and rather large-headed. ♂ unmistakably blue, orange and white. Upperparts, including wings and tail, bright blue with some black on wing-coverts, tertials and primaries. Blue extends onto face, to throat- and breast-sides. Supraloral spot extends over eye as narrow supercilium, variable but often silvery-white. Underparts largely clean white, with narrow white throat and distinctive large orange-red flank patches. ♀ appears 'open-faced' with very prominent eye in pale eye-ring, and distinctive narrow white throat; upperparts olive-brown, greyer on rump and blue-grey on tail. Imm. ♂ (also defends territory) much like ♀, but has brighter orange flanks, bluer tail and often some blue in wings. Juv. heavily spotted pale brown, but also has blue tail. **BP** Bill black; eyes, especially of ♀, appear large, irides black; long tarsi black. **Vo** Calls frequently, a soft *tuc tuc-tuc* or *heet katt-katt* (first note high-pitched, latter two hard). Song a rapid, cheerful: *hichuri churiririchurochii* or *hyoro-hyurururip*. **TN** Formerly *Tarsiger cyanurus*. Himalayan population (Orange-flanked Bush Robin) formerly included in this taxon.

SIBERIAN BLUE ROBIN

bochaiensis

ad ♂

ad ♀

juv

WHITE-BROWED ROBIN

formosana

ad ♂

JOHNSTONE'S ROBIN

ad ♂

ad ♀

RED-FLANKED BLUETAIL

ad ♂

ad ♀

SWINHOE'S ROBIN CTKJR
Luscinia sibilans

L 13–14cm. **SD** Taiga zone from Yenisei to Sea of Okhotsk, Kamchatka, Sakhalin and apparently S Kurils (but not Hokkaido); winters SE China and SE Asia. Fairly common migrant Korea (may breed in N), very rare offshore Japan, Hokkaido and Taiwan. Monotypic. **HH** Damp well-wooded or forested areas, both deciduous and coniferous, typically in dense undergrowth, often in gullies or near streams, to 1,200m. Commonly skulks in dense vegetation, but also ascends trees. Stance rather erect; shivers tail and rear body. **ID** Rather plain brown upperparts, with bright rufous uppertail-coverts and tail (resembling Japanese Robin but deeper, less bright, see p.414). Pale supraloral and eye-ring. Underparts greyish-white with greyish-brown scaling on throat-sides, chest and flanks, where brownest. **BP** Bill black; eyes appear large in rather plain face, irides black; long tarsi reddish-pink. **Vo** A low *tuc-tuc*. Song a strong, erratic rattling trill, more drawn-out than Japanese Robin, varied trilling phrases with falling intonation: *shu-rurururu, hin-rururun, hyururururururu, hichochochocho*. Similar to Siberian Blue Robin, but less powerful. **AN** Swinhoe's Red-tailed Robin.

ORIENTAL MAGPIE-ROBIN CT
Copsychus saularis

L 19–21cm; WT 29–41g. **SD** Ranges from India to SE Asia and Indonesia, with *C. s. prosthopellus* across S & E China north to Zhejiang; an established exotic in Taiwan now increasing in urban areas. Common (native) resident on Kinmen. **HH** Open and secondary forests, urban parks and gardens, and villages, where conspicuous; also coastal mangrove. **ID** ♂ boldly pied, with glossy black hood and upperparts to tail. Wings glossy black, with white scapulars, secondary-coverts and secondaries, and white outer tail; underparts white. ♀ less contrasting, with grey hood, and duller grey-black back, wings and tail, also with white (as ♂). Underparts white, with buff-orange wash on flanks. **BP** Bill somewhat long, fine, black; eyes dark brown; long tarsi dark-grey/black. **Vo** A long, plaintive *swee-ee*, and in alarm a harsh *chr-r*. Song, given from perch or in flight, a varied warbling, mixing churrs, whistles and mimicry: *suiii-suuuh-swiit-swer-swiit-siiuh*.

WHITE-RUMPED SHAMA T
Copsychus malabaricus

L 21–28cm; WT 31–42g. **SD** Natural range India to Indonesia, but introduced and breeds locally on Taiwan (race uncertain). **HH** Secondary forest, urban parks and gardens. **ID** ♂ has glossy black hood, mantle, wings and long graduated tail (up to 7cm), with orange lower breast, flanks and vent, and white rump and outer rectrices. ♀ has grey hood and mantle, is less richly coloured below and has shorter tail. **BP** Bill black; eyes black; tarsi dull pink. **Vo** A scolding *tschak* and *krr-krr*. Song is beautiful (hence a common cagebird), a series of strong, thrush-like, fluty phrases incorporating mimicry.

PRZEVALSKI'S REDSTART C
Phoenicurus alaschanicus

L 14–16cm. **SD** Breeds NC China, but in non-breeding season (Oct–Mar) reaches east to Hebei and Beijing; rather rare winter visitor to NE China. Monotypic. **HH** Favours rocky and scrub-covered hillsides above 3,300m in summer, but descends to *c*.2,000m in winter. **ID** A strongly marked black, orange and grey redstart. ♂ has crown, nape and neck-sides pale to mid grey; mantle, back, rump and tail all strongly rufous-orange; scapulars and wings black, with strong white scapular patch, white primary bases and in tertials. Underparts rufous-orange from chin to vent, whiter on central belly. ♀ plain, dull earth brown, slightly more buff on underparts; wings blackish-brown with pale tips to coverts, secondaries and tertials; rump and tail-sides orange, central rectrices blackish-brown. **BP** Bill black; narrow white eye-ring and black eyes; tarsi black. **Vo** Unknown. **AN** Alashan Redstart.

BLACK REDSTART CTKJ
Phoenicurus ochruros

L 14–115cm; WT 12–20g. **SD** Europe and montane regions from Caucasus and C Asia east to Yenisei and C China. Winters from S Europe and N Africa to India. *P. (o.) rufiventris* (**Eastern Black Redstart**, may warrant specific status), a rare winter visitor to E China, including Hebei and Shandong, and accidental Japan, Korea and Taiwan. **HH** Generally in areas with sparse vegetation; including open rocky areas, cliffs, rock slides and old buildings. **ID** Strongly dimorphic. ♂ generally sooty-black, blackest on face and chest, greyer on mantle with brown cast to wings; grey on forehead (white in ♂ Common Redstart, see p.420). Underparts and tail bright brick orange. Tail especially distinctive, with orange rump and sides to black-centred tail. ♀ much plainer, warm mid brown above and paler below, especially on chin where nearly white, very similar to Daurian Redstart (p.420), but both sexes of *P. (o.) rufiventris* lack white wing patches (though other extralimital subspecies possess them); darker overall than Common. Rump and tail resemble ♂ but less bright. **BP** Bill fine, black; eye of ♀ especially prominent in plain face, irides black; long tarsi black. **Vo** A sharp *tsip*, harder *tuc* and combination of hard tacking and thin, almost whistled, notes: *gatt hee* or *gap gap hee*. Song comprises repeated phrases of short jingling notes with buzzy crackling and whistles: *tuwiduwiwiwi drrr-drrr-duwiidu*.

SWINHOE'S ROBIN

ad

ad ♂

ORIENTAL
MAGPIE-ROBIN

prosthopellus

ad ♀

WHITE-RUMPED SHAMA

ad ♀

ad ♂

ad ♂

1st-win ♂

BLACK
REDSTART

rufiventris

ad ♂

PRZEVALSKI'S
REDSTART

ad ♀

ad ♀

COMMON REDSTART JR
Phoenicurus phoenicurus

L 14cm; WT 11–23g. **SD** Widespread in Europe and east in narrowing range as far as Lake Baikal and extreme NW China; winters in Africa and Arabia. *P. p. phoenicurus* accidental offshore Japan (Oct/Nov) and C Kuril Is. **HH** Open woodland, parks and large gardens. **ID** ♂ bright; combination of black, grey and orange. Upperparts range from white on forehead to mid ash-grey on crown, nape, mantle and back; rump, uppertail-coverts and tail-sides bright rusty-red, tail itself blackish-brown. Face, from above bill to ear-coverts, chin and throat sooty-black; breast brick orange becoming white on belly, flanks and vent. Wings plain, blackish-brown, lacking white. ♀ plainer, with mid brown face, head and upperparts; rump and tail bright rusty-red; off-white throat, breast and belly washed dull orange-brown (closely resembles ♀ Black Redstart but paler, see p.418). **BP** Bill fine, black; eyes dark brown; tarsi dark brown. **Vo** A strong, liquid, somewhat Willow Warbler-like, plaintive *hueet* commonly repeated, or in combination as *hueet-hueet hueet-tick*, and harder, scolding *tchak*.

WHITE-THROATED REDSTART C
Phoenicurus schisticeps

L 15cm; WT 15–17g. **SD** E Himalayas across C China northeast to Shaanxi. Accidental further northeast to Beidaihe. Monotypic. **HH** Generally at 2,400–4,300m in summer, in subalpine scrub and conifers, descending to 1,400m in winter, in meadows, scrub-covered areas and dry park-like forest. **ID** ♂ is black, white, and red. Upperparts range from dark slate-blue on crown and nape, brighter blue on forehead, to dark greyish-black on mantle; lower back and rump orange-red. Wings black with white scapulars and fringes to tertials; tail also black, with rufous bases to outer feathers. Underparts black on face, with white throat spot, otherwise deep red from breast to undertail-coverts. ♀ earth-brown where ♂ is black, dark brown wings with white (pattern as ♂); underparts pale brown with white throat, rump and outer tail-feathers orange, tail blackish-brown. **BP** Bill fine, black; eyes black; tarsi dark brownish-black. **Vo** A drawn-out *zieh* and a high, dry rattle.

DAURIAN REDSTART CTKJR
Phoenicurus auroreus

L 14–15cm; WT 11–20g. **SD** Fairly common E Asian species: *P. a. auroreus* from just west of Lake Baikal and Mongolia east to Amur Estuary, Sakhalin, south to NE China and Korea, wintering to C & S Korea, Japan south of Hokkaido, Taiwan and coastal SE China (Oct–Apr); *P. a. leucopterus* from E Himalayas to C & E China, wintering to NE India and Indochina. **HH** Favours open hillsides, very open forest with rocky areas; in winter at woodland edges, agricultural margins, in parks and large gardens. **ID** Strongly dimorphic. Ad. ♂ largely black and orange, with silver-grey forehead and crown-sides, grey crown and nape, jet black face and chin. Mantle and wings largely black but some feathers tipped brown, and prominent white patch on secondaries and tertials; lower back, rump and tail-sides orange, central rectrices black. Imm. ♂ resembles ad., but less cleanly marked. ♀ largely brown and orange, much plainer, warm mid brown above, paler below with conspicuous white wing patch and orange rump and tail-sides. **BP** Bill fine, black; eye of ♀ especially prominent on plain face, irides black; long tarsi black. **Vo** Call combines high sharp notes with harder notes; typically gives strong double *hit wheet*, *heett katt* or *tuc tuc peet* similar to Red-flanked Bluetail, but straighter and louder.

GÜLDENSTÄDT'S REDSTART C
Phoenicurus erythrogastrus

L 18cm; WT 21–29g. **SD** Mountains of Caucasus to Tibetan plateau: *P. e. grandis* occurs in NW & C China, and NE of L Baikal, with some females wintering northeast to Hebei. **HH** Breeds in rocky alpine zone above 3,600m; winters at lower elevations on rocky, scrubby hillsides down to 1,500m. **ID** A large redstart. ♂ distinctive in black, white and red; has white cap, black face, throat, mantle, and back. Wings black with very extensive white patch on coverts and secondaries. Underparts, rump and tail deep orange-red. ♀ very pale sandy-brown, greyer on head and upperparts, buffier on underparts, wings dark brown with pale fringes to lesser and median coverts; rump and outer tail-feathers orange-red, central rectrices blackish-brown. **BP** Bill black; eyes black; tarsi black. **Vo** A loud *tyeet-teck-teck*, hard *drrrt* and *tak-tak-tak*. **AN** White-winged Redstart.

COMMON REDSTART

phoenicurus

ad ♂

ad ♀

1st-win ♂

ad ♂

WHITE-THROATED REDSTART

ad ♂

ad ♂

ad ♀

ad ♂ *auroreus*

DAURIAN REDSTART

ad ♀
auroreus

1st-win ♂
auroreus

ad ♂
auroreus

ad ♂
leucopterus

GÜLDENSTÄDT'S REDSTART

grandis

ad ♂

ad ♂

ad ♀

WHITE-BELLIED REDSTART C
Hodgsonius phaenicuroides

L 18–19cm; WT 22–25g. **SD** Himalayas to C China, wintering to N India, Thailand and N Indochina. Isolated population in mountains of Hebei and Beijing referable to *H. p. ichangensis*. **HH** Breeds in thickets and bushy areas close to forest near treeline. In winter descends lower, when found on bush- or scrub-covered hillsides to 1,300m. **ID** Large, dark, long-tailed redstart with a rufous-based tail and white belly. ♂ head, mantle, breast and wing-coverts slate-blue; wings blackish-blue with white spots on black alula; tail long, graduated, black with orange-brown basal outer half; lower breast, belly and vent off-white. ♀ sooty-brown above, paler yellowish-buff below, whiter on belly; tail brown with more rufous-brown basal outer half. **BP** Bill black; eyes black; tarsi pinkish-brown. **Vo** A hard *tuk* or *chuck* and grating *chack*, and in alarm *tsiep tsiep tek tek*. Song a melancholic whistled phrase, *teuuh-tiyou-tuh*, the middle note longer and higher than the others. **AN** Hodgson's Shortwing.

PLUMBEOUS REDSTART CTK
Rhyacornis fuliginosa

L 12–13cm; WT 13–23g. **SD** *R. f. fuliginosa* from Himalayas and SE Asia, across China, northeast to Hebei and Shandong; *R. f. affinis* restricted to Taiwan. Reported Korea. **HH** Restricted to mid- and high-elevation forests, along rivers and streams, at 1,000–4,300m (200–2,500m on Taiwan), with some altitudinal migration. Singles or pairs highly territorial along fast-flowing rivers, typically on boulders (like dipper or forktails). Often pumps and fans tail. Individuals seen near coasts of Taiwan presumed to be migrants or vagrants. **ID** Small, perky, robin-like bird. ♂ (both races) plain dull slate-grey, with deep chestnut tail, uppertail-coverts and undertail-coverts. ♀ (*R. f. fuliginosa*) paler grey, with two prominent white wingbars; white underparts from chin to vent, scaled extensively grey from chin to belly; white crescentic patch on uppertail-coverts contrasts with all greyish-black tail. ♀ (*R. f. affinis*) similar, but mid grey, with two faint white wingbars; grey underparts scaled white from chin to breast; rump and tail pattern the same. Juv. resembles ♀, but brown with buff spotting on upperparts. In flight, round-tipped tail is commonly fanned. **BP** Bill short, fine, dark grey; eyes large, irides black; tarsi brown or greyish-brown. **Vo** Hard clicks and sharp, rising *ziet ziet*. Song, given in brief song-flight or from rock in stream, a strong ratcheting phrase of shrill whistles and warbles with jaunty rhythm, recalling accentors, rising in pitch and often repeated: *striiii-triiii-triiii-tríííih*.

WHITE-CAPPED REDSTART CTK
Chaimarrornis leucocephalus

L 18–19cm; WT 24–42g. **SD** Mountains of C Asia, Himalayas and SE Asia east to Zhejiang, with eastern birds migratory and wintering in S & SE China. Monotypic. Records from Taiwan and Korea are puzzling, and perhaps relate to escaped cagebirds (those from Taiwan involving breeding), but an altitudinal migrant and migratory in easternmost range, thus genuine vagrancy possible. **HH** Stays close to rushing streams and rivers in rugged terrain. Bobs on landing, and fans, flicks and wags tail. **ID** Large, black and rufous chat. Ad. has foreparts to lower breast, and back, black, with crown and nape brilliant white; lower back, rump, tail and underparts rusty-red; wings short, black; long tail has rounded tip and black terminal band. Juv. less cleanly marked than ad., with black-scaled cap, greyer rather than black upperparts, and buffish-grey flecking on lower breast/belly. **BP** Bill black; eyes black; tarsi black. **Vo** A long, sharply whistled *teeeet* or *shviiit*. Song consists of repeated melancholic whistled phrases: *tieu-yieu-yieu-yieu*. **AN** River Redstart.

WHITE-TAILED ROBIN TJ
Myiomela leucura

L 17–19cm; WT 24–30g. **SD** Himalayas to SE Asia. In region, *M. l. montium* is resident on Taiwan. Accidental offshore Japan, though possibly escaped cagebird. **HH** On Taiwan, favours shady areas of mature and secondary forest, bamboo and plantations. Found in mid- and upper-elevation forests (*c*.1,000–2,500m), but descends to *c*.200m in winter. **ID** Large blue-black bird recalling Blue-and-white Flycatcher (see p.440). ♂ generally black, but has shining blue forehead, scapulars and wing-coverts, and outer fringes to flight-feathers. Black tail is long, round-tipped, with prominent white 'lozenges' at base of lateral (but not outermost) feathers, revealed as tail is fanned or flicked, and in flight. ♀ mostly dark olive- to earth brown, with diffuse white throat, and has similar tail pattern to ♂, but perches less obviously and does not fan tail as frequently. **BP** Bill dark grey; eyes black; tarsi black. **Vo** A hard *chut*, low *tuc-tuc-tuc* and a high-pitched, straight whistled *sweet*. Song a simple trilled warble, repeated 3–4 times: *tsi-weddleuu*, the phrase having a rather melancholy ring (similar to Blue-and-white Flycatcher song). Last phrase in series is sometimes a doh, ray, me-like *tsidiri*, ending on a high, barely audible note. **AN** White-tailed Blue Robin.

WHITE-BELLIED REDSTART

ichangensis

ad ♂

ad ♀

PLUMBEOUS REDSTART

ad ♂ *fuliginosa*

ad ♀ *affinis*

ad ♂ *fuliginosa*

ad ♀ *fuliginosa*

ad ♂ *affinis*

WHITE-CAPPED REDSTART

ad ♂

ad ♀

ad ♂

WHITE-TAILED ROBIN

montium

ad ♂

ad ♀

LITTLE FORKTAIL CT
Enicurus scouleri

L 12–14cm; WT 11–19g. **SD** From Pamirs, Altai and SW Tien Shan across China to continental coast north to Zhejiang, and mountains of Taiwan. Monotypic. **HH** Alongside white-water sections of cold, fast-flowing streams, rivers and waterfalls. Frequently bobs, flicks wings and fans tail. In contrast to Plumbeous Redstart (p.422), tail is constantly 'shivered' open and shut, without clear rhythm. More often found in montane areas (500m to >2,500m), where Plumbeous Redstart is thinly distributed, but pairs found to as low as 200m. **ID** A small, perky, pied bird, unusually small, plump and short-tailed for a forktail. Foreparts, including breast, mantle and scapulars, largely black, except extensive white forehead. Lower back, rump and belly bright white, with dark bar separating white rump from white lower back. Wings black with white bar on coverts and short, notched black tail has broad white outer feathers. **BP** Bill somewhat short, thick, black; eyes black; tarsi very pale flesh pink. **Vo** Song/call a loud, thin, *ts-youeee*, penetrating through sound of rushing water.

SLATY-BACKED FORKTAIL C
Enicurus schistaceus

L 22–25cm; WT 26–38g. **SD** C Himalayas to SE Asia and S China east to SE coasts. Monotypic. **HH** Montane forest streams (400–1,800m). **ID** Typical forktail, with largely pied plumage and long tail. Easily distinguished from other species by slate-grey crown, nape, mantle and scapulars. Upperparts largely grey, but contrastingly white on forehead, lower back and rump. Wings black with white bar on greater coverts and white patch at base of primaries; long tail is black, banded and tipped with white. Face, ear-coverts and chin black, underparts otherwise white. **BP** Bill black; eyes black; tarsi pale flesh pink. **Vo** A high, thin *teenk*, shrill *seet*, rasping *chaat* or squeaky *weeng*.

WHITE-CROWNED FORKTAIL C
Enicurus leschenaulti

L 25–28cm; WT 27–53g. **SD** N India to Indonesia and across S China, mostly south of Changjiang River: *E. l. sinensis* is race in S & SE China. **HH** Clear montane streams to 1,400m. **ID** A large, pied forktail, similar to Slaty-backed, but has black hood reaching upper breast, white forehead and forecrown, black mantle and scapulars, white back, white greater coverts and white tips to tertials. **BP** Bill black; eyes dark pink; tarsi pale flesh pink. **Vo** Calls include *tseee* or *tseee chit-chit-chit* in flight, and a dry, slurred *gzuweet*.

SPOTTED FORKTAIL C
Enicurus maculatus

L 25–27cm; WT 34–48g. **SD** Afghanistan and Himalayas to Vietnam and across S China; *E. m. bacatus* reaches east into region to Fujian. **HH** Like other forktails, but found along narrower, fast-flowing montane streams, generally above 1,200m in region. **ID** Large forktail, very similar to White-crowned, but in addition to white forehead and forecrown, black mantle and scapulars are heavily spotted and scalloped with white, as is lower black area of chest. Black tail has very narrow white bands and broad white tip. **BP** Bill black; eyes dark brown; tarsi pale flesh-pink. **Vo** A rasping *tseeek*, penetrating *tjeet*, disyllabic *juwee*, creaky *cheek-chik-chik-chik-chik* and buzzy *zhih-zhih-zhih....*

LITTLE FORKTAIL

ad

juv

SLATY-BACKED FORKTAIL

ad

juv

WHITE-CROWNED FORKTAIL

sinensis

ad

SPOTTED FORKTAIL

bacatus

ad

WHINCHAT   J
Saxicola rubetra

L 12–15cm; WT 13–26g. **SD** W Palearctic species that is widespread in drier parts of Europe, W Russia east to CW Siberia and Yenisei River, wintering in Africa. Accidental Japan. Monotypic. **ID** Small brown chat with prominent supercilia. Upperparts dark brown with paler stripes on crown and fringes to feathers of mantle and back, rump buffy-orange heavily streaked with black, tail dark blackish-brown with distinct white flashes at sides of base. Wings blackish-brown with narrow white patch on scapulars and square white patch on greater coverts. ♂ has contrasting face, with long pale supercilium from lores to rear of ear-coverts and dark blackish-brown ear-coverts. Chin/throat white, but has strong orange wash from chin to lower breast, rest of underparts white. Non-breeding ♂ resembles ♀, which is somewhat duller than breeding ♂, with weaker face pattern, paler brown cheeks and light streaking on neck-sides and upper breast; differs from ♀ Siberian Stonechat in having darker ear-coverts, pale supercilium, more defined eyestripe and more distinct malar, but beware 1st-winter ♂ Siberian Stonechat. **BP** Bill fine, sharply pointed, grey/horn; eyes dark brown; long tarsi pink. **Vo** Call combines short whistled *phew* and hard clicking: *phew-tak phew-tak-tak….*

SIBERIAN STONECHAT   CTKJR
Saxicola maurus

L 12–13cm; WT 13–17g. **SD** Urals to Japan, with much subspecific variation. In region, *S. m. stejnegeri* ranges in summer (Apr–Aug) from Baikal across Yakutia to inland Chukotka (but not N or E tundra, Koryakia or Kamchatka), Sea of Okhotsk, Sakhalin and N Japan south to Korea, E Mongolia and NE China; migrates through E China and visits Taiwan, wintering in SE China and SE Asia. *S. m. przewalskii* of Tibet to C China winters as far east as E China. **HH** Favours dry steppe, open grassland with bushes, damp meadows and agricultural land, in lowlands and hills. Perches prominently on bushes and grass stems. **ID** Strongly dimorphic. ♂ breeding has strongly contrasting pattern of largely black and white with some orange; black hood, upperparts and tail, with bold, broad white neck and scapular patches and large clear white uppertail-coverts/rump patch; underparts clean white with small, bright orange upper chest patch. Winter ♂ resembles ♀, but retains only ghost of black hood, particularly on face. ♀ rather drab plain sandy orange-brown, with grey-brown head, mantle and wings, more orange-brown on underparts with plain orange-buff rump, tail

blackish; white scapular patch less prominent than in ♂. In flight, short black tail, white (or pale) rump patch and white scapular patches unmistakable. **BP** Bill fine, short, black; eyes black; tarsi black. **Vo** Call a hard stony *ja ja*, *jat* or *hit*, like pebbles being knocked together, given in combination with sharp whistle: *wist jat-jat*. Song a clear, but rather thin, weak and rather formless chattering: *hii-hyoro-hiri-hii* or *hiichu hichii chii pii chochii*. **TN** Formerly within Eurasian Stonechat *S. torquatus*.

PIED BUSHCHAT   J
Saxicola caprata

L 13–14cm; WT 14–26g. **SD** Iran to New Guinea, reaching as close to region as Philippines. Accidental in winter to S Japan (Jan–Mar); subspecies uncertain. **HH** Occurs near habitation in damp meadows and grasslands, often near water. **ID** ♂ smart, almost entirely glossy black, with white only on wing-coverts, rump, lower belly and vent; 1st-winter ♂ less black, feathers narrowly fringed buff. ♀ plain, mid to dark brown above, more rufous sandy-brown below, with rusty, rufous-orange rump. **BP** Bill fine, black; eyes dark brown; tarsi black. **Vo** A plaintive *hweet* and hard, insistent *chek-chek*, similar to that of Northern Wheatear.

GREY BUSHCHAT   CTKJ
Saxicola ferreus

L 14–15cm; WT 14–16g. **SD** Himalayas to N SE Asia. Mostly resident or local migrant across China south of Changjiang to continental coast, some wintering south of main range. Accidental offshore Japan, offshore Korea, and Taiwan. Monotypic. **HH** Dry, open areas with scrub, also agricultural land. **ID** A small pale chat recalling Siberian Stonechat in structure, but longer tailed. ♂ grey, with prominent white supercilia, white chin and throat contrasting with black mask; upperparts mid-grey streaked blacker, white bar on lesser coverts (partly obscured at rest but conspicuous in flight), pale grey rump and outer tail-feathers, with black central rectrices, and underparts plain mid-grey. ♀ warm orange-brown, with white chin/throat, orange-buff supercilia contrast slightly with darker brown crown and ear-coverts, mantle orange-brown streaked black, rump and outer tail-feathers plain orange-brown, central tail black, and wings blackish-brown with orange-brown outer fringes to flight-feathers but no white wing patch. Underparts dull orange-brown, whiter on belly. **BP** Bill, fine, black; eyes black; tarsi dark-grey. **Vo** A soft *zizz*, insect-like *jijijijijit* and harder *jahi jahi* and *tak-tak-tak* notes.

WHINCHAT

SIBERIAN STONECHAT

stejnegeri

ad ♂
worn

1st-win

ad ♂

PIED
BUSHCHAT

ad ♀

1st-win

ad ♂

GREY BUSHCHAT

ad ♀

ad ♂

ad ♀

ISABELLINE WHEATEAR CTKJ
Oenanthe isabellina

L 16–17cm; WT 21–39g. **SD** SE Europe across C Asia to upper Amur River, Mongolia, and Inner Mongolia east to W Jilin, China. Accidental migrant mainly in spring (Apr/May) to coastal China (Beidaihe), offshore Korea and offshore Japan. Monotypic. **HH** Favours stony plateaus, arid sandy areas with rocks on plains, and in hills in steppes and semi-deserts. **ID** Large, upright, sandy, grey-brown and buff wheatear lacking contrast between head, mantle and wings; head and bill rather large, tail rather short. Face plain, relieved by buff supercilia. Black lores and alula diagnostic, in combination with dark-centred median coverts always paler than alula (unlike similar plumage of Northern Wheatear). Ad. ♂ generally has blacker lores than ♀ or 1st-winter ♂, but sexes otherwise similar. Rump, uppertail-coverts and basal half of outer tail white, rest of tail blackish-brown; tail is shorter than in Desert Wheatear with more extensive white at base and sides. **BP** Black bill appears rather large; eyes black; long tarsi black. **Vo** A hard *chack* or *chek-chek*, and whistled *wiiu*. Song, given during short display flight with tail fanned, comprises repetitive, short sharp clear whistles mixed with harsher notes in rambling series: *vi-vi-vi-vew-vew-vew-vuy-vuy-vuy...* and includes mimicry of other passerines.

NORTHERN WHEATEAR CTKJR
Oenanthe oenanthe

L 14–17cm; WT 18–33g. **SD** Very wide range, from W Europe across Russia (except northernmost tundra), including Yakutia and Chukotka and across Bering Strait to N & C Alaska, but absent from Kamchatka and S Russian Far East. Also across N China; winters SW Asia and Africa. Accidentals have reached Jiangsu and Hebei (China), and Japan (Hokkaido to Tsushima, and Ogasawara and Iwo Is), also reported Korea. Subspecies in region is *O. o. oenanthe*. **HH** Highly terrestrial; prefers dry open grassy areas and open stony ground from steppe to tundra; on migration also agricultural land with short vegetation. **ID** Pale grey or grey-brown wheatear with short tail, long legs, strongly contrasting plumage and much white in tail. Very upright, bobs and flicks wings frequently. Typically flies fast and low. ♂ has pale ash-grey or blue-grey upperparts from forecrown to lower back with narrow white forehead band; lores and ear-coverts black; rump, uppertail-coverts and two-thirds of basal part of outer tail-feathers white, central rectrices and band across tip black, wings also black. Underparts off-white with variable peach wash to throat and breast. In autumn, upperparts tinged brownish and black mask is narrower and slightly duller. ♀ browner overall, lacks mask, instead has pale supercilium above dark eyestripe, wings strongly contrast with back, tail pattern as ♂ but duller (larger Isabelline Wheatear is very similar, but lacks contrast between wings and body). Juv. is more richly coloured above and below, with russet-brown upperparts and greater/median coverts, and rich warm-orangey-buff wash over much of underparts (Isabelline not quite as richly coloured as young Northern, but very similar, though lacks contrast between wings and body). **BP** Bill black; eyes black; tarsi black. **Vo** Sharp whistles, *wheeet*, and hard *tchak tchak* notes. Song, commonly given from rise in ground or exposed rock, occasionally in flight, is rapid and varied, including whistles, trills and squeaks interspersed with call notes.

PIED WHEATEAR CKJ
Oenanthe pleschanka

L 14.5–16cm; WT 16–22g. **SD** Ranges from SE Europe to NE China, east to Hebei. Accidental Japan and Korea. Monotypic. **HH** Dry open habitats, generally steppes or semi-desert areas and on migration appears in cultivated areas. **ID** ♂ breeding boldly patterned black, grey and white. In summer, crown and nape grey-white, mantle and upper back black, lower back, rump and most of tail-sides white (though outer feathers have black edges), central rectrices and narrow band across tail tip black. Black face, chin and throat are connected to black upperparts, creamy white breast, belly and vent, with buff wash to breast-sides. Winter ♂ retains black face and throat, but upperparts brown, wings blackish-brown with buff fringes to feathers. ♀ drabber, brown and white; upperparts and breast pale to mid brown, wing-coverts and flight-feathers blackish, contrasting with paler upperparts; has faint pale supercilia, clean white rump, tail pattern like ♂ and white belly. 1st-winter (both sexes) has pale fringes to feathers of mantle, back and scapulars, whilst ♂ has faintly indicated dusky throat patch. **BP** Bill fine, sharply pointed, black; eyes black; long tarsi black. **Vo** A harsh *tschak*, nasal *chep* and buzzy *brsche*.

DESERT WHEATEAR TJ
Oenanthe deserti

L 14–15cm; WT 15–34g. **SD** N Africa and Middle East to W & N China and Mongolia, wintering from NW India to Africa. *O. d. deserti* is an accidental visitor (Oct–Mar) to Japan and Taiwan. **HH** Arid, stony deserts with rocks and scattered low vegetation, in lowland plains and hills. **ID** Pale sandy-brown wheatear with considerable contrast between pale body and black wings and tail; tail all black, contrasting with white or buff tail base and rump. Rest of upperparts from forehead pale sandy-brown. Underparts off-white with buff wash strongest on breast. ♂ breeding has black face, chin/throat and neck-sides; wings black with brown fringes to some coverts, secondaries and tertials. Winter ♂ has black face/bib scaled orange-buff (white in 1st-winter) and wing-feathers broadly fringed orange-buff. ♀ breeding paler and greyer-brown overall, note especially pale lesser wing-coverts, has rufous-tinged cheeks, lacks black bib, and has brownish-black tail and wings; wing-coverts and tertials have black centres. Winter ♀ has paler wings; all feathers more extensively fringed orange-buff, alula black but does not contrast with rest of wing (see Isabelline Wheatear). **BP** Bill black; eyes dark brown; tarsi black. **Vo** A whistled *swii*, hard clicking *tsak* and muffled, rattled *tk-tk-tk*.

ISABELLINE WHEATEAR

ad ♀ spring

ad ♂ win

NORTHERN WHEATEAR

oenanthe

1st-win

ad ♀ spring

ad ♂ spring

PIED WHEATEAR

1st-win

ad ♀ sum

ad ♂ win

ad ♂ spring

DESERT WHEATEAR

deserti

1st-win ♀

ad ♂ win

RUFOUS-TAILED ROCK THRUSH   CJ
Monticola saxatilis

L 16–19cm; WT 40–65g. **SD** SW Europe to C Asia and Transbaikalia, winters in sub-Saharan Africa and S Asia; ranges east to Inner Mongolia and Hebei, China. Monotypic. Accidental Japan and E China. Monotypic. **HH** Favours arid montane slopes and dry hilly steppe with sparse vegetation. **ID** ♂ has greyish-blue head, neck, mantle and back, with white across middle of back; wings dark blackish-brown with grey fringes/tips to most feathers, tail blackish-brown with redstart-like orange-red uppertail-coverts and fringes to outer tail-feathers. Chin/throat and face grey-blue, sharply delineated from warm rusty-orange of entire underparts. ♀ duller greyish-brown, but shares tail pattern; upperparts grey-brown with buff spotting and scaling, underparts grey-brown on face, buff-brown below with orange wash to chest and dark scalloping over much of rest. Reddish tail distinguishes ♀ from ♀ Blue Rock Thrush. **BP** Bill fine, sharply pointed, quite long, grey/horn; eyes dark brown; tarsi black. **Vo** Varied including a hard *chack-chack*, shrike-like *ks-chrrr*, whistled *feweet* and thin *he-he hee*. **AN** White-backed Rock Thrush.

BLUE ROCK THRUSH   CTKJR
Monticola solitarius

L 20–23cm; WT 37–70g. **SD** SW Europe and NW Africa to SE Asia and coastal E Asia, wintering to Africa and Indonesia. *M. (s.) philippensis* is a summer visitor to NE China, Korea, parts of Russian Far East, Sakhalin, S Kuril Is and Hokkaido; resident in much of Japan south to Nansei Shoto, and parts of E China, and a common winter visitor to Taiwan and SE China. Some breed on rocky coasts of Taiwan. *M. s. pandoo* of C & E China is accidental to offshore Korea, Japan and Taiwan. Species merits splitting as **Blue Rock Thrush** (including *M. s. pandoo*) and **Red-bellied Rock Thrush** *M. philippensis*. **HH** Locally common in rocky steppe, rocky coasts, cliffs, crags, even harbours and coastal towns, where often perches prominently. **ID** ♂ has deep blue hood, upperparts, rump and thighs, blackish-brown wings with blue outer fringes, and blue-grey tail. *M. (s.) philippensis* has solid, deep chestnut underparts, whereas more wide-ranging continental *pandoo* is smaller and has all-blue underparts (though birds with restricted chestnut on vent reported). Imm. ♂ *philippensis* resembles ad., but has narrow scaling over blue and chestnut areas. ♀ is drabber, grey-brown with blue-grey cast to upperparts; underparts are spotted/scaled pale grey-brown on face, throat and upper breast, and barred on lower chest, belly and vent. **BP** Bill strong, rather long, black; eyes black; tarsi black. **Vo** Varied but unremarkable range of harsh calls: *tak-tak*, *ka-tchuc-tchuc*, *hee* or *chin*. Song (both sexes), given from perch or in flight, comprises sweeter, melodious whistled notes in short series of varied phrases: *hee choicho peechiyo, hiyochee pee pipipi chuu* or *tju-sri tjurr-titi wuchi-trr-trrt-tri*, and includes mimicry of other passerines.

CHESTNUT-BELLIED ROCK THRUSH   C
Monticola rufiventris

L 21–23cm; WT 48–61g. **SD** Pakistan through northern SE Asia to SE China, north to Fujian. Monotypic. **HH** Breeds in moist montane coniferous and oak forests at 1,000–3,000m, but descends to lowland forested and rocky areas in winter. **ID** ♂ deep blue, black and orange-red. Crown, nape, shoulders, rump and tail-sides deep shining blue; mantle blue-grey, wings black with extensive blue fringes to most feathers, lores, ear-coverts, throat and wings black; underparts from upper breast to vent bright orange-red. ♀ like large White-throated Rock Thrush, rather dull grey-brown, heavily scaled on breast and flanks, with prominent pale buff/white crescent on rear ear-coverts. **BP** Bill fine, sharply pointed, black; eyes large, irides dark brown; tarsi black. **Vo** Variable, includes *quock*, harsh jay-like and higher *tick* notes, and squirrel-like *chhrrr* in alarm. Song, from treetop or in flight, is a series of sweet, pleasant but subdued warbles: *teetatewleedee-tweet tew* or *twew-twi-er tre-twi teedle-desh*, also incorporating piercing whistles and upslurred notes. **AN** Chestnut-breasted Rock Thrush.

WHITE-THROATED ROCK THRUSH   CTKJR
Monticola gularis

L 16–19cm; WT 32–37g. **SD** Breeds only in NE China, Russia from Lake Baikal to Russian Far East, and N & C Korea, wintering to SE China and SE Asia. Uncommon migrant along Chinese coast, scarce in Korea, accidental but annual in Japan, also recorded Sakhalin. Monotypic. **HH** Montane mixed and coniferous forests with rocks and crags (above 200m in Amurland, 300–1,650m in the other parts of eastern Russia and over 1,500m in China), on migration also on coasts and in coastal forests. **ID** ♂ is a beautiful mix of cobalt-blue, black and orange; has shining deep blue crown, shoulders and tail-sides, a black mask, mantle and wings (latter with conspicuous white patch), and orange lores, malar and neck-sides. Underparts from breast to vent, also back and rump, bright rufous-orange; white throat patch noticeable when face-on. ♀ generally plainer, cooler brown, but has black-centred, buff-fringed feathers on back giving barred or scaled appearance, and black-fringed pale buff feathers to chest and flanks; malar, rear ear-coverts and throat distinctly white. Very upright stance. **BP** Bill short, thick, black, faintly yellow at gape; eyes large, irides black; tarsi pinkish-brown. **Vo** A soft *tsip* or *tseep* in flight, and harsh *chak* or *tack-tack*; song comprises various beautiful drawn-out rising whistles *swee wee lalee lu lu lu*. **AN** White-breasted Rock Thrush.

RUFOUS-TAILED ROCK THRUSH

1st-year ♂

ad ♀ sum

ad ♂ sum

ad ♂ *philippensis*

ad ♂ *philippensis*

ad ♂ sum

ad ♂ sum

BLUE ROCK THRUSH

ad ♂ *pandoo*

ad ♀ *pandoo*

CHESTNUT-BELLIED ROCK THRUSH

juv/1st-year ♂

ad ♂ sum

ad ♂ sum

ad ♂ sum

ad ♀ sum

ad ♂ sum

ad ♀ sum
Chestnut-bellied

1st-year ♂

ad ♂ sum

WHITE-THROATED ROCK THRUSH

BROWN-CHESTED JUNGLE FLYCATCHER   C
Rhinomyias brunneatus

L 15cm; WT 14–22g. **SD** *R. b. brunneatus* is a summer visitor to S & E China north to Fujian and Zhejiang, thus potential vagrant further N and E. Winters Malay Peninsula. **HH** Uncommon in broadleaf evergreen forest, plantations and bamboo thickets, favouring forest edge, from lowlands to 1,200m. **ID** Slightly larger and plumper than other brown or grey flycatchers of E Asia, but least distinguished. Ad. has mid-brown on upperparts, wings and tail. Face mostly plain brown, with buff supraloral line. White on chin/throat with some fine black scaling at sides, pale greyish-brown in band across breast and flanks, otherwise white below. Wings rather short, thus tail appears somewhat longer than *Muscicapa*. Juv. like ad., but upperparts scaly with buff tips, whilst wing-coverts and tertials have rufous-brown tips. **BP** Bill appears rather long and hook-tipped, dark grey above, pale dull yellow below, with rictal bristles at base; eyes large with buff eye-ring, irides dark brown; tarsi brownish-pink. **Vo** Call a harsh churring *trrrr*. Song a series of loud descending whistles.

SPOTTED FLYCATCHER   J
Muscicapa striata

L 13.5–14.5cm; WT 11.2–25.3g. **SD** Main range spans Europe to Lake Baikal. *M. s. neumanni* occurs east to W Transbaikalia; *M. s. mongola* east to SE Transbaikalia and N Mongolia. Winters E & S Africa. Vagrants have reached Bering Sea islands and offshore Japan. **HH** Open woodland and forest edge, also gardens. Upright stance on perch, from where sallies to catch insects; frequently flicks tail. **ID** Upperparts ash-grey, palest on forehead, with noticeable fine black streaking on forehead and crown. Weak eye-ring, faintly pale lores and faint dusky malar stripe, face otherwise plain grey-brown. Wings dark grey-brown with narrow pale fringes to coverts, inner secondaries and tertials. Tail grey-brown with pale outer fringes to outer feathers. Underparts have distinct pale greyish-brown streaks on white breast and flanks. Closely resembles Grey-streaked Flycatcher, but wings shorter, reaching about halfway to tail tip. Head less rounded and profile more elongated than other grey/brown flycatchers. **BP** Bill fine, black; eyes not as large as Grey-streaked, irides black; tarsi black. **Vo** Call either a short *seep* or harsh *tek-tek*, slightly longer *zee zeet-eet*, harder *eez-tuk-tuk* or rattling alarm *ch-r-r-r-r-rer*. Song comprises a more squeaky sequence of high-pitched notes: *sip-sip-see-sitti-se-see*.

GREY-STREAKED FLYCATCHER   CTKJR
Muscicapa griseisticta

L 12.5–14cm; WT 15.1–17.4g. **SD** E Asian breeding endemic: S & C Kamchatka, Kuril Is, N Sea of Okhotsk, Sakhalin, S Russian Far East, N Korea and adjacent NE China; scarce migrant in Hokkaido. Winters Indonesia and Philippines. On migration, fairly common Korea, offshore Japan and Nansei Shoto (uncommon main islands of Japan), Taiwan and E China. Monotypic. **HH** Generally in mature mixed broadleaf forest, but also larch forest, in hills and lowlands. Wider range of wooded and open habitats on migration. **ID** A rather plain *Muscicapa* with noticeably long wings; primary extension noticeably longer than in Dark-sided or Asian Brown Flycatchers (p.434), primaries reach close to tail tip. Closely resembles Spotted but more heavily streaked below, and has almost plain (not distinctly streaked) crown. Upperparts entirely dark ash-grey brown from crown to tail (Spotted has pale grey streaks on forehead and crown). Face plain, ash grey-brown, with prominent whitish lores and narrow eye-ring. White chin, throat and broad submoustachial stripe extend as neck bar. Dark malar streak broadens and merges into dark grey-brown streaking on white breast, sides and flanks. Clear, narrow dark streaking on white lower breast/upper belly distinguishes it from similar Dark-sided and Spotted. Undertail-coverts white. Flight-feathers blackish-brown, with narrow off-white fringes to tertials; narrow white fringes to greater coverts form a single distinct pale wingbar. **BP** Bill blackish, slightly longer and thicker than in Dark-sided; eyes large (but less so than Asian Brown), with indistinct pale eye-ring, irides black; tarsi blackish-grey. **Vo** Call a whispery, thin *heest*, slightly rising, also a thin *tsuii*, *chii* or *speet-teet-teet*; in alarm a plaintive *tsr tsr*. Song, typically given from high dead branch, a formless high twittering. **AN** Grey-spotted Flycatcher.

DARK-SIDED FLYCATCHER   CTKJR
Muscicapa sibirica

L 13–14cm; WT 8.5–12g. **SD** Summer visitor (May–Oct) to Himalayas and S China, also east to Japan and Kamchatka: *M. s. sibirica* ranges from upper Ob and Transbaikalia east along N Sea of Okhotsk to S Kamchatka and Kuril Is, south through Russian Far East, Sakhalin, N Japan, NE China and N Korea; migrant Taiwan. Winters in S China and SE Asia. **HH** Generally in montane regions, in mixed taiga with predominance of conifers, but also mature mixed broadleaf forest, in undergrowth and mid levels. On migration and in winter occurs in lowland habitats ranging from urban parks to various types of woodland. **ID** Dark-sided, Grey-streaked and Asian Brown Flycatchers all have a rather large-headed appearance and very upright stance. Dark-sided is a medium-sized, rather plain flycatcher, with shorter wings than Grey-streaked, the primaries reaching only mid tail. Upperparts dark grey-brown from crown to tail. Face plain, dusky with pale grey lores (less distinct than Asian Brown, see p.434), thus pale eye-ring is more prominent. Pale submoustachial indistinct, but dark malar contrasts strongly with white chin/throat and pale or white partial collar. Underparts dirty white, with dusky grey-brown wash across breast and onto flanks, with broad, diffuse streaks on centre and sides of breast/flanks, and greyish-brown spots on undertail-coverts (occasionally lacking). Underparts darker than Asian Brown or Grey-streaked, but variable and confusing. Pale fringes to greater coverts form single pale buffy-brown wingbar; pale fringes to tertials form a prominent wing-panel. Primary length intermediate between Asian Brown and Grey-streaked, exposed primaries longer, by 15–20%, than tertials. **BP** Bill shorter and less deep-based than Grey-streaked, shorter than Asian Brown, and blackish base to lower mandible slightly paler; eyes large, with pale eye-ring, sooty to mid-grey lores, irides black; tarsi black. **Vo** Call a thin *tsuii* or *chii* and downslurred *feeeer*; flight call a thin straight *siht*. Song comprises a weak series of high-pitched notes, trills and whistles that is quiet, easily overlooked and very similar to Asian Brown, and usually includes *tsichiriri* or *tsee-tsee-tsee* notes. **AN** Sooty Flycatcher; Siberian Flycatcher.

BROWN-CHESTED JUNGLE FLYCATCHER

brunneatus

ad

SPOTTED FLYCATCHER

mongola

ad worn

ad fresh

GREY-STREAKED FLYCATCHER

ad worn

ad fresh

DARK-SIDED FLYCATCHER

sibirica

1st-win

ad

ASIAN BROWN FLYCATCHER CTKJR
Muscicapa dauurica

L 12–14cm; WT 7.8–16g. **SD** C Siberia to Japan. *M. d. dauurica* occurs across Russian taiga from Yenisei River, Transbaikalia east to Amurland and Ussuriland, Sakhalin, S Kuril Is, Japan, NE China and N & C Korea. Winters uncommonly in Taiwan, and outside region in S China, SE Asia, Philippines and Indonesia. On migration fairly common through Korea, Japan, Taiwan and E China. **HH** Generally in mature mixed broadleaf forest, in temperate zone, and larches and birches in southern boreal region. Wider range of wooded and open habitats on migration. **ID** Small, rather plain flycatcher, lacking prominent field marks (see Dark-sided and Grey-streaked, p.432), but with large head and short tail. Upperparts entirely plain greyish-brown from crown to tail. Face very plain, pale grey-brown, with prominent pale lores and narrow band above bill, and distinct eye-ring. White of chin/throat blends subtly into grey-brown cheeks, but may show indistinct malar. Lacks bold malar and half-collar of Dark-sided. Underparts off-white with grey-brown wash to breast-sides and flanks, but far less heavy than Dark-sided; sometimes faintly streaked on breast; undertail-coverts unmarked white. Wings plainer than either Dark-sided or Grey-streaked, with only very indistinct whitish wingbar and faint wing-panel formed by pale tertial fringes (though more prominent in 1st-winter). Primary length shorter than other related flycatchers, exposed primaries shorter (*c.*80–90%) than tertials. **BP** Bill blackish, with prominent pale base to lower mandible, longer than Dark-sided; eyes large (largest of grey/brown flycatchers), with pale grey lores, narrow but prominent off-white eye-ring, and black irides; tarsi black. **Vo** Call a quite hard *trrr* followed by *dit tit it* notes, or some *chuck* notes; also a thin *tsui* or *chee*. Flight call a thin *siht* sounding identical to Dark-sided Flycatcher. Song subdued and thin but varied, a drawn-out jumble of complex chatters, warbles and chuckles, *tsii-chiriri-chopiriri* at a high pitch that is easily overlooked, slower than Dark-sided. **TN** Formerly *M. latirostris*.

FERRUGINOUS FLYCATCHER CTKJ
Muscicapa ferruginea

L 12–13cm; WT 9–16.7g. **SD** Summer visitor to Himalayas and C & S China, wintering Hainan and Philippines to Malay Peninsula, Borneo, Sumatra and Java. On migration in SE China north to Fujian and Zhejiang. Summer visitor to Taiwan, near-annual vagrant to S Japan (Mar/Apr), accidental Korea. Monotypic. **HH** Shady forest clearings and wooded stream sides, but prefers to breed in primary montane forest (1,200–2,400m) in Taiwan. **ID** Small, warm-coloured *Muscicapa*. Broad head of ad. is grey, with pale buff lores and prominent white eye-ring, chin/throat and neck-sides white, bordered by narrow dark malar; underparts dull to bright orange-brown, brightest on flanks and vent, belly whitish. Upperparts warm mid brown on mantle and back; rump, uppertail-coverts, and outer tail-feathers warm orange-brown, central rectrices and tail tip dark brown. Wings blackish-brown with orange fringes to greater coverts forming distinct orange wingbar, tertials and secondaries also fringed orange. Juv. like ad., but has spotted grey

head and large pale orange-buff spots on mantle and underparts. **BP** Bill short, dark grey with yellow base to lower mandible; eyes appear large, with white eye-ring, irides black; tarsi black. **Vo** Very high, sibilant *tsiiiii*, ending abruptly. Song harsh and shrill consisting of high-pitched notes, *tsit-tittu-tittu*, followed by 2 or 3 high whistles.

EUROPEAN PIED FLYCATCHER J
Ficedula hypoleuca

L 12–14cm; WT 9.7–22.2g. **SD** Breeds from SW Europe to Urals and Siberia. *F. h. sibirica* breeds in W & SC Siberia east to upper Yenisei. Winters in WC Africa. Accidental Japan (Oct). **HH** Mature open deciduous and mixed forests. **ID** ♂ a boldly pied *Ficedula*, glossy black on head, face and entire upperparts including wings and tail. Forehead white or very narrowly divided into two small round patches on forehead, and underparts bright white from chin to vent. Closed black wings have broad white patch on greater coverts connected to broad white fringes of tertials (similar extralimital Collared Flycatcher *F. albicollis* slightly larger, more robust with larger bill, ♂ has broader white forehead patch, narrow white hind-collar, broader and more prominent white bar on closed wing). *F. h. sibirica* also has grey morph, in which head, face and mantle are dark grey, rather than black. ♀ warm brown on head, face and upperparts, whitish on chin, but pale ash-brown on underparts; wings show similar but less boldly marked panel than ♂. Differs from ♀ Collared (which has paler mantle and rump, more grey than Pied) by whiter tips to greater and median coverts and broad white patch on primary-coverts. Collared also has larger white patch extending from wingbar and almost reaching edge of wing. **BP** Bill black; eyes black; tarsi black. **Vo** A soft tongue-clicking *teck*, a more metallic and more repetitive alarm note *pik pik pik*, and when excited *whee-tic*.

YELLOW-RUMPED FLYCATCHER CTKJR
Ficedula zanthopygia

L 13–13.5cm; WT 9–15.5g. **SD** Restricted to E & NE China, E Transbaikalia, S Russian Far East and Korea. Winters in peninsular SE Asia and Sumatra. Rare migrant to offshore Japan, common Korea, also occurs Taiwan (accidental) and E & S China. Monotypic. **HH** Generally in mature mixed broadleaf forest, often near water. Wider range of habitats (including woodland, parks and scrub) on migration. **ID** Small, very bright *Ficedula*. ♂ largely black, white and yellow. Black from crown to tail, with broad white supercilia from above lores to behind eye, prominent white wing patch (on inner greater coverts) and white tertial fringes. Entire underparts, and lower back and rump, bright lemon-yellow. ♀ plainer, olive-green and yellow, lacking contrast; upperparts plain olive-green, underparts off-white, wings largely olive with white greater covert bar and tertial fringes forming fairly prominent patch on folded wing, and prominent bright lemon-yellow rump. In flight, white innerwing patch and white tertial fringes of ♂ contrast with black wings and yellow back/rump; ♀ in flight has prominent rump and less distinct wing patches (see Narcissus, p.436). **BP** Bill black; eyes black, ♀ has narrow white eye-ring; tarsi black. **Vo** Call resembles Taiga Flycatcher, but deeper, harder rattling *trrrt*; also a more rippling *prrrip-prrrip*. Song similar to Narcissus, but shorter, more thrush-like whistling: *tji tji tjiririri…* **AN** Tricoloured Flycatcher.

ASIAN BROWN
FLYCATCHER

dauurica

FERRUGINOUS FLYCATCHER

ad worn

ad fresh

ad

EUROPEAN PIED
FLYCATCHER

sibirica

ad ♀

ad ♂

YELLOW-RUMPED
FLYCATCHER

ad ♀

ad ♂

NARCISSUS FLYCATCHER CTKJR
Ficedula narcissina

L 13–13.5cm; WT 11–12g. **SD** Range-restricted NE Asian flycatcher. Breeds only in Ussuriland, Sakhalin, S Kuril Is and Japan (Hokkaido to S Kyushu). Scarce migrant S Korea and Taiwan; winters Borneo. Monotypic. **HH** Mature mixed broadleaf deciduous forest with some conifers in north, and broadleaf evergreen forest in south; prefers shady areas. On migration appears in wide range of habitats including urban parks. **ID** Small, very colourful flycatcher. Ad. ♂ black from forehead to tail, with long yellow supercilium broadening noticeably behind eye; wings black, with small but prominent white innerwing patch (tertials lack white or present only at bases); lower back and rump bright rich yellow, the rump feathers often erected prominently in display. Underparts grade from bright, deep orange on chin/throat to bright yellow on breast and flanks, and white on lower belly/vent. 1st-winter ♂ resembles ad. but frequently has much of nape and mantle dark olive- or greyish-green, often bordered black, and has greyer mantle and wings. ♀ plain olive-brown or olive-green (closely resembling ♀ Blue-and-white Flycatcher, see p.440), with faint pale supercilium before eye, lacks yellow rump patch and white wing patch of Yellow-rumped (p.434); rump dull green or yellowish-olive, uppertail-coverts and tail fringes somewhat richer brown; wing-coverts and tertials darker grey than ♀ Blue-and-white, from which best separated on size. Underparts off-white; breast-sides and flanks brownish-white, throat and belly slightly yellow. In flight, ♂ has small white innerwing patch and bright yellow rump; ♀ plain (see Yellow-rumped). **BP** Bill slate-blue or black; eyes black; tarsi slate-blue to bluish-grey. **Vo** A soft *tink-tink* and bubbling, rising *brrrut* (thicker in tone than Ryukyu or Mugimaki Flycatchers). On migration, a series of plaintive upslurred whistles like a pheasant chick, *puee puee puee*, or series of downslurred plaintive *piu piu piu* notes interspersed with quite deep tutting *tchuk* notes. Highly vocal and attractive songster; song extremely varied and including deft mimicry. Song a 3–4-part series of warbled or whistled notes; the motif *cho-tee-cho-turr* with second syllable higher is common; and a Chinese Bamboo Partridge-like phrase *pi-pe-poi* is frequently incorporated. ♂ gives an urgent subsong full of metallic *zink* and whistled notes in defence of territory.

CHINESE FLYCATCHER CTKJ
Ficedula elisae

L 13–13.5cm. **SD** Most restricted range of the colourful NE Asian flycatchers; breeds only in Hebei and Shaanxi, E China; accidental Taiwan and W Honshu, Japan; reported Korea. Winters south to peninsular SE Asia. Monotypic. **HH** Broadleaf forest, preferring shady areas. **ID** Resembles Narcissus but has shorter wings. Furthermore, ♂ has dull greenish-olive head, but bright yellow supraloral stripe to just above eye and narrow eye-ring (lacks broadening yellow supercilium of Narcissus); dark

greyish-olive mantle and scapulars; rump pale or bright yellow; wings greyish-black with larger white coverts patch than Narcissus; tail dark greyish-black; underparts plain yellow (lacks orange throat of Narcissus). 1st-spring ♂ has been erroneously described as full species ('*F. beijingnica*'), but song analysis has confirmed identity as *F. elisae*. ♀ has olive-green upperparts, pale yellow patch above lores and dull yellow or yellowish-buff underparts. **BP** Bill black; eyes black; tarsi black. **Vo** A low *tok tok tok*, purring *tchook tchook* and sharp *tek tek*. Song awaits transcription. **TN** Formerly within Narcissus Flycatcher. **AN** Elisa's Flycatcher; Green-backed Flycatcher.

RYUKYU FLYCATCHER TKJ
Ficedula owstoni

L 13–13.5cm. **SD** Endemic to Nansei Shoto; Tanega-shima and Amami south to Yaeyama Is where largely resident, but reported from Taiwan north to Kuroshima (off Kyushu), and even Korea. Monotypic. **HH** Primary and secondary broadleaf evergreen subtropical forest (even gardens on Yaeyama Is), preferring shady areas. **ID** Similar to Narcissus, but slightly smaller with shorter wings and primary projection. Ad. ♂ is less black than Narcissus, with dark olive-green crown, mantle and upper back, and orange-yellow lower back and rump, fairly prominent but small wing patch (formed by white tertial fringes); supercilium less golden-yellow than Narcissus and underparts pale yellow, lacking orange-tinged throat of Narcissus. 1st-year ♂ (may breed and sing in this plumage), closely resembles ♀; rather plain olive-brown, with blackish- or greenish-brown crown, upper ear-coverts and eyestripe, lacks supercilium; face and underparts to mid belly yellowish-brown (lacking strong orange suffusion of ad. ♂ Narcissus, and paler than 1st-year ♂ Narcissus), vent and lower belly off-white; mantle, nape and lower ear-coverts dark moss-green. Wings blackish-brown, with paler brown fringes to coverts and tertials. In 1st-year ♂, first feature to develop is white on tertial webs. ♀ perhaps inseparable from ♀ Narcissus, but upperparts tinged green (somewhat browner in Narcissus), imm. frequently has yellowish eye-ring. **BP** Bill black, but heavier than Narcissus, may appear slightly upturned; eyes black; tarsi dark lead grey. **Vo** A piping, downslurred but insistent *piu...piu...piu...* sometimes interspersed with a low *tuc tuc* or low, strong *chuc* recalling Dusky Warbler. Similar to some Narcissus calls, but lacks latter's distinctive *prrit*. Song comprises short, simple phrases (closer to Yellow-rumped than Narcissus) – a rather constant three-part warbling phrase, with intervals of 1–2 seconds between each part: *chur-lee.... tridlee... chidlee....* (lacks mimicry of Narcissus). Like Narcissus, has an aggressive subsong given near ground in presence of intruders, a complex series of metallic churrs and rattles, some very high-pitched. **TN** Formerly within Narcissus Flycatcher.

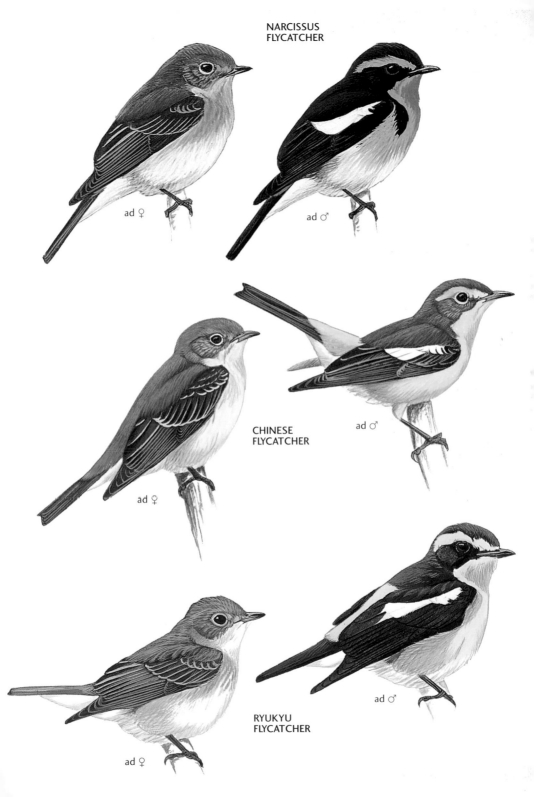

NARCISSUS
FLYCATCHER

ad ♀

ad ♂

CHINESE
FLYCATCHER

ad ♂

ad ♀

RYUKYU
FLYCATCHER

ad ♀

ad ♂

MUGIMAKI FLYCATCHER CTKJR
Ficedula mugimaki

L 12.5–13.5cm; WT 9.5–12g. **SD** Breeds from Tomsk and Krasnoyarsk east to S Russian Far East, Sakhalin, NE China and N Korea. Winters in S China, SE Asia, Malaysia, Indonesia and Philippines. Scarce migrant offshore Japan (rare main islands), common Korea, also occurs Taiwan and E China. Monotypic. **HH** Favours rather shady areas of mature mixed broadleaf forest, damp montane and lowland taiga, and a wider range of habitats on migration including conifers. **ID** Small, brightly marked flycatcher, appearing plump, round-headed and short-billed. Ad. ♂ largely black from face to tail, with short white streak above and behind eye, bright orange from chin to lower breast; belly white. Black wings have large patch formed by white median coverts, and narrow white fringes to tertials and basal sides to tail, making it very distinctive in flight. Juv. resembles ad., but is dark grey instead of black and orange of underparts less bright. ♀ less contrasting, mid-brown where ♂ is black, with narrow greater covert bar and tertial fringes, lacks white in tail base; underparts washed pale orange. **BP** Bill short, black; eyes large, irides black; tarsi grey or black. **Vo** *Beerirri* calls similar to, but softer than, Narcissus, also a rattling *turrt*, low *chuck* and soft *tyu*. Song consists of loud trills.

SLATY-BACKED FLYCATCHER CT
Ficedula hodgsonii

L 13–13.5cm; WT 8.5–11g. **SD** Himalayas to C China, also northeast to Shanxi and Beijing, and has reached coast (Happy I.). Winters E India to SE Asia. Monotypic. **HH** Dense, moist forest at high altitudes (>2,000m) in summer, lower in winter; and generally in middle or lower strata of forest. **ID** Small, plain, two-tone flycatcher. ♂ dull slaty-blue (lacks iridescence) on face and upperparts from crown to rump, black on lores and cheeks, with rufous-orange chin to belly, paler on lower belly and vent. Wings uniform slaty-blue, tail blue-black with white bases to outer feathers. ♀ dull, ashy- or olive-brown above, with distinct pale eye-ring and supraloral line; lacks white in tail base, but has very slightly rufous uppertail-coverts; somewhat paler greyish-buff on underparts. **BP** Bill black; eyes black; tarsi dark grey. **Vo** A hard *tchat*, *chink* or *chink-it*, deep rattled *terrht* or *tchrt*; song is a short, rapid descending series of flute-like whistles: *per-ip-it-u* or *per-ip-it-tu*. **AN** Rusty-breasted Blue Flycatcher.

RUFOUS-GORGETED FLYCATCHER C
Ficedula strophiata

L 13–14.5cm; WT 10–15g. **SD** From Himalayas to C China and locally in SE Asia, wintering to S China and SE Asia. *F. s. strophiata* occurs closest to region. Accidental northeast to Beidaihe. **HH** Broadleaf and mixed montane forests (1,000–3,000m), where favours lower levels, but descends in winter. **ID** Small, distinctively marked flycatcher. ♂ smart with black face and chin, narrow white forehead over lores to eye; crown and upperparts including wings dark olive-brown, uppertail-coverts and most of tail black, with prominent white flashes at base of outer feathers. Dark grey borders to black face, particularly on ear-coverts, neck-sides and

breast, offset by bright orange-red patch in centre of throat, rest of underparts brownish-buff to pale buff, vent white. ♀ very similar, but less sharply marked. Juv. heavily spotted, like young European Robin. **BP** Bill black; eyes large, irides black; tarsi dark grey. **Vo** Various calls including a high-pitched, upslurred *pee-tweet*, hard *tchuk-tchuktchuk*, sharp *pink* and harsh *trrt*. Song a thin *zwi-chirr rri* or *zreet-creet-creet-chirt-chirt*. **AN** Orange-gorgeted Flycatcher.

TAIGA FLYCATCHER CTKJR
Ficedula albicilla

L 11–13cm; WT 8–14g. **SD** Breeds across Russian taiga from Urals to Yakutia, inland Chukotka, south through Koryakia and Kamchatka. Migrates through E Asia to winter from E India to S China and N SE Asia. Uncommon migrant Korea and E China, scarce migrant and rare winter visitor Japan (Oct–May); locally not uncommon winter visitor to coastal Taiwan. Monotypic. **HH** Breeds in mixed deciduous and spruce taiga. On migration/winter found more widely in woodland, orchards, parks and forest edge – favours clearings, rides and glades. **ID** Small *Ficedula* with a large head and typically cocked tail. Crown brown giving 'capped' appearance, otherwise upperparts plain ash-brown with dark brown wings (some pale fringes to coverts and tertials), black or dark blackish-brown tail with prominent white basal patches, and uppertail-coverts blacker than tail in both sexes. ♂ breeding has warm rusty-orange on very restricted area of chin and throat, bordered broadly at sides and below with grey, grading into off-white underparts with pale peach suffusion. Face plain, lores, malar and ear-coverts grey. Acquires ♀-like winter plumage. 1st-summer ♂ already resembles ad. ♀ lacks red throat, being more uniform grey-buff below with small white throat patch. **BP** Bill quite broad-based, almost entirely black, deeper than Red-breasted; eyes prominent in plain face, irides dark brown; tarsi black. **Vo** A rolling, but hard fast, dry trilling or clicking *trrrt* or *trrrrr* (faster than Red-breasted), recalls a quiet Mistle Thrush. Also a softer *tic* and hard *tzit*. Song consists of a fast phrase of trills and high whistled notes, with characteristic trilled *r* sounds. **TN** Formerly within Red-breasted Flycatcher.

RED-BREASTED FLYCATCHER TKJ
Ficedula parva

L 11–13cm; WT 8.5–11.5g. **SD** Breeds from S Scandinavia across C & E Europe to Ukraine and Caucasus, wintering in Pakistan and India. Accidental Taiwan, Korea and Japan. Monotypic. **HH** Breeds in mixed deciduous and spruce forests. On migration/winter found more widely in woodland, orchards, parks and forest edge – favours clearings, rides and glades. **ID** Small *Ficedula*, closely resembles Taiga but wings marginally shorter, and uppertail-coverts paler than or concolorous with tail. ♂ breeding has extensive warm rusty-orange throat and upper breast, merging into white belly. Retains red bib throughout winter. 1st-summer ♂ resembles ♀, but may defend territory. ♀ lacks red throat patch, being more creamy buff below with a larger, more diffuse white throat. **BP** Bill quite broad-based, black above with reddish-brown lower mandible tipped darker (compare Taiga); eyes large and prominent on plain face, narrow off-white eye-ring, irides dark brown; tarsi black. **Vo** A slurred, rattled *serrrt* or wren-like *drrrrrr* (softer and slower than Taiga). Song opens rhythmically, but terminates in characteristic series of descending notes: *diu-tvi-diu-tvi*.

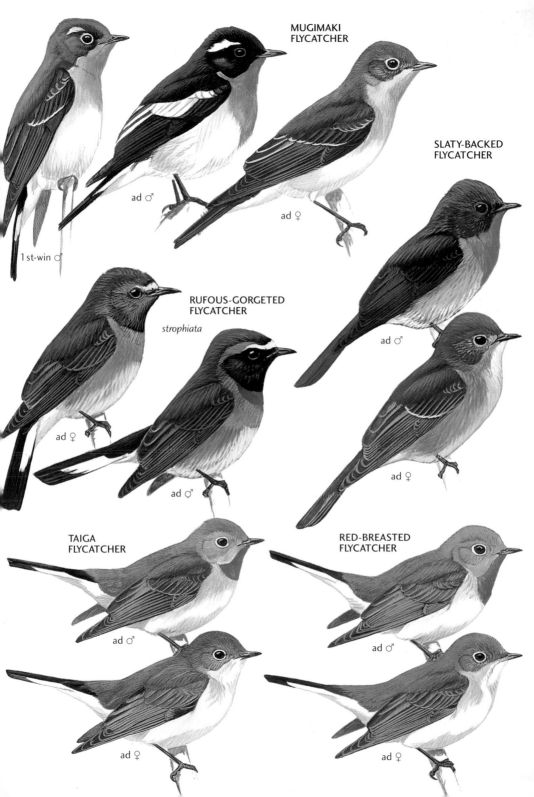

MUGIMAKI FLYCATCHER

1st-win ♂

ad ♂

ad ♀

SLATY-BACKED FLYCATCHER

ad ♂

ad ♀

RUFOUS-GORGETED FLYCATCHER

strophiata

ad ♀

ad ♂

TAIGA FLYCATCHER

ad ♂

ad ♀

RED-BREASTED FLYCATCHER

ad ♂

ad ♀

SNOWY-BROWED FLYCATCHER   T
Ficedula hyperythra

L 11–13cm; WT 6–10g. **SD** Himalayas to Indonesia. *F. h. innexa* endemic to Taiwan. **HH** Shady forest, bamboo and scrub at mid to upper elevations (*c*.1,000–2,000m in Taiwan). Descends (as low as *c*.200m) in winter. Forages low in forest, including on ground. **ID** A small, short-tailed flycatcher somewhat recalling Red-flanked Bluetail. ♂ *F. h. innexa* has deep blue head and face, and ash-blue upperparts; wings black with blue fringes to flight-feathers and blue coverts; rump blue, short, square-tipped tail brown with ash-blue edges. Forehead and stripe back to eye snowy white, lores black. Underparts peach/orange on chin/throat, brighter orange on neck-sides, olive on flanks, white from breast to vent. ♀ dark ash-grey on head with yellowish-brown face; mantle, scapulars and back ash-grey; rump and tail brown, wings also dark brown with olive-grey fringes to wing-coverts and orange-brown outer fringes to flight-feathers. Lores pale buff-grey, chin/throat pale buff, underparts white with orange-buff wash to neck-sides and chestnut on flanks. **BP** Bill fine, black; eyes appear large in plain face, irides black; tarsi pale grey to pink. **Vo** Call is a thin upslurred *chee* or *seep* and oft-repeated softer or thinner *sip*. High-pitched song a rather quiet, descending series of evenly spaced notes, *tsit-tsip-tsee-tswiii...* or *chee-chee-chee*. **AN** Thicket Flycatcher; Rufous-breasted Blue Flycatcher.

SLATY-BLUE FLYCATCHER   C
Ficedula tricolor

L 12.5–13cm; WT 6–10g. **SD** W Himalayas to S China, wintering just to south of breeding range or at lower elevations. *F. t. diversa* breeds WC China, wintering to S China, and N Thailand, Laos and Vietnam. Has strayed into region, to Hebei (Beidaihe). **HH** Typically in subalpine evergreen montane forest and shrub above 1,500m, but lower in winter when found in undergrowth and scrub. **ID** A small, dark flycatcher. ♂ *F. t. diversa* is dull slate-blue from crown to rump, black on forehead, face, wings and tail (with white basal patches recalling larger Blue-and-white Flycatcher). Off-white to grey-buff throat merges with dull buff breast and flanks. ♀ olive-brown above, with pale eye-ring and lores, throat pale buff, underparts brownish-buff, rump and tail rufous-brown. **BP** Bill black; eyes large, irides dark brown; tarsi black. **Vo** A rapid ticking *tek-tek-tek* and in alarm a distinctive *ee-tick* or *chreet-tic-tic*.

BLUE-AND-WHITE FLYCATCHER   CTKJR
Cyanoptila cyanomelana

L 16–17cm; WT *c*.25g. **SD** E Asian breeding endemic: *C. c. cumatilis* a summer visitor (Apr–Oct) to restricted range in NE China, N Korea and S Russian Far East (accidental Kyushu); *C. c. cyanomelana* summer visitor to S Kuril Is, S Korea and Japan from Hokkaido south. Migrant through E China and Taiwan. Winters in Philippines, SE Asia and Borneo, with some in S Taiwan. **HH** Lowlands to *c*.1,200m in forested mountains, generally in mature mixed broadleaf forest with dense undergrowth, often near streams, rivers or waterfalls; on migration also in parks. **ID** A large flycatcher, ad. ♂ is bright, whilst ♀ is especially plain. Ad. ♂ has shining deep blue upperparts from crown to tail including wing-coverts, with blue outer fringes to tertials, secondaries and primary bases; forehead may appear silvery blue. Face, throat, breast and sides black, belly, vent and sides to tail base pure white. 1st-winter/1st-year ♂ resembles ♀, but has blue wings, rump and tail. ♀ generally warm, pale to mid brown above, rather rufous on rump and tail, with rufous-brown fringes to tertials and inner flight-feathers, and a very plain face; olive-grey on neck and breast, and off-white belly and vent. ♂ *C. c. cumatilis* has turquoise or azure-blue crown and upperparts (cobalt-blue in *C. c. cyanomelana*) with tinge of green, and deep greenish-blue or aquamarine face to breast and flanks (instead of black), throat has greenish gloss; neck and mantle lack purple tinge of *C. c. cyanomelana*, flanks pale grey, and has white centre to tail base (rather than sides). ♀ *cumatilis* closely resembles ♀ *cyanomelana* but is generally darker or more rufous-brown with less white on throat. **BP** Bill black; eyes large, appearing particularly so in plain face of ♀/juv., irides black; tarsi black. **Vo** A strong *tchuck* or *chik chuk*. A persistent songster, typically from high perch, giving a varied, short fluty warble with a slightly cracked and melancholic tone: *hi-hwi-pipipipi tsi tsi tsi* or *piii hii hii piii chuichui*, ending with a soft *jitt jitt*, or *hii-rii-rii-chichin*. ♀ occasionally gives short version of ♂'s song.

ASIAN VERDITER FLYCATCHER   CTK
Eumyias thalassinus

L 15–17cm; WT 15–20g. **SD** India to S China, SE Asia and Indonesia, with some of race *E. t. thalassinus* wintering in SE China north to Fujian. Accidental Taiwan and vagrant Korea. **HH** Favours open mixed lowland and low montane forest, including pines, often near habitation. **ID** A large, rather plain turquoise flycatcher, with a longish tail and upright stance. ♂ entirely greenish-blue except black lores, black ventral streaking, and black wings and tail (heavily fringed turquoise). ♀ similar to ♂ but has dusky-grey lores. Juv. duller, grey-brown with turquoise wash, and buff and black spotting/scaling. **BP** Bill blackish-grey; eyes large, irides black; tarsi black. **Vo** A short, plaintive *pseeut* and longer, drier *tze-ju-jui*. Song, from high perch, starts hesitantly then runs down scale, *p'p'pwe...p'p'pwe... pe-tititi-wu-pitititi-weu*, ending abruptly.

SNOWY-BROWED FLYCATCHER

innexa

ad ♀

ad ♂

SLATY-BLUE FLYCATCHER

diversa

ad ♀

ad ♂

ad ♀
cyanomelana

ad ♂
cyanomelana

BLUE-AND-WHITE FLYCATCHER

ad ♀
cumatilis

ad ♂
cumatilis

ad ♂

ad ♀

ASIAN VERDITER FLYCATCHER

thalassinus

HILL BLUE FLYCATCHER K
Cyornis banyumas

L 14–15.5cm; WT 14–17g. **SD** E Himalayas to Indonesia, with some local and altitudinal migration. Accidental offshore Korea; subspecies probably *C. b. whitei*. **HH** Favours shady areas of primary and secondary broadleaf forest. **ID** Medium-sized, blue, orange and white (♂) or grey, brown and buff (♀) flycatcher. ♂ has largely deep dark blue upperparts, with bright pale blue forehead band and blue-black face; exposed fringes to blackish flight- and tail-feathers also blue. Orange on chin, throat and flanks, white on belly and vent. ♀, which has pale buff eye-ring and lores, is generally greyish-brown where ♂ is blue, but has dull rufous rump and brown tail. Pale orange-buff on throat, pale orange on breast, grading to white on vent. **BP** Bill black; eyes dark brown; legs violet-grey. **Vo** Calls range from a soft *tac* to a hard *chek-chek* and scolding *trrrt-trrt-trrt*. **TN** Formerly within Large Blue Flycatcher *C. magnirostris*. **AN** Large-billed Blue Flycatcher.

BLUE-THROATED BLUE FLYCATCHER C
Cyornis rubeculoides

L 14–15cm; WT 10–20g. **SD** Himalayas to SE Asia, also W & C China. Vagrant northeast into the region to Hebei; race is *C. r. glaucicomans*. **ID** ♂ has all-blue hood including face and chin; upperparts, wings and tail also blue, black only on lores and primaries. Bright orange chest, with white belly and vent. ♀ has earth brown or grey-brown upperparts, face and chin, with buff eye-ring and lores, and orange throat to breast, white below. **BP** Bill black; eyes dark brown; tarsi pink to dark blue-grey. **Vo** A soft *tac* or rasping *trrt-trrt* in alarm.

FUJIAN NILTAVA C
Niltava davidi

L 18cm. **SD** S China and N SE Asia. In region, has restricted range in montane Fujian. Accidental northeast to Hebei (Beidaihe). Monotypic. **HH** Favours dense broadleaf evergreen forest under-growth on middle montane slopes, but on migration occurs at lower elevations and in scrub and gardens. **ID** Medium-sized, upright, somewhat large-headed flycatcher resembling Vivid Niltava. ♂ has black forehead, face, ear-coverts, chin and throat, with bright, shiny blue patch on neck-sides, with clear demarcation between black throat and bright orange-red underparts. Upperparts bright blue, brightest on forecrown, rump and tail; scapulars shining blue, wings blue-black. ♀ has pale ashy-brown head, breast, mantle and scapulars, with rufous-brown tail and wings; face rather plain, relieved by pale loral spot and eye-ring, narrow white chin patch, broader white fore-collar below throat, and small blue patch on neck-sides; central belly and ventral region white. **BP** Bill strong, black with prominent rictal bristles; eyes appear large on plain face, irides black; tarsi black. **Vo** Metallic *tit tit tit* notes and a harsh rattle *trrt trrt trrt tit tit trrt trrt*.

VIVID NILTAVA TJ
Niltava vivida

L 18–19cm; WT 17g. **SD** NE India to SE Asia. *N. (v.) vivida*, **Taiwan Niltava**, endemic to Taiwan, may be separate species. Accidental in Nansei Shoto (Okinawa and Yonaguni-jima). **HH** Mid-elevation evergreen and mixed deciduous forest at *c*.1,000–2,500m in Taiwan, where favours middle strata and canopy. Descends (as low as 200m) in winter. **ID** Medium to large, upright, somewhat large-headed flycatcher with high-peaked crown and long, round-tipped tail. ♂ has shining deep blue hood and upperparts including tail, darker blackish-blue wings, black or blue-black lores and face, and deep orange upper breast to vent; young ♂ has speckled brown head and dull grey-blue upperparts, wings and tail. ♀ superficially resembles ♀ Blue-and-white (p.440), but has rather greyer head with buffy-orange forehead, lores, face and throat; underparts plain grey; upperparts grey-brown, paler on rump, blackish-brown tail has rufous fringes to outer feathers, but rounded tip (square in Blue-and-white). **BP** Bill slightly hook-tipped, black with prominent rictal bristles (lacking in Blue-and-white); eyes appear large on plain face, irides dark brown; tarsi black. **Vo** Call is a clear whistled *yiyou-yiyou* and a hard, penetrating *pit* with a metallic ring. Song is a short series of 1 to 3 high-pitched whistles with a twittering flourish of higher and then lower notes: *tseeoo tyüdü tswee-prrit*, the opening, wavering, note similar to the whistled *tiLÜI* of Siberian Rubythroat.

GREY-HEADED CANARY-FLYCATCHER CTR
Culicicapa ceylonensis

L 12–13cm; WT 6–9g. **SD** Indian subcontinent to SE Asia and Indonesia. *C. c. calochrysea* occurs to SW, C & S China. Vagrants have reached Hebei, Taiwan and Russia. **HH** Favours dry broad-leaf forests where very active at lower and mid levels in pursuit of insects. **ID** Very distinctive small flycatcher with unique combination of grey hood and chest, bright green upperparts including wings and tail, and yellow belly and vent. Stance rather upright, rear crown peak accentuated by crest-like feathers. **BP** Bill black above, pinkish below; eyes large, with narrow white eye-ring, irides black; tarsi dark pink. **Vo** Rather vocal year-round, even singing in winter. Calls include a *chap*, sharp *kitwik kitwik* and soft *pit pit pit* notes. Squeaky, high-pitched song is a loud five-note phrase: *tit-titu-wheeee*. **AN** Grey-headed Flycatcher.

ad ♀

ad ♂

**HILL BLUE
FLYCATCHER**

whitei

ad ♀

ad ♂

**BLUE-THROATED
BLUE FLYCATCHER**

glaucicomans

**FUJIAN
NILTAVA**

ad ♀

ad ♂

ad ♀

ad ♂

**VIVID
NILTAVA**

vivida

ad

**GREY-HEADED
CANARY-FLYCATCHER**

calochrysea

BROWN DIPPER *Cinclus pallasii*   CTKJR

L 21–23cm; WT 76.3–88.7g. **SD** From mountains of C Asia and Himalayas across China to SE Asia, northeast to Kamchatka, S Russian Far East and Japan. *C. p. pallasii* is locally common in suitable habitat on Taiwan, in E & NE China, Korea, Japan (except most of Nansei Shoto), Sakhalin and E & NE Russia to Kamchatka. **HH** Largely resident and strongly territorial along fast-flowing montane streams and rivers (to *c.*1,800m in region); in winter descends to ice-free areas including river mouths, but true migration unknown. **ID** A plump, uniformly dark, chocolate-brown bird of rivers; size of a small thrush, but resembles a large wren due to its dark plumage, short rounded wings and short tail. Behaviour distinctive: it bobs on waterside rocks, flicking tail upwards. Flight very fast, direct and low over water. **BP** Bill short, thrush-like, black; irides dark brown, but white nictitating membrane commonly flicked over eye; its stout legs and large, oversized, feet are dark grey/black. **Vo** Flight call a repetitive, hard *dzzeet* or *cheet*, or buzzing *zzit zzit*; song a very loud and extremely varied rich warbling, *chityiijoijoi*, including trills, rattles and buzzing notes audible even above the sound of fast water. **AN** Pallas's Dipper.

ORANGE-BELLIED LEAFBIRD   CT
Chloropsis hardwickii

L 17–19.8cm. **SD** Himalayas east through S & SE China. *C. h. melliana* occurs in S China east to Fujian and Zhejiang. Escapee (?) on Taiwan. **HH** Forages in deciduous to broadleaf evergreen forest canopy; generally above 1,000m. **ID** Colourful but rather cryptic, medium-sized forest bird. ♂ has crown to rump largely apple green; scapulars, coverts, primaries and rectrices deep cobalt-blue or black, secondaries deep green. Mask black, bib deep indigo, with azure moustachial streak, underparts from chest to vent orange. ♀ less brightly marked, largely green with blue moustachial streak and dull yellow-orange band on central belly to vent. Juv. entirely bright green. **BP** Bill slightly arched, dark grey; eyes dark brown; tarsi grey. **Vo** Very varied including sharp *chittick* take-off call, mewing *chair*, wheezing *frease*, and rapid twittering *ti-ti-tsyi-tsyi-tsyi-tsyi-tsyi*. Song is long, sustained and melodious, incorporating many varied phrases and mimicry.

PLAIN FLOWERPECKER *Dicaeum concolor*   CT

L 8–9cm; WT 4–6g. **SD** India to SE Asia. *D. c. olivaceum* in S China; *D. c. uchidai* resident in Taiwan. **HH** Cultivation, secondary and hill forests (*c.*150–500m in Taiwan); almost invariably found feeding in mistletoe clumps or on other fleshy fruits, or calling from exposed perch. **ID** Tiny, very active plain bird, with very short tail; forages from flowers and on fruits and insects. Upperparts olive-green from crown to tail. Underparts pale greyish-yellow; pale lores and supraloral area from base of bill to eye. **BP** Bill slender, arched, black (pink in young); large eyes prominent in plain face, irides orange-brown; short tarsi black. **Vo** A high, slightly hoarse, sharp *tiek* or *tit*, recalling call of Winter Wren.

FIRE-BREASTED FLOWERPECKER   CT
Dicaeum ignipectum

L 9cm; WT 4–8g. **SD** Himalayas to SE Asia and S China. *D. i. ignipectum*, C & S China east to Fujian; *D. i. formosum* on Taiwan. **HH** Montane deciduous forests, generally above 800m to over 2,500m. In Taiwan, small numbers descend to *c.*500m in winter. Usually in treetops, where favours mistletoes. **ID** Tiny, very active bird, with very short tail, forages from flowers and on fruits and insects. ♂ has deep blue head, face, upperparts, wings and tail; underparts very colourful, with white chin/throat and red upper breast, pale yellow lower breast, belly, flanks and vent, with black streak on centre of breast and belly. ♀ dark olive-green from crown to tail, underparts pale buffy-yellow, brightest on flanks; differs from Plain Flowerpecker in darker face and darker upperparts, with paler rump and warmer yellow-buff flanks. **BP** Bill fine, arched, black; eyes prominent in plain face, irides black; short tarsi black. **Vo** Call is a very high pitched *tzip* often rapidly repeated in series. Song is a rather sunbird-like short zippy phrase of very high notes, *tissit-tissit-tissit*, ending in a trill. **AN** Buff-bellied Flowerpecker.

SCARLET-BACKED FLOWERPECKER   C
Dicaeum cruentatum

L 9cm; WT 7–8g. **SD** SE Asia to SE China, with *D. c. cruentatum* ranging east to Fujian. **HH** Gardens and secondary growth with flowers and mistletoes, below 1,000m. **ID** Very small, active flowerpecker. ♂ has blue-black face, wings and tail, and is scarlet from forehead to uppertail-coverts; underparts creamy white with grey flanks. ♀ is plain olive-green, including face (no pale loral and supraloral patch), somewhat brighter on wings; rump and uppertail-coverts bright scarlet, tail black; underparts creamy white. Juv. like ♀ but lacks scarlet rump, at most has orange tinge. **BP** Bill fine, arched, dark grey (pale orange in juv.); eyes black; short tarsi black. **Vo** A hard, metallic *tchik-tchik-tchik*, higher *chizee* and thinner *tsi*. Song a dry buzzy rattle and thin repetitive *tissit tissit tissit*.

FORK-TAILED SUNBIRD *Aethopyga christinae*   C

L 10cm; WT 4.7–5.9g. **SD** Laos, Vietnam and C & E China, with *A. c. latouchii* east to Fujian. **HH** Visits flowering shrubs and trees in town gardens, woods and forests. **ID** A small sunbird, the only one in region, with bright maroon, green and yellow plumage. ♂ has deep metallic green crown, nape and malar band (extending to neck-sides), uppertail-coverts and tail; mantle and wings bright green, rump pale yellow, supercilium, ear-coverts, chin, throat and breast deep maroon. Pale yellow belly, flanks and vent. Wings short; short tail has white outer corners and two elongated plumes. ♀ is plainer, olive-green on head and face, brighter on rest of upperparts, with a yellow rump; chin to vent pale lemon-yellow, tail green with white tips to outer feathers, lacks extended plumes. **BP** Bill long, fine, arched, black; eyes black; slender tarsi black. **Vo** A sharp, loud and high-pitched *twis* and nervous *ts-wi'i'i'i'i'i'i'i*.

BROWN DIPPER

(not to scale)

pallasii

juv

ad

melliana

ad ♀

ad ♂

**PLAIN
FLOWERPECKER**

ad *uchidai*

ad *olivaceum*

**ORANGE-BELLIED
LEAFBIRD**

(not to scale)

**FIRE-BREASTED
FLOWERPECKER**

ad ♂
formosum

ad ♂
ignipectum

ad ♀

**SCARLET-BACKED
FLOWERPECKER**

cruentatum

ad ♂

juv

ad ♀

ad ♂

latouchii

ad ♀

**FORK-TAILED
SUNBIRD**

HOUSE SPARROW   CTKJR
Passer domesticus

L 14–16cm; WT 24–38g. **SD** W Europe almost to Sea of Okhotsk. Widely introduced elsewhere in world including N America. Race in region *R. d. domesticus* occurs in Amurland, Yakutsk, Khabarovsk, N & NE China locally in Sakhalin and Kamchatka, and NE Chukotka. Accidental Japan (Hegura, Honshu and Hokkaido, Apr–Oct) and Taiwan (possibly ship-assisted); reported Korea. **HH** Rarely found away from villages, towns or cities. Gregarious, forms flocks in non-breeding season. **ID** Large, heavy-set grey and brown sparrow, with a large head. ♂ has unique combination of dull grey crown, black lores and extensive black bib onto chest, broad chestnut eyestripe and nape. Mantle brown, streaked heavily with black, lower back and rump plain grey-brown, rather short tail is blackish-brown with brown outer fringes. Wings brown, with prominent white patch on median coverts. ♀ rather dull, overall ashy grey-brown with very pale supercilium (see ♀ Russet Sparrow, which is brighter and neater). **BP** Bill stout, blackish-grey in breeding ♂, yellowish in non-breeding ♂, and yellow below with grey culmen in ♀ and young; eyes black; tarsi pinkish-grey. **Vo** A hard *chup*, similar to Eurasian Tree; several birds together maintain a steady chattering. Song a complex, but monotonous, series of chirps: *tsii-tsuriri-pyo-pip-riri* or *chilp chev chilp chelp chyurp*.

RUSSET SPARROW   CTKJR
Passer rutilans

L 13–14.5cm. **SD** Afghanistan to SE Asia. *P. r. rutilans* resident throughout much of its southern range in S & E China and Taiwan (rare and declining), also summer visitor north to Shandong, Korea, from C Honshu to Hokkaido and S Sakhalin and S Kuril Is. Locally common offshore Korea, and individuals seen near coast of Taiwan presumably migrants. Winter visitor from C Honshu south to Kyushu. **HH** Breeds in mature forested areas, riparian woodland, also around rural villages; in winter found in agricultural land, often in large flocks, but only rarely mixes with Eurasian Tree Sparrow. **ID** Paler, neater, brighter than Eurasian Tree; sexes dissimilar. ♂ rich rufous-brown on crown, nape, mantle and scapulars, streaked black on mantle; wings black with broad brown fringes to flight-feathers, white tips to coverts forming clear double wingbar; rump and tail grey-brown, tail shorter than Eurasian Tree. Chin and stripe from bill to eye black, cheeks off-white; underparts off-white on breast to grey on flanks and vent. Winter ♂ less richly coloured and has pale patch on rear crown-sides. ♀ plainer, sandy-brown with a prominent broad supercilium contrasting with dark forecrown and dark eyestripe; upperparts brown with black and cream stripes on chestnut mantle; underparts dark sandy-brown. Closed wing shows white median coverts bar, pale fringes to greater coverts and pale panel at base of primaries. Differs from ♀ House in having more contrasting face pattern and darker, more chestnut-brown upperparts with dark streaking on mantle and scapulars. **BP** Bill more slender and pointed than Eurasian, greyish-black in summer ♂, with yellow/horn base to lower mandible in winter ♂ and ♀; eyes dark brown; legs dull pink. **Vo** Varied, including a typical *Passer* hard *chup* note and high *cheet*. Song a hurried *chee tsuri pyo piriri*. **AN** Cinnamon Sparrow.

EURASIAN TREE SPARROW   CTKJR
Passer montanus

L 14–15cm; WT 18–27g. **SD** W Europe to E Russia and E China. *P. m. montanus* in SE & E China, Taiwan, Korea and Japan to Sakhalin, Kamchatka and Commander Is, and throughout continental region; summer visitor to areas west of Sea of Okhotsk. **HH** Around habitation, from dense urban areas to rural villages, in gardens, parks and agricultural land; may nest in loose colonies and commonly flocks in winter; essentially occupies niche of House Sparrow in most of E Asia. **ID** Neatly marked, brown and white sparrow (sexes alike). Cap and nape chestnut; cheeks, throat and neck-sides white, with contrasting black lores, chin and cheek spot forming diagnostic head pattern, with complete off-white or grey neck-band. Upperparts warm brown with dark streaking on mantle; wings brown with white fringes to coverts forming two narrow longish bars, primaries and secondaries black, fringed warm brown; rump and tail grey-brown. Underparts off-white with buff-washed flanks. Young lack black chin and ear-coverts patch. **BP** Bill stubby, black in summer, yellow-horn at base of lower mandible in winter; eyes black; legs dull pink. **Vo** Commonly gives an abrupt *chun chun*, also a cheery, nasal *twuweet*, and dry rattled *tett-ett-ett* in flight. Song is little more than a hurried series of calls.

ROCK SPARROW   C
Petronia petronia

L 14–17cm; WT 26–35g. **SD** NW Africa to Mongolia. *P. p. brevirostris* from Altai and Mongolia to SE Transbaikalia and adjacent Inner Mongolia, China. **HH** Particularly associated with arid montane regions with cliffs and rockslides, or montane steppe, but also around habitation in such areas. **ID** A rather pale, sandy-brown, heavily streaked and heavily built sparrow with a striking head pattern (sexes alike). Head boldly striped, pale crown has broad dark lateral stripes, long broad, pale supercilia, and browner ear-coverts. Pale sandy grey-brown overall, with heavy brown streaking on mantle; wings and tail mid brown with paler fringes. Unique distinguishing features are lemon-yellow spot in centre of breast, white patch at base of primaries and white spots in tips of tail-feathers. In flight, wings broad, tail short. **BP** Bill deep, dull grey above, horn below; eyes dark brown; tarsi brownish-pink. **Vo** A short, nasal *vuee*, shrill, rising *tchuveee*, hard rattling *tee-turrrr* and drawn-out *sle-veeit*. **AN** Rock Petronia.

HOUSE SPARROW

domesticus

ad ♀

ad ♂

RUSSET SPARROW

rutilans

ad ♀

ad ♂

**EURASIAN
TREE SPARROW**

montanus

juv

ad

ROCK SPARROW

brevirostris

ad

LESSER MASKED WEAVER — TJ
Ploceus intermedius

L 13cm; WT 17.3–26.8g. **SD** Natural range Africa, from Ethiopia to Namibia, but introduced and now established in Taiwan and C Honshu. **HH** In region, favours open woodland, agricultural land and parks. **ID** ♂ has black mask and forehead, with contrasting pale eyes; rear crown, neck and underparts bright yellow, upperparts yellow with black streaking on mantle and back, wings black with broad yellow fringes to coverts and flight-feathers; tail short, black with yellow fringes to feathers. ♀ (and non-breeding ♂) has yellowish olive-green upperparts from forehead to rump, with prominent yellow supercilium and pale eyes, mantle heavily streaked black, wings as ♂ but slightly duller, underparts dull yellow. **BP** Bill strong, long and conical, black; eyes pale yellow to white (♂), pale yellow (♀) or dark brown (juv.); tarsi grey. **Vo** Mixes liquid and nasal calls into an accelerating series of notes for song.

NORTHERN RED BISHOP — J
Euplectes franciscanus

L 11cm; WT 11.5–22g. **SD** Natural range Africa, from Senegal to Kenya, but a common cagebird and small local populations sometimes viable in warmer regions of Japan. **ID** A small, large-headed finch, with short wings and tail. Tail frequently flicked open. ♂ breeding is unmistakable, with black cap and cheeks, black belly, brown wings and tail, and otherwise bright red plumage. ♀ (and non-breeding ♂) plainer, mid brown above with pale streaking, and pale sandy-buff below, face plain, cap dark. **BP** Bill deep, stubby, black in breeding ♂, pink with grey culmen in non-breeding ♂, and ♀; eyes dark brown; tarsi dull yellowish-brown. **Vo** Dry chipping notes. Song a tuneless dry chipping and trilling. **AN** Orange Bishop.

ORANGE-CHEEKED WAXBILL — TJ
Estrilda melpoda

L 10cm; WT 6–9.6g. **SD** Natural range tropical W Africa, from Senegambia to DR Congo, but popular cagebird and introduced in Japan where feral populations established in C Honshu (presumably *E. m. melpoda*). **HH** Dry grassland and open waste ground. **ID** A small brown waxbill with red bill and orange cheeks. Crown and nape grey, mantle and wings warm chestnut, rump deep red, tail black. Underparts from chin to vent off-white. **BP** Bill deep-based, short, waxy red; eyes black; tarsi black. **Vo** A sibilant twittering: *shirii shirii shi shi shii shii.*

BLACK-RUMPED WAXBILL — J
Estrilda troglodytes

L 10cm; WT 6.5–8.7g. **SD** Natural range Africa, from Senegal to Ethiopia and Kenya; frequent in cagebird trade and introduced to Japan, where local breeder. **HH** Favours dry vegetation, grasslands and edges of reedbeds. **ID** A pale, neat waxbill; rather pale plain brown above, with prominent black rump and tail; underparts very pale whitish-buff with broad red stripe through eye and variable pink wash or red stripe on centre of belly. **BP** Bill stout, bright pink (ad.) or black (juv.); eyes black; tarsi dark grey. **Vo** A loud metallic *chip*; song includes both loud chipping and long upslurred *soyiiiii* notes.

COMMON WAXBILL — T
Estrilda astrild

L 10–11cm; WT 7–10g. **SD** Widespread sub-Saharan African species; popular cagebird. Introduced and breeding locally on Taiwan. **HH** Favours woodland edge and grassy habitats. **ID** A small, rather plain brown waxbill. Ad. has broad red stripe through eye and pink wash on central belly. Cheeks pale buff, underparts pale brown, vent black, and much of upper- and underparts finely barred brown. Juv. paler with less pink on underparts. **BP** Bill stout, bright pink (ad.) or black (juv.); eyes black; tarsi dark grey. **Vo** Short dry squeaking and chipping notes, and a somewhat buzzy *tzep*, flocks usually maintain a steady twittering. Song consists of paired phrases, *tikatik-wheez tikatik wheez* or *tcher-tcher-preee*, rising in pitch to end.

LESSER MASKED WEAVER

juv

ad ♂

ad ♀

ad ♂ sum

**NORTHERN
RED BISHOP**

ad ♂
moulting
to sum

ad ♀ /
ad ♂ win

**ORANGE-CHEEKED
WAXBILL**

ad

juv

**COMMON
WAXBILL**

ad

juv

**BLACK-RUMPED
WAXBILL**

ad

juv

PLATE 211 : MUNIAS I

RED AVADAVAT TJ
Amandava amandava

L 9–11cm; WT 10g. **SD** Natural range S & SE Asia, but introduced populations established in S Japan and Taiwan (race uncertain). **HH** Dry grasslands, including reedbed fringes, scrub and agricultural fields, where moves in restless flocks. **ID** A small dark red (♂) or pale sandy-brown (♀) finch. ♂ has deep red forehead, crown, cheeks, rump and uppertail-coverts, wings and short tail are black, with small silver spots on wing-coverts, tips of tertials and outer tail-feathers; lores and around eye black; underparts orange-red from chin to belly, with white spots on breast-sides and flanks, vent black. ♀ upperparts dull grey-brown, but bright red on rump and uppertail-coverts; wings and tail like ♂ but fewer spots on coverts, face plain grey with black lores and around eyes, white chin and throat, buff flanks and belly, vent paler. Juv. lacks red and has buff double wingbars. **BP** Bill sharply pointed, orange; eyes red; tarsi pink. **Vo** Calls weak and thin, *teei* or *zsi*, in flight a short, soft *chick-chick*. Song comprises a series of call-like notes, but includes a motif like an unoiled sign creaking in wind, and ends in a short trill.

INDIAN SILVERBILL T
Lonchura malabarica

L 11–12cm. **SD** Natural range Saudi Arabia to Sri Lanka, also a common cagebird that has been introduced and is well established on Taiwan. **HH** Dry grassland and cultivation. **ID** Small, very pale brown finch with blackish-brown wings, and white lower back and rump contrasting with long, blackish-brown tail. Upperparts of ad. pale sandy-brown. Underparts, from lores and chin to vent creamy white with buff flanks. Juv. has more extensive buff over most of underparts and on rump. Wings and tail blackish-brown, scapulars, coverts and tertials pale sandy-brown. **BP** Bill large, silvery blue-grey; eyes black; tarsi brownish-pink. **Vo** Higher, more ringing flight call, *chit-it*, than Scaly-breasted Munia. **AN** White-throated Silverbill.

WHITE-RUMPED MUNIA CTJ
Lonchura striata

L 10–11cm. **SD** Resident across S & SE Asia; *L. s. swinhoei* across much of S & E China north to Zhejiang, and Taiwan; residents in Nansei Shoto, perhaps natural colonists. **HH** Gardens, agricultural croplands, scrub and forest edge, in small noisy flocks. May prefer more wooded areas than Scaly-breasted, though they often occur together. **ID** Small brown finch with a white lower back and rather pointed tail. Dark, almost black upper breast and head impart hooded appearance. Mostly dark brown with fine, off-white; streaking on head, breast, mantle and back; lower back white, rump plain brown, pointed tail is black. Wings dark blackish-brown with paler fringes to many feathers. Underparts off-white. **BP** Bill large, upper mandible blackish-grey, lower mandible pale blue-grey; eyes black; tarsi dark grey. **Vo** A thin twittering *chee chee*, *peet peet* and high-pitched, rolling *prree* or *breeet* flight call.

SCALY-BREASTED MUNIA CTKJ
Lonchura punctulata

L 10–12cm. **SD** Widespread resident of S & SE Asia. *L. p. topela* in much of S & E China north to Zhejiang, and Taiwan. Birds in S Japan may be natural colonists or derive from feral stock. Reported Korea. **HH** Dry grasslands, gardens, cultivated fields and scrub, in small to very large flocks, often with other species. **ID** A small, rather drab finch. Ad. largely mid brown, becoming blackish-brown on chin, throat and tail. Underparts off-white with brown centres to most feathers affording distinctive scaled appearance; belly and vent off-white. Juv. plain brown above, pale buff below. **BP** Bill large, dark grey upper mandible with paler culmen, lower mandible pale grey; eyes black; tarsi dark grey. **Vo** A plaintive, slightly nasal *weh weh* and a *pee yu*. **AN** Spotted Munia; Nutmeg Mannikin; Spice Finch.

RED AVADAVAT

ad ♂

ad ♀

INDIAN SILVERBILL

juv

ad

WHITE-RUMPED
MUNIA

ad

swinhoei

SCALY-BREASTED
MUNIA

topela

ad

juv

BLACK-HEADED MUNIA   TJ
Lonchura malacca

L 10–12cm. **SD** Widespread in S & SE Asia and across S China; *L. m. formosana* is a scarce local resident on Taiwan. Some populations may be introduced, and perhaps involve various other subspecies; *L. (m.) atricapilla* **Chestnut Munia** of NE India to Yunnan possibly merits specific status, introduced with established populations in Okinawa, Osaka and Tokyo. **HH** Scrub, grasslands and ricefields in single-species flocks, or mixed with Scaly-breasted (p.450). **ID** A small finch with dark plumage and very large bill. Ad. hood black, as is central belly to vent. Rest of plumage plain dark chocolate-brown from neck to tail, wings, lower breast and flanks (nominate *L. m. malacca* and some other races have white flanks). Ad. *L. m. formosana* has grey crown and nape (not black), dark brow and face-sides, and more extensive black on belly than *L. (m.) atricapilla*. Juv. paler brown and lacks black hood. **BP** Bill massive, pale blue-grey; eyes black; tarsi dark grey. **Vo** A three-noted *chirp chirp chirp* in flight, and high-pitched reedy *pwi-pwi*. Song comprises quiet musical notes interspersed with long whistles and bill-clicking.

WHITE-HEADED MUNIA   J
Lonchura maja

L 10cm. **SD** Resident from southern SE Asia to Bali. Introduced to Japan where feral populations established on Honshu, around Tokyo area. Monotypic. **HH** Frequents ricefields and reedbeds, often flocking with Scaly-breasted Munias (p.450). **ID** A small, white-headed munia. The entire head is white, bordered with buff on sides and back of neck; plumage otherwise chestnut-brown. **BP** Bill massive at base, tapering to sharp point, blue-grey; black eyes appear large in white face; tarsi dark blue-grey. **Vo** A high-pitched piping *pee-eet*. Song a repetitive, tinkling *weeeeee heeheeheeheehee* interspersed with bill-clicking.

JAVA SPARROW   CTJ
Lonchura oryzivora

Vulnerable. L 12.5–16cm. **SD** Natural range Java and Bali, where threatened by cagebird trade. Introduced widely around the world and frequent escapees have established local populations in E coastal China, from Shanghai to Fujian, also S Japan and Taiwan. Monotypic. **HH** Urban and rural areas, in gardens and croplands. **ID** Stocky grey finch with massive bill and distinctive face pattern. Ad. has black head with large white cheek patch, upper- and underparts largely grey, pinkish-grey on belly, white on vent and black on tail. Juv. dull, greyish-brown above, greyish-buff below, with dusky-brown cap. **BP** Bill massive, blunt, red; eye-ring red, eyes black; tarsi pink. **Vo** Hard *tup*, *t'luk* or *ch'luk* notes. Song: bell-like soft chattering notes accelerate into a trill interspersed with both higher and deeper notes, ending with a prolonged whistled *ti-tui*.

EASTERN PARADISE WHYDAH   J
Vidua paradisaea

L ♂ 38cm; WT 15.8–29.3g. **SD** Natural range Sudan to Natal, but it is an established exotic in C Honshu. Monotypic. **HH** Dry ricefields and other cultivation with associated grassland or reeds. **ID** ♂ breeding has black face, cap and throat, black mantle, back and wings, and extremely elongated tapering tail; cheeks and breast golden-yellow to chestnut and rest of underparts yellowish-buff to off-white. ♀ and non-breeding ♂ are generally nondescript mid-brown with dark blackish-brown streaking on upperparts and breast, but with striking head pattern. Black crown with buff central crown-stripe, prominent white supercilia and black eyestripes extending as two black crescents, one behind eye, the other bordering ear-coverts. **BP** Bill short, deep-based, black; eyes black; tarsi pink. **Vo** Combination of hard chipping, nasal, squeaking and chattering notes. **AN** Long-tailed Paradise Wydah.

BLACK-HEADED MUNIA

formosana

juv

ad ♂

ad ♂ sum

EASTERN PARADISE WHYDAH

ad ♂ win

WHITE-HEADED MUNIA

ad

juv

JAVA SPARROW

juv

ad

ALPINE ACCENTOR CTKJR
Prunella collaris

L 15–19cm; WT 39–45g. **SD** Local and largely sedentary in high mountains from SW Europe and N Africa to Himalayas and E Asia. *P. c. erythropygia* in SE Siberia, NE China and N Korea, with isolated population on Honshu; some winter around W Yellow Sea. *P. c. fennelli* endemic to Taiwan. **HH** In summer, generally above treeline (usually above 1,800m, but typically above 3,300m in Taiwan), on rocky crags and alpine meadows, but also on coastal cliffs and rocky islands of Russian Far East. Descends in winter, when found around rocky areas in forest and along roads, and may also vacate breeding range. Typically confiding, moving quietly and methodically across snow-free banks, or near streams and damp areas in winter; alone or in family parties. **ID** A small, grey and brown thrush-like bird. Larger than other accentors. *P. c. erythropygia* has distinctly grey-hooded appearance with fine silver spotting on chin, and white crescents above and below eyes; upperparts rather dark, streaked chestnut and black; scapulars chestnut, wings black and rufous with wingbars formed by white tips to coverts; rump and uppertail-coverts rufous, tail blackish-rufous with white tip. Grey chest, rufous/chestnut belly, flanks and vent, with dark grey chevrons on flanks and lower belly. *P. c. fennelli* greyer on head and paler below. **BP** Bill strong; tip and upper mandible black, lower mandible and base pale yellow; eyes dark reddish-brown; tarsi dark pinkish-grey or brown. **Vo** Short *kyon* or *kyo*, lark-like *drip*, *truiririp* or *tschirr*, and loud chirruping flight call. ♂ song, given from rock or in short display flight, a soft, but powerful, musical warbling, *kyichichi-jiri-jiri* or *chuchuri churuchurubyurirupitti* and *kyorikyori, kyorikyorichi*. ♀ song is a simpler series of trills.

SIBERIAN ACCENTOR CTKJR
Prunella montanella

L 13–15cm; WT 17.5g. **SD** Extreme NE Europe to Chukotka; *P. m. badia* in NE Siberian taiga from lower Lena River to Anadyr River of Chukotka, south to N Kamchatka and W Sea of Okhotsk. *P. m. montanella* breeds patchily across southern latitudes, reaching region in montane forests of E Siberia from Baikal to Amurland and Sikhote-Alin. Winters (Oct–Mar) south to Korea and NE China; rare in winter in Japan. **HH** Favours forests and edges of subarctic and taiga zones, typically in valleys with tangled riverside vegetation. In winter in thickets, shrubs near streams, sometimes in dry forest and grassland on hillsides; forages on ground. Can be rather secretive and quiet. **ID** May appear generally drab brown at first glance, but has distinctive mask-like head pattern with black of crown, lores and ear-coverts contrasting with long, broad yellow-ochre supercilium and yellow-ochre ear-spot; chin, throat, breast and flanks dull ochre-yellow, with diffuse brown streaking on flanks; belly and undertail-coverts buff to off-white. Neck-sides grey,

upperparts warm brown streaked black; wings brown with single prominent, white, greater covert bar, rump brownish-grey, tail blackish-brown. *P. m. badia* smaller and darker, with richer chestnut mantle, deeper ochre underparts and more strongly rufous flank streaks. **BP** Bill fine, sharply pointed, black with ochre base to lower mandible; eyes dark brown; tarsi pinkish-brown or orange-brown. **Vo** A trisyllabic, insect-like *ti-ti-ti*; *tiriri* or *see-see-see*, whilst song is a penetrating, high-pitched *chirichiriri* given from a bush or treetop, resembling Japanese Accentor.

BROWN ACCENTOR C
Prunella fulvescens

L 14–15cm; WT 20g. **SD** Mountains of C Asia east to Altai, S Siberia east to China. Subspecies in region is *P. f. dahurica*. **HH** Open, dry, rocky/stony, alpine and subalpine slopes with thickets and open scree. Moves seasonally up and down mountains, descending to below treeline, even into villages. **ID** A rather distinctive bird with a dark greyish- and blackish-brown crown, prominent off-white supercilium and very dark, blackish-brown ear-coverts; creamy white on throat. Recalls Siberian Accentor but paler and greyer. Upperparts rather cold, dark greyish-brown, streaked darker on crown and back, plain greyish-brown rump and greyish-brown tail; underparts unstreaked buff or ochre. **BP** Bill fine, sharply pointed, black; eyes yellowish-brown; tarsi yellowish- to reddish-brown. **Vo** A bunting-like *ziet-ziet-ziet*. Song a fairly loud trilling warble, given from elevated perch, rarely in flight: *tuk-tileep-tilee-tileep-tileep*.

JAPANESE ACCENTOR KJR
Prunella rubida

L 14–16cm; WT 19g. **SD** Essentially restricted to Japan, but also in S Sakhalin and S Kuril Is. Summer visitor to N of range (e.g. Sakhalin, Kurils, Hokkaido and high mountains of Honshu). Resident elsewhere in Honshu and on Shikoku. Altitudinal and latitudinal migrant, descending lower and to south in winter, winters from C Honshu west to NE Kyushu. Reported Korea. Monotypic. **HH** In summer favours stone pines and dense thickets of Rowan near treeline, in winter also in deciduous forests at middle elevations, often near rivers/streams, in thickets and scrub. **ID** A dark accentor (closely resembles extralimital Dunnock *P. modularis*). Crown brown, nape greyish-brown, mantle and wings dark chestnut-brown with heavy blackish-brown and paler brown streaking; underparts, head and face dark grey, but face 'frosted' paler grey, flanks and vent have some rufous streaking. **BP** Bill fine, sharply pointed, blackish-grey; eyes dark reddish-brown; tarsi brownish-pink. **Vo** A thin *tsuririri* or *chiririri*. Song a rapid, clear and tinkling *tsee-tsee-syuu tsee-tsee-syuu; chiri chirichiri* or *chuririri chuuchii hirihichiririri*, typically from top of stone pine.

ALPINE ACCENTOR

ad *fennelli*

ad
erythropygia

juv
erythropygia

SIBERIAN ACCENTOR

montanella

1st-win

BROWN ACCENTOR

dahurica

ad

ad sum

JAPANESE ACCENTOR

ad

FOREST WAGTAIL
Dendronanthus indicus
CTKJR

L 16–18cm; WT 14–17g. **SD** Breeds through C, E & NE China, Korea, northeast to Amurland, winters S China and SE Asia; rare, annual migrant Japan (some breed in SW), scarce Korea, Taiwan (Sep–May). Monotypic. **HH** Breeds around open glades and paths, near streams/rivers, in broadleaf evergreen and deciduous forest, to 1,500m. Winter in woodland, orchards and plantations. Well-camouflaged against leaf litter; when flushed may alight quickly on ground or in trees. **ID** Distinctive brown, black and white wagtail with long slender tail equal in length to body. Upperparts drab olive-brown, with prominent white supercilia, black wing-coverts with broad yellowish-white fringes forming double wingbar; long tail brown with white outer rectrices. Underparts whitish with central black breast patch extending as narrow collar, and second black bar on sides of lower breast. In flight, distinctive black median coverts bordered above and below with white, and black wingtips unmistakable. On ground commonly, and distinctively, sways rear and tail from side to side. **BP** Bill grey above, pinkish below; eyes dark brown; legs pink. **Vo** A single *pink* or double metallic *pink-pink*, occasionally a very quiet *tsip* while foraging. Song a simple repetitive disyllabic *tsi-fee*, or a husky *chuchupii chuchupii* recalling Eastern Great Tit.

WESTERN YELLOW WAGTAIL
Motacilla flava
KJR

L 16–18cm; WT 11.2–26.4g. **SD** Ranges from W Europe to the Kolyma; winters south to sub-Saharan Africa, S & SE Asia. In region, *M. f. thunbergi* 'plexa' (or *M. f. plexa*) breeds from Lena east to Kolyma River. Taxonomy subject to much debate;some lump all subspecies, others split into two (as here), three, or more species. **HH** Wet meadows, wetland margins and grassy swamps. **ID** Ad. breeding has plain olive-green upperparts, blackish wings with white fringes to coverts (narrow wingbars in flight), long, slender, blackish tail with white outer feathers. Pale yellow chin, brighter yellow from throat to belly, slaty blue-grey crown and nape, with slight white supercilium behind eye; lores and cheeks black. Ad. non-breeding has grey-brown upperparts, off-white to buff flanks, pale yellow lower vent and undertail-coverts, and white supercilium does not wrap around ear-coverts as in Citrine Wagtail. Tail approximately same length as body. **BP** Bill black (summer), or dark grey above, horn below (winter); eyes dark brown; tarsi black. **Vo** A thin buzzy *tsweep*, *bizi* or *zi*. Song, from perch, comprises very short series of twitters *srii-sriiit*.

EASTERN YELLOW WAGTAIL
Motacilla tschutschensis
CTKJR

L 16–18cm; WT 11.2–26.4g. **SD** Breeds to Alaska, and to Amurland and NE China. Common migrant throughout E of region. Various subspecies and forms recorded: 'Alaska Wagtail' *M. t. tschutschensis* of N American range, Chukotka and N Kamchatka, winters in E Asia and Indonesia; 'Kamchatka' *M. t. tschutschensis* (form 'simillima') of C & S Kamchatka, Commander and N Kuril Is, winters from Philippines to N Australia; 'Green-headed'

M. t. taivana (sometimes treated specifically) of SE Siberia, Vilyuy and Lena River basins to Sea of Okhotsk, Amurland, Sakhalin and N Hokkaido, winters from Taiwan and SE China to Philippines and N Australia; and 'Manchurian' *M. t. macronyx* of Transbaikalia to N & NE China, Amurland and Ussuriland, winters in SE China (scarce Taiwan) and SE Asia. Some or all may represent species, or all may be included with Western Yellow Wagtail. **HH** Wet meadows, grassy swamps and wetland margins, often with Red-throated Pipit (see p.464). **ID** Variable, but all breeding ads. share: uniform olive-green upperparts, blackish wings with white fringes to coverts (narrow wingbars in flight) and long, slender, blackish tail with other feathers (approximately same length as body). All non-breeding ads. extremely similar (sometimes inseparable), sharing: brown upperparts, off-white to buff flanks, some with pale yellow lower vent and undertail-coverts; white supercilium does not wrap around ear-coverts as in Citrine Wagtail. 'Green-headed' has bright yellow underparts from chin to belly, broad yellow supercilia, olive-green face, crown and upperparts, and prominent white covert fringes. 'Kamchatka' is bright yellow below, blue-grey crown, cheeks and nape; prominent white supercilia. In winter, particularly buffy on flanks; supercilia whitish. 'Alaska' has dark blue-grey crown and nape, white supercilia and throat, and dull grey wash to chest. 'Manchurian' has crown and cheeks greyish-black, no supercilia, white chin, and bright yellow throat to belly. **BP** All subspecies: bill black (summer), or dark grey above, horn below (winter); eyes dark brown; tarsi black. **Vo** A thin buzzy *tsweep*, *bizi* or *zi*. Song, from perch, a very short series of twittered notes *srii-sriiit*.

CITRINE WAGTAIL *Motacilla citreola*
CTKJ

L 16.5–20cm; WT 18–25g. **SD** Breeds from N Europe to Himalayas, and C China, *M. c. citreola* east to Transbaikalia, Mongolia and NE China. Migrates through E China, wintering in S & SE Asia, S & SE China. Accidental or scarce migrant in Korea, Japan, Taiwan. **HH** Breeds in damp grasslands, wetland margins, and wet tundra. On migration/winter usually at wetlands. **ID** Large, plain, long-tailed wagtail. ♂ breeding unmistakable due to bright yellow head and underparts, becoming whiter towards vent. Broad blackish bar on hindneck, grey mantle, black and white wings, long black tail with white outer feathers. ♀ is rather plain, but upperparts generally grey, with uniform grey crown, nape and mantle; face yellow, pale yellow on supercilium (which continues around ear-coverts) and malar region, darker grey on cheeks; underparts off-white with grey-washed flanks. Juv. resembles ♀, but has brownish-grey upperparts, lacks yellow wash on face, supercilia (with dark border above) and ear-coverts surround buff, forehead and lores also buff. 1st-winter much greyer than either yellow wagtail, with white supercilia and ear-coverts surround, pale lores and white undertail-coverts. In flight, black wings and pale grey coverts with white fringes forming broad double wingbar, distinctive in all plumages. **BP** Bill all black (summer), or all dark (winter); eyes black; tarsi black. **Vo** Call like Eastern Yellow, but slightly sweeter, *tsuili*, a loud, ripping *tsreep*, also *biju biju*, *ju ju* and *chui*. Song, in flight or from perch, consists of short simple phrases repeated after several seconds.

FOREST WAGTAIL

ad

WESTERN YELLOW WAGTAIL

thunbergi

ad ♂

ad ♀

EASTERN YELLOW WAGTAIL

ad ♂ spring
macronyx

ad ♂ spring
taivana

ad ♀
macronyx

ad ♀
tschutschensis

ad ♂ spring
tschutschensis

CITRINE WAGTAIL

citreola

1st-win

ad ♀ sum

ad ♂ sum

GREY WAGTAIL
Motacilla cinerea

CTKJR

L 17–20cm; WT 14–22g. **SD** Widespread, from W Europe to NE Russia, wintering from N Africa to Philippines and Indonesia. *M. c. robusta* in C & S Kamchatka, Kuril Is, Japan, Amurland, NE China and Korea; *M. c. melanope* from Urals to Sea of Okhotsk and Koryakia, also Transbaikalia to C & E China, and Taiwan. Winters Japan, Korea, Taiwan, across China south of Changjiang River, and in SE Asia. **HH** Generally along fast-flowing rocky streams and rivers; also wetlands and coasts in winter. **ID** The largest and longest-tailed wagtail in region (tail longer than body). Upperparts plain dark grey, wings black, long tail black with white outer feathers. ♂ breeding has prominent white supercilium, black chin and throat, and white malar/submoustachial stripe contrasting with dark grey ear-coverts; underparts bright lemon yellow from chest to vent; ♂ non-breeding has white chin and throat, and buffy supercilium. ♀ has less distinct supercilium, white chin/throat mottled black, and generally paler underparts, white suffused pale yellow, particularly on vent/undertail-coverts. Non-breeding ♀ paler with more buff on breast. Juv. is like non-breeding ♀, but dark parts more olive, pale parts buffier. In flight, grey back, black wings with single white wingbar, bright yellow rump and very long tail diagnostic. **BP** Bill black (♂) or dark grey (♀); eyes dark brown; legs flesh-pink. **Vo** Variable, hard metallic *tzit, tzit-tzit, tzit-zee* or *chichin chichin*, or sharp *chittick*; stronger, clearer and higher than White Wagtail. Song, given in display flight ending in parachuting to ground, is long, complex and sibilant, with various phrases, including *si-si-si* and *see-see-swee*, usually ending in a trill: *tzii-tz-tzi-tzi-tzee-ree-ree-ree*.

WHITE WAGTAIL
Motacilla alba

CTKJR

L 16.5–18cm; WT 17.6–24.6g. **SD** W Europe to E Asia and extreme W Alaska, wintering south from Africa to SE Asia and Philippines. Various races in region: '**Baikal Wagtail**' *M. a. baicalensis* from Lake Baikal to Amurland, wintering in E China and SE Asia; '**East Siberian**' *M. a. ocularis* from Yenisei to W Alaska, also Kamchatka and Sea of Okhotsk, wintering to E & SE Asia; '**Black-backed**' *M. a. lugens* (formerly treated as a separate species) of Kamchatka, Kuril Is, Japan, Russian Far East, NE Korean islands to Amur Estuary and Sakhalin, wintering mainly south from C Japan (few further N) through Ryukyus, also Korea, to Taiwan and S China; '**Amur**' *M. a. leucopsis* from NE & E China, Korea, S Russian Far East, also SW Japan (scarce) and Taiwan, wintering to E & SE Asia. Various forms occur together on migration. In addition, '**Masked**' *M. a. personata* of C Asia is a rare visitor to Japan, and the largely sedentary '**Himalayan**' *M. a. alboides* has also been reported

from Japan. **HH** Common, occurs in wide range of habitats from riversides and cultivated land to coasts. **ID** Four distinctive subspecies occur in region, all overall 'pied', with grey or black backs, largely white wings, and long black tails with white outer feathers. ♂ *M. a. leucopsis* is black from mid crown to rump, with white forehead, face and chin, and large black chest patches (especially in breeding plumage) unconnected to black of neck/mantle. ♀ is grey on crown and upperparts with small black chest patch. Both sexes have plain white face from forehead and chin to ear-coverts, with prominent black eyes. *M. a. baicalensis* closely recalls *M. a. leucopsis*, but ♂'s mantle is grey (not black) and black chest patch extends higher onto throat and, in point, to neck-sides. *M. a. ocularis* (♂ most similar to winter-plumaged Black-backed), has mantle and rump grey, black of crown broad at rear and connected to black eyeline; chin, throat and chest patch black, but do not connect to mantle or neck patch. ♂ *M. a. lugens* has black upperparts from mid crown to rump, black eyestripe from bill to nape, broad black throat patch contrasting with white chin and connected to black of mantle; wing-coverts very broadly fringed white. Winter ♂ has grey not black mantle and reduced black chest patch (though still connected to nape). ♀ has grey upperparts, with dark grey/blackish eyeline meeting dark grey nape, and narrow black chest band also connected to collar. In flight, wings almost all white, with white flight-feathers and coverts, but scapulars and tips to outer primaries black in ad. **BP** Bill black; eyes black; legs black. **Vo** Variable, but clearly bisyllabic *chuchun chuchun, chichin chichin* or *chizzick chizzick*. Song is either simple *tsutsu tsutsu zuizui* or very much more complex, fast chirping series of slurred warbled notes.

JAPANESE WAGTAIL
Motacilla grandis

CTKJR

L 21–23cm. **SD** Regional endemic restricted to Japan (from Hokkaido to Kyushu) and Korea, with some seasonal altitudinal and latitudinal movements; accidental S Russian Far East, Taiwan, Fujian and Hebei. Monotypic. **HH** Locally common and largely sedentary, typically along fast-flowing cold-water montane streams and rivers, and broad shingle-bedded lowland rivers; occasionally at ponds and lakes and in coastal areas. **ID** Ad. has black hood, mantle, scapulars and broad chest patch, with white forehead, supercilium and narrow chin patch. ♂ is typically jet black, but ♀ appreciably paler greyish-black above. Juv. grey where ad. is black and may lack white supercilia. At rest has extensive white wing patch, white-fringed tertials and very long black tail with white outer feathers. In flight, wings largely white but has extensive black tips to primaries, and black lesser primary-coverts. **BP** Bill black; eyes dark brown; legs black. **Vo** Calls distinctively buzzy, *jiji jiji, bjee* or *bjeen*. Song complex, somewhat similar to White Wagtail's, transcribed as *chichii joijui* and *tsutsuchiiichiii juu juu* and, more fully, as *tz tzui tztzui, pitz pitz tztzui pitz pitz bitz, bitzeen bitz bitzeen bitz bitzeen, tztzui tzigi chigi jijijiji*.

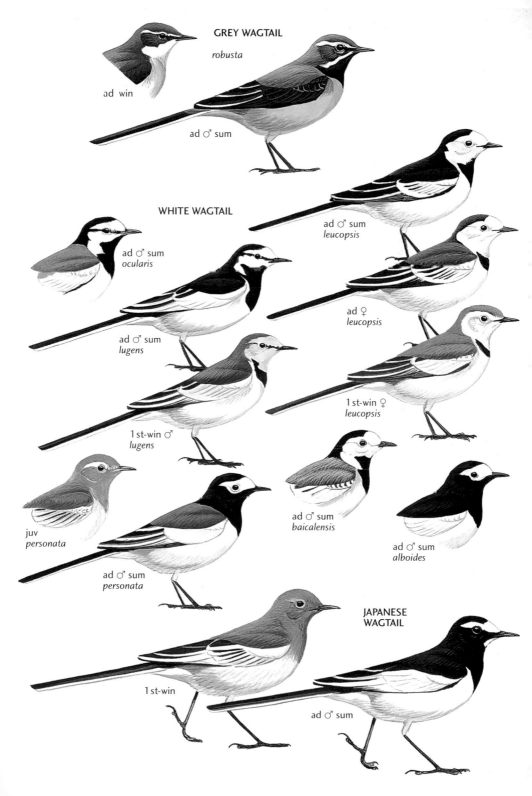

GREY WAGTAIL

robusta

ad win

ad ♂ sum

WHITE WAGTAIL

ad ♂ sum
oculis

ad ♂ sum
leucopsis

ad ♂ sum
lugens

ad ♀
leucopsis

1st-win ♂
lugens

1st-win ♀
leucopsis

juv
personata

ad ♂ sum
baicalensis

ad ♂ sum
alboides

ad ♂ sum
personata

**JAPANESE
WAGTAIL**

1st-win

ad ♂ sum

RICHARD'S PIPIT
CTKJR
Anthus richardi

L 17–18cm; WT 21–40g. **SD** Summer visitor across C Asia to the Lena, Sea of Okhotsk, S Russian Far East and E China; wintering S & SE China and Taiwan, south to SE Asia. Fairly common migrant (Apr/May; Sep/Oct) Korea and rare offshore Japan, and rare winter visitor S Japan. Three races involved (probably indistinguishable in field): *A. r. dauricus* of N Mongolia to middle Lena River and Amurland; *A. r. ussuriensis* of Shandong to NE China and possibly Korea, wintering in C & E China and SE Asia; and *A. r. sinensis* of C & E China, wintering to SE Asia. **HH** Typically in rather arid habitats; steppe, semi-desert, grassland and cultivated areas. **ID** Largest pipit in region. Rather bulky, deep-chested; upright stance, long tail, stout legs and pale, well-streaked appearance all distinctive (but see Blyth's Pipit). Overall sandy-brown, with mid-brown upperparts, crown, mantle and scapulars streaked darker. Prominent pale supercilia, lores, chin and throat; narrow, blackish malar stripe broadens to form triangular patch then merges with breast streaking. Underparts buffy-brown, especially on upper chest and flanks, otherwise off-white; narrow dark streaking on lower neck and upper breast. Median coverts have blackish centres and broad buff/whitish fringes, tertials and secondaries blackish fringed buff. 1st-winter cleaner, with narrow white fringes to median and greater coverts and tertials. In deeply undulating flight, long tail very conspicuous, tail-sides white; when flushed typically gains height and distance quickly. Walks and runs strongly. **BP** Bill longish, stout, grey above, pink below, somewhat heavy and arched; eyes dark brown; rather long tarsi pink, hindclaw very long. **Vo** Loud, harsh, rather sparrow-like and explosive *reep*, *chree-eep* or *schreep*, commonly given on take-off and in flight. During high, circular display flight, gives monotonous, buzzy song *tree-ee-aya-eeya tree-eeya-eeya-eeya-eeya* or *tzwee-tzwee-tzwee-tzwee-tzwee-tzwee-tzwee*.

BLYTH'S PIPIT
CTKJ
Anthus godlewskii

L 15–17cm; WT 17–30.5g. **SD** Breeds from S Russia to NE China and winters south of Himalayas in India. In region, migrates through NE China, a rare migrant (Mar–May; Sep/Oct) through Korea and accidental Japan. Monotypic. **HH** Typically on dry montane slopes, stony steppes and arid cultivated habitats, though may also occur in swampier habitats on migration. **ID** Slightly smaller and more compact than very similar Richard's Pipit, from which best distinguished by size, less upright stance, shorter tail and shorter legs; finer, more sharply pointed bill, and vocalisations. Overall pale like Richard's, but plumages of both variable. Median coverts with squarer blackish centres and broader pale fringes, form more prominent wingbars than in Richard's, but only in ads. Upperparts more neatly streaked, and underparts more extensive and plainer buff than Richard's. **BP** Bill grey above, pale pink below; eyes dark brown; tarsi pale or yellowish-pink, tarsi shorter and hindclaw shorter and more arched than Richard's. **Vo** Variable; distinctive wheezy or buzzing *spzeeu* or *pscheeo* (very similar to Richard's, but higher, softer, more shrill and descending), and a mellow,

low *chip-chup*, shorter and higher pitched than Richard's. Song, typically during display flight, is unpipit-like *zret zret zret...sri-sri-serererererereleee-ueh*. Alarm a harsh *bzrp* or *brzi*.

MEADOW PIPIT
J
Anthus pratensis

L 14.5–15cm; WT 14.5–22g. **SD** Ranges across cooler regions of Europe and Russia east to the Ob; *A. p. pratensis* closest to region, winters in S Europe and N Africa to Middle East, also Caucasus and C Asia. Accidental offshore and S Japan (autumn and winter). **HH** Damp boggy areas, wet grasslands, damp meadows and tundra. **ID** A small, dark, olive-brown pipit. Dark olive crown and mantle, narrowly streaked black, rump plainer and unstreaked, tail blackish-brown with white outer feathers. Face plain, with only faint supercilium, narrow pale eye-ring, ear-coverts mid olive-brown, chin/throat off-white with narrow lateral throat-stripe extending as gorget of dark streaks as far as flanks, but breast streaking weaker, and flank streaking stronger than Tree or Olive-backed (p.462); belly buff. **BP** Bill fine, blackish-brown; eyes dark brown; tarsi yellowish-brown, hind-claw longer than hindtoe, and only slightly curved (longer than Tree or Olive-backed). **Vo** Similar to Buff-bellied: a thin, sibilant *sip-sip-sip*, though slightly more squeaky than Buff-bellied, and a more insistent *spip-spip-spip* or *ist ist ist*.

TREE PIPIT
KJR
Anthus trivialis

L 14–15cm; WT 15–39g. **SD** Widespread Palearctic breeder, from NW Europe across Russia to Kolyma River, wintering S Europe, sub-Saharan Africa and India. *A. t. sibiricus* breeds into NE Russia, but most migrate well west of our region, though has strayed to Japan and Korea on migration/winter. **HH** Typically in woodland clearings and edge, heaths and grassland; wintering in similar habitats and more open savanna, also open areas with isolated trees. Sings from perch in trees, and when flushed commonly lands in trees. **ID** Small to medium-sized pipit resembling Olive-backed (p.462), but overall browner-buff (lacks Olive-backed's distinctive olive-green cast), smaller, and has less striking face pattern, with indistinct, or much-reduced, uniform buff-coloured supercilium, narrower and least distinct before eye, strongest above eye, lacking dark border above, affording open-faced 'surprised' look; generally (but not always) lacks prominent creamy spot and black crescent at rear of ear-coverts of Olive-backed. Upperparts brownish-tinged with distinct, broad dark streaking on mantle (more heavily streaked than Olive-backed, including on crown). Median coverts have black centres with 'tooth' extending to break off-white fringes (often palest feature of species). Strong buff tinge to white of breast and flanks, heavily streaked breast, changing abruptly to narrower, elongated dusky streaks on flanks. **BP** Bill strong, appears somewhat long, rather heavy-based and culmen slightly decurved at tip, dark brown above, pale horn/pink at base; eyes dark brown; tarsi pale or pinkish. **Vo** A somewhat drawn-out yellow wagtail-like *tseep*, *teez* or *spihz*, and *bzzt*, also a very similar *duii* to Olive-backed (much variation and overlap); alarm *syt*. Song highly varied, but involves repeated phrases and slurring trills, *seep-seep-seep-seea-seea-seea*, given in parachuting display flight to low bush or treetop.

RICHARD'S PIPIT

1st-win

juv

ad *sinensis*

ad Richard's

ad Blyth's

BLYTH'S PIPIT

1st-win

ad

ad Meadow

MEADOW PIPIT

pratensis

ad

1st-win

ad Tree

ad

TREE PIPIT

1st-win

sibiricus

OLIVE-BACKED PIPIT CTKJR
Anthus hodgsoni

L 15–17cm; WT 17–26.3g. **SD** Breeds from Himalayas and Urals to Japan: *A. h. yunnanensis* N Siberia, Yakutia to W Chukotka, Sea of Okhotsk, S Kamchatka, Russian Far East, Sakhalin, Hokkaido, also N Mongolia and N Manchuria, winters in S & SE Asia; *A. h. hodgsoni* from Himalayas to S Mongolia, Korea and Honshu, winters to SE Asia. In region a common migrant, winter visitor and/or summer visitor. **HH** Typically breeds at taiga edge in N and in riparian woodland and montane (sometimes lowland) forest further S. Winters in drier woodlands (often in lowlands), parks and grasslands near forest. Behaviour much as Tree Pipit. **ID** Medium-sized pipit, overall rather bright olive-green and off-white (see Tree Pipit, p.460). Head rounded and well marked, more contrasting than Tree; face pattern distinctive, with prominent two-tone supercilium, buffish before eye, pale cream behind eye, bordered above by narrow blackish line; olive ear-covert patch has distinct white spot (like extension of supercilium), above blackish crescent at rear. Chin/throat whitish, split by narrow black malar stripe, merging into broad black streaks on breast, narrowing only slightly on flanks. Upperparts mid to dark greenish-olive, the mantle olive-green and unstreaked, or with faint narrow dark streaking in spring; wings and tail have greenish fringes, wings with double clear wingbars, and median coverts have dark, rounded centres with unbroken pale, often rather creamy white, fringes. Combination of clear supercilia and wingbars strongly support identification as Olive-backed. *A. h. hodgsoni* has broader stripes on back, scapulars, breast and flanks than *A. h. yunnanensis*, and upper body appears more uniform olive with less black. *A. h. hodgsoni* breeding in Japan have less distinct streaking on back/scapulars, and thus are closer to *A. h. yunnanensis*. Primaries extend further beyond tertial tips in Japanese breeding population than elsewhere. **BP** Bill short, narrow-based, with fine, sharp tip, dark grey above, dark horn below; eyes dark brown; tarsi pale or yellowish-pink. **Vo** Call, commonly given at take-off and in flight, a thin *dzzt*, *dsst*, *tseez*, *zii* or *duii*, similar to Tree (usually inseparable); alarm *sit*. Song is fast, light, varied and complex, somewhat skylark-like, *tsui-tsui-choi-choi zizizi*, but delivered from treetops in forest, or in descending display flight ending on treetop; song frequently terminates in calls.

PECHORA PIPIT CTKJR
Anthus gustavi

L 14cm; WT 19.8–26g. **SD** NE European Russia to Chukotka and NE China: *A. g. gustavi* north of range to Chukotka and Kamchatka; *A. g. stejnegeri* Commander Is; *A. g. menzbieri* S Ussuriland and NE China (sometimes split as **Menzbier's Pipit** *A. menzbieri*). Winters in Philippines and Indonesia. Migrates (May; Sep/Oct) through NE & E coastal China; scarce migrant Korea, rare Japan and Taiwan. Movement of races unclear, but Menzbier's has occurred in Shandong and off W Korea, nominate in Jiangsu. **HH** In summer, in lowland bushy tundra and swampy taiga, also beside streams with willows; on migration/in winter in wet grassland, along trails in forest, as well as grassy forest edges. Skulking. **ID** A rather dark pipit, closely resembling 1st-winter Red-throated (RTP, p.464), but generally darker, more heavily streaked (easily separated on breeding grounds using voice, but migrants often silent). Face pattern slightly more distinct, with subtle supercilium from bill to just rear of eye, blackish loral stripe from eye to bill (absent in RTP). Upperparts warmer and more boldly marked than RTP, ear-coverts brown, crown and nape brown with sharp, well-defined, narrow blackish streaks (duller nape in RTP), mantle brown-olive with more prominent, contrasting, black and white 'braces'; whitish fringes to median and greater coverts form double wingbars (slightly bolder than RTP), greater coverts usually blacker than RTP, creating contrasting wing pattern. Underparts distinctive with creamy white chin/throat, clear yellowish-buff wash on breast/flanks, and whitish belly, giving white/buff/white pattern to underparts; narrow malar does not reach bill, but joins dark neck patch and merges into heavy breast and flank streaking (underparts yellower in Menzbier's or has buff wash to breast, white on rest of underparts). Fresh young RTP can be well washed buffy below, but less clean-cut tripartite pattern than Pechora. Unique amongst pipits in having tertials noticeably shorter than primaries, with 2–3 primary tips extending slightly beyond tertials; especially useful in separating 1st-winter from RTP, in which all primaries cloaked by tertials. In flight, prominent 'braces' and wingbars. *A. g. stejnegeri* is larger and darker than nominate. Menzbier's is more extensively black above, with narrower, paler fringes, deeper yellow-buff underparts and deeper buff undertail-coverts; wings shorter. **BP** Bill short, thickish, grey above, pink below and at base; eyes dark brown; tarsi pale pinkish. **Vo** Usually silent, even when flushed (whereas RTP almost always gives high, thin *speee* or *seet* when flushed), but abrupt, hard 1- or 2-syllable clicking call may be given on take-off or in flight: *pwit*, *chi*, *chu* or *tsep*; calls of migrants (mostly nocturnal) vary from squeaky but emphatic *dit dit dit* to more Grey Wagtail-like hard *tsip*. Song, usually in display flight but also from perch, is variable, hard and mechanical, mixing dry trills with double call-note, commencing *tsivi-tsivi*. *A. g. stejnegeri*: *zi-zi-zi-zi-zeerrrrrrrr-zezezezeze*. Menzbier's Pipit song is a low, buzzing *cheep-cheecheecheechzhzhzhzh cheenee*.

ROSY PIPIT CK
Anthus roseatus

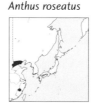

L 15–16.5cm; WT 17–25g. **SD** Afghanistan and W Himalayas to S & EC China, east to Hebei, and reaches coastal China, wintering south of breeding range to SE Asia. Accidental Korea. Monotypic. **HH** Breeds in montane alpine zone, but winters at lower altitudes, frequenting grasslands and ricefields, where typically skulks near streams. **ID** Small pipit with distinct breeding and non-breeding plumages. Resembles Meadow (p.460) and Buff-bellied (p.464) Pipits, but has more strongly streaked mantle and bolder supercilium (pale and broad with black brow above); lores and ear-coverts dark. Ad. breeding has rather grey head, with pink supercilium, distinct black eyestripe and bold black malar; underparts rather plain with pale pink flush, flanks lightly streaked dark. Non-breeding has buffy-pink supercilium, mantle greyer, boldly streaked black; underparts lack pink with narrow black malar merging into more extensive spotting on breast-sides to flanks with buff wash. **BP** Bill short, fine; grey above, horn/pink below; eyes dark brown; tarsi pale or pinkish. **Vo** A weak *tseep* or *seep-seep*. Song (in flight) a twittering during ascent, and pleasant musical *tsuli tsuli tsuli* in descent.

OLIVE-BACKED PIPIT

1st-win
yunnanensis

ad *hodgsoni*

ad
yunnanensis

ad Olive-backed
yunnanensis

PECHORA PIPIT

ad *gustavi*

ad *menzbieri*

ad Pechora
gustavi

ROSY PIPIT

ad win Rosy

1st-win

ad sum

RED-THROATED PIPIT *Anthus cervinus*   CTKJR

L 14–15cm; WT 16.4–29.3g. **SD** N Fennoscandia to Beringia (including extreme W Alaska): *A. c. cervinus* breeds N Yakutia, Chukotka, Kamchatka, Commander Is and N Sea of Okhotsk. Winters Africa, S & SE Asia, in region from SW Kyushu and S Nansei Shoto, coastal China from Changjiang, and Taiwan. **HH** Breeds on montane and coastal tundra with willows and streams. During winter confiding, often accompanies yellow wagtails, favours wet grasslands and marshy edges, often where animals have trampled. **ID** Small to medium-sized pipit, compact with rather short tail, and a plain face. Resembles Buff-bellied, but upperparts sandy-brown with more distinct black streaking. Two parallel pale lines ('braces') on mantle (cf. also Pechora). Extent of red individually and seasonally variable; summer ♂ brightest, pale orange to deep brick red, extending across face, head and upper breast, forming distinct hood in some, more diffuse with paler orange wash to belly and flanks in others; breast-sides and flanks heavily streaked (streaks do not narrow on flanks as in Tree or Olive-backed). 1st-winter more contrasting, with distinct supercilia, bolder mantle streaking, and thin, bold lateral throat-stripes extending into more extensive streaking on breast/flanks. Unlike similar Pechora, tertials cover primaries. **BP** Bill short, fine, grey at tip, yellowish-horn at base; eyes dark brown; tarsi pinkish. **Vo** When flushed typically gives clear, thin *teeze*, a yellow wagtail-like *pssiih*, or *chuirii*; alarm a harder *chyup*. Display flight song a mixture of dry rattling buzzing sounds and thin drawn-out notes: *chu-chu-chu swee-swee-swee-swee psiu psiu psiiu siirrrr wi-wi-wi-wi tswee-tswee-tswee-tswee*.

BUFF-BELLIED PIPIT   CTKJR
Anthus rubescens

L 14–17cm; WT 21g. **SD** N America, E Siberia south to Sakhalin. American and Asian races perhaps specifically distinct: **American Pipit** *A. r. rubescens* N America, perhaps across Bering Strait to E Asia and may winter in Russian Far East and Japan; **Buff-bellied Pipit** *A. r. japonicus* E Asia from SE Taimyr and Lake Baikal through Yakutia, Chukotka, Kamchatka, Commander Is, Sakhalin and Kuril Is, wintering south to Japan, Korea, coastal E & S China, to NE India and SE Asia. Along with Olive-backed, one of commonest pipits in E Asia. **HH** Common migrant throughout region. In summer, rocky alpine and subalpine tundra; on migration/in winter damp grasslands, fields, wetland and woodland edges. **ID** Rather drab, grey-brown, lacking distinguishing features. Face weakly marked, with pale supercilia, plain pale lores, cream chin/throat with prominent black malar forming black neck mark, and merging with breast and flanks streaks. Upperparts dull, indistinctly streaked; underparts buff with variable streaking from breast to flanks. Ad. breeding orange-buff or pale rufous-cinnamon below (pink-toned in Water), with lighter black streaking; upperparts grey-brown. Non-breeding browner above with dark streaking, underparts whiter with much heavier black streaking and more distinct lateral throat-stripe; winter *A. r. japonicus* has stronger, darker streaking on underparts than *A. r. rubescens*. *A. r. japonicus* averages slightly darker above, more heavily streaked

below, with bolder dark malar, and brighter, fleshier legs. *A. r. japonicus* also resembles Meadow (p.460) but has greyer-brown, only faintly streaked, upperparts and only lightly streaked crown (crown streaking prominent in Meadow). **BP** Bill blackish-brown, grey-tipped, horn-coloured base (shorter and finer than Water); eyes dark brown; tarsi dull, pale yellowish-pink (*japonicus*) or dark blackish-brown (*rubescens*). **Vo** A sharp, clear *pit*; *pi-pit* or squeaky *tseep* or *speep-eep* recalling Meadow, sometimes rapidly repeated, *si-si-si-si-sif*, and in alarm a rising *pwisp*. Song, in flight or from perch, a prolonged series of high notes given in discrete phrases: *treeu-treeu-treeu, pleetrr pleetrr pleetrr, chwee-chwee, tsip-tsip-tsiru*.

WATER PIPIT *Anthus spinoletta*   CTKJ

L 15–17.5cm; WT 18.7–23g. **SD** SW Europe to Transbaikalia and C China, wintering N Africa, Middle East, NW India and S China. Status and range in E Asia of *A. s. blakistoni* unclear, but accidental to coastal E China, Taiwan and S Korea, and reported S Japan. **HH** Breeds in rocky alpine areas with short vegetation. Winters at wetlands, from coastal saltmarshes to inland marshes, rivers, lakes and wet fields. **ID** Medium-sized, pale, somewhat colourful pipit. Ad. breeding crown, face and nape largely grey, crown lightly streaked black, broad creamy supercilium behind eye and pale crescent below it, lores largely dark, breaking eye-ring; mantle dark greyish-brown streaked lightly black, wings blackish-brown, tail long, blackish-brown with prominent white outer feathers. Underparts creamy, even peach, or buff from chin to lower breast/flanks very narrow dark lateral throat streak not linked to fine streaking on breast-sides and flanks (less heavily streaked below than Rosy or Buff-bellied), belly and vent white. Non-breeding more streaked, head less grey, with less distinct supercilium, lacks extensive dark malar, and has fine brown rather than coarse dark underparts streaking. Has less distinct brown breast streaking than Asian Buff-bellied and generally cleaner white underparts. Lacks buffy wash to underparts of American Buff-bellied. **BP** Bill fine, sharp, largely dark grey or blackish-brown, pink at base of lower mandible; eyes small, black; tarsi dark grey or blackish-brown. **Vo** A single or double, thin *weest*, also *psri* and *pheet*.

UPLAND PIPIT *Anthus sylvanus*   C

L 17cm; WT 18.7g. **SD** Disjunct range in Himalayas and across S & SE China south of the Changjiang including Fujian and Zhejiang. Monotypic. **HH** Grassy areas with bushes in hills, above 500m in region; occurs in similar, lower, habitats in winter. **ID** Large pipit, resembling Richard's (p.460), but warmer brown, and more heavily streaked below. Crown to mantle warm mid brown with fine to broad blacker streaking; wings blackish-brown with broad brown fringes to coverts and flight-feathers, tail long, blackish-brown; prominent supercilium is clean white, ear-coverts pale brown, with buff malar, and narrow black lateral throat-stripe from bill joining narrow streaking on neck-sides. Underparts off-white, with warm buff wash on flanks, fine dark streaking on chest and belly, and somewhat broader streaking on flanks. **BP** Bill short, thick, grey above, pink below; eyes black; tarsi brownish-pink. **Vo** A high-pitched, somewhat bunting-like *zip zip*, or sparrow-like *chirp*. Song varied, partly recalling Eastern Great Tit, partly some buntings: a repetitive *seetyu-seetyu* and a monotonous *weeeee tch weeee tch*.

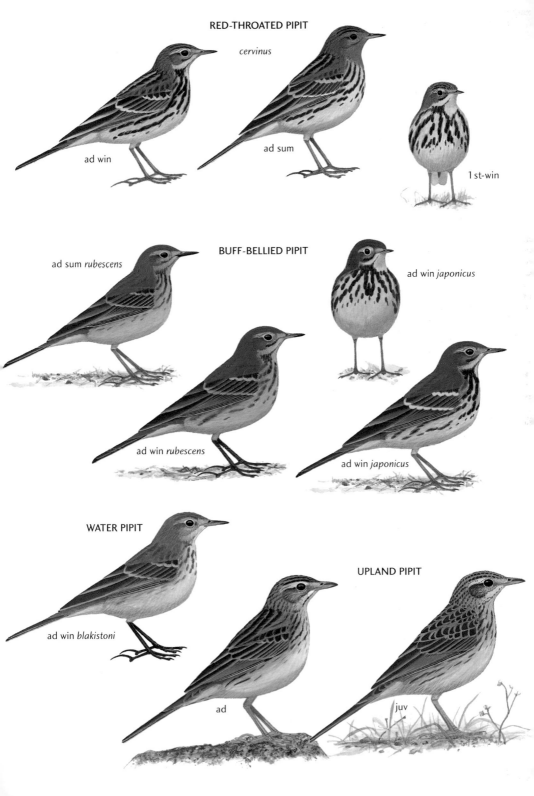

RED-THROATED PIPIT

cervinus

ad win

ad sum

1st-win

ad sum *rubescens*

BUFF-BELLIED PIPIT

ad win *japonicus*

ad win *rubescens*

ad win *japonicus*

WATER PIPIT

UPLAND PIPIT

ad win *blakistoni*

ad

juv

COMMON CHAFFINCH CKJ
Fringilla coelebs

L 14–16cm; WT 18–29g. **SD** W Europe and forested zone of S Siberia east to Angara River. Some winter east NE China, Hebei and Liaoning; accidental in Japan, and reported Korea. Race in region is *F. c. coelebs*. **HH** Gardens, parks, mixed and deciduous forests, where typically feeds on ground. Migratory in some areas, forming flocks in winter. **ID** A rather colourful sparrow-sized finch. ♂ has blue-grey crown and nape, brown cheeks and mantle, olive lower back and rump, and dull, pinkish-buff underparts. Wings black, with blue-grey shoulders and broad white wing-panel and wingbar; tail long, notched, blackish with white outer feathers. ♀ is rather plain greyish-brown, with very plain face, but has white shoulder patch and white tips to wing-coverts forming second wingbar. **BP** Bill rather thin, conical and sharply pointed, grey; eyes black; tarsi dull grey-brown. **Vo** A sharp, explosive *spink, fink* or *chink*, in flight a soft *yupp* or *chup* recalling Brambling's flight call. Song an accelerating, descending series of cheerful musical notes, ending in a trisyllabic flourish: *chink-chink-chink-tee-tee-tee-terree-erree-erree-chissee chu-ee-oo*.

BRAMBLING CTKJR
Fringilla montifringilla

L 14–16cm; WT 17–30g. **SD** Widespread and common, from Scandinavia to E Russia, across Yakutia, Chukotka, south through Koryakia and Kamchatka, Sea of Okhotsk to Amurland and Sakhalin. Winters (Oct–Apr) commonly, even abundantly, but erratically throughout Japan and from S Russian Far East, NE China and Korea to S China; rare Taiwan. Monotypic. **HH** In summer, mature taiga, particularly where there are birches amongst conifers; in winter forests, woodlands and agricultural land. **ID** A largely orange and black finch. ♂ in breeding plumage has black head, mantle and back, orange throat, breast and shoulders, white rump and black tail. Orange from chin to breast and flanks; belly to vent white, with some black streaking on flanks. Orange scapulars, tips of greater coverts and outer fringes of secondaries. Winter ♂ paler, with browner head and mantle, and orange of underparts less bright. ♀ resembles winter ♂, but has dusky grey-brown ear-coverts with broad grey crescent on head-sides to rear of ear-coverts; breast, scapulars and greater coverts all tinged orange or rusty-buff. In flight, relatively long-winged, orange and black wing pattern and contrasting white rump and black tail distinctive. Tail quite long and obviously notched. **BP** Bill conical, pointed, dull horn-coloured with dark grey tip (♀ lacks darker tip); eyes black; legs dull pink. **Vo** A strong nasal *dzwee* and in flight an oft-repeated soft *chup, yeck* or *tk-tk-tk*. Song rather monotonous and slow, but incorporates distinctive buzzing *rrrrrhew* and strong nasal *byeen*.

YELLOW-FRONTED CANARY T
Serinus mozambicus

L 11–13cm. **SD** Natural range Senegal to S Africa. Common in cagebird trade; escapees established in Taiwan (race unknown), with some evidence of seasonal migration there. **HH** Agricultural areas with scrub and open woodland. **ID** A small, yellow-green finch with bright yellow wingbars, rump and tail-sides. ♂ has broad yellow forecrown and supercilium, black malar dividing bright yellow cheeks from throat, and bright yellow underparts. Upperparts olive-green from crown and nape to mantle and back, mantle and scapulars streaked blackish-olive, rump yellow. ♀ duller and paler. **BP** Bill stout, sharply conical, pinkish-grey; eyes black; tarsi blackish-pink. **Vo** Song prolonged, but consists of short interrupted phrases: *si-yu sisi si yu*.

ORIENTAL GREENFINCH CTKJR
Carduelis sinica

L 13–16cm; WT 17–22g. **SD** Widespread and common resident across C, S & E China, Korea, S Russian Far East, Japan, and a summer visitor to Sakhalin, through Kuril Is to S & C Kamchatka. Winters as far south as Taiwan (rare). Resident on Kinmen. Five races in region: *C. s. ussuriensis* NE China, Korea, Russian Far East and S Sea of Okhotsk region; *C. s. kawarahiba* of Kamchatka, Kuril Is, Sakhalin and Hokkaido, wintering to Japan, SE China and Taiwan; *C. s. minor* largely resident in Hokkaido, Honshu and Izu Is, Kyushu and Tsushima; *C. s. kittlitzi* resident in Ogasawara and Iwo Is; *C. s. sinica* C & E China south to N & C Vietnam. **HH** Mixed coniferous and deciduous forests and agricultural areas in summer; flocks in woodland, forest edge, riverine scrub, agricultural land and gardens in winter. **ID** Large, dark, olive-brown finch with prominent, bright yellow wing and tail patches. ♂ has olive-grey head, black around eye and bill, and grey nape; upperparts dark olive-brown; underparts slightly paler olive-brown. ♀ is browner, and young birds duller with heavy streaking above and below. Wing-feathers mostly black, but yellow bases to primaries form bold yellow flash on closed wing, and has grey-white fringes to tertials; rump grey-brown, tail black with yellow vent and basal tail-sides. Tail rather broad with shallow notch. In flight shows broad yellow wingbar; flight typically finch-like, undulating deeply, but display flight is loose-winged and slow, somewhat butterfly-like. *C. s. kittlitzi* is smaller, generally darker olive-green, with little yellow on underparts except vent, and larger bill than *C. s. minor*, whilst *C. s. kawarahiba* is much larger with broader white tertial fringes. **BP** Bill strong, conical, pink; eyes black; tarsi pink. **Vo** A very nasal *djeeeen, djuwee* or *dzweee*. Song a slightly sweet chattering comprising *kirr korr* and *kirikiri-korokoro-been* notes given from high perch or in slow song-flight. Song is interspersed with or preceded by calls. **AN** Grey-capped Greenfinch.

COMMON CHAFFINCH

coelebs

ad ♂

ad ♀

ad ♀

BRAMBLING

ad ♀ win

ad ♂ sum

ad ♀ win

ad ♂ win

YELLOW-FRONTED CANARY

ad ♂

ad ♂ *kawarahiba*

ORIENTAL GREENFINCH

ad ♀ *kawarahiba*

ad ♂ *kawarahiba*

ad ♂ *ussuriensis*

EURASIAN SISKIN
CTKJR
Carduelis spinus

L 11–13cm; WT 11–18g. **SD** Widespread, from Scandinavia to N Sea of Okhotsk, maritime SE Russia, Sakhalin, NE China, S Kuril Is, and perhaps Hokkaido and C Kamchatka; winter visitor (Oct–May) Japan, Korea and E & SE China; rare Taiwan. Monotypic. **HH** Mixed taiga (particularly birch and alder); in winter flocks favour conifers and alders, riparian woodland, parks and gardens. Irruptive, winter numbers vary greatly between years. **ID** Small yellow-green finch with bright yellow wing, rump and tail patches. ♂ has black forehead and chin, yellow face with olive-green ear-coverts, grey nape, olive-green mantle streaked black, yellow rump, black tail with yellow bases to outer feathers; shallowly notched tail is narrower and shorter than in Oriental Greenfinch (p.466). Underparts yellow, whiter on belly and flanks, with dark grey streaking on flanks. Wings black with broad yellow tips to coverts and outer fringes to flight-feathers. ♀ duller, lacks black forehead and chin, less yellow below. Note extralimital Pine Siskin *C. pinus* is a migrant to Alaska and could appear amoung Eurasian Siskin in NE Asia; differs in colder grey plumage with heavy blackish-brown streaking; yellow only on greater coverts and outer fringes to primaries (♂), or white bar on greater coverts and only slight yellow fringes to primaries (♀). **BP** Bill fine, sharply conical, pink with grey culmen and tip; eyes black; tarsi blackish-pink. **Vo** Calls varied, rising or falling, either thin *tsuu-ee*, *chueen* (clearer than Oriental Greenfinch) or *tilu;* or hard, dry rattling *tet-tet*. Song prolonged, rambling and complex, many thin twittering notes interspersed with trills: *tseen jukuku tsupyee...*

COMMON REDPOLL
CTKJR
Carduelis flammea

L 11–14cm; WT 12–16g. **SD** Widespread across northern N America, and from Scandinavia to Bering Strait; introduced New Zealand. *C. f. flammea* a summer visitor to extreme N Russia, resident across taiga of Yakutia and Chukotka to Kamchatka, Sakhalin and S Russian Far East, and winter visitor (Nov–Mar) to Kuril Is, Japan south to C Honshu, NE China to Shandong and Jiangsu, and Korea (rare). **HH** Tundra, forest-tundra and mixed coniferous taiga, favours birches and alders, especially in winter also in hill and lowland forest, riparian woodland, parks and gardens. Irruptive, commonly flocks, but winter numbers vary greatly between years. **ID** Small, grey-brown, rather streaky finch. ♂ has deep red forehead, black lores and black chin spot. Upperparts greyish-brown, rather heavily streaked dark grey-brown, rump plainer grey-brown, wings and tail black, with white tips to coverts forming double wingbar, white fringes to tertials and outer margins of secondaries. Underparts off-white washed buff and variably streaked, sometimes heavily on flanks, whiter on lower belly and vent with broad, diffuse, dusky streaks on undertail-coverts. Lower face, throat, breast and rump of ♂ variably washed pale to deep pink, brightening during winter. ♀ lacks pink. Juv. lacks red forehead and black on face/throat, and is slightly darker and more heavily streaked, but this plumage rarely seen south of breeding range. Tail dull black with obvious notch. Flight buoyant, undulating; flocks active, 'dancing' in air. **BP** Bill fine, sharply conical, yellow/horn; eyes black; tarsi dark grey/black. **Vo** Calls frequently; nasal *dsooee* when perching, flight notes recall Eurasian Siskin, but longer and harder *ju ju chueen* or reverberating, chattering *che che che che* or *chett-chett-chett* with hard metallic edge. Song, given during slow, circling and undulating display flight comprises 2–3 hard flight calls interspersed with a dry, wheezy reeling *zzzrrrrrrrr*. **AN** Mealy Redpoll.

ARCTIC REDPOLL
CKJR
Carduelis hornemanni

L 12–16cm; WT 10–16g. **SD** High-latitude Holarctic. In Eurasia, *C. h. exilipes* ('**Coues's Redpoll**') from N Scandinavia to Chukotka and Bering Strait. Scarce to rare winter visitor (Oct–Mar) south to NE China (has reached Beidaihe), very rare N Japan; accidental N Korea. **HH** Tundra, forest-tundra, tundra-edge and stunted taiga (birch and willow thickets). **ID** Paler, and less streaked than Common. Shares similar deep red forehead, black lores and chin, but mantle paler grey-brown with whitish fringes, or almost white with darker brown streaks; lower back and rump unstreaked white (♂ with variable pale pink tone); underparts only lightly streaked on flanks and almost entirely pure white; vent and undertail-coverts white, unstreaked (longest feathers may have dark central shaft-streak). Overall pale, especially on lower back/rump. ♂ suffused pink on cheeks and breast, though may have buff wash to breast rather than pink. ♀ lacks pink flush. Wings whiter on secondaries and coverts, with prominent white wingbar and tertial fringes. Obviously notched tail is blackish with pale outer fringes, contrasting strongly with white rump. Common is highly variable, thus complicating identification. In flight, white wingbars, white vent and unstreaked rump are all useful features. **BP** Bill stubby, short, deep, with straight culmen, pale yellow/horn; eyes black; tarsi dark grey/black. **Vo** Very similar to Common, but slightly softer and higher. **AN** Hoary Redpoll.

EURASIAN GOLDFINCH
J
Carduelis carduelis

L 12–15cm; WT 14–19g. **SD** W Europe to mountains of C Asia. Accidental in region only in S & W Japan (Feb, May, Oct). *C. c. carduelis* ('**European Goldfinch**') and *C. c. caniceps* ('**Grey-crowned Goldfinch**'), perhaps separate species, both recorded Japan; records of *C. c. carduelis* probably escapes. **HH** Mixed deciduous and coniferous forests, forest edge, parks and gardens. **ID** *C. c. caniceps* colourful, with bright red face, black lores, grey-white head-sides and underparts, darker grey mantle and flanks, black wings with yellow bases to primaries, secondaries and outer tertials, forming broad golden-yellow band, white fringes and tips to tertials, white rump and short, notched black tail with white spots near tip. *C. c. carduelis* has deeper, more extensive bright red forehead and chin, black crown and hind-collar, white head-sides and underparts, dark grey-brown mantle and breast-sides, blacker wings with broad golden-yellow band, and only white tips to tertials. **BP** Bill sharply conical, pale pinkish-horn; eyes black; tarsi dull pink. **Vo** A musical tinkling *kiri kiri* or *tickelitt*; flight call a lilting, liquid *tilili* or *tulilit*. Song a more protracted version of calls, including high tinkling notes, stalling buzzing notes and chirps.

EURASIAN SISKIN

juv

ad ♂

ad ♂

ad ♀

ad ♀ win

COMMON REDPOLL

flammea

juv

ad ♂ win

ad ♂ win

ad ♂ sum

EURASIAN GOLDFINCH

ARCTIC REDPOLL

exilipes

ad ♂ sum

ad ♂ sum

ad *carduelis*

ad ♂ win

juv

ad *caniceps*

PLATE 221 : ROSY FINCHES AND ROSEFINCHES

ASIAN ROSY FINCH CKJR
Leucosticte arctoa

L 16cm. **SD** Ranges from Lake Baikal to S Yakutia, Koryakia and Kamchatka, probably Chukotka, and N Kuril Is, wintering locally (Nov–Apr) to S Russian Far East, NE China, Hokkaido (where may breed) and N & C Honshu, sometimes south to Kyushu; scarce N Korea. *L. a. brunneonucha* in most of region; *L. a. gigliolii* in SW Yakutia. **HH** In summer, rocky tundra or mountains above treeline; in winter often in flocks in low mountains and forests, also lowlands and snow-free coasts and headlands. **ID** Large, rather dull finch with few distinguishing features, but rather long-winged. ♂ rather black, particularly on forehead, face, underparts, wings and tail. Pale brown crown and nape grade into brown mantle streaked dark grey, lower back and rump dark brown. Underparts blackish-brown spotted pale grey on throat and breast, with raspberry-pink flanks; scapulars and wing-coverts fringed and tipped brighter pink; undertail-coverts also pink. ♀ has pink restricted to scapulars, covert tips and flanks. Flight strong but undulating, wings long, tail long and notched. **BP** Bill yellow/ horn with dark grey tip; eyes dark brown; tarsi dark grey/black. **Vo** Sparrow-like *iyuu* and a dry *peut*. Song, given from exposed rock, a range of twittering notes. **AN** Arctic Rosy Finch.

GREY-CROWNED ROSY FINCH R
Leucosticte tephrocotis

L 16cm; WT 26g. **SD** Primarily western N America including Aleutians and Bering Sea islands: *L. t. maxima* is a common endemic on Commander Is off Kamchatka, and straggler to E Kamchatka. **HH** Open tundra, rocky outcrops, rocky shores and headlands. **ID** Breeding ♂ has black forehead/throat and very dark brown mantle and breast, wings blackish-brown with pink in scapulars, coverts and fringes to flight-feathers, also flanks and rump; face, cheeks and nape grey. ♀ is less contrasting. Flight strong, buoyant, undulating, with very pale flight-feathers and long, notched tail. **BP** Bill black (♂) or yellow with dark tip (♀/young); eyes black; tarsi black. **Vo** A soft *cheew* or buzzy *jeerf*, and in flight a husky chirping. Song a slow series of descending whistles: *jeew jeew jeew*.

MONGOLIAN FINCH C
Eremopsaltria mongolica

L 11–14cm; WT 18–26g. **SD** E Turkey to Mongolia, barely reaching region in Inner Mongolia. Possible elsewhere in NE Asia as vagrant. Monotypic. **HH** Semi-arid and arid scrub, stony desert and mountains. **ID** Pale, large-headed, rather plain finch of arid country. ♂ has pale grey-brown upperparts, with pink-washed lores, face-sides and breast; underparts white. Wings black with extensive white coverts with pink tinge; white bases to secondaries, pink fringes to primaries, tail short, notched, white at base with pink wash on lower back/rump, black at tip. ♀/winter ♂ duller, lacking pink. **BP** Bill stout with rounded tip, pale horn; eyes prominent in very plain face, irides black; tarsi brownish-pink. **Vo** A rising *tew-vweet*, slightly nasal *vzheen*, clear *vee-tyu*, and melancholy

piu. Brief song a slowly repeated, quite loud, fluting *viit-vüah... vreyah*. **TN** Formerly *Bucanetes mongolicus*.

LONG-TAILED ROSEFINCH CKJR
Uragus sibiricus

L 13–18cm. **SD** Resident over much of S Siberia, from Kazakhstan to SE Russia and NE China. Fairly common summer visitor (May–Sep) S Sakhalin, Hokkaido and S Kurils, wintering (Oct–Apr) in Japan south of Hokkaido and Korea, when found alone or in family parties. *U. s. sibiricus* NE China; *U. s. ussuriensis* NE Manchuria, Amurland, Ussuriland and Korea; *U. s. sanguinolentus* Sakhalin, S Kurils, Hokkaido and N Honshu, wintering further south in Japan. **HH** Forest and wetland fringes, riparian scrub, open woodland and agricultural land with bushes or trees. **ID** Small-bodied but long-tailed. ♂ deep pink with silvery crown and cheeks, mantle dark pink with black streaking, lower back and rump unstreaked deep pink, and tail black with white outer rectrices. Forehead and lores dark pink, underparts deep pink, silvery on flanks, greyer on belly. Wings largely black, with two prominent wingbars and a pale panel formed by white fringes to tertials. Winter ♂ duller, less pink, browner above. ♀ plainer brown with very plain face, but wing pattern similar to ♂. First-winter similar. Flight rather weak and fluttering. **BP** Bill stubby, pale greyish; eye of ♂ set in dark patch, but that of ♀ prominent in plain face resembling Common Rosefinch, irides black; tarsi dull dark pink. **Vo** Soft, plaintive fluty, commonly double or triple whistles: *hwit-hwot, fee fee, pee popo* and distinctive ringing *stip* audible at some distance. Song a hurried *churu churu chee fee fee fee*.

COMMON ROSEFINCH CTKJR
Carpodacus erythrinus

L 13–15cm; WT 19–27g. **SD** Common from Scandinavia to Chukotka, Kamchatka, S Russian Far East and Sakhalin, NE China and N Korea. Migrates through E China and Korea; rare Japan, accidental Taiwan. Winters S & SE China. *C. e. erythrinus* ranges east to Lena and Kolyma basins; *C. e. grebnitskii* NE China and N Korea to Sakhalin, NE Siberia and Kamchatka. **HH** In summer, forest, thickets and woodland edge, often favours deciduous stands (e.g. birches); in winter/on migration in scrub and lowland woodland. **ID** Medium-sized, rather plain finch. ♂ has deep red head, chin/throat and breast, pinkish-white on rest of underparts; reddish-grey mantle and scapulars, bright red rump; wings and tail blackish with broad grey-brown fringes, tinged pink. Young ♂ has red confined to head, rest of body grey-brown like ♀. ♀ very plain, especially face, with no distinguishing features, except dark eye in pale unmarked face. Upperparts dark olive-brown streaked darker; underparts pale grey-brown with diffuse streaking on breast and flanks. Flight strong but undulating, tail rather long and notched. **BP** Bill stout with rounded tip, dull grey; eyes black; tarsi greyish-pink. **Vo** A strongly whistled *vüi* or *chüi*, or *chooee-ee* commonly given in flight, including by migrants overhead; also a quiet, short *zik* before take-off and in flight. Song a simple, sweetly whistled combination of call-like notes: *vidyew-vui vidyew-vui vidyew-vidyew-vui*, typically given from high song post or treetop. **AN** Scarlet Rosefinch.

ASIAN ROSY FINCH

ad ♂
gigliolii

ad ♂
brunneonucha

ad ♀
brunneonucha

ad ♂ win

maxima

GREY-CROWNED ROSY FINCH

MONGOLIAN FINCH

ad ♀ win

ad ♀

ad ♂

ad ♂ sum

ad ♂
grebnitskii

COMMON ROSEFINCH

LONG-TAILED ROSEFINCH

sanguinolentus

ad ♂
erythrinus

ad ♀ sum

ad ♂ sum

ad ♂
grebnitskii

ad ♂ win

ad ♀/juv

CHINESE BEAUTIFUL ROSEFINCH C
Carpodacus davidianus

L 15cm. **SD** Himalayas northeast across W & C China, just reaching Inner Mongolia and Hebei. Monotypic. **HH** Normally at high altitudes (above 3,600m) in juniper scrub, dwarf oak and rhododendrons but descends in winter when may appear beyond usual range. **ID** Medium-sized, rather chunky, dark brown and bright reddish-pink rosefinch. ♂ has largely mid-brown crown to lower back and scapulars (dark-centred feathers with pale brown fringes). Broad supercilium is deep red before eye, becoming paler pink at rear; eyestripe brown, face fairly bright, deep pinkish-red (lores, chin to ear-coverts darkest), breast to flanks deep pink, rump pale whitish-pink, and belly and vent off-white. ♀ lacks pink, is plainer brown above with dark streaking, pale supercilium above dark ocular area, and underparts pale with weak streaking on flanks. Wings and tail blackish-brown with paler brown fringes to coverts and tertials. Tail rather square-ended, lacking distinct fork. **BP** Bill stout with pointed tip, dull grey; eyes dark brown; tarsi greyish-pink. **Vo** Call a staccato, metallic *tsink tsink*. **TN** Formerly within Beautiful Rosefinch *C. pulcherrimus*.

VINACEOUS ROSEFINCH T
Carpodacus vinaceus

L 13–16cm. **SD** Locally in Himalayas and W China; *C. v. formosanus* occurs in upper montane areas of Taiwan. **HH** Scrub, bamboo thickets, roadside vegetation and forest edge at *c.*2,000–3,500m. **ID** ♂ deep wine-red with silvery stripe from above eye broadening on headsides; blackish-red forehead, face and chin; wing-coverts and flight-feathers black with white-tipped tertials, tail quite long and notched, black, but tinged deep wine-red. ♀ dull brown, streaked blackish-brown on mantle; underparts paler dull brown with narrow blackish streaks on throat, somewhat broader less clear streaking on flanks; resembles very dark Common Rosefinch (p.470), with short dark, off-white-tipped tertials. **BP** Bill sharply pointed, grey; eyes black; tarsi dark brownish-pink. **Vo** Calls include a sharp, dry *ziht ziht*, a plaintive *pee-u* and a repetitive, nasal *flee-yuu*.

PALLAS'S ROSEFINCH CKJR
Carpodacus roseus

L 15–17cm. **SD** Breeds C & S Siberia, from Yenisei to Yakutia, south to Amur Estuary and N Sakhalin, also N Korea. Winters (Oct–Apr) southeast to S Russian Far East, Hokkaido, N & C Honshu, Korea and NE & E China: *C. r. roseus* C & E Siberia to Kolyma mountains, wintering to NE China and Korea; *C. r. portenkoi* Sakhalin winters in N Japan. **HH** During summer found in upper montane taiga, but in winter generally scarce and rather erratic, occurring only locally within broad range, favouring forest edges, scrub and agricultural land with trees, also wooded parks. Often in family parties or small flocks in winter. **ID** Heavy-set finch with rather large head and prominent eye. ♂ generally dark raspberry-pink with forehead and chin/throat covered in silver spots. Mantle and back dark pink streaked black; underparts deep pink grading to white on belly and vent; lower back and rump plain pink; wings black with pink scapulars, white fringes to wing-coverts forming double wingbar, and white-fringed tertials and pale pink outer fringes to primaries and secondaries; notched tail black, also with pale pinkish outer fringes. Young ♂ resembles ♀, but has variable pink on head, breast and rump. ♀ generally brown with rather heavily streaked upperparts and warm rusty-brown tone to breast, with fine dark streaking; crown and nape greyish-brown, wings have pale fringes to coverts and flight-feathers, and dark pink rump is unstreaked. **BP** Bill short, deep, dull grey; eyes small, irides black; legs pinkish-grey, feet grey. **Vo** A soft whistled *fee*, a metallic *tsuiii*, a quite loud, discordant, bunting-like *dzih* and strong *chek-chek*, slightly deeper and softer than Hawfinch; the song includes repeated whistled call notes.

PINE GROSBEAK CKJR
Pinicola enucleator

L 19–24cm; WT 47–64g. **SD** A high-latitude, often high-altitude species of N America and Eurasia, from Scandinavia to Chukotka. Mainly resident, but some move to lower altitudes, even lower latitudes, in winter, when found south to NE China (Heilongjiang and Jilin): *P. e. kamtschatkensis* occurs across Siberia from Yenisei to Kamchatka, wintering to N China, and presumably this subspecies accidental Korea; *P. e. sakhalinensis* occurs in Sakhalin, Kuril Is and Hokkaido. **HH** Scarce, favours taiga-forest, preferably deciduous areas with alder and birch, but also low pine forest and alpine stone pines in summer, visiting similar habitats at lower elevations in winter, when also visits berry-bearing shrubs/trees even in towns/villages, sometimes with waxwings. **ID** Very large, heavy-set finch with rather small rounded head. ♂ generally dark pink with black lores and eye patch, and mantle, back and wings largely black, but mantle-feathers fringed deep pink, scapulars with grey, coverts with white, forming double white wingbar; rump red with some black scaling, and notched, rather long tail is black. Underparts deep reddish-pink grading to dark grey on flanks, belly and vent. Juv. ♂ orange on head, breast and rump. ♀ has yellow-green head and breast, upperparts and much of underparts grey. Flight slow and undulating. **BP** Bill large and stubby, greyish-black; eyes black; tarsi black. *P. e. kamtschatkensis* generally identical to *P. e. sakhalinensis* but bill is smaller, shorter and less deep. **Vo** Calls are soft, somewhat bullfinch-like, sweetly whistled: *pee-u*, *pyüru pyüru* or *pyou-you lee*. Song is soft, fast and high-pitched, *pyuru pyuru pyuro ruriri*, similar to pattern of Red-flanked Bluetail.

CHINESE BEAUTIFUL
ROSEFINCH

ad ♀

ad ♂

VINACEOUS ROSEFINCH

formosanus

ad ♀

ad ♂

PALLAS'S ROSEFINCH

roseus

ad ♂

ad ♀

PINE GROSBEAK

kamtschatkensis

ad ♀

1st-sum ♂

ad ♂

COMMON CROSSBILL CKJR
Loxia curvirostra

L 15–17cm; WT 35–50g. **SD** Common in boreal Holarctic, across much of N America, in Eurasia from W Europe to S Russian Far East, Sakhalin and S Kuril Is, also Himalayas to W China, and NE China and N & C Korea. Summer visitor to some regions, and winters (Oct–May) south to Hokkaido, Honshu, Korea and parts of E China to Zhejiang: *L. c. japonica* in NE China, Russian Far East, Japan, Korea, Sakhalin, S Kuril Is, wintering to Japan and China; *L. c. tianschanica* of C Asia and W China winters to Liaoning and Hebei. **HH** Coniferous taiga- and temperate forests, favouring fir rather than pine, and in winter in mature mixed forest often in mountains, sometimes in lowlands. Irruptive and partly migratory species, varying greatly in numbers between years within broad winter range, from some parts of which it may be absent in any given winter. **ID** Large colourful finch, rather stocky with largish rounded head. ♂ variably brick or orange-red on head, underparts, mantle and rump, mantle with some brown, wings and tail black. ♂ *L. c. japonica* bright and often shows white on vent. ♀ olive to yellowish-green, rather greyer on head and face, brighter green on rump, with dark wings and tail; juv. resembles ♀, but heavily streaked on underparts and from crown to back; 1st-year ♂ greenish-brown with orange or rusty tinge to back and underparts. In flight, appears stocky and bull-headed, tail clearly notched; flight deeply undulating and often vocal. **BP** Bill deep, stubby (though size variable) with unusual crossed tips (shared only with next species), blackish-grey; eyes black; tarsi dark grey/black, pinkish-grey in young. **Vo** Calls, given frequently in flight and from treetop perch, are deep, explosive, even metallic, *chup chup chup*, *jip jip* or *glipp-glipp-glipp*. Song a loud mixture of trills and twittering: *chu chui chu pyuu pee pee*. **AN** Red Crossbill.

TWO-BARRED CROSSBILL CKJR
Loxia leucoptera

L 14–16cm; WT 25–38g. **SD** Common Holarctic species, ranging across much of northern N America and Eurasia: *L. l. fasciata* from N Scandinavia to Sea of Okhotsk, occurring sympatrically with Common Crossbill over much of range but less extensively distributed in S Russian Far East (absent Sakhalin), NE China; winters south to Liaoning and Hebei. **HH** Deciduous taiga, especially larch, birch and rowan, but also fir and pine. Winter range in E Asia poorly known, but like Common Crossbill can be irruptive in winter (sometimes with Common), to N Japan and Korea. **ID** Large colourful finch with prominent wing pattern, and thinner bill, smaller head and longer tail than Common Crossbill. ♂ has bright raspberry-red head, underparts and mantle; wings, lower back and tail black, but wing-coverts broadly fringed white, forming very prominent double wingbar, with broad white tips to tertials. ♀ dull olive-green, streaked darker greyish-brown on crown and, especially, mantle and scapulars, and underparts rather paler yellow-olive, streaked dark grey; rump greenish-yellow. Juv. resembles ♀, but more heavily streaked, and 1st-year ♂ is pale, orange-red. In flight, resembles Common, but appears more slender, all plumages show clear white wingbars, tail clearly notched; flight deeply

undulating; different flight call. **BP** Bill longer, more slender than Common, but similarly scissor-like, blackish-grey; eyes black; tarsi dark grey/black. **Vo** Softer and higher pitched than Common: *chip-chipp-chipp* or *glib glib*, scratchier and somewhat redpoll-like, and a discordant piping *tviiht*. Song varied and twittering, more like Eurasian Siskin than Common Crossbill. **AN** White-winged Crossbill.

BROWN BULLFINCH CT
Pyrrhula nipalensis

L 15–17cm. **SD** Resident, across Himalayas to SE Asia. *P. n. ricketti* S & SE China north to Fujian; *P. n. uchidai* local resident of mid- to upper-elevation slopes on Taiwan. **HH** Subalpine forests above 2,000m in summer, but lower on Taiwan (*c.*500–2,500m). **ID** Plain brown finch with long tail. ♂ has entire plumage drab grey-brown; blackish on lores, around eye, chin and forehead, with narrow white streak below eye, and off-white behind eye. Cheeks, face-sides to throat and nape grey. Lower back black, white rump band narrow, tail long and glossy black with notched tip, graduated with shorter feathers overlying longer ones, and having oddly crinkled appearance. Wings largely glossy black, with grey-brown scapulars and broad off-white band on greater coverts. ♀ (and juv.) lacks black on face and is brown in wings not white. In flight, appears buff-brown with narrow rump patch and grey-white wingbar on median coverts. **BP** Bill stubby, black at tip, pale blue-grey at base; eyes black; tarsi dull pink. **Vo** Differs greatly from other bullfinches: a quiet *veh* contact call and hard, rolled, flight call *tyerrlip*. Overhead flocks sound like Lapland or Snow Buntings. Song a sweet, wavering and slightly rising *tweert* or *per-lee* given from treetops.

BEAVAN'S BULLFINCH CT
Pyrrhula erythaca

L 15–17cm. **SD** Resident from Himalayas to C & NE China: *P. e. erythaca* N Hebei; *P. (e.) owstoni* on Taiwan, may be specifically distinct (as **Owston's Bullfinch**). **HH** Mixed and coniferous subalpine forests and alpine zone; usually above 2,000m on Taiwan. **ID** Larger, more richly coloured bird than Brown Bullfinch, with larger white rump patch. Ad. has distinctive black mask from forehead to chin, extending to point behind eye, bordered above by white. Glossy blue-black wings with broad pale grey band formed by greater coverts, and broad white rump, white vent and glossy blue-black tail with crinkled, graduated feathers like Brown. ♂ has pale grey crown, nape, mantle, scapulars and greater coverts, and warm orange-buff breast and pale grey or off-white underparts. ♀ largely grey from crown to lower back, underparts greyish-white with orange-buff wash to flanks, grey back and greater coverts. In flight, grey-white wingbar and white rump conspicuous; ♂ closely resembles Brown, but lacks black lower back, and ♀ much greyer, with greyer wingbar. Subspecies very similar, but *P. e. owstoni* less colourful, less orange-buff and more dull pinkish-grey on underparts. **BP** Bill stubby, grey; eyes black; legs dull pink. **Vo** Calls variable, typically a soft whistled Eurasian Bullfinch-like *soo-ee*, but *P. e. owstoni* gives soft, thin *yifu yifu* and nasal *swik wu*, and in flight a bullfinch-like *pu* call. **AN** Grey-headed Bullfinch.

COMMON CROSSBILL

japonica

ad ♂

ad ♀

juv

TWO-BARRED CROSSBILL

fasciata

ad ♂

ad ♀

BROWN BULLFINCH

ricketti

ad ♂

juv

ad ♂ *erythaca*

BEAVAN'S BULLFINCH

ad ♀ *erythaca*

ad ♂ *owstoni*

juv *erythaca*

EURASIAN BULLFINCH   CKJR
Pyrrhula pyrrhula

L 14–18cm; WT 27–38g. **SD** Widespread, from W Europe to Kamchatka south to Japan and NE China. Chiefly resident, but northern birds winter to NE & N China and Korea; in Japan resident in Hokkaido, N & C Honshu, but rare winter visitor to S Kyushu. *P. p. pyrrhula* N Europe to Lake Baikal and S Yakutia; *P. p. cassinii* Kamchatka and Koryakia, around Sea of Okhotsk, wintering to Japan and NE China; *P. p. rosacea* Sakhalin, Russian Far East, NE China, and winters E China, Korea and Japan (sometimes considered synonymous with *griseiventris*). Two taxa perhaps warrant specific status: *P. (p.) cineracea* (**Baikal Bullfinch**) east to N Mongolia and Transbaikalia; *P. (p.) griseiventris* (**Grey-bellied Bullfinch**) in Kuril Is and Hokkaido. **HH** Fairly common in montane and lowland mixed and coniferous forests in summer, also woodland edge, scrub and gardens in winter. **ID** Plump, bull-necked finch with prominent white rump. ♂'s of all races have glossy black cap, black lores and eye patch and very narrow black bib, mid to dark grey upperparts, black wings with narrow white lesser covert bar, and broad white greater covert bar, vent and rump, tail black. *P. (p.) griseiventris*, the common form in Japan and Kuril Is, is larger than other races and dark grey above, with pale to mid grey upperparts, and entirely mid to dark grey underparts, only cheeks and throat bright pink. *P. p. rosacea*, the common form in winter in Korea is pinker, that of cheeks less intense than *P. (p.) griseiventris*, contrasting less with pinkish-brown underparts, mantle brownish-grey and has white spots in outer tail. *P. p. cassinii* is pinkest race, colour extending from cheeks over entire underparts, only lower belly and vent white; more extensive white in outer tail. *P. (p.) cineracea* lacks pink. ♀ shares same pattern of black and white, but upperparts plain grey-brown, and underparts grey (pinky-brown in *rosacea*). Juv. resembles ♀, but lacks black cap. **BP** Bill short, stubby, dark greyish-black; eyes small, black; tarsi dull dark pink. **Vo** Soft, melancholic, repetitive piping: *teu teu teu* or downslurred *pheew pheew pheew*. Song a softly whistled *fiyo fiyo fee*. Readily investigates whistled rendition of call or song.

HAWFINCH   CTKJR
Coccothraustes coccothraustes

L 16–18cm; WT 46–70g. **SD** W Europe to Russian Far East, Sakhalin, S Kamchatka, Kuril Is and Japan. Resident NE China and locally N Korea; winters Korea, Japan south of Hokkaido, E China south to Fujian; rare Taiwan. *C. c. coccothraustes* S Siberia to Transbaikalia and N Mongolia; *C. c. japonicus* Sakhalin, Kamchatka, N & C Japan; winters in S Japan, SE China and straggles to Taiwan. **HH** Widespread but nowhere common (though often very common on migration), in montane and lowland deciduous and mixed forests in summer, also woodlands, parks and gardens in winter. **ID** Large, compact, bull-headed but short-tailed finch. Sexes very similar, but ♂ somewhat brighter. Head warm orange-brown, with black lores, eye surround, narrow forehead bar and chin patch; grey nape and neck-sides. Upperparts dark brown, mid brown on rump; underparts buffy-brown, but white on vent. Scapulars and coverts brown, with broad white bar on greater coverts, primaries and secondaries glossy blue-black, the inner primaries with strangely crinkled broad spatulate tips; rump and uppertail-coverts brown, short, slightly notched tail black with broad white tip. In deeply undulating flight, wings and tail flash much white – showing a white panel at base of primaries, broad bar across coverts and broad tail tip. **BP** Bill deep and massive, grey at base, blackish at tip, and horn-coloured in winter/juv.; eyes small, irides mid brown; tarsi pink. **Vo** Distinctive explosive *tsi, tzick* or *tic* given at rest, or repeatedly in flight. Song comprises various harsh chattering sounds: *tsutsutsu chuu-pip-pip-tsiriri.*

CHINESE GROSBEAK   CTKJR
Eophona migratoria

L 15–19cm; WT c.50g. **SD** Migrant to E & NE China to Amurland: *E. m. sowerbyi* NE China; *E. m. migratoria* NE China, Ussuriland, Amurland and Korea. Winters C Honshu to SE China; rare Taiwan, has reached Sakhalin. **HH** Montane and lowland deciduous taiga-forest or woodlands and orchards in summer, also parks and gardens in winter. **ID** Large, highly vocal and boldly marked finch, superficially resembling Japanese Grosbeak, but smaller. ♂ has fuller black cap, extending to ear-coverts and chin; upperparts brown, pale grey-brown on rump; underparts pinkish-brown, strongly washed rufous or cinnamon on flanks and vent. Wings glossy blue-black, with small white patch at base of primaries and all-white primary and secondary tips; tail glossy black and deeply notched. In flight, narrow white primary bar, trailing edge and wingtips, and pale grey rump contrasting with black tail diagnostic. ♀ lacks black hood with duller upperparts and less glossy black wings and tail. **BP** Bill large, dull yellow with blackish tip and bluish base (Japanese has a much yellower bill); eyes red-brown; tarsi dull pink. **Vo** A loud *tek-tek* or *dek-dek*. Song a loudly whistled *chee chee choree kirichoo*, similar to Japanese. **AN** Yellow-billed Grosbeak.

JAPANESE GROSBEAK   CTKJR
Eophona personata

L 18–23cm; WT c.80g. **SD** Partly migratory, partially resident E Asian endemic. *E. p. magnirostris* is a summer visitor to NE China, Ussuriland and Sakhalin, and perhaps N Korea, migrates through E China, wintering to S China. *E. p. personata* breeds Hokkaido and N Honshu, year-round and wintering through rest of main Japanese archipelago; accidental Taiwan. **HH** Fairly common, in montane and lowland deciduous taiga forest, also mixed lowland broadleaf forest, in summer, and woodland, parks and gardens in winter. **ID** Very large, highly vocal, boldly marked finch. Ad. readily recognised by very large bill, glossy black face and cap, and generally mid grey plumage, lightly tinged brown on tertials, grey on vent. Wings glossy black to tip, with bold white patch midway along primaries; long tail is glossy black and deeply notched. Juv. has black lores and margin to bill, but otherwise head, back and underparts brownish-grey, with same wing and tail pattern as ad. **BP** Bill massive, bright rich yellow; eyes dark brown; tarsi brownish-pink. **Vo** A deep, hard *tuk-tuk*. Simple song comprises 4–5 strongly whistled notes, rising and falling and ending in a longer note: *tsuki-hi-hoshi.*

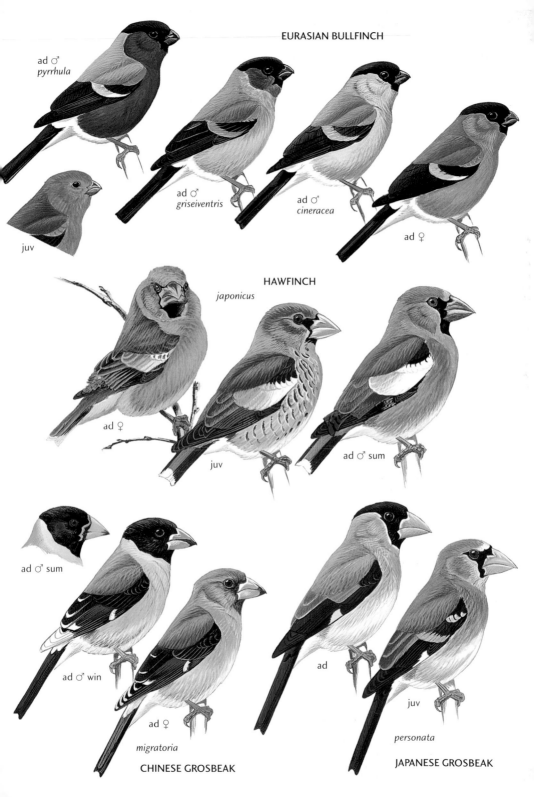

EURASIAN BULLFINCH

ad ♂
pyrrhula

juv

ad ♂
griseiventris

ad ♂
cineracea

ad ♀

HAWFINCH

japonicus

ad ♀

juv

ad ♂ sum

ad ♂ sum

ad ♂ win

ad ♀

migratoria

CHINESE GROSBEAK

ad

juv

personata

JAPANESE GROSBEAK

ORANGE-CROWNED WARBLER   R
Vermivora celata

L 13cm; WT 9g. **SD** N America, wintering south to C America, and breeding north to Alaska. Taiga form *V. c. celata* of W Alaska has strayed almost annually to St Lawrence I, and is accidental to Chukotski Peninsula. *V. c. lutescens* of W Canada and SC Alaska, has also strayed twice to Gambell, so possible in NE Asia. **ID** A small, rather dull, grey-green warbler with very fine bill, dark eyestripe and yellow undertail-coverts in all plumages. Upperparts dull grey-green, with narrow dark eyeline and pale yellow supercilium. Underparts pale yellow with blurred olive streaks on chest, and brightest yellow on undertail-coverts. Often has narrow pale line at leading edge of wing. *V. c. lutescens* brighter and paler yellow than *V. c. celata*. **BP** Bill very fine, dark grey; eyes black, appearing quite large with pale yellow (♂) or white (♀/juv.) crescents above and below; tarsi dark. **Vo** A sharp, flat *chip*, and in flight a short, rising *seet*.

YELLOW-RUMPED WARBLER   R
Dendroica coronata

L 12–14cm; WT 12g. **SD** Widespread in N America, wintering to Mexico and C America. One population, '**Myrtle Warbler**' *D. c. hooveri*, breeds as far northwest as W Alaska. Accidental (Sept) to Bering Sea islands and Chukotka. **HH** Open woodland; typically perches conspicuously and rather upright. **ID** Distinctive, with yellow patches on rump and breast-sides in all plumages. ♂ breeding blue-grey, blacker on ear-coverts, white on throat ('Myrtle' only) and underparts, with black streaking on mantle, across chest and sides; yellow on crown, broken white supercilium, with two white wing-covert bars. ♀ is like a dull ♂, with dark ear-coverts, whitish throat, but browner-grey mantle and chest. 1st-winter browner than ad. ♀ with dark ear-coverts bordered above by narrow dark eyeline and below by pale off-white or buff-white throat wrapping around rear of ear-coverts ('Myrtle' only). **BP** Bill slender, black; prominent white crescents above and below black eyes; tarsi black. **Vo** Quite vocal, including on migration. A low, flat *chep* or *tep* and in flight a clear, rising *svit*.

The related **Yellow** and **Blackpoll Warblers** are both common American species with similar migration patterns that breed as far northwest as W Alaska, and are overdue vagrants to NE Asia. Particularly likely in autumn – many other N American warblers yet to be recorded in Asia have occurred at Gambell, St Lawrence I. (USA), within *c*.70km of Chukotka (see Appendix 2).

YELLOW WARBLER   Extralimital
Dendroica petechia

L 12–13cm; WT 9.5g. **SD** Has reached St Lawrence I. several times. *D. p. banksi* winters in C America and N South America and migrates to breed in Alaska and NW Canada, making it a likely overshoot into the region. **ID** Ad. bright yellowish-green

with distinctive yellow fringes to tail-feathers; all have bright yellow face with beady dark eye prominent in plain face. ♂ has rufous-brown streaks on breast and flanks, ♀ lacks streaking but has white-fringed tertials, and young are drab greyish-yellow with white-fringed tertials. Short-tailed (shorter than Orange-crowned and much shorter than Wilson's). **Vo** A clear, loud *chip* or soft *tsep*, and in flight a clear trilling *tzip*.

BLACKPOLL WARBLER   Extralimital
Dendroica striata

L 14cm; WT 13g. **SD** Has reached St Lawrence and breeds in W Alaska. Monotypic. **ID** Ad. greyish or yellowish-grey. Breeding ♂ has black cap and white cheeks, two broad white wingbars and heavy black streaking on white underparts. Non-breeding ♂/♀ more yellowish-grey, 1st-winter birds grey on neck-sides and streaked above. All ages have conspicuous white undertail-coverts. In flight, two white wingbars and prominent white spots on outer tail-feathers. **BP** Tarsi dull to bright yellowish-flesh. **Vo** A clear, sharp *chip* and in flight a sharp buzzing *tzzz*.

NORTHERN WATERTHRUSH   R
Seiurus noveboracensis

L 12.5–15.5cm; WT 18g. **SD** Breeds as close to Asia as W Alaska. Accidental to St Lawrence and to Chukotski Peninsula. Monotypic. **HH** Wooded areas with fresh water, often in boggy areas dominated by low willows; typically forages on ground at water's edge. Commonly crouches and bobs rear part of body. **ID** A rather dark, long-legged, short-tailed, pipit-like bird. Generally very dark brown, with prominent, long, pale (often yellowish-white) supercilium narrowing behind eye, above a dark eyeline and dusky ear-coverts; pale throat and pale yellowish-white underparts heavily streaked on breast, with long dark streaks extending on flanks. Pale areas (supercilium and underparts) whitish in some, washed yellow in others. **BP** Bill dark grey with pinkish sides; eyes black; tarsi pink. **Vo** A loud, sharp *chink* and in flight a high buzzy *zzip*. Calls are reminiscent of *Motacilla* wagtails.

WILSON'S WARBLER   J
Wilsonia pusilla

L 11–12.5cm; WT 8g. **SD** Breeds over much of N America, wintering from Mexico south. *W. p. pileolata* breeds as close to Asia as W Alaska. **HH** Mainly riparian thickets, often of willows and alders, usually with dense ground cover. Accidental to Japan (Hegura-jima). **ID** Very small, very active, bright yellow bird, the size of Japanese White-eye, with rounded wings and a long, thin tail. Ad. bright, plain olive-green above and yellow below. Face very plain, yellowish-olive, with distinct black cap (♂), or forehead patch (♀), yellow forehead and head-sides in ♂ and dull green crown in imm. ♀. **BP** Bill very fine, short, grey-pink; eyes appear large on plain yellow/olive face, irides black; tarsi dull pink. **Vo** A flat *timp*, and in flight a clear, abrupt *tilk*.

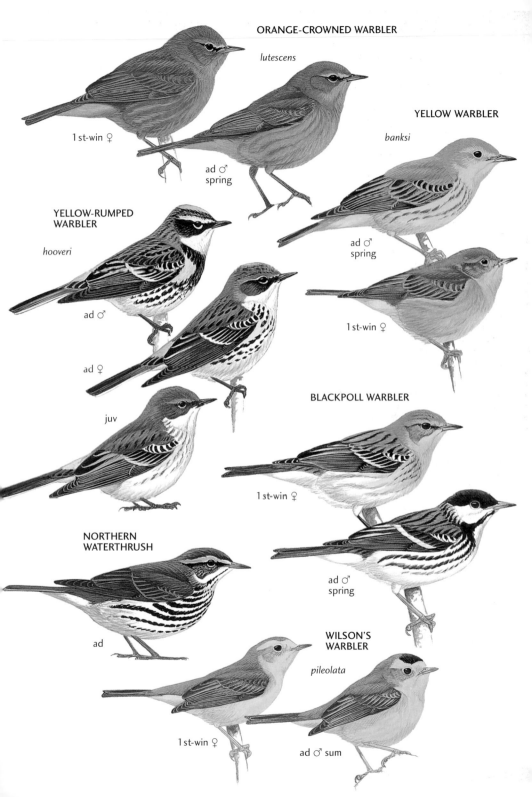

ORANGE-CROWNED WARBLER

lutescens

1 st-win ♀

ad ♂ spring

YELLOW WARBLER

banksi

ad ♂ spring

1 st-win ♀

YELLOW-RUMPED WARBLER

hooveri

ad ♂

ad ♀

juv

BLACKPOLL WARBLER

1 st-win ♀

ad ♂ spring

NORTHERN WATERTHRUSH

ad

WILSON'S WARBLER

pileolata

1 st-win ♀

ad ♂ sum

RUSTY BLACKBIRD R
Euphagus carolinus

L 23cm; WT 60g. **SD** Widespread across N America: western race *E. c. carolinus* breeds as far north as Alaska. Has reached St Lawrence (Aug–Sep) and accidental in Chukotka. **HH** In normal range, occurs in wooded swamps and wet alder and willow thickets adjoining tundra. **ID** Rather slender-billed with a long, full tail, resembling cross between starling and thrush. Breeding ♂ is black with dull blue gloss on head and breast, greener gloss (difficult to see) on mantle and wings. Non-breeding ♂ is blue-black with rusty feathering on head, nape, mantle and flanks (but much variation in its extent), and similar rusty fringes to wing-feathers including tertials (lacking in very similar extralimital Brewer's Blackbird *E. cyanocephalus*), with variable barring below. ♀ rather dull ashy grey-brown in breeding season, generally rufous or rusty-buff in non-breeding season, with prominent pale supercilium and black eye patch; rufous back and fringes to wing-feathers contrast with plain grey rump. **BP** Thin, slightly decurved bill is rather sharply pointed, dark grey; eyes lemon-yellow in breeding ♂, otherwise white or yellowish-white, contrasting with dark face or eye patch; tarsi black. **Vo** A soft *chuck*.

WESTERN MEADOWLARK R
Sturnella neglecta

L 24cm; WT 97g. **SD** Widespread across W & C N America, reaching Pacific NW, vagrant to Alaska, and accidental Chukotka. Monotypic. **HH** In normal range occurs in open grassy areas. **ID** Rather stout, long-billed, long-legged and short-tailed bird. Ad. breeding (Feb–Aug) has characteristic yellow underparts. Upperparts mid brown with pale brown fringes to most feathers; supercilium broad, yellow before eye, buff behind it, ear-coverts, mantle, back, wings, rump and tail brown; short tail barred dark brown and buff, and has white in outer 2–3 feathers. Yellow malar (white in near-identical, extralimital Eastern Meadowlark *S. magna*), chin and throat, black collar broadest in centre, yellow from breast to belly, with blackish-brown streaking on off-white flanks, vent off-white. Non-breeding (Sep–Jan) perhaps slightly duller, paler grey-brown, with less contrasting head pattern. **BP** Bill long, sharply pointed, blue-grey with dark culmen, more straw-coloured in non-breeding plumage; eyes dark brown; tarsi pink. **Vo** A sharp *chuck* or dull rattled *vidididididididi*, and in flight *veeet* or *rrink*.

CRESTED BUNTING CTJ
Melophus lathami

L 16–18cm. **SD** Himalayas and India across N SE Asia and S & E China north to Zhejiang; accidental Taiwan and offshore Japan. Monotypic. **HH** Dry grassy areas in hills. **ID** A large, dark, bunting with prominent pointed crest in both sexes. ♂ largely black with rufous-brown wings (greater coverts and fringes to flight-feathers), vent and tail, with black rump/uppertail-coverts and tail tip. ♀ is brown, pale-throated, darkest on chest with narrow blackish streaks, and has warm dull rufous-brown wings and fringes to dark-brown tail. **BP** Bill large, dusky greyish-pink; eyes large, dark brown; tarsi dull brownish-pink. **Vo** A hard *chi*.

SLATY BUNTING CT
Latoucheornis siemsseni

L 13cm. **SD** Breeds in C China, wintering disjunctly to south, including Zhejiang and Fujian in E; unconfirmed record from Penghu I. (Taiwan). Monotypic. **HH** Scrub and second growth. **ID** Small, very distinctive dark bunting. ♂ has deep slate head, upperparts, wings and tail, and breast and flanks, and white belly, vent and outer tail-feathers. ♀ is dark brown, warm rufous-brown on crown, nape and breast, paler on ear-coverts and chin; mantle heavily streaked blackish, wings blackish-brown fringed paler brown, forming double wingbar and pale fringes to flight-feathers; rump scalloped grey, tail long, dark brown with white outer feathers. **BP** Bill grey; eyes black; tarsi brownish-pink. **Vo** A sharp *zick*; song is high-pitched and likened to *Parus* tits. **AN** Chinese Blue Bunting.

RUSTY BLACKBIRD

carolinus

ad ♂ sum

ad ♀

WESTERN MEADOWLARK

ad

CRESTED BUNTING

ad ♂

ad ♀

ad ♂

ad ♀

SLATY BUNTING

YELLOWHAMMER   CTKJ
Emberiza citrinella

L 15–18cm; WT 25–36g. **SD** Widespread, from British Isles to W & C Siberia to Lake Baikal. Race closest to region is *E. c. erythrogenys*; winters in Transcaucasus and C Asia. Accidental in winter NE China and Japan; unconfirmed records Taiwan and Korea. **HH** Forest-steppe, in thickets, woodland edge and agricultural land with scattered trees. **ID** Large, long-tailed yellow-headed bunting. ♂ has essentially yellow head and underparts, with dusky patch at rear of crown abutting rear ear-coverts. Upperparts brown with heavy black streaking on mantle, bright rufous on rump. Rufous-brown also in band across breast and streaking on flanks. Tail long, notched, often flicked, showing white-fringed outer feathers. ♀ duller, with mainly brown head, somewhat yellower on supercilium, malar and throat, but also has rufous-brown rump and streaking on belly; palest ♀ very similar to ♀ Pine Bunting, but never white on belly. Yellowhammer x Pine Bunting hybrids represent a potential identification pitfall. **BP** Bill stubby, grey; eyes pale brown; tarsi greyish-pink. **Vo** Calls include *tsui, tsik-tsik* and a liquid, clicking *pt..pt..pt.. pittlitt*. Song a cheerful but monotonous *chichi chichi chichi chichi chichi jyueeee*.

PINE BUNTING   CKJR
Emberiza leucocephalos

L 16–18cm; WT 24–35g. **SD** Summers across much of Siberia from E Urals to Magadan and Amur Estuary, Sakhalin and Kuril Is, also NE China; winters through S Russian Far East, adjacent NE China, Hokkaido to Kyushu, and Korea, vagrant south as far as Jiangsu, China. Race in region is *E. l. leucocephalos* **HH** Uncommon and local, in open mixed deciduous forest, woodland edge and agricultural land with trees. **ID** A large, dark, rufous bunting. Breeding ♂ has distinctive head pattern of white crown and ear-coverts (both bordered black), broad chestnut supercilia, chestnut chin/throat and head-sides. Nape grey, mantle rufous-brown streaked black, lower back, rump and uppertail-coverts plain rufous-brown, tail blackish-brown with white outer feathers. Underparts rufous-brown, with narrow grey-white collar and neck patch separating chestnut throat from rufous breast; belly and vent white; wings blackish-brown with rufous-brown fringes to most feathers. Winter ♂ duller, greyer brown above and below, with only 'ghost' of summer head pattern, though cheek patch still prominent. ♀ (see similar Yellowhammer) is generally greyish-brown, with paler supercilia, off-white malar bordered above by black moustachial and below by dark lateral throat-stripe which extends as heavy dark grey streaking on breast and sides; lacks yellowish tones of Yellowhammer. 1st-winter resembles ad. of respective sex. **BP** Bill sharp-pointed, blue-grey with dark culmen; eyes black; tarsi brownish-pink. **Vo** Mostly identical to Yellowhammer, but also gives hard *tsick* and downslurred *chüeh*. Song very similar to Yellowhammer: *jui jui jui jueeen*.

GODLEWSKI'S BUNTING   C
Emberiza godlewskii

L 17cm. **SD** NE India across China to Mongolia. *E. g. godlewskii* Mongolia and Transbaikalia; *E. g. omissa* in C & N China east to S Heilongjiang. **HH** Dry habitats, from fields to rocky hills with ravines, scrub and forest edge. **ID** A distinctly grey-headed bunting. ♂ has mid-grey hood with grey crown bordered by chestnut lateral crown-stripes, black lores, chestnut stripe behind eye and bordering ear-coverts, and black moustachial. Upperparts mid brown with black streaking on mantle and back, rump plain rufous-brown, tail blackish-brown with white outer fringes to outer rectrices. ♀ has cream or off-white hood, with grey-streaked crown-sides, black lores and moustachial, but less distinct chestnut eyestripe and border to ear-coverts than ♂; upperparts paler, sandier brown. **BP** Bill grey; eyes black; tarsi brownish-pink. **Vo** A hard *pett pett* and thin *tziii*. Song, which commences with high-pitched *tsitt* notes, is long with twittering phrases. **Vo** Formerly within Rock Bunting *E. cia*.

MEADOW BUNTING   CTKJR
Emberiza cioides

L 15–18cm. **SD** Widespread from Tien Shan and Altai to Japan. In region, occurs from Fujian to NE China, and Russian Far East to Sakhalin and S Kuril Is, also Korea and Japan. Resident throughout most of range, but summer visitor to Hokkaido, S Kuril Is and Sakhalin; accidental Taiwan. *E. c. weigoldi* in NE China, Russian Far East and N Korea; *E. c. castaneiceps* in E China and Korea, wintering to S China and Taiwan; *E. c. ciopsis* in Japan (northern birds moving south to winter). **HH** Fairly common, in open wooded areas, thickets, cleared forests and agricultural areas in lowlands and low hills. **ID** A large, dark, warm rufous-brown bunting. ♂ has distinctive pied head, with white supercilium joining collar which runs to chin, black lores and ear-coverts, white malar and black lateral throat-stripe, and crown is dark chestnut bordered laterally with black. Upperparts warm, dark, rufous-brown from crown to tail, with some black streaking on mantle and back, but plain rufous on lower back, rump and uppertail-coverts; tail blackish and rufous-brown with white outer fringes to outer feathers. Chin and throat white, upper breast orange-brown, then plain dark rufous-brown grading to off-white on belly. ♀ generally paler, sandy- rather than rufous-brown, with broad white supercilium, mainly white chin and throat, and brown ear-coverts, otherwise like washed-out ♂. Races somewhat variable; *E. c. castaneiceps* is smallest and darkest, with less streaked upperparts and chestnut auriculars, whilst *E. c. weigoldi* is bright, rich chestnut, and *E. c. ciopsis* has almost black-and-white head pattern, in particular with black ear-coverts (unlike others), and distinctly streaked upperparts. **BP** Bill dark grey; eyes black; tarsi brownish-pink. **Vo** Call differs from other common buntings of region in being uttered in series, often three, *zit-zit-zit* notes (sometimes 2–4), not singly. However, Elegant regularly gives double or triple *tsip* notes. Song *cho-pizt-two-chirr* or *tsui chocho tsuryi cho tsupitchi richipi*.

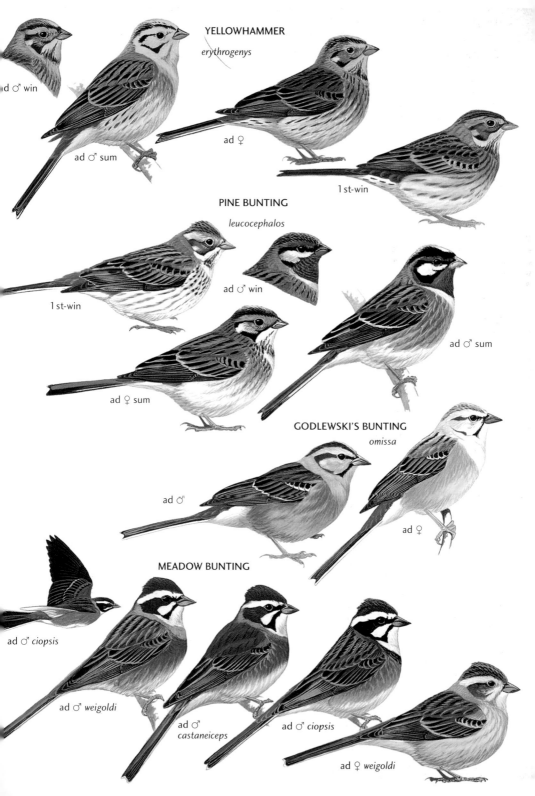

YELLOWHAMMER

erythrogenys

d ♂ win

ad ♂ sum

ad ♀

1st-win

PINE BUNTING

leucocephalos

1st-win

ad ♂ win

ad ♀ sum

ad ♂ sum

GODLEWSKI'S BUNTING

omissa

ad ♂

ad ♀

MEADOW BUNTING

ad ♂ *ciopsis*

ad ♂ *weigoldi*

ad ♂ *castaneiceps*

ad ♂ *ciopsis*

ad ♀ *weigoldi*

JANKOWSKI'S BUNTING CKR
Emberiza jankowskii

Vulnerable. L 16–17cm. **SD** Extremely limited breeding range includes NE China (SE Heilongjiang and Jilin), adjacent S Russian Far East (now extinct), and extreme NE N Korea (current status unknown). Winters Liaoning, Hebei and Inner Mongolia. Monotypic. **HH** Scarce, local and declining with very narrow habitat requirements: open, low-altitude sandy areas with grasses, scrub or bushes, in ecotone between deciduous forest and steppe. **ID** A large pale, rufous bunting somewhat like Meadow Bunting. ♂ like pale Meadow Bunting, with plain chestnut forehead, crown and nape, prominent white supercilia, but grey ear-coverts, dark brown lateral throat-stripe and off-white underparts with pale orange-brown wash to flanks and unique dark brown central belly patch. Rufous-brown with black streaking on mantle, plain rufous-brown on lower back, rump and uppertail-coverts; tail blackish and rufous-brown with white outer fringes to outer feathers; face black on lores, white on malar, chin and throat. Wings have white fringes to lesser and greater coverts, forming narrow but conspicuous double wing-bars, and rufous-buff fringes to secondaries and tertials. ♀ less cleanly marked on head, is paler, sandier brown streaked black on mantle, lacks central belly patch, and dark lateral throat-stripe merges into streaking on sides of upper breast. **BP** Bill grey; eyes black; tarsi brownish-pink. **Vo** A soft single or double *tsik-tsik*, thin *hsiu* and explosive *sstlitt* in alarm; song a short buzzy trill *chu-chu cha-cha cheee*. **AN** Rufous-backed Bunting.

GREY-NECKED BUNTING J
Emberiza buchanani

L 15–17cm; WT 17–26g. **SD** Caucasus to S Altai, wintering in S Asia. Accidental Japan (Oct); race uncertain. **HH** Arid areas, on rocky slopes and rocky areas with sparse vegetation. **ID** Rather plain bunting, resembling Ortolan. ♂ has dull blue-grey head, face and neck, distinctive pale buff malar, grey lateral throat-stripe and whitish throat; underparts warm rufous-buff (lacks Ortolan's grey upper breast), whiter on belly; upperparts, wings and tail mid brown with darker streaking and white outer tail-feathers. ♀ is less clearly marked, and 1st-winter and juv. only have slight grey cast to head and buffish underparts, with light streaking on mantle and breast. **BP** Bill pinkish-orange; eye-ring white, eyes black; legs dull pink to brownish-pink. **Vo** A rather strong *chi*.

ORTOLAN BUNTING Emberiza hortulana J

L 15–17cm. **SD** Warm regions of W Europe across S Russia to Irkutsk. Accidental Japan (Oct). Monotypic. **HH** Open dry habitats, hilly areas and upland steppes with thickets. **ID** A rather large bunting. ♂ has olive-grey head, neck and breast, yellow malar stripe, chin and throat patch. Upperparts brown with heavy dark streaking, underparts rufous-brown on belly. Lesser wing-coverts blackish in Ortolan (grey in very similar Grey-necked). ♀ lacks olive-grey of ♂, being brown, with whitish-ochre throat, breast and belly, and brown streaking on breast (unmarked in ♀ Grey-necked). 1st-winter and juv. heavily streaked on mantle, breast and malar, with yellow tinge to buff submoustachial stripe and chin/throat. **BP** Bill pink; eyes paler, prominent, with narrow white ring, irides dark brown; tarsi pink or brownish-pink. **Vo** Varied, including a soft, clicking *plett* or *tsupitt tsupitt*, harder *chu* and disyllabic *sli-e*. Song: *chichi juijui chee*.

TRISTRAM'S BUNTING CTKJR
Emberiza tristrami

L 14–15cm. **SD** Uncommon. Breeds in NE China, N & C Korea, and adjacent S Russian Far East to Amur River. Winters in S & SE China and N SE Asia. Fairly common migrant Korea, rare offshore Japan; accidental Taiwan. Monotypic. **HH** Typically in shady areas of mixed taiga-forest, particularly beneath stands of fir; shy and nervous. **ID** Distinctive, small but rather stocky, compact bunting with a very heavily striped head. ♂ has black head and chin, with grey-white coronal stripe, supercilium and malar, and a white spot on rear ear-coverts. Greyish-brown on mantle with black streaking, rufous-brown to chestnut on lower back, rump and tail; underparts white, with extensive dull rufous-brown wash on breast and flanks with dark streaking. ♀ less contrasting, greyish-brown instead of black on head, but shares similar striped pattern, with buff supercilium and ear-coverts, and has whitish chin/throat with dark lateral throat-stripe. **BP** Sharply pointed bill is grey above and at tip, pink below; eyes black; tarsi deep pink. **Vo** A slightly metallic *tsip* resembling Rustic and Elegant buntings.

CHESTNUT-EARED BUNTING CTKJR
Emberiza fucata

L 15–16cm. **SD** Wide-ranging but uncommon, summering from Lake Baikal to SE Russian Far East, NE China, Korea and Japan, also Himalayas and SE China, wintering in S & SE China, rarely Taiwan, and SE Asia. In Japan, summer visitor in Kyushu and from C Honshu north, winter visitor from C Honshu south, resident in coastal E China. *E. f. fucata* breeds NE China, Korea, Russian Far East to Sakhalin and Japan; *E. f. kuatunensis* in SE China (Fujian and Zhejiang). **HH** Open grassy habitats with thickets, including rank meadows and wetland fringes; winters in open agricultural land. **ID** A large well-patterned bunting with distinctive chestnut ear patch (see Little Bunting) and double breast-band. ♂ has grey crown and nape, bright chestnut ear-coverts, and contrasting face pattern of black moustachial, white malar and chin/throat, and black lateral throat-stripe extending as necklace of black streaks. Mantle grey-brown with black streaking, scapulars, lower back, rump and uppertail-coverts plain rufous-brown, tail blackish-brown. Underparts white on chin/throat with black, white and chestnut breast-bands, then warm orange-brown wash on sides grading to white on belly; wings blackish-brown with mid brown feather fringes. ♀ less cleanly marked, but has chestnut ear-coverts patch, lacks contrast of ♂, but also has dark lateral throat-stripe extending as streaking on breast and flanks. **BP** Bill grey above, pinkish-grey below; prominent narrow white eye-ring, irides black; tarsi brownish-pink. **Vo** A rather spitting, staccato *pit* or *pt*. Song recalls Meadow Bunting, but is deeper, shorter, weaker, less clear, ending in three-part phrase *chip chip chil-ri-wit chi chi tsiririri* or *che chitsu chirinju*. **AN** Grey-headed Bunting.

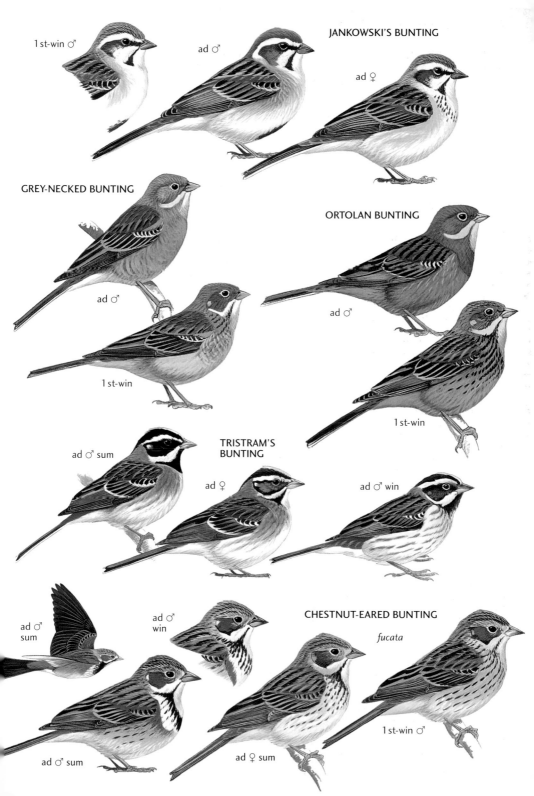

1 st-win ♂

ad ♂

JANKOWSKI'S BUNTING

ad ♀

GREY-NECKED BUNTING

ad ♂

1 st-win

ORTOLAN BUNTING

ad ♂

1 st-win

ad ♂ sum

TRISTRAM'S BUNTING

ad ♀

ad ♂ win

ad ♂ sum

ad ♂ win

CHESTNUT-EARED BUNTING

fucata

1 st-win ♂

ad ♂ sum

ad ♀ sum

LITTLE BUNTING CTKJR
Emberiza pusilla

L 12–14cm; WT 13–19g. **SD** Widespread in summer from N Scandinavia to Yakutia and Chukotka, south through Koryakia to NE Sea of Okhotsk; also NE China. In winter, found across S Siberia, C Asia and much of China south of the Chiangjiang. On migration occurs widely in E China, an uncommon migrant Taiwan and Korea, rare offshore Japan; also recorded in winter in S Japan and Taiwan. Monotypic. **HH** In summer, favours taiga and forest-tundra habitats and in winter/on migration found in scrub, woodland fringes and dry agricultural land with tall grasses or reeds. **ID** A small, neatly compact bunting with distinctly chestnut face (see Chestnut-eared Bunting, p.484), and striped head. ♂ has mostly chestnut head, face and lores with black lateral crown-stripes, narrow eyestripe and outer border to ear-coverts, a narrow moustachial and lateral throat-stripes merging into prominent but narrow black streaking on sides of otherwise clean white breast and flanks. Upperparts greyish-brown with black streaking on mantle and scapulars; wings also blackish-brown, with white tips to coverts forming faint wingbars, and rufous fringes to secondaries. ♀ and 1st-winter ♂ resemble ad. ♂, but lateral crown-stripes less clear and face less distinctly chestnut, but still fairly bright. **BP** Bill rather small, grey; distinctive narrow white eye-ring around black eye; tarsi dull orange/pink. **Vo** A short, clipped, slightly metallic *zip*, somewhat wetter sounding than other buntings. Short song comprises mixture of clear notes and rasping rolling sounds in long phrases *tee-tee-tee-teerch*.

YELLOW-BROWED BUNTING CTKJR
Emberiza chrysophrys

L 14–16cm. **SD** Uncommon to rare in limited breeding range, from Lake Baikal towards Sea of Okhotsk, but not reaching its shores; winters in E & SE China south of the Changjiang; rare Taiwan. Migrates via E & NE China; uncommon Korea and rare offshore Japan. Monotypic. **HH** Taiga forest, at woodland edges, in thickets and also, on migration, agricultural land with trees. **ID** A small compact bunting with unique head pattern; superficially resembles Tristram's Bunting, but less dark overall and less striped on head. ♂ has black head, but grey nape (not chestnut), a broad yellow supercilium from just above eye becoming white at rear, white spot on otherwise black ear-coverts, white malar, black lateral throat-stripe and narrow black streaking on chin, breast-sides and flanks; underparts otherwise off-white with brown-washed flanks. Upperparts brown on neck and mantle with heavy streaking, plain mid brown lower back and rump, and brownish-black tail with white outer feathers. White-tipped coverts form two narrow wingbars. ♀ has less boldly marked head, but longer supercilium is yellow from bill base to above eye, then white behind. **BP** Bill grey above and at tip, pinkish-grey below; eyes black; tarsi pink. **Vo** A thin, metallic *tzip*, recalling Little Bunting, but less 'wet'. Short clear song: *hwee tee-tee tu-tu-tu*.

RUSTIC BUNTING CTKJR
Emberiza rustica

L 13–15.5cm; WT 17–22g. **SD** Wide-ranging, breeding from Scandinavia to N Sakhalin, Sea of Okhotsk, Yakutia, Chukotka and Kamchatka. In region, winters (Oct–May) throughout Japan south of Hokkaido, Korea (fairly common) and E China; accidental Taiwan. Monotypic. **HH** In summer, favours taiga forest and forest edge habitats, riparian thickets and swampy woodlands. In winter found in dry lowland woodland, forest on lower slopes, riverine scrub and margins of agricultural land with rank vegetation. **ID** A smallish compact bunting with distinct crest and striking head pattern. Breeding ♂ has black head, with erectile crest, white lateral crown-stripe from above eye to nape, and white chin and throat with black lateral throat-stripe. Upperparts chestnut on hindneck, brown on back but heavily streaked black, lower back and rump dull chestnut, tail long, blackish with white outer feathers. Underparts white, with chestnut band on upper breast and chestnut streaking on flanks. Winter ♂ less contrasting, generally lacking black on head, but does have black outline to brownish ear-coverts, and crown still appears crested, whilst dark lateral throat-stripe blends into chestnut of upper breast. ♀ resembles winter ♂, but paler and less boldly marked, lacking any black on ear-coverts. White-tipped coverts form two narrow wingbars, most prominent in summer ♂. **BP** Bill grey above and at tip, pink at base in summer ♂, pinkish-grey in ♀/winter ♂; eyes black; tarsi dull pink. **Vo** A strong *tsip* or *fuchip*. Song a complex, strong, far-carrying melodious *pyo-hyorori-kyururu*, lacking sharp or hard sounds.

ELEGANT BUNTING CTKJR
Emberiza elegans

L 15–16cm. **SD** C & NE China, Korea, and S Russian Far East; winter visitor to Japan south of Hokkaido and coastal SE China south to Burma. *E. e. elegans* breeds NE China (south to Hebei), S Russian Far East, Korea and Tsushima, wintering to S & E China, Japan and Taiwan. **HH** Open deciduous woodland, town parks, and woodland and forest edges with tall grasses, bordering overgrown agricultural land, often on low hillsides. **ID** Extremely attractive. Breeding ♂ has black crown and short erectile crest, black mask from supraloral to ear-coverts and a black chest shield; remainder of head is bright lemon-yellow. Grey half-collar on nape, mantle pale brown heavily streaked rufous-brown, rump greyish-brown, and tail ash-grey and blackish-brown. Underparts largely white with chestnut streaking on sides; wings blackish-brown with white fringes to coverts, brown fringes to flight-feathers and broad rusty-brown fringes to tertials. Winter ♂ is similar, but with duller, less contrasting head pattern. ♀ less contrasting, lacking black on head or underparts, but has distinct yellow wash to supercilium, head-sides and chin/throat; lack of malar stripe makes separation from ♀ Rustic easy. **BP** Rather stout bill, grey above, pinkish-grey below; eyes black; tarsi pink. **Vo** A clear, somewhat moist-sounding *tsi* or *tsi-ti tsip*. Song a long, slurred trilling *chichi churichuri churiri chichi*. **AN** Yellow-throated Bunting.

LITTLE BUNTING

ad ♂ sum

ad win

YELLOW-BROWED BUNTING

ad ♂

1st-win ♀

ad ♀

RUSTIC BUNTING

ad ♂ sum

ad ♀ sum

ad ♂

1st-win

ELEGANT BUNTING

elegans

ad ♂

ad ♀

YELLOW-BREASTED BUNTING   CTKJR
Emberiza aureola

L 14–16cm; WT 17–26g. **SD** Breeds from Scandinavia to Chukotka, Kamchatka and Kuril Is, Hokkaido south as far as NE China (May–Sep); winters in S China and SE Asia; rare winter visitor Taiwan. Once abundant, now greatly reduced in numbers in much of E Asia due to trapping for food. *E. a. aureola* from Finland to Kolyma River; *E. a. ornata* NE Mongolia, NE China to Kamchatka, Anadyr, Kuril Is, Sakhalin, and Hokkaido (where almost extinct). **HH** Breeds in wetland fringes, in reedbeds and tall grass, rank meadows and riparian grasslands; in winter also in dry agricultural land with scrub, grasses and reeds. **ID** A quite large, somewhat stout, attractive bunting. Breeding ♂ very boldly marked, dark chocolate-brown above, from rear crown to rump, bright yellow below, with broad black mask, chocolate-brown chest-band and streaks on flanks, conspicuous white scapulars and bold white wingbar. Winter/imm. ♂ lack mask and chest-band and are greyer on ear-coverts bordered narrowly below with black. ♀ resembles winter ♂, but has broad pale supercilium, dark border to ear-coverts, prominent white median covert bar and brighter yellow underparts. Juv. heavily streaked on breast and flanks. Yellow-breasted lacks chestnut coloration on rump of Chestnut Bunting. **BP** Bill heavy, pink in breeding ♂, otherwise grey above and at tip, pink below and at base; eyes black; tarsi pink. **Vo** A soft *tip-tip* or *tzip*. Song a loud, clearly whistled, ascending series of notes until final one, which descends: *tsyu tsui tee-e tee-e tsee-tee*; *fee hyo hyui chui churee* or *filyou-filyou-filee-filee-filee-tyou-tyou*.

CHESTNUT BUNTING   CTKJR
Emberiza rutila

L 12–15cm. **SD** Breeds Lake Baikal to Sea of Okhotsk, also NE China; winters in S China and SE Asia. Common or fairly common on migration through E China, Taiwan and Korea, rare offshore Japan. Monotypic. **HH** In summer occurs in taiga zone where favours low thickets, coniferous and mixed forests; and in winter/on migration found in scrub, woodland fringes and dry agricultural land with trees/shrubs. **ID** A very attractive and distinctively colourful small bunting. Breeding ♂ has bright chestnut hood, mantle, scapulars, wing-coverts, back and rump, with bright yellow underparts, becoming grey only on sides; wings blackish-brown, broadly fringed chestnut on coverts, secondaries and tertials; tail blackish-brown. Winter ♂ has chestnut hood and mantle less bright and flecked with yellow. ♀ recalls ♀ Yellow-breasted, but smaller; head to back greyish-brown with heavy streaking on mantle and scapulars, lower back, rump and uppertail-coverts quite bright chestnut, tail blackish with white outer feathers. Face rather nondescript with weak yellowish-buff supercilium, and malar, throat and underparts also yellowish-buff with dark lateral throat-stripe merging into fine streaking on flanks. **BP** Bill delicate, grey; eyes black; tarsi dull pink. **Vo** A short, hard *tsip* or *zit*. Song a loud *töi-töi-tööi-öö-öi-öi-see-see-see* or *chui chui chichichi chui chui chuchuchuu*.

BLACK-HEADED BUNTING   CTKJ
Emberiza melanocephala

L 15–18cm; WT 24–33g. **SD** Occurs west of region, from SE Europe to Sea of Azov, wintering in India, but has occurred regularly in recent years in offshore Korea, offshore and S Japan, and Fujian, China, perhaps suggesting an undiscovered closer breeding population, with records from Laos, Thailand, Borneo and Hong Kong strengthening this supposition. Monotypic. **HH** In native range, occurs in open steppe with bushes; as vagrant found in thickets and woodland edge near agricultural land. **ID** A large, bulky, bull-headed bunting. ♂ breeding unmistakable, with black head, face and neck-sides contrasting with rufous-brown or chestnut mantle and lower back, and bright yellow throat and underparts. Non-breeding ♂ has pale buff fringes to feathers of head and mantle. ♀ (very closely recalls ♀ Red-headed Bunting) has plain sandy-brown head, breast and upperparts, strong yellow tinge to throat, breast, belly and undertail-coverts, and narrow streaking on sides. Both sexes have white fringes to wing-coverts, forming two wingbars, and broad pale fringes to tertials. Most E Asian records involve ♀s or juvs., which are best separated from Red-headed by rufous fringes to mantle and back, slightly stronger contrast between grey-buff ear-coverts and paler off-white chin/throat, and more uniformly yellow underparts. **BP** Large bill rather long, bulbous, sharply pointed, with straight culmen, dark grey; eyes brown often with white eyering; tarsi quite bright pink. **Vo** Varied and similar to other large buntings, strong *jüp*, *zrt*, *tsit* and clicking *ptr'r'r*.

RED-HEADED BUNTING   CKJ
Emberiza bruniceps

L 16–17cm; WT 18–31g. **SD** Occurs west of region in C Asia and Kazakhstan, but has reached region several times, in coastal China (Happy I.), offshore Korea and offshore Japan. Monotypic. **HH** In native range occurs in dry open steppe with hills and thickets; as vagrant found in thickets and woodland edge near agricultural land. **ID** A large, bulky, bull-headed bunting. ♂ breeding unmistakable, with bright chestnut head and chest, bright yellow underparts, yellow-olive upperparts and yellow rump; mantle and rump streaked black and wings black with white fringes to all feathers. Non-breeding ♂ has less clearly marked chestnut head and breast, yellow malar stripe and yellow flecks in chestnut. ♀ (often inseparable from ♀ Black-headed) is plain sandy-brown on head, breast and upperparts, yellowish on flanks and belly, bright yellow on vent and greenish-yellow on rump/uppertail-coverts; wings blackish, the feathers fringed buff. **BP** Large bill rather sharply pointed, with straight culmen, but shorter more triangular than Black-headed, dark grey; eyes, prominent in plain face, irides black; tarsi quite bright pink. **Vo** Calls indistinguishable from Black-headed.

YELLOW-BREASTED BUNTING

ornata

ad ♂ win

ad ♂ sum

ad ♀

imm/1st-win

CHESTNUT BUNTING

ad ♂

ad ♀

1st-sum ♂

1st-win ♀

RED-HEADED BUNTING

BLACK-HEADED BUNTING

ad ♂ win

ad ♂ win

ad ♂ sum

ad ♂ sum

ad ♀

ad ♀

JAPANESE YELLOW BUNTING   CTKJR
Emberiza sulphurata

Vulnerable. L 13–14cm. **SD** Range-restricted E Asian endemic, found in summer (May–Sep) C to N Honshu, Japan. Winters primarily in Philippines (rarely perhaps in Fujian and Taiwan); scarce migrant Korea. Monotypic. **HH** Scarce and local, in mixed, but mainly deciduous, forest at mid elevations (600–1,500m) in summer; and in forest (to 1,500m), woodland fringes, grassland, scrub and agricultural land in winter. **ID** Yellow and olive-green bunting superficially resembling larger *personata* Black-faced, but is cleaner, quieter, less active, with plainer face and more prominent wingbars. ♂ has head, face, nape and neck-sides plain greyish-olive; lores black contrasting with narrow but noticeable white eye-ring. Upperparts greyish-olive with black-streaked mantle and scapulars, lower back and rump plain yellowish-olive, tail dark greyish-olive with white outer feathers. Underparts pale yellowish-green, with some streaking on flanks. ♀ lacks black lores, but has white eye-ring and yellowish wash to creamy-buff underparts from chin to vent, with some contrast in malar region. Prominent white tips to median and greater coverts form distinct double wingbars (slightly less distinct in ♀). **BP** Bill blue-grey; eyes particularly noticeable due to white crescents above and below, irides black; tarsi dull pink. **Vo** Call a metallic *tsip*, softer than Black-faced but very similar to Tristram's and Chestnut. Song recalls Black-faced, but softer and faster: *chip-in chin-chin chee-chee-chee che-rui* or *chichon pipi chiichii chiichii*.

BLACK-FACED BUNTING   CTKJR
Emberiza spodocephala

L 13–16cm. **SD** Wide-ranging across E Russia, from Yenisei to Lena, and to Sea of Okhotsk south through S Russian Far East, N & C Korea, Japan and C China. Winters from C Japan south also S Korea, S & E China, Taiwan and SE Asia. *E. s. spodocephala* breeds E Siberia and NE China to N Sakhalin and N & C Korea, and winters E & S China and Taiwan; *E. (s.) personata* **Masked Bunting** breeds S Sakhalin, S Kuril Is and N & C Japan, wintering in S Japan and E & S coastal China (elevated to full species by Russian authors); *E. s. sordida* in EC China, moving to S & E China and Taiwan; rare offshore Japan. **HH** The common bunting of E Asia, occurring in mixed, but mainly deciduous, forest of lowlands and river valleys, to mid-elevation mountains in summer, and in forest, woodland fringes, scrub, parks, gardens and agricultural land in winter, particularly in areas with dwarf bamboo ground cover. **ID** All races share double wingbar, white outer tail-feathers, and calls and nervous behaviour, often flicking tail. *E. s. spodocephala*, the common continental form, is a rather dull, dark bunting, somewhat resembling Grey Bunting, but less dark and with white in tail. ♂ head and breast dark greenish-grey or greyish-olive (even blackish-grey), with extensive black lores and narrow black chin; upperparts dark brown with black and grey streaking, rump plain brown, tail darker with white outer feathers; wings blackish with dark brown fringes to all feathers; underparts pale yellow on belly, browner and streaked on flanks. ♀ lacks black of lores and chin, and dark-hooded appearance, instead has dull grey-brown supercilium, broad pale malar bordered darker on moustachial and lateral throat-stripe, breast dusky olive-brown heavily streaked, and rest of underparts off-white. *E. (s.) personata*, the common form in Japan, is much brighter. ♂ has same black mask, but crown, nape and ear-coverts are more olive, underparts bright yellow, with dark olive-green neck-sides and streaking on breast-sides/flanks, upperparts paler brown and wings also paler brown than nominate. White fringes to lesser and greater coverts form faint wingbars. ♀ resembles nominate ♀, but supercilium, malar and underparts washed variably dull to pale yellow. ♂ *E. s. sordida* is plainer, with dark olive-green hood sharply demarcated from bright yellow underparts; ♀ resembles ♀ *E. s. personata*, but has browner crown and ear-coverts, with buff not yellow supercilium. **BP** Bill greyish-pink with grey tip and culmen; eyes black; tarsi dull brownish-pink. **Vo** A hard, forced, slightly discordant, *tsip* or *jit*. Song a slow *tsip-chee-tree phirrr* or *choppiichott pii chiriri*.

GREY BUNTING   CTKJR
Emberiza variabilis

L 14–17cm. **SD** Range-restricted E Asian endemic, from S Kamchatka to Nansei Shoto. *E. v. variabilis* summer visitor from C Honshu north through Hokkaido, S Sakhalin, Kuril Is; winter visitor from N Honshu to Nansei Shoto, rarely Korea (especially far S), and accidental E China and Taiwan (south to Lanyu). *E. v. musica* in S Kamchatka; winters Japan. **HH** Uncommon and local, favouring mixed deciduous and coniferous forests of mid elevations, with undergrowth and thickets of dwarf bamboo, preferring dense shady cover where quite skulking; in winter in shady forest, woodland and parks. **ID** A rather plain, dark grey bunting, with no white in tail. ♂ mostly plain slate-grey from head to tail, with black-streaked mantle, black wing-feathers with broad slate-grey fringes to wing-coverts and tertials. Imm. ♂ has face and underparts like dull ♂, upperparts like ♀. ♀ is generally dark grey-brown, with grey-brown ear-coverts bordered below by blackish moustachial stripe, pale grey malar and dark brown lateral throat-stripe; chin/throat whitish. Underparts dark dusky-brown with quite heavy darker streaking; upperparts dark brown with black streaking, but rump plain dark rufous-brown (particularly noticeable in flight), and tail all brown. **BP** Bill pink with grey culmen and tip; eyes black; tarsi pink. **Vo** A fine, thin *tsi*. Song slow, slightly sweet and somewhat flycatcher-like, with 3–5 distinct notes: *swee swee chi-chi-chi*; also *huiiii tsi-tsi tsu-chi hee hee* or *fee choichoi*. **AN** Japanese Grey Bunting.

JAPANESE YELLOW BUNTING

ad ♂

ad ♀

1st-win ♀

BLACK-FACED BUNTING

ad ♂
spodocephala

ad ♀
spodocephala

ad ♂
spodocephala

ad ♂ *sordida*

ad ♂
personata

ad ♀
personata

GREY BUNTING

variabilis

ad ♂

ad ♀

1st-win ♂

PALLAS'S REED BUNTING
CTKJR

Emberiza pallasi

L 12–15cm; WT 12–16g. **SD** C & E Siberia, from Yenisei to Chukotka, ranging further N in summer than Common Reed Bunting to edge of tundra. Winters in Ussuriland, Korea and E coastal China; accidental Taiwan. Very rare winter visitor S Japan. *E. p. minor* breeds from N & E Transbaikalia to maritime Russian Far East; *E. p. polaris* occurs in N Siberia, Yakutia, Chukotka and Koryakia south to N Kamchatka. **HH** Locally common in thickets along rivers and in tundra, also around marshes, wetland fringes, reedbeds and dry grasslands; in winter, also in dry agricultural land with scrub, grasses and reeds. **ID** Smaller, paler, more delicate than similar Common Reed Bunting, with small head and fine bill. ♂ breeding has glossy black hood broken by broad white malar stripe; nape, neck-sides and underparts clear white, unstreaked (or with some very fine streaking on flanks and thin dusky streaks on breast-sides) and lacking grey on sides; upperparts pale tawny-brown with contrasting black streaking on mantle and back, the lower back, rump and uppertail-coverts sandy or pale buff, and long tail ashy-black with white outer feathers. Wings lack rufous on lesser coverts of Common Reed, instead has pale blue-grey lesser wing-coverts and sandy fringes to rest of wing-feathers. Winter ♂ lacks black hood, has prominent buff supercilium and malar, and dusky lateral throat-stripe; underparts pale buff, but remain unstreaked. Lacks pale median forecrown-stripe of Common Reed. ♀ resembles winter ♂, but has narrow distinct lateral throat-stripe and whiter underparts, still with some faint dusky streaking. **BP** Bill short, sharply pointed with straight culmen, lower mandible much slighter than Common Reed, black in breeding ♂, otherwise greyish-black with mostly dull pinkish-horn lower mandible (see Common Reed, which is entirely dark or grey); eyes, appears small, irides black; tarsi dull brownish-orange. **Vo** Somewhat sparrow-like *chep* and *tschirp*, as well as more typical bunting-like *tsip* and occasional Common Reed-like *dsiu*. Song a plain and monotonous series of rasping *srih-srih-srih-srih* notes.

JAPANESE REED BUNTING
CKJR

Emberiza yessoensis

L 14–15cm. **SD** E Asian endemic. *E. y. continentalis* restricted to small area of NE China (Heilongjiang) and extreme S Russian Far East in Ussuriland, migrates through NE China to winter in E coastal China (Jiangsu to Fujian); reported Taiwan. *E. y. yessoensis* very local in N & E Honshu and Kyushu, Japan. In winter, very local in S & W Honshu, Korea. **HH** Very locally common in marshy habitats, wetland fringes with bushes, reedbeds, and in some regions tall-grass meadows; in winter, also open agricultural land near water and particularly coastal marshes. **ID** The most richly coloured of the three reed buntings. ♂ breeding has unbroken, glossy black hood extending only to lower neck at front; nape, neck-sides, upperparts, wings, breast-sides and flanks warm rufous/orange-brown; mantle heavily streaked

black and white, lesser wing-coverts show some blue-grey, otherwise wings mostly rufous with prominent black tertials. Winter ♂ retains strong 'ghost' of black hood, with prominent orange-buff supercilium. 1st-winter ♂ resembles ad. ♀ from Dec onwards, head darkens Feb onwards. ♀ resembles winter ♂, but has distinct orange-buff malar and narrow distinct lateral throat-stripe contrasting with off-white throat. Lower back, rump and uppertail-coverts warm rufous-brown in both sexes, readily distinguishing them from Common Reed or Pallas's Reed. **BP** Bill rather sharply pointed with straight culmen, lower mandible heavier than Pallas's Reed, black in breeding ♂, grey above and at tip, bright pink below, especially at base, in winter ♂/♀; eyes black; tarsi quite bright pink. **Vo** A hard *tsu tsu cho* or *chi*. Song like a repetitive Meadow Bunting: *chui-tsui-chirin*, *choppi churiri pi* or *cho pichu piichuu picho*. **AN** Ochre-rumped Bunting.

COMMON REED BUNTING
CTKJR

Emberiza schoeniclus

L 13–19cm; WT 16–25cm. **SD** Wide range, from W Europe to Japan. Summer visitor (May–Sep) across taiga to S Russian Far East, Sakhalin, Hokkaido, Kuril Is, S Kamchatka and NE China. Winters from Honshu and S Korea south through E & SE China (Oct–Apr); accidental Taiwan. *E. s. parvirostris* breeds across Siberia to C Yakutia; *E. s. pyrrhulina* breeds in Transbaikalia to NE China and N Japan, Kuril Is and Kamchatka, wintering to Japan, Korea and E China; *E. s. pallidior* breeds SC Siberia to Lake Baikal, also wintering in SE China. **HH** Locally common in wetland fringes, reedbeds, thickets and tall grasses, in rank meadows and riparian grasslands, where may be found with Yellow-breasted Bunting; winters in similar habitats, but also in dry agricultural land with scrub, grasses and reeds. **ID** A large, long-tailed bunting with a black hood; largest of the three reed buntings. ♂ breeding has glossy black hood to breast, broken only by bold white malar stripe; nape, neck-sides and underparts white, somewhat grey-white at sides; upperparts pale rufous-brown heavily streaked black on mantle, back and scapulars, but lower back, rump and uppertail-coverts ashy grey-brown, tail long, ashy-black with white outer feathers. Winter ♂ duller, lacks black hood, but has weak median forecrown-stripe, pale ash-brown supercilium, brown lores, crown and ear-coverts, and black lateral throat-stripe forming prominent neck patch and merging into narrow streaking on flanks. Often has stronger grey tone to rump. ♀ resembles winter ♂, but lateral throat-stripe less prominent and supercilium continues to base of bill. **BP** Bill heavy, culmen slightly arched, lower mandible robust, greyish-black; eyes black; tarsi dull brownish-orange. **Vo** A rather plaintive, sibilant *siooo*, harsher *bjee* and *chuiin*. Song a rising *chii chuichui chi chui jurin* or *shree-shree-teeree-teeree*.

PALLAS'S REED BUNTING

ad ♂ win

ad ♂ sum

polaris

ad ♀

1st-win ♀

JAPANESE REED BUNTING

ad ♂ win

yessoensis

ad ♂ sum

1st-win ♀

ad ♀

ad ♂ sum
parvirostris

COMMON REED BUNTING

ad ♂ sum
pyrrhulina

ad ♂ win
pallidior

ad ♂ win
parvirostris

ad ♀ sum
parvirostris

LAPLAND BUNTING   CTKJR
Calcarius lapponicus

L 15–17cm; WT 20–28g. **SD** Widespread summer visitor across high-latitude Holarctic. Uncommon winter visitor (Oct–Mar) to N Japan, Korea, NE & E China. *C. l. lapponicus* breeds N Europe to C Chukotka and N Anadyr; *C. l. kamtschaticus* in S Anadyr, Koryakia, Kamchatka and N Sea of Okhotsk, wintering to E Asia including Japan; *C. l. coloratus* of Commander Is, wintering to E China; *C. l. alascensis* of E Chukotka, to Alaska and Canada, wintering N America. **HH** Summers on open tundra, particularly with dwarf birches or willow thickets, and winters on beaches and dunes, or open agricultural land. Commonly adopts hunched posture resembling larks. Usually solitary, in pairs or small parties in winter, occasionally in larger flocks. **ID** Stocky, medium-sized bunting with large head. ♂ breeding has black head and chest, and flanks streaking, white supercilia curving down neck- to breast-sides, and nape/hindneck bright chestnut. Upperparts brown streaked black. ♀ is buff on breast and flanks, with white belly. Wings quite long – reach beyond uppertail-coverts – with long primary projection, brownish-black with distinctive rufous fringes to greater coverts and tertials in all plumages; tail long, notched, with white outer feathers. Winter ♂ has plainer face, usually with 'ghost' of black summer plumage on throat and chest, retains chestnut nape and has chestnut on wing-coverts and tertials. Non-breeding ♀ generally pale brown on upperparts and breast, wing-coverts and tertials warm brown; face pale with dark corners to ear-coverts, white malar and dark lateral throat-stripe. **BP** Bill yellowish-horn with grey tip; eyes black; tarsi blackish-brown, hindclaw longer than hindtoe. **Vo** A distinctive dry rattled *prrrt* (like Snow Bunting) and whistled *tleuw*, or run together as *ticketick-tleuw*. Also a soft *pee-dle* on breeding grounds. Song, commonly delivered from perch on rock or tall tundra umbellifer, and occasionally during high display flight, a husky, jingling *freew didi freer di fridi fideew.* **AN** Lapland Longspur.

SNOW BUNTING   CTKJR
Calcarius nivalis

L 15–18cm; WT 28–50g. **SD** High-latitude Holarctic. Summer visitor to shores of Arctic Ocean east to Bering Strait, EC Kamchatka and Commander Is. Uncommon and local winter visitor to coastal W Kamchatka, Sakhalin, S Russian Far East and Hokkaido, also E Inner Mongolia and N Heilongjiang; vagrant to Kyushu, Korea, and Hebei. *C. n. vlasowae* across Eurasia to NE Siberia; *C. n. townsendi* on Commander Is and W Aleutians. **HH** Summers on open rocky tundra, coastal tundra and mountains; winters on open ground, along snow-free beaches,

riverbanks and grassy capes. In winter in restless flocks, fluttering in 'dancing' flight, then dropping suddenly to settle. **ID** Large, unmistakably black and white bunting with long wings reaching beyond uppertail-coverts. ♂ breeding jet black and white; head, entire underparts, lower back, rump and much of wings white, only primary tips black. Mantle, scapulars, primaries, secondaries and tertials black, as is short, notched tail. Winter ♂ has warm orange-brown crown and ear-coverts, back mottled brown and black, and warm brown breast-sides and flanks. ♀ breeding resembles winter ♂ but has dusky grey-black streaks on mantle, is not as cleanly black and white, and head less clean white with dusky streaking, especially on crown and ear-coverts; winter ♀ like winter ♂. Flight buoyant and flickering on long wings, may circle repeatedly before settling; wings largely white with extensive black tips; tail largely black with broad white outer panel and base. **BP** Bill rather small, black in summer ♂, yellow with grey tip in ♀/winter ♂; black eyes prominent in plain face; tarsi black. **Vo** Varied, similar to Lapland Bunting, but more liquid, rippling *prrrp*, rattled *tiriririt* and softer *tew* or *pyeuw*, often given in flight; also a jarring *jrrt* given occasionally while feeding. Song given from exposed rock or raised ground a clear, twittering *swito-süvee-vitüta-süveh* or *hudidi feet feet feew hudidi feet feet feew hudidi*. **TN** Formerly *Plectrophenax nivalis.*

MCKAY'S BUNTING   R
Calcarius hyperboreus

L 15–18cm; WT 54g. **SD** Remote Bering Sea islands, wintering on Alaskan coast; regular migrant at Gambell, St Lawrence I. (Aug to Oct), just *c*.70km from Russian coast. Accidental, or perhaps more likely largely overlooked, scarce winter visitor to Russian coast. In N America, has strayed south of Alaska to British Columbia, Washington and Oregon, so could occur south to N Japan. Monotypic. **HH** Barren tundra and shores. **ID** Very similar to Snow Bunting, but has more white in wings and tail, and less black. ♂ breeding almost all white, lacking prominent black back of breeding Snow, with black restricted to tertials, primary tips and tips of central rectrices. Winter ♂ lacks brown-streaked back of winter Snow, but has some orange tones to crown, ear-coverts and neck-sides. ♀ and juv. have whiter head, paler less heavily streaked mantle and back, more extensive white in wings and tail than corresponding plumages of Snow. In flight, almost entire wing appears white, with only extreme tips of primaries and tertials black, and tail also almost all white, with a little black near tips of central feathers. **BP** Bill black (both sexes breeding) or straw-yellow (winter both sexes); eyes black; tarsi black. **Vo** A hard *chi cha*, but essentially calls and song identical to Snow Bunting. **TN** Formerly *Plectrophenax hyperboreus.*

♂ win

ad ♀ sum

ad ♂ sum

LAPLAND BUNTING

kamtschaticus

1st-win ♀

SNOW BUNTING

vlasowac

ad ♀ win

ad ♀ sum

ad ♂ sum

ad ♂ sum

ad ♂ sum

juv

ad ♂ win

ad ♂ sum

MCKAY'S BUNTING

ad ♀ sum

FOX SPARROW JR
Passerella iliaca

L 16–19cm; WT 32g. **SD** Very widespread polytypic N American species (split pending, probably into four species). Two populations, **Sooty Fox Sparrow** P. (i.) *unalaschensis* group and **Red Fox Sparrow** P. i. *zaboria*, migrate to coastal Alaska and Aleutians. Sooty Fox is accidental in E Russia, on Chukotski Peninsula, Wrangel I., Anadyr and Commander Is, and Japan in winter. **HH** Favours brushy habitat or lush tundra with sheltered gullies. **ID** Records so far have involved Sooty Fox of Aleutians and Alaska, which is a large, dark brown sparrow, with plain head, face and upperparts (unstreaked), wings and tail. Tail tinged warmer brown or reddish-brown. Underparts white with extensive, heavy dark streaking or mottling on throat, breast and flanks, somewhat whiter with less streaking on belly. Red Fox is overdue, as it nests closest to Chukotka; it has reddish back, cheeks and underparts streaking. **BP** Bill rather large, short, straw yellow or orange-yellow base to lower mandible, with dark grey culmen; eyes dark brown; tarsi dull brownish-pink. **Vo** A loud, low *tschup*, also described as *chi*, *cha* or *check*, a sharp *smack* or *chap*, and in flight a sharp, rising *seeep*. **TN** Formerly *Zonotrichia iliaca*.

SONG SPARROW JR
Melospiza melodia

L 13–16cm; WT 20g. **SD** Very widespread N American species, wintering south as far as Mexico. Nearly 40 subspecies recognised of which several occur on islands off Alaska west along Aleutian chain, iincluding *M. m. maxima* and *M. m. sanaka*. Accidental NE Russia and Japan. **HH** Low vegetation with brush. **ID** Fairly large, round-headed and long-tailed sparrow, with short, rounded wings; rather coarsely streaked, thus superficially recalls various ♀ *Emberiza* but, like similar Savannah Sparrow (see p.498), lacks streaking in outer tail-feathers. Larger, darker NW races, most likely to stray to region. Ad. generally dull grey with heavy brown streaks often converging to central breast spot; crown-sides, ear-coverts and lateral throat-stripe brown, mantle and breast heavily streaked brown, wings and tail largely plain dark brown. Supercilium buffish-grey lacking yellow. Juv. even more heavily streaked with colder, darker blackish-brown. **BP** Bill sharply pointed, grey; eyes dark brown; tarsi dull greyish-pink. **Vo** A husky *jimp*, hard, high *tik* (in alarm) and in flight a thin, level *seeet*.

WHITE-CROWNED SPARROW KJR
Zonotrichia leucophrys

L 16–18.5cm; WT 29g. **SD** Widespread N American species breeding in far north and wintering in southern states and Mexico; *Z. l. gambelii* migrates as far north as W Alaska. Accidental (Oct–May) in NE Russia (Wrangel I.), Japan and Korea. **HH** Favours brushy and weedy, not wooded, habitats, on agricultural land and coastal areas. **ID** Ad. breeding unmistakable, with narrow white crown bordered by broad black lateral crown-stripes, grey lores, white supercilium broad behind eye to nape, and black eyestripe from eye to nape. Face, chin/throat and breast plain grey, flanks brown; upperparts grey-brown with black streaking, rump grey-brown, tail rather long and plain brown. Wings brown with two white bars formed by tips to coverts, and rufous fringes to tertials. 1st-winter far more likely to occur in Asia, providing most records at Gambell, St Lawrence I. Head pattern much weaker than ad., with brown crown, dull grey supercilium, brown eyestripe from eye, grey to grey-brown face and throat separated by dark brown lateral throat-stripe, nape streaked and grades into dull grey-brown back streaked blackish-brown; underparts dusky grey-brown, browner on flanks. **BP** Bill pale pink or pale yellow; eyes black with white crescents above and below; tarsi yellowish-brown. **Vo** A rather strong, sharp *pink* or metallic *pzit* and in flight a high, thin, rising *seeet*.

GOLDEN-CROWNED SPARROW JR
Zonotrichia atricapilla

L 18cm; WT 29g. **SD** Migrant through western N America, wintering in S USA, breeding in Canada northwest to W Alaska. Accidental E Russia, on Chukotski Peninsula, Wrangel I., and Japan. Monotypic. **HH** Breeds in willow and alder thickets bordering alpine and arctic tundra; winters in brushy areas. **ID** Large, plain-faced sparrow with long tail and two-toned bill. Ad. breeding generally similar to White-crowned, but has golden-yellow forecrown, whiter at rear, and broad black band across head-sides to eye and bill, face plain grey, underparts dull grey with rufous-brown wash to flanks. Non-breeding ad. has reduced black and yellow on head, and is more grey. 1st-winter has little or no black on head, and only small area of often dingy yellow on forehead. **BP** Bill yellowish-horn below, grey above in ad., all grey in young; eyes black; tarsi flesh pink. **Vo** Call a sharp, flat *pink* and in flight a thin *seet*.

DARK-EYED JUNCO R
Junco hyemalis

L 13–16cm; WT 19g. **SD** Occurs across breadth of N America, with many recognisable races, one of which, **Slate-coloured Junco** J. (h.) *hyemalis* migrates as far northwest as C Alaska to breed. Vagrant to St Lawrence I. (Sep–Oct). Accidental to E Russia, on Chukotski Peninsula and Wrangel I. **HH** Breeds in coniferous taiga or boreal forest, wintering in open woodland and brushy areas. **ID** Rather bunting-like in overall proportions. ♂ slate-grey on head, upperparts, breast and flanks; white on belly and vent. ♀/juv. slightly paler grey on face, breast, back and rump, with brown wash to crown, mantle and flanks. Wings rather rounded, slate-grey in ♂, greyish with brown fringes to most feathers in ♀; tail long, blackish-grey with distinct white outer feathers. **BP** Bill small and rather conical, pink; eyes appear small, but quite prominent in plain face, irides black; tarsi dull, pale fleshy-pink. **Vo** A hard clicking *tik* or *check*, high-pitched *stip* and in flight a buzzy *tzeet*, but when flushed gives a series of twittering notes *tsititit tit*.

FOX SPARROW

unalaschensis

ad

SONG SPARROW

sanaka

ad ♂

WHITE-CROWNED SPARROW

gambelii

ad sum

1st-win

GOLDEN-CROWNED SPARROW

ad ♂

1st-win

DARK-EYED JUNCO

hyemalis

1st-win

ad ♂

SAVANNAH SPARROW KJR
Passerculus sandwichensis

L 14–16cm; WT 20g. **SD** Widespread N American species. *P. s. athinus* breeds regularly on E Chukotski Peninsula and Alaska. Accidental elsewhere in NE Russia (e.g. Wrangel I.), Korea, and rare but regular winter visitor to Japan. **HH** Grassy areas including grass-covered dunes, agricultural areas, fallow fields, open grasslands and coasts. **ID** A small, neat sparrow, with clean streaking and characteristic yellow patch on lores (limited or lacking in some imms.). Recalls several ♀ buntings, but lacks white outer tail-feathers. Face striped, with pale buff supercilium becoming prominently yellow in front of eye, dark eyestripe broadening at rear, brown ear-coverts, blackish moustachial and lateral throat-stripe. Narrow whitish lateral crown-stripe. Upperparts grey-brown with dark, blackish streaking; underparts off-white with heavily streaked breast and flanks, clean white belly and undertail-coverts. Wings brown with white tips and fringes to black tertials. In flight, wings quite long, tail quite short, square or slightly notched, outer feathers may appear pale. Quite similar to slightly larger Yellow-browed Bunting, but yellow confined to supraloral area, rather than extending behind eye, and Yellow-browed has stronger, better-defined, contrasting black, white and yellow head pattern, especially in ad. ♂ breeding plumage, but to lesser extent in ad. ♀, making confusion unlikely. **BP** Bill small, sharply pointed, dull pink with grey culmen and tip; eyes black with narrow white eye-ring; tarsi dull pink. **Vo** A sharp, slightly metallic *tsui, ji* or *tsip* and in flight a thin, weakly descending *tsiw*. **TN** Formerly *Ammodramus sandwichensis*.

AMERICAN TREE SPARROW R
Spizella arborea

L 16cm; WT 20g. **SD** Occurs over breadth of N America; *S. a. ochracea* breeds as far northwest as W Alaska. Accidental to St Lawrence I. and to E Russia, on Chukotski Peninsula and Wrangel I. **HH** Breeds in brush and scrub, and winters in weedy areas with brush. **ID** A small sparrow with rather large head. All have rufous crown and eyeline, but otherwise grey head. Neck grey, mantle rufous-brown with black streaking, plain grey rump and tail lacking white outer feathers. White throat grades into grey-buff breast/belly with rufous wash at sides and dark spot in centre of breast. Two white wingbars and rufous fringes to most wing-feathers. Juv. as ad. **BP** Bill small, dark grey above, yellow below; eye with broken narrow white ring rather prominent, irides black; tarsi quite dull pink. **Vo** A thin *seet*, unusual jingling *teedleoo*, and in flight a sharp *tsiiw*.

CHIPPING SPARROW R
Spizella passerina

L 14cm; WT 12g. **SD** Wide-ranging N American species; *S. p. passerina* breeds as far northwest as Alaska. Presumably it is this subspecies that has reached islands off Alaskan coast, Bering Sea (e.g. St Lawrence Aug–Sep) and, in region, is accidental in NE Russia. **HH** During breeding season favours open woodland, but on migration/in winter open areas with brush and scrub. **ID** Somewhat plain, grey-rumped sparrow, size of Savannah. Ad. breeding distinctive due to chestnut cap, broad white supercilium, narrow black eyestripe, white chin/throat, and plain grey neck, breast and belly; upperparts mid brown streaked heavily blackish-brown, with two prominent white wingbars. Non-breeding less cleanly marked on head, in particular, crown is blackish-brown and supercilium buff, but retains strong eyestripe. **BP** Bill small, dark grey above, pale greyish-horn below (breeding) or more extensively pink (non-breeding); eyes black; tarsi pink. **Vo** A sharp *chip* and in flight a thin, rising *tsiis*.

RED-CRESTED CARDINAL TJ
Paroaria coronata

L 18–19cm. **SD** Natural range E Bolivia to C Argentina and S Brazil, but common in cagebird trade, and feral population established in C Honshu, Japan and reported Taiwan. Monotypic. **ID** Large finch-like bird with unmistakable plumage. Entire head, erectile crest, face and throat bright red, upperparts including wings and long tail grey, and underparts white. **BP** Bill pale whitish-pink; eyes brown; tarsi black. **Vo** Song consists of thrush-like whistled phrases: *tsilewp-jewp tsilewp-jewp...* or *dew-dewe- duh-dew-diuh*.

SAVANNAH SPARROW

athinus

ad

ad ♂

ad

AMERICAN TREE SPARROW

ochracea

ad ♂ win

CHIPPING SPARROW

passerina

ad ♂ sum

ad ♂

RED-CRESTED CARDINAL

ad

Species from E Asia that have either become extinct (†) or are presumed to have become extinct (†?) in recent times.

†SPECTACLED CORMORANT R
Phalacrocorax perspicillatus

L >75cm. **SD** Endemic to Commander Is, Russia; *perhaps* also in coastal Kamchatka and outer Aleutians. Common, when discovered by G. W. Steller on Bering I in 1741; described by Pallas only in 1811 and last specimen taken 1850. Extirpated due to human consumption. **ID** Large, short-winged, probably flightless cormorant; black, with loose facial plumes, red gular/gape patch and white eye-ring. **BP** Bill and legs stout, grey or dark grey.

†?CRESTED SHELDUCK *Tadorna cristata* CKJR

L 60–63cm. **SD** Occurred in Ussiriland, NE China and Korea, visited Japan. Earliest known specimen taken in 1877, last in 1916; described by Nagamichi Kuroda in 1917. Last reliable sighting (1♂2♀) on islands S of Vladivostok, Russia, May 1964, but unsubstantiated reports from NE China in 1970s and 1980s, and in 1990 from S China. **HH** Coastal wetlands and rivers; resident or local migrant. **ID** A striking, dark shelduck with black cap extending to nape. ♂ greyish-white face and neck, black breast and collar, greyish-brown mantle, wing-coverts and underparts, dull orange undertail-coverts; ♀ similar, with white face, 'spectacles' and breast. In flight, dark with white underwing-coverts. **BP** Bill waxy red; eyes dark brown; tarsi orange.

†?ESKIMO CURLEW *Numenius borealis* R

L 29–34cm. **SD** Before 1850, abundant breeding species of high-Arctic Canadian tundra and the Russian Chukchi Peninsula (until the 1880s), migrated to winter as far south as S Argentina. Since 1960s, unsubstantiated reports only from Americas. Decline due to over-hunting and loss of locust prey on migration route. **HH** Bred on Arctic tundra, visited coastal wetlands, staging on prairies on migration and grassland in winter. **ID** Marginally larger than Little Curlew (p.176), but darker, with buff-cinnamon underparts more heavily marked with black chevrons, longer wings (primaries extending well beyond tail), shorter legs, and distinctly cinnamon underwing. **BP** Bill short, strongly arched; legs short, grey.

†BONIN WOODPIGEON *Columba versicolor* J

L 45cm. **SD** Endemic Nakoudo-jima and Chichi-jima, Ogasawara Is. First discovered on Chichi-jima, Jun 1827, by Beechey, collected on only three further occasions: two on Chichi-jima, May 1828 (Kittlitz), and one on Nakoudo-jima, 15 Sep 1889 (Holst). Not recorded since. Cause of extinction unclear. **HH** Inhabited subtropical forests; diet fruits, seeds and buds. **ID** Larger than extant Black Woodpigeon (p.246). Dark grey, darkest, almost slate-grey, on wings and tail, with maroon, or green and purple gloss to back and rump, green tinge to flanks, with paler crescentic patch on neck/breast. **BP** Bill and cere dull yellow; legs red.

†RYUKYU WOODPIGEON *Columba jouyi* J

L 45cm. **SD** Endemic Okinawa, Kerama and Daito Is, Japan. Last record 1904 (Okinawa) and 1936 (Daito Is). Cause of extinction unclear. **HH** Inhabited subtropical forests; diet fruits, seeds and buds. **ID** More uniformly black than close relative Bonin Woodpigeon, with purple-glossed head, green-glossed neck and underparts, and distinctive silvery crescent on upper back. **BP** Bill and cere dark blue; legs red.

†MIYAKO KINGFISHER *Halcyon miyakoensis* J

SD Previously considered endemic taxon, but collected only once, Miyako I., Japan (1887), an island quite unlikely to have supported such a species. At best, perhaps a local subspecies of Micronesian Kingfisher *H. cinnamominus miyakoensis*, but almost certainly only accidental, ship-assisted, or collected elsewhere. **ID** Largely brownish-orange, with blue-green stripe from bill base across face to back; wings, back and tail blue-green with cobalt-blue rump and uppertail-coverts. **BP** Specimen lacks bill sheath, hence colour unknown; legs red.

†KITTLITZ'S THRUSH *Zoothera terrestris* J

SD Endemic Ogasawara Is, Japan. Quite common when discovered in 1828 by Baron F. H. von Kittlitz – the only naturalist to observe it. Missed there, surprisingly, by Beechey in 1827, and never found again. Kittlitz collected four specimens on Chichi-jima in Jun 1828. Probably eradicated by introduced mammals. **HH** Inhabited subtropical woodland; ground-dwelling and perhaps ground-nesting. **ID** A small, brown *Zoothera* with dark-streaked mantle and white areas on the throat, belly and vent. **BP** Bill dark grey; legs dull ochre.

†BONIN GROSBEAK *Chaunoproctus ferreorostris* J

L 18.5cm. **SD** Endemic Ogasawara Is, Japan. Discovered by Beechey in 1827 and observed by Kittlitz in 1828. Specimens vary in size, perhaps indicating there were two different island populations. Searched for unsuccessfully during 1850s, none found by naturalists after Kittlitz. Like Kittlitz's Thrush, probably exterminated by introduced mammals. **HH** Inhabited subtropical forest, where encountered alone or in pairs on forest floor. **ID** Large, dimorphic, grosbeak-like bird, brown with red forehead, supercilia, lower face and neck (♂), or plain brown (♀). **BP** Massive bill was dark grey; legs dull brownish-pink. **Vo** A soft, pure, high piping call.

SPECTACLED CORMORANT

CRESTED SHELDUCK

ad ♂

ad ♀

ESKIMO CURLEW

RYUKYU
WOODPIGEON

MIYAKO KINGFISHER

BONIN
WOODPIGEON

ad ♂

ad ♀

KITTLITZ'S THRUSH

BONIN GROSBEAK

APPENDIX 1. THE STATUS OF THE BIRDS OF EAST ASIA

Generalised status information is given for each species in all of the regions covered by the guide:

C = Chinese coastal provinces from Fujian northwards and northeast China

T = Taiwan, including its insular territories such as Lanyu, Penghu, Matsu and Kinmen islands, but as the latter two are adjacent to the continental coast, status of some species there is very different from that on mainland Taiwan, thus in such cases additional status information for Matsu/Kinmen is given in parentheses

K = The Korean Peninsula and its associated islands

J = The Japanese archipelago including the Ogasawara and Bonin Islands

R = The Russian Far East, from east of Lake Baikal to the Bering Strait, north to the Arctic Ocean, and including Sakhalin, Chukotka, Kamchatka and the Kuril Islands

Species are classified as:
R = resident, **B** = breeds, **M** = migrant, **W** = winters. To these status codes, qualifiers (1–4 and ?) are appended indicating relative status, in relation to the likelihood of seeing the species in appropriate habitat at the appropriate season.

Breeders (approximately):
B1 = abundant (100,000–1 million pairs)
B2 = common (10,000–100,000 pairs)
B3 = scarce (100–10,000 pairs)
B4 = rare (fewer than 100 pairs)
B? = breeding suspected, not confirmed

Migrants and winter visitors (annual):
M or W1 = abundant (10,000 plus individuals)
M or W2 = common (1,000–10,000 individuals)
M or W3 = scarce (50–1,000 individuals)
M or W4 = rare (fewer than 50 individuals)

In addition, **V** = vagrant or accidental visitor from outside the region (for potential vagrants likely to reach the region in the future see Appendix 2). **V*** = approximately 1–5 records to date, **V**** = approximately 6–20 records to date, **V*** = more than 20 records, perhaps an annual visitor, **??** = unpublished, unconfirmed, or uncertain record or uncertain identification, **I** = introduced and established, **Esc?** = recorded, but possibly escapee, released or introduced, but not yet established. **†** = extinct (previously native).

These status codes have been assessed using a range of published sources; they should be taken only as general guides to the relative status of a species within a particular range, not as an absolute indicator of population size. Publication has been taken as the basis for records for the region. That many records go unpublished is unfortunate, as it makes a true understanding of status difficult, if not impossible, to ascertain.

✓	Common name	C	T	K	J	R
	Hazel Grouse	RB2		RB3	RB3	RB1
	Siberian Grouse	RB4/ V***				RB2
	Black-billed Capercaillie	RB4				RB1
	Eurasian Black Grouse	RB3		RB3		RB2
	Rock Ptarmigan	RB4			RB3	RB1
	Willow Ptarmigan	RB4				RB1
	Chukar Partridge	RB2				
	Chinese Francolin	RB2				
	Daurian Partridge	RB2-3				RB2
	Japanese Quail	BM2	V*/M4	BM3W4	BM3W4	RB2
	Blue-breasted Quail	RB3	RB4			
	Taiwan Hill Partridge		RB3			
	White-necklaced Partridge	RB3-4				
	Chinese Bamboo Partridge	RB2	RB2		IRB2	

✓	Common name	C	T	K	J	R
	Cabot's Tragopan	RB4				
	Koklass Pheasant	RB3				
	Silver Pheasant	RB2				
	Swinhoe's Pheasant		RB3			
	Brown Eared-Pheasant	RB4				
	Elliot's Pheasant	RB3-4				
	Mikado Pheasant		RB3			
	Copper Pheasant				ERB3	
	Reeves's Pheasant	ERB4 (†?)				
	Ring-necked Pheasant	RB2-3	RB3	RB2	IRB2	RB2
	Japanese Green Pheasant				ERB2	
	Indian Peafowl		Esc		IRB3	
	Lesser Whistling Duck	RB4	V**		(RB†)V**	
	Swan Goose	RBW3	V*	MW3	(W3†)V*	B3
	Bean Goose	W2	W4	W1	W2	B2

✓	Common name	C	T	K	J	R
	Greylag Goose	RBW4	V**	V**	V*	B3-4?
	Greater White-fronted Goose		W4	W1	W1	B2
	Lesser White-fronted Goose	W3-4	V**	W4	W4	B3
	Bar-headed Goose	V*?		V***	V**	V***
	Emperor Goose			V***	V***	RB3
	Snow Goose	V**		W4	W4	B3
	Ross's Goose					V***
	Lesser Canada Goose	V*	V***	V*	W4	(B†)V*/M4
	Greater Canada Goose	IB3-4			IB3-4/V***	
	Brent Goose	W4	V***	W3-4	MW3	B2
	Mute Swan	B4W3-4	V***	MW4	V***/IRB3	V*
	Black Swan				IB3	
	Whooper Swan	W3	V**	W2	W1	B2W2
	Trumpeter Swan				V**	V**
	Tundra Swan	W2-3	V**	W2	W1	B2-3
	Common Shelduck	BW3	W4	W1-2	W3	V*
	Ruddy Shelduck	RB3W3	V*/W4	W2-3	V*/W4	B3
	Mandarin Duck	RB3W4	RB4W4	RB3	RB3W2	B3
	Cotton Pygmy-Goose	RB4	V**		V***	
	Gadwall	RBW3	W4	B?R?W2	B4R?W3	B3
	Falcated Duck	RB?W3	W4	W2	W2	B2
	Eurasian Wigeon	RB3W2	W2-3	W2	W1	B1
	American Wigeon		V*	W4	W4	B4
	Black Duck			V***	?V***	
	Mallard	RBW2	W3	RB3W1	RB2W1	B2
	Philippine Duck		V**		V***	
	Eastern Spot-billed Duck	RBW2	RB3W2	RB3W1	RB2W1	B3
	Blue-winged Teal				V***	
	Northern Shoveler	RBW2	W2	RB?W2	RB3?W1	B2
	Northern Pintail	W2	W2	W1	W1	B1
	Garganey	MW3	M3W4	M2	RB?M3	B2
	Baikal Teal	B4W2-3	W4	W1	W3	B1-2
	Eurasian Teal	RB2W2	W1	W1	W1	B1
	Green-winged Teal	V***	V***	V**	V*	V*
	Red-crested Pochard	W3	V***	V**	V*	
	Canvasback	V***	V***	V***	V*	V***
	Redhead			V***	V***	V***
	Common Pochard	W2	W4	MW1	aB4W1-2	B3
	Baer's Pochard	RBW4	W4	V*	V*	B4
	Ferruginous Duck	V**	V***	V**	V**	
	Ring-necked Duck				V**	
	Tufted Duck	RBW2	WM2	W2	W2	B1
	Greater Scaup	W4	WM3	W1-2	W1	B1
	Lesser Scaup				V**	V**

✓	Common name	C	T	K	J	R
	Steller's Eider	W4			W4	B3W3
	Spectacled Eider					BW3
	King Eider				V***	B2W2
	Common Eider				V***	B2W2
	Harlequin Duck	V**		W3	B4W2-3	B2W2
	Surf Scoter				V*	B4/V*
	White-winged Scoter	M3		W2-3	W2	B1W1
	Black Scoter	W4V**		W3	W1-2	B1W1
	Long-tailed Duck	W4V**		W4	W2-3	B1W1
	Bufflehead				W4?	W4
	Common Goldeneye	RB4W3	V**	W2	W1-2	B1W1
	Barrow's Goldeneye			V***	V***	V**
	Smew	W3	V**	W2-3	W2	B2
	Hooded Merganser				V***	
	Goosander	RBW3	V*	RB4W2	W1-2	B1
	Red-breasted Merganser	RB3W	WM4	W2-3	W2	B1W2
	Scaly-sided Merganser	RBW4	V*	W3-4	V*/W4	B3
	Red-throated Diver	MW4	V** (W3-4)	W3	W2-3	B2W2
	Black-throated Diver	V**	V**	W3	W3	B2-3W2
	Pacific Diver	V**		W3	W2-3	B2-3W2
	Great Northern Diver				V***	V***
	White-billed Diver			W4	W4	?
	Laysan Albatross	V***	V*		WRB3-4	?
	Black-footed Albatross	V**?	V*/W4		RB4M3	M4
	Short-tailed Albatross	V**?	V**	V***	RB3M3	M4
	Wandering Albatross				V***	
	Northern Fulmar	MW3		MW4?	M2	BM1
	Cape Petrel				V***	
	Providence Petrel				M3	M4
	Juan Fernandez Petrel				V***	
	White-necked Petrel				M4	
	Kermadec Petrel		V***		M3	
	Hawaiian Petrel				V**	
	Mottled Petrel				V**	M3
	Black-winged Petrel				V***	
	Bonin Petrel	V**	V**	V***	BM3	M?
	Stejneger's Petrel				M4	
	Tahiti Petrel		V***			
	Streaked Shearwater	BMW3	M2	BM1W4	BM1W3	BM2?
	Kiritimati Shearwater				V*	
	Wedge-tailed Shearwater	V**	M2-3		BM2	
	Buller's Shearwater				M4	V***
	Tropical Shearwater				BM3W3	
	Manx Shearwater				V***	
	Flesh-footed Shearwater	?	?	M3-4	M3	V*/M4

503

✓	Common name	C	T	K	J	R
	Sooty Shearwater	BM3W3	V**	??	M1-2	M1-2
	Short-tailed Shearwater		M3-4	M3	M1-2	M1-2?
	Bulwer's Petrel	BM3	B4M2		BM3	
	Wilson's Storm-petrel				V*	
	Band-rumped Storm-petrel				BM2-3	V*/M4
	Swinhoe's Storm-petrel	BM3	BM2-3	BM2-3	BM2-3	V***
	Leach's Storm-petrel	V***			BM3	BM2
	Tristram's Storm-petrel		V***		BM3	
	Matsudaira's Storm-petrel		V*		BM4	
	Fork-tailed Storm-petrel		V***		MW3	BM2
	Little Grebe	RB3M2	RB3	RB3M2	RB3M2	RB3M2
	Red-necked Grebe	BM2W3?		W3	BM4W2	BM2
	Great Crested Grebe	BM3W2	M4(W3)	BM3W2	BM3W2	BM2
	Slavonian Grebe	M3W4	V**	W3	W2	BM2-3?
	Black-necked Grebe	BM3W2	V*/W4	W2-3	W2	BM3?
	Painted Stork	R4(†?)				
	Black Stork	BM3-4	MW4	BM3-4W4	V**	BM3-4
	White Stork	V***				
	Oriental Stork	BM3W3?	MW4	B?W3-4	(†)(l)V*	BM4
	Sacred Ibis		IB3			
	Black-headed Ibis	W4	V*/W4	V**	V*	V***
	Crested Ibis	RB4	V***	†/V***	†	†/V**
	Glossy Ibis	V***	V***		I?/V***	
	Eurasian Spoonbill	BM4W3	W4	W3	W3	BM4
	Black-faced Spoonbill	BM3W3?	W3	BM3W4	W4	V***
	Eurasian Bittern	BM3W3?	W4	W3-4	B4W4	BM3-4
	Yellow Bittern	RBM2	RB2	BM2-3	RBM3	BM3-4
	Schrenck's Bittern	BM3	M4	BM3-4	BM4	BM3
	Cinnamon Bittern	BM3	RB3	V**	RB3	V**
	Black Bittern	RB2-3	V*	V**	V**	
	Striated Heron	RBM2	RB4WM3	BM3	RBM2	BM3?
	White-eared Night Heron	RB4				
	Japanese Night Heron	M4	M4	V*/M4	BMW4	V**
	Malayan Night Heron		RB3MW3		RB?3	
	Black-crowned Night Heron	BMW1	RB1	RB2W3	RB2M2	
	Rufous Night Heron		V***		RB†	
	Chinese Pond Heron	RBM2	M3	B4M3	M3	B3
	Eastern Cattle Egret	RBM1	BM1W1	BM2-3W4	BM2W3	M4
	Grey Heron	RBM1	W2	RB3W2	RB1W1	BM1
	Great Blue Heron					V***
	Purple Heron	RBM3	MW3-4	M4	RB4M3	M4?
	Great White Egret	BM2	W2	BM2W2	RB2	BM2
	Intermediate Egret	BM2W2	W2B4	BM3W4	BM2W4	B4
	Pied Heron		V***			
	White-faced Heron	V***	V***			
	Little Egret	RBM1	RB1	RBM1-2	RB1-2	

✓	Common name	C	T	K	J	R
	Pacific Reef Egret	M4	RB3	RB3W3-4	RB2-3	
	Swinhoe's Egret	BM3	M3	BM3	M4	BM4
	Red-tailed Tropicbird		V**		BM4	
	White-tailed Tropicbird		V***		(B?)M4	
	Christmas Is. Frigatebird	V***			V***	
	Great Frigatebird	V**	V**	V***	V**	
	Lesser Frigatebird	V**	M4	V**	V*	V**
	Great White Pelican	W4	(l)V***		(l)V***	
	Spot-billed Pelican	(E)	V***/?	V***	V***/?	
	Dalmatian Pelican	W4	V*/M4	?	V**	
	Masked Booby	?	BM4	M4/V*	BM4	
	Red-footed Booby	V**	BM4		BM4	V***
	Brown Booby	M4	BM2	V*/M4	BM2W2	
	Double-crested Cormorant					V***
	Great Cormorant	W1-2?	W2	RB3W1-2	RB2W1	BM2?
	Temminck's Cormorant	BM3W4	W4	BM3W2	BM2W2	BM2
	Pelagic Cormorant	B4W3	V***	BM3W3	BM2W2	(R)BM2
	Red-faced Cormorant	V***?			RB4W4	RB3
	Pied Falconet	RB4				
	Lesser Kestrel	BM3	V***	??	V**	
	Eurasian Kestrel	RBM2	W2	RBM3	RBM2	BM2
	Amur Falcon	BM3	V***	B3M4	M4?	BM3
	Merlin	W3	V*	W3	W3	BM3
	Northern Hobby	BM3W2?	M4	BM3-4	BM3	BM2?
	Saker Falcon	BM3		V**		
	Gyr Falcon	V**			W4	RB2
	Peregrine Falcon	RBW2-3	B4MW3	RB3	RB3	BM2
	Osprey	BW2	W3	B4?W4	RBM3	BM2?
	Black Baza	BM3	V***	??		
	Oriental Honey Buzzard	BM2	B4M2W3	BM2	BM2	BM2
	Black-shouldered Kite	RB3-4	RB4		V***	
	Black-eared Kite	RBM1	RB3	BM2W2?	RB1-2	BM1
	Brahminy Kite	RB4(†?)	V***			
	White-bellied Sea Eagle	RB3	V***			
	Pallas's Sea Eagle	RB3				
	White-tailed Sea Eagle	RB3W2	V*/W4	RB4W3	RB4W2	RB2
	Bald Eagle					V***
	Steller's Sea Eagle	W4	??	W3-4	W2-3	RB2-3
	Bearded Vulture	V***		V***		
	Monk Vulture	W4	V***	W2-3	V**	
	Short-toed Snake Eagle	V*				
	Crested Serpent Eagle	RB2	RB2	V***		
	Ryukyu Serpent Eagle				RB3-4	
	Western Marsh Harrier	V***		??	V***	
	Eastern Marsh Harrier	W2	MW3	MW3	B3MW3	BM2
	Hen Harrier	BM2	MW4	W3	W3	BM2?

✓	Common name	C	T	K	J	R
	Northern Harrier				V***	?
	Pallid Harrier	V***				
	Pied Harrier	BM2	V*	M3	M4	BM2
	Montagu's Harrier	V***			V***?	
	Crested Goshawk	RB3	RB2			
	Chinese Sparrowhawk	BM2	M1	BM2	M2-3	
	Japanese Sparrowhawk	BM2-3 W3	MW3	BM2-3	BM2-3 (RB3)	BM2-3?
	Besra	RB2	RB2			?
	Eurasian Sparrowhawk	BM2W2	WM4	RB3W3	RB2W2	BM2
	Northern Goshawk	BM2?	WM4	RB4W3	RB3	BM2?
	Grey-faced Buzzard	BM2	M1-2W3	B?M2-3	B2M1 W3	BM2
	Eastern Buzzard	BMW2	MW4	W2-3	RB3MW2	BM2
	Rough-legged Buzzard	M3	V**	W3-4	W4	RB3
	Upland Buzzard	BM3W4	V**	MW4	V*	
	Indian Black Eagle	RB4	RB3			
	Greater Spotted Eagle	BM3W3?	MW4	W4	V*	BM3
	Steppe Eagle	BM3		V**	V***	
	Eastern Imperial Eagle	MW4	V**	W4	V**	
	Golden Eagle	RB3 & V**		RB4W4	RB3	RB3
	Bonelli's Eagle	RB3				
	Booted Eagle	B4?/V**		V***		
	Mountain Hawk Eagle	RB3(RB4)	RB4	V**	RB3	
	Great Bustard	BM3W4		W4	V**	BM3?
	Little Bustard				V***	
	Swinhoe's Rail	BM4		V*/MW4	B?W4	BM4
	Red-legged Crake		V***			
	Slaty-legged Crake		RB3		RB3	
	Okinawa Rail				RB3	
	Slaty-breasted Rail	RB2-3?	RB3			
	Eastern Water Rail	BMW3	MW4	W3	BMW3	BM3
	Brown Crake	RB3	(RB4)			
	White-breasted Waterhen	RB2	RB2	BM3	RB3M3	V**
	Baillon's Crake	BM2	V*	M3	BM3W4	BM2
	Ruddy-breasted Crake	RB(BM)2	RB2-3	BM2-3	RBM2	B4
	Band-bellied Crake	BM3	V***	B?M4?	V***	BM3
	White-browed Crake		V***		†	
	Watercock	BM3	BM3-4	BM3	(B)M4	V***
	Purple Swamphen		V***			
	Common Moorhen	RB(BM)1	RB1	BM2	RB(BM)1	BM1
	Common Coot	BM2W2	W3	BM2W2	BM2W2	BM2
	Demoiselle Crane	BM3	V***	V**	V**	
	Siberian Crane	M4		V**	V**	BM3-4
	Sandhill Crane	V**		V**	W4	B3
	White-naped Crane	BM3	V***	MW2	(B)W3	BM3
	Common Crane	BM3	V**	MW4	W4	BM3

✓	Common name	C	T	K	J	R
	Hooded Crane	RB3W2	V**	M1W2	W1-2	RB3
	Red-crowned Crane	RB3	V***	W3	RB3-4	RB3
	Small Buttonquail	RB4	RB4			
	Yellow-legged Buttonquail	RB (BM)3	V***	M3-4		BM3
	Barred Buttonquail	RB2	RB2		RB3	
	American Black Oystercatcher					V***
	Eurasian Oystercatcher	B3MW2	MW4	RB3	W3-4	BM3
	Ibisbill	RB4				
	Black-winged Stilt	B3M2	RB3 MW2	BM3	RB3M2 3 (V***)	B3 4M4
	Pied Avocet	M1W1B3	MW3	W4	W4	B4
	Northern Lapwing	B?M1W1	W3	W3	W2-3	BM2
	Grey-headed Lapwing	B3M2	MW4	M3-4	RB3MW3	V*
	Sociable Plover	V***				
	Pacific Golden Plover	M2	W2	M3	MW2	BM2
	American Golden Plover				V**	BM4/V*
	Grey Plover	M1W2	MW2-3	MW2	MW2	BM2
	Common Ringed Plover	V**	V**	V**	MW3-4	BM2
	Semipalmated Plover				V***	BM3-4
	Long-billed Plover	RBM3W3	W4	RB3	RB3	BM3
	Little Ringed Plover	BM2(RB2)	RB3MW2	BM2	BM2W3	BM2
	Killdeer					V***
	Kentish Plover	BM2(RB2)	RB3MW1	B4M2	RB2	BM3
	Lesser Sand Plover	MW2	W3M2	M2	MW2	BM2
	Greater Sand Plover	M1W3?	M2W3	M4	M3	V***
	Oriental Plover	M4	M4	M4	M4	
	Eurasian Dotterel	M4		V***	M4	BM3
	Pheasant-tailed Jacana	RB3	RB4M4	BM4	(B) M4/V*	V***
	Greater Painted Snipe	RB3	RB2MW3	B4MW4	RB3	V***
	Eurasian Woodcock	BM2MW3	MW3-4	WM3	BM2MW3	BM2
	Amami Woodcock				RB3	
	Jack Snipe	MW4	V*	V***	V*/W4	BM3
	Solitary Snipe	RB3MW4		MW3	W3	RB3M2
	Latham's Snipe	M4	M4	M4	BM2-3	B3M2
	Pin-tailed Snipe	M2W3	MW3	M3	MW3-4	BM2
	Swinhoe's Snipe	M2-3	MW3	M3	MW3-4	BM3?
	Common Snipe	BM2MW2	MW2	M2W3	MW2	BM1-2
	Short-billed Dowitcher				V***	
	Long-billed Dowitcher	V**	MW4	V**	MW4	BM2
	Asian Dowitcher	BM4M2	M4	M3	M4	BM3
	Eastern Black-tailed Godwit	B?M2	MW3	M2	M3	BM2
	Hudsonian Godwit				V***	
	Bar-tailed Godwit	M1W3	MW3	M2	M2	BM2
	Little Curlew	M1	M3	M4	M4	BM2-3
	Whimbrel	M2W3	M2W3	M2	M2W3?	B2M1

✓	Common name	C	T	K	J	R
	Bristle-thighed Curlew				V***	V***
	Slender-billed Curlew				(†?)V***	
	Eurasian Curlew	B?M1W1	M2W3	MW2	M2W3?	BM2
	Far Eastern Curlew	M1B4?	MW3-4	M2	M3	BM3
	Spotted Redshank	M1W2	MW3	M2	M2	BM2
	Common Redshank	M1W2	MW2	MW2-3	BM3MW2	BM3
	Marsh Sandpiper	B3M1W2	MW2-3	M3	M3	BM3
	Common Greenshank	M1W2	MW2	M2W4	MW2	BM2
	Nordmann's Greenshank	M4	M4	M4	M4	BM3
	Greater Yellowlegs			V***	V**	V***
	Lesser Yellowlegs		V**		V*	V***
	Green Sandpiper	M2W2	M2W3	M2W3	M2W3	BM2
	Wood Sandpiper	BM1W3	MW2	M2	M2W3	BM1-2
	Grey-tailed Tattler	M3	M2W3	M2	M2W3-4	BM2
	Wandering Tattler	V***	V***	V***	V*	BM3-4
	Terek Sandpiper	M2	M2W4	M2	M2	BM2
	Common Sandpiper	BMW2	MW2	B3M2W3	B3MW2	BM1-2
	Spotted Sandpiper				V***	V***
	Ruddy Turnstone	M2W3	M2W2-3	M2-3	M2W3	BM2
	Black Turnstone				V***	V***
	Great Knot	M1W3-4	M2W4	M1	M2-3	B2M1
	Red Knot	M1W3	MW3-4	M1	M3-4	B2M1
	Sanderling	M2W2	MW3	MW3	M2W2-3	BM2
	Semipalmated Sandpiper		V***		??	B3
	Western Sandpiper	V***	V***	V***	V**	B3
	Red-necked Stint	M1W2	WM1-2	M1-2	M1W2-3	BM1
	Little Stint	M4	V*	V**	M4	BM2
	Temminck's Stint	M2W3	MW3	M3	MW3	MB2
	Long-toed Stint	M2-3	M2W2-3	M2-3	M3W3-4	BM2
	Least Sandpiper		V***		??	V***
	White-rumped Sandpiper	V***			V***	V***
	Baird's Sandpiper	V***	V***		V**	BM3-4
	Pectoral Sandpiper	V**	V*	V*/M4	M3-4	BM3
	Sharp-tailed Sandpiper	M1	M2	M2	M2	BM2-3
	Curlew Sandpiper	M1	M3W4	M3-4	M3	BM2
	Rock Sandpiper	V***			W4	BM3
	Stilt Sandpiper		V***		V**	V***
	Dunlin	M1W1	MW1-2	M1-2 W3	M1-2 W2-3	BM2
	Spoon-billed Sandpiper	M4	M4	M4	M4	B3-4M3
	Broad-billed Sandpiper	M3	M3W4	M3	M3	BM3
	Buff-breasted Sandpiper		V***	V***	V**	B3-4M4
	Ruff	M3W4	M3W4	M3	M3	BM2

✓	Common name	C	T	K	J	R
	Wilson's Phalarope			V***	V**	??
	Red-necked Phalarope	M2-3	M1-2	M1-2	M1-2	BM1
	Red Phalarope	V***	V*	V**	M3	BM1-2
	Oriental Pratincole	BM3	BM3	B?M3-4	BM3	BM4(V*)
	Black-tailed Gull	BW2-3	W3M2 (B4)	RB2	RB2	BM2
	Common Gull	B3?W3	W4	W2	W2	BM2
	Ring-billed Gull			??	V***	
	California Gull				V***	
	Glaucous-winged Gull	V**		W4	W3	BM3
	Glaucous Gull	W4	V***	W3	W1-2	BM1-2?
	Iceland Gull				V***	V**
	Thayer's Gull			W4	W4	V***
	Vega Gull	W2-3	W3	W2	W1-2	BM2
	American Herring Gull	V**		V**	V*	
	Mongolian Gull	BM3 W3	WM3	B3MW2-3	WM3-4	BM3
	Slaty-backed Gull	W2-3	W4	W2-3	RB2W1-2	BM1-2
	Heuglin's Gull	W3	MW3	MW3	MW3	?
	Pallas's Gull	MW4	V**	V**	V*	
	Relict Gull	M2-3 W2-3		W4	V***	?
	Brown-headed Gull	V***	V***		V***	
	Black-headed Gull	MW1-2?	M2W1-2	MW1-2	M1W2	BM1-2
	Slender-billed Gull	V***		V***	V***	
	Bonaparte's Gull				V***	V***
	Saunders's Gull	BMW3	MW3	B2MW2	W2-3	BM3
	Laughing Gull				V***	
	Franklin's Gull	V***	V***		V***	
	Little Gull	BM3(V*)	V***	??	V**	BM3
	Ross's Gull	V***		??	V*/W4	BM3
	Ivory Gull			V***	V**	BM3-4
	Sabine's Gull			V***	V**	BM3
	Black-legged Kittiwake	V*	V**	W2-3	W2	BM2
	Red-legged Kittiwake				V*/W4	BM3
	Caspian Tern	MW3	MW3	M4	MW4	BM3
	Lesser Crested Tern				V***	
	Sandwich Tern		V***			
	Chinese Crested Tern	BM4/V**	V** (BM4)			
	Greater Crested Tern	BM3	BM2	V***	BM3	
	Roseate Tern	BM3	BM2-3		BM3	
	Black-naped Tern	BM3	BM2		BM3	
	Common Tern	BM2W3	M2	M2	M2	BM2
	Arctic Tern				V**	BM2
	Little Tern	BM2	B3MW3	BM3	BM3	V***
	Aleutian Tern	M3?	M4	V***	V*	BM3?
	Grey-backed Tern				B†?V*	

✓	Common name	C	T	K	J	R
	Bridled Tern	BM3	BM1-2	V***	BM4(V*)	
	Sooty Tern	V**	M4	V***	BM3-4 (V**)	
	Gull-billed Tern	BM3	M3/W4	V*	M3-4	
	Whiskered Tern	BM2	MW2-3	M3-4	M3W4	BM3
	White-winged Tern	BM3W3	M2-3W3	M3-4	M3	BM3
	Black Tern	V**	V**	??	V**	
	Brown Noddy		BM3		BM3	
	Black Noddy				V**	
	Blue Noddy				V***	
	White Tern				V**	
	South Polar Skua		V*/M4	M3	M3	
	Pomarine Skua	M4	M3-4	MW3	MW2-3	BM2-3
	Arctic Skua	?	M4	V*	MW2-3	BM2-3
	Long-tailed Skua	M3	M3-4	M3	M3	BM3
	Little Auk				V***	B4M4
	Brünnich's Guillemot			V***	W2	RB1-2
	Common Guillemot		V***	W4	B4W2-3	BM2
	Black Guillemot					BM4
	Pigeon Guillemot			??	W3	BMW3?
	Spectacled Guillemot			B?V*?	B4W3	BM2-3
	Long-billed Murrelet	W4		W4	W4	BM3-4
	Kittlitz's Murrelet				??	BMW4
	Ancient Murrelet	B3W2	V**	B3W2	B3?W2	B2W2
	Japanese Murrelet	V***	V***	B4M3	BW3	V***
	Cassin's Auklet				??	V***
	Parakeet Auklet				V*/W4	RB2
	Least Auklet			V**	W2-3	RB1-2
	Whiskered Auklet			V***	V*	BM4
	Crested Auklet	V***			W2	RB3-4
	Rhinoceros Auklet	V***		B?W3-4	B1-2W2	RB2
	Tufted Puffin				B4W3	B1-2W2
	Horned Puffin				W4	BW2-3
	Pallas's Sandgrouse	B3W3		V**	V**	??
	Rock Dove	RB4(I2)	I2	I2	I2	I2
	Hill Pigeon	RB3		RB3-4		RB3-4
	Stock Dove	V*		V***	V**	
	Ashy Woodpigeon		RB2			
	Black Woodpigeon	RB4(†?)	V**	RB3	RB3	V*
	Oriental Turtle Dove	RBM2	RB2	RB2	RB2/BM2	BM2
	Red Turtle Dove	RB2	RB2	V**	W4	V***
	Spotted Dove	RB2	RB2	V***		
	Eurasian Collared Dove	RB2-3	Esc?	RB4	IRB3/V***	
	Barred Cuckoo Dove	RB2-3	V***/Esc?			
	Philippine Cuckoo Dove		RB3			

✓	Common name	C	T	K	J	R
	Emerald Dove	?	RB3		RB3	
	Orange-breasted Green Pigeon		V***			
	White-bellied Green Pigeon	RB3-4 M4	RB2-3	V*	RB2-3	V***
	Taiwan Green Pigeon		RB3-4		V**	
	Ryukyu Green Pigeon		V***		RB2	
	Black-chinned Fruit Dove		RB4		V***	
	Tanimbar Corella		IB4			
	Sulphur-crested Cockatoo		IB4			
	White Cockatoo		IB4			
	Alexandrine Parakeet				IB3	
	Rose-ringed Parakeet		I4		IB3	
	Red-breasted Parakeet				IB3	
	Red Lory		IB4			
	Budgerigar		I4	I4	IB3	
	Chestnut-winged Cuckoo	BM3	M4	V**	V**	
	Plaintive Cuckoo	BM2	V***			
	Large Hawk Cuckoo	BM3	BM3	V***	V***	
	Northern Hawk Cuckoo	BM3	V***	BM3	BM3	BM3
	Indian Cuckoo	BM2?	V***(M4)	BM3	V*(M4?)	BM2?
	Eurasian Cuckoo	BM2	V*/M4	BM2-3	BM2-3	BM2
	Himalayan Cuckoo	BM2	BM2?			
	Oriental Cuckoo	BM2	BM2	BM3	BM2-3	BM2
	Lesser Cuckoo	BM2	BM4	BM4	BM2-3	BM2
	Asian Drongo Cuckoo	BM3	V***	V***	V***	
	Asian Koel	BM2	V*/M4 (B3)	??	V*	
	Greater Coucal	RBM3	(RB3)			
	Lesser Coucal	BM2	RB2	V***	V**	
	Eastern Grass Owl	RB3	RR4	V***	V***	
	Mountain Scops Owl	RB2	RB2			
	Collared Scops Owl	RB3	RB2	RB3-4	RB2-3	RB3-4?
	Oriental Scops Owl	BM2	M3-4	BM2-3	BM2-3	BM3
	Elegant Scops Owl	RB2-3			RB2	
	Snowy Owl	W4		V***	W4	RB3
	Eurasian Eagle Owl	RB3		RB3	RB4 (V***)	RB3
	Blakiston's Fish Owl	RB4		?	RB3-4	RB3
	Tawny Fish Owl	RB4	RB4			
	Brown Wood Owl	RB4	RB3-4			
	Chinese Tawny Owl	RB3	RB3	RB3		
	Ural Owl	RB2		RB3	RB2	RB2
	Great Grey Owl	RB4				RB3-4
	Northern Hawk Owl	W4		RB4		RB3
	Eurasian Pygmy Owl	RB4				RB3?
	Collared Owlet	RB2	RB2			
	Asian Barred Owlet	RB2				

✓	Common name	C	T	K	J	R
	Little Owl	RBM3	V***/ Esc?	RBM4	V**/ Esc?	RB2-3
	Tengmalm's Owl	RB3			RB3-4	RB3
	Brown Hawk Owl	BM2	RBM3	BM2	RBM2	?
	Northern Long-eared Owl	RBM2-3	W4	B?W3-4	RBM3	RB2-3
	Short-eared Owl	RB2-3?	W3	W3-4	W3	RB2-3
	Grey Nightjar	BM2-3	M4	BM3	BM2-3	BM2-3
	Savanna Nightjar	BM2	RB3			
	Himalayan Swiftlet		V***	V**	V***	
	White-throated Needletail	BM3?	M3-4	BM3	BM3	BM2-3?
	Silver-backed Needletail		BM3W4?			
	Common Swift	BM3		??	??	BM3?
	Pacific Swift	RB3-4 M2	BM3	BM2-3	BM2	BM2
	House Swift	RB2	RB2	BM4	RBM3	
	Rufous Hummingbird					V***
	Red-headed Trogon	RB3				
	Dollarbird	BM3	M3-4	B3M2-3	BM3-4	BM3
	Ruddy Kingfisher	BM3	M3-4	BM3	BM3RB2	V***
	White-throated Kingfisher	RB2-3	V** (RB4)		V***	
	Black-capped Kingfisher	BM2-3	MW4	BM2-3	M4	V***
	Collared Kingfisher	V**	V***		V**	
	Oriental Dwarf Kingfisher		V***			
	Common Kingfisher	BM2-3 RB2	RB2	RB3M3	RB2M3	BM2-3
	Blyth's Kingfisher	RB4				
	Crested Kingfisher	RB3		B†/V*	RB3	V**?
	Pied Kingfisher	RB2?	(RB4)			
	Blue-tailed Bee-eater	BM3	(BM3)			
	Rainbow Bee-eater		V***		V***	
	Blue-throated Bee-eater	RB3-4?	(V***)			
	Hoopoe	BM2?	M3(RB3)	BM2-3	M3	BM2-3
	Great Barbet	RB3				
	Taiwan Barbet		RB2			
	Eurasian Wryneck	BMW3?	MW4	M3	BMW3	BM2?
	Speckled Piculet	RB2-3?				
	Rufous-bellied Woodpecker	BM3		B?M4	V***	BM3?
	Japanese Pygmy Woodpecker	RB3		RB2-3	RB2	RB2
	Grey-capped Pygmy Woodpecker	RB3	RB2	RB3		RB3
	Lesser Spotted Woodpecker	RB3		RB4	RB4	RB3
	Rufous Woodpecker	RB3				
	Pryer's Woodpecker				RB4	
	White-backed Woodpecker	RB3	RB3	RB3	RB3-4	RB3
	Great Spotted Woodpecker	RB2 M3-4		RB2-3	RB2M4	RB2 M3-4

✓	Common name	C	T	K	J	R
	Eurasian Three-toed Woodpecker	RB3-4		RB4?	RB4?	RB3
	White-bellied Woodpecker		†(S) RB4(N)	†		
	Black Woodpecker	RB3		RB4	RB3	RB2
	Lesser Yellow-naped Woodpecker	RB4				
	Greater Yellow-naped Woodpecker	RB4				
	Japanese Woodpecker				RB2	
	Grey-headed Woodpecker	RB3	RB3	RB2-3	RB3	RB2-3
	Pale-headed Woodpecker	RB4				
	Bay Woodpecker	RB2				
	Hooded Pitta				V***	
	Fairy Pitta	BM3	BM3	BM3		
	Blue-winged Pitta		V***/ Esc?			
	Large Woodshrike	RB3				
	White-breasted Woodswallow		V***	V***		
	Large Cuckoo-shrike	RB3	RB3			
	Black-winged Cuckoo-shrike	BM2	M4	V***	V***	
	Swinhoe's Minivet	BM3	??			
	Ashy Minivet	BM2M2	M3	BM2M2	BM3M3	BM2-3
	Ryukyu Minivet		V*	??	RB3	
	Grey-chinned Minivet	RB2	RB2			
	Long-tailed Minivet	RB3?	V***			
	Scarlet Minivet	RB2				
	Tiger Shrike	BM3	V*	BM3M3	BM4M4	BM3
	Bull-headed Shrike	BM2	V*	RB2	RB2(W2)	BM2-3
	Brown Shrike	BM2W3	MW1	BM2M2	BM3M3	BM2-3
	Red-backed Shrike	?		??	V***	
	Isabelline Shrike				V***	
	Long-tailed Shrike	RBM2	RB2-3	V**?	V*/W4	
	Great Grey Shrike	W3		W4	W4	RB/BM3
	Chinese Grey Shrike	BM3	V**	B4W4	W4(V*)	BM3
	Maroon Oriole		RB3			
	Black-naped Oriole	BM2	B4M4	BM2M2	M3	BM2
	Black Drongo	BM2	RB2	M3	M3	V**
	Ashy Drongo	BM2	MW4	M4	M4	
	Bronzed Drongo		RB2			
	Hair-crested Drongo	BM3	M4	V**	V**	V***
	Black-naped Monarch		RB2			
	Asian Paradise Flycatcher	BM3	M4/V*	B4?/ V**	??	B4?
	Japanese Paradise Flycatcher	M3-4	M3RB4	BM3M3	R/BM3	V***
	Siberian Jay	RB2-3				RB2-3
	Eurasian Jay	RB2(M3)	RB2-3	RB2	RB2(M3)	RB2(M3)
	Lidth's Jay				RB2-3	

✓	Common name	C	T	K	J	R
	Azure-winged Magpie	RB2	IB4	RB2	RB3	RB3?
	Taiwan Blue Magpie		RB2-3			
	Red-billed Blue Magpie	RB2	IB4			
	Grey Treepie	RB2	RB2			
	Common Magpie	RB2	IB2-3	RB2	I?RB3(V**)	RB2
	Spotted Nutcracker	RB2	RB2-3	RB3	RB3	RB2
	Red-billed Chough	RB3		V***		
	Western Jackdaw				V***	
	Daurian Jackdaw	RB2W3	V**	B3-4W3	W4	RB2
	House Crow		V***		V***	
	Rook	RBW3	V*	W3	W3	RB3?
	Carrion Crow	RB2	V**	RB2?	RB2	
	Collared Crow	RB3	V*** (RB3)			
	Large-billed Crow	RB2	RB2-3	RB2	RB1-2M2	RB2
	Northern Raven	W4		W4	W4	RB3?
	Bohemian Waxwing	W3(V**)	V**	W3	W2-3	BM2
	Japanese Waxwing	BMW3	V*	W3	WM2-3	BM3
	Northern Great Tit					RB2
	Eastern Great Tit	RB2		RB2	RB2	RB2
	Southern Great Tit	RB2			RB2	
	Green-backed Tit		RB2			
	Yellow Tit		RB3			
	Yellow-bellied Tit	RB2		V***		
	Yellow-cheeked Tit	RB3	Esc			
	Coal Tit	RB2	RB2	RB2	RBM2	RB2
	Azure Tit	RB3		??	V***	RB3?
	Varied Tit	RB3	RB3	RB2	RB2	?
	Marsh Tit	RB2		RB2	RB2	RB2
	Willow Tit	RB2		V*?	RB2-3	RB2
	Songar Tit	RB3				
	Siberian Tit					RB2
	Sultan Tit	RB3				
	Yellow-browed Tit	RB2-3				
	Chinese Penduline Tit	RB2-3	V**	WM3	W3	B4
	Grey-throated Martin		RB2-3		V***	
	Sand Martin	BM2	M4	M3	BM2-3W4	BM2
	Pale Martin	BM3?	V**			
	Tree Swallow					V***
	Purple Martin					V***
	Barn Swallow	BM1-2	B2M1W4	BM1-2	BM1-2W3	BM2
	Pacific Swallow		RB2		RB2	
	Eurasian Crag Martin	BM2-3?		??		
	Northern House Martin	BM3W4		M4/V*	M4/V*	?
	Asian House Martin	BM2	RBM2	B2	BM2 W3-4	BM2

✓	Common name	C	T	K	J	R
	Red-rumped Swallow	BM2-3?	M4	BM2-3	BM2-3W4	BM2-3
	Striated Swallow		RB2			
	Cliff Swallow					V***
	Long-tailed Tit	RB2		RB2	RB2	RB2
	Black-throated Tit	RB2	RB1-2			
	Bimaculated Lark			??	V**	
	Mongolian Lark	BM2W3			V***	
	Greater Short-toed Lark	BM3W3	V**	MW4	MW4	
	Asian Short-toed Lark	BM3W3	V**	M3	V*(MW4)	
	Crested Lark	RB3		RB3-4		
	Eurasian Skylark	BM2	MW4	RB2W2	RB2W2	BM2
	Oriental Skylark	RB3	RB2		V***	
	Shore Lark	BMW3			WM3-4	W?
	Zitting Cisticola	RB2	RB2	RB2	RB2	V***
	Golden-headed Cisticola	RB3	RB2			
	Chinese Hill Warbler	RB3		RB3		
	Striated Prinia	RB2	RB3			
	Hill Prinia	RB2				
	Yellow-bellied Prinia	RB2	RB2			
	Plain Prinia	RB2	RB1-2		V***	
	Common Tailorbird	RB2				
	Collared Finchbill	RB2	RB2-3			
	Red-whiskered Bulbul	RB2	Esc		IB3	
	Brown-breasted Bulbul	RB2				
	Chinese Bulbul	RB2	RB1	M4	RB2-3	
	Taiwan Bulbul		RB2			
	Sooty-headed Bulbul	RB2				
	Mountain Bulbul	RB3				
	Brown-eared Bulbul	RB?2-3?	RB1-2/V*	RB2	RB1-2	?
	Chestnut Bulbul	RB2?	V***			
	Himalayan Black Bulbul	RB2	RB1		V***	
	Marsh Grassbird	RB3-4		??	RB3	BM3-4?
	Asian Stubtail	BM2W3	MW3	BM2	BM2W3-4	BM3
	Japanese Bush Warbler	BM2	MW4	BM2	RB2	BM2
	Korean Bush Warbler	BM2	WM1	BM2	MW4	BM2
	Brownish-flanked Bush Warbler	RB2	RB2			
	Yellowish-bellied Bush Warbler	RB3	RB1-2			
	Siberian Bush Warbler	BM3-4		??	??	?
	Chinese Bush Warbler	BM3-4				BM3-4?
	Brown Bush Warbler	RB3				
	Taiwan Bush Warbler		RB2			
	Russet Bush Warbler	RB3				
	Lanceolated Warbler	B2M3	M3	B3M3	B2M3	BM2?
	Pallas's Grasshopper Warbler	BM2-3	M4	M3	V*/M4	BM2-3

✓	Common name	C	T	K	J	R
	Middendorff's Warbler	M3	MW3	(B?)M3	BW3-4	BM2
	Pleske's Warbler	B4M3-4	WM4	BM3M3	BM3	BM3
	Gray's Warbler	BM3M3	MW3	M3	BM3 M3-4	BM2-3?
	Thick-billed Warbler	BM3	SR	M3-4	V*	BM3
	Oriental Reed Warbler	BM2W2	M2W3	BM2	BM2	BM2
	Black-browed Reed Warbler	BM2	M4 W3-4	BM3-4	BM2	BM2
	Streaked Reed Warbler	BMW3	V***		V***	
	Blunt-winged Warbler	BM3				
	Manchurian Reed Warbler	BM2-3	??	BKP		
	Paddyfield Warbler		V**	??	V***	
	Eurasian Reed Warbler	V***				
	Blyth's Reed Warbler				V***	
	Booted Warbler				V***	
	Willow Warbler			V***	V*/M4	BM3-4
	Siberian Chiffchaff	V***		V**	(MW)V*	V***
	Wood Warbler				V***	
	Dusky Warbler	BM2	MW3	BM2M3	MW3-4	BM2
	Tickell's Leaf Warbler	V***		V***	V***	
	Buff-throated Warbler	BM3W3				
	Yellow-streaked Warbler	BM3W3				
	Radde's Warbler	BM3M3	M4	BM3M3	MW4/V*	BM3
	Chinese Leaf Warbler	BM3				
	Pallas's Leaf Warbler	BM2	M3W4	BM2 M2-3	M4	BM2
	Yellow-browed Warbler	BM2W2	MW2	M3	MW3-4	BM2
	Hume's Warbler	BM3		??	??	
	Arctic Warbler	BM2 W2-3	MW2	M2	BM2M 2W4	BM2
	Pale-legged Leaf Warbler	BM2	V*?/ M4?	B3M3	M4	BM2
	Sakhalin Leaf Warbler	M4	M3	M4?	BM2-3	BM2-3
	Large-billed Leaf Warbler	B3-4?				
	Greenish Warbler		V***	??	V***	
	Two-barred Greenish Warbler	BM2-3	V***	V*	V*	BM2-3
	Eastern Crowned Warbler	BM2 M2-3	M3	BM2 M2	BM2 M2	BM2
	Ijima's Warbler		M4W4 ?V**		BM3 M3	
	Claudia's Leaf Warbler	BM3				
	Hartert's Warbler	BM3 (V***)				
	Kloss's Leaf Warbler	BM3				
	Sulphur-breasted Leaf Warbler	BM3	SRP			
	White-spectacled Warbler	RB3				
	Alström's Warbler	BM3				
	Bianchi's Warbler	RB3 (V***)				

✓	Common name	C	T	K	J	R
	Chestnut-crowned Warbler	BM3 W3				
	Lesser Whitethroat	V***	V***	V***	V***	V***
	Desert Lesser Whitethroat	V***				
	Rufous-faced Warbler	RB2	RB2			
	Black-necklaced Scimitar-babbler		RB2			
	Grey-sided Scimitar-babbler	RB2-3				
	Taiwan Scimitar-Babbler		RB2			
	Streak-breasted Scimitar-Babbler	RB2				
	Taiwan Wren-Babbler		RB2			
	Pygmy Wren-Babbler	RB2-3				
	Rufous-capped Babbler	RB2	RB2			
	Chinese Babax	RB3				
	Masked Laughingthrush	RB2				
	Rufous-crowned Laughingthrush		RB3			
	Lesser Necklaced Laughingthrush	RB2?				
	Greater Necklaced Laughingthrush	RB2?				
	Black-throated Laughingthrush		IB4			
	Plain Laughingthrush	RB2-3				
	Moustached Laughingthrush	RB3			IB3-4	
	Rusty Laughingthrush		RB2-3			
	Buffy Laughingthrush	RB2				
	Chinese Hwamei	RB2	IB3-4		IB3	
	Taiwan Hwamei		RB3			
	White-browed Laughingthrush	RB2-3				
	White-whiskered Laughingthrush		RB2			
	Red-tailed Laughingthrush	RB3				
	Steere's Liocichla		RB2			
	Red-billed Leiothrix	RB2	Esc		IB3	
	White-browed Shrike-babbler	RB2				
	Green Shrike-babbler	RB3				
	Taiwan Barwing		RB3			
	Taiwan Fulvetta		RB2			
	Grey-hooded Fulvetta	RB2				
	Dusky Fulvetta	RB2-3	RB2			
	Grey-cheeked Fulvetta	RB1-2	RB1			
	Taiwan Sibia		RB2			
	Indochinese Yuhina	RB2				
	Taiwan Yuhina		RB1			
	Black-chinned Yuhina	RB2				

✓	Common name	C	T	K	J	R
	White-bellied Erpornis	RB2	RB2			
	Bearded Tit	RB3W4		V***	V***	
	Spot-breasted Parrotbill	RB3				
	Vinous-throated Parrotbill	RB2	RB2	RB2	V***	RB2
	Golden Parrotbill	RB3	RB3			
	Short-tailed Parrotbill	RB3				
	Grey-headed Parrotbill	RB2				
	Northern Parrotbill					RB3
	Reed Parrotbill	RB3				
	Chestnut-flanked White-eye	BM2	B?M2	M4	BM2	
	Japanese White-eye	RB2	RB1	RB3M3	RB2	B3?
	Lowland White-eye		RB2			
	Bonin Honeyeater				RB2	
	Flamecrest		RB2			
	Goldcrest	BM2W3	MW4	W2-3?	RB2W3	BM2
	Ruby-crowned Kinglet					V***
	Winter Wren	RB2	RB2-3	RB2?	RB2-3	BM2?
	Eurasian Nuthatch	RB2	RB2	RB2	RB2	RB2
	Chinese Nuthatch	RB3		RB3 (M3-4)		
	Wallcreeper	RB3				
	Eurasian Treecreeper	RB2-3		B?W3	RB2-4	RB2-3
	Asian Glossy Starling		IB3(V***)		V***	
	Great Myna		IB4			
	Crested Myna	RB2	RB4 (RB3)		(IB3) V**	
	Javan Myna		IB1 2		IB3	
	Jungle Myna		IB3-4			
	Bank Myna		IB4		IB3?	
	Common Myna		IB2		IB3	
	Black-collared Starling	RB2	IB3			
	Asian Pied Starling		IB4		IB3	
	Daurian Starling	BM2-3	M4	M3	M3-4	BM2-3
	Chestnut-cheeked Starling	M3	M3W4	M4	BM2-3	BM3?
	White-shouldered Starling	RB2-3	MW2	V**?	W4	V***
	Chestnut-tailed Starling		IB4(V***)			
	Rosy Starling	V***	V***	??	V***	
	Red-billed Starling	RB2	MW3	M4/V*	MW3-4	
	White-cheeked Starling	BM2W3	MW2	RB2	RB2W3	BM2
	Common Starling	M4	MW4	MW 3-4	WM 3-4	B?/V***
	Taiwan Whistling Thrush		RB2			
	Blue Whistling Thrush	RB 2-3?	V*** (RB3)			
	Orange-headed Thrush	RBM3	Esc	??		
	Siberian Thrush	BM3	V*	M3	BM3M3	BM3?
	White's Thrush	RBW2-3	RB4/WM2	RBW2-3	RBW2-3	BM2

✓	Common name	C	T	K	J	R
	Amami Thrush				RB3	
	Varied Thrush					V***
	Grey-cheeked Thrush				V***	B2-3
	Swainson's Thrush					V***
	Hermit Thrush					V***
	Grey-backed Thrush	BM2 M2-3	M4	BM2 M2	M4	BM2
	Japanese Thrush	BM3M3	M4	M4	BM3M3	V**
	Eurasian Blackbird	RB2-3	MW4 (MW2)	MW4	V*	
	Island Thrush		RB3-4			
	Grey-sided Thrush	BM3	V***			
	Eyebrowed Thrush	BM2 M2-3	M(W)2	M2	M2-3W3	BM2
	Pale Thrush	MW2	MW2	BM3?	BM4 MW2	BM2
	Brown-headed Thrush	MW3	MW2	M4	BM2 W2-3	BM2
	Izu Thrush				RB3	
	Black-throated Thrush	V*	??	V*	V*	
	Red-throated Thrush	MW2-3	V***	V**	V***	
	Naumann's Thrush	MW3-4	MW3	M3W3	MW4	BM2
	Dusky Thrush	W2	MW3	M2W3?	W1-2	RB2
	Fieldfare	V***			V**	B3?/V***
	Redwing			V***	V**	B2?
	Song Thrush				V***	
	Chinese Thrush	B4MW3		??		
	Mistle Thrush				V**	
	Green Cochoa	RB4				
	Lesser Shortwing	RB3				
	White-browed Shortwing	RB3			??/Esc	
	Taiwan Shortwing		RB2			
	Bluethroat	BM3M3	MW4	WM3-4	WM4	BM3
	European Robin	V***		V***	V***	
	Japanese Robin	MW3-4	MW4	V*/M4	BM2-3	BM2-3
	Ryukyu Robin		V**		RB2-3	
	Siberian Rubythroat	BM2W2	MW2	BM2	BM2W3	BM2
	Siberian Blue Robin	BM2-3M3	V**	BM2 M3	BM2-3M3	BM2-3
	White-browed Robin		RB3			
	Johnstone's Robin		RB2			
	Red-flanked Bluetail	BM2 W2	MW3	MW2?	BM2 W2	BM2
	Swinhoe's Robin	M2-3	V*	M2-3	M4	BM2
	Oriental Magpie-Robin	RB1-2	IB3-4 (RB2)			
	White-rumped Shama		IB4			
	Przevalski's Redstart	V***/M4				
	Black Redstart		V*	V**	V**	V**

✓	Common name	C	T	K	J	R
	Common Redstart				V***	V***
	White-throated Redstart	RB3 (V***)				
	Daurian Redstart	BM2-3W2	MW2	RB2	W2	BM2?
	Güldenstädt's Redstart	W3-4				
	White-bellied Redstart	BM3-4				
	Plumbeous Redstart	RB2-3	RB2	??		
	White-capped Redstart	RB2-3	V** or IB4	??		
	White-tailed Robin		RB2		I?V***	
	Little Forktail	RB3	RB3			
	Slaty-backed Forktail	RB3				
	White-crowned Forktail	RB3				
	Spotted Forktail	RB2?				
	Whinchat				V***	
	Siberian Stonechat	BM2 M2-3	MW3	BM2 M2-3	BM2 M2-3	BM2
	Pied Bushchat				V**	
	Grey Bushchat	RB2-3	M4	V***	V**	
	Isabelline Wheatear	B3	V***	V***	V**	
	Northern Wheatear	BM3	V***	??	V**	BM2-3
	Pied Wheatear	BM2-3		V***	V**	
	Desert Wheatear		V***		V**	
	Rufous-tailed Rock Thrush	V***			V***	
	Blue Rock Thrush	BM2 W2	RB3 MW2	BM2	RB2	BM2-3?
	Chestnut-bellied Rock Thrush	BM3 W3				
	White-throated Rock Thrush	BM3 M3	V**	M4	V*	BM3
	Brown-chested Jungle Flycatcher	BM3				
	Spotted Flycatcher				V***	
	Grey-streaked Flycatcher	BM3 M2-3	M2-3	M2-3	M2-3	BM2-3?
	Dark-sided Flycatcher	BM2 M2	M3	BM2-3M3	BM2-3M3	BM2
	Asian Brown Flycatcher	BM2 M2	M3W4	BM2 M2	BM2-3M2-3	BM2
	Ferruginous Flycatcher	BM2-3	BM2	??	V*	
	European Pied Flycatcher				V***	
	Yellow-rumped Flycatcher	BM2 M3-4	V*	BM2 M2	M4	BM2?
	Narcissus Flycatcher	BM3M3	M3	M3	BM2	BM2-3
	Chinese Flycatcher	BM3M3	V**	??	V***	
	Ryukyu Flycatcher		??	??	RB3	
	Mugimaki Flycatcher	BM2-3M2-3	M4	BM2-3M2	M4	BM2-3
	Slaty-backed Flycatcher	BM4	V***			
	Rufous-gorgetted Flycatcher	V***				

✓	Common name	C	T	K	J	R
	Taiga Flycatcher	M3	MW3	M3	M3W4	BM3?
	Red-breasted Flycatcher		V***	V**	V**	
	Snowy-browed Flycatcher	?	RB2			
	Slaty-blue Flycatcher	V***				
	Blue-and-white Flycatcher	BM2	M3W4	BM2 M2	BM2 M2	BM 2-3?
	Asian Verditer Flycatcher	W3-4	V**	V***		
	Hill Blue Flycatcher			V***		
	Blue-throated Blue Flycatcher	V***				
	Fujian Niltava	V***				
	Vivid Niltava		RB2		V***	
	Grey-headed Canary-Flycatcher	V**	V***			V***
	Brown Dipper	RB3-4	RB3	RB2	RB2	RB2
	Orange-bellied Leafbird	RB2	Esc			
	Plain Flowerpecker	RB3	RB3			
	Fire-breasted Flowerpecker	RB2-3	RB2			
	Scarlet-backed Flowerpecker	RB3?				
	Fork-tailed Sunbird	RB2-3				
	House Sparrow	RB2-3	V***	??	V**	RB2-3
	Russet Sparrow	RB2 (BM2)	RB4	BM2 W2	BM2 MW2	BM2
	Eurasian Tree Sparrow	RB1-2?	RB1	RB1-2	RB1-2	RB2
	Rock Sparrow	RB3				
	Lesser Masked Weaver		IB4		IB3	
	Northern Red Bishop				IB3	
	Orange-cheeked Waxbill		IB3-4		IB3	
	Black-rumped Waxbill				IB3	
	Common Waxbill		IB3-4			
	Red Avadavat		IB3-4		IB3	
	Indian Silverbill		IB3			
	White-rumped Munia	RB2	RB2		IB3 (RB3?)	
	Scaly-breasted Munia	RB2	RB2	??	IB3 (RB3?)	
	Black-headed Munia		RB3		IB3	
	White-headed Munia		IB3		IB3	
	Java Sparrow	IB3	IB4		IB3-4	
	Eastern Paradise Whydah				IB3	
	Alpine Accentor	RB3W3	RB3	W3	RB3	RB3
	Siberian Accentor	W3	V***	W3	W4	RB3
	Brown Accentor	RB3				
	Japanese Accentor			??	BM/ RB2-3	BM3
	Forest Wagtail	BM3M4	MW4	BM3M4	M4	BM3
	Western Yellow Wagtail		?	?		BM2
	Eastern Yellow Wagtail	BM2	MW2	M2-3	BM2W3	BM2

✓	Common name	C	T	K	J	R
	Citrine Wagtail	BM3 M3-4	V*/M4	V**	V*	?
	Grey Wagtail	BM2W2	MW2	RBM2	RBM2	BM2
	White Wagtail	BM2 MW2	MW2 RB3	RB2/ MW2	RB3/ MW2	BM2
	Japanese Wagtail	V***	V***	RB3-4	RB2-3	V***
	Richard's Pipit	B3MW3	MW3	M2(W?)	MW3	BM2
	Blyth's Pipit	BM3	V***	M4	V*	
	Meadow Pipit				V**	
	Tree Pipit			V**	V**	BM2?
	Olive-backed Pipit	BM2W2	MW2	?	BM2W2	BM2
	Pechora Pipit	BM3M3	M3	M3	M4	BM2
	Rosy Pipit	BM3			V**	
	Red-throated Pipit	MW2-3	MW2	MW2-3	MW3	BM2
	Buff-bellied Pipit	W2	MW3	W2	W2	BM2
	Water Pipit	MW3	V***	V**	??	
	Upland Pipit	RB3				
	Common Chaffinch	W3-4			V***	
	Brambling	W2-3	MW3	W1-2	W1-2	BM2W2
	Yellow-fronted Canary		IB4			
	Oriental Greenfinch	BM/ RB2	MW3 (RB3)	RB2 W2	RB2 W2	BM2
	Eurasian Siskin	BM2	MW3	W2	BM3?W2	BM2
	Common Redpoll	W2	V***/I?	W4	W2	BM2
	Arctic Redpoll	W3-4		V***	V**	BMW3
	Eurasian Goldfinch				V***	
	Asian Rosy Finch	W3-4		W4	B?W3M4	RB3
	Grey-crowned Rosy Finch					RR3?
	Mongolian Finch	V***				
	Long-tailed Rosefinch	RB2-3		BM3?W3	BM2-3W3	BM2-3
	Common Rosefinch	BM2W3	V**	M2	M4	BM2
	Chinese Beautiful Rosefinch	RB3				
	Vinaceous Rosefinch		RB2			
	Pallas's Rosefinch	MW3-4		MW3-4	MW3-4	BM3W3
	Pine Grosbeak	W3-4		V***	RB3	RB3
	Common Crossbill	BM2-3W3-4		BM2-3W3-4?	BM2-3W3-4?	BM2
	Two-barred Crossbill	BM3 W3-4		V***	V**	BM3-4
	Brown Bullfinch	RB2-3	RB2-3			
	Beavan's Bullfinch	RB2-3	RB2			
	Eurasian Bullfinch	RB2		W2-3?	W2-3	RB2
	Hawfinch	RB3	M4	W2-3?	BM3 ?W2-3	BM2-3
	Chinese Grosbeak	BM3W3	M4	M3	MW4	BM3
	Japanese Grosbeak	BM2-3W2	M4	B?M4	BM2 RB2	BM2-3
	Orange-crowned Warbler					V***
	Yellow-rumped Warbler					V***

✓	Common name	C	T	K	J	R
	Northern Waterthrush					V***
	Wilson's Warbler				V***	
	Rusty Blackbird					V***
	Western Meadowlark					V***
	Crested Bunting	RB3	V***		V***	
	Slaty Bunting	RB3	??			
	Yellowhammer	V***	??	??	V**	
	Pine Bunting	RB3 (V***)	V***	MW3-4	MW3-4	RB3
	Godlewski's Bunting	RB2-3				
	Meadow Bunting	RB2	M4/V*	RB2-3?	BM2 RB2	BM2
	Jankowski's Bunting	BM3 W3		BM4/ V***		Ex
	Grey-necked Bunting				V***	
	Ortolan Bunting				V***	
	Tristram's Bunting	BM2W3	M3-4	M2	M3-4	BM2
	Chestnut-eared Bunting	BM3 W3	M4	BM3 W3	BM3 W2-3	BM2
	Little Bunting	BM3W2	M3W4	MW3-4	MW3-4	BM2
	Yellow-browed Bunting	W2-3	M3-4	M2-3	M4	?
	Rustic Bunting	W2-3	M3-4	W2-3?	W2-3	BM2
	Elegant Bunting	BM2-3W3-4	M3-4W4	BM2-3W2	BM3 W2	BM2?
	Yellow-breasted Bunting	BM2-3M3	M4	M3	BM3-4M3	BM2
	Chestnut Bunting	M2	M3-4	M2	M3-4	?
	Black-headed Bunting	V**	V**	V***	V*	
	Red-headed Bunting	V***		V***	V**	
	Japanese Yellow Bunting	WM3-4	M3W4	M3	BM3-4M3	V***
	Black-faced Bunting	RB2W2	MW2-3	RB2W2	RB2W2	RB2
	Grey Bunting	V***	V**	M1	BM3 WM3	BM3
	Pallas's Reed Bunting	W3	V*	W3	V**(W4)	BM3W3
	Japanese Reed Bunting	BM3 MW4		W3-4	BM3 WM4	BM3
	Common Reed Bunting	BM2W2	V*	W2	BM2W2	BM2
	Lapland Bunting	W3	V***	W3-4	W3-4M4	BM2-3
	Snow Bunting	V***	V***	V***	W3-4M4	BM2W3
	McKay's Bunting					V***
	Fox Sparrow				V***	V**
	Song Sparrow				V***	V***
	White-crowned Sparrow			V***	V**	V***
	Golden-crowned Sparrow				V***	V***
	Dark-eyed Junco					V***
	Savannah Sparrow			V**	W4	B4/V*
	American Tree Sparrow					V***
	Chipping Sparrow					V***
	Red-crested Cardinal		Esc		IB3	

APPENDIX 2. SPECIES LIKELY TO REACH EAST ASIA AS VAGRANTS

Vagrants may reach the East Asian region from various directions, particularly North America (1), from North and Central Asia (Siberia and northern and northwest China, from Mongolia and perhaps from further west in central Asia) (2), South China (3), from Southeast Asia and Indonesia (4), and the northern or southern Pacific, or even other oceans (5).

The best understood are perhaps those from North America (1). Considerable research has been conducted on islands of the Bering Sea, particularly in recent years at Gambell, St Lawrence Island. Adults of a number of species may overshoot on spring migration on headings from southern wintering grounds towards western Alaska and thus reach northeast Siberia; in addition, reverse migration of immature birds in their first autumn may take many North American migrants out in to the Bering Sea, a number of which have reached Gambell in August, September and October (1G). Consideration of potential vagrancy from North and Central Asia (2) and Southern China (3) has relied on maps in Mackinnon and Phillipps (2000), although not always reliable, they remain the best available. Pelagic seabirds are renowned for extraordinary wanderings, and a number (perhaps many more than listed here) are potential vagrants to the western North Pacific (5).

The potential for vagrancy into the region from the south involves both northbound migrants in spring, reverse migration of immatures in autumn, and also local/regional species making minor range adjustments bringing them just into the region. The completely unpredictable direction is from the far west - which has already brought the Eurasian Pied Flycatcher twice to Japan - involving strays very far from their normal ranges. It is impossible to imagine what might appear next from this direction. The regular movement of typhoons from the Pacific towards the East Asian coastline means that a number of pelagic species from the central and northern Pacific are very likely to occur in East Asian waters, perhaps most likely off Japan, or the Kuril Islands, or in the western Bering Sea off the Commander Islands.

Species	Scientific Name	Origin
Western Capercaillie	Tetrao urogallus	2
Common Quail	Coturnix coturnix	2
Shy Albatross	Thalassarche cauta	5
Murphy's Petrel	Pterodroma ultima	5
Phoenix Petrel	Pterodroma alba	5
Herald Petrel	Pterodroma arminjoniana	5
Cook's Petrel	Pterodroma cookii	5
Pink-footed Shearwater	Puffinus creatopus	5
Great Shearwater	Puffinus gravis	5
White-faced Storm-petrel	Pelagodroma marina	5
Lesser Adjutant	Leptopius javanicus	3
Javan Pond Heron	Ardeola javanicus	4
Red-billed Tropicbird	Phaethon aethereus	5
Shikra	Accipiter badius	3
Corn Crake	Crex crex	2
Little Crake	Porzana parva	2
Black-necked Crane	Grus nigricollis	3
Asian Emerald Cuckoo	Chrysococcyx maculatus	3
Northern Saw-whet Owl	Aegolius acadius	1G
European Bee-eater	Merops apiaster	2
Olive-sided Flycatcher	Contopus cooperi	1G
Least Flycatcher	Empidonax minimus	1G
Pacific Slope Flycatcher	Empidonax difficilis	1G

Species	Scientific Name	Origin
Rosy Minivet	Pericrocotus roseus	3
Burmese Shrike	Lanius colluroides	3
Warbling Vireo	Vireo gilvus	1G
Savi's Warbler	Locustella luscinoides	2
Clamorous Reed Warbler	Acrocephalus stentoreus	2
Sulphur-bellied Warbler	Phylloscopus griseolus	2
Barred Warbler	Sylvia nisoria	2
Desert Warbler	Sylvia nana	2
Red-breasted Nuthatch	Sitta canadensis	1G
Little Pied Flycatcher	Ficedula westermanni	3
Hainan Blue Flycatcher	Cyornis hainanus	3
Pale Blue Flycatcher	Cyornis unicolor	3
Small Niltava	Niltava macgrigoriae	3
Tawny Pipit	Anthus campestris	2
Purple Finch	Carpodacus purpureus	1G
Tennessee Warbler	Vermivora peregrina	1G
Nashville Warbler	Vermivora ruficapilla	1G
Magnolia Warbler	Dendroica magnolia	1G
Townsend's Warbler	Dendroica townsendi	1G
American Redstart	Setophaga ruticilla	1G
MacGillivray's Warbler	Oporornis tolmiei	1G
Bullock's Oriole	Icterus bullockii	1G
Lincoln's Sparrow	Melospiza licolnii	1G

INDEX